Basic College Mathematics

Julie Miller

Molly O'Neill

Nancy Hyde

 **Higher Education**

Boston Burr Ridge, IL   Dubuque, IA   Madison, WI   New York   San Francisco   St. Louis
Bangkc Bogotá Caracas   Kuala Lumpur   Lisbon   London   Madrid   Mexico City
Milan Montreal   New Delhi   Santiago   Seoul   Singapore   Sydney   Taipei   Toronto

Higher Education

BASIC COLLEGE MATHEMATICS

Published by McGraw-Hill, a business unit of The McGraw-Hill Companies, Inc., 1221 Avenue of the Americas, New York, NY 10020. Copyright © 2007 by The McGraw-Hill Companies, Inc. All rights reserved. No part of this publication may be reproduced or distributed in any form or by any means, or stored in a database or retrieval system, without the prior written consent of The McGraw-Hill Companies, Inc., including, but not limited to, in any network or other electronic storage or transmission, or broadcast for distance learning.

Some ancillaries, including electronic and print components, may not be available to customers outside the United States.

This book is printed on acid-free paper.
Printed in China

4 5 6 7 8 9 0 CTP/CTP 0 9 8

ISBN-13 978-0-07-302318-2
ISBN-10 0-07-302318-3

ISBN-13 978-0-07-302319-9 (Annotated Instructor's Edition)
ISBN-10 0-07-302319-1

Publisher: *Elizabeth J. Haefele*
Sponsoring Editor: *David Millage*
Senior Developmental Editor: *Erin Brown*
Marketing Manager: *Barbara Owca*
Project Manager: *Jodi Rhomberg*
Senior Production Supervisor: *Sherry L. Kane*
Lead Media Project Manager: *Stacy A. Patch*
Media Producer: *Amber M. Huebner*
Designer: *Laurie B. Janssen*
Interior Designer: *Rokusek Design*
(USE) Cover Image: *© Index Stock*
Lead Photo Research Coordinator: *Carrie K. Burger*
Photo Research: *Pam Carley/Sound Reach*
Supplement Producer: *Melissa M. Leick*
Compositor: *Techbooks/GTS, York, PA*
Typeface: *10/12 TimesTen Roman*
Printer: *CTPS*

Photo Credits
Chapter 1 Opener: © Richard Hutchings/PhotoEdit; p. 5: © SMSU-WP Photo/Vickie Driskell; p. 21: © Tom Carter/PhotoEdit; p. 32: © Bettmann/Corbis; p. 48: © Jeff Greenberg/PhotoEd; p. 54: © Mary Kate Denny/PhotoEdit; p. 65: © Richard Clark; p. 73: © Vol. 165/Corbis; p. 82: © Siede Pre/Getty RF; p. 83: © Greg Flume/NewSport/Corbis; Chapter 2 Opener: © Vol. 26/PhotoDisc/Getty; p. 10. (left to right) © Fred Prouser/Reuters/Corbis; © Jessica Rinaldi/Reuters/Corbis; © Shannon Stapletn/Reuters/Corbis; © Lisa O'Connor/Zuma/Corbis; © Tobias Schwarz/Reuters/Corbis; p. 126: © NASA; 138: © David Young-Wolff/Photo Edit; 149: © James Prince/Photo Researchers, Inc.; p. 156: © P.P.A. Exprer/Photo Researchers; Chapter 3 Opener: © PhotoDisc RF/Getty; p. 186: © Vol. 85/PhotoDisc/Gettyp. 196: © Darrel Gulin/Getty; p. 207: (top to bottom) © Daniel J. Catt; © James R. Fisher/Photo Researche1, Inc.; © Steve Maslowski/Photo Researchers, Inc.; © Steve Maslowski/Photo Researchers, Inc.; p. 216: © Kim Sayer/Corbis; p. 217: © Hans Georg Roth/Corbis; p. 225: © Molly O'Neill; Chapter 4 Opener: © A/Wide World Photos; p. 242: © EP4 PhotoDisc/Getty; 259: © Corbis; p. 276: © Corbis; p. 285: © Vol. 1 PhooDisc/Getty; p. 296: © Corbis; p. 311 © Digital Vision RF; Chapter 5 Opener: ©Vol. 6/PhotoDisc/Getty; p. 31 © Vol.77/ PhotoDisc/Getty; p. 329: Christina Pahnke/New Sport/CORBIS; p. 345: © Linda Waymire; Chater 6 Opener: © Vol. 1/PhotoDisc/Getty; p. 378: © Vol. 76/PhotoDisc/Getty; p. 390: © Tom Carter/Photo Researchers, Inc.; p. 396: © Galen Rowell/Corbis; p. 427: © Wild Views/Digital Vision; p. 428: © Bill Aron/PhotEdit; Chapter 7 Opener: © Duomo/Corbis; p. 458: © Vol. 1/PhotoDisc/Getty; p. 462: © Rachel EpsteiPhotoEdit; p. 474: © Tony Freeman/PhotoEdit; p. 482: © Bill Aron/PhotoEdit; Chapter 8 Opener: © Kyko Hamada/GettyImages; p. 542: © Linda Waymire; Chapter 9 Opener: © Hal Lott/Corbis; Chapter 10pener: © Vol. 34/PhotoDisc/Getty; Chapter 11 Opener: © Felicia Martinez/PhotoEdit

Library of Congress Cataloging-in-Publication Data

Miller, Julie, 1962–
      Basic college mathematics / Julie Miller, Molly O'Neill, Nancy Hyde. —st ed.
         p.   cm.
      Includes index.
      ISBN 978-0-07-302318-2 — 0-07-302318-3 (acid-free paper)
      1. Mathematics—Textbooks. I. O'Neill, Molly, 1953–. II. Hyde, Nancy. II. Title.

QA37.3.M55   2007
510—dc22   2005054026

www.mhhe.com

Contents

Brief Contents

Nancy Hyde has been a full time faculty member of the Mathematics Department at Broward Community College for 24 years. During this time she has taught the full spectrum of courses from developmental math through differential equations. She received a bachelor of science degree in math education from Florida State University and a master's degree in math education from Florida Atlantic University. She has conducted workshops and seminars for both students and teachers on the use of technology in the classroom. In addition to this textbook, she has authored a graphing calculator supplement for College Algebra.

NANCY HYDE

"I grew up in Brevard County, Florida, with my father working at Cape Canaveral. I was always excited by mathematics and physics in relation to the space program. As I studied higher levels of mathematics I became more intrigued by its abstract nature and infinite possibilities. It is enjoyable and rewarding to convey this perspective to students while helping them to understand mathematics."

—Nancy Hyde

Preface

From the Authors

First and foremost, we would like to thank the students and colleagues who have helped us prepare this text. The content and organization are based on a wealth of resources. In addition to an accumulation of our own notes and experiences as teachers, we recognize the influence of colleagues at Daytona Beach Community College and Broward Community College, as well as fellow presenters and attendees of national mathematics conferences and meetings. Perhaps our single greatest source of inspiration has been our students, who ask good, probing questions every day and challenge us to find new and better ways to convey mathematical concepts. We gratefully acknowledge the part that each has played in the writing of this book.

In designing the framework for this text, the time we have spent with our students has proved especially invaluable. Over the years we have observed that students struggle consistently with certain topics. We have also come to know the influence of forces beyond the math, particularly motivational issues. An awareness of the various pitfalls has enabled us to tailor pedagogy and techniques that directly address students' needs and promote their success. Those techniques and pedagogy are outlined here.

Active Classroom

First, we believe students retain more of what they learn when they are actively engaged in the classroom. Consequently, as we wrote each section of text, we also wrote accompanying worksheets called **Classroom Activities** to foster accountability and to encourage classroom participation. Classroom Activities resemble the examples that students encounter in the textbook. The activities can be assigned to individual students or to pairs or groups of students. Most of the activities have been tested in the classroom with our own students. In one class in particular, the introduction of Classroom Activities transformed a group of "clock watchers" into students who literally had to be ushered out of the classroom so that the next class could come in. The activities can be found in the *Instructor's Resource Manual*, which is available through MathZone.

Conceptual Support

While we believe students must practice basic skills to be successful in any mathematics class, we also believe concepts are important. To this end, we have included **Concept Connections** questions and homework exercises that ask students to **"interpret the meaning in the context of the problem."** These questions make students stop and think, so they can process what they learn. In this way, students will learn underlying concepts. They will also form an understanding of what their answers mean in the contexts of the problems they solve.

We have also included special exercises within the section-ending homework exercises called **Number Sense and Estimation** (see page 276, for example). The goal of these exercises is to help students sharpen their reasoning skills. We want to encourage students to get into the habit of asking themselves whether the answers they produce are reasonable. If students can develop this habit, they will become stronger independent thinkers and more confident problem solvers.

Writing Style

Many students believe that reading a mathematics text is an exercise in futility. However, students who take the time to read the text and the features within the margins may cast that notion aside. In particular, the **Tips** and **Avoiding Mistakes** boxes should prove especially enlightening. They offer the types of insights and hints that are usually only revealed during classroom lecture. On the whole, students should be very comfortable with the reading level, as the language and tone are consistent with those used daily within our own developmental mathematics classes.

Real-World Applications

Another critical component of the text is the inclusion of **contemporary real-world examples and applications**. We based examples and applications on information that students encounter daily when they turn on the news, read a magazine, or surf the World Wide Web. We incorporated data for students to answer mathematical questions based on data in tables and graphs. When students encounter facts or information that is meaningful to them, they will relate better to the material and remember more of what they learn.

Study Skills

Many students in this course lack the basic study skills needed to be successful. Therefore, at the beginning of every set of homework exercises, we included a set of **Study Skills Exercises**. The exercises focus on one of nine areas: learning about the course, using the text, taking notes, completing homework assignments, test taking, time management, learning styles, preparing for a final exam, and defining **key terms**. Through completion of these exercises, students will be in a better position to pass the class and adopt techniques that will benefit them throughout their academic careers.

Language of Mathematics

Finally, for students to succeed in mathematics, they must be able to understand its language and notation. We place special emphasis on the skill of translating mathematical notation to English expressions and vice versa through **Translating Expressions Exercises**. These appear intermittently throughout the text. We also include key terms in the homework exercises and ask students to define these terms.

While we have made every effort to fine-tune this textbook to serve the needs of all students, we acknowledge that no textbook can satisfy every student's needs entirely. However, we do trust that the thoughtfully designed pedagogy and contents of this textbook offer any willing student the opportunity to achieve success, opening the door to a wider world of possibilities.

Listening to Students' and Instructors' Concerns

Our editorial staff has amassed the results of reviewer questionnaires, user diaries, focus groups, and symposia. We have consulted with a nine-member panel of basic mathematics instructors and their students on the development of this book. In addition, we have read hundreds of pages of reviews from instructors across the country. At McGraw-Hill symposia, faculty from across the United States gathered to discuss issues and trends in developmental mathematics. These efforts have involved hundreds of faculty and have explored issues such as content, readability, and even the aesthetics of page layout.

What Sets This Book Apart?

While this textbook offers complete coverage of the basic college mathematics curriculum, there are several concepts that receive special emphasis.

Order of Operations

The rules for the order of operations first appear in Section 1.7 Exponents and the Order of Operations. We offer repeated exposure to this topic, including it in the following sections:

Section 2.4   Multiplication of Fractions and Applications
Section 2.5   Division of Fractions and Applications
Section 3.1   Addition and Subtraction of Like Fractions
Section 3.3   Addition and Subtraction of Unlike Fractions
Section 3.5   Order of Operations and Applications of Fractions
Section 4.6   Order of Operations and Applications of Decimals
Section 10.5   Order of Operations and Scientific Notation

By exposing students to the order of operations repeatedly, we hope to reinforce their understanding and boost retention.

Variables

The term *variable* is defined in Chapter 1 (Section 1.5), when the formula for the area of a rectangle is presented. In this context, the student learns to substitute numerical values for variables within a formula and then performs the order of operations. This skill is interspersed throughout the early part of the text, as variables are used in applications. Formulas are then used again in detail in the geometry chapter (Chapter 8) before the formal application of the concept occurs in Chapter 11. By this practice, students may be "eased" into the concept, with early and repeated exposure increasing their familiarity and comfort.

Geometry

An introduction to geometry is covered formally in Chapter 8. However, some geometry topics such as perimeter and area are introduced early as applications of operations on whole numbers, fractions, and decimals. In this way, students have repeated exposure to geometry formulas and will be more likely to remember them in the long term.

Suggestions Welcome!

Many features of this book, and many refinements in writing, illustrations, and content, came about because of suggestions and questions from instructors and their students. We invite your comments with regard to this textbook as we work to further shape and refine its contents.

Julie Miller	Molly O'Neill	Nancy Hyde
millerj@dbcc.edu	oneillm@dbcc.edu	hyde_n@firn.edu

Acknowledgments and Reviewers

The development of this textbook would never have been possible without the creative ideas and constructive feedback offered by many reviewers. We are especially thankful to the following instructors for their valuable feedback and careful review of the manuscript.

Board of Advisers

Allan Brinkman, *Cleveland State University*
Mary Deas, *Johnson County Community College*
Vivian Dennis-Monzingo, *Eastfield College*
Linda Franko, *Cuyahoga Community College*
Shelbra Jones, *Wake Technical Community College*
Chris Kolaczewski-Ferris, *University of Akron*
Joanne Peeples, *El Paso Community College*
Jordan Neus, *Suffolk County Community College–Brentwood*
Susan Santolucito, *Delgado Community College*

Manuscript Reviewers

Rosalie Abraham, *Community College at Jacksonville*
Marwan Abu-Sawwa, *Florida Community College at Jacksonville*
Darla Aguilar, *Pima Community College*
Khadija Ahmed, *Monroe County Community College*
Anthony Aikens, *South Georgia Technical College*
Sheila Anderson, *Housatonic Community College*
Eugene J. D. Bowen, *Brookdale Community College*
Patricia Bower, *Mt. San Antonio College*
Jerome Brown, *Harford Community College*
Connie Buller, *Metropolitan Community College*
Susan Caldiero, *Cosumnes River College*
Carol Caponigro, *Lake Michigan College*
Judy Connell, *Lanier Technical College*
June Decker, *Three Rivers Community College*
Sue Duff, *Guilford Technical Community College*
Jay Faircloth, *Moultrie Technical College*
Thomas Geil, *Milwaukee Area Technical College*
Naomi Gibbs, *Pitt Community College*
David Gillette, *Chemeketa Community College*
Tania Giordani, *Columbia College*
Cynthia B. Gubitose, *Southern Connecticut State University*
Joseph Guiciardi, *CCAC–Boyce*
Richard Hobbs, *Mission College*
Joe Howe, *St. Charles Community College*
Juan Carlos Jiménez, *Springfield Technical Community College*
Jackie King, *Community College of Denver*
Lynette King, *Gadsden State Community College*

Alan Kunkle, *Ivy Tech State College*

Jeanine M. Lewis, *Aims Community College*

Jacqueline J. Lindquist, *Central Lakes College*

Sharon Louvier, *Lee College*

Diane Martling, *William Rainey Harper College*

Val Mohanakumar, *Hillsborough Community College*

Tammy Payton, *North Idaho College*

Faith Peters, *Miami-Dade College*

Mary Rack, *Johnson County Community College*

Manuel Rodriguez, *DeVry University*

Mohammad Sharifian, *Compton Community College*

Moshen Shirani, *Tennessee State University*

Mark A. Shore, *Allegany College of Maryland*

Cathy Singleton, *Glenville State College*

Dennis Stramiello, *Nassau Community College*

Sharon Testone, *Onondaga Community College*

Alexis Thurman, *County College of Morris*

Stephen Toner, *Victor Valley Community College*

Patrick Wagener, *Los Medanos College*

Claire Wladis, *Borough of Manhattan Community College*

Abbas Zadegan, *Florida Memorial College*

Special thanks go to Carrie Green for preparing the Instructor's Solutions Manual and the Student's Solutions Manual, to Yolanda Davis and Patricia Jayne for their appearance in and work on the video series, and to Lauri Semarne for her work in ensuring accuracy. Further thanks go to Ethel Wheland for preparing the instructors notes.

Finally, we are forever grateful to the many people behind the scenes at McGraw-Hill, our publishing family. To Erin Brown, our lifeline on this project, without you we'd be lost. To Liz Haefele, your passion for excellence has been a constant inspiration. To Michael Lange and David Dietz, thanks for your vision and input and for being there all these years. To Barb Owca and David Millage, we marvel at your creative ideas in a world that's forever changing. To Jeff Huettman and Amber Huebner for your awesome work with the technology and to Jodi Rhomberg for her support and keen attention to detail during production.

Most importantly, we give special thanks to all the students and instructors who use *Basic College Mathematics* in their classes.

Julie Miller   Molly O'Neill   Nancy Hyde

A COMMITMENT TO ACCURACY

You have a right to expect an accurate textbook, and McGraw-Hill invests considerable time and effort to make sure that we deliver one. Listed below are the many steps we take to make sure this happens.

OUR ACCURACY VERIFICATION PROCESS

First Round

Step 1: Numerous **college math instructors** review the manuscript and report on any errors that they may find, and the authors make these corrections in their final manuscript.

Second Round

Step 2: Once the manuscript has been typeset, the **authors** check their manuscript against the first page proofs to ensure that all illustrations, graphs, examples, exercises, solutions, and answers have been correctly laid out on the pages, and that all notation is correctly used.

Step 3: An outside, **professional mathematician** works through every example and exercise in the page proofs to verify the accuracy of the answers.

Step 4: A **proofreader** adds a triple layer of accuracy assurance in the first pages by hunting for errors, then a second, corrected round of page proofs is produced.

Third Round

Step 5: The **author team** reviews the second round of page proofs for two reasons: 1) to make certain that any previous corrections were properly made, and 2) to look for any errors they might have missed on the first round.

Step 6: A **second proofreader** is added to the project to examine the new round of page proofs to double check the author team's work and to lend a fresh, critical eye to the book before the third round of paging.

Fourth Round

Step 7: A **third proofreader** inspects the third round of page proofs to verify that all previous corrections have been properly made and that there are no new or remaining errors.

Step 8: Meanwhile, in partnership with **independent mathematicians,** the text accuracy is verified from a variety of fresh perspectives:
- The **test bank author** checks for consistency and accuracy as they prepare the computerized test item file.
- The **solutions manual author** works every single exercise and verifies their answers, reporting any errors to the publisher.
- A **consulting group of mathematicians,** who write material for the text's MathZone site, notifies the publisher of any errors they encounter in the page proofs.
- A video production company employing **expert math instructors** for the text's videos will alert the publisher of any errors they might find in the page proofs.

Final Round

Step 9: The **project manager,** who has overseen the book from the beginning, performs a **fourth proofread** of the textbook during the printing process, providing a final accuracy review.

⇒   What results is a mathematics textbook that is as accurate and error-free as is humanly possible, and our authors and publishing staff are confident that our many layers of quality assurance have produced textbooks that are the leaders of the industry for their integrity and correctness.

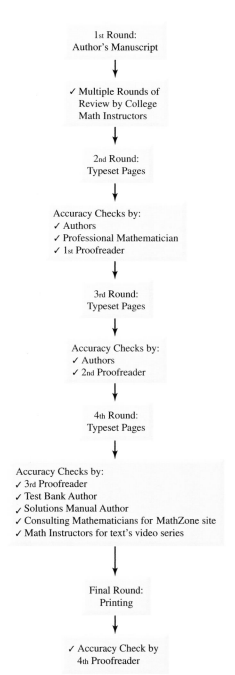

1st Round:
Author's Manuscript

↓

✓ Multiple Rounds of
Review by College
Math Instructors

↓

2nd Round:
Typeset Pages

↓

Accuracy Checks by:
✓ Authors
✓ Professional Mathematician
✓ 1st Proofreader

↓

3rd Round:
Typeset Pages

↓

Accuracy Checks by:
✓ Authors
✓ 2nd Proofreader

↓

4th Round:
Typeset Pages

↓

Accuracy Checks by:
✓ 3rd Proofreader
✓ Test Bank Author
✓ Solutions Manual Author
✓ Consulting Mathematicians for MathZone site
✓ Math Instructors for text's video series

↓

Final Round:
Printing

↓

✓ Accuracy Check by
4th Proofreader

Guided Tour

Chapter Opener

Each chapter opens with an application relating to an exercise presented in the chapter. Section titles are clearly listed for easy reference.

Decimals

4

4.1 Decimal Notation and Rounding
4.2 Addition and Subtraction of Decimals
4.3 Multiplication of Decimals
4.4 Division of Decimals
4.5 Fractions as Decimals
4.6 Order of Operations and Applications of Decimals

Chapter 4 is devoted to the study of decimal numbers. We begin with a discussion of place values and then perform addition, subtraction, multiplication, and division. The applications of decimal numbers are far-reaching. For example, in 1993, Becky and Keith Dilley were the proud parents of sextuplets (six children from a multiple birth). The children, now in their teen years, weighed just over 2 lb each at birth. See Exercise 56 of Section 4.2 to determine the total birth weight of the Dilley sextuplets.

231

---

chapter 2 preview

The exercises in this chapter preview contain concepts that have not yet been presented. These exercises are provided for students who want to compare their levels of understanding before and after studying the chapter. Alternatively, you may prefer to work these exercises when the chapter is completed and before taking the exam.

Section 2.1

1. Identify the fractions as proper or improper.

   a. $\frac{4}{5}$ b. $\frac{16}{8}$ c. $\frac{15}{15}$

2. Write $4\frac{3}{8}$ as an improper fraction.

3. Write $\frac{39}{7}$ as a mixed number.

4. Write a fraction that represents the shaded area.

   a.   b.

5. There are 8 different brands of wine on the wine list at a restaurant. Of these wines, 5 are red wines and the rest are white wines. What fraction represents the white wines?

Section 2.2

6. Is the number 1092 divisible by 2, 3, or 5? Explain your answers.

7. List all the factors of 45.

8. Write the prime factorization of 630.

Section 2.3

9. Which of the following fractions is simplified to lowest terms?

   a. $\frac{16}{25}$ b. $\frac{12}{14}$ c. $\frac{30}{15}$

Section 2.4

For Exercises 10–12, multiply and simplify the answer to lowest terms.

10. $\frac{9}{13} \times \frac{39}{27}$ 11. $\left(\frac{7}{12}\right)\left(\frac{6}{35}\right)$ 12. $12 \cdot \frac{77}{84}$

13. Find the area of the triangle.

$\frac{10}{3}$ cm

3 cm

Section 2.5

For Exercises 14–16, divide. Simplify the answer to lowest terms.

14. $\frac{64}{21} \div 8$ 15. $\frac{33}{20} \div \frac{44}{15}$ 16. $\frac{3}{0}$

17. George painted $\frac{2}{3}$ of a wall that measures 18 ft by 10 ft. How much area did he paint?

18. If a recipe requires $2\frac{1}{2}$ cups of flour for one batch of cookies, how many batches can be made from 10 cups of flour?

Section 2.6

For Exercises 19–21, multiply or divide the mixed numbers. Write the answer as a mixed number or a whole number.

19. $1\frac{3}{4} \cdot 9\frac{5}{7}$ 20. $8\frac{5}{6} \div 2\frac{1}{2}$ 21. $4\frac{1}{3} \cdot 2\frac{7}{10} \div 1\frac{4}{5}$

Chapter Preview

A Chapter Preview appears at the beginning of each chapter. It contains exercises, grouped by section. The exercises are based on topics not yet presented, offering students an opportunity to compare levels of understanding before and after studying the chapter.

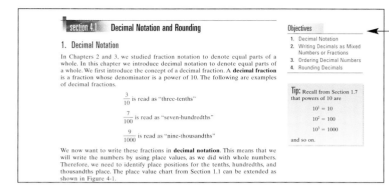

Objectives

A list of important learning objectives is provided at the beginning of each section. Each objective corresponds to a heading within the section, making it easy for students to locate topics as they study or as they work through homework exercises.

Concept Connections

Students can test their understanding of what they have read by completing the Concept Connections that appear in the margins. These questions test how well students grasp concepts. Students can check their responses by referring to answers at the bottom of the page.

Skill Practice Exercises

Every worked example is paired with a Skill Practice Exercise. These exercises appear in the margin directly beside the worked examples and offer students an immediate opportunity to work problems that mirror the examples. Students can then check their work by referring to the answers at the bottom of the page.

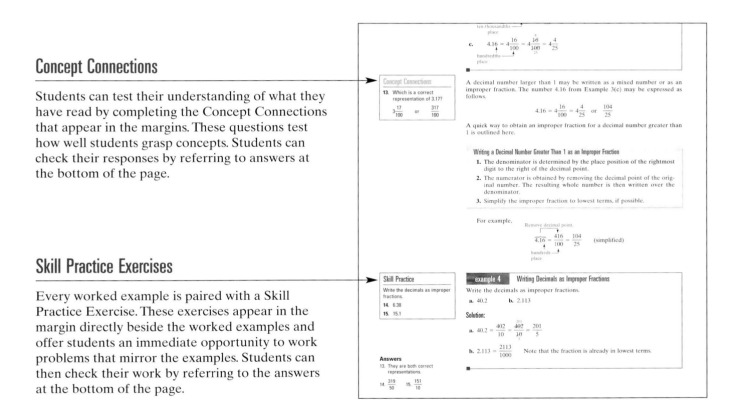

Avoiding Mistakes

Through notes labeled Avoiding Mistakes students are alerted to common errors and are shown methods to avoid them.

Tips

Tip boxes appear throughout the text and offer helpful hints and insight.

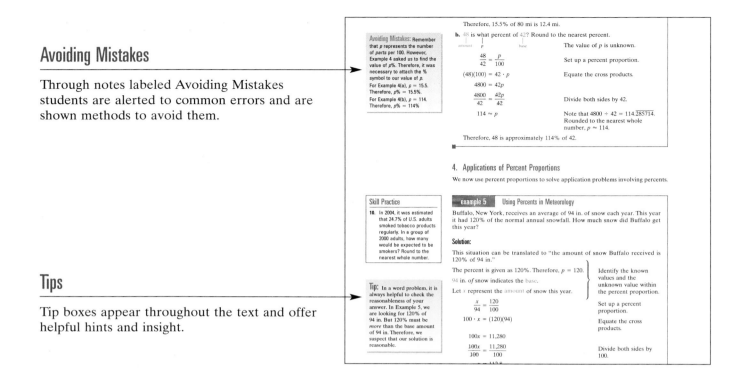

Therefore, 15.5% of 80 mi is 12.4 mi.

b. 48 is what percent of 42? Round to the nearest percent.

amount p base

The value of p is unknown.

$$\frac{48}{42} = \frac{p}{100}$$

Set up a percent proportion.

$$(48)(100) = 42 \cdot p$$

Equate the cross products.

$$4800 = 42p$$

$$\frac{4800}{42} = \frac{42p}{42}$$

Divide both sides by 42.

$$114 \approx p$$

Note that $4800 \div 42 = 114.\overline{285714}$. Rounded to the nearest whole number, $p \approx 114$.

Therefore, 48 is approximately 114% of 42.

Avoiding Mistakes: Remember that p represents the number of *parts* per 100. However, Example 4 asked us to find the value of $p\%$. Therefore, it was necessary to attach the % symbol to our value of p. For Example 4(a), $p = 15.5$. Therefore, $p\% = 15.5\%$. For Example 4(b), $p = 114$. Therefore, $p\% = 114\%$

4. Applications of Percent Proportions

We now use percent proportions to solve application problems involving percents.

Skill Practice

10. In 2004, it was estimated that 24.7% of U.S. adults smoked tobacco products regularly. In a group of 2000 adults, how many would be expected to be smokers? Round to the nearest whole number.

Tip: In a word problem, it is always helpful to check the reasonableness of your answer. In Example 5, we are looking for 120% of 94 in. But 120% must be *more* than the base amount of 94 in. Therefore, we suspect that our solution is reasonable.

example 5   Using Percents in Meteorology

Buffalo, New York, receives an average of 94 in. of snow each year. This year it had 120% of the normal annual snowfall. How much snow did Buffalo get this year?

Solution:

This situation can be translated to "the amount of snow Buffalo received is 120% of 94 in."

The percent is given as 120%. Therefore, $p = 120$.

94 in. *of* snow indicates the base.

Let x represent the amount of snow this year.

Identify the known values and the unknown value within the percent proportion.

$$\frac{x}{94} = \frac{120}{100}$$

Set up a percent proportion.

$$100 \cdot x = (120)(94)$$

Equate the cross products.

$$100x = 11,280$$

$$\frac{100x}{100} = \frac{11,280}{100}$$

Divide both sides by 100.

Worked Examples

Examples are set off in boxes and organized so that students can easily follow the solutions. Explanations appear beside each step and color-coding is used, where appropriate. For additional step-by-step instruction, students can run the "e-Professors" in MathZone. The e-Professors are based on worked examples from the text and use the solution methodologies presented in the text.

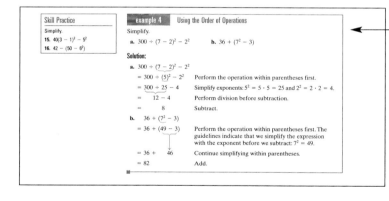

Skill Practice

Simplify.

15. $40(3 - 1)^2 - 5^2$

16. $42 - (50 - 6^2)$

example 4   Using the Order of Operations

Simplify.

a. $300 \div (7 - 2)^2 - 2^2$ **b.** $36 + (7^2 - 3)$

Solution:

a. $300 \div (7 - 2)^2 - 2^2$

$= 300 \div (5)^2 - 2^2$ Perform the operation within parentheses first.

$= 300 \div 25 - 4$ Simplify exponents: $5^2 = 5 \cdot 5 = 25$ and $2^2 = 2 \cdot 2 = 4$.

$= 12 - 4$ Perform division before subtraction.

$= 8$ Subtract.

b. $36 + (7^2 - 3)$

$= 36 + (49 - 3)$ Perform the operation within parentheses first. The guidelines indicate that we simplify the expression with the exponent before we subtract: $7^2 = 49$.

$= 36 + 46$ Continue simplifying within parentheses.

$= 82$ Add.

Midchapter Review

Midchapter Reviews are provided to help solidify the foundation of concepts learned in the beginning of a chapter before expanding to new ideas presented later in the chapter.

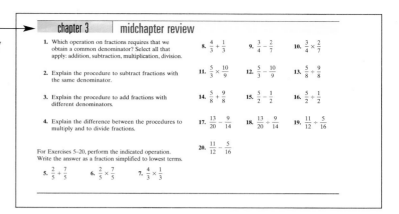

Instructor Note (AIE only)

Throughout each section of the Annotated Instructor's Edition (AIE), notes to the Instructor can be found in the margins. The notes may assist with lecture preparation in that they point out items that tend to confuse students, or lead students to err.

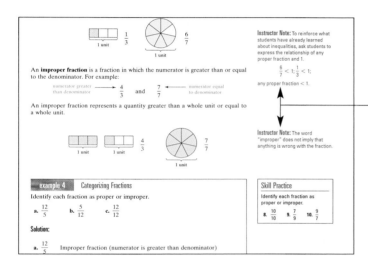

References to Classroom Activities (AIE only)

References are made to Classroom Activities at the beginning of each set of Practice Exercises in the AIE. The activities may be found in the *Instructor's Resource Manual*, which is available through MathZone, and can be used during lecture, or assigned for additional practice.

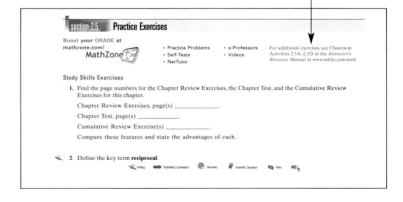

Practice Exercises

A variety of problem types appear in the section-ending Practice Exercises. Problem types are clearly labeled with either a heading or an icon for easy identification. References to MathZone are also found at the beginning of the Practice Exercises to remind students and instructors that additional help and practice problems are available. The core exercises for each section are organized by section objective. General references to examples are provided for blocks of core exercises. **Mixed Exercises** are also provided in some sections where no reference to objectives or examples is offered.

Icon Key

The following key has been prepared for easy identification of "themed" exercises appearing within the Practice Exercises.

Student Edition and AIE
Exercises Keyed to Video
Calculator Exercises

AIE only
Writing
Translating Expressions
Geometry
Number Sense and Estimation NS&E

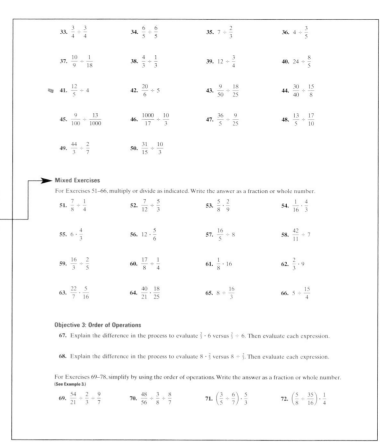

33. $\frac{3}{4} \div \frac{3}{4}$ 34. $\frac{6}{5} \div \frac{6}{5}$ 35. $7 \div \frac{2}{3}$ 36. $4 \div \frac{3}{5}$

37. $\frac{10}{9} \div \frac{1}{18}$ 38. $\frac{4}{3} \div \frac{1}{3}$ 39. $12 \div \frac{3}{4}$ 40. $24 \div \frac{8}{5}$

41. $\frac{12}{5} \div 4$ 42. $\frac{20}{6} \div 5$ 43. $\frac{9}{50} \div \frac{18}{25}$ 44. $\frac{30}{40} \div \frac{15}{8}$

45. $\frac{9}{100} \div \frac{13}{1000}$ 46. $\frac{1000}{17} \div \frac{10}{3}$ 47. $\frac{36}{5} \div \frac{9}{25}$ 48. $\frac{13}{5} \div \frac{17}{10}$

49. $\frac{44}{3} \div \frac{2}{7}$ 50. $\frac{31}{15} \div \frac{10}{3}$

Mixed Exercises

For Exercises 51–66, multiply or divide as indicated. Write the answer as a fraction or whole number.

51. $\frac{7}{8} \div \frac{1}{4}$ 52. $\frac{7}{12} \div \frac{5}{3}$ 53. $\frac{5}{8} \cdot \frac{2}{9}$ 54. $\frac{1}{16} \cdot \frac{4}{3}$

55. $6 \cdot \frac{4}{3}$ 56. $12 \cdot \frac{5}{6}$ 57. $\frac{16}{5} \div 8$ 58. $\frac{42}{11} \div 7$

59. $\frac{16}{3} \div \frac{2}{5}$ 60. $\frac{17}{8} \div \frac{1}{4}$ 61. $\frac{1}{8} \cdot 16$ 62. $\frac{2}{3} \cdot 9$

63. $\frac{22}{7} \cdot \frac{5}{16}$ 64. $\frac{40}{21} \cdot \frac{18}{25}$ 65. $8 \div \frac{16}{3}$ 66. $5 \div \frac{15}{4}$

Objective 3: Order of Operations

67. Explain the difference in the process to evaluate $\frac{2}{3} \cdot 6$ versus $\frac{2}{3} \div 6$. Then evaluate each expression.

68. Explain the difference in the process to evaluate $8 \cdot \frac{2}{3}$ versus $8 \div \frac{2}{3}$. Then evaluate each expression.

For Exercises 69–78, simplify by using the order of operations. Write the answer as a fraction or whole number. (See Example 3.)

69. $\frac{54}{21} \div \frac{2}{3} \cdot \frac{9}{7}$ 70. $\frac{48}{56} \div \frac{3}{8} \div \frac{8}{7}$ 71. $\left(\frac{3}{5} \div \frac{6}{7}\right) \cdot \frac{5}{3}$ 72. $\left(\frac{5}{8} \div \frac{35}{16}\right) \cdot \frac{1}{4}$

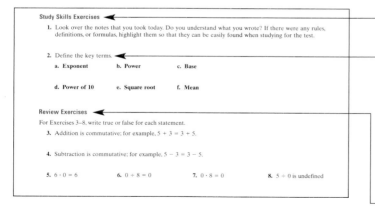

Study Skills Exercises

1. Look over the notes that you took today. Do you understand what you wrote? If there were any rules, definitions, or formulas, highlight them so that they can be easily found when studying for the test.

2. Define the key terms.
   a. Exponent   b. Power   c. Base
   d. Power of 10   e. Square root   f. Mean

Review Exercises

For Exercises 3–8, write true or false for each statement.

3. Addition is commutative; for example, $5 + 3 = 3 + 5$.

4. Subtraction is commutative; for example, $5 - 3 = 3 - 5$.

5. $6 \cdot 0 = 6$ 6. $0 \div 8 = 0$ 7. $0 \cdot 8 = 0$ 8. $5 \div 0$ is undefined

Study Skills Exercises appear at the beginning of the exercise set. They are designed to help students learn techniques to improve their study habits, including exam preparation, note taking, and time management.

In the Practice Exercises, where appropriate, students are asked to define the **Key Terms** that are presented in the section. Assigning these exercises will help students to develop and expand their mathematical vocabulary.

Review Exercises also appear at the start of the Practice Exercises. The purpose of the Review Exercises is to help students retain their knowledge of concepts previously learned.

Writing Exercises offer students an opportunity to conceptualize and communicate their understanding of arithmetic. These, along with the **Translating Expressions Exercises** enable students to strengthen their command of mathematical language and notation and improve their reading and writing skills.

Review Exercises

For Exercises 2–11, translate the English phrase into a mathematical statement and simplify.

2. 89 decreased by 66   **3.** 71 increased by 14   **4.** 16 more than 42

5. Twice 14   **6.** The difference of 93 and 79   **7.** Subtract 32 from 102

8. Divide 12 into 60   **9.** The product of 10 and 13   **10.** The total of 12, 14, and 15

11. The quotient of 24 and 6

Objective 1: Problem-Solving Strategies

12. In your own words, list the guidelines or strategy that you would use to solve an application problem.

For Exercises 13–16, write two or more key words or phrases that represent the given operation. Answers may vary.

13. Addition   **14.** Multiplication   **15.** Subtraction   **16.** Division

Objective 2: Applications Involving One Operation

17. A graphing calculator screen consists of an array of rectangular dots called *pixels*. If the screen has 96 rows of pixels and 126 pixels in each row, how many pixels are in the whole screen? (See Example 3.)

18. The floor of a rectangular room has 62 rows of tile with 38 tiles in each row. How many total tiles are there?

51. If you wanted to line the outside of a garden with a decorative border, would you need to know the area of the garden or the perimeter of the garden?

52. If you wanted to know how much sod to lay down within a rectangular backyard, would you need to know the area of the yard or the perimeter of the yard?

53. A homeowner wants to fence her rectangular backyard. The yard is 75 ft by 90 ft. If fencing costs $5 per foot, how much will it cost to fence the yard?

54. Alexis wants to buy molding for a room that is 12 ft by 11 ft. No molding is needed for the doorway which measures 3 ft. See the figure. If molding costs $2 per foot, how much money will it cost? (See Example 7.)

11 ft
3 ft
12 ft   12 ft
11 ft

55. What is the cost to carpet the room whose dimensions are shown in the figure? Assume that carpeting costs $34 per square yard and that there is no waste.

6 yd
5 yd

56. What is the cost to tile the room whose dimensions are shown in the figure? Assume that tile costs $3 per square foot.

12 ft
20 ft

Geometry Exercises appear throughout the Practice Exercises and encourage students to review and apply geometry concepts.

Calculator Exercises signify situations where a calculator would provide assistance for time-consuming calculations. These exercises were carefully designed to demonstrate the types of situations where a calculator is a handy tool rather than a "crutch."

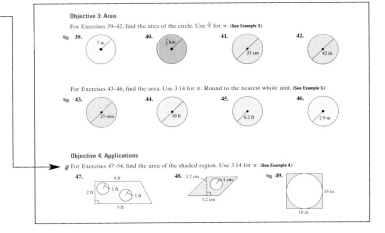

Objective 3: Area

For Exercises 39–42, find the area of the circle. Use $\frac{22}{7}$ for π. (See Example 3.)

39. 7 m   **40.** $\frac{2}{3}$ km   **41.** 21 cm   **42.** 42 in.

For Exercises 43–46, find the area. Use 3.14 for π. Round to the nearest whole unit. (See Example 3.)

43. 25 mm   **44.** 10 ft   **45.** 6.2 ft   **46.** 2.9 m

Objective 4: Applications

For Exercises 47–54, find the area of the shaded region. Use 3.14 for π. (See Example 4.)

47.   4 ft   **48.** 3.2 cm   1 cm   **49.**
2 ft 1 ft 1 ft
5 ft   5.2 cm   16 in.
16 in.

Exercises Keyed to Video are labeled with an icon to help students and instructors identify those exercises for which accompanying video instruction is available.

57. Ling has three jobs. He works for a lawn maintenance service 4 days a week. He also tutors math and works as a waiter on weekends. His hourly wage and the number of hours for each job are given for a 1-week period. How much money did Ling earn for the week?

	Hourly Wage	Number of Hours
Tutor	$30/hr	4
Waiter	10/hr	16
Lawn maintenance	8/hr	30

58. An electrician, a plumber, a mason, and a carpenter work at a certain construction site. The hourly wage and the number of hours each person worked are summarized in the table. What was the total amount paid for all four workers?

	Hourly Wage	Number of Hours
Electrician	$36/hr	18
Plumber	28/hr	15
Mason	26/hr	24
Carpenter	22/hr	48

Number Sense and Estimation exercises test students' ability to reason with numbers. The exercises often ask students to determine whether the answers they produce make sense in light of the problems they are asked to solve.

Objective 4: Applications of Decimal Division

When multiplying or dividing decimals, it is important to place the decimal point correctly. For Exercises 75–78, determine whether you think the number is reasonable or unreasonable. If the number is unreasonable, move the decimal point to a position that makes more sense.

75. Steve computed the gas mileage for his Honda Civic to be 3.2 miles per gallon.

76. The sale price of a new refrigerator is $96.0.

77. Mickey makes $8.50 per hour. He estimates his weekly paycheck to be $3400.

78. Jason works in a legal office. He computes the average annual income for the attorneys in his office to be $1400 per year.

For Exercises 79–84, solve the application. Check to see if your answers are reasonable.

79. A membership at a health club costs $560 per year. The club has a payment plan in which a member can pay $50 down and the rest in 12 equal payments. How much is each payment? **(See Example 9.)**

Applications based on real-world facts and figures motivate students and enable them to hone their problem-solving skills.

80. Brooke owes $39,628.68 on the mortgage for her house. If her monthly payment is $695.24, how many months does she still need to pay? How many years is this?

81. It is reported that on average 42,000 tennis balls are used and 650 matches are played at the Wimbledon tennis tournament each year. On average, how many tennis balls are used per match? Round to the nearest whole unit.

82. A package of dental floss contains 100 yd of floss. If Patty uses floss once a day and it lasts for 230 days, approximately how long is each piece that she uses? Round the answer to the nearest tenth of a yard.

83. In baseball the batting average is found by dividing the number of hits by the number of times a batter was at bat. Babe Ruth was at bat 8399 times and had 2873 hits. What was his batting average? Round to the thousandths place. **(See Example 10.)**

84. Ty Cobb was at bat 11,434 times and had 4189 hits, giving him the all time best batting average. Find his average. Round to the thousandths place. (Refer to Exercise 83.)

Expanding Your Skills, found near the end of most Practice Exercises, challenge students' knowledge of the concepts presented.

Expanding Your Skills

Sometimes an expression will have parentheses within parentheses. This is called *nested parentheses*. Often different shapes such as ( ), [ ], or { } are used to make it easier to match up the pairs of parentheses, for example,

$$\{300 - 4[4 + (5 + 2)^2] + 8\} - 31$$

It is important to note that the symbols ( ), [ ], or { } all represent parentheses and are used for grouping. When nested parentheses occur, simplify the innermost set first. Then work your way out. For example, simplify

$$\{300 - 4[4 + (5 + 2)^2] + 8\} - 31$$

The solution is

$$\{300 - 4[4 + (5 + 2)^2] + 8\} - 31$$
$$= \{300 - 4[4 + (7)^2] + 8\} - 31 \qquad \text{Simplify within the innermost parentheses first ( ).}$$
$$= \{300 - 4[4 + 49] + 8\} - 31 \qquad \text{Simplify the exponent.}$$
$$= \{300 - 4[53] + 8\} - 31 \qquad \text{Simplify within the next innermost parentheses [ ].}$$
$$= \{300 - 212 + 8\} - 31 \qquad \text{Multiply before adding.}$$
$$= \{88 + 8\} - 31 \qquad \text{Subtract and add in order from left to right within the parentheses { }.}$$
$$= 96 - 31 \qquad \text{Simplify within the parentheses { }.}$$
$$= 65 \qquad \text{Simplify.}$$

For Exercises 97–100, simplify the expressions with nested parentheses.

97. $3[4 + (6 - 3)^2] - 15$

98. $2[5(4 - 1) + 3] \div 6$

99. $5\{21 - [3^2 - (4 - 2)]\}$

100. $4\{18 - [(10 - 8) + 2^3]\}$

Optional Calculator Connections are located at the end of the Practice Exercises and appear intermittently. They can be implemented at the instructor's discretion depending on the amount of emphasis placed on the calculator in the course. The Calculator Connections display keystrokes and include a set of exercises that provide an opportunity for students to apply the skill introduced.

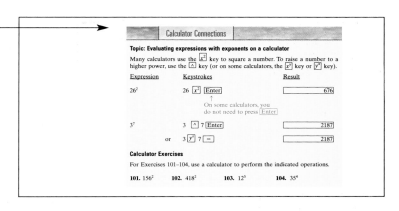

End-of-Chapter Summary and Exercises

The **Summary**, located at the end of each chapter, outlines key concepts for each section and illustrates those concepts with examples.

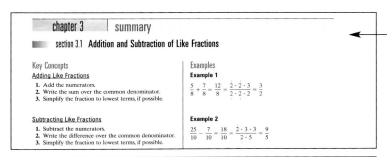

Following the Summary is a set of **Review Exercises** that are organized by section. A **Chapter Test** appears after each set of Review Exercises. Chapters 2–11 also include **Cumulative Reviews** that follow the Chapter Tests. These end-of-chapter materials provide students with ample opportunity to prepare for quizzes or exams.

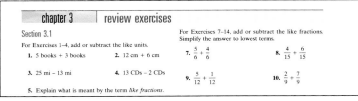

SUPPLEMENTS

For the Instructor

Instructor's Resource Manual

The *Instructor's Resource Manual* (IRM), written by the authors, is a printable electronic supplement available through MathZone. The IRM includes discovery-based classroom activities, worksheets for drill-and-practice, materials for a student portfolio, and some tips for implementing successful cooperative learning. Numerous classroom activities are available for each section of text and can be used as a complement to lecture or can be assigned for work outside of class. The activities are designed for group or individual work and take about 5–10 minutes each. With increasing demands on faculty schedules, these ready-made lessons offer a convenient means for both full-time and adjunct faculty to promote active learning in the classroom.

 www.mathzone.com

McGraw-Hill's **MathZone** is a complete, online tutorial and course management system for mathematics and statistics, designed for greater ease of use than any other system available. Available with selected McGraw-Hill texts, the system allows instructors to **create and share courses and assignments** with colleagues and adjuncts with only a few clicks of the mouse. All assignments, questions, e-Professors, online tutoring, and video lectures are directly tied to **text-specific** materials.

MathZone courses are customized to your textbook, but you can **edit** questions and algorithms, **import** your own content, and **create** announcements and due dates for assignments.

MathZone has **automatic grading** and reporting of easy-to-assign algorithmically generated homework, quizzing, and testing. All student activity within MathZone is automatically recorded and available to you through a **fully integrated grade book** that can be downloaded to Excel.

MathZone offers:
- **Practice exercises** based on the text and generated in an unlimited number for as much practice as needed to master any topic you study.
- **Videos** of classroom instructors giving lectures and showing you how to solve exercises from the text.
- **e-Professors** to take you through animated, step-by-step instructors (delivered via on-screen text and synchronized audio) for solving problems in the book, allowing you to digest each step at your own pace.
- **NetTutor,** which offers live, personalized tutoring via the Internet.

Instructor's Testing and Resource CD

This cross-platform CD-ROM provides a wealth of resources for the instructor. Among the supplements featured on the CD-ROM is a **computerized test bank** utilizing Brownstone Diploma ® algorithm-based testing software to create customized exams quickly. This user-friendly program enables instructors to search for questions by topic, format, or difficulty level; to edit existing questions or to add new ones; and to scramble questions and answer keys for multiple versions of a single test. Hundreds of text-specific open-ended and multiple-choice questions are included in the question bank. Sample chapter tests are also provided.

ALEKS (**A**ssessment and **LE**arning in **K**nowledge **S**paces) is an artificial intelligence-based system for mathematics learning, available over the web 24/7. Using unique adaptive questioning, ALEKS accurately assesses what topics each student knows and then determines exactly what each student is ready to learn next. ALEKS interacts with the students much as a skilled human tutor would, moving between explanation and practice as needed, correcting and analyzing errors, defining terms, changing topics on request, and helping them master the course content more quickly and easily. Moreover, the new ALEKS 3.0 now links to text-specific videos, multimedia tutorials, and textbook pages in PDF format. ALEKS also offers a robust classroom management system that allows instructors to monitor and direct student progress toward mastery of curricular goals. See www.highed.aleks.com.

Miller/O'Neill/Hyde Video Lectures on Digital Video Disk (DVD)

In the videos, qualified instructors work through selected problems from the textbook, following the solution methodology employed in the text. The video series is available on DVD or online as an assignable element of MathZone (see next page). The DVDs are closed-captioned for the hearing impaired, subtitled in Spanish, and meet the Americans with Disabilities Act Standards for Accessible Design. Instructors may use them as resources in a learning center, for online courses, and/or to provide extra help for students who require extra practice.

Annotated Instructor's Edition

In the *Annotated Instructor's Edition* (*AIE*), **answers to all exercises and tests appear adjacent to each exercise**, in a color used *only* for annotations. The *AIE* also contains **Instructor Notes** that appear in the margin. The notes may assist with lecture preparation. Also found in the *AIE* are icons within the Practice Exercises that serve to guide instructors in their preparation of homework assignments and lessons.

Instructor's Solutions Manual

The *Instructor's Solutions Manual* provides comprehensive, worked-out solutions to all exercises in the Chapter Previews; the Practice Exercises; the Midchapter Reviews; the end-of-chapter Review Exercises; the Chapter Tests; and the Cumulative Review Exercises.

For the Student

 **www.mathzone.com**

McGraw-Hill's MathZone is a powerful web-based tutorial for homework, quizzing, testing, and multimedia instruction. Also available in CD-ROM format, MathZone offers:

Practice exercises based on the text and generated in an unlimited quantity for as much practice as needed to master any objective

Video clips of classroom instructors showing how to solve exercises from the text, step-by-step

e-Professor animations that take the student through step-by-step instructions, delivered on-screen and narrated by a teacher on audio, for solving exercises from the textbook; the user controls the pace of the explanations and can review as needed

NetTutor, which offers personalized instruction by live tutors familiar with the textbook's objectives and problem-solving methods

Every assignment, exercise, video lecture, and e-Professor is derived from the textbook.

Student's Solutions Manual

The *Student's Solutions Manual* provides comprehensive, worked-out solutions to the odd-numbered exercises in the Chapter Previews, the Practice Exercise sets; the Midchapter Reviews, the end-of-chapter Review Exercises, the Chapter Tests, and the Cumulative Review Exercises.

Video Lectures on Digital Video Disk (DVD)

The video series is based on exercises from the textbook. Each presenter works through selected problems, following the solution methodology employed in the text. The video series is available on DVD or online as part of MathZone. The DVDs are closed-captioned for the hearing impaired, subtitled in Spanish, and meet the Americans with Disabilities Act Standards for Accessible Design.

NetTutor

Available through MathZone, NetTutor is a revolutionary system that enables students to interact with a live tutor over the Web. NetTutor's Web-based, graphical chat capabilities enable students and tutors to use mathematical notation and even to draw graphs as they work through a problem together. Students can also submit questions and receive answers, browse previously answered questions, and view previous sessions. Tutors are familiar with the textbook's objectives and problem-solving styles.

Whole Numbers

1

Chapter 1 begins with adding, subtracting, multiplying, and dividing whole numbers. We also include rounding, estimating, and applying whole numbers in a variety of real-world situations. For example, in Exercise 43 in Section 1.4, the total sales for five top-selling candy bars is given. After rounding each value to the nearest million, we find that consumers spent over $150 million on these items.

Brand	Sales ($)
M&Ms	97,404,576
Hershey's Milk Chocolate	81,296,784
Reese's Peanut Butter Cups	54,391,268
Snickers	53,695,428
KitKat	38,168,580

chapter 1 | preview

The exercises in this chapter preview contain concepts that have not yet been presented. These exercises are provided for students who want to compare their levels of understanding before and after studying the chapter. Alternatively, you may prefer to work these exercises when the chapter is completed and before taking the exam.

Section 1.1

1. For the number 6,873,129 identify the place value of the underlined digit.

2. Write the number in standard form: five million, two hundred three thousand, fifty-one

3. Write the following inequality in words:
$130 < 244$

Section 1.2

For Exercises 4–5, add.

4. $73 + 41$

5. $71 + 4 + 81 + 106$

Section 1.3

For Exercises 6–7, subtract. Check by using addition.

6. $284 - 171$

7. $\begin{array}{r} 1001 \\ -235 \\ \hline \end{array}$

Section 1.4

8. Approximate the perimeter of the triangle by first rounding the numbers to the hundreds place.

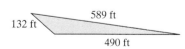

132 ft   589 ft   490 ft

For Exercises 9–10, estimate the sum or difference by first rounding the numbers to the tens place.

9. $682 + 249$

10. $768 - 241$

Sections 1.5 and 1.6

For Exercises 11–16, multiply or divide as indicated.

11. $31 \cdot 8$

12. $12\overline{)1032}$

13. $737 \div 7$

14. $\begin{array}{r} 409 \\ \times\ 228 \\ \hline \end{array}$

15. $\dfrac{0}{61}$

16. $0\overline{)341}$

17. Find the area of the rectangle.

28 m
5 m

Section 1.7

18. Write the repeated multiplication in exponential notation. Do not evaluate.

a. $7 \cdot 7 \cdot 7 \cdot 7 \cdot 7 \cdot 7$

b. $3 \cdot 3 \cdot 3 \cdot 3 \cdot 10 \cdot 10 \cdot 10$

19. Simplify the expression, using the order of operations: $14 - 2(20 \div 5)$

20. Herman collects snow globes. He purchased 4 in the past year and paid $25, $30, $19, and $22. What is the average price per globe?

Section 1.8

21. When migrating, a hawk travels a distance of 9445 mi while a swallow travels 9258 mi. Determine how much farther the hawk travels.

22. Liz is taking a natural herb in capsule form. She purchased 4 bottles containing 30 capsules each. The directions state that she can take either 2 or 3 capsules per day.

a. How many days will the capsules last if Liz takes 3 per day?

b. How many days will the capsules last if she takes only 2 per day?

c. How many more days can she take the herb if she takes only 2 per day?

section 1.1   Introduction to Whole Numbers

1. Place Value

Numbers provide the foundation that is used in mathematics. We begin this chapter by discussing how numbers are represented and named. All numbers in our numbering system are composed from the **digits** 0, 1, 2, 3, 4, 5, 6, 7, 8, and 9. In mathematics, the numbers 0, 1, 2, 3, 4, 5, 6, 7, 8, 9, 10, 11, 12, ... are called the *whole numbers*. (The three dots are called *ellipses* and indicate that the list goes on indefinitely.)

For large numbers, commas are used to separate digits into groups of three called **periods**. For example, the number of live births in the United States in a recent year was 4,058,614 (Source: *The World Almanac*). Numbers written in this way are said to be in **standard form**. The position of each digit within a number determines the place value of the digit. To interpret the number of births in the United States, refer to the place value chart (Figure 1-1).

Objectives

1. Place Value
2. Standard Notation and Expanded Notation
3. Writing Numbers in Words
4. The Number Line and Order

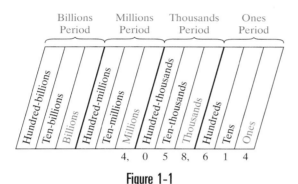

Figure 1-1

Concept Connections

1. Explain the difference between the two 3s in the number 303.

The digit 5 in the number 4,058,614 represents 5 ten-thousands because it is in the ten-thousands place. The digit 4 at the left represents 4 millions, whereas the digit 4 on the right represents 4 ones.

example 1   Determining Place Value

Determine the place value of the digit 2 in each number.

a. 417,216,900   **b.** 724   **c.** 502,000,700

Solution:

a. 417,216,900   hundred-thousands

b. 724   tens

c. 502,000,700   millions

Skill Practice

Determine the place value of the digit 4 in each number.

2. 547,098,632

3. 1,659,984,036

example 2   Determining Place Value

Mount Everest, the highest mountain on earth, is 29,035 feet (ft) tall. Give the place value for each digit in this number.

Answers

1. First 3 (on the left) represents 3 hundreds, while the second 3 (on the right) represents 3 ones.
2. Ten-millions
3. Thousands

Solution:

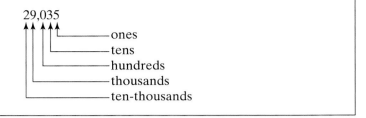

2. Standard Notation and Expanded Notation

A number can also be written in an expanded form by writing each digit with its place value units. For example, the number 287 can be written as

$$287 = 2 \text{ hundreds} + 8 \text{ tens} + 7 \text{ ones}$$

This is called **expanded form.**

example 3   Converting Standard Form to Expanded Form

Convert to expanded form.

a. 4,672

b. 257,016

Solution:

a. 4,672   4 thousands + 6 hundreds + 7 tens + 2 ones

b. 257,016   2 hundred-thousands + 5 ten-thousands + 7 thousands + 1 ten + 6 ones

example 4   Converting Expanded Form to Standard Form

Convert to standard form.

a. 2 hundreds + 5 tens + 9 ones

b. 1 thousand + 2 tens + 5 ones

Solution:

a. 2 hundreds + 5 tens + 9 ones = 259

b. Each place position from the thousands place to the ones place must contain a digit. In this problem, there is no reference to the hundreds place digit. Therefore, we assume 0 hundreds. Thus,

$$1 \text{ thousand} + 0 \text{ hundreds} + 2 \text{ tens} + 5 \text{ ones} = 1,025$$

3. Writing Numbers in Words

The word names of some two-digit numbers appear with a hyphen while others do not. For example:

Number	Number Name
12	twelve
68	sixty-eight
40	forty
42	forty-two

Concept Connections

9. Write the name of a two-digit number that is not hyphenated. Write the name of a two-digit number that is hyphenated.

To write a three-digit or larger number, begin at the leftmost group of digits. The number named in that group is followed by the period name, followed by a comma. Then the next period is named, and so on.

---

example 5   Writing a Number in Words

Write the number 621,417,325 in words.

Solution:

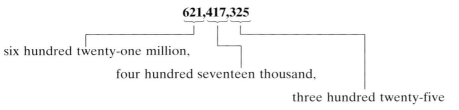

Skill Practice

10. Write the number 1,450,327,214 in words.

Notice from Example 5 that when naming numbers, the name of the ones period is not attached to the last group of digits. Also note that for whole numbers, the word *and* should not appear in word names. For example, the number 405 should be written as four hundred five.

---

example 6   Writing a Number in Words

Write the number 1,206,427,200 in words.

Solution:

$$1,206,427,200$$

One billion, two hundred six million, four hundred twenty-seven thousand, two hundred

Skill Practice

11. Write the number 401,207 in words.

Answers

9. For example: fourteen
   For example: fifty-six
10. One billion, four hundred fifty million, three hundred twenty-seven thousand, two hundred fourteen
11. Four hundred one thousand, two hundred seven

---

example 7   Writing a Number in Standard Form

Write the number in standard form.

Six million, forty-six thousand, nine hundred three

Solution:

6,046,903

We have seen several examples of writing a number in standard form, in expanded form, and in words. Standard form is the most concise representation. Also note that when we write a four-digit number in standard form, the comma is often omitted. For example, the number 4,389 is often written as 4389.

4. The Number Line and Order

Whole numbers can be visualized as equally spaced points on a line called a *number line* (Figure 1-2).

Figure 1-2

The whole numbers begin at 0 and are ordered from left to right by increasing value.

A number is graphed on a number line by placing a dot at the corresponding point. For any two numbers graphed on a number line, the number to the left is less than the number to the right. Similarly, a number to the right is greater than the number to the left. In mathematics, the symbol $<$ is used to denote "is less than," and the symbol $>$ means "is greater than." Therefore,

$3 < 5$ means   3 is less than 5
$5 > 3$ means   5 is greater than 3

---

example 8   Determining Order Between Two Numbers

Fill in the blank with the symbol $<$ or $>$.

a. 4 $\boxed{\phantom{<}}$ 10   **b.** 7 $\boxed{\phantom{<}}$ 0   **c.** 82 $\boxed{\phantom{<}}$ 30

Solution:

a. 4 $\boxed{<}$ 10

b. 7 $\boxed{>}$ 0

c. 82 $\boxed{>}$ 30

To visualize the numbers 82 and 30 on the number line, it may be necessary to use a different scale. Rather than setting equally spaced marks in units of 1, we can use units of 10. The number 82 must be somewhere between 80 and 90 on the number line.

---

section 1.1   Practice Exercises

Boost *your* **GRADE** at
mathzone.com!

MathZone+x

- Practice Problems
- Self-Tests
- NetTutor

- e-Professors
- Videos

Study Skills Exercises

In this text we provide skills for you to enhance your learning experience. Each set of practice exercises begins with an activity that focuses on one of eight areas: learning about your course, using your text, taking notes, doing homework, taking an exam (test and math anxiety), managing your time, recognizing your learning style, and studying for the final exam.

Each activity requires only a few minutes and will help you to pass this class and become a better math student. Many of these skills can be carried over to other disciplines and help you to become a model college student.

1. To begin, write down the following information.

   a. Instructor's name

   c. Instructor's telephone number

   e. Instructor's office hours

   g. The room number in which the class meets

   b. Instructor's office number

   d. Instructor's email address

   f. Days of the week that the class meets

   h. Is there a lab requirement for this course? If so, how often and what is the location of the lab?

2. Define the key terms.

   a. Digit   **b. Standard form**   **c. Periods**   **d. Expanded form**

Objective 1: Place Value

3. Name the place values for each of the digits in the number 8,213,457. **(See page 3, Figure 1-1.)**

4. Name the place values for each of the digits in the number 103,596.

For Exercises 5–24, determine the place value for each underlined digit. **(See Example 1.)**

   5. 3$\underline{2}$1

   6. 6$\underline{8}$9

   7. $\underline{2}$14

   8. 73$\underline{8}$

   9. 8,$\underline{7}$10

   10. 2,$\underline{2}$93

   11. $\underline{1}$,430

   12. $\underline{3}$,101

   13. $\underline{4}$52,723

   14. $\underline{6}$55,878

   15. $\underline{1}$,023,676,207

   16. $\underline{3}$,111,901,211

   17. 2$\underline{2}$,422

   18. 5$\underline{8}$,106

   19. 51,0$\underline{3}$3,201

   20. 93,9$\underline{7}$1,224

21. The number of U.S. travelers abroad in a recent year was $\underline{1}$0,677,881. **(See Example 2.)**

22. The area of Lake Superior is 3_1_,820 mi².

23. For a recent year, the total number of U.S. $1 bills in circulation was _7_,653,468,440.

24. For a certain flight, the cruising altitude of a commercial jet is 31,000 ft.

Objective 2: Standard Notation and Expanded Notation

For Exercises 25–34, convert the numbers to expanded form. **(See Example 3.)**

25. 58   **26.** 71   **27.** 539   **28.** 382

29. 503   **30.** 809   **31.** 10,241   **32.** 20,873

33. 2,006,004   **34.** 5,001,009

For Exercises 35–42, convert the numbers to standard form. **(See Example 4.)**

35. 5 hundreds + 2 tens + 4 ones   **36.** 3 hundreds + 1 ten + 8 ones

37. 1 hundred + 5 tens   **38.** 6 hundreds + 2 tens

39. 1 thousand + 9 hundreds + 6 ones   **40.** 4 thousands + 2 hundreds + 1 one

41. 8 ten-thousands + 5 thousands + 7 ones   **42.** 2 ten-thousands + 6 thousands + 2 ones

43. Write your favorite three-digit number in both standard form and expanded form.

44. Write your favorite four-digit number in both standard form and expanded form.

45. Name the first four periods of a number (from right to left).

46. Name the first four place values of a number (from right to left).

Objective 3: Writing Numbers in Words

For Exercises 47–56, write the number in words. **(See Examples 5 and 6.)**

47. 241   **48.** 327   **49.** 603   **50.** 108

51. The Shuowen jiezi dictionary, an ancient Chinese dictionary that dates back to the year 100, contained 9,535 characters. Write the number 9,535 in words.

52. Researchers calculate that about 590,712 stone blocks were used to construct the Great Pyramid. Write the number 590,712 in words.

53. 31,530   **54.** 52,160   **55.** 100,234   **56.** 400,199

57. Mt. McKinley in Alaska is 20,320 ft high. Write the number 20,320 in words.

58. There are 1,800 seats in the Regal Champlain Theater in Plattsburgh, New York. Write the number 1,800 in words.

59. Interstate I-75 is 1,377 miles (mi) long. Write the number 1,377 in words.

60. In the United States, there are approximately 60,000,000 cats living in households. Write the number 60,000,000 in words.

For Exercises 61–66, convert the number to standard form. **(See Example 7.)**

61. Six thousand, five

62. Four thousand, four

63. Six hundred seventy-two thousand

64. Two hundred forty-eight thousand

65. One million, four hundred eighty-four thousand, two hundred fifty

66. Two million, six hundred forty-seven thousand, five hundred twenty

Objective 4: The Number Line and Order

For Exercises 67–68, graph the numbers on the number line.

67. a. 6   **b.** 13   **c.** 8   **d.** 1

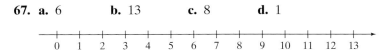

68. a. 5   **b.** 3   **c.** 11   **d.** 9

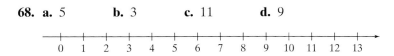

69. On a number line, what number is 4 units to the right of 6?

70. On a number line, what number is 8 units to the left of 11?

71. On a number line, what number is 3 units to the left of 7?

72. On a number line, what number is 5 units to the right of 0?

For Exercises 73–76, translate the inequality to words.

73. $8 > 2$ **74.** $6 < 11$ **75.** $3 < 7$ **76.** $14 > 12$

For Exercises 77–88, insert the appropriate inequality. Choose from $<$ or $>$. **(See Example 8.)**

77. $6 \,\square\, 11$ **78.** $14 \,\square\, 13$ **79.** $21 \,\square\, 18$ **80.** $5 \,\square\, 7$

81. $3 \,\square\, 7$ **82.** $14 \,\square\, 24$ **83.** $95 \,\square\, 89$ **84.** $28 \,\square\, 30$

85. $0 \,\square\, 3$ **86.** $8 \,\square\, 0$ **87.** $90 \,\square\, 91$ **88.** $48 \,\square\, 47$

Expanding Your Skills

89. Answer true or false. The number 12 is a digit.

90. Answer true or false. The number 26 is a digit.

91. What is the greatest two-digit number?

92. What is the greatest three-digit number?

93. What is the greatest whole number?

94. What is the least whole number?

95. How many zeros are there in the number ten million?

96. How many zeros are there in the number one hundred billion?

97. What is the greatest three-digit number that can be formed from the digits 6, 9, and 4? Use each digit only once.

98. What is the greatest three-digit number that can be formed from the digits 0, 4, and 8? Use each digit only once.

section 1.2   Addition of Whole Numbers

1. Addition of Whole Numbers Using the Number Line

We use addition of whole numbers to represent an increase in quantity. For example, suppose Jonas types 5 pages of a report before lunch. Later in the afternoon he types 3 more pages. The total number of pages that he typed is found by adding 5 and 3.

$$5 \text{ pages} + 3 \text{ pages} = 8 \text{ pages}$$

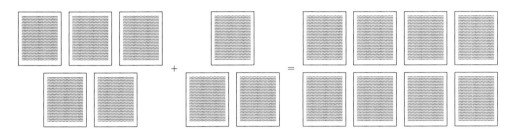

The result of an addition problem is called the **sum**, and the numbers being added are called **addends**. Thus,

$$5 + 3 = 8$$
addends   sum

The number line is a useful tool to visualize the operation of addition. To add 5 and 3 on a number line, begin at 0 and move 5 units to the right. Then move an additional 3 units to the right. The final location indicates the sum.

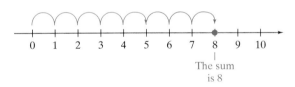

The sum
is 8

You can use a number line to find the sum of any pair of digits. The sums for all possible pairs of one-digit numbers should be memorized (see Exercise 9). Memorizing these basic addition facts will make it easier for you to add larger numbers.

2. Addition of Whole Numbers

To add whole numbers, line up the numbers vertically by place value. Then add the digits in the corresponding place positions.

example 1	Adding Whole Numbers

Add.

$$24 + 61$$

Objectives

1. Addition of Whole Numbers Using the Number Line
2. Addition of Whole Numbers
3. Properties of Addition
4. Translations and Applications Involving Addition
5. Perimeter

Concept Connections

1. Identify the addends and the sum.

    $$3 + 7 + 12 = 22$$

Skill Practice

2. Add. $\begin{array}{r} 47 \\ + 32 \\ \hline \end{array}$

Answers

1. Addends: 3, 7, and 12; sum: 22
2. 79

Solution:

$$24 = 2 \text{ tens} + 4 \text{ ones}$$
$$\underline{+\ 61 = 6 \text{ tens} + 1 \text{ one}}$$
$$85 = 8 \text{ tens} + 5 \text{ ones}$$

example 2   Adding Whole Numbers

Add.

$$261 + 28$$

Solution:

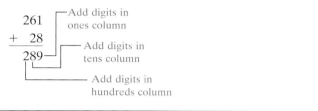

Sometimes when adding numbers, the sum of the digits in a given place position is greater than 9. If this occurs, we must do what is called *carrying* or *regrouping*. Example 3 illustrates this process.

example 3   Adding Whole Numbers with Carrying

Add.

$$35 + 48$$

Solution:

$$35 = 3 \text{ tens} + \ 5 \text{ ones}$$
$$\underline{+\ 48 = 4 \text{ tens} + \ 8 \text{ ones}}$$
$$7 \text{ tens} + 13 \text{ ones} \longleftarrow$$

The sum of the digits in the ones place exceeds 9. But 13 ones is the same as 1 ten and 3 ones. We can *carry* 1 ten to the tens column while leaving the 3 ones in the ones column. Notice that we placed the carried digit above the tens column.

$$\overset{1 \text{ ten}}{35} = 3 \text{ tens} + 5 \text{ ones} \longleftarrow$$
$$\underline{+\ 48 = 4 \text{ tens} + 8 \text{ ones}}$$
$$83 = 8 \text{ tens} + 3 \text{ ones}$$

The sum is 83.

example 4   Adding Whole Numbers With Carrying

Add.

$$458 + 67$$

Solution:

$$\overset{\overset{1}{}}{458}$$
$$+\ \ 67$$
$$\overline{5}$$

Add the digits in the ones column: $8 + 7 = 15$. Write 5 in the ones column, and carry the 1 to the tens column.

$$\overset{\overset{1\ 1}{}}{458}$$
$$+\ \ 67$$
$$\overline{25}$$

Add the digits in the tens column (including the carry): $1 + 5 + 6 = 12$. Write the 2 in the tens column, and carry the 1 to the hundreds column.

$$\overset{\overset{1\ 1}{}}{458}$$
$$+\ \ 67$$
$$\overline{525}$$

Add the digits in the hundreds column.

The sum is 525.

■—————————————————————

Addition of numbers may include more than two addends.

example 5   Adding Whole Numbers

Add.

$$21{,}076 + 84{,}158 + 2419$$

Solution:

$$\overset{\overset{1\ \ \ 1\ 2}{}}{21{,}076}$$
$$84{,}158$$
$$+\ \ \ 2{,}419$$
$$\overline{107{,}653}$$

In this example, the sum of the digits in the ones column is 23. Therefore, we write the 3 and carry the 2.

■—————————————————————

3. Properties of Addition

We present three properties of addition that you may already have discovered.

Addition Property of 0

The sum of any number and 0 is that number.

Examples: $5 + 0 = 5$

                              $0 + 2 = 2$

Commutative Property of Addition

Changing the order of two addends does not affect the sum.

Example: $5 + 7$ is equivalent to $7 + 5$

Skill Practice

6. Add.

   657
   + 89

7. Add.

   $3087 + 25{,}686$

Skill Practice

8. Add.

   57,296
    4,089
   + 9,762

Answers

6. 746
7. 28,773
8. 71,147

In mathematics we use parentheses ( ) as grouping symbols. To add more than two numbers, we can group them and then add. For example:

$(2 + 3) + 8$ Parentheses indicate that $2 + 3$ is added first, and the result is added to 8.

$= 5 + 8$

$= 13$

$2 + (3 + 8)$ Parentheses indicate that $3 + 8$ is added first, and the result is added to 2.

$= 2 + 11$

$= 13$

Associative Property of Addition

The manner in which addends are grouped does not affect the sum.

Example: $(1 + 7) + 3$ is equivalent to $1 + (7 + 3)$

example 6   Applying the Properties of Addition

a. Rewrite $9 + 6$, using the commutative property of addition.

b. Rewrite $(15 + 9) + 5$, using the associative property of addition.

Solution:

a. $9 + 6 = 6 + 9$ Change the order of the addends.

b. $(15 + 9) + 5 = 15 + (9 + 5)$ Change the grouping of the addends.

4. Translations and Applications Involving Addition

In the English language, there are many different words and phrases that imply addition. A partial list is given in Table 1-1.

table 1-1

Word/Phrase	Example	In Symbols
Sum	The sum of 6 and 2	$6 + 2$
Added to	3 added to 8	$8 + 3$
Increased by	7 increased by 2	$7 + 2$
More than	10 more than 6	$6 + 10$
Plus	8 plus 3	$8 + 3$
Total of	The total of 9 and 6	$9 + 6$

example 7   Translating an English Phrase to a Mathematical Statement

Translate each phrase to an equivalent mathematical statement and simplify.

a. 12 added to 109   b. The sum of 1386 and 376

Solution:

a. 109 + 12

$$
\begin{array}{r}
\overset{1}{1}09 \\
+\ 12 \\
\hline
121
\end{array}
$$

b. 1386 + 376

$$
\begin{array}{r}
\overset{1\ 1}{1}386 \\
+\ 376 \\
\hline
1762
\end{array}
$$

Skill Practice

Translate and simplify.

12. 50 more than 80

13. 12 increased by 14

14. The sum of 10, 20, and 30

Addition of whole numbers is sometimes necessary to solve application problems.

example 8   Solving an Application Problem

Carlita works as a waitress at El Pinto restaurant in Albuquerque, New Mexico. Her tips for the last five nights were $30, $18, $66, $102, and $45. Find the total amount she made in tips.

Solution:

To find the total, we add.

$$
\begin{array}{r}
\overset{1\ 2}{\$\ 30} \\
18 \\
66 \\
102 \\
+\ 45 \\
\hline
\$261
\end{array}
$$

Carlita made $261 in tips.

Skill Practice

15. Talita received test scores of 92, 100, 84, and 96 on her first four math tests. She also earned 8 points of extra credit. How many total points did she earn?

Tables and graphs are often used to summarize information in an organized manner. Examples 9 and 10 demonstrate the interpretation of these tools.

example 9   Solving an Application Problem Involving a Table

The following table gives the top five most-visited websites for a recent month.

Website	Number of Visitors
AOL Time Warner Network	97,995
MSN-Microsoft sites	89,819
Yahoo! sites	83,433
Google sites	37,460
Terra Lycos	36,173

Find the total number of visitors to the top five most-visited websites.

Answers
12. 80 + 50; 130
13. 12 + 14; 26
14. 10 + 20 + 30; 60
15. 380

Skill Practice

16. The table gives the number of gold, silver, and bronze medals won in the 2002 Winter Olympics for selected countries. Find the total number of medals won by Canada.

	Gold	Silver	Bronze
Germany	12	16	7
United States	10	13	11
Norway	11	7	6
Canada	6	3	8

Skill Practice

17. Samira's monthly expenses are summarized in the graph. Find the sum of her expenses.

Monthly Budget

Utilities $170
Food $300
Car $340
Other $250
Rent $660

Solution:

$$
\begin{array}{r}
\scriptstyle 3\ 3 2\ 2 2 \\
97,995 \\
89,819 \\
83,433 \\
37,460 \\
+\ \ 36,173 \\
\hline
344,880
\end{array}
$$

There were 344,880 combined visitors to these websites.

---

example 10   Solving an Application Problem Involving a Graph

The graph in Figure 1-3 gives the number of new AIDS cases in the United States for the years 2000, 2001, and 2002. The red bars in the graph represent the values for the number of women (aged 13 and older). The blue bars in the graph represent the values for the number of men (aged 13 and older). (Source: Centers for Disease Control.)

Find the total number of new AIDS cases for women in the United States in the years 2000–2002.

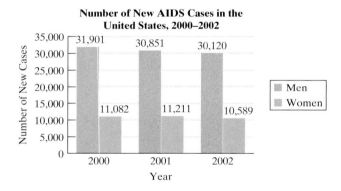

Figure 1-3

Solution:

We need to find the number of new AIDS cases for women only. Therefore, add the values corresponding to the red bars in the graph.

$$
\begin{array}{r}
\scriptstyle 1\ 1 \\
11,082 \\
11,211 \\
+\ 10,589 \\
\hline
32,882
\end{array}
$$

There were 32,882 new AIDS cases attributed to women in the years 2000–2002.

Answers

16. 17 medals
17. $1720

5. Perimeter

One special application of addition is to find the perimeter of a polygon. A **polygon** is a flat figure formed by line segments connected at their ends.

Familiar figures such as triangles, rectangles, and squares are examples of polygons. See Figure 1-4.

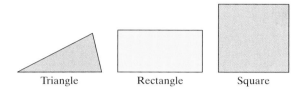

Figure 1-4

The **perimeter** of any polygon is the distance around the outside of the figure. To find the perimeter, add the lengths of the sides.

---

 example 11   Finding Perimeter

Find the perimeter of the triangle.

Solution:

The perimeter is the sum of the lengths of the sides.

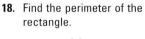

$$
\begin{array}{r}
\overset{1}{18}\text{ in.} \\
24\text{ in.} \\
+\ 30\text{ in.} \\
\hline
72\text{ in.}
\end{array}
$$

The perimeter is 72 inches (in.).

■

Skill Practice

18. Find the perimeter of the rectangle.

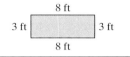

 **example 12   Finding Perimeter**

A paving company wants to edge the perimeter of a parking lot with concrete curbing. Find the perimeter of the parking lot.

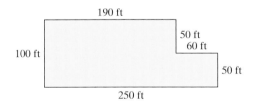

Solution:

The perimeter is the sum of the lengths of the sides.

$$
\begin{array}{r}
\overset{3}{190}\text{ ft} \\
50\text{ ft} \\
60\text{ ft} \\
50\text{ ft} \\
250\text{ ft} \\
+\ 100\text{ ft} \\
\hline
700\text{ ft}
\end{array}
$$

The distance around the parking lot (the perimeter) is 700 ft.

■

Skill Practice

19. Find the perimeter of the garden.

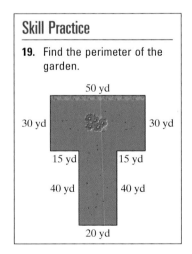

Answers
18. 22 ft
19. 240 yd

section 1.2   Practice Exercises

**Boost *your* GRADE at
mathzone.com!**

MathZone

- Practice Problems
- Self-Tests
- NetTutor

- e-Professors
- Videos

Study Skills Exercises

1. Taking 12 credit-hours is the equivalent of a full-time job. Often students try to work too many hours while taking classes at school.

   a. Write down how many hours you work per week and the number of credit-hours you are taking this term.

   Number of hours worked per week _____

   Number of credit hours this term _____

   b. The table gives a recommended limit on the number of hours you should work based on the number of credit-hours you are taking at school. (Keep in mind that other responsibilities in your life such as your family might also make it necessary to limit your hours at work even more.) How do your numbers from part (a) compare to those in the table? Are you working too many hours?

Number of Credit-hours	Maximum Number of Hours of Work per Week
3	40
6	30
9	20
12	10
15	0

2. Define the key terms.

   a. Sum   **b. Addends**   **c. Polygon**   **d. Perimeter**

Review Exercises

For Exercises 3–8, write the number in the form indicated.

3. Convert the number 351 to expanded form.

4. Write the number 351 in words.

5. Convert the number 107 to expanded form.

6. Write the given number in standard form: two thousand, four

7. Write the given number in standard form: four thousand, twelve

8. Convert the given number to standard form: 6 thousands + 2 hundreds + 6 ones

Objective 1: Addition of Whole Numbers Using the Number Line

9. Fill out the chart. Use the number line if necessary.

+	0	1	2	3	4	5	6	7	8	9
0										
1										
2										
3										
4										
5										
6										
7										
8										
9										

For Exercises 10–15, identify the addends and the sum.

10. $5 + 9 = 14$

11. $2 + 8 = 10$

12. $12 + 5 = 17$

13. $11 + 10 = 21$

14. $1 + 13 + 4 = 18$

15. $5 + 8 + 2 = 15$

Objective 2: Addition of Whole Numbers

For Exercises 16–31, add. **(See Examples 1 and 2.)**

16. $\begin{array}{r} 42 \\ + 33 \\ \hline \end{array}$

17. $\begin{array}{r} 21 \\ + 53 \\ \hline \end{array}$

18. $\begin{array}{r} 39 \\ + 20 \\ \hline \end{array}$

19. $\begin{array}{r} 15 \\ + 43 \\ \hline \end{array}$

20. $\begin{array}{r} 12 \\ 15 \\ + 32 \\ \hline \end{array}$

21. $\begin{array}{r} 10 \\ 8 \\ + 30 \\ \hline \end{array}$

22. $\begin{array}{r} 7 \\ 21 \\ + 10 \\ \hline \end{array}$

23. $\begin{array}{r} 6 \\ 11 \\ + 2 \\ \hline \end{array}$

24. $341 + 225$

25. $407 + 181$

26. $890 + 107$

27. $444 + 354$

28. $4 + 13 + 102$

29. $11 + 221 + 5$

30. $31 + 7 + 430$

31. $24 + 14 + 160$

For Exercises 32–51, add the whole numbers with carrying. **(See Examples 3–5.)**

32. $\begin{array}{r} 76 \\ + 45 \\ \hline \end{array}$

33. $\begin{array}{r} 25 \\ + 59 \\ \hline \end{array}$

34. $\begin{array}{r} 87 \\ + 24 \\ \hline \end{array}$

35. $\begin{array}{r} 38 \\ + 77 \\ \hline \end{array}$

36. $\begin{array}{r} 658 \\ + 231 \\ \hline \end{array}$

37. $\begin{array}{r} 642 \\ + 295 \\ \hline \end{array}$

38. $\begin{array}{r} 152 \\ + 549 \\ \hline \end{array}$

39. $\begin{array}{r} 462 \\ + 388 \\ \hline \end{array}$

40. $15 + 5 + 9$ **41.** $2 + 31 + 8$ **42.** $14 + 9 + 17$ **43.** $7 + 18 + 4$

44. $79 + 112 + 12$ **45.** $62 + 907 + 34$ **46.** $331 + 422 + 76$ **47.** $87 + 119 + 630$

48. $4980 + 10{,}223$ **49.** $23{,}112 + 892$ **50.** $8721 + 3212$ **51.** $12{,}333 + 788$

Objective 3: Properties of Addition

For Exercises 52–55, rewrite the addition problem, using the commutative property of addition. **(See Example 6.)**

52. $12 + 6 = \square + \square$ **53.** $30 + 21 = \square + \square$ **54.** $101 + 44 = \square + \square$ **55.** $8 + 13 = \square + \square$

For Exercises 56–59, rewrite the addition problem using the associative property of addition, by inserting a pair of parentheses.

56. $(4 + 8) + 13 = 4 + 8 + 13$ **57.** $(23 + 9) + 10 = 23 + 9 + 10$

58. $7 + (12 + 8) = 7 + 12 + 8$ **59.** $41 + (3 + 22) = 41 + 3 + 22$

60. Explain the difference between the commutative and the associative properties of addition.

61. Explain the addition property of 0. Then simplify the expressions.

    a. $423 + 0$ **b.** $0 + 25$ **c.** $\begin{array}{r} 67 \\ + \ 0 \\ \hline \end{array}$

Objective 4: Translations and Applications Involving Addition

For Exercises 62–71, translate the English phrase into a mathematical statement and simplify. **(See Example 7.)**

62. The sum of 13 and 7   **63.** The sum of 100 and 42   **64.** 45 added to 7

65. 81 added to 23   **66.** 5 more than 18   **67.** 2 more than 76

68. 1523 increased by 90   **69.** 1320 increased by 448   **70.** The total of 5, 39, and 81

71. The total of 78, 12, and 22

For Exercises 72–77, write an English phrase from the mathematical statement. Answers may vary.

72. $54 + 24$ **73.** $33 + 15$ **74.** $12 + 88$ **75.** $70 + 15$

76. $4 + 23 + 77$ **77.** $11 + 41 + 53$

78. The attendance at a high school play during one weekend was as follows: 103 on Friday, 112 on Saturday, and 61 at the Sunday matinee. What was the total attendance?

79. To schedule enough drivers for an upcoming week, a local pizza shop manager recorded the number of deliveries each day from the previous week: 38, 54, 44, 61, 97, 103, 124. What was the total number of deliveries for the week?
(See Example 8.)

80. Three top television shows entertained the following number of viewers in one week: 27,300,000 for *CSI*, 20,800,000 for *Survivor*, and 19,900,000 for *ER*. Find the sum of the viewers for these shows.

81. To travel from Houston to Corpus Christi, a salesperson must stop in San Antonio. If it is 195 mi from Houston to San Antonio and 228 mi from San Antonio to Corpus Christi, how far will she travel on this trip?

82. Nora earned $43,000 last year. This year her salary was increased by $2500. What is her present salary?

83. The number of participants in the Special Olympics increased by 1,205,655 since it began in 1968 with 1000 athletes. How many athletes are presently participating?

84. The table gives the number of desks and chairs delivered each quarter to an office supply store. What is the total number of desks delivered for the year?

	Chairs	Desks
March	220	115
June	185	104
September	201	93
December	198	111

85. A portion of Jonathan's checking account register is shown. What is the total amount of the four checks written?
(See Example 9.)

Check No.	Description	Credit	Debit	Balance
1871	Electric bill		$60	$180
1872	Groceries		52	128
1873	Department store		75	53
	Payroll	$1256		1309
1874	Restaurant		58	1251
	Transfer from savings	150		1401

86. The graph displays the number of public school teachers in the United States. Find the number of elementary school teachers (include prekindergarten and kindergarten).

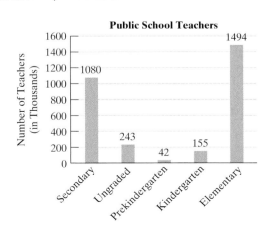

Public School Teachers

87. The staff for U.S. public schools is categorized in the graph. Determine the number of staff other than teachers.

Number of Public School Staff

Teachers 2,997,741
Counselors 100,052
Supervisors 45,934
Aides 675,038

88. The pie graph shows the costs incurred in managing Sub-World sandwich shop for one month. From this information, determine the total cost for one month.

Sub-World Monthly Expenses

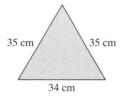

Food $7329
Labor $9560
Nonfood items $1248
Overhead $3500

89. The Student Career Experience Program is a program that places students in government jobs. The chart displays the number of participants during 2006 in the top six agencies. Find the total number of participants in the program. **(See Example 10.)**

Student Career Experience Program, Top Six Employing Agencies (Fiscal Year 2006)

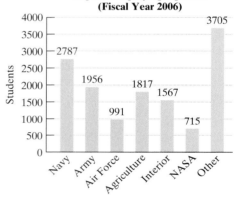

Navy 2787, Army 1956, Air Force 991, Agriculture 1817, Interior 1567, NASA 715, Other 3705

Objective 5: Perimeter

For Exercises 90–97, find the perimeter. **(See Example 11.)**

90.

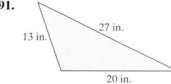

35 cm   35 cm
34 cm

91.

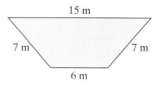

13 in.   27 in.
20 in.

92. Find the perimeter of an NBA basketball court.

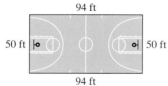

94 ft
50 ft   50 ft
94 ft

93.

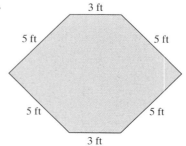

15 m
7 m   7 m
6 m

94.

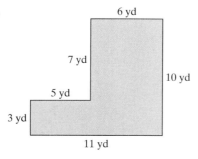

6 yd
7 yd
10 yd
5 yd
3 yd
11 yd

95.

3 ft
5 ft   5 ft
5 ft   5 ft
3 ft

96.

21 m 20 m
21 m 18 m
11 m 19 m

97. A major league baseball diamond is in the shape of a square. Find the distance a batter must run if he hits a home run. **(See Example 12.)**

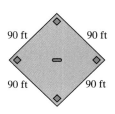

90 ft 90 ft
90 ft 90 ft

Calculator Connections

Topic: Adding on a calculator

The following keystrokes demonstrate the procedure to add numbers on a calculator. The $\boxed{\text{Enter}}$ key (or, on some calculators, the $\boxed{=}$ key or $\boxed{\text{Exe}}$ key) tells the calculator to complete the calculation. Notice that commas used in large numbers are not entered into the calculator.

Expression	Keystrokes	Result
92,406 + 83,168	92406 $\boxed{+}$ 83168 $\boxed{\text{Enter}}$	175574

Your calculator may use the $\boxed{=}$ key or $\boxed{\text{Exe}}$ key instead.

Calculator Exercise

For Exercises 98–101, add by using a calculator.

98. 9,084,037 + 452,903

99. 899,382 + 9406

100.
```
   45,418
   81,990
    9,063
+  56,309
```

101.
```
  9,300,050
  7,803,513
  3,480,009
+   907,822
```

102. The number of viewers for four television programs for a selected week is given in the table. What is the total number of viewers?

Program	Number of Viewers
ABC premier event	17,457,000
American Idol	17,164,000
CSI	17,004,000
Law and Order	15,717,000

103. The number of votes tallied for the leading Presidential candidates for the 2004 election is given in the table. Find the total number of votes for these three candidates.

Candidate	Number of Votes
Nader	411,304
Kerry	59,028,109
Bush	62,040,606

section 1.3   Subtraction of Whole Numbers

1. Introduction to Subtraction

Jeremy bought a case of 12 sodas, and on a hot afternoon he drank 3 of the sodas. We can use the operation of subtraction to find the number of sodas remaining.

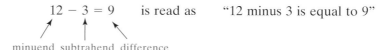

12 sodas − 3 sodas   =   9 sodas

The symbol − between two numbers is a subtraction sign, and the result of a subtraction is called the **difference**. The number being subtracted (in this case, 3) is called the **subtrahend**. The number 12 from which 3 is subtracted is called the **minuend**.

$$12 - 3 = 9 \quad \text{is read as} \quad \text{"12 minus 3 is equal to 9"}$$

minuend subtrahend difference

Subtraction is the reverse operation of addition. To find the number of sodas that remain after Jeremy takes 3 sodas away from 12 sodas, we ask the following question:

"3 added to what number equals 12?"

That is,

$$12 - 3 = ? \quad \text{is equivalent to} \quad ? + 3 = 12$$

Subtraction can also be visualized on the number line. To evaluate $7 - 4$, start from the point on the number line corresponding to the minuend (7 in this case). Then move to the *left* 4 units. The resulting position on the number line is the difference.

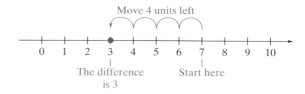

To check the result, we can use addition.

$$7 - 4 = 3 \quad \text{because} \quad 3 + 4 = 7$$

Skill Practice

Subtract. Check by using addition.

1. $11 - 5$ **2.** $8 - 0$

3. $7 - 2$ **4.** $5 - 5$

Answers

1. 6   2. 8   3. 5   4. 0

example 1   Subtracting Whole Numbers

Subtract and check the answer, using addition.

a. $8 - 2$ **b.** $10 - 6$ **c.** $5 - 0$ **d.** $3 - 3$

Solution:

a. $8 - 2 = 6$ because $6 + 2 = 8$

b. $10 - 6 = 4$ because $4 + 6 = 10$

c. $5 - 0 = 5$ because $5 + 0 = 5$

d. $3 - 3 = 0$ because $0 + 3 = 3$

■

2. Subtraction of Whole Numbers

When subtracting large numbers, it is usually more convenient to write the numbers vertically. We write the minuend on top and the subtrahend below it. Starting from the ones column, we subtract digits having corresponding place values.

example 2	Subtracting Whole Numbers Without Borrowing

Subtract and check the answer by using addition.

a. $\begin{array}{r} 976 \\ -\ 124 \\ \hline \end{array}$ **b.** $\begin{array}{r} 2498 \\ -\ 197 \\ \hline \end{array}$

Solution:

a. $\begin{array}{r} 976 \\ -\ 124 \\ \hline 852 \end{array}$ Check: $\begin{array}{r} 852 \\ +\ 124 \\ \hline 976 \end{array}$ ✓

        ──Subtract the ones column digits
        ──Subtract the tens column digits
        ──Subtract the hundreds column digits

b. $\begin{array}{r} 2498 \\ -\ 197 \\ \hline 2301 \end{array}$ Check: $\begin{array}{r} 2301 \\ +\ 197 \\ \hline 2498 \end{array}$ ✓

■

When a digit in the subtrahend is larger than the corresponding digit in the minuend, we must "regroup" or borrow a value from the column to the left.

$\begin{array}{l} 92 = 9 \text{ tens} + 2 \text{ ones} \\ -\ 74 = 7 \text{ tens} + 4 \text{ ones} \\ \hline \end{array}$

In the ones column, we cannot take 4 away from 2. We will regroup by borrowing 1 ten from the minuend. Furthermore, 1 ten = 10 ones.

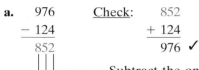

We now have 12 ones in the minuend.

$\begin{array}{l} \overset{8\ 12}{9\ 2} = 9 \text{ tens} + 12 \text{ ones} \\ -\ 7\ 4 = 7 \text{ tens} + \ 4 \text{ ones} \\ \hline 1\ 8 = 1 \text{ ten} + \ 8 \text{ ones} \end{array}$

Tip: The process of *borrowing* in subtraction is the reverse operation of *carrying* in addition.

example 3	Subtracting Whole Numbers With Borrowing

Subtract and check the result with addition.

a. $\begin{array}{r} 134,616 \\ -\ 53,438 \\ \hline \end{array}$ **b.** $500 - 247$

Skill Practice

Subtract. Check by using addition.

5. $\begin{array}{r} 472 \\ -\ 261 \\ \hline \end{array}$ **6.** $\begin{array}{r} 3947 \\ -\ 137 \\ \hline \end{array}$

Concept Connections

7. Which subtraction (a or b) requires borrowing?

   a. $\begin{array}{r} 76 \\ -\ 24 \\ \hline \end{array}$ **b.** $\begin{array}{r} 76 \\ -\ 49 \\ \hline \end{array}$

Answers

5. 211   6. 3810   7. b

Solution:

a.
$$\overset{\overset{0\ 16}{}}{1\ 3\ 4,6\ \cancel{1}\ \cancel{6}}$$
$$-\ \ \ 5\ 3,4\ 3\ 8$$
$$\overline{8}$$

In the ones place, 8 is greater than 6. In the minuend, we borrow 1 ten from the tens place.

$$\overset{\overset{5\ \cancel{0}\ 16}{10}}{1\ 3\ 4,\cancel{6}\ \cancel{1}\ \cancel{6}}$$
$$-\ \ \ 5\ 3,4\ 3\ 8$$
$$\overline{7\ 8}$$

In the tens place, 3 is greater than 0. In the minuend, we borrow 1 hundred from the hundreds place.

$$\overset{\overset{0\ 13\ \ \ \ 5\ \cancel{0}\ 16}{10}}{\cancel{1}\ \cancel{3}\ 4,\cancel{6}\ \cancel{1}\ \cancel{6}}$$
$$-\ \ \ 5\ 3,4\ 3\ 8$$
$$\overline{8\ 1,1\ 7\ 8}$$

In the ten-thousands place, 5 is greater than 3. We borrow 1 hundred-thousand from the hundred-thousands place.

Check:
$$\overset{1\ \ \ \ 1\ 1}{81,178}$$
$$+\ 53,438$$
$$\overline{134,616\ \checkmark}$$

b.
$$500$$
$$-\ 247$$

In the ones place, 7 is greater than 0. We try to borrow 1 ten from the tens place. However, the tens place digit is 0. Therefore we must first borrow from the hundreds place.

$$\overset{\overset{4\ 10}{}}{\cancel{5}\ \cancel{0}\ 0}$$
$$-\ 2\ 4\ 7$$

$$\overset{\overset{4\ \cancel{10}\ 10}{9}}{\cancel{5}\ \cancel{0}\ \cancel{0}}$$ ←—Now we can borrow 1 ten to add to the ones place.
$$-\ 2\ 4\ 7$$
$$\overline{2\ 5\ 3}$$ Subtract.

Check:
$$\overset{1\ 1}{253}$$
$$+\ 247$$
$$\overline{500\ \checkmark}$$

3. Translations and Applications Involving Subtraction

In applications of mathematics, several words and phrases imply subtraction. A partial list is provided in Table 1-2.

table 1-2

Word/Phrase	Example	In Symbols
Minus	15 minus 10	15 − 10
Difference	The difference of 10 and 2	10 − 2
Decreased by	9 decreased by 1	9 − 1
Less than	5 less than 12	12 − 5
Subtract . . . from	Subtract 3 from 8	8 − 3

In Table 1-2, make a note of the last two entries. The phrases *less than* and *subtract . . . from* imply a specific order in which the subtraction is performed. In both cases, begin with the second number listed and subtract the first number listed.

---

example 4   Translating an English Phrase to a Mathematical Statement

Translate the English phrase to a mathematical statement and simplify.

a. The difference of 150 and 38

b. 30 subtracted from 82

Solution:

a. From Table 1-2, the *difference* of 150 and 38 implies that the first number (150) is the minuend and the second number (38) is the subtrahend. Therefore, we have $150 - 38$.

$$\begin{array}{r} 1\overset{4}{\cancel{5}}\overset{10}{0} \\ -\ \ 38 \\ \hline 112 \end{array}$$

b. The phrase "30 subtracted from 82" implies that 30 is taken away from 82. Therefore, we must start with 82 as the minuend and subtract 30. We have $82 - 30$.

$$\begin{array}{r} 82 \\ -\ 30 \\ \hline 52 \end{array}$$

Skill Practice

Translate the English phrase into a mathematical statement and simplify.

11. Twelve decreased by eight

12. Subtract three from nine.

In Section 1.2 we saw that the operation of addition is commutative. That is, the order in which two numbers are added does not affect the sum. This is *not* true for subtraction. For example, $82 - 30$ is not equal to $30 - 82$. The symbol $\neq$ means "is not equal to." Thus, $82 - 30 \neq 30 - 82$.

Most applications of subtraction generally fall into two categories.

1. The first type is phrased as a subtraction problem in which the minuend and subtrahend are given.

Example: Shawn has $52 and then spends $40. How much money does he have left? (In this problem, we subtract $40 from $52.)

$$\$52 - \$40 = \$12$$

2. The second type is phrased as an addition problem with a missing addend.

Example: Maria received 72 points on her last math test, but needed 90 points to receive an A. How many more points would she have needed to earn an A? (In this problem, the addition problem can be translated to subtraction.)

$$72 + ? = 90 \quad \text{is equivalent to} \quad 90 - 72 = ?$$

Because $90 - 72 = 18$, Maria would have needed 18 more points.

Skill Practice

13. The temperature at 1:00 P.M. in Denver was 47°F. Three hours later, the temperature was 34°F. By how much did the temperature drop?

---

example 5   Solving an Application Problem

A biology class started with 35 students. By mid-semester, 7 students had dropped. How many students are still in the class?

Solution:

$$35 - 7 = 28 \qquad \text{There are 28 students still in the class.}$$

Skill Practice

14. Teresa earned test scores of 98, 84, and 90 on her first three exams. How many points must she score on the fourth exam to earn a total of 360 points?

---

example 6   Solving an Application Problem

A surveyor knows that the perimeter of the lot shown is 620 ft. Find the length of the missing side. See Figure 1-5.

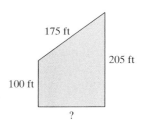

175 ft

205 ft

100 ft

?

Figure 1-5

Solution:

Recall that the perimeter of a polygon is the sum of the lengths of its sides. The sum of the three known sides in Figure 1-5 is 480 ft:

$$
\begin{array}{r}
\overset{1}{1}00 \\
175 \\
+\ 205 \\
\hline
480
\end{array}
$$

This value plus the length of the fourth side equals the perimeter: 480 ft + ? = 620 ft. Equivalently, we can subtract 480 ft from the perimeter to find the length of the missing side:

$$620 \text{ ft} - 480 \text{ ft} = ?$$

$$
\begin{array}{r}
\overset{5\ \ 12}{\cancel{6}2\,0} \\
-\ 4\,8\,0 \\
\hline
1\,4\,0
\end{array}
$$

The missing side is 140 ft long.

A third application of subtraction is to compute a change (increase or decrease) in an amount.

Skill Practice

15. At Houston Community College, the total enrollment for the fall semester in 2000 was 49,520 students. In 2001, the fall semester enrollment was 53,565.

a. Has the enrollment increased or decreased?

b. Determine the amount of increase or decrease.

---

example 7   Solving an Application Problem

The number of reported robberies in the United States has fluctuated each year as shown in the graph.

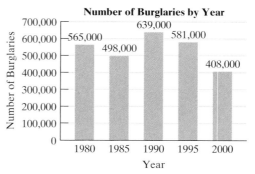

Answers

13. 13°F   14. 88 points
15. a. Increased   b. 4045

a. Find the increase in the number of reported robberies from the year 1985 to 1990.

b. Find the decrease in the number of reported robberies from the year 1995 to 2000.

Solution:

For the purpose of finding an amount of increase or decrease, we will subtract the smaller number from the larger number.

a. Because the number of robberies went *up* from 1985 to 1990, there was an *increase*. To find the amount of the increase, we subtract the smaller number from the larger number.

$$\begin{array}{r} \overset{5\ 13}{6\ \cancel{3}\ 9,0\ 0\ 0} \\ -\ 4\ 9\ 8,0\ 0\ 0 \\ \hline 1\ 4\ 1,0\ 0\ 0 \end{array}$$

From 1985 to 1990, there was an increase of 141,000 reported robberies in the United States.

b. Because the number of robberies went *down* from 1995 to 2000, there was a *decrease*. To find the amount of the decrease, we subtract the smaller number from the larger number.

$$\begin{array}{r} \overset{7\ 11}{5\ 8\ \cancel{1},0\ 0\ 0} \\ -\ 4\ 0\ 8,0\ 0\ 0 \\ \hline 1\ 7\ 3,0\ 0\ 0 \end{array}$$

From 1995 to 2000 there was a decrease of 173,000 reported robberies in the United States.

section 1.3   Practice Exercises

Boost *your* GRADE at mathzone.com!

MathZone

- Practice Problems
- Self-Tests
- NetTutor
- e-Professors
- Videos

Study Skills Exercises

1. It is very important to attend class every day. Math is cumulative in nature, and you must master the material learned in the previous class to understand today's lesson. Because this is so important, many instructors tie attendance into the final grade. Write down the attendance policy for your class.

2. Define the key terms.

    a. Difference   **b. Subtrahend**   **c. Minuend**

Review Exercises

For Exercises 3–5, add.

3. $330 + 821$

4.
$$\begin{array}{r} 782 \\ 21 \\ + 1046 \\ \hline \end{array}$$

5.
$$\begin{array}{r} 46 \\ 804 \\ + 49 \\ \hline \end{array}$$

6. Circle the true statement:
$14 > 21, 14 < 21$

7. Circle the true statement:
$0 < 10, 0 > 10$

8. Write the inequality in words:
$22 < 25$

Objective 1: Introduction to Subtraction

For Exercises 9–14, identify the minuend, subtrahend, and the difference.

9. $12 - 8 = 4$

10. $6 - 1 = 5$

11. $21 - 12 = 9$

12. $32 - 2 = 30$

13.
$$\begin{array}{r} 9 \\ - 6 \\ \hline 3 \end{array}$$

14.
$$\begin{array}{r} 17 \\ - 3 \\ \hline 14 \end{array}$$

For Exercises 15–18, write the subtraction problem as a related addition problem. For example, $19 - 6 = 13$ can be written as $13 + 6 = 19$.

15. $27 - 9 = 18$

16. $20 - 8 = 12$

17. $102 - 75 = 27$

18. $211 - 45 = 166$

For Exercises 19–24, subtract, then check the answer by using addition. **(See Example 1.)**

19. $8 - 3$ Check: $\square + 3 = 8$

20. $7 - 2$ Check: $\square + 2 = 7$

21. $4 - 1$ Check: $\square + 1 = 4$

22. $9 - 1$ Check: $\square + 1 = 9$

23. $6 - 0$ Check: $\square + 0 = 6$

24. $3 - 0$ Check: $\square + 0 = 3$

Objective 2: Subtraction of Whole Numbers

For Exercises 25–38, subtract and check the answer by using addition. **(See Example 2.)**

25.
$$\begin{array}{r} 68 \\ - 23 \\ \hline \end{array}$$

26.
$$\begin{array}{r} 54 \\ - 31 \\ \hline \end{array}$$

27.
$$\begin{array}{r} 88 \\ - 27 \\ \hline \end{array}$$

28.
$$\begin{array}{r} 75 \\ - 50 \\ \hline \end{array}$$

29.
$$\begin{array}{r} 1347 \\ - 221 \\ \hline \end{array}$$

30.
$$\begin{array}{r} 4865 \\ - 713 \\ \hline \end{array}$$

31.
$$\begin{array}{r} 1525 \\ - 1204 \\ \hline \end{array}$$

32.
$$\begin{array}{r} 8843 \\ - 5612 \\ \hline \end{array}$$

33. $12,806 - 2802$

34. $12,771 - 1240$

35. $14,356 - 13,253$

36. $34,550 - 31,450$

37. $95,432 - 61,101$

38. $80,529 - 20,117$

For Exercises 39–62, subtract the whole numbers involving borrowing. **(See Example 3.)**

39. 76
        $-\,59$

40. 64
        $-\,48$

41. 87
        $-\,38$

42. 94
        $-\,75$

43. 240
        $-\,136$

44. 360
        $-\,225$

45. 710
        $-\,189$

46. 850
        $-\,303$

47. 4350
        $-\,4327$

48. 7293
        $-\,7255$

49. 6002
        $-\,1238$

50. 3000
        $-\,2356$

51. 10,425
        $-\;9,122$

52. 23,901
        $-\;8,164$

53. 62,088
        $-\,59,871$

54. 32,112
        $-\,28,334$

55. $470 - 92$

56. $674 - 89$

57. $3709 - 2987$

58. $8052 - 2788$

59. $32,439 - 1498$

60. $21,335 - 4123$

61. $8,007,234 - 2,345,115$

62. $3,045,567 - 1,871,495$

Objective 3: Translations and Applications Involving Subtraction

For Exercises 63–72, translate the English phrase into a mathematical statement and simplify. **(See Example 4.)**

63. 78 minus 23

64. 45 minus 17

65. 78 decreased by 6

66. 50 decreased by 12

67. Subtract 100 from 422.

68. Subtract 42 from 89.

69. 72 less than 1090

70. 60 less than 3111

71. The difference of 50 and 13

72. The difference of 405 and 103

For Exercises 73–76, write an English phrase for the mathematical statement. (Answers will vary.)

73. $93 - 27$

74. $80 - 20$

75. $165 - 85$

76. $171 - 42$

77. Use the expression $7 - 4$ to explain why subtraction is not commutative.

78. Is subtraction associative? Use the numbers 10, 6, 2 to explain.

79. A $50 bill was used to purchase $17 worth of gasoline. Find the amount of change received. **(See Example 5.)**

80. There are 55 DVDs to shelve one evening at a video rental store. If Jason puts away 39 before leaving for the day, how many are left for Patty to handle?

81. The songwriting team of John Lennon and Paul McCartney had 118 chart hits while Mick Jagger and Keith Richards had 63. How many more chart hits did Lennon and McCartney have than Jagger and Richards?

82. In 2005 it was estimated that the urban population of China was about 536 million people while the urban population of India was about 313 million people. What was the difference in the populations of China and India in 2005?

83. In landscaping a yard, Lily would like 26 plants for a border. If she has 18 plants in her truck, how many more will she need to finish the job?

84. A collection is taken to buy flowers for a co-worker who is in the hospital. If $30 has been collected and the flower arrangement costs $43, how much more needs to be collected?

 85. At the time of John Elway's retirement from football, his total passing yardage was 51,475 yd. Brett Favre had 42,285 yd as of 2002. How many more yards would Favre need to reach Elway's total?

86. The musical *Cats* has the record of being the longest-running musical on Broadway with 7485 performances. *The Phantom of the Opera* had performed 6231 times as of January 1, 2003. How many more performances must there be of *Phantom* to equal the record?

For Exercises 87 and 88, for each figure find the missing length.

87. The perimeter of the triangle is 39 m.

14 m / 12 m / ?

88. The perimeter of the figure is 547 cm.

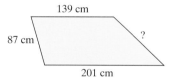

139 cm
87 cm ?
201 cm

89. A homeowner knows that the perimeter of his backyard is 56 yd. Find the length of the missing side. (**See Example 6.**)

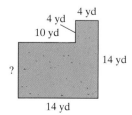

4 yd 4 yd
10 yd
14 yd
?
14 yd

90. Barbara has 15 ft of molding to install in her bathroom, as shown in the figure. What is the missing length? *Note:* There will be no molding by the tub or door.

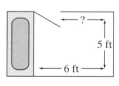

?
5 ft
6 ft

For Exercises 91–94, use the information from the graph on page 33. (**See Example 7.**)

 91. What is the difference in the number of marriages for the 20-year period from 1980 to 2000?

92. Of the years presented in the graph, which year had the greatest number of marriages? Which year had the least?

93. What is the difference in the number of marriages between the year having the greatest and the year having the least?

94. Between which two 5-year periods did the greatest increase in the number of marriages occur? What is the increase?

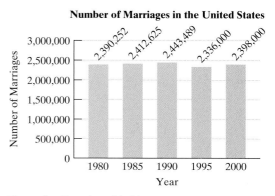

Number of Marriages in the United States

Figure for Exercises 91–94

Calculator Connections

Topic: Subtracting on a calculator

To subtract numbers on a calculator, use the subtraction key $\boxed{-}$. Do not confuse the subtraction key with the $\boxed{(-)}$ key. The $\boxed{(-)}$ is presented later to enter negative numbers.

Expression	Keystrokes	Result
345,899 − 43,018	345899 $\boxed{-}$ 43018 $\boxed{\text{Enter}}$	302881

Calculator Exercises

For Exercises 95 and 96, subtract by using a calculator.

95.   4,905,620
        − 458,318

96.   953,400,415
        − 56,341,902

For Exercises 97–100, refer to the table showing the land area for five states.

State	Land Area (mi²)
Rhode Island	1,045
Tennessee	41,217
West Virginia	24,078
Wisconsin	54,310
Colorado	103,718

97. Find the difference in land area between Colorado and Wisconsin.

98. Find the difference in land area between Tennessee and West Virginia.

99. Find the difference in land area between the state with the greatest land area and the state with the least land area.

100. How much more land area does Wisconsin have than Tennessee?

Objectives

1. Rounding
2. Estimation
3. Using Estimation in Applications

Concept Connections

1. Is the number 82 closer to 80 or to 90? Round 82 to the nearest ten.
2. Is the number 65 closer to 60 or to 70? Round the number to the nearest ten.

section 1.4   Rounding and Estimating

1. Rounding

Rounding a whole number is a common practice when we do not require an exact value. For example, a recent enrollment figure for the College of DuPage in Glyn Ellyn, Illinois, was 29,423 students. We might round this number to the nearest thousand and say that there were approximately 29,000 students. In mathematics we use the symbol $\approx$ to read "is approximately equal to." Hence $29,423 \approx 29,000$.

A number line is a helpful tool to understand rounding. For example, the number 48 is closer to 50 than it is to 40. Therefore, 48 rounded to the nearest ten is 50.

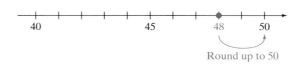

The number 43, on the other hand, is closer to 40 than to 50. Therefore, 43 rounded to the nearest ten is 40.

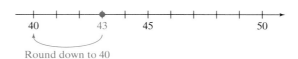

The number 45 is halfway between 40 and 50. In such a case, our convention will be to round *up* to the next-larger ten.

The decision to round up or down to a given place value is determined by the digit to the *right* of the given place value. The following steps outline the procedure.

Rounding Whole Numbers

1. Identify the digit one position to the right of the given place value.
2. If the digit in step 1 is a 5 or greater, add 1 to the digit in the given place value. Then replace each digit to the right of the given place value by 0.
3. If the digit in step 1 is less than 5, replace it and each digit to its right by 0. Note that in this case, the digit in the original given place value does not change.

Skill Practice

3. Round 12,461 to the nearest thousand.

Answers

1. Closer to 80; 80
2. The number 65 is the same distance from 60 and 70; round up to 70.
3. 12,000

example 1   Rounding a Whole Number

Round 3741 to the nearest hundred.

Solution:

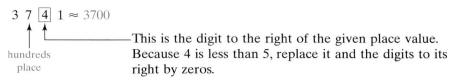

3 7 4 1 ≈ 3700

hundreds place

This is the digit to the right of the given place value. Because 4 is less than 5, replace it and the digits to its right by zeros.

The number 3700 is 3741 rounded to the nearest hundred.

◼—

Example 1 could also have been solved by drawing a number line. Use the part of a number line showing multiples of 100 on either side of 3741.

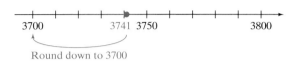

Round down to 3700

---

example 2   Rounding a Whole Number

Round 1,790,641 to the nearest hundred-thousand.

Solution:

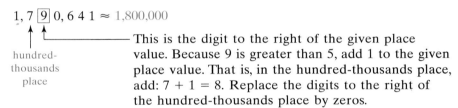

1,7 9 0,6 4 1 ≈ 1,800,000

hundred-thousands place

This is the digit to the right of the given place value. Because 9 is greater than 5, add 1 to the given place value. That is, in the hundred-thousands place, add: 7 + 1 = 8. Replace the digits to the right of the hundred-thousands place by zeros.

The number 1,800,000 is 1,790,641 rounded to the nearest hundred-thousand.

◼—

Skill Practice

4. Round 147,316 to the nearest ten-thousand.

---

example 3   Rounding a Whole Number

Round 1503 to the nearest thousand.

Solution:

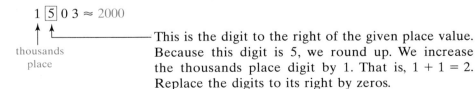

1 5 0 3 ≈ 2000

thousands place

This is the digit to the right of the given place value. Because this digit is 5, we round up. We increase the thousands place digit by 1. That is, 1 + 1 = 2. Replace the digits to its right by zeros.

The number 2000 is 1503 rounded to the nearest thousand.

◼—

Skill Practice

5. Round 7,521,460 to the nearest million.

---

example 4   Rounding a Whole Number

Round the number 24,961 to the hundreds place.

Skill Practice

6. Round 39,823 to the nearest thousand.

Answers

4. 150,000   5. 8,000,000   6. 40,000

Avoiding Mistakes:

When rounding numbers, the digit in the original given place value either increases or remains the same. It never decreases.

Solution:

$$2\,4,9\,\boxed{6}\,1 \approx 25{,}000$$

This value is greater than 5. Therefore, add 1 to the hundreds place digit. Replace the digits to the right of the hundreds place with 0.

$$\overset{+1}{24},900 \qquad \text{which equals} \qquad 25{,}000$$

The number 25,000 is 24,961 rounded to the nearest hundred.

2. Estimation

We use the process of rounding to estimate the result of numerical calculations. For example, to estimate the following sum, we can round each addend to the nearest ten.

31	rounds to $\longrightarrow$	30
12	rounds to $\longrightarrow$	10
+ 49	rounds to $\longrightarrow$	+ 50
		90

The estimated sum is 90 (the actual sum is 92).

Skill Practice

7. Estimate the sum by rounding each number to the nearest hundred.

   $3162 + 4931 + 2206$

---

example 5   Estimating a Sum

Estimate the sum by rounding to the nearest thousand.

$$6109 + 976 + 4842 + 11{,}619$$

Solution:

6,109	rounds to $\longrightarrow$	$\overset{1}{6},000$
976	rounds to $\longrightarrow$	1,000
4,842	rounds to $\longrightarrow$	5,000
+ 11,619	rounds to $\longrightarrow$	+ 12,000
		24,000

The estimated sum is 24,000 (the actual sum is 23,546).

Skill Practice

8. Estimate the difference by rounding each number to the nearest million.

   $35{,}264{,}000 - 21{,}906{,}210$

---

example 6   Estimating a Difference

Estimate the difference $4817 - 2106$ by rounding each number to the nearest hundred.

Solution:

4817	rounds to $\longrightarrow$	4800
− 2106	rounds to $\longrightarrow$	− 2100
		2700

The estimated difference is 2700 (the actual difference is 2711).

Answers

7. 10,300
8. 13,000,000

3. Using Estimation in Applications

example 7   **Estimating a Sum in an Application**

A driver for a delivery service must drive from Chicago, Illinois, to Dallas, Texas, and make several stops on the way. The driver follows the route given on the map. Estimate the total mileage by rounding each distance to the nearest ten miles.

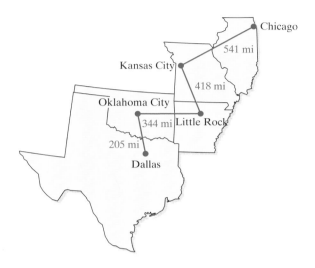

Solution:

541	rounds to ⟶	$\overset{1\ 1}{540}$
418	rounds to ⟶	420
344	rounds to ⟶	340
+ 205	rounds to ⟶	+ 210
		1510

The driver traveled approximately 1510 mi.

■

example 8   **Estimating a Difference in an Application**

In a recent year, the U.S. Census Bureau reported that the number of males over the age of 18 was 100,994,367. The same year, the number of females over 18 was 108,133,727. Round each value to the nearest million. Estimate how many more females over 18 there were than males over 18.

Solution:

The number of males was approximately 101,000,000. The number of females was approximately 108,000,000.

$$
\begin{array}{r}
108,000,000 \\
- \ 101,000,000 \\
\hline
7,000,000
\end{array}
$$

There were approximately 7 million more women over age 18 in the United States than men.

■

Skill Practice

9. The number of births in the United States for the years 1999, 2000, and 2001 is given in the chart. (Source: National Center for Health Statistics.) Round the values to the nearest million. Estimate the total number of births for these three years.

Skill Practice

10. In a recent year, there were 135,073,000 persons over the age of 16 employed in the United States. During the same year, there were 6,742,000 persons over the age of 16 who were unemployed. Approximate each value to the nearest million. Use these values to approximate how many more people were employed than unemployed.

Answers

9. 1999: 4,000,000;
   2000: 4,000,000;
   2001: 4,000,000;
   total: 12,000,000 births
10. Employed: 135,000,000;
    unemployed: 7,000,000;
    difference: 128,000,000

section 1.4   Practice Exercises

Study Skills Exercises

1. Purchase a three-ring binder for your math notes and homework. Use section dividers to separate each chapter that you cover in the text. Keep your homework and notes in the appropriate section. What other course materials might you keep organized in your notebook?

2. Define the key term **rounding**.

Review Exercises

For Exercises 3–5, add or subtract as indicated.

3.   59
     $-$ 33

4.   130
     $-$ 98

5.   4009
     $+$ 998

6.   12,033
     $+$ 23,441

7. Determine the place value of the digit 6 in the number 1,860,432.

8. Determine the place value of the digit 4 in the number 1,860,432.

Objective 1: Rounding

9. Explain how to round a whole number to the hundreds place.

10. Explain how to round a whole number to the tens place.

For Exercises 11–32, round each number to the given place value. **(See Examples 1–4.)**

11. 342; tens

12. 834; tens

13. 725; tens

14. 445; tens

15. 9384; hundreds

16. 8363; hundreds

17. 8539; hundreds

18. 9817; hundreds

19. 9982; hundreds

20. 7974; hundreds

21. 2578; thousands

22. 3511; thousands

23. 34,992; thousands

24. 76,831; thousands

25. 109,337; thousands

26. 437,208; thousands   **27.** 489,090; ten-thousands   **28.** 388,725; ten-thousands

29. In the first five months after its release, the movie *Harry Potter and the Sorcerer's Stone* grossed $317,093,502. Round this number to the millions place.

30. The year 1999 saw the highest number of computer viruses. There were 26,193. Round this number to the nearest thousand.

31. The largest English dictionary contains 21,543 words. Round this number to the thousands place.

32. A shopping center in Edmonton, Alberta, Canada, covers an area of 492,000 square meters (m²). Round this number to the hundred-thousands place.

Objective 2: Estimation

For Exercises 33–36, estimate the sum by first rounding each number to the nearest ten. (See Example 5.)

33.   57
         82
       + 21

34.   33
         78
       + 41

35.   41
         12
      + 129

36.   29
         73
      + 113

For Exercises 37–40, estimate the difference by first rounding each number to the nearest hundred. (See Example 6.)

37.   898
        − 422

38.   731
        − 584

39.   412
        − 252

40.   771
        − 544

Objective 3: Using Estimation in Applications

41. The number of women in the 40–44 age group who gave birth in 1981 is 23,326. By 2001 this number increased to 92,813. Round each value to the nearest thousand to estimate how many more women in the 40–44 age group gave birth in 2001.

42. The number of women in the 45–49 age group who gave birth in 1981 is 1190. By 2001 this number increased to 4844. Round each value to the nearest thousand to estimate how many more women in the 45–49 age group gave birth in 2001.

43. The table shows the sales of the top five candy bars for the year 2001.

Brand	Manufacturer	Sales ($)
M&Ms	Mars	97,404,576
Hershey's Milk Chocolate	Hershey Chocolate	81,296,784
Reese's Peanut Butter Cups	Hershey Chocolate	54,391,268
Snickers	Mars	53,695,428
KitKat	Hershey Chocolate	38,168,580

Round the sales to the nearest million to estimate the total sales brought in by the Mars company. (See Example 7.)

44. Refer to the chart in Exercise 43. Round the sales to the nearest million to estimate the total sales brought in by the Hershey Chocolate Company.

For Exercises 45–48, use the given table.

45. Round the revenue to the nearest hundred-thousand to estimate the total revenue for the years 1996 through 1999.

46. Round the revenue to the nearest hundred-thousand to estimate the total revenue for the years 2000 through 2003.

47. a. Determine the year with the greatest revenue. Round this revenue to the nearest hundred-thousand.

   b. Determine the year with the least revenue. Round this revenue to the nearest hundred-thousand.

Beach Parking Revenue for Daytona Beach, Florida	
Year	Revenue
1994	$3,603,462
1995	3,152,743
1996	3,499,468
1997	3,257,846
1998	3,235,061
1999	3,514,777
2000	3,316,897
2001	3,272,028
2002	3,360,289
2003	3,470,295

Table for Exercises 45–48

48. Estimate the difference between the year with the greatest revenue and the year with the least revenue.

For Exercises 49–52, use the graph provided.

49. Determine the state with the greatest number of students enrolled in grades 6–12. Round this number to the nearest thousand.

50. Determine the state with the least number of students enrolled in grades 6–12. Round this number to the nearest thousand.

51. Use the information in Exercises 49 and 50 to estimate the difference between the number of students in the state with the highest enrollment and that of the lowest enrollment. (See Example 8.)

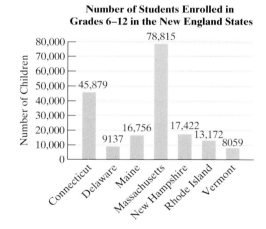

Figure for Exercises 49–52

52. Estimate the total number of students enrolled in grades 6–12 in the New England states by first rounding the number of students to the thousands place.

53. If you were to estimate the following sum, what place value would you round to and why?

$$389,220 + 2988 + 12,824 + 101,333$$

54. Identify the place value that you would round to when estimating the answer to the following problem. Then round the values and estimate the answer.

$$4208 - 932 + 1294$$

Expanding Your Skills

For Exercises 55–58, round the numbers to estimate the perimeter of each figure. (Answers may vary.)

55.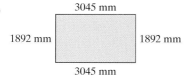

56.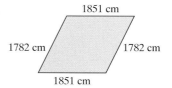

57.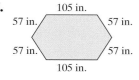

58.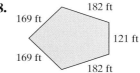

chapter 1 | midchapter review

For Exercises 1–4, add or subtract.

1. $35 + 998$

2. $6723 - 3342$

3. $590 - 489$

4. $9110 + 432$

For Exercises 5–8, determine the place value for the underlined digit.

5. $2\underline{3},981$

6. $8\underline{7}39$

7. $\underline{7}83,870$

8. $\underline{3},324,921$

For Exercises 9–10, estimate the answer by rounding as indicated.

9. $23,981 - 8739$; thousands

10. $783,870 + 3,324,921$; hundred-thousands

For Exercises 11–14, use your knowledge of adding and subtracting to find the number.

11. Find a number that when decreased by 23 yields 245.

12. Find a number that when increased by 23 yields 245.

13. Find a number that when increased by 51 yields 72.

14. Find a number that when decreased by 51 yields 72.

Concept Connections

1. How can multiplication be used to compute the sum $4 + 4 + 4 + 4 + 4 + 4 + 4$?

Concept Connections

2. In the product 4(8), explain why the parentheses are needed to indicate multiplication.

Skill Practice

Identify the factors and the product.

3. $3 \times 11 = 33$
4. $2 \times 5 \times 8 = 80$

Answers

1. 7×4
2. Without the parentheses, the expression would look like the number 48.
3. Factors: 3 and 11; product: 33
4. Factors 2, 5, and 8; product: 80

---

section 1.5   Multiplication of Whole Numbers

1. Introduction to Multiplication

Suppose that Carmen buys three cartons of eggs to prepare a large family brunch. If there are 12 eggs per carton, then the total number of eggs can be found by adding three 12s.

12 eggs

12 eggs

$\underline{+\ 12 \text{ eggs}}$
36 eggs

When each addend in a sum is the same, we have what is called *repeated* addition. Repeated addition is also called **multiplication**. We use the multiplication sign $\times$ to express repeated addition more concisely.

$$12 + 12 + 12 \quad \text{is equal to} \quad 3 \times 12$$

The expression 3×12 is read "3 times 12" to signify that the number 12 is added 3 times. The numbers 3 and 12 are called **factors**, and the number 36 is called the **product**.

The symbol $\cdot$ may also be used to denote multiplication such as in the expression $3 \cdot 12 = 36$. Two factors written adjacent to each other with no other operator between them also implies multiplication. The quantity $2y$, for example, is understood to be 2 times y. If we use this notation to multiply two numbers, parentheses are used to group one or both factors. For example,

$$3(12) = 36 \qquad (3)12 = 36 \qquad \text{and} \qquad (3)(12) = 36$$

all represent the product of 3 and 12.

> **Tip:** In the expression 3(12), the parentheses are necessary because two adjacent factors written together with no grouping symbol would look like the number 312.

The products of one-digit numbers such as $4 \times 5 = 20$ and $2 \times 7 = 14$ are basic facts. All products of one-digit numbers should be memorized (see Exercise 6).

example 1   Identifying Factors and Products

Identify the factors and the product.

a. $6 \times 3 = 18$ **b.** $5 \times 2 \times 7 = 70$

Solution:

a. Factors: 6, 3; product: 18   **b.** Factors: 5, 2, 7; product: 70

2. Properties of Multiplication

Recall from Section 1.2 that the order in which two numbers are added does not affect the sum. The same is true for multiplication. This is stated formally as the *commutative property of multiplication.*

Commutative Property of Multiplication

Changing the order of two factors does not affect the product.

Example: 2×5 is equivalent to 5×2

The following rectangular arrays help us visualize the commutative property of multiplication.

$2 \times 5 = 10$ 2 rows of 5

$5 \times 2 = 10$ 5 rows of 2

Multiplication is also an associative operation.

Associative Property of Multiplication

The manner in which factors are grouped under multiplication does not affect the product.

Example: $(3 \times 5) \times 2$ is equivalent to $3 \times (5 \times 2)$

---

example 2   Applying Properties of Multiplication

a. Rewrite the expression 3×9, using the commutative property of multiplication. Then find the product.

b. Rewrite the expression $(4 \times 2) \times 3$, using the associative property of multiplication. Then find the product.

Solution:

a. $3 \times 9 = 9 \times 3$. The product is 27.

b. $(4 \times 2) \times 3 = 4 \times (2 \times 3)$.

To find the product, we have

$$4 \times (2 \times 3)$$
$$= 4 \times (6)$$
$$= 24$$

The product is 24.

Skill Practice

5. Rewrite the expression 6×5, using the commutative property of multiplication. Then find the product.

6. Rewrite the expression $3 \times (1 \times 7)$, using the associative property of multiplication. Then find the product.

Two other important properties of multiplication involve factors of 0 and factors of 1.

Answers

5. 5×6; product is 30
6. $(3 \times 1) \times 7$; product is 21

Multiplication Property of 0

The product of any number and 0 is 0.

Examples:
$$5 \times 0 = 0$$
$$0 \times 12 = 0$$

The product $5 \times 0 = 0$ can easily be understood by writing the product as repeated addition.

$$\underbrace{0 + 0 + 0 + 0 + 0}_{\text{add 0 five times}} = 0$$

Multiplication Property of 1

The product of any number and 1 is that number.

Examples:
$$1 \times 4 = 4$$
$$3 \times 1 = 3$$

These examples can be understood by considering rectangular arrays.

$1 \times 4 = 4$ one row of 4

$3 \times 1 = 3$ three rows of 1

The last property of multiplication involves both addition and multiplication. First consider the expression $2(4 + 3)$. By performing the operation within parentheses first, we have

$$2(4 + 3) = 2(7) = 14$$

We get the same result by multiplying 2 times each addend within the parentheses:

$$2(4 + 3) = (2 \times 4) + (2 \times 3) = 8 + 6 = 14$$

This result illustrates the **distributive property of multiplication over addition** (sometimes we simply say *distributive property* for short).

Skill Practice

Apply the distributive property and simplify.

7. $2(6 + 4)$

8. $5(0 + 8)$

example 3   Applying the Distributive Property of Multiplication Over Addition

Apply the distributive property and simplify.

a. $3(4 + 8)$ **b.** $7(3 + 0)$

Solution:

a. $3(4 + 8) = (3 \times 4) + (3 \times 8) = 12 + 24 = 36$

b. $7(3 + 0) = (7 \times 3) + (7 \times 0) = 21 + 0 = 21$

Answers

7. $(2 \times 6) + (2 \times 4)$; 20
8. $(5 \times 0) + (5 \times 8)$; 40

3. Multiplying Many-Digit Whole Numbers

When multiplying numbers with several digits, it is sometimes necessary to carry. To see why, consider the product 3×29. By writing the factors in expanded form, we can apply the distributive property. In this way we see that 3 is multiplied by both 20 and 9.

$$3 \times 29 = 3(20 + 9) = (3 \times 20) + (3 \times 9)$$
$$= 60 + 27$$
$$= 6 \text{ tens} + 2 \text{ tens} + 7 \text{ ones}$$
$$= 8 \text{ tens} + 7 \text{ ones}$$
$$= 87$$

Now we will multiply 29×3 in vertical form.

Multiply $3 \times 9 = 27$. Write the 7 in the ones column and carry the 2.

Multiply 3×2 tens $= 6$ tens. Add the carry: 6 tens + 2 tens = 8 tens. Write the 8 in the tens place.

---

example 4   Multiplying a Many-Digit Number by a One-Digit Number

Multiply.

$$\begin{array}{r} 368 \\ \times\ 5 \end{array}$$

Solution:

Using the distributive property, we have

$$5(300 + 60 + 8) = 1500 + 300 + 40 = 1840$$

This can be written vertically as:

$$\begin{array}{r} 368 \\ \times\ 5 \\ \hline 40 \\ 300 \\ +\ 1500 \\ \hline 1840 \end{array}$$

Multiply 5×8.
Multiply 5×60.
Multiply 5×300.
Add.

The numbers 40, 300, and 1500 are called *partial sums*. The product of 386 and 5 is found by adding the partial sums. The product is 1840.

■

The solution to Example 4 can also be found by using a shorter form of multiplication. We outline the procedure:

$$\begin{array}{r} \overset{4}{3}68 \\ \times\quad 5 \\ \hline 0 \end{array}$$

Multiply $5 \times 8 = 40$. Write the 0 in the ones place and carry the 4.

$$
\begin{array}{r}
{\scriptstyle 3\,4} \\
368 \\
\times \quad 5 \\
\hline
40
\end{array}
$$

Multiply 5×6 tens $= 300$. Add the carry. $300 + 4$ tens $= 340$. Write the 4 in the tens place and carry the 3.

$$
\begin{array}{r}
{\scriptstyle 3\,4} \\
368 \\
\times \quad 5 \\
\hline
1840
\end{array}
$$

Multiply 5×3 hundreds $= 1500$. Add the carry. $1500 + 3$ hundreds $= 1800$. Write the 8 in the hundreds place and the 1 in the thousands place.

The next example demonstrates the process to multiply two factors with many digits.

Skill Practice

10. Multiply.

$$
\begin{array}{r}
59 \\
\times\ 26 \\
\hline
\end{array}
$$

---

example 5   Multiplying a Many-Digit Number by a Many-Digit Number

Multiply:

$$
\begin{array}{r}
72 \\
\times\ 83 \\
\hline
\end{array}
$$

Solution:

Writing the problem vertically and computing the partial sums, we have

$$
\begin{array}{r}
72 \\
\times\ 83 \\
\hline
216 \\
+\ 5760 \\
\hline
5976
\end{array}
$$

Multiply 3×72.
Multiply 80×72.
Add.

The product is 5976.

The procedure to use the short form of multiplication is as follows.

Step 1:

$$
\begin{array}{r}
72 \\
\times\ 83 \\
\hline
216
\end{array}
$$

Multiply 3×2. Write the product, 6, in the ones column.
Multiply 3×7 tens $= 210$. Write the 1 in the tens column.
Write the 2 in the hundreds column.

Step 2:

$$
\begin{array}{r}
{\scriptstyle 1} \\
72 \\
\times\ 83 \\
\hline
216 \\
5760
\end{array}
$$

Multiply 8 tens $\times 2 = 160$. Write the 0 in the ones column and the 6 in the tens column. Carry the 1.
Multiply 8 tens $\times 7$ tens $= 5600$.
Add the carry: $5600 + 1$ hundred $= 5700$.
Write the 7 in the hundreds place. Write the 5 in the thousands place.

Step 3:

$$
\begin{array}{r}
{\scriptstyle 1} \\
72 \\
\times\ 83 \\
\hline
216 \\
+\ 5760 \\
\hline
5976
\end{array}
$$

Add.

Answer

10. 1534

example 6	Multiplying Two Multidigit Whole Numbers

Use the short-form procedure to compute 368×497.

Solution:

$$
\begin{array}{r}
2\,3 \\
6\,7 \\
4\,5 \\
368 \\
\times\ 497 \\
\hline
2576 \\
33120 \\
+\ 147200 \\
\hline
182{,}896
\end{array}
$$

Skill Practice

11. Multiply.

$$
\begin{array}{r}
274 \\
\times\ 586
\end{array}
$$

4. Estimating Products by Rounding

A special pattern occurs when one or more factors in a product ends in zero. Consider the following products:

$$
\begin{aligned}
12 \times 20 &= 240 & 120 \times 20 &= 2400 \\
12 \times 200 &= 2400 & 1200 \times 20 &= 24{,}000 \\
12 \times 2000 &= 24{,}000 & 12{,}000 \times 20 &= 240{,}000
\end{aligned}
$$

Notice in each case the product is $12 \times 2 = 24$ followed by the total number of zeros from each factor. Consider the product 1200×20.

$$
\begin{array}{r|l}
12 & 00 \\
\times\ 2 & 0 \\
\hline
24 & 000
\end{array}
$$
Shift the numbers 1200 and 20 so that the zeros appear to the right of the multiplication process. Multiply $12 \times 2 = 24$.
Write the product 24 followed by the total number of zeros from each factor.

Concept Connections

12. How many zeros would the product $3000 \times 400{,}000$ have?

example 7	Estimating a Product

Estimate the product 795×4060 by rounding 795 to the nearest hundred and 4060 to the nearest thousand.

Solution:

$$
\begin{array}{lcl}
795 & \text{rounds to} \longrightarrow & 800 \\
4060 & \text{rounds to} \longrightarrow & 4000
\end{array}
\qquad
\begin{array}{r|l}
8 & 00 \\
\times\ 4 & 000 \\
\hline
32 & 00000
\end{array}
$$

The product is approximately 3,200,000.

Skill Practice

13. Estimate the product 421×869 by rounding each factor to the nearest hundred.

example 8	Estimating a Product in an Application

For a trip from Atlanta to Los Angeles, the average cost of a plane ticket was $495. If the plane carried 218 passengers, estimate the total revenue for the airline. (*Hint*: Round each number to the hundreds place and find the product.)

Skill Practice

14. A small newspaper has 16,850 subscribers. Each subscription costs $149 per year. Estimate the revenue for the year by rounding the number of subscriptions to the nearest thousand and the cost to the nearest ten.

Answers

11. 160,564   12. 8 zeros
13. 360,000   14. $2,550,000

Solution:

$495   rounds to ⟶ $ 5 | 00

218   rounds to ⟶   × 2 | 00

$10 | 0000

The airline received approximately $100,000 in revenue.

5. Translations and Applications Involving Multiplication

In English there are many different words that imply multiplication. A partial list is given in Table 1-3.

table 1-3

Word/Phrase	Example	In Symbols
Product	The product of 4 and 7	4×7
Times	8 times 4	8×4
Multiply … by…	Multiply 6 by 3	6×3

Multiplication may also be warranted in applications involving unit rates. In Example 8, we multiplied the cost per customer ($495) by the number of customers (218). The value $495 is a unit rate because it gives the cost per one customer (per one unit).

Skill Practice

15. Gail can type 65 words per minute. How many words can she type in 45 minutes?

example 9   Solving an Application Involving Multiplication

The average weekly income for production workers is $489. How much does a production worker make in 1 year (assume 52 weeks in 1 year).

Solution:

The value $489 per week is a unit rate. The total earnings for 1 year are given by $489 × 52.

$$
\begin{array}{r}
{}^{4\,4} \\
{}^{1\,1} \\
489 \\
\times\ 52 \\
\hline
978 \\
+\ 24450 \\
\hline
25428
\end{array}
$$

The yearly earnings are $25,428.

Tip: This product can be estimated quickly by rounding the factors.

489   rounds to ⟶   5 | 00

52   rounds to ⟶   × 5 | 0

25 | 000

The total yearly income is approximately $25,000. Estimating gives a quick approximation of a product. Furthermore, it also checks for the reasonableness of our exact product. In this case $25,000 is close to our exact value of $25,428.

Answer

15. 2925 words

6. Area of a Rectangle

Another application of multiplication of whole numbers lies in finding the area of a region. **Area** measures the amount of surface contained within the region. For example, a square that is 1 in. by 1 in. occupies an area of 1 square inch, denoted as 1 in^2. Similarly, a square that is 1 centimeter (cm) by 1 cm occupies an area of 1 square centimeter. This is denoted by 1 cm^2.

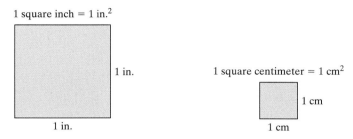

The units of square inches and square centimeters (in.^{2} and cm^2) are called *square units*. For larger regions, we measure the number of square units occupied in that region. For example, the region in Figure 1-6 occupies 6 cm^2.

Figure 1-6

The 3-cm by 2-cm region in Figure 1-6 suggests that to find the **area of a rectangle**, multiply the length by the width. If the area is represented by A, the length is represented by l, and the width is represented by w, then we have

$$\text{Area of rectangle} = (\text{length}) \times (\text{width})$$

$$A = l \times w$$

The letters A, l, and w are called **variables** because their values *vary* as they are replaced by different numbers.

---

example 10   Finding the Area of a Rectangle

Find the area and perimeter of the rectangle.

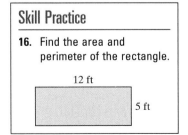

Solution:

Area:

$A = l \times w$

$A = (7 \text{ yd}) \times (4 \text{ yd})$

$\quad = 28 \text{ yd}^2$

Recall from Section 1.2 that the perimeter of a polygon is the sum of the lengths of the sides. In a rectangle the opposite sides are equal in length. Thus,

Perimeter:

$P = 7 \text{ yd} + 4 \text{ yd} + 7 \text{ yd} + 4 \text{ yd}$

$\quad = 22 \text{ yd}$

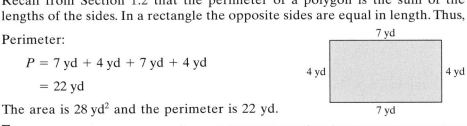

The area is 28 yd^2 and the perimeter is 22 yd.

Skill Practice

16. Find the area and perimeter of the rectangle.

12 ft

5 ft

Avoiding Mistakes: Notice that area is measured in square units (such as yd^2) and perimeter is measured in units of length (such as yd). It is important to apply the correct units of measurement.

Answer

16. Area: 60 ft^2; perimeter: 34 ft

example 11   Finding Area in an Application

The state of Wyoming is approximately the shape of a rectangle (Figure 1-7). Its length is 355 mi and its width is 276 mi. Approximate the total area of Wyoming by rounding the length and width to the nearest ten.

WY   276 mi

355 mi

Figure 1-7

Solution:

$$
\begin{array}{r}
\overset{1}{\overset{4}{}}\ \\
36\,|\,0 \\
\times\ 28\,|\,0 \\
\hline
288 \\
720 \\
\hline
1008\,|\,00
\end{array}
$$

355   rounds to ⟶
276   rounds to ⟶

The area of Wyoming is approximately 100,800 mi².

Answer

17. 13,244 ft²

Study Tips

1. List the materials that you need to bring to class every day (for example, paper, pencil, etc.).

2. Define the key terms.

   a. Multiplication   **b. Factor**   **c. Product**

   d. Distributive property of multiplication over addition

   e. Area   **f. Area of a rectangle**   **g. Variable**

Review Exercises

For Exercises 3–5, estimate the answer by rounding to the indicated place value.

3. $869,240 + 34,921 + 108,332$; ten-thousands

4. $907,801 - 413,560$; hundred-thousands

5. $8821 - 3401$; hundreds

Objective 1: Introduction to Multiplication

6. Fill out the chart of multiplication facts.

$\times$	0	1	2	3	4	5	6	7	8	9
0										
1										
2										
3										
4										
5										
6										
7										
8										
9										

For Exercises 7–10, write the repeated addition as multiplication and simplify.

7. $5 + 5 + 5 + 5 + 5 + 5$

8. $2 + 2 + 2 + 2 + 2 + 2 + 2 + 2 + 2$

9. $9 + 9 + 9$

10. $7 + 7 + 7 + 7$

For Exercises 11–16, identify the factors and the product. **(See Example 1.)**

11. $6 \times 4 = 24$

12. $5 \times 8 = 40$

13. $13 \times 42 = 546$

14. $26 \times 9 = 234$

15. $3 \cdot 5 \cdot 2 = 30$

16. $4 \cdot 1 \cdot 8 = 32$

17. Write the product of 5 and 12, using three different notations. (Answers may vary.)

18. Write the product of 23 and 14, using three different notations. (Answers may vary.)

Objective 2: Properties of Multiplication

For Exercises 19–24, match the property with the expression.

19. $8 \times 1 = 8$

a. Commutative property of multiplication

20. $6 \cdot 13 = 13 \cdot 6$

b. Associative property of multiplication

21. $2(6 + 12) = 2 \cdot 6 + 2 \cdot 12$

c. Multiplication property of 0

22. $5 \cdot (3 \cdot 2) = (5 \cdot 3) \cdot 2$

d. Multiplication property of 1

23. $0 \times 4 = 0$

e. Distributive property of multiplication over addition

24. $7(14) = 14(7)$

For Exercises 25–30, rewrite the expression, using the indicated property. **(See Examples 2–3.)**

25. 14×8; commutative property of multiplication

26. 3×9; commutative property of multiplication

27. $6 \times (2 \times 10)$; associative property of multiplication

28. $(4 \times 15) \times 5$; associative property of multiplication

29. $5(7 + 4)$; distributive property of multiplication over addition

30. $3(2 + 6)$; distributive property of multiplication over addition

Objective 3: Multiplying Many-Digit Whole Numbers

For Exercises 31–60, multiply. **(See Examples 4–6.)**

31. $\begin{array}{r} 24 \\ \times\ 6 \\ \hline \end{array}$

32. $\begin{array}{r} 18 \\ \times\ 5 \\ \hline \end{array}$

33. $\begin{array}{r} 26 \\ \times\ 2 \\ \hline \end{array}$

34. $\begin{array}{r} 71 \\ \times\ 3 \\ \hline \end{array}$

35. $\begin{array}{r} 131 \\ \times\ 5 \\ \hline \end{array}$

36. $\begin{array}{r} 720 \\ \times\ 3 \\ \hline \end{array}$

37. $\begin{array}{r} 344 \\ \times\ 4 \\ \hline \end{array}$

38. $\begin{array}{r} 105 \\ \times\ 9 \\ \hline \end{array}$

39. $\begin{array}{r} 1410 \\ \times\ 8 \\ \hline \end{array}$

40. $\begin{array}{r} 2016 \\ \times\ 6 \\ \hline \end{array}$

41. $\begin{array}{r} 3312 \\ \times\ 7 \\ \hline \end{array}$

42. $\begin{array}{r} 4801 \\ \times\ 5 \\ \hline \end{array}$

43. $\begin{array}{r} 42{,}014 \\ \times\ 9 \\ \hline \end{array}$

44. $\begin{array}{r} 51{,}006 \\ \times\ 8 \\ \hline \end{array}$

45. $\begin{array}{r} 32 \\ \times\ 14 \\ \hline \end{array}$

46. $\begin{array}{r} 41 \\ \times\ 21 \\ \hline \end{array}$

47. 68 · 24   **48.** 55 · 41   **49.** 72 · 12   **50.** 13 · 46

51. (143)(17)   **52.** (722)(28)   **53.** (349)(19)   **54.** (512)(31)

55.   151   **56.**   703   **57.**   222   **58.**   387
     × 127   × 146   × 841   × 506

59.   3532   **60.**   2810
     × 6014   × 1039

Objective 4: Estimating Products by Rounding

For Exercises 61–68, multiply the numbers, using the method found on page 47. **(See Example 7.)**

61.   600   **62.**   900   **63.**   3000   **64.**   4000
     × 40   × 50   × 700   × 400

65.   8000   **66.**   1000   **67.** 90,000   **68.** 50,000
     × 9000   × 2000   × 400   × 6000

For Exercises 69–72, estimate the product by first rounding the number to the indicated place value.

69. 11,784 × 5201; thousands place   **70.** 45,046 × 7812; thousands place

71. 82,941 × 29,740; ten-thousands place   **72.** 630,229 × 71,907; ten-thousands place

73. Suppose a hotel room costs $189 per night. Round this number to the nearest hundred to estimate the cost for a five-night stay. **(See Example 8.)**

74. The science department of Comstock High School must purchase a set of calculators for a class. If the cost of one calculator is $129, estimate the cost of 28 calculators by rounding the numbers to the nearest tens place.

75. The average price for a ticket to see U2 during their 2001 Elevation tour was $78 (rounded). If a concert stadium seats 10,256 fans, estimate the amount of money made during that performance by rounding the number of seats to the nearest ten-thousand.

76. A breakfast buffet at a local restaurant serves 48 people. Estimate the maximum revenue for one week (7 days) if the price of a breakfast is $12.

Objective 5: Translation and Applications Involving Multiplication

77. If it takes 15 pounds (lb) of grapes to make 1 gallon (gal) of red wine, how many pounds of grapes would be needed for 25 gal? **(See Example 9.)**

78. Dustin's monthly homeowner's association dues are $82. How much will he pay the association in 1 year (12 months)?

79. It costs about $45 for a cat to have a medical exam. If a humane society has 37 cats, find the cost of medical exams for their cats.

80. A can of Coke contains 12 fluid ounces (fl oz). Find the number of ounces in a case of Coke containing 12 cans.

81. PaperWorld shipped 115 cases of copy paper to a business. There are 5 reams of paper in each case and 500 sheets of paper in each ream. Find the number of sheets of paper delivered to the business.

82. A dietary supplement bar has 14 grams (g) of protein. If Kathleen eats 2 bars a day for 6 days, how many grams of protein will she get from this supplement?

83. Tylee's car gets 31 miles per gallon (mpg) on the highway. How many miles can he travel if he has a full tank of gas (12 gal)?

84. Sherica manages a small business called Pizza Express. She has 23 employees who work an average of 32 hours (hr) per week. How many hours of work does Sherica have to schedule each week?

Objective 6: Area of a Rectangle

For Exercises 85–88, find the area of the rectangle. **(See Example 10.)**

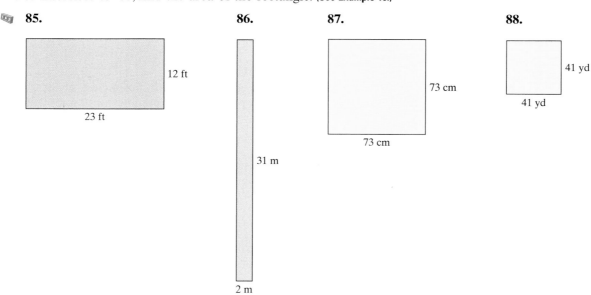

85.

12 ft

23 ft

86.

31 m

2 m

87.

73 cm

73 cm

88.

41 yd

41 yd

89. Mr. Beckwith wants to paint his garage door that is 8 ft by 16 ft. To decide how much paint to buy, he must find the area of the door. What is the area of the door? **(See Example 11.)**

90. Mrs. Patel must have her driveway pressure-cleaned. The cost of the job depends on the area of the driveway. Find the area of the driveway given that it is a rectangle with a length of 80 ft and a width of 12 ft.

91. The front of a building has windows that are 44 in. by 58 in.

  a. Find the area of one window.

  b. If the building has 3 floors and each floor has 14 windows, how many windows are there?

  c. What is the total area of all the windows?

92. A Monopoly game board is 19 in. by 19 in. Find the area of the game board.

Calculator Connections

Topic: Multiplying on a calculator

To multiply numbers on a calculator, use the $\boxed{\times}$ key.

Expression	Keystrokes	Result
38,319 × 1561	38319 $\boxed{\times}$ 1561 $\boxed{\text{Enter}}$	59815959

Calculator Exercises

For Exercises 93–96, solve the problem. Use a calculator to perform the calculations.

93. The United States consumes approximately 21,000,000 barrels (bbl) of oil per day. How much does it consume in 1 year?

94. The average time to commute to work for people living in Washington state is 26 minutes (min) (round trip 52 min). How much time does a person spend commuting to and from work if the person works 5 days a week for 50 weeks per year?

95. The cost of tuition, room, and board for Yale University for the 2004–2005 school year was reported as $41,470 per year (Source: *International Herald Tribune*). What would be the total cost for 1300 students to go to school at Yale for 4 years? (Assume that the yearly rate does not change.)

96. The average cost of tuition to attend a public 4-year university is $5295 per year. What would be the total cost of tuition for 20,000 students to attend for 4 years?

section 1.6   Division of Whole Numbers

1. Introduction to Division

Suppose 12 pieces of pizza are to be divided evenly among 4 children (Figure 1-8). The number of pieces that each child would receive is given by $12 \div 4$, read "12 divided by 4."

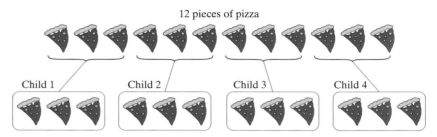

Figure 1-8

The process of separating 12 pieces of pizza among 4 children is called **division**. The statement $12 \div 4 = 3$ indicates that each child receives 3 pieces of pizza. The number 12 is called the **dividend**. It represents the number to be divided. The number 4 is called the **divisor**, and it represents the number of groups. The result of the division (in this case 3) is called the **quotient**. It represents the number of items in each group.

Division may be represented in several ways. For example, the following are all equivalent statements.

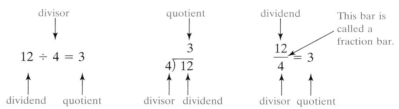

Recall that subtraction is the reverse operation of addition. In the same way, division is the reverse operation of multiplication. For example, we say $12 \div 4 = 3$ because $3 \times 4 = 12$.

---

example 1   Identifying the Dividend, Divisor, and Quotient

Simplify each expression. Then identify the dividend, divisor, and quotient.

a. $48 \div 6$ **b.** $9\overline{)36}$ **c.** $\dfrac{63}{7}$

Solution:

a. $48 \div 6 = 8$ because $8 \times 6 = 48$
The dividend is 48, the divisor is 6, and the quotient is 8.

b. $9\overline{)36}$ (quotient 4)   because $4 \times 9 = 36$

The dividend is 36, the divisor is 9, and the quotient is 4.

c. $\dfrac{63}{7} = 9$ because $9 \times 7 = 63$

The dividend is 63, the divisor is 7, and the quotient is 9.

2. Properties of Division

Example 2 illustrates some of the interesting properties of division.

Dividing Whole Numbers

Divide.

a. $8 \div 8$ **b.** $\dfrac{6}{6}$ **c.** $5 \div 1$

d. $1\overline{)7}$ **e.** $0 \div 6$ **f.** $\dfrac{0}{4}$

Solution:

a. $8 \div 8 = 1$ because $1 \times 8 = 8$

b. $\dfrac{6}{6} = 1$ because $1 \times 6 = 6$

c. $5 \div 1 = 5$ because $5 \times 1 = 5$

d. $\begin{array}{r} 7 \\ 1\overline{)7} \end{array}$ because $7 \times 1 = 7$

e. $0 \div 6 = 0$ because $0 \times 6 = 0$

f. $\dfrac{0}{4} = 0$ because $0 \times 4 = 0$

Skill Practice

Divide.

4. $3\overline{)3}$ **5.** $5 \div 5$

6. $\dfrac{4}{1}$ **7.** $8 \div 1$

8. $\dfrac{0}{7}$ **9.** $3\overline{)0}$

Properties of Division

1. Any number divided by itself is 1. See Examples 2(a) and 2(b).
2. Any number divided by 1 is the number itself. See Examples 2(c) and 2(d).
3. Zero divided by any nonzero number is zero. See Examples 2(e) and 2(f).

Note that $0 \div 6 = 0$. However, reversing the dividend and divisor produces an undefined quotient. That is,

$$6 \div 0 \text{ is undefined}$$

This is so because there is no number that when multiplied by 0, will produce a product of 6.

You should also note that unlike addition and multiplication, division is neither commutative nor associative. In other words, reversing the order of the dividend and divisor may produce a different quotient. Similarly, changing the manner in which numbers are grouped with division may affect the outcome. See Exercises 31 and 32.

Concept Connections

10. Which expression is undefined?

    $\dfrac{5}{0}$ or $\dfrac{0}{5}$

11. Which expression is equal to zero?

    $0\overline{)4}$ or $4\overline{)0}$

3. Long Division

To divide larger numbers we use a process called **long division**. This process uses a series of estimates to find the quotient. We illustrate long division in Example 3.

Answers

4. 1   5. 1   6. 4   7. 8

8. 0   9. 0   10. $\dfrac{5}{0}$ 11. $4\overline{)0}$

Skill Practice

12. Divide.

$8\overline{)136}$

---

example 3   Using Long Division

Divide.

$$7\overline{)161}$$

Solution:

Estimate $7\overline{)161}$ by first estimating $7\overline{)16}$ and writing the result in the tens place of the quotient. Since $7 \times 2 = 14$, there are at least 2 sevens in 16.

$$\begin{array}{r} 2 \\ 7\overline{)161} \\ -140 \\ \hline 21 \end{array}$$

The 2 in the tens place represents 20 in the quotient.
←Multiply 7×20 and write the result under the dividend.
Subtract 140. We see that our estimate leaves 21.

Repeat the process. Now divide $7\overline{)21}$ and write the result in the ones place of the quotient.

$$\begin{array}{r} 23 \\ 7\overline{)161} \\ -140 \\ \hline 21 \\ -21 \\ \hline 0 \end{array}$$

← Multiply 7×3.
Subtract.

The quotient is 23.   Check: $\begin{array}{r} 23 \\ \times 7 \\ \hline 161 \end{array}$ ✔

We can streamline the process of long division by "bringing down" digits of the dividend one at a time.

Skill Practice

13. Divide.

$2891 \div 7$

---

example 4   Using Long Division

Divide.

$$6138 \div 9$$

Solution:

$$\begin{array}{r} 682 \\ 9\overline{)6138} \\ -54 \\ \hline 73 \\ -72 \\ \hline 18 \\ -18 \\ \hline 0 \end{array}$$

$9 \times 6 = 54$ and subtract.
Bring down the 3.
$9 \times 8 = 72$ and subtract.
Bring down the 8.
$9 \times 2 = 18$ and subtract.

The quotient is 682.   Check: $\begin{array}{r} {}^{7\,1} \\ 682 \\ \times\ 9 \\ \hline 6138 \end{array}$ ✔

Answers
12. 17
13. 413

In many instances, quotients do not come out evenly. For example, suppose we had 13 pieces of pizza to distribute among 4 children (Figure 1-9).

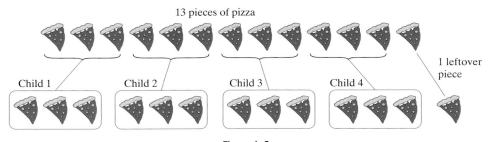

Figure 1-9

The mathematical term given to the "leftover" piece is called the **remainder**. The division process may be written as

$$\begin{array}{r} 3\text{ R1} \\ 4\overline{)13} \\ -12 \\ \hline 1 \end{array}$$

The remainder is written next to the 3.

The **whole part of the quotient** is 3, and the remainder is 1. Notice that the remainder is written next to the whole part of the quotient.

We can check a division problem that has a remainder. To do so, multiply the divisor by the whole part of the quotient and then add the remainder. The result must equal the dividend. That is,

(Divisor)(whole part of quotient) + remainder = dividend

Thus,

$$(4)(3) + 1 \stackrel{?}{=} 13$$
$$12 + 1 \stackrel{?}{=} 13$$
$$13 = 13 \ ✔$$

---

example 5   Using Long Division

Divide.

$$1595 \div 6$$

Solution:

$$\begin{array}{r} 265\text{ R5} \\ 6\overline{)1595} \\ -12 \\ \hline 39 \\ -36 \\ \hline 35 \\ -30 \\ \hline 5 \end{array}$$

$6 \times 2 = 12$ and subtract.
Bring down the 9.
$6 \times 6 = 36$ and subtract.
Bring down the 5.
$6 \times 5 = 30$ and subtract.
The remainder is 5.

To check, verify that $6 \times 265 + 5 = 1595.$ ✔

■

Skill Practice

14. Divide.
    $1482 \div 5$

Answer
14. 296 R2

4. Dividing by a Many-Digit Divisor

When the divisor has more than one digit, we still use a series of estimations to find the quotient.

---

example 6   Dividing by a Two-Digit Number

Divide.

$$32\overline{)1259}$$

Solution:

To estimate the leading digit of the quotient, estimate the number of times 30 will go into 125. Since $30 \cdot 4 = 120$, our estimate is 4.

$$\begin{array}{r} 4 \\ 32\overline{)1259} \\ -128 \end{array}$$

$32 \times 4 = 128$ is too big. We cannot subtract 128 from 125. Revise the estimate in the quotient to 3.

$$\begin{array}{r} 3 \\ 32\overline{)1259} \\ -96\downarrow \\ \hline 299 \end{array}$$

$32 \times 3 = 96$ and subtract.

Bring down the 9.

Now estimate the number of times 30 will go into 299. Because $30 \times 9 = 270$, our estimate is 9.

$$\begin{array}{r} 39 \ \text{R}11 \\ 32\overline{)1259} \\ -96\downarrow \\ \hline 299 \\ -288 \\ \hline 11 \end{array}$$

$32 \times 9 = 288$ and subtract.

The remainder is 11.

To check, verify that $32 \times 39 + 11 = 1259$. ✔

---

example 7   Dividing by a Many-Digit Number

Divide.

$$\frac{82{,}705}{602}$$

Solution:

$$\begin{array}{r} 137 \ \text{R}231 \\ 602\overline{)82{,}705} \\ -602\downarrow \\ \hline 2250 \\ -1806\downarrow \\ \hline 4445 \\ -4214 \\ \hline 231 \end{array}$$

$602 \times 1 = 602$ and subtract.

Bring down the 0.

$602 \times 3 = 1806$ and subtract.

Bring down the 5.

$602 \times 7 = 4214$ and subtract.

The remainder is 231.

To check, verify that $602 \times 137 + 231 = 82{,}705$. ✔

5. Translations and Applications Involving Division

Several words and phrases imply division. A partial list is given in Table 1-4.

table 1-4

Word/Phrase	Example	In Symbols		
Divide	Divide 12 by 3	$12 \div 3$ or	$\dfrac{12}{3}$ or	$3\overline{)12}$
Quotient	The quotient of 20 and 2	$20 \div 2$ or	$\dfrac{20}{2}$ or	$2\overline{)20}$
Per	110 mi per 2 hr	$110 \div 2$ or	$\dfrac{110}{2}$ or	$2\overline{)110}$
Divides into	4 divides into 28	$28 \div 4$ or	$\dfrac{28}{4}$ or	$4\overline{)28}$
Divided, or shared equally among	64 shared equally among 4	$64 \div 4$ or	$\dfrac{64}{4}$ or	$4\overline{)64}$

---

example 8   Solving an Application Involving Division

A painting business employs 3 painters. The business collects $1950 for painting a house. If all painters are paid equally, how much does each person make?

Solution:

This is an example where $1950 is shared equally among 3 people. Therefore, we divide.

```
      650
  3)1950
   -18↓       3 × 6 = 18 and subtract.
    15|       Bring down the 5.
   -15↓       3 × 5 = 15 and subtract.
    00        Bring down the 0. The remainder is 0.
```

Each painter makes $650.

Skill Practice

17. Four players play Hearts with a standard 52-card deck of cards. If the cards are equally distributed, how many cards does each player get?

---

example 9   Solving an Application Involving Division with Estimation

Elaine and Max drove from South Bend, Indiana, to Bonita Springs, Florida. The total driving distance was 1089 mi, and the driving time was approximately 20 hr. Estimate the average speed by rounding the distance to the nearest hundred.

Skill Practice

18. A college has budgeted $4800 to buy graphing calculators. Each calculator costs $119. Estimate the number of calculators that the college can buy by rounding the cost to the nearest ten.

Answers

17. 13 cards   18. 40 calculators

Solution:

1089 mi rounds to 1100 mi. The speed is represented by 1100 mi per 20 hr, or

$$\begin{array}{r} 55 \\ 20\overline{)1100} \\ -100\downarrow \\ \hline 100 \\ -100 \\ \hline 0 \end{array}$$

$20 \times 5 = 100$ and subtract.

Bring down the 0.

$20 \times 5 = 100$ and subtract.

Max and Elaine averaged approximately 55 miles per hour (mph).

Skill Practice

19. The cost for four different types of pastry at a French bakery is shown in the graph.

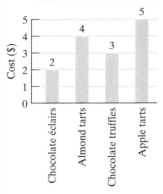

Cost for Selected Pastries

Melissa has $360 to spend on desserts.

a. If she spends all the money on chocolate éclairs, how many can she buy?

b. If she spends all the money on apple tarts, how many can she buy?

example 10   Solving an Application Involving Division

The chart in Figure 1-10 depicts the number of calories burned per hour for selected activities.

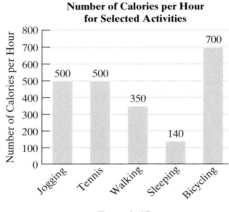

Figure 1-10

a. Janie wants to burn 3500 calories per week exercising. For how many hours must she jog?

b. For how many hours must Janie bicycle to burn 3500 calories?

Solution:

a. The total number of calories must be distributed in 500-calorie increments. Thus, the number of hours required is given by $3500 \div 500$.

$$\begin{array}{r} 7 \\ 500\overline{)3500} \\ -3500 \\ \hline 0 \end{array}$$

Janie requires 7 hr of jogging to burn 3500 calories.

b. 3500 calories must be distributed in 700-calorie increments. The number of hours required is given by $3500 \div 700$.

$$\begin{array}{r} 5 \\ 700\overline{)3500} \\ -3500 \\ \hline 0 \end{array}$$

Janie requires 5 hr of bicycling to burn 3500 calories.

Answers

19. a. 180 chocolate éclairs
   b. 72 apple tarts

section 1.6   Practice Exercises

Study Skills Exercises

1. In your next math class, take notes by drawing a vertical line about three-fourths of the way across the paper, as shown. On the left side, write down what your instructor puts on the board or overhead. On the right side, make your own comments about important words, procedures, or questions that you have.

2. Define the key terms.

   a. Division   **b. Dividend**   **c. Divisor**   **d. Quotient**

   e. Long division   **f. Remainder**   **g. Whole part of the quotient**

Review Exercises

For Exercises 3–10, add, subtract, or multiply as indicated.

3. $48 \cdot 103$ **4.** $678 - 83$ **5.** $1008 + 245$ **6.** $14(220)$

7. 5230×127 **8.** $789(25)$ **9.** $4890 - 3988$ **10.** $38{,}002 + 3902$

Objective 1: Introduction to Division

For Exercises 11–16, simplify each expression. Then identify the dividend, divisor, and quotient. **(See Example 1.)**

11. $72 \div 8$ **12.** $32 \div 4$ **13.** $8\overline{)64}$

14. $5\overline{)35}$ **15.** $\dfrac{45}{9}$ **16.** $\dfrac{20}{5}$

Objective 2: Properties of Division

17. In your own words, explain the difference between dividing a number by zero and dividing zero by a number.

18. Explain what happens when a number is either divided or multiplied by 1.

For Exercises 19–30, use the properties of division to simplify the expression, if possible. **(See Example 2.)**

19. $15 \div 1$

20. $21\overline{)21}$

21. $0 \div 10$

22. $\dfrac{0}{3}$

23. $0\overline{)9}$

24. $4 \div 0$

25. $\dfrac{20}{20}$

26. $1\overline{)9}$

27. $\dfrac{16}{0}$

28. $\dfrac{5}{1}$

29. $8\overline{)0}$

30. $13 \div 13$

31. Show that $6 \div 3 = 2$ but $3 \div 6 \neq 2$ by using multiplication to check.

32. Show that division is not associative, using the numbers 36, 12, and 3.

Objective 3: Long Division

33. Explain the process for checking a division problem when there is no remainder.

34. Show how checking by multiplication can help us remember that $0 \div 5 = 0$ and that $5 \div 0$ is undefined.

For Exercises 35–46, divide and check by multiplying. **(See Examples 3 and 4.)**

35. $78 \div 6$

36. $364 \div 7$

37. $5\overline{)205}$

38. $8\overline{)152}$

39. $\dfrac{972}{2}$

40. $\dfrac{582}{6}$

41. $1227 \div 3$

42. $236 \div 4$

43. $4\overline{)3808}$

44. $3\overline{)2895}$

45. $\dfrac{4932}{6}$

46. $\dfrac{3619}{7}$

For Exercises 47–54, check the following division problems. If it does not check, find the correct answer.

47. $4\overline{)224}^{\,56}$

48. $7\overline{)574}^{\,82}$

49. $761 \div 3 = 253 \text{ R2}$

50. $604 \div 5 = 120 \text{ R4}$

51. $\dfrac{1021}{9} = 113 \text{ R4}$

52. $\dfrac{1311}{6} = 218 \text{ R3}$

53. $8\overline{)203}^{\,25 \text{ R6}}$

54. $7\overline{)821}^{\,117 \text{ R5}}$

For Exercises 55–70, divide and check the answer. **(See Example 5.)**

55. $61 \div 8$

56. $89 \div 3$

57. $9\overline{)92}$

58. $5\overline{)74}$

59. $\dfrac{55}{2}$

60. $\dfrac{49}{3}$

61. $593 \div 3$

62. $801 \div 4$

63. $\dfrac{382}{9}$

64. $\dfrac{428}{8}$

65. $3115 \div 2$

66. $4715 \div 6$

67. $6014 \div 8$

68. $9013 \div 7$

69. $6\overline{)5012}$

70. $2\overline{)1101}$

Objective 4: Dividing by a Many-Digit Divisor

For Exercises 71–82, divide. **(See Examples 6 and 7.)**

71. $9110 \div 19$

72. $3505 \div 13$

73. $24\overline{)1051}$

74. $41\overline{)8104}$

75. $\dfrac{8008}{26}$

76. $\dfrac{9180}{15}$

77. $68{,}012 \div 54$

78. $92{,}013 \div 35$

79. $69{,}712 \div 304$

80. $51{,}107 \div 221$

81. $114\overline{)34{,}428}$

82. $421\overline{)87{,}989}$

Objective 5: Translations and Applications Involving Division

For Exercises 83–88, for each English sentence, write a mathematical expression and simplify.

83. Find the quotient of 497 and 71.

84. Find the quotient of 1890 and 45.

85. Divide 877 by 14.

86. Divide 722 by 53.

87. Divide 6 into 42.

88. Divide 9 into 108.

89. There are 392 students signed up for Anatomy 101. If each classroom can hold 28 students, find the number of classrooms that are needed. **(See Example 8.)**

90. A wedding reception is planned to take place in the fellowship hall of a church. The bride anticipates 120 guests, and each table will seat 8 people. How many tables should be set up for the reception to accommodate all the guests?

91. A case of tomato sauce contains 32 cans. If a grocer has 168 cans, how many cases can he fill completely? How many cans will be left over?

92. Austin has $425 to spend on dining room chairs. If each chair costs $52, does he have enough to purchase 8 chairs? If so, will he have any money left over?

93. Pauline drove 312 mi in 6 hr. Find Pauline's average speed (in miles per hour).

94. A house cleaning company charges $144 to clean a 3-room apartment. At this rate, how much does it cost to clean 1 room?

95. If it takes 2200 lb of grapes to make 100 gal of white wine, how many pounds are needed for 1 gal?

96. There are 7280 acres of ferns in Florida that are owned by 260 farmers. Find the average size of each farm.

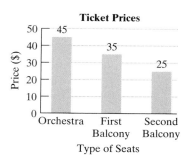

Ticket Prices

Figure for Exercise 97

97. A group of 18 people go to a concert. Ticket prices are given in the chart. If the group has $450, can they all attend the concert? If so, which type of seats can they buy?   **(See Example 10.)**

98. The chart gives the average annual income for four professions: teacher, professor, CEO, and programmer. Find the monthly income for each of the four professions.

99. Suppose Genny can type 1234 words in 22 min. Round each number to estimate her rate in words per minute. **(See Example 9.)**

100. On a trip to California from Illinois, Lavu drove 2780 mi. The gas tank in his car allows him to travel 405 mi. Round each number to the hundreds place to estimate the number of tanks of gas needed for the trip.

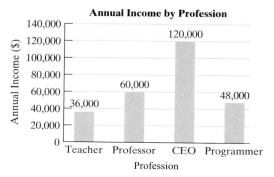

Annual Income by Profession

Figure for Exercise 98

Calculator Connections

Topic: Dividing on a calculator

To divide numbers on a calculator, use the ⊟ key.

Expression	Keystrokes	Result
2,449,216 ÷ 6248	2449216 ⊟ 6248 Enter	392

Calculator Exercises

For Exercises 101–104, use a calculator to divide.

101. 3,437,226 ÷ 14,689

102. 982)‾625,534‾

103. The budget for the U.S. federal government for 2006 is approximately $2160 billion dollars. How much could the government spend each month and still stay within its budget?

104. At a weigh station, a truck carrying 96 crates weighs in at 34,080 lb. If the truck weighs 9600 lb when empty, how much does each crate weigh?

section 1.7   Exponents and the Order of Operations

Objectives

1. Exponents
2. Square Roots
3. Order of Operations
4. Computing a Mean (Average)

1. Exponents

Thus far in the text we have learned to add, subtract, multiply, and divide whole numbers. We now present the concept of an **exponent** to represent repeated multiplication. For example, the product

exponent

$$3 \cdot 3 \cdot 3 \cdot 3 \cdot 3 \quad \text{can be written as} \quad 3^5$$

base

The expression 3^5 is written in exponential form. The exponent, or **power**, is 5 and represents the number of times the **base**, 3, is multiplied. The expression 3^5 is read as "three to the fifth power." Other expressions in exponential form are shown below.

5^2	is read as	"five squared" or "five to the second power"
5^3	is read as	"five cubed" or "five to the third power"
5^4	is read as	"five to the fourth power"
5^5	is read as	"five to the fifth power"

Tip: The expression $5^1 = 5$. Any number without an exponent explicitly written is assumed to be to the first power.

Exponential form is a shortcut notation for repeated multiplication. However, to simplify an expression in exponential form, we often write out the individual factors.

Concept Connections

1. Write the expression in exponential form: $2 \cdot 2 \cdot 2$.

2. How would you read the expression 7^3?

example 1   Evaluating Exponential Expressions

Evaluate.

a. 6^2 **b.** 5^3 **c.** 2^4

Solution:

a. $6^2 = 6 \cdot 6$ The exponent, 2, indicates the number of times the
      $= 36$ base, 6, is multiplied.

b. $5^3 = 5 \cdot 5 \cdot 5$ When three factors are multiplied, we can group the first
      $= (\underline{5 \cdot 5}) \cdot 5$ two factors and perform the multiplication.

      $= (25) \cdot 5$ Then multiply the product of the first two factors by
                          the last factor.
      $= 125$

c. $2^4 = 2 \cdot 2 \cdot 2 \cdot 2$

      $= (\underline{2 \cdot 2}) \cdot 2 \cdot 2$ Group the first two factors.

      $= 4 \cdot 2 \cdot 2$ Multiply the first two factors.

      $= (\underline{4 \cdot 2}) \cdot 2$ Multiply the product by the next factor to the right.

      $= 8 \cdot 2$

      $= 16$

Skill Practice

Evaluate.

3. 8^2 **4.** 4^3 **5.** 2^5

Answers

1. 2^3
2. "Seven cubed" or "seven to the third power"
3. 64   4. 64   5. 32

One important application of exponents lies in recognizing **powers of 10**, that is, 10 raised to a whole-number power. For example, consider the following expressions.

$$10^1 = 10$$

$$10^2 = 10 \cdot 10 = 100$$

$$10^3 = 10 \cdot 10 \cdot 10 = 1000$$

$$10^4 = 10 \cdot 10 \cdot 10 \cdot 10 = 10{,}000$$

$$10^5 = 10 \cdot 10 \cdot 10 \cdot 10 \cdot 10 = 100{,}000$$

$$10^6 = 10 \cdot 10 \cdot 10 \cdot 10 \cdot 10 \cdot 10 = 1{,}000{,}000$$

From these examples, we see that a power of 10 results in a 1 followed by several zeros. The number of zeros is the same as the exponent on the base of 10.

Skill Practice

Evaluate.

6. 10^8

7. 10^{13}

2. Square Roots

To square a number means that we multiply the base times itself. For example, $5^2 = 5 \cdot 5 = 25$.

To find a positive **square root** of a number means that we reverse the process of squaring. For example, finding the square root of 25 is equivalent to asking, What positive number when squared, equals 25? The symbol $\sqrt{}$, (called a *radical sign*) is used to denote the positive square root of a number. Therefore, $\sqrt{25}$ is the positive number that when squared, equals 25. Thus, $\sqrt{25} = 5$ because $(5)^2 = 25$.

Skill Practice

Find the square roots.

8. $\sqrt{4}$

9. $\sqrt{100}$

10. $\sqrt{81}$

example 2   Evaluating Square Roots

Find the square roots.

a. $\sqrt{9}$ **b.** $\sqrt{64}$ **c.** $\sqrt{1}$ **d.** $\sqrt{0}$

Solution:

a. $\sqrt{9} = 3$ because $(3)^2 = 3 \cdot 3 = 9$

b. $\sqrt{64} = 8$ because $(8)^2 = 8 \cdot 8 = 64$

c. $\sqrt{1} = 1$ because $(1)^2 = 1 \cdot 1 = 1$

d. $\sqrt{0} = 0$ because $(0)^2 = 0 \cdot 0 = 0$

Tip: To simplify square roots, it is advisable to become familiar with the following squares and square roots.

$$0^2 = 0 \longrightarrow \sqrt{0} = 0 \qquad 7^2 = 49 \longrightarrow \sqrt{49} = 7$$

$$1^2 = 1 \longrightarrow \sqrt{1} = 1 \qquad 8^2 = 64 \longrightarrow \sqrt{64} = 8$$

$$2^2 = 4 \longrightarrow \sqrt{4} = 2 \qquad 9^2 = 81 \longrightarrow \sqrt{81} = 9$$

$$3^2 = 9 \longrightarrow \sqrt{9} = 3 \qquad 10^2 = 100 \longrightarrow \sqrt{100} = 10$$

$$4^2 = 16 \longrightarrow \sqrt{16} = 4 \qquad 11^2 = 121 \longrightarrow \sqrt{121} = 11$$

$$5^2 = 25 \longrightarrow \sqrt{25} = 5 \qquad 12^2 = 144 \longrightarrow \sqrt{144} = 12$$

$$6^2 = 36 \longrightarrow \sqrt{36} = 6 \qquad 13^2 = 169 \longrightarrow \sqrt{169} = 13$$

Answers

6. 100,000,000

7. 10,000,000,000,000

8. 2

9. 10

10. 9

3. Order of Operations

A numerical expression may contain more than one operation. For example, the following expression contains both multiplication and subtraction.

$$18 - 5(2)$$

The order in which the multiplication and subtraction are performed will affect the overall outcome.

<u>Multiplying first yields</u>

$18 - 5(2) = 18 - 10$

$= 8$ (correct)

<u>Subtracting first yields</u>

$18 - 5(2) = 13(2)$

$= 26$ (incorrect)

To avoid confusion, mathematicians have outlined the proper order of operations. In particular, multiplication is performed before addition or subtraction. The guidelines for the order of operations are given below. These rules must be followed in all cases.

Order of Operations

1. Perform all operations inside parentheses first.
2. Simplify any expressions containing exponents or square roots.
3. Perform multiplication or division in the order that they appear from left to right.
4. Perform addition or subtraction in the order that they appear from left to right.

Concept Connections

11. For the expression $20 - 9 \div 3 + 1$, would you subtract first, divide first, or add first?

---

example 3   Using the Order of Operations

Simplify.

a. $15 - 10 \div 2 + 3$ **b.** $(5 - 2) \cdot 7 - 1$ **c.** $\sqrt{100} - 2^3$

Solution:

a. $15 - \underbrace{10 \div 2} + 3$

$= \underbrace{15 - 5} + 3$ Perform the division $10 \div 2$ first.

$=\quad 10 + 3$ Perform addition and subtraction from left to right.

$=\quad\quad 13$ Add.

b. $\underbrace{(5 - 2)} \cdot 7 - 1$

$= \underbrace{(3) \cdot 7} - 1$ Perform the operation inside parentheses first.

$=\quad 21 - 1$ Perform multiplication before subtraction.

$=\quad 20$ Subtract.

c. $\sqrt{100} - 2^3$

$= 10 - 8$ Simplify any expressions with exponents or square roots. Note that $\sqrt{100} = 10$, and $2^3 = 2 \cdot 2 \cdot 2 = 8$.

$= 2$ Subtract.

Skill Practice

Simplify.

12. $18 + 6 \div 2 - 4$

13. $(20 - 4) \div 2 + 1$

14. $2^3 - \sqrt{25}$

Answers

11. Divide first
12. 17
13. 9
14. 3

example 4   Using the Order of Operations

Simplify.

  a. $300 \div (7-2)^2 - 2^2$ **b.** $36 + (7^2 - 3)$

Solution:

  a. $300 \div (7-2)^2 - 2^2$

$= 300 \div (5)^2 - 2^2$ Perform the operation within parentheses first.

$= 300 \div 25 - 4$ Simplify exponents: $5^2 = 5 \cdot 5 = 25$ and $2^2 = 2 \cdot 2 = 4$.

$=\quad 12 - 4$ Perform division before subtraction.

$=\quad\quad 8$ Subtract.

  b. $36 + (7^2 - 3)$

$= 36 + (49 - 3)$ Perform the operation within parentheses first. The guidelines indicate that we simplify the expression with the exponent before we subtract: $7^2 = 49$.

$= 36 + \quad 46$ Continue simplifying within parentheses.

$= 82$ Add.

4. Computing a Mean (Average)

The order of operations must be used when we compute an average. The technical term for the average of a list of numbers is the **mean** of the numbers. To find the mean of a set of numbers, first compute the sum of the values. Then divide the sum by the number of values. This is represented by the formula

$$\text{Mean} = \frac{\text{sum of the values}}{\text{number of values}}$$

example 5   Computing a Mean

Six drivers were stopped for speeding in a 25-mph zone. The speed for each driver (in miles per hour) is given below. Find the mean speed.

45, 38, 42, 50, 41, 54

Solution:

$$\text{Mean speed} = \frac{45 + 38 + 42 + 50 + 41 + 54}{6}$$

$$= \frac{270}{6}$$ Add the values in the list first.

$$= 45$$ Divide.

$$\begin{array}{r} 45 \\ 6\overline{)270} \\ -24 \\ \hline 30 \\ -30 \\ \hline 0 \end{array}$$

The mean speed is 45 mph.

section 1.7   Practice Exercises

Study Skills Exercises

1. Look over the notes that you took today. Do you understand what you wrote? If there were any rules, definitions, or formulas, highlight them so that they can be easily found when studying for the test.

2. Define the key terms.

    a. Exponent   **b. Power**   **c. Base**

    d. Power of 10   **e. Square root**   **f. Mean**

Review Exercises

For Exercises 3–8, write true or false for each statement.

3. Addition is commutative; for example, $5 + 3 = 3 + 5$.

4. Subtraction is commutative; for example, $5 - 3 = 3 - 5$.

5. $6 \cdot 0 = 6$ **6.** $0 \div 8 = 0$ **7.** $0 \cdot 8 = 0$ **8.** $5 \div 0$ is undefined

Objective 1: Exponents

9. Write an exponential expression with 9 as the base and 4 as the exponent.

10. Write an exponential expression with 3 as the base and 8 as the exponent.

11. Write an exponential expression with 7 as the exponent and 2 as the base.

12. Write an exponential expression with 5 as the exponent and 6 as the base.

For Exercises 13–16, write the repeated multiplication in exponential form. Do not simplify.

13. $3 \cdot 3 \cdot 3 \cdot 3 \cdot 3 \cdot 3$ **14.** $7 \cdot 7 \cdot 7 \cdot 7$ **15.** $4 \cdot 4 \cdot 4 \cdot 4 \cdot 2 \cdot 2 \cdot 2$ **16.** $5 \cdot 5 \cdot 5 \cdot 10 \cdot 10 \cdot 10$

For Exercises 17–20, expand the exponential expression as a repeated multiplication. Do not simplify.

17. 8^4 **18.** 2^6 **19.** 4^8 **20.** 6^2

For Exercises 21–36, evaluate the exponential expressions. **(See Example 1.)**

21. 2^3

22. 4^2

23. 3^2

24. 5^2

25. 3^3

26. 11^2

27. 5^3

28. 4^3

29. 2^5

30. 6^3

31. 3^4

32. 5^4

33. 1^2

34. 1^3

35. 1^4

36. 1^5

37. Explain what happens when the number 1 is raised to any power. **(See Exercises 33–36.)**

For Exercises 38–41, evaluate the powers of 10.

38. 10^2

39. 10^3

40. 10^4

41. 10^5

42. Explain how to get 10^9 *without* doing the repeated multiplication. **(See Exercises 38–41.)**

Objective 2: Square Roots

For Exercises 43–50, evaluate the square roots. **(See Example 2.)**

43. $\sqrt{4}$

44. $\sqrt{9}$

45. $\sqrt{36}$

46. $\sqrt{81}$

47. $\sqrt{100}$

48. $\sqrt{49}$

49. $\sqrt{0}$

50. $\sqrt{16}$

Objective 3: Order of Operations

51. Does the order of operations indicate that addition is always performed before subtraction? Explain.

52. Does the order of operations indicate that multiplication is always performed before division? Explain.

For Exercises 53–84, simplify using the order of operations. **(See Examples 3–4.)**

53. $6 + 10 \cdot 2$

54. $4 + 3 \cdot 7$

55. $10 - 3^2$

56. $11 - 2^2$

57. $(10 - 3)^2$

58. $(11 - 2)^2$

59. $36 \div 2 \div 6$

60. $48 \div 4 \div 2$

61. $15 - (5 + 8)$

62. $41 - (13 + 8)$

63. $(13 - 2) \cdot 5$

64. $(8 + 4) \cdot 6$

65. $4 + 12 \div 3$

66. $9 + 15 \div \sqrt{25}$

67. $30 \div 2 \cdot \sqrt{9}$

68. $55 \div 11 \cdot 5$

69. $7^2 - 5^2$

70. $3^3 - 2^3$

71. $(7 - 5)^2$

72. $(3 - 2)^3$

73. $\sqrt{81} + 2(9 - 1)$ **74.** $\sqrt{121} + 3(8 - 3)$ **75.** $36 \div (2^2 + 5)$ **76.** $42 \div (3^2 - 2)$

77. $80 - (20 \div 4) + 6$ **78.** $120 - (48 \div 8) - 40$ **79.** $(43 - 26) \cdot 2 - 4^2$ **80.** $(51 - 48) \cdot 3 + 7^2$

81. $(18 - 5) - (23 - \sqrt{100})$ **82.** $(\sqrt{36} + 11) - (31 - 16)$ **83.** $80 \div (9^2 - 7 \cdot 11)^2$

84. $108 \div (3^3 - 6 \cdot 4)^2$

Objective 4: Computing a Mean (Average)

For Exercises 85–90, find the mean (average) of each set of numbers. **(See Example 5.)**

85. 5, 8, 4, 3, 10   **86.** 7, 2, 4, 8, 1, 2   **87.** 32, 41, 68, 51

88. 19, 21, 18, 21, 16   **89.** 105, 114, 123, 101, 100, 111   **90.** 1480, 1102, 1032, 1002

91. Neelah took 6 quizzes and received the following scores: 19, 20, 18, 19, 18, 14. Find her quiz average.

92. Shawn's scores on his last 4 tests were 83, 95, 87, and 91. What is his test average?

93. At a certain grocery store, Jessie notices that the price of bananas varies from week to week. During a 3-week period she buys bananas for 33¢ per pound, 39¢ per pound, and 42¢ per pound. What does Jessie pay on average per pound?

94. On a trip, Stephen had his car washed 4 times and paid $7, $10, $8, and $7. What was the average amount spent per wash?

95. The monthly rainfall for Seattle, Washington, is given in the table. All values are in millimeters (mm).

	Jan.	Feb.	Mar.	Apr.	May	Jun.	Jul.	Aug.	Sep.	Oct.	Nov.	Dec.
Rainfall	122	94	80	52	47	40	15	21	44	90	118	123

Find the average monthly rainfall for the winter months of November, December, and January.

96. The monthly snowfall for Alpena, Michigan, is given in the table. All values are in inches.

	Jan.	Feb.	Mar.	Apr.	May	Jun.	Jul.	Aug.	Sep.	Oct.	Nov.	Dec.
Snowfall	22	16	13	5	1	0	0	0	0	1	9	20

Find the average monthly snowfall for the winter months of November, December, January, February, and March.

Expanding Your Skills

Sometimes an expression will have parentheses within parentheses. This is called *nested parentheses*. Often different shapes such as ( ), [ ], or { } are used to make it easier to match up the pairs of parentheses, for example,

$$\{300 - 4[4 + (5 + 2)^2] + 8\} - 31$$

It is important to note that the symbols ( ), [ ], or { } all represent parentheses and are used for grouping. When nested parentheses occur, simplify the innermost set first. Then work your way out. For example, simplify

$$\{300 - 4[4 + (5 + 2)^2] + 8\} - 31$$

The solution is

$\{300 - 4[4 + (5 + 2)^2] + 8\} - 31$

$= \{300 - 4[4 + (7)^2] + 8\} - 31$ Simplify within the innermost parentheses first ( ).

$= \{300 - 4[4 + 49] + 8\} - 31$ Simplify the exponent.

$= \{300 - 4[53] + 8\} - 31$ Simplify within the next innermost parentheses [ ].

$= \{300 - 212 + 8\} - 31$ Multiply before adding.

$= \{88 + 8\} - 31$ Subtract and add in order from left to right within the parentheses { }.

$= 96 - 31$ Simplify within the parentheses { }.

$= 65$ Simplify.

For Exercises 97–100, simplify the expressions with nested parentheses.

97. $3[4 + (6 - 3)^2] - 15$ **98.** $2[5(4 - 1) + 3] \div 6$

99. $5\{21 - [3^2 - (4 - 2)]\}$ **100.** $4\{18 - [(10 - 8) + 2^3]\}$

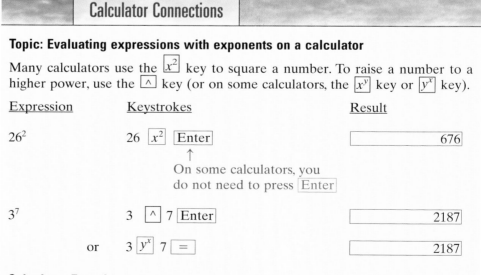

Calculator Connections

Topic: Evaluating expressions with exponents on a calculator

Many calculators use the $\boxed{x^2}$ key to square a number. To raise a number to a higher power, use the $\boxed{\wedge}$ key (or on some calculators, the $\boxed{x^y}$ key or $\boxed{y^x}$ key).

Expression	Keystrokes	Result
26^2	26 $\boxed{x^2}$ $\boxed{\text{Enter}}$	676
	↑	
	On some calculators, you do not need to press $\boxed{\text{Enter}}$	
3^7	3 $\boxed{\wedge}$ 7 $\boxed{\text{Enter}}$	2187
	or 3 $\boxed{y^x}$ 7 $\boxed{=}$	2187

Calculator Exercises

For Exercises 101–104, use a calculator to perform the indicated operations.

101. 156^2 **102.** 418^2 **103.** 12^5 **104.** 35^4

For Exercises 105–110, simplify the expressions by using the order of operations. For each step use the calculator to simplify the given operation.

105. $8126 - 54{,}978 \div 561$

106. $92{,}168 + 6954 \times 29$

107. $(3548 - 3291)^2$

108. $(7500 \div 625)^3$

109. $\dfrac{89{,}880}{384 + 2184}$

110. $\dfrac{54{,}137}{3393 - 2134}$

---

section 1.8   Problem-Solving Strategies

1. Problem-Solving Strategies

In this section we offer additional practice with applications of whole numbers. Keep in mind that all word problems are different and that there is no magic "trick" to solve an application problem. However, we can offer the following guidelines.

Objectives

1. Problem-Solving Strategies
2. Applications Involving One Operation
3. Applications Involving Multiple Operations

Guidelines for Problem Solving

1. Read the problem carefully and familiarize yourself with the situation. If possible, draw a diagram or write down an appropriate formula. Sometimes you may be able to estimate a reasonable answer.
2. Write down what information is given and what must be found.
3. Form a strategy. Identify what mathematical operation applies (addition, subtraction, multiplication, or division). Sometimes a combination of operations is necessary.
4. Perform the mathematical operations to solve for the unknown.
5. Check the answer. If the answer is reasonable and checks, state the answer in words.

2. Applications Involving One Operation

We illustrate these guidelines with a variety of examples. To assist with step 3 where we must identify an appropriate mathematical operation, we summarize some of our key words and phrases. See Table 1-5.

table 1-5

Operation	Key Word or Phrase
Addition	Sum, added to, increased by, more than, plus, total of
Subtraction	Difference, minus, decreased by, less, subtract
Multiplication	Product, times, multiply
Division	Quotient, divide, per, shared equally

Concept Connections

1. The odometer of a car read 24,316 mi last year. This year the reading is 37,134. How many miles was the car driven during the year?

example 1   Solving a Travel Application

Kent travels from Columbus, Ohio, to Indianapolis, Indiana, and then on to Springfield, Illinois. The total distance he drives is 351 mi. The distance between Columbus and Indianapolis is 168 mi. Find the distance between Indianapolis and Springfield.

Solution:

Familiarize and draw a picture.

Given:	In this case, we know the total distance and one of the parts.
Find:	Find the second distance (between Indianapolis and Springfield).
Operation:	This problem can be phrased as an addition problem with a missing addend or as an equivalent subtraction problem.

$$168 + ? = 351 \quad \text{or} \quad 351 - 168 = ?$$

Subtracting yields

$$351 - 168 = 183$$

The distance between Indianapolis and Springfield is 183 mi.

Tip: The answer is reasonable because the sum of the distances equals the total distance.

$$183 \text{ mi } + 168 \text{ mi } = 351 \text{ mi}$$

Skill Practice

Refer to the chart in Example 2.

2. Find the total number of trucks sold during this 3-month period.

3. How many more vehicles were sold in September than in August?

example 2   Solving a Sales Application

A used car business keeps records of vehicle sales by type of vehicle and month.

	July	August	September
Cars	23	28	32
Trucks	13	8	10
SUVs	15	18	21

Find the total number of vehicles sold in July.

Solution:

In this problem, we can use the chart as our diagram. We are looking for the total number of vehicles sold in July, so we might highlight the July column.

	July	August	September
Cars	23	28	32
Trucks	13	8	10
SUVs	15	18	21

| Given: | The number of each type of vehicle sold in July (highlighted in red). |
| Find: | The total number of vehicles sold in July. |

Answers

1. 12,818 mi
2. 31 trucks
3. 9

Operation: The word *total* indicates addition.

$$\text{Total} = 23 + 13 + 15$$
$$= 51$$

There were 51 vehicles sold in July.

---

example 3   Solving a Business Application

A ream of paper holds 500 sheets. Gail purchases 24 reams for her office. How many sheets of paper is this?

Solution:

Familiarize and draw a picture.

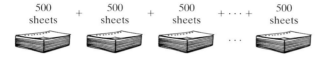

This situation calls for repeated addition. Therefore, we will multiply.

$$24(500) = 12,000$$

There are 12,000 sheets of paper.

> **Skill Practice**
>
> **4.** One page of print in a book contains 48 lines of text. How many lines of text are in one chapter containing 21 pages?

> **Tip:** We might also reason that each pair of 2 reams produces 1000 sheets of paper. Since there are 12 pairs of 2 in 24, there must be 12 thousand sheets of paper.

---

example 4   Solving a Consumer Application

A 5-speed Jeep Cherokee gets 23 mpg (miles per gallon) on the highway. How many gallons of gas would be required for a 667-mi drive from El Paso to Dallas?

Solution:

Familiarize and draw a picture.

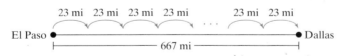

Given: The total distance, 667 mi, and the gas mileage, 23 mpg.

Find: How many increments of 23 mi would be required for the trip?

This is a situation where 667 mi must be divided into 23-mi increments. Use the operation of division.

$$\begin{array}{r} 29 \\ 23\overline{)667} \\ -46 \\ \hline 207 \\ -207 \\ \hline 0 \end{array}$$

The drive from El Paso to Dallas in a Jeep Cherokee will require 29 gal of gas.

> **Skill Practice**
>
> **5.** A vat of flour at a food distributor holds 580 lb of flour. How many 5-lb bags of flour can be filled from one vat?

> **Tip:** The solution to Example 4 can be checked by multiplication. Twenty-nine gallons of gas at 23 mpg produces
>
> $$(29 \text{ gal})(23 \text{ mi/gal}) = 667 \text{ mi}$$

3. Applications Involving Multiple Operations

Sometimes more than one operation is needed to solve an application problem.

Answers
4. 1008 lines
5. 116 bags

Skill Practice

6. Danielle buys a new entertainment center with a new plasma television for $4240. She pays $1000 down, and the rest is paid off in equal monthly payments over 2 years. Find Danielle's monthly payment.

example 5   Solving a Consumer Application

Jorge bought a car for $18,340. He paid $2500 down and then paid the rest in equal monthly payments over a 4-year period. Find the amount of Jorge's monthly payment (not including interest).

Solution:

Familiarize and draw a picture.

Given: total price: $18,340
          down payment: $2500

          payment plan: 4 years
          (48 months)

Find: monthly payment

$18,340   Original cost of car

$-2,500   Minus down payment

$15,840   Amount to be paid off

Divide payments over 4 years (48 months)

Operations:

1. The amount of the loan to be paid off is equal to the original cost of the car minus the down payment. We use subtraction:

$$\begin{array}{r} \$18,340 \\ -\ 2,500 \\ \hline \$15,840 \end{array}$$

2. This money is distributed in equal payments over a 4-year period. Because there are 12 months in 1 year, there are $4 \times 12 = 48$ months in a 4-year period. To distribute $15,840 among 48 equal payments, we divide.

$$\begin{array}{r} 330 \\ 48\overline{)15,840} \\ -144 \\ \hline 144 \\ -144 \\ \hline 00 \end{array}$$

Jorge's monthly payments will be $330.

Tip: The solution to Example 5 can be checked by multiplication. Forty-eight payments of $330 each amount to 48($330) = $15,840. This added to the down payment totals $18,340 as desired.

Skill Practice

7. Taylor makes $18 per hour for the first 40 hr worked each week. His overtime rate is $27 per hour for hours exceeding the normal 40-hr workweek. If his total salary for one week is $963, determine the number of hours of overtime worked.

example 6   Solving a Travel Application

Linda must drive from Clayton to Oakley. She can travel directly from Clayton to Oakley on a mountain road, but will only average 40 mph. On the route through Pearson, she travels on highways and can average 60 mph. Which route will take less time?

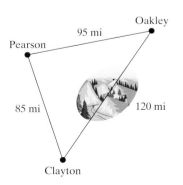

Oakley

95 mi

Pearson

85 mi

120 mi

Clayton

Answers

6. $135 per month
7. 9 hr

Solution:

Read and familiarize: A map is presented in the problem.

Given: The distance for each route and the speed traveled along each route.

Find: Find the time required for each route. Then compare the times to determine which will take less time.

Operations:

1. First note that the total distance of the route through Pearson is found by using addition.

$$85 \text{ mi} + 95 \text{ mi} = 180 \text{ mi}$$

2. The speed of the vehicle gives us an increment of distance traveled per hour. Therefore, the time of travel equals the total distance divided by the speed.

From Clayton to Oakley through the mountains, we divide 120 mi by 40-mph increments to determine the number of hours.

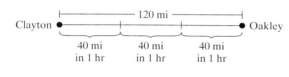

$$\text{Time} = \frac{120 \text{ mi}}{40 \text{ mph}} = 3 \text{ hr}$$

From Clayton to Oakley through Pearson, we divide 180 mi by 60-mph increments to determine the number of hours.

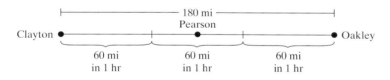

$$\text{Time} = \frac{180 \text{ mi}}{60 \text{ mph}} = 3 \text{ hr}$$

Therefore, each route takes the same amount of time, 3 hr.

---

example 7   Solving a Construction Application

A rancher must fence the corral shown in Figure 1-11. However, no fencing is required on the side adjacent to the barn. If fencing costs $4 per foot, what is the total cost?

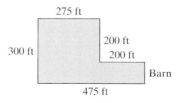

Figure 1-11

Skill Practice

8. Alain wants to put molding around the base of the room shown in the figure. No molding is needed where the door, closet, and bathroom are located. Find the total cost if molding is $2 per foot.

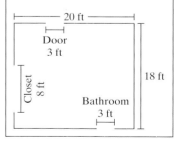

Answer

8. $124

Solution:

Read and familiarize: A figure is provided.

Strategy

With some application problems, it helps to work backward from your final goal. In this case our final goal is to find the total cost. However, to find the total cost, we must first find the total distance to be fenced. To find the total distance, we add the lengths of the sides that are being fenced.

$$
\begin{array}{r}
\overset{\scriptstyle 1\ 1}{275}\text{ ft} \\
200\text{ ft} \\
200\text{ ft} \\
475\text{ ft} \\
+\ 300\text{ ft} \\
\hline
1450\text{ ft}
\end{array}
$$

Therefore,

$$
\begin{pmatrix}\text{Total cost} \\ \text{of fencing}\end{pmatrix} = \begin{pmatrix}\text{total} \\ \text{distance} \\ \text{in feet}\end{pmatrix}\begin{pmatrix}\text{cost} \\ \text{per foot}\end{pmatrix}
$$

$$
= (1450\text{ ft})(\$4\text{ per ft})
$$

$$
= \$5800
$$

The total cost of fencing is $5800.

---

section 1.8   Practice Exercises

Boost *your* GRADE at mathzone.com!

MathZone

- Practice Problems
- Self-Tests
- NetTutor
- e-Professors
- Videos

Study Skills Exercise

1. Sometimes you may run into a problem with homework, or you find that you are having trouble keeping up with the pace of the class. A tutor can be a good resource. Answer the following questions.

   a. Does your college offer tutoring?

   b. Is it free?

   c. Where would you go to sign up for a tutor?

Review Exercises

For Exercises 2–11, translate the English phrase into a mathematical statement and simplify.

2. 89 decreased by 66

3. 71 increased by 14

4. 16 more than 42

5. Twice 14

6. The difference of 93 and 79

7. Subtract 32 from 102

8. Divide 12 into 60

9. The product of 10 and 13

10. The total of 12, 14, and 15

11. The quotient of 24 and 6

Objective 1: Problem-Solving Strategies

12. In your own words, list the guidelines or strategy that you would use to solve an application problem.

For Exercises 13–16, write two or more key words or phrases that represent the given operation. Answers may vary.

13. Addition

14. Multiplication

15. Subtraction

16. Division

Objective 2: Applications Involving One Operation

17. A graphing calculator screen consists of an array of rectangular dots called *pixels*. If the screen has 96 rows of pixels and 126 pixels in each row, how many pixels are in the whole screen? **(See Example 3.)**

18. The floor of a rectangular room has 62 rows of tile with 38 tiles in each row. How many total tiles are there?

19. The Honda Hybrid gets 60 miles per gallon (mpg) in stop-and-go traffic. How many gallons will it use in 540 mi of stop-and-go driving?

20. A couple travels an average speed of 52 mph for a cross-country trip. If the couple drove 1352 mi, how many hours was the trip?

21. White Mountain Peak in California is 14,246 ft high. Denali in Alaska is 20,320 ft high. How much higher is Denali than White Mountain Peak? **(See Example 1.)**

22. In a recent year, *Reader's Digest* was the best-selling U.S. magazine with 12,212,000 yearly subscriptions. *Sports Illustrated* was 15th overall and had 3,252,900 yearly subscriptions. How many more subscriptions did *Reader's Digest* have than *Sports Illustrated*?

23. Jeannette has two children who each attended college in Boston. Her son Ricardo attended Bunker Hill Community College where the yearly tuition and fees came to $2600. Her daughter Ricki attended M.I.T. where the yearly tuition and fees totaled $26,960. If Jeannette paid the full amount for both children to go to school, what was her total expense for tuition and fees for one year? **(See Example 2.)**

24. Clyde and Mason each leave a rest area on the Florida Turnpike. Clyde travels north and Mason travels south. After 2 hr, Clyde has gone 138 mi and Mason, who ran into heavy traffic, traveled only 96 mi. How far apart are they?

25. The Honda Insight gets 66 mpg on the highway. How many miles can it go on 20 gal?

26. A 3 credit-hour class at a certain college meets 3 hr per week. If a semester is 16 weeks long, how many hours will the class meet during the semester?

27. At one time, Tidewater Community College in Virginia had 3000 students who registered for Beginning Algebra. If the average class size is 25 students, how many Beginning Algebra classes will the college have to offer? **(See Example 4.)**

28. Eight people are to share equally in an inheritance of $84,480. How much money will each person receive?

29. A movie theater has 70 rows and 45 seats in a row. What is the maximum seating capacity?

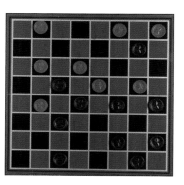

30. A square checkerboard has 8 boxes per row and 8 rows. What is the total number of boxes?

Objective 3: Applications Involving Multiple Operations

31. The balance in Gina's checking account is $278. If she writes checks for $82, $59, and $101, how much will be left over?

32. The balance in Jose's checking account is $3455. If he write checks for $587, $36, and $156, how much will be left over?

33. A community college bought 72 new computers and 6 new printers for a computer lab. If computers were purchased for $2118 each and the printers for $256 each, what was the total bill (not including tax)?

34. Tickets to the San Diego Zoo in California cost $14 for children aged 3–11 and $21 for adults. How much money is required to buy tickets for a class of 33 children and 6 adult chaperones?

35. A discount music store buys used CDs from its customers for $3. Furthermore, a customer can buy any used CD in the store for $8. Latayne sells 16 CDs.

    a. How much money does she receive by selling the 16 CDs?

    b. How many CDs can she then purchase with the money?

36. Shevona earns $8 per hour and works a 40-hr workweek. At the end of the week, she cashes her paycheck and then buys two tickets to a Janet Jackson concert.

    a. How much money is her paycheck worth?

    b. If the concert tickets cost $64 each, how much money does she have left over from her paycheck after buying the tickets?

37. During his 13-year career with the Chicago Bulls, Michael Jordan scored 12,192 field goals (worth 2 points each). He scored 581 three-point shots and 7327 free-throws (worth 1 point each). How many total points did he score during his career with the Bulls?

38. A.J. is a manager for a surf shop. One month he bought 80 T-shirts that cost $6 each, 35 bathing suits that cost $12 each, and 20 pairs of men's shorts that cost $18 each. How much money did he spend?

39. In a recent year, Atlanta's Hartsfield Airport was the busiest airport in the world with an estimated 75,858,500 passengers. In the same year, Chicago's O'Hare Airport was the second busiest with 67,448,000 passengers.

    a. What was the difference between the numbers of passengers traveling through Hartsfield Airport and O'Hare Airport?

    b. What was the total number of passengers for the two airports?

40. Recently, the American Medical Association reported that there were 618,233 male doctors and 195,537 female doctors in the United States.

    a. What is the difference between the number of male doctors and the number of female doctors?

    b. What is the total number of doctors?

41. On a map, each inch represents 60 mi.

    a. If Las Vegas and Salt Lake City are approximately 6 in. apart on the map, what is the actual distance between the cities?

    b. If Madison, Wisconsin, and Dallas, Texas, are approximately 840 mi apart, how many inches would this represent on the map?

42. On a map, each inch represents 40 mi.

   a. If Wichita, Kansas, and Des Moines, Iowa, are approximately 8 in. apart on the map, what is the actual distance between the cities?

   b. If Seattle, Washington and Sacramento, California are approximately 600 mi apart, how many inches would this represent on the map?

43. A textbook company ships books in boxes containing a maximum of 12 books. If a bookstore orders 1250 books, how many boxes can be filled completely? How many books will be left over?

44. A farmer sells eggs in containers holding a dozen eggs. If he has 4257 eggs, how many containers will be filled completely? How many eggs will be left over?

45. Marc pays for an $84 dinner with $20 bills.

   a. How many bills must he use?

   b. How much change will he receive?

46. Shawn buys 3 CDs for a total of $54 and pays with $10 bills.

   a. How many bills must he use?

   b. How much change will he receive?

47. Jackson purchased a car for $16,540. He paid $2500 down and paid the rest in equal monthly payments over a 36-month period. How much were his monthly payments? **(See Example 5.)**

48. Lucio purchased a refrigerator for $1170. He paid $150 at the time of purchase and then paid off the rest in equal monthly payments over 1 year. How much was his monthly payment?

49. Monika must drive from Watertown to Utica. She can travel directly from Watertown to Utica on a small county road, but will only average 40 mph. On the route through Syracuse, she travels on highways and can average 60 mph. Which route will take less time? **(See Example 6.)**

Figure for Exercise 49

50. It takes Rex 4 hr to travel from Oklahoma City to Fort Smith. If the distance between Oklahoma City and Fort Smith is 180 mi, what is his average speed (in miles per hour).

51. If you wanted to line the outside of a garden with a decorative border, would you need to know the area of the garden or the perimeter of the garden?

52. If you wanted to know how much sod to lay down within a rectangular backyard, would you need to know the area of the yard or the perimeter of the yard?

53. A homeowner wants to fence her rectangular backyard. The yard is 75 ft by 90 ft. If fencing costs $5 per foot, how much will it cost to fence the yard?

54. Alexis wants to buy molding for a room that is 12 ft by 11 ft. No molding is needed for the doorway which measures 3 ft. See the figure. If molding costs $2 per foot, how much money will it cost? **(See Example 7.)**

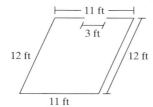

55. What is the cost to carpet the room whose dimensions are shown in the figure? Assume that carpeting costs $34 per square yard and that there is no waste.

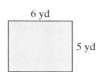

56. What is the cost to tile the room whose dimensions are shown in the figure? Assume that tile costs $3 per square foot.

57. Ling has three jobs. He works for a lawn maintenance service 4 days a week. He also tutors math and works as a waiter on weekends. His hourly wage and the number of hours for each job are given for a 1-week period. How much money did Ling earn for the week?

	Hourly Wage	Number of Hours
Tutor	$30/hr	4
Waiter	10/hr	16
Lawn maintenance	8/hr	30

58. An electrician, a plumber, a mason, and a carpenter work at a certain construction site. The hourly wage and the number of hours each person worked are summarized in the table. What was the total amount paid for all four workers?

	Hourly Wage	Number of Hours
Electrician	$36/hr	18
Plumber	28/hr	15
Mason	26/hr	24
Carpenter	22/hr	48

chapter 1 | summary

section 1.1 Introduction to Whole Numbers

Key Concepts

The place value for each **digit** of a number is shown in the chart.

	Billions Period			Millions Period			Thousands Period			Ones Period		
	Hundred-billions	Ten-billions	Billions	Hundred-millions	Ten-millions	Millions	Hundred-thousands	Ten-thousands	Thousands	Hundreds	Tens	Ones
						3,	4	0	9,	1	1	2

Numbers can be written in different forms, for example:

<u>Standard Form</u>: 3,409,112

<u>Expanded Form</u>: 3 millions + 4 hundred-thousands + 9 thousands + 1 hundred + 1 ten + 2 ones

<u>Words</u>: three million, four hundred nine thousand, one hundred twelve

The order of the whole numbers can be visualized by placement on a number line.

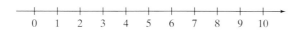

Examples

Example 1

The digit 9 in the number 24,891,321 is in the ten-thousands place.

Example 2

The standard form of the number forty-one million, three thousand, fifty-six is 41,003,056.

Example 3

The expanded form of the number 76,903 is 7 ten-thousands + 6 thousands + 9 hundreds + 3 ones.

Example 4

In words the number 2504 is two thousand, five hundred four.

Example 5

To show that $8 > 4$, note the placement on the number line: 8 is to the right of 4.

▬ section 1.2 Addition of Whole Numbers

Key Concepts

The **sum** is the result of adding numbers called **addends**.

Addition is performed with and without carrying.

<u>Addition Property of Zero</u>:
The sum of any number and zero is that number.

<u>Commutative Property of Addition</u>:
Changing the order of the addends does not affect the sum.

<u>Associative Property of Addition</u>:
The manner in which the addends are grouped does not affect the sum.

There are several words and phrases that indicate addition, such as *sum, added to, increased by, more than, plus,* and *total of.*

The **perimeter** of a **polygon** is the distance around the outside of the figure. To find perimeter, take the sum of the lengths of all sides of the figure.

Examples

Example 1
For $2 + 7 = 9$, the addends are 2 and 7, and the sum is 9.

Example 2

$$
\begin{array}{r} 23 \\ +\,41 \\ \hline 64 \end{array}
\qquad
\begin{array}{r} \overset{1\ 1}{189} \\ +\,76 \\ \hline 265 \end{array}
$$

Example 3

$16 + 0 = 16$

Example 4

$3 + 12 = 12 + 3$

Example 5

$2 + (19 + 3) = (2 + 19) + 3$

Example 6
The sum of 6 and 18 translates to $6 + 18$.

Example 7
The expression $5 + 4$ can be translated as 5 increased by 4, or 4 more than 5.

Example 8
The perimeter is found by adding the lengths of all sides.

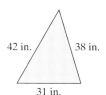

Perimeter = 42 in. + 38 in. + 31 in. = 111 in.

section 1.3 Subtraction of Whole Numbers

Key Concepts

The **difference** is the result of subtracting the **subtrahend** from the **minuend**.

Subtracting numbers with and without borrowing.

There are several words and phrases that indicate subtraction, such as *minus, difference, decreased by, less than,* and *subtract from.*

Example 5 is an application involving subtraction.

Examples

Example 1

For $19 - 13 = 6$, the minuend is 19, the subtrahend is 13, and the difference is 6.

Example 2

$$
\begin{array}{r} 398 \\ -227 \\ \hline 171 \end{array}
\qquad
\begin{array}{r} \overset{9}{\underset{}{\cancel{2}}}\,\overset{1}{\cancel{0}}\,4 \\ -\ 8\,8 \\ \hline 1\,1\,6 \end{array}
$$

Example 3

The difference of 15 and 7 translates to $15 - 7$.

Example 4

The expression $31 - 20$ can be translated to 31 decreased by 20, or subtract 20 from 31.

Example 5

Henry has to drive to his in-law's house 185 mi away. If he drives 105 mi today, how many miles will he have to drive tomorrow?

Solution:
$$105 + \boxed{?} = 185$$
$$185 - 105 = \boxed{80}$$

He must drive 80 mi tomorrow.

section 1.4 Rounding and Estimating

Key Concepts

To **round a number**, follow these steps.

1. Identify the digit one position to the right of the given place value.
2. If the digit in step 1 is a 5 or greater, add 1 to the digit in the given place value. Then replace each digit to the right of the given place value by 0.
3. If the digit in step 1 is less than 5, replace it and each digit to its right by 0.

Round to estimate sums and differences.

Examples

Example 1

Round each number to the indicated place.

  a. 4942; hundreds place
  b. 3712; thousands place
  c. 135; tens place

Solution:

  a. 4900   **b.** 4000   **c.** 140

Example 2

Round to the thousands place to estimate the sum: $3929 + 2528 + 5452$.

Solution: $4000 + 3000 + 5000 = 12{,}000$

The sum is approximately 12,000.

section 1.5 Multiplication of Whole Numbers

Key Concepts

Multiplication is repeated addition.

The **product** is the result of multiplying **factors**.

<u>Commutative Property of Multiplication</u>:

Changing the order of the factors does not affect the product.

<u>Associative Property of Multiplication</u>:

The manner in which the factors are grouped does not affect the product.

<u>Multiplication Property of 0</u>:

The product of any number and 0 is 0.

<u>Multiplication Property of 1</u>:

The product of any real number and 1 is that number.

<u>Distributive Property of Multiplication Over Addition</u>:

$a(b + c) = (a \times b) + (a \times c)$

Multiplying whole numbers.

Estimating products by rounding.

There are several words and phrases that indicate multiplication, such as *product*, *times*, and *multiply by*.

The **area of a rectangle** with length l and width w is given by $A = l \cdot w$.

Examples

Example 1

$16 + 16 + 16 + 16 = 4 \times 16 = 64$

Example 2

For $3 \times 13 \times 2 = 78$ the factors are 3, 13, and 2, and the product is 78.

Example 3

$4(7) = 7(4)$

Example 4

$6 \times (5 \times 7) = (6 \times 5) \times 7$

Example 5

$43 \times 0 = 0$

Example 6

$290 \times 1 = 290$

Example 7

$5 \times (4 + 8) = (5 \times 4) + (5 \times 8)$

Example 8

$$3 \times 14 = 42 \qquad 7(4) = 28$$

$$
\begin{array}{r}
312 \\
\times\ 23 \\
\hline
936 \\
6240 \\
\hline
7176
\end{array}
$$

Example 9

$3102 \times 698 \approx 3000 \times 700 = 2{,}100{,}000$

Example 10

The product of 25 and 3 translates to 25(3).

Example 11

The expression $78 \cdot 12$ can be translated to 78 times 12, or 78 multiplied by 12.

Example 12

Find the area of the rectangle.

23 cm

70 cm

<u>Solution</u>:

$$A = (23 \text{ cm}) \cdot (70 \text{ cm}) = 1610 \text{ cm}^2$$

section 1.6 Division of Whole Numbers

Key Concepts

A **quotient** is the result of dividing the **dividend** by the **divisor**.

<u>Properties of Division</u>:
1. Any number divided by itself is 1.
2. Any number divided by 1 is the number itself.
3. Zero divided by any nonzero number is zero.

Note: A number divided by zero is undefined.

Long division, with and without a **remainder**.

There are several words and phrases that indicate division, such as *divide, quotient, per, divides into,* and *shared equally*.

Example 7 is an application of division.

Examples

Example 1

For $36 \div 4 = 9$, the dividend is 36, the divisor is 4, and the quotient is 9.

Example 2

1. $13 \div 13 = 1$

2.
$$\begin{array}{r} 37 \\ 1\overline{)37} \end{array}$$

3. $\dfrac{0}{2} = 0$

Example 3

$\dfrac{2}{0}$ is undefined

Example 4

$$\begin{array}{r} 263 \\ 3\overline{)789} \\ -6 \\ \hline 18 \\ -18 \\ \hline 09 \\ -9 \\ \hline 0 \end{array}$$

$$\begin{array}{r} 41 \text{ R } 12 \\ 21\overline{)873} \\ -84 \\ \hline 33 \\ -21 \\ \hline 12 \end{array}$$

Example 5

The quotient of 72 and 9 translates to $72 \div 9$.

Example 6

The expression $4\overline{)84}$ can be translated to 84 divided by 4, or 4 divides into 84.

Example 7

Rwanda has an area of 10,169 mi^2 and a population of 7,398,000 people. The population density of Rwanda is the number of people per square mile. Round the numbers to the nearest ten-thousand to estimate the population density of Rwanda.

<u>Solution</u>: $7,400,000 \div 10,000 = 740$

There are approximately 740 people per square mile.

section 1.7   Exponents and Order of Operations

Key Concepts

A number raised to an **exponent** represents repeated multiplication.

For 6^3, 6 is the **base** and 3 is the exponent or **power**.

The **square root** of 16 is 4 because $4^2 = 16$. That is, $\sqrt{16} = 4$.

Order of Operations
1. Perform all operations inside parentheses first.
2. Simplify any expressions containing exponents or square roots.
3. Perform multiplication or division in the order that they appear from left to right.
4. Perform addition or subtraction in the order that they appear from left to right.

Powers of 10 can be expressed as the number 1 followed by zeros. The number of zeros is the same as the exponent on the base of 10.

$10^1 = 10$

$10^2 = 100$

$10^3 = 1000$ and so on.

The **mean** is the average of a set of numbers. To find the mean, add all the values and divide by the number of values.

Examples

Example 1

$9^4 = 9 \cdot 9 \cdot 9 \cdot 9 = 6561$

Example 2

$\sqrt{49} = 7$

Example 3

$19 - 32 \div 2^4 + 21$

$= 19 - 32 \div 16 + 21$

$= 19 - 2 + 21$

$= 17 + 21$

$= 38$

Example 4

$(17 - 12)^2 - \sqrt{16}(10 - 2 \cdot 4)$

$= (17 - 12)^2 - 4(10 - 2 \cdot 4)$

$= (5)^2 - 4(10 - 8)$

$= (5)^2 - 4(2)$

$= 25 - 4(2)$

$= 25 - 8$

$= 17$

Example 5

$10^5 = 10,000$ 1 followed by 5 zeros

Example 6

Find the mean of Michael's scores from his homework assignments.

40, 41, 48, 38, 42, 43

Solution:

$$\frac{40 + 41 + 48 + 38 + 42 + 43}{6} = \frac{252}{6} = 42$$

The mean is 42.

section 1.8 Problem-Solving Strategies

Key Concepts

Guidelines for Problem Solving

1. Read the problem carefully. Draw a diagram or write an appropriate formula. Estimate a reasonable answer.
2. Write down what information is given and what must be found.
3. Form a strategy. Identify what mathematical operation or operations apply.
4. Perform the mathematical operations to solve for the unknown.
5. Check the answer.

Examples

Example 1

Nolan received a doctor's bill for $984. His insurance will pay $200, and the balance can be paid in 4 equal monthly payments. How much will each payment be?

Solution:

To find the amount not paid by insurance, subtract $200 from the total bill.
$$984 - 200 = 784$$
To find Nolan's 4 equal payments, divide the amount not covered by insurance by 4.
$$784 \div 4 = 196$$
Nolan must make 4 payments of $196 each.

chapter 1 | review exercises

Section 1.1

For Exercises 1–2, determine the place value for each underlined digit.

1. 1̲0,024

2. 8̲21,811

For Exercises 3–4, convert the numbers to standard form.

3. 9 ten-thousands + 2 thousands + 4 tens + 6 ones

4. 5 hundred-thousands + 3 thousands + 1 hundred + 6 tens

For Exercises 5–6, convert the numbers to expanded form.

5. 3,400,820

6. 30,554

For Exercises 7–8, write the numbers in words.

7. 245

8. 30,861

For Exercises 9–10, write the numbers in standard form.

9. Three thousand, six-hundred two

10. Eight hundred thousand, thirty-nine

For Exercises 11–12, place the numbers on the number line.

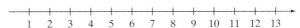

11. 2

12. 7

For Exercises 13–14, determine if the inequality is true or false.

13. $3 < 10$

14. $10 > 12$

Section 1.2

For Exercises 15–16, identify the addends and the sum.

15. $105 + 119 = 224$

16.
$$\begin{array}{r} 53 \\ + 21 \\ \hline 74 \end{array}$$

For Exercises 17–20, add.

17. $18 + 24 + 29$

18. $27 + 9 + 18$

19.
$$\begin{array}{r} 8403 \\ + 9007 \\ \hline \end{array}$$

20.
$$\begin{array}{r} 68{,}421 \\ + 2{,}221 \\ \hline \end{array}$$

21. For each of the mathematical statements, identify the property used. Choose from the commutative property or the associative property.

  a. $6 + (8 + 2) = (8 + 2) + 6$

  b. $6 + (8 + 2) = (6 + 8) + 2$

  c. $6 + (8 + 2) = 6 + (2 + 8)$

For Exercises 22–25, translate the English phrase to a mathematical statement and simplify.

22. The sum of 403 and 79

23. 92 added to 44

24. 7 more than 36

25. 23 increased by 6

26. The chart gives the number of cars sold by three dealerships during one week.

	Honda	Ford	Toyota
Bob's Discount Auto	23	21	34
AA Auto	31	25	40
Car World	33	20	22

  a. What is the total number of cars sold by AA Auto?

  b. What is the total number of Fords sold by these three dealerships?

27. The bar graph represents the distribution of the U.S. population by age group for a recent year. Determine the number of seniors (aged 60 and over).

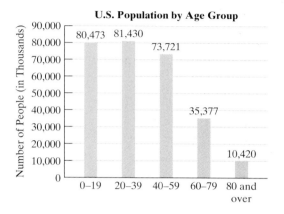

28. Find the perimeter of the figure.

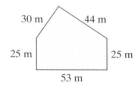

Section 1.3

For Exercises 29–30, identify the minuend, subtrahend, and difference.

29. $14 - 8 = 6$

30.
$$\begin{array}{r} 102 \\ - 78 \\ \hline 24 \end{array}$$

For Exercises 31–32, subtract and check your answer by addition.

31.
$$\begin{array}{r} 37 \\ - 11 \\ \hline \end{array}$$
Check: $\boxed{} + 11 = 37$

32.
$$\begin{array}{r} 61 \\ - 41 \\ \hline \end{array}$$
Check: $\boxed{} + 41 = 61$

For Exercises 33–36, subtract.

33.   2005
        − 1884

34. 1389 − 299

35. 86,000 − 54,981

36. 67,000 − 32,812

For Exercises 37–40, translate the English phrase into a mathematical statement and simplify.

37. 38 minus 31

38. 111 decreased by 15

39. Subtract 42 from 251

40. The difference of 90 and 52

41. There were 95,191,761 tons of watermelons and 23,299,323 tons of cantaloupes produced in 2006. What is the difference between the weight of the watermelons and the weight of the cantaloupes?

42. Tiger Woods earned $7,392,188 from the PGA tours as of 2002. If Phil Mickelson earned $4,311,971, find the difference in their winnings.

43. The graph gives the estimated number of overseas visitors (in thousands) for five cities in the United States for a recent year. What is the difference between the number of visitors to New York and the number of visitors to Orlando?

Overseas Visitors in U.S. Cities

Section 1.4

For Exercises 44–45, round each number to the given place value.

44. 5,234,446; millions

45. 9,332,945; ten-thousands

For Exercises 46–47, estimate the sum by rounding to the indicated place value.

46. 894,004 − 123,883; hundred-thousands

47. 330 + 489 + 123 + 571; hundreds

48. In 2004, the population of Russia was 144,112,353, and the population of Japan was 127,295,333. Estimate the difference in their populations by rounding to the nearest million.

49. The state of Missouri has two dams: Fort Peck with a volume of 96,050 cubic meters (m^3) and Oahe with a volume of 66,517 m^3. Round the numbers to the nearest thousand to estimate the total volume of these two dams.

Section 1.5

For Exercises 50–51, identify the factors and the product.

50. $32 \cdot 12 = 384$

51. $33 \times 40 = 1320$

52. Indicate whether the statement is equal to the product of 8 and 13.

a. 8(13)   **b.** (8) · 13   **c.** (8) + (13)

For Exercises 53–57, for each property listed, choose an expression from the right column that demonstrates the property.

53. Associative property   **a.** 3(4) = 4(3)
of multiplication

54. Distributive property   **b.** 19 × 1 = 19
of multiplication
over addition

55. Multiplication property of 0

c. $(1 \cdot 8) \cdot 3 = 1 \cdot (8 \cdot 3)$

56. Commutative property of multiplication

d. $0 \cdot 29 = 0$

57. Multiplication property of 1

e. $4(3 + 1) = 4 \cdot 3 + 4 \cdot 1$

For Exercises 58–60, multiply.

58. 142
    $\times\ 43$

59. $(1024)(51)$

60. 6000
    $\times\ 500$

61. A discussion group needs to purchase books that are accompanied by a workbook. The price of the book is $26, and the workbook costs an additional $13. If there are 11 members in the group, how much will it cost the group to purchase both the text and workbook for each student?

62. Orcas, or killer whales, eat 551 lb of food a day. If Sea World has two adult killer whales, how much food will they eat in one week?

Section 1.6

For Exercises 63–64, perform the division. Then identify the divisor, dividend, and quotient.

63. $42 \div 6$

64. $4\overline{)52}$

For Exercises 65–68, use the properties of division to simplify the expression, if possible.

65. $3 \div 1$

66. $3 \div 3$

67. $3 \div 0$

68. $0 \div 3$

69. Explain how you check a division problem if there is no remainder.

70. Explain how you check a division problem if there is a remainder.

For Exercises 71–73, divide and check the answer.

71. $348 \div 6$

72. $11\overline{)458}$

73. $\dfrac{1043}{20}$

For Exercises 74–75, write the English phrase as a mathematical expression and simplify.

74. The quotient of 72 and 4

75. 108 divided by 9

76. Quinita has 105 photographs that she wants to divide equally among herself and three siblings. How many photos will each person receive? How many photos will be left over?

77. Ashley has $60 to spend on souvenirs at a surf shop. The prices of several souvenirs are given in the chart.

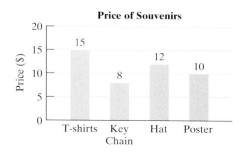

Price of Souvenirs

a. How many souvenirs can Ashley buy if she chooses all T-shirts?

b. How many souvenirs can Ashley buy if she chooses all hats?

Section 1.7

For Exercises 78–79, write the repeated multiplication in exponential form. Do not simplify.

78. $8 \cdot 8 \cdot 8 \cdot 8 \cdot 8$

79. $2 \cdot 2 \cdot 2 \cdot 2 \cdot 5 \cdot 5 \cdot 5$

For Exercises 80–83, evaluate the exponential expressions.

80. 5^3

81. 4^4

82. 1^7

83. 10^6

For Exercises 84–85, evaluate the square roots.

84. $\sqrt{64}$

85. $\sqrt{144}$

For Exercises 86–89, evaluate the expression using the order of operations.

86. $14 \div 7 \cdot 4 - 1$

87. $2 + 3 \cdot 12 \div 2 - \sqrt{25}$

88. $6^2 - 4^2 + (9 - 7)^3$

89. $26 - 2(10 - 1) + (3 + 4 \cdot 11)$

90. Find the mean for the set of numbers 7, 6, 12, 5, 7, 6, 13.

91. Carolyn's electric bills for the past 5 months have been $80, $78, $101, $92, and $94. Find her average monthly charge.

92. The table shows the number of homes sold by a realty company in the last 6 months. Determine the average number of houses sold per month for these 6 months.

Month	Number of Houses
May	6
June	9
July	11
August	13
September	5
October	4

Section 1.8

93. The Cincinnati Zoo houses about 17,000 animals that represent 750 species. The San Diego Zoo has 4000 animals representing 800 species.

a. Which zoo has the most animals? How many more animals does it have?

b. Which zoo has the most species? How many more species does it have?

94. Doris drives her son to extracurricular activities each week. She drives 5 mi round trip to baseball practice 3 times a week and 6 mi round trip to piano lessons once a week.

a. How many miles does she drive in one week to get her child to his activities?

b. Approximately how many miles does she travel during a school year consisting of 10 months (there are approximately 4 weeks per month)?

95. At one point in his baseball career, Alex Rodriquez signed a contract for $252,000,000 for a 9-year period between 2001 and 2010. Suppose federal taxes amount to $75,600,000 for the contract. After taxes, how much will Alex receive per year?

96. Aletha wants to buy plants for a rectangular garden in her backyard that measures 12 ft by 8 ft. She wants to divide the garden into 2-square-foot (2 ft^2) areas, one for each plant.

a. How many plants should Aletha buy?

b. If the plants cost $3 each, how much will it cost Aletha for the plants?

c. If she puts a fence around the perimeter of the garden that costs $2 per foot, how much will it cost for the fence?

d. What will be Aletha's total cost for this garden?

chapter 1 | test

1. Determine the place value for the underlined digit.

  a. 4<u>9</u>2   **b.** 2<u>3</u>,441   **c.** <u>2</u>,340,711   **d.** 340,5<u>9</u>2

2. Fill in the table with either the word name for the number or the number in standard form.

	Population	
State / Province	**Standard Form**	**Word Name**
a. Kentucky		Four million, sixty-five thousand
b. Texas	21,325,000	
c. Pennsylvania	12,287,000	
d. New Brunswick, Canada		Seven hundred twenty-nine thousand
e. Ontario, Canada	11,410,000	

3. Translate the phrase by writing the numbers in standard form and inserting the appropriate inequality. Choose from $<$ or $>$.

  a. Fourteen is greater than six.

  b. Seventy-two is less than eighty-one.

For Exercises 4–17, perform the indicated operation.

4. $\begin{array}{r} 51 \\ +\ 78 \\ \hline \end{array}$ **5.** $\begin{array}{r} 82 \\ \times\ 4 \\ \hline \end{array}$

6. $\begin{array}{r} 154 \\ -\ 41 \\ \hline \end{array}$ **7.** $4\overline{)908}$

8. $58 \cdot 49$ **9.** $149 + 298$

10. $324 \div 15$ **11.** $3002 - 2456$

12. $10{,}984 - 2881$ **13.** $\dfrac{840}{42}$

14. $(500{,}000)(3000)$

15. $34 + 89 + 191 + 22$

16. $403(0)$ **17.** $0\overline{)16}$

18. For each of the mathematical statements, identify the property used. Choose from the commutative property of multiplication and the associative property of multiplication. Explain your answer.

  a. $(11 \cdot 6) \cdot 3 = 11 \cdot (6 \cdot 3)$

  b. $(11 \cdot 6) \cdot 3 = 3 \cdot (11 \cdot 6)$

19. Round each number to the indicated place value.

  a. 4850; hundreds   **b.** 12,493; thousands

  c. 7,963,126; hundred-thousands

20. The attendance to the Van Gogh and Gauguin exhibit in Chicago was 690,951. The exhibit moved to Amsterdam, and the attendance was 739,117. Round the numbers to the ten-thousands place to estimate the total attendance of this exhibit.

For Exercises 21–23, simplify, using the order of operations.

21. $8^2 \div 2^4$ **22.** $26 \cdot \sqrt{4} - 4(8 - 1)$

23. $36 \div 3(14 - 10)$

24. Brittany and Jennifer are taking an online course in business management. Brittany has taken 6 quizzes worth 30 points each and received the following scores: 29, 28, 24, 27, 30, and 30. Jennifer has only taken 5 quizzes so far, and her scores are 30, 30, 29, 28, and 28. At this point in the course, which student has a higher average?

25. The number of foreign adoptions by U.S. citizens rose by 862 children from the year 2001 to 2002. If there were 19,237 foreign adoptions in 2001, how many were there in 2002?

26. The use of the cell phone has grown every year for the past 13 years. See the figure.

    a. Find the change in the number of phones used from 2001 to 2002.

    b. Of the years presented in the chart, between which two years was the increase the greatest?

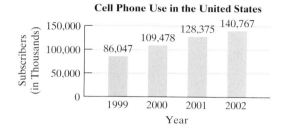

27. The table gives the number of calls to three fire departments during a selected number of weeks. Find the number of calls per week of each department to determine which department is the busiest.

	Number of Calls	Time Period (Number of Weeks)
North Side Fire Department	80	16
South Side Fire Department	72	18
East Side Fire Department	84	28

28. Find the perimeter of the figure.

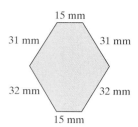

29. Find the perimeter and the area of the rectangle.

30. Round to the nearest hundred to estimate the area of the rectangle.

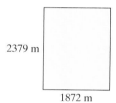

Fractions and Mixed Numbers: Multiplication and Division

2

In this chapter we study the concept of a fraction and a mixed number. We learn how to simplify fractions by reducing to lowest terms. Multiplication and division of fractions and mixed numbers are also presented along with a variety of applications. For example, suppose Ricardo buys a house for $240,000. The bank requires $\frac{1}{10}$ of the cost of the house as a down payment. Ricardo's mother pays $\frac{2}{3}$ of the down payment as a gift. Find out how much Ricardo and his mother each pay toward the down payment of the house in Exercise 93 in Section 2.5.

chapter 2   preview

The exercises in this chapter preview contain concepts that have not yet been presented. These exercises are provided for students who want to compare their levels of understanding before and after studying the chapter. Alternatively, you may prefer to work these exercises when the chapter is completed and before taking the exam.

Section 2.1

1. Identify the fractions as proper or improper.

   a. $\dfrac{4}{5}$ b. $\dfrac{16}{8}$ c. $\dfrac{15}{15}$

2. Write $4\frac{3}{5}$ as an improper fraction.

3. Write $\frac{39}{7}$ as a mixed number.

4. Write a fraction that represents the shaded area.

   a.   b.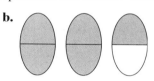

5. There are 8 different brands of wine on the wine list at a restaurant. Of these wines, 5 are red wines and the rest are white wines. What fraction represents the white wines?

Section 2.2

6. Is the number 1092 divisible by 2, 3, or 5? Explain your answers.

7. List all the factors of 45.

8. Write the prime factorization of 630.

Section 2.3

9. Which of the following fractions is simplified to lowest terms?

   a. $\dfrac{16}{25}$ b. $\dfrac{12}{14}$ c. $\dfrac{30}{15}$

Section 2.4

For Exercises 10–12, multiply and simplify the answer to lowest terms.

10. $\dfrac{9}{13} \times \dfrac{39}{27}$ 11. $\left(\dfrac{7}{12}\right)\left(\dfrac{6}{35}\right)$ 12. $12 \cdot \dfrac{77}{84}$

13. Find the area of the triangle.

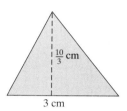

$\frac{10}{3}$ cm

3 cm

Section 2.5

For Exercises 14–16, divide. Simplify the answer to lowest terms.

14. $\dfrac{64}{21} \div 8$ 15. $\dfrac{33}{20} \div \dfrac{44}{15}$ 16. $\dfrac{3}{0}$

17. George painted $\frac{2}{3}$ of a wall that measures 18 ft by 10 ft. How much area did he paint?

18. If a recipe requires $2\frac{1}{2}$ cups of flour for one batch of cookies, how many batches can be made from 10 cups of flour?

Section 2.6

For Exercises 19–21, multiply or divide the mixed numbers. Write the answer as a mixed number or a whole number.

19. $1\dfrac{3}{4} \cdot 9\dfrac{5}{7}$ 20. $8\dfrac{5}{6} \div 2\dfrac{1}{2}$ 21. $4\dfrac{1}{3} \cdot 2\dfrac{7}{10} \div 1\dfrac{4}{5}$

section 2.1   Introduction to Fractions and Mixed Numbers

1. Definition of a Fraction

In Chapter 1, we studied operations on whole numbers. In this chapter we work with numbers that represent part of a whole. When a whole unit is divided into equal parts, we call the parts **fractions** of a whole. For example, the pie in Figure 2-1 is divided into 5 equal parts. One-fifth ($\frac{1}{5}$) of the pie has been eaten, and four-fifths ($\frac{4}{5}$) of the pie remains.

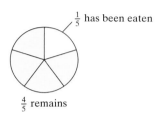

$\frac{1}{5}$ has been eaten

$\frac{4}{5}$ remains

Figure 2-1

> **Tip:** The fraction $\frac{a}{b}$ may also be written as a/b. However, we discourage the use of the "slanted" fraction bar. In later applications of algebra, the slanted fraction bar can cause confusion.

A fraction is written in the form $\frac{a}{b}$, where a and b are whole numbers and $b \neq 0$. In the fraction $\frac{5}{8}$, the "top" number, 5, is called the *numerator*. The bottom number, 8, is called the *denominator*.

$$\text{numerator} \longrightarrow \frac{5}{8} \longleftarrow \text{denominator}$$

example 1   Identifying the Numerator and Denominator of a Fraction

For each fraction, identify the numerator and denominator.

a. $\frac{3}{5}$ **b.** $\frac{1}{8}$ **c.** $\frac{8}{1}$

Solution:

a. $\frac{3}{5}$ The numerator is 3. The denominator is 5.

b. $\frac{1}{8}$ The numerator is 1. The denominator is 8.

c. $\frac{8}{1}$ The numerator is 8. The denominator is 1.

Skill Practice

Identify the numerator and denominator.

1. $\frac{4}{11}$ **2.** $\frac{0}{5}$ **3.** $\frac{6}{1}$

The **denominator** of a fraction denotes the number of equal pieces into which a whole unit is divided. The **numerator** denotes the number of pieces being considered.

Answers

1. Numerator: 4, denominator: 11
2. Numerator: 0, denominator: 5
3. Numerator: 6, denominator: 1

For example, the garden in Figure 2-2 is divided into 10 equal parts. Three sections contain tomato plants. Therefore, $\frac{3}{10}$ of the garden contains tomato plants.

$\frac{3}{10}$ tomato plants

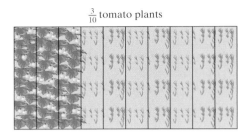

Figure 2-2

example 2   Writing Fractions

Write a fraction for the shaded portion and a fraction for the unshaded portion of the figure.

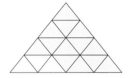

Solution:

Shaded portion: $\dfrac{13}{16}$ ◄——— 13 pieces are shaded.
                                    ◄——— The triangle is divided into 16 equal pieces.

Unshaded portion: $\dfrac{3}{16}$ ◄——— 3 pieces are not shaded.
                                      ◄——— The triangle is divided into 16 equal pieces.

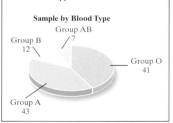
example 3   Writing Fractions

What portion of the group of celebrities shown below is female?

Solution:

The group is divided among 5 members. Therefore, the denominator is 5. There are 2 women being considered. Thus, $\frac{2}{5}$ of the group is female.

In Section 1.6 we learned that fractions represent division. For example, note that the fraction $\frac{5}{1} = 5 \div 1 = 5$. In general, a fraction of the form $\frac{n}{1} = n$. This implies that any whole number may be written as a fraction by writing the whole number over 1.

Further recall that for $a \neq 0$, $0 \div a = 0$ and $a \div 0$ is undefined. Therefore, a fraction of the form $\frac{0}{a} = 0$ and $\frac{a}{0}$ is undefined. For example,

$$\frac{0}{5} = 0 \text{ whereas } \frac{5}{0} \text{ is undefined}$$

2. Proper and Improper Fractions

If the numerator is less than the denominator in a fraction, then the fraction is called a **proper fraction**. Furthermore, a proper fraction represents a number less than 1 whole unit. The following are proper fractions.

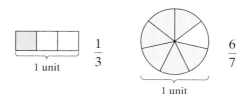

An **improper fraction** is a fraction in which the numerator is greater than or equal to the denominator. For example:

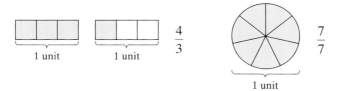

An improper fraction represents a quantity greater than a whole unit or equal to a whole unit.

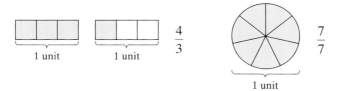

---

example 4   Categorizing Fractions

Identify each fraction as proper or improper.

a. $\frac{12}{5}$ **b.** $\frac{5}{12}$ **c.** $\frac{12}{12}$

Solution:

a. $\frac{12}{5}$ Improper fraction (numerator is greater than denominator)

b. $\frac{5}{12}$ Proper fraction (numerator is less than denominator)

c. $\frac{12}{12}$ Improper fraction (numerator is equal to denominator)

Skill Practice

Identify each fraction as proper or improper.

8. $\frac{10}{10}$ **9.** $\frac{7}{9}$ **10.** $\frac{9}{7}$

Answers
6. Undefined   7. 0
8. Improper   9. Proper
10. Improper

Skill Practice

11. Write an improper fraction representing the shaded area.

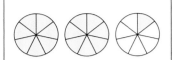

Avoiding Mistakes: In Example 5, each whole unit is divided into 8 pieces. Therefore the screw is $\frac{11}{8}$ in., not $\frac{11}{16}$ in.

 **Writing Improper Fractions**

Write an improper fraction to represent the fractional part of an inch for the screw shown in the figure.

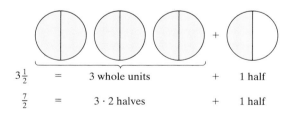

Solution:

Each 1-in. unit is divided into 8 parts, and the screw extends for 11 parts. Therefore, the screw is $\frac{11}{8}$ in.

---

3. Mixed Numbers

Sometimes a mixed number is used instead of an improper fraction to denote a quantity greater than one whole. For example, suppose a typist typed $\frac{9}{4}$ pages of a report. We would be more likely to say that the typist typed $2\frac{1}{4}$ pages (read as "two and one-fourth pages"). The number $2\frac{1}{4}$ is called a *mixed number* and represents 2 wholes plus $\frac{1}{4}$ of a whole.

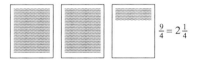

In general, a **mixed number** is a sum of a whole number and a fractional part of a whole. However, by convention the plus sign is left out.

$$3\frac{1}{2} \quad \text{means} \quad 3 + \frac{1}{2}$$

Suppose we want to change a mixed number to an improper fraction. From Figure 2-3, we see that the mixed number $3\frac{1}{2}$ is the same as $\frac{7}{2}$.

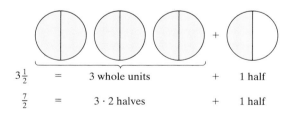

$3\frac{1}{2}$	=	3 whole units	+	1 half
$\frac{7}{2}$	=	$3 \cdot 2$ halves	+	1 half

Figure 2-3

This process to convert a mixed number to an improper fraction can be summarized as follows.

Changing a Mixed Number to an Improper Fraction

1. Multiply the whole number by the denominator.
2. Add the result to the numerator.
3. Write the result from step 2 over the denominator.

Answer

11. $\frac{15}{7}$

For example,

Multiply the whole number by the denominator.   Add the numerator.

$$3\frac{1}{2} = \frac{3 \times 2 + 1}{2} = \frac{7}{2}$$

Write the result over the denominator.

---

example 6   Converting Mixed Numbers to Fractions

Convert the mixed number to an improper fraction.

a. $7\frac{1}{4}$ **b.** $8\frac{2}{5}$

Solution:

a. $7\frac{1}{4} = \dfrac{7 \times 4 + 1}{4}$ **b.** $8\frac{2}{5} = \dfrac{8 \times 5 + 2}{5}$

$\phantom{7\frac{1}{4}} = \dfrac{28 + 1}{4}$ $\phantom{8\frac{2}{5}} = \dfrac{40 + 2}{5}$

$\phantom{7\frac{1}{4}} = \dfrac{29}{4}$ $\phantom{8\frac{2}{5}} = \dfrac{42}{5}$

Skill Practice

Convert the mixed number to an improper fraction.

12. $10\frac{5}{8}$ **13.** $15\frac{1}{2}$

Now suppose we want to convert an improper fraction to a mixed number. In Figure 2-4, the improper fraction $\frac{13}{5}$ represents 13 slices of pie where each slice is $\frac{1}{5}$ of a whole pie. If we divide the 13 pieces into groups of 5, we make 2 whole pies with 3 pieces left over. Thus,

$$\frac{13}{5} = 2\frac{3}{5}$$

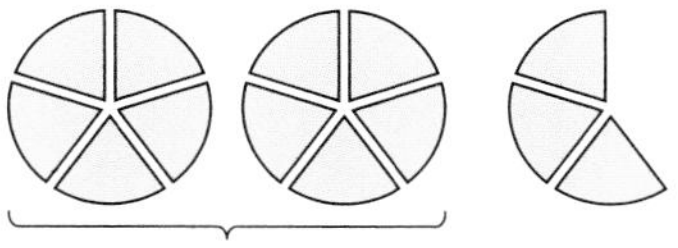

13 pieces =   2 groups of 5   + 3 left over

$\frac{13}{5}$ =   2   + $\frac{3}{5}$

Figure 2-4

This process can be accomplished by division.

$$\frac{13}{5} \longrightarrow \begin{array}{r} 2 \\ 5\overline{)13} \\ -10 \\ \hline 3 \end{array} \quad 2\frac{3}{5} \begin{array}{l} \text{remainder} \\ \\ \text{divisor} \end{array}$$

Changing an Improper Fraction to a Mixed Number

1. Divide the numerator by the denominator to obtain the quotient and remainder.

2. The mixed number is then given by

$$\text{Quotient} + \frac{\text{remainder}}{\text{divisor}}$$

Answers

12. $\dfrac{85}{8}$ 13. $\dfrac{31}{2}$

Skill Practice

Convert the improper fraction to a mixed number.

14. $\dfrac{14}{5}$　**15.** $\dfrac{95}{22}$

example 7　Converting an Improper Fraction to a Mixed Number

Convert to a mixed number.

a. $\dfrac{25}{6}$　　**b.** $\dfrac{162}{41}$

Solution:

a. $\dfrac{25}{6}$ ⟶ $\begin{array}{r} 4 \\ 6\overline{)25} \\ -24 \\ \hline 1 \end{array}$　$4\dfrac{1}{6}$

remainder
divisor

b. $\dfrac{162}{41}$ ⟶ $\begin{array}{r} 3 \\ 41\overline{)162} \\ -123 \\ \hline 39 \end{array}$　$3\dfrac{39}{41}$

remainder
divisor

The process to convert an improper fraction to a mixed number indicates that the result of a division problem can be written as a mixed number.

Skill Practice

Divide and write the quotient as a mixed number.

16. $5967 \div 41$

example 8　Writing a Quotient as a Mixed Number

Divide. Write the quotient as a mixed number.

$$28\overline{)4217}$$

Solution:

$\begin{array}{r} 150 \\ 28\overline{)4217} \\ -28 \\ \hline 141 \\ -140 \\ \hline 17 \\ -0 \\ \hline 17 \end{array}$　$150\dfrac{17}{28}$

remainder
divisor

4. Fractions and the Number Line

Fractions can be visualized on a number line. For example, to graph the fraction $\frac{3}{4}$, divide the distance between 0 and 1 into 4 equal parts. To plot the number $\frac{3}{4}$, start at 0 and count over 3 parts.

Answers

14. $2\dfrac{4}{5}$　　15. $4\dfrac{7}{22}$　　16. $145\dfrac{22}{41}$

example 9   Plotting Fractions on a Number Line

Plot the point on the number line corresponding to each fraction.

a. $\dfrac{1}{2}$ **b.** $\dfrac{5}{6}$ **c.** $\dfrac{21}{5}$

Solution:

a. $\dfrac{1}{2}$ Divide the distance between 0 and 1 into 2 equal parts.

b. $\dfrac{5}{6}$ Divide the distance between 0 and 1 into 6 equal parts.

c. $\dfrac{21}{5} = 4\dfrac{1}{5}$ Write $\frac{21}{5}$ as a mixed number.

Thus, $\frac{21}{5} = 4\frac{1}{5}$ is located one-fifth of the way between 4 and 5 on the number line.

Skill Practice

Plot the numbers on a number line.

17. $\dfrac{4}{5}$ **18.** $\dfrac{1}{3}$

19. $\dfrac{13}{4}$ **20.** $\dfrac{20}{7}$

Answers

17.

18.

19.

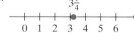

20.

---

section 2.1   Practice Exercises

Boost *your* GRADE at mathzone.com!

- Practice Problems
- Self-Tests
- NetTutor
- e-Professors
- Videos

Study Skills Exercises

1. After doing a section of homework, check the odd answers in the back of the text. Choose a method to identify the exercises that you got wrong or had trouble with (i.e., circle the number or put a star by the number). List some reasons why it is important to label these problems.

2. Define the key terms.

a. Fraction   **b. Numerator**   **c. Denominator**

d. Proper fraction   **e. Improper fraction**   **f. Mixed number**

Objective 1: Definition of a Fraction

For Exercises 3–10, write a fraction that represents the shaded area. **(See Example 2.)**

3.

4.

 **5.**

6.

7.

8.

9.

10.

11. Write a fraction to represent the portion of gas in a gas tank represented by the gauge.

12. Write a fraction that represents the portion of medicine left in the bottle.

For Exercises 13–16, identify the numerator and the denominator for each fraction.
(See Example 1.)

13. $\dfrac{2}{3}$

14. $\dfrac{8}{9}$

15. $\dfrac{12}{11}$

16. $\dfrac{1}{2}$

For Exercises 17–24, write the fraction as a division problem and simplify, if possible.

17. $\dfrac{6}{1}$

18. $\dfrac{9}{1}$

19. $\dfrac{2}{2}$

20. $\dfrac{8}{8}$

 21. $\dfrac{0}{3}$

22. $\dfrac{0}{7}$

23. $\dfrac{2}{0}$

24. $\dfrac{11}{0}$

25. What fraction of the umbrellas is yellow?
(See Example 3.)

26. Write a fraction representing the boats in the marina that are sailboats.

27. A class has 21 children—11 girls and 10 boys. What fraction of the class is made up of boys?

28. A restaurant has 33 tables. Of the total, 10 are reserved for customers who smoke, and 23 are for nonsmokers. Write a fraction representing the tables reserved for the smokers.

Objective 2: Proper and Improper Fractions

For Exercises 29–34, label the fraction as proper or improper. **(See Example 4.)**

29. $\dfrac{7}{8}$

30. $\dfrac{2}{3}$

31. $\dfrac{10}{10}$

32. $\dfrac{3}{3}$

33. $\dfrac{7}{2}$

34. $\dfrac{21}{20}$

For Exercises 35–38, write an improper fraction for the shaded portion of each group of figures. **(See Example 5.)**

35.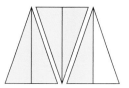

36.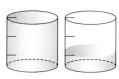

37.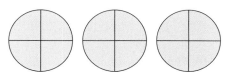

38.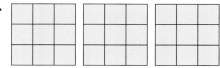

Objective 3: Mixed Numbers

For Exercises 39–40, write an improper fraction and a mixed number for the shaded portion of each group of figures.

39.

40.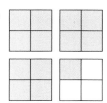

41. Write an improper fraction and a mixed number that represent the fraction of the length of the ribbon.

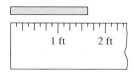

42. Write an improper fraction and a mixed number that represent the fraction of the amount of sugar needed for a batch of cookies as indicated in the figure.

For Exercises 43–58, convert the mixed number to an improper fraction. **(See Example 6.)**

43. $1\dfrac{3}{4}$

44. $6\dfrac{1}{3}$

45. $4\dfrac{2}{9}$

46. $3\dfrac{1}{5}$

47. $3\dfrac{3}{7}$

48. $8\dfrac{2}{3}$

49. $7\dfrac{1}{4}$

50. $10\dfrac{3}{5}$

51. $11\dfrac{5}{12}$

52. $12\dfrac{1}{6}$

53. $21\dfrac{3}{8}$

54. $15\dfrac{1}{2}$

55. $8\dfrac{5}{16}$

56. $7\dfrac{4}{15}$

57. $13\dfrac{9}{20}$

58. $9\dfrac{17}{18}$

59. How many eighths are in $2\dfrac{3}{8}$?

60. How many fifths are in $2\dfrac{3}{5}$?

61. How many fourths are in $1\dfrac{3}{4}$?

62. How many thirds are in $5\dfrac{2}{3}$?

For Exercises 63–78, convert the improper fraction to a mixed number. **(See Example 7.)**

63. $\dfrac{37}{8}$

64. $\dfrac{13}{7}$

65. $\dfrac{39}{5}$

66. $\dfrac{19}{4}$

67. $\dfrac{27}{10}$

68. $\dfrac{43}{18}$

69. $\dfrac{52}{9}$

70. $\dfrac{67}{12}$

71. $\dfrac{133}{11}$

72. $\dfrac{51}{10}$

73. $\dfrac{23}{6}$

74. $\dfrac{115}{7}$

75. $\dfrac{68}{9}$

76. $\dfrac{40}{17}$

77. $\dfrac{65}{8}$

78. $\dfrac{101}{15}$

For Exercises 79–84, divide. Write the quotient as a mixed number. **(See Example 8.)**

79. $7\overline{)309}$

80. $4\overline{)921}$

81. $5281 \div 5$

82. $7213 \div 8$

83. $8913 \div 11$

84. $4257 \div 23$

Objective 4: Fractions and the Number Line

For Exercises 85–94, plot the fraction on the number line. (**See Example 9.**)

85. $\frac{3}{4}$

86. $\frac{1}{2}$

87. $\frac{1}{3}$

88. $\frac{1}{5}$

89. $\frac{2}{3}$

90. $\frac{5}{6}$

91. $\frac{7}{6}$

92. $\frac{7}{5}$

93. $\frac{5}{3}$

94. $\frac{3}{2}$

Expanding Your Skills

95. True or false? Whole numbers can be written both as proper and improper fractions.

96. True or false? Suppose m and n are whole numbers where $m > n$. Then $\frac{m}{n}$ is an improper fraction.

97. True or false? Suppose m and n are whole numbers where $m > n$. Then $\frac{n}{m}$ is a proper fraction.

98. True or false? Suppose m and n are whole numbers where $m > n$. Then $\frac{n}{3m}$ is a proper fraction.

Calculator Connections

Topic: Converting mixed numbers to improper fractions

Calculator Exercises

For Exercises 99–102, convert the mixed number to an improper fraction. Use a calculator to help you with the calculations.

99. $21\frac{39}{407}$

100. $184\frac{17}{91}$

101. $48\frac{23}{112}$

102. $25\frac{59}{73}$

Objectives

1. Factors and Factorizations
2. Divisibility Rules
3. Prime and Composite Numbers
4. Prime Factorization
5. Identifying All Factors of a Whole Number

Skill Practice

Find four different factorizations of the given number.

1. 18   **2.** 30

section 2.2   **Prime Numbers and Factorization**

1. Factors and Factorizations

Recall from Section 1.5 that two numbers multiplied to form a product are called factors. For example, $2 \cdot 3 = 6$ indicates that 2 and 3 are factors of 6. Likewise, because $1 \cdot 6 = 6$, the numbers 1 and 6 are factors of 6. In general, a **factor** of a number n is a nonzero whole number that divides evenly into n.

The products $2 \cdot 3$ and $1 \cdot 6$ are called factorizations of 6. In general, a **factorization** of a number n is a product of factors that equals n.

example 1   Finding Factorizations of a Number

Find four different factorizations of 12.

Solution:

$$12 = \begin{cases} 1 \cdot 12 \\ 2 \cdot 6 \\ 3 \cdot 4 \\ 2 \cdot 2 \cdot 3 \leftarrow \end{cases}$$

Tip: Notice that a factorization may include more than two factors.

2. Divisibility Rules

The number 20 is said to be divisible by 5 because 5 divides evenly into 20. To determine whether one number is divisible by another, we can perform the division and note whether the remainder is zero. However, there are several rules by which we can quickly determine whether a number is divisible by 2, 3, 5, or 10. These are called divisibility rules.

Divisibility Rules for 2, 3, 5, and 10

- *Divisibility by 2.* A whole number is divisible by 2 if it is an even number. That is, the ones-place digit is 0, 2, 4, 6, or 8.
  Examples: 26 and 384

- *Divisibility by 5.* A whole number is divisible by 5 if its ones-place digit is 5 or 0.
  Examples: 45 and 260

- *Divisibility by 10.* A whole number is divisible by 10 if its ones-place digit is 0.
  Examples: 30 and 170

- *Divisibility by 3.* A whole number is divisible by 3 if the sum of its digits is divisible by 3.
  Example: 312   (sum of digits is $3 + 1 + 2 = 6$ which is divisible by 3)

We address other divisibility rules for 4, 6, 8, and 9 in the Expanding Your Skills portion of the exercises. However, these divisibility rules are harder to remember, and it is often easier simply to perform division to test for divisibility.

Answers

1. For example: $1 \cdot 18, 2 \cdot 9, 6 \cdot 3,$
   $2 \cdot 3 \cdot 3$
2. For example: $1 \cdot 30, 2 \cdot 15, 3 \cdot 10,$
   $5 \cdot 6, 2 \cdot 3 \cdot 5$

example 2   Applying the Divisibility Rules

Determine whether the given number is divisible by 2, 3, 5, or 10.

a. 624   **b.** 82   **c.** 720

Solution:

Test for divisibility

a. 624   By 2:   Yes.   The number 624 is even.
        By 3:   Yes.   The sum $6 + 2 + 4 = 12$ is divisible by 3.
        By 5:   No.   The ones-place digit is not 5 or 0.
        By 10: No.   The ones-place digit is not 0.

b. 82   By 2:   Yes.   The number 82 is even.
        By 3:   No.   The sum $8 + 2 = 10$ is not divisible by 3.
        By 5:   No.   The ones-place digit is not 5 or 0.
        By 10: No.   The ones-place digit is not 0.

c. 720   By 2:   Yes.   The number 720 is even.
        By 3:   Yes.   The sum $7 + 2 + 0 = 9$ is divisible by 3.
        By 5:   Yes.   The ones-place digit is 0.
        By 10: Yes.   The ones-place digit is 0.

Tip: When in doubt about divisibility, you can check by performing the division. For instance, in Example 2(a), we can verify that 624 is divisible by 3.

$$\begin{array}{r} 208 \\ 3\overline{)624} \\ -6 \\ \hline 24 \\ -24 \\ \hline 0 \end{array}$$

3. Prime and Composite Numbers

Two important classifications of whole numbers are prime numbers and composite numbers.

Definition of Prime and Composite Numbers

- A **prime number** is a whole number greater than 1 that has only two factors (itself and 1).

- A **composite number** is a whole number greater than 1 that is not prime. That is, a composite number will have at least one factor other than 1 and the number itself.

Note: The whole numbers 0 and 1 are neither prime nor composite.

example 3   Identifying Prime and Composite Numbers

Determine whether the number is prime, composite, or neither.

a. 19   **b.** 51   **c.** 1

Tip: The number 2 is the only even prime number.

Solution:

a. The number 19 is prime because its only factors are 1 and 19.

b. The number 51 is composite because $3 \cdot 17 = 51$. That is, 51 has factors other than 1 and 51.

c. The number 1 is neither prime nor composite by definition.

Prime numbers are used in a variety of skills in mathematics. We advise you to become familiar with the first several prime numbers: 2, 3, 5, 7, 11, 13, 17, 19, 23, 29, . . .

4. Prime Factorization

In Example 1 we found four factorizations of 12.

$$1 \cdot 12$$
$$2 \cdot 6$$
$$3 \cdot 4$$
$$2 \cdot 2 \cdot 3$$

The last factorization $2 \cdot 2 \cdot 3$ consists of only prime-number factors. Therefore, we say $2 \cdot 2 \cdot 3$ is the prime factorization of 12. Note that the order in which we write the factors within a factorization does not affect its product (this is so because multiplication is commutative). Therefore, the products $2 \cdot 2 \cdot 3$, $2 \cdot 3 \cdot 2$, and $3 \cdot 2 \cdot 2$ are all equivalent and all represent the prime factorization of 12.

> ### Prime Factorization
>
> The **prime factorization** of a number is the factorization in which every factor is a prime number.
>
> *Note:* The order in which the factors are written does not affect the product.

Prime factorizations of numbers will be particularly helpful when we add, subtract, multiply, divide, and simplify fractions to lowest terms.

Concept Connections

9. Is the product $2 \cdot 3 \cdot 10$ the prime factorization of 60? Explain.

10. Which is the prime factorization of 70?

    $2 \cdot 5 \cdot 7$ or $2 \cdot 7 \cdot 5$
    or $5 \cdot 7 \cdot 2$

Concept Connections

11. How would you write the factorization,
    $2 \cdot 2 \cdot 2 \cdot 2 \cdot 3 \cdot 5 \cdot 5$
    using exponents?

Avoiding Mistakes: Make sure that the end of each branch is a prime number.

---

example 4   Determining the Prime Factorization of a Number

Find the prime factorization of 220.

Solution:

One method to factor a whole number is to make a factor tree. Begin by determining any two numbers that when multiplied equal 220. Then continue factoring each factor until the branches "end" in prime numbers.

```
      220
     /   \
   10     22
  / \    / \
 5  2   2   11  ←—Branches end in prime numbers.
```

> **Tip:** The prime factorization from Example 4 can also be expressed by using exponents as $2^2 \cdot 5 \cdot 11$.

Therefore, the prime factorization of 220 is $2 \cdot 2 \cdot 5 \cdot 11$.

■

In Example 4, note that the result of a prime factorization does not depend on the original two-number factorization. Similarly, the order in which the factors are written does not affect the product, for example,

$$220 = 2 \cdot 2 \cdot 5 \cdot 11 \qquad 220 = 2 \cdot 2 \cdot 5 \cdot 11 \qquad 220 = 11 \cdot 2 \cdot 2 \cdot 5$$

> **Tip:** You can check the prime factorization of any number by multiplying the factors.

Answers

9. No. The factor 10 is not a prime number. The prime factorization of 60 is $2 \cdot 2 \cdot 3 \cdot 5$.

10. All three factorizations represent the prime factorization. The order in which we write the factors does not affect the product.

11. $2^4 \cdot 3 \cdot 5^2$

Another technique to find the prime factorization of a number is to divide the number by the smallest known prime factor. Then divide the quotient by its smallest known prime factor. Continue dividing in this fashion until the quotient is a prime number. The prime factorization is the product of divisors and the final quotient. For example,

the last quotient is prime ──────────→   11

5 is the smallest prime factor of 55 ──→ $5\overline{)55}$

2 is the smallest prime factor of 110 ──→ $2\overline{)110}$

2 is the smallest prime factor of 220 ──→ $2\overline{)220}$

Therefore, the prime factorization of 220 is $2 \cdot 2 \cdot 5 \cdot 11$ or $2^2 \cdot 5 \cdot 11$.

---

example 5   Determining Prime Factorizations

Find the prime factorization.

a. 198   **b.** 153

Solution:

a.

$$11$$
$$3\overline{)33}$$
$$3\overline{)99}$$ ←── The sum of the digits
$$2\overline{)198}$$ $9 + 9 = 18$ is divisible by 3.

Because 198 is even, we → know it is divisible by 2.

The prime factorization of 198 is $2 \cdot 3 \cdot 3 \cdot 11$ or $2 \cdot 3^2 \cdot 11$.

b.

$$17$$
$$3\overline{)51}$$
$$3\overline{)153}$$

The prime factorization of 153 is $3 \cdot 3 \cdot 17$ or $3^2 \cdot 17$.

Skill Practice

Find the prime factorization of the given number.

12. 168   **13.** 990

5. Identifying All Factors of a Whole Number

Sometimes it is necessary to identify all factors (both prime and other) of a number. Take the number 30, for example. A list of all factors of 30 is a list of all whole numbers that divide evenly into 30.

Factors of 30:   1, 2, 3, 5, 6, 10, 15, and 30

---

example 6   Listing All Factors of a Number

List all factors of 36.

Solution:

Begin by listing all the two-number factorizations of 36. This can be accomplished by systematically dividing 36 by 1, 2, 3, and so on. Notice, however, that after the product $6 \cdot 6$, the two-number factorizations are repetitious, and we can stop the process.

Skill Practice

List all the factors of the given number.

14. 45   **15.** 52

Answers

12. $2 \cdot 2 \cdot 2 \cdot 3 \cdot 7$ or $2^3 \cdot 3 \cdot 7$
13. $2 \cdot 3 \cdot 3 \cdot 5 \cdot 11$ or $2 \cdot 3^2 \cdot 5 \cdot 11$
14. 1, 3, 5, 9, 15, 45
15. 1, 2, 4, 13, 26, 52

Tip: When listing a set of factors, it is not necessary to write the numbers in any specified order. However, in general we list the factors in order from smallest to largest.

$1 \cdot 36$
$2 \cdot 18$
$3 \cdot 12$
$4 \cdot 9$
$6 \cdot 6$
$9 \cdot 4$
$12 \cdot 3$
$18 \cdot 2$
$36 \cdot 1$

} These products are repetitious of the factorizations above. Therefore, we can stop at $6 \cdot 6$.

The list of all factors of 36 consists of the individual factors in the products. The factors are 1, 2, 3, 4, 6, 9, 12, 18, 36.

---

section 2.2   Practice Exercises

Boost *your* GRADE at mathzone.com!

MathZone

- Practice Problems
- Self-Tests
- NetTutor
- e-Professors
- Videos

Study Skills Exercises

1. A rule of thumb is that 2 to 3 hours of study time per week is needed for each 1 hour per week of class time. Based on the number of hours you are in class this semester, how many hours per week should you be studying?

2. Define the key terms.

   a. **Factor**   b. **Factorization**   c. **Prime number**

   d. **Composite number**   e. **Prime factorization**

Review Exercises

For Exercises 3–5, write two fractions, one representing the shaded area and one representing the unshaded area.

3.
   0   1

4.

5.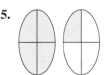

6. Write a fraction with numerator 6 and denominator 5. Is this fraction proper or improper?

7. Write a fraction with denominator 12 and numerator 7. Is this fraction proper or improper?

8. Write a fraction with denominator 6 and numerator 6. Is this fraction proper or improper?

9. Write the improper fraction $\frac{23}{5}$ as a mixed number.   **10.** Write the mixed number $6\frac{2}{7}$ as an improper fraction.

Objective 1: Factors and Factorization

For Exercises 11–16, find two different factorizations of each number. (Answers may vary.) **(See Example 1.)**

11. 8   **12.** 20   **13.** 24   **14.** 14

15. 32   **16.** 54

17. Find two factors whose product is the number in the top row and whose sum is the number in the bottom row. The first column is done for you as an example.

Product	36	42	30	15	81
Factor	12				
Factor	3				
Sum	15	13	31	16	30

18. Find two factors whose product is the number in the top row and whose difference is the number in the bottom row. The first column is done for you as an example.

Product	36	42	45	72	24
Factor	9				
Factor	4				
Difference	5	1	12	14	5

Objective 2: Divisibility Rules

19. State the divisibility rule for dividing by 2.   **20.** State the divisibility rule for dividing by 10.

21. State the divisibility rule for dividing by 3.   **22.** State the divisibility rule for dividing by 5.

For Exercises 23–32, determine if the number is divisible by

a. 2   **b.** 3   **c.** 5   **d.** 10 **(See Example 2.)**

23. 45   **24.** 100   **25.** 72   **26.** 57

27. 108   **28.** 1040   **29.** 3140   **30.** 2115

31. 137   **32.** 241

33. Ms. Haefele has 28 students in her class. Can she distribute a package of 84 candies evenly to her students?

34. Mr. Dietz has 22 students in an algebra class. He has 110 sheets of graph paper. Can he distribute the graph paper evenly among his students?

Objective 3: Prime and Composite Numbers

35. Are there any whole numbers that are not prime or composite? If so, list them.

36. True or false? The square of any prime number is also a prime number.

37. True or false? All odd numbers are prime.   **38.** True or false? All even numbers are composite.

39. One method for finding prime numbers is the *sieve of Eratosthenes*. The natural numbers from 2 to 50 are shown in the table. Start at the number 2 (the smallest prime number). Leave the number 2 and cross out every second number after the number 2. This will eliminate all numbers that are multiples of 2. Then go back to the beginning of the chart and leave the number 3, but cross out every third number after the number 3 (thus eliminating the multiples of 3). Begin at the next open number and continue this process. The numbers that remain are prime numbers. Use this process to find the prime numbers less than 50.

	2	3	4	5	6	7	8	9	10
11	12	13	14	15	16	17	18	19	20
21	22	23	24	25	26	27	28	29	30
31	32	33	34	35	36	37	38	39	40
41	42	43	44	45	46	47	48	49	50

40. Use the sieve of Eratosthenes to find the prime numbers less than 80.

	2	3	4	5	6	7	8	9	10
11	12	13	14	15	16	17	18	19	20
21	22	23	24	25	26	27	28	29	30
31	32	33	34	35	36	37	38	39	40
41	42	43	44	45	46	47	48	49	50
51	52	53	54	55	56	57	58	59	60
61	62	63	64	65	66	67	68	69	70
71	72	73	74	75	76	77	78	79	80

For Exercises 41–56, determine whether the number is prime, composite, or neither. **(See Example 3.)**

41. 7   **42.** 17   **43.** 10   **44.** 21

45. 51   **46.** 57   **47.** 23   **48.** 31

49. 1   **50.** 0   **51.** 121   **52.** 69

53. 19   **54.** 29   **55.** 39   **56.** 49

Objective 4: Prime Factorization

For Exercises 57–60, determine whether or not the factorization represents the prime factorization. If not, explain why.

57. $36 = 2 \cdot 2 \cdot 9$ **58.** $48 = 2 \cdot 3 \cdot 8$ **59.** $210 = 5 \cdot 2 \cdot 7 \cdot 3$ **60.** $126 = 3 \cdot 7 \cdot 3 \cdot 2$

For Exercises 61–72, find the prime factorization. **(See Examples 4 and 5.)**

61. 70 **62.** 495 **63.** 260 **64.** 175

65. 147 **66.** 102 **67.** 138 **68.** 231

69. 616 **70.** 364 **71.** 47 **72.** 41

Objective 5: Identifying All Factors of a Whole Number

For Exercises 73–80, list all the factors of the number. **(See Example 6.)**

73. 12 **74.** 18 **75.** 32 **76.** 55

77. 81 **78.** 60 **79.** 48 **80.** 72

Expanding Your Skills

For Exercises 81–84, determine whether the number is divisible by 4. Use the following divisibility rule: A whole number is divisible by 4 if the number formed by its last two digits is divisible by 4.

81. 230 **82.** 1046 **83.** 4616 **84.** 10,264

For Exercises 85–88, determine whether the number is divisible by 8. Use the following divisibility rule: A whole number is divisible by 8 if the number formed by its last three digits is divisible by 8.

85. 1032 **86.** 2520 **87.** 17,126 **88.** 25,058

For Exercises 89–92, determine whether the number is divisible by 9. Use the following divisibility rule: A whole number is divisible by 9 if the sum of its digits is divisible by 9.

89. 396 **90.** 414 **91.** 8453 **92.** 1587

For Exercises 93–96, determine whether the number is divisible by 6. Use the following divisibility rule: A whole number is divisible by 6 if it is divisible by both 2 and 3 (use the divisibility rules for 2 and 3 together).

93. 522 **94.** 546 **95.** 5917 **96.** 6394

Objectives

1. Equivalent Fractions
2. Simplifying Fractions to Lowest Terms
3. Simplifying Fractions by Powers of 10
4. Applications of Simplifying Fractions

1. Equivalent Fractions

The fractions $\frac{3}{6}, \frac{2}{4}$, and $\frac{1}{2}$ all represent the same portion of a whole. See Figure 2-5. Therefore, we say that the fractions are *equivalent*.

$$\frac{3}{6} \quad = \quad \frac{2}{4} \quad = \quad \frac{1}{2}$$

Figure 2-5

> **Avoiding Mistakes:** The test to determine whether the two fractions are equivalent is not the same process as multiplying fractions. Multiplying of fractions is covered in Section 2.4.

One method to show that two fractions are equivalent is to calculate their cross products. For example, to show that $\frac{3}{6} = \frac{2}{4}$, we have

$$\frac{3}{6} \underset{\nwarrow \nearrow}{\times} \frac{2}{4}$$

$$3 \times 4 \overset{?}{=} 6 \times 2$$

$$12 = 12 \qquad \text{Yes. The fractions are equivalent.}$$

Skill Practice

Fill in the blank ☐ with = or ≠.

1. $\dfrac{13}{24}$ ☐ $\dfrac{6}{11}$

2. $\dfrac{9}{4}$ ☐ $\dfrac{54}{24}$

---

example 1   **Determining Whether Two Fractions Are Equivalent**

Fill in the blank ☐ with = or ≠.

a. $\dfrac{18}{39}$ ☐ $\dfrac{6}{13}$ b. $\dfrac{5}{7}$ ☐ $\dfrac{7}{9}$

Solution:

a. $\dfrac{18}{39} \underset{\nwarrow \nearrow}{\times} \dfrac{6}{13}$

$18 \times 13 \overset{?}{=} 39 \times 6$

$234 = 234$

Therefore, $\dfrac{18}{39} \boxed{=} \dfrac{6}{13}$.

b. $\dfrac{5}{7} \underset{\nwarrow \nearrow}{\times} \dfrac{7}{9}$

$5 \times 9 \overset{?}{=} 7 \times 7$

$45 \neq 49$

Therefore, $\dfrac{5}{7} \boxed{\neq} \dfrac{7}{9}$.

---

2. Simplifying Fractions to Lowest Terms

In Figure 2-5 we see that $\frac{3}{6}, \frac{2}{4}$, and $\frac{1}{2}$ all represent equal quantities. However, the fraction $\frac{1}{2}$ is said to be in **lowest terms** because the numerator and denominator share no common factor other than 1.

To simplify a fraction to lowest terms, we use the following important principle.

> ### The Fundamental Principle of Fractions
>
> Consider the fraction $\dfrac{a}{b}$ and the nonzero number c. Then
>
> $$\frac{a}{b} = \frac{a \div c}{b \div c}$$

Answers

1. ≠
2. =

The fundamental principle of fractions indicates that dividing both the numerator and the denominator by the same nonzero number results in an equivalent fraction. For example, the numerator and denominator of the fraction $\frac{3}{6}$ both share a common factor of 3. To simplify $\frac{3}{6}$, we will divide both the numerator and denominator by the common factor 3.

$$\frac{3}{6} = \frac{3 \div 3}{6 \div 3} = \frac{1}{2}$$

Before applying the fundamental principle of fractions, it is helpful to write the prime factorization of both the numerator and the denominator. This will allow us to find the common factors. For example, to simplify $\frac{35}{42}$ to lowest terms, begin by writing

$$\frac{35}{42} = \frac{5 \cdot 7}{2 \cdot 3 \cdot 7}$$

In this form, it is clear that 7 is the common factor. Use the fundamental principle of fractions to divide the numerator and denominator by 7:

$$\frac{5 \cdot 7 \div 7}{2 \cdot 3 \cdot 7 \div 7} = \frac{5}{2 \cdot 3} = \frac{5}{6}$$

Because there are no other common factors, we say that $\frac{5}{6}$ is simplified to lowest terms.

A shorthand notation of writing the division by 7 is to strike out the common factors with "cancel lines." Then replace the common factor of 7 by the new common factor of 1.

$$\frac{35}{42} = \frac{5 \cdot 7}{2 \cdot 3 \cdot 7} = \frac{5 \cdot \overset{1}{\cancel{7}}}{2 \cdot 3 \cdot \underset{1}{\cancel{7}}} = \frac{5 \cdot 1}{2 \cdot 3 \cdot 1} = \frac{5}{6}$$

Tip: Simplifying a fraction is also called reducing a fraction to lowest terms. For example, the simplified (or reduced) form of $\frac{35}{42}$ is $\frac{5}{6}$.

---

example 2	**Simplifying Fractions to Lowest Terms**

Simplify to lowest terms. Write the answer as a fraction or whole number.

a. $\dfrac{30}{20}$ **b.** $\dfrac{8}{24}$ **c.** $\dfrac{110}{99}$ **d.** $\dfrac{75}{25}$

Solution:

a. $\dfrac{30}{20} = \dfrac{2 \cdot 3 \cdot \overset{1}{\cancel{5}}}{\underset{1}{\cancel{2}} \cdot 2 \cdot \underset{1}{\cancel{5}}} = \dfrac{1 \cdot 3 \cdot 1}{1 \cdot 2 \cdot 1} = \dfrac{3}{2}$

b. $\dfrac{8}{24} = \dfrac{\overset{1}{\cancel{2}} \cdot \overset{1}{\cancel{2}} \cdot \overset{1}{\cancel{2}}}{\underset{1}{\cancel{2}} \cdot \underset{1}{\cancel{2}} \cdot \underset{1}{\cancel{2}} \cdot 3} = \dfrac{1 \cdot 1 \cdot 1}{1 \cdot 1 \cdot 1 \cdot 3} = \dfrac{1}{3}$

c. $\dfrac{110}{99} = \dfrac{2 \cdot 5 \cdot \overset{1}{\cancel{11}}}{3 \cdot 3 \cdot \underset{1}{\cancel{11}}} = \dfrac{2 \cdot 5}{3 \cdot 3} = \dfrac{10}{9}$

d. $\dfrac{75}{25} = \dfrac{3 \cdot \overset{1}{\cancel{5}} \cdot \overset{1}{\cancel{5}}}{\underset{1}{\cancel{5}} \cdot \underset{1}{\cancel{5}}} = \dfrac{3 \cdot 1 \cdot 1}{1 \cdot 1} = \dfrac{3}{1} = 3$

Avoiding Mistakes: Don't forget to write the 1's when you strike out common factors. This is particularly important when all the factors in the numerator "cancel."

Tip: Recall that any fraction of the form $\frac{n}{1} = n$. Therefore, $\frac{3}{1}$ equals the whole number 3.

Skill Practice

Simplify to lowest terms.

3. $\dfrac{15}{35}$ **4.** $\dfrac{48}{12}$

5. $\dfrac{14}{84}$ **6.** $\dfrac{26}{195}$

Another method to simplify a fraction to lowest terms is to identify the greatest factor shared by both the numerator and denominator (called the *greatest*

Answers

3. $\dfrac{3}{7}$ 4. 4   5. $\dfrac{1}{6}$ 6. $\dfrac{2}{15}$

common factor). For example, with the fraction $\frac{48}{32}$, you might notice that 16 is the greatest number that divides evenly into both the numerator and the denominator. Therefore, we can factor the numerator and denominator by using 16 as one of the factors. Then we divide the numerator and denominator by 16.

$$\frac{48}{32} = \frac{3 \cdot 16}{2 \cdot 16} = \frac{3 \cdot \cancel{16}}{2 \cdot \cancel{16}} = \frac{3}{2}$$

The drawback to this method is that students don't always divide by the *greatest* common factor from the numerator and denominator. For example, suppose we had incorrectly thought that 8 was the greatest number that divides into the numerator and denominator. We show that our final answer is not simplified completely.

$$\frac{48}{32} = \frac{6 \cdot 8}{4 \cdot 8} = \frac{6 \cdot \cancel{8}}{4 \cdot \cancel{8}} = \frac{6}{4} \longleftarrow \quad \text{6 and 4 still share a} \\ \longleftarrow \quad \text{common factor of 2.}$$

The fraction $\frac{6}{4}$ is only *partially simplified* because we did not divide by the *greatest* common factor. To complete the simplification, we must reduce again by dividing the numerator and denominator by the common factor of 2.

$$\frac{6}{4} = \frac{2 \cdot 3}{2 \cdot 2} = \frac{3}{2}$$

3. Simplifying Fractions by Powers of 10

Consider the fraction $\frac{50}{70}$. Both the numerator and the denominator are divisible by 10 because the ones-place digit ends in 0. To simplify this fraction, we can divide both numerator and denominator by 10.

$$\frac{50}{70} = \frac{5 \cdot \cancel{10}}{7 \cdot \cancel{10}} = \frac{5}{7}$$

Notice that dividing numerator and denominator by 10 has the effect of eliminating the 0 in the ones place from each number.

$$\frac{5\cancel{0}}{7\cancel{0}} = \frac{5}{7}$$

This is a quick way to simplify, or partially simplify, a fraction when the numerator and denominator share a common factor of 10, 100, 1000, and so on. This is demonstrated in Example 3.

Skill Practice

Simplify to lowest terms by first reducing by 10, 100, or 1000.

8. $\dfrac{630}{190}$ **9.** $\dfrac{1300}{52,000}$

10. $\dfrac{21,000}{35,000}$

example 3   Simplifying Fractions by 10, 100, and 1000

Simplify each fraction to lowest terms by first reducing by 10, 100, or 1000. Write the answer as a fraction.

a. $\dfrac{170}{30}$ **b.** $\dfrac{2500}{7500}$ **c.** $\dfrac{5000}{130,000}$

Solution:

a. $\dfrac{170}{30} = \dfrac{17\cancel{0}}{3\cancel{0}}$ Both 170 and 30 are divisible by 10. "Strike through" one zero.

$= \dfrac{17}{3}$ The fraction $\frac{17}{3}$ is simplified completely.

Answers

7. The fraction $\frac{39}{52}$ can be simplified further to $\frac{3}{4}$.

8. $\dfrac{63}{19}$ 9. $\dfrac{1}{40}$ 10. $\dfrac{3}{5}$

b. $\dfrac{2500}{7500} = \dfrac{25\cancel{00}}{75\cancel{00}}$ Both 2500 and 7500 are divisible by 100. Strike through two zeros.

$= \dfrac{25}{75}$ The fraction is partially simplified.

$= \dfrac{\overset{1}{\cancel{5}} \cdot \overset{1}{\cancel{5}}}{3 \cdot \underset{1}{\cancel{5}} \cdot \underset{1}{\cancel{5}}}$ Factor and simplify further.

$= \dfrac{1}{3}$

c. $\dfrac{5000}{130{,}000} = \dfrac{5\cancel{000}}{130{,}\cancel{000}}$ The numbers 5000 and 130,000 are both divisible by 1000. Strike through three zeros.

$= \dfrac{5}{130}$

$= \dfrac{\overset{1}{\cancel{5}}}{2 \cdot \underset{1}{\cancel{5}} \cdot 13}$ Factor and simplify further.

$= \dfrac{1}{26}$

---

4. Applications of Simplifying Fractions

example 4   Simplifying Fractions in an Application

Madeleine got 28 out of 35 problems correct on an algebra exam. David got 27 out of 45 questions correct on a different algebra exam.

  a. What fractional part of the exam did each student answer correctly?

  b. Which student performed better?

Solution:

  a. Fractional part correct for Madeleine:

$$\dfrac{28}{35} \quad \text{or equivalently} \quad \dfrac{28}{35} = \dfrac{2 \cdot 2 \cdot \overset{1}{\cancel{7}}}{5 \cdot \underset{1}{\cancel{7}}} = \dfrac{4}{5}$$

    Fractional part correct for David:

$$\dfrac{27}{45} \quad \text{or equivalently} \quad \dfrac{27}{45} = \dfrac{\overset{1}{\cancel{3}} \cdot \overset{1}{\cancel{3}} \cdot 3}{\underset{1}{\cancel{3}} \cdot \underset{1}{\cancel{3}} \cdot 5} = \dfrac{3}{5}$$

  b. From the simplified form of each fraction, we see that Madeleine performed better because $\frac{4}{5} > \frac{3}{5}$. That is, 4 parts out of 5 is greater than 3 parts out of 5. This is also easily verified on a number line.

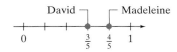

Concept Connections

11. How many zeros may be eliminated from the numerator and denominator of the fraction $\frac{430{,}000}{154{,}000{,}000}$?

Skill Practice

12. Joanne planted 77 seeds in her garden and 55 sprouted. Geoff planted 140 seeds and 80 sprouted.

  a. What fractional part of the seeds sprouted for Joanne and what part sprouted for Geoff?

  b. For which person did a greater portion of seeds sprout?

Answers

11. Four zeros; the numerator and denominator are both divisible by 10,000.
12. **a.** Joanne: $\frac{5}{7}$; Geoff: $\frac{4}{7}$
    b. Joanne had a greater portion of seeds sprout.

section 2.3   Practice Exercises

**Boost *your* GRADE at
mathzone.com!**

MathZone

- Practice Problems
- Self-Tests
- NetTutor
- e-Professors
- Videos

Study Skills Exercises

1. Write down the page number(s) for the Midchapter Review for this chapter. _____ Explain the purpose of the Midchapter Review.

2. Define the key term **lowest terms**.

Review Exercises

For Exercises 3–9, write the prime factorization for each number.

3. 145   **4.** 114   **5.** 92   **6.** 153

7. 85   **8.** 120   **9.** 195

Objective 1: Equivalent Fractions

For Exercises 10–13, shade the second figure so that it expresses a fraction equivalent to the first figure.

10.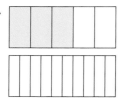

11.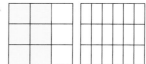

12.

13.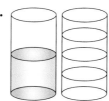

14. True or false? The fractions $\frac{1}{3}$ and $\frac{3}{1}$ are equivalent.   **15.** True or false? The fractions $\frac{4}{5}$ and $\frac{5}{4}$ are equivalent.

16. In your own words, explain the concept of equivalent fractions.

For Exercises 17–24, determine if the fractions are equivalent. Then fill in the blank with either $=$ or $\neq$.
(See Example 1.)

17. $\frac{2}{3} \square \frac{3}{5}$ **18.** $\frac{1}{4} \square \frac{2}{9}$ **19.** $\frac{1}{2} \square \frac{3}{6}$ **20.** $\frac{6}{16} \square \frac{3}{8}$

21. $\dfrac{12}{16} \square \dfrac{3}{4}$ **22.** $\dfrac{4}{5} \square \dfrac{12}{15}$ **23.** $\dfrac{8}{9} \square \dfrac{20}{27}$ **24.** $\dfrac{5}{6} \square \dfrac{12}{18}$

Objective 2: Simplifying Fractions to Lowest Terms

For Exercises 25–54, simplify the fraction to lowest terms. Write the answer as a fraction or a whole number. **(See Example 2.)**

25. $\dfrac{12}{24}$ **26.** $\dfrac{15}{18}$ **27.** $\dfrac{6}{18}$ **28.** $\dfrac{21}{24}$

29. $\dfrac{36}{20}$ **30.** $\dfrac{49}{42}$ **31.** $\dfrac{15}{12}$ **32.** $\dfrac{30}{25}$

33. $\dfrac{20}{25}$ **34.** $\dfrac{8}{16}$ **35.** $\dfrac{14}{14}$ **36.** $\dfrac{8}{8}$

37. $\dfrac{50}{25}$ **38.** $\dfrac{24}{6}$ **39.** $\dfrac{9}{9}$ **40.** $\dfrac{2}{2}$

41. $\dfrac{105}{140}$ **42.** $\dfrac{84}{126}$ **43.** $\dfrac{33}{11}$ **44.** $\dfrac{65}{5}$

45. $\dfrac{77}{110}$ **46.** $\dfrac{85}{153}$ **47.** $\dfrac{130}{150}$ **48.** $\dfrac{70}{120}$

49. $\dfrac{385}{195}$ **50.** $\dfrac{39}{130}$ **51.** $\dfrac{34}{85}$ **52.** $\dfrac{69}{92}$

53. $\dfrac{145}{58}$ **54.** $\dfrac{114}{95}$

Objective 3: Simplifying Fractions by Powers of 10

For Exercises 55–62, simplify to lowest terms by first reducing the powers of 10. **(See Example 3.)**

55. $\dfrac{120}{160}$ **56.** $\dfrac{720}{800}$ **57.** $\dfrac{3000}{1800}$ **58.** $\dfrac{2000}{1500}$

59. $\dfrac{42,000}{22,000}$ **60.** $\dfrac{50,000}{65,000}$ **61.** $\dfrac{5100}{30,000}$ **62.** $\dfrac{9800}{28,000}$

Objective 4: Application of Simplifying Fractions

63. Aundrea tossed a coin 48 times and heads came up 20 times. What fractional part of the tosses came up heads? What fractional part came up tails?

64. **a.** What fraction of the alphabet is made up of vowels? (Include the letter y as a vowel, not a consonant.)

 b. What fraction of the alphabet is made up of consonants?

65. Of the 88 constellations that can be seen in the night sky, 12 are associated with astrological horoscopes. The names of as many as 36 constellations are associated with animals or mythical creatures.

 a. Of the 88 constellations, what fraction is associated with horoscopes?

 b. What fraction of the constellations have names associated with animals or mythical creatures?

66. At Pizza Company, Lee made 70 pizzas one day. There were 105 pizzas sold that day. What fraction of the pizzas did Lee make?

67. Jonathan and Jared sell candy bars for a fund-raiser. Jonathan sold 25 of his 35 candy bars, and Jared sold 24 of his 28 candy bars. **(See Example 4.)**

 a. What fractional part of his total number of candy bars did each boy sell?

 b. Which boy sold the greater fractional part?

68. Lynette and Lisa are taking online courses. Lisa has completed 14 out of 16 assignments in her course while Lynette has completed 15 out of 24 assignments.

 a. What fractional part of her total number of assignments did each woman complete?

 b. Which woman has completed more of her course?

69. Raymond read 720 pages of a 792-page book. His roommate, Travis, read 540 pages from a 660-page book.

 a. What fractional part of the book did each person read?

 b. Which of the roommates read a greater fraction of his book?

70. Mr. Bishop and Ms. Waymire both gave exams today. By mid-afternoon, Mr. Bishop had finished grading 16 out of 36 exams, and Ms. Waymire had finished grading 15 out of 27 exams.

    a. What fractional part of her total has Ms. Waymire completed?

    b. What fractional part of his total has Mr. Bishop completed?

Expanding Your Skills

71. Write three fractions equivalent to $\frac{3}{4}$.

72. Write three fractions equivalent to $\frac{1}{3}$.

73. Write three fractions equivalent to $\frac{12}{18}$.

74. Write three fractions equivalent to $\frac{80}{100}$.

---

chapter 2 | midchapter review

1. Write a definition of an improper fraction.

2. Write a definition of a proper fraction.

3. Of the fractions $\frac{2}{3}$ and $\frac{3}{2}$, which one is proper and which one is improper?

4. Draw a figure that represents the fraction $\frac{2}{3}$. Answers may vary.

5. Draw a figure that represents the fraction $\frac{3}{2}$. Answers may vary.

6. List all the factors of the given number.

    a. 16   **b.** 63

    c. 40   **d.** 30

7. Write the prime factorization of each number.

    a. 16   **b.** 63

    c. 40   **d.** 30

For Exercises 8–13, simplify the fraction to lowest terms and write the answer as a fraction.

8. $\dfrac{16}{30}$ **9.** $\dfrac{20}{30}$ **10.** $\dfrac{20}{16}$

11. $\dfrac{30}{24}$ **12.** $\dfrac{24}{16}$ **13.** $\dfrac{20}{24}$

For Exercises 14–16, simplify the fraction and write the answer as a mixed number.

14. $\dfrac{30}{12}$ **15.** $\dfrac{22}{8}$ **16.** $\dfrac{12}{9}$

Objectives

1. Multiplication of Fractions
2. Fractions and the Order of Operations
3. Area of a Triangle
4. Applications of Multiplying Fractions

Concept Connections

1. What fraction is $\frac{1}{2}$ of $\frac{1}{4}$ of a whole?

---

section 2.4　Multiplication of Fractions and Applications

1. Multiplication of Fractions

Suppose Elija takes $\frac{1}{3}$ of a cake and then gives $\frac{1}{2}$ of this portion to his friend Max. Max gets $\frac{1}{2}$ of $\frac{1}{3}$ of the cake. This is equivalent to the expression $\frac{1}{2} \cdot \frac{1}{3}$. See Figure 2-6.

Elija takes $\frac{1}{3}$

Max gets $\frac{1}{2}$ of $\frac{1}{3} = \frac{1}{6}$

Figure 2-6

From the illustration, the product $\frac{1}{2} \cdot \frac{1}{3} = \frac{1}{6}$. Notice that the product $\frac{1}{6}$ is found by multiplying the numerators and multiplying the denominators. This is true in general to multiply fractions.

Multiplication of Fractions

To multiply fractions, write the product of the numerators over the product of the denominators. Then simplify the resulting fraction, if possible.

The rule for multiplying fractions can be expressed symbolically as follows.

$$\frac{a}{b} \cdot \frac{c}{d} = \frac{a \cdot c}{b \cdot d} \qquad \text{provided } b \text{ and } d \text{ are not equal to } 0$$

Skill Practice

Multiply. Write the answer as a fraction.

2. $\frac{2}{3} \cdot \frac{5}{9}$　**3.** $\frac{7}{12} \times 11$

---

example 1　Multiplying Fractions

Multiply.

a. $\frac{2}{5} \cdot \frac{4}{7}$　　**b.** $\frac{8}{3} \times 5$

Solution:

a. $\frac{2}{5} \cdot \frac{4}{7} = \frac{2 \cdot 4}{5 \cdot 7} = \frac{8}{35}$ ← Multiply the numerators.
← Multiply the denominators.

Notice that the product $\frac{8}{35}$ is simplified completely because there are no common factors shared by 8 and 35.

b. $\frac{8}{3} \times 5 = \frac{8}{3} \times \frac{5}{1}$　　First write the whole number as a fraction.

$= \frac{8 \times 5}{3 \times 1}$　　Multiply the numerators. Multiply the denominators.

$= \frac{40}{3}$　　The product is not reducible because there are no common factors shared by 40 and 3.

Answers

1. $\frac{1}{8}$　2. $\frac{10}{27}$　3. $\frac{77}{12}$

Example 2 illustrates a case where the product of fractions must be simplified.

---

example 2   Multiplying and Simplifying Fractions

Multiply the fractions and simplify if possible.

$$\frac{4}{30} \cdot \frac{5}{14}$$

Solution:

$$\frac{4}{30} \cdot \frac{5}{14} = \frac{4 \cdot 5}{30 \cdot 14}$$ Multiply the numerators. Multiply the denominators.

$$= \frac{20}{420}$$ The fraction $\frac{20}{420}$ is not simplified.
To simplify, write the prime factorization of 20 and 420.

$$\begin{array}{cc} & 7 \\ & 5\overline{)35} \\ 5 & 3\overline{)105} \\ 2\overline{)10} & 2\overline{)210} \\ 2\overline{)20} & 2\overline{)420} \end{array}$$

$$20 = 2 \cdot 2 \cdot 5 \qquad 420 = 2 \cdot 2 \cdot 3 \cdot 5 \cdot 7$$

$$\frac{20}{420} = \frac{\overset{1}{2} \cdot \overset{1}{2} \cdot \overset{1}{\cancel{5}}}{\underset{1}{2} \cdot \underset{1}{2} \cdot 3 \cdot \underset{1}{\cancel{5}} \cdot 7}$$

$$= \frac{1 \cdot 1 \cdot 1}{1 \cdot 1 \cdot 3 \cdot 1 \cdot 7}$$

$$= \frac{1}{21}$$

It is often easier to factor the numerator and denominator of a fraction *before* multiplying. Consider the product from Example 2. The numbers 4, 5, 30, and 14 can be factored more easily than the larger numbers of 20 and 420.

$$\frac{4}{30} \cdot \frac{5}{14} = \frac{2 \cdot 2}{3 \cdot 2 \cdot 5} \cdot \frac{5}{2 \cdot 7}$$ First factor the numbers in the original fractions.

$$= \frac{\overset{1}{2} \cdot \overset{1}{2}}{3 \cdot 2 \cdot \underset{1}{\cancel{5}}} \cdot \frac{\overset{1}{\cancel{5}}}{2 \cdot 7}$$ We can simplify *before* multiplying and obtain the same result.

$$= \frac{1}{21}$$

As a general rule, this method is used most often in the text.

---

example 3   Multiplying and Simplifying Fractions

Multiply and simplify.

$$\frac{10}{18} \times \frac{21}{55}$$

Skill Practice

Multiply and simplify.

6. $\dfrac{6}{25} \times \dfrac{15}{18}$ **7.** $\dfrac{7}{8} \cdot \dfrac{4}{9}$

Solution:

$$\frac{10}{18} \times \frac{21}{55} = \frac{2 \cdot 5}{2 \cdot 3 \cdot 3} \times \frac{3 \cdot 7}{5 \cdot 11} \qquad \text{Factor the numerators and denominators.}$$

$$= \frac{\overset{1}{2} \cdot \overset{1}{5}}{\underset{1}{2} \cdot \underset{1}{3} \cdot 3} \times \frac{\overset{1}{3} \cdot 7}{\underset{1}{5} \cdot 11} \qquad \text{Simplify.}$$

$$= \frac{7}{33} \qquad \text{Multiply.}$$

Sometimes you may recognize factors common to both the numerator and the denominator without writing the prime factorizations first. In such a case, you may opt to simplify without factoring the individual numerators or denominators. For example, consider the product:

$$\frac{10}{18} \times \frac{21}{55} = \frac{\overset{2}{\cancel{10}}}{\underset{6}{18}} \times \frac{\overset{7}{\cancel{21}}}{\underset{11}{\cancel{55}}} \qquad \begin{array}{l} \text{10 and 55 share a common factor of 5.} \\ \text{18 and 21 share a common factor of 3.} \end{array}$$

$$= \frac{\overset{1}{\overset{2}{\cancel{10}}}}{\underset{6}{\underset{3}{18}}} \times \frac{\overset{7}{21}}{\underset{11}{55}} \qquad \begin{array}{l} \text{We can simplify further because 2} \\ \text{and 6 share a common factor of 2.} \end{array}$$

$$= \frac{7}{33}$$

Skill Practice

Multiply and simplify. Write the answer as a fraction.

8. $9\left(\dfrac{2}{15}\right)$

9. $\left(\dfrac{3}{16}\right)\left(\dfrac{4}{9}\right)\left(\dfrac{6}{5}\right)$

example 4   Multiplying and Simplifying Fractions

Multiply and simplify. Write the answer as a fraction.

a. $6\left(\dfrac{3}{8}\right)$ **b.** $\dfrac{21}{25} \cdot \dfrac{65}{24} \cdot \dfrac{15}{39}$

Solution:

a. $6\left(\dfrac{3}{8}\right) = \dfrac{6}{1} \cdot \dfrac{3}{8}$ Write the whole number as a fraction.

$$= \frac{2 \cdot \overset{1}{3}}{1} \cdot \frac{3}{2 \cdot 2 \cdot \underset{1}{2}} \qquad \begin{array}{l} \text{Factor the numerators and denominators.} \\ \text{Simplify.} \end{array}$$

$$= \frac{9}{4} \qquad \text{Multiply.}$$

b. $\dfrac{21}{25} \cdot \dfrac{65}{24} \cdot \dfrac{15}{39} = \dfrac{3 \cdot 7}{5 \cdot 5} \cdot \dfrac{5 \cdot 13}{2 \cdot 2 \cdot 2 \cdot 3} \cdot \dfrac{3 \cdot 5}{3 \cdot 13}$ Factor.

$$= \frac{\overset{1}{3} \cdot 7}{\underset{1}{5} \cdot \underset{1}{5}} \cdot \frac{\overset{1}{5} \cdot \overset{1}{13}}{2 \cdot 2 \cdot 2 \cdot \underset{1}{3}} \cdot \frac{\overset{1}{3} \cdot \overset{1}{5}}{\underset{1}{3} \cdot \underset{1}{13}} \qquad \text{Simplify.}$$

$$= \frac{7}{8} \qquad \text{Multiply.}$$

2. Fractions and the Order of Operations

In Section 1.7 we learned to recognize powers of 10. These are $10^1 = 10$, $10^2 = 100$, and so on. In this section, we learn to recognize the **powers of one-tenth**, that is, $\frac{1}{10}$ raised to a whole-number power. For example, consider the following expressions.

Concept Connections

10. Evaluate 10^2 and $\left(\frac{1}{10}\right)^2$

11. Evaluate 10^5 and $\left(\frac{1}{10}\right)^5$.

$$\left(\frac{1}{10}\right)^1 = \frac{1}{10}$$

$$\left(\frac{1}{10}\right)^2 = \frac{1}{10} \cdot \frac{1}{10} = \frac{1}{100}$$

$$\left(\frac{1}{10}\right)^3 = \frac{1}{10} \cdot \frac{1}{10} \cdot \frac{1}{10} = \frac{1}{1000}$$

$$\left(\frac{1}{10}\right)^4 = \frac{1}{10} \cdot \frac{1}{10} \cdot \frac{1}{10} \cdot \frac{1}{10} = \frac{1}{10,000}$$

$$\left(\frac{1}{10}\right)^5 = \frac{1}{10} \cdot \frac{1}{10} \cdot \frac{1}{10} \cdot \frac{1}{10} \cdot \frac{1}{10} = \frac{1}{100,000}$$

$$\left(\frac{1}{10}\right)^6 = \frac{1}{10} \cdot \frac{1}{10} \cdot \frac{1}{10} \cdot \frac{1}{10} \cdot \frac{1}{10} \cdot \frac{1}{10} = \frac{1}{1,000,000}$$

From these examples, we see that a power of one-tenth results in a fraction with a 1 in the numerator. The denominator has a 1 followed by the same number of zeros as the exponent on the base of $\frac{1}{10}$.

For problems with more than one operation, the order of operations must still be considered.

---

example 5   **Simplifying Expressions**

Simplify.

a. $\left(\frac{2}{5}\right)^3$ **b.** $\left(\frac{2}{15} \cdot \frac{3}{4}\right)^2$

Skill Practice

Simplify. Write the answer as a fraction.

12. $\left(\frac{4}{3}\right)^2$ **13.** $\left(\frac{6}{5} \cdot \frac{1}{12}\right)^3$

Solution:

a. $\left(\frac{2}{5}\right)^3 = \frac{2}{5} \cdot \frac{2}{5} \cdot \frac{2}{5}$ With an exponent of 3, multiply 3 factors of the base.

$\phantom{\left(\frac{2}{5}\right)^3} = \frac{2 \cdot 2 \cdot 2}{5 \cdot 5 \cdot 5}$ Multiply the numerators. Multiply the denominators.

$\phantom{\left(\frac{2}{5}\right)^3} = \frac{8}{125}$

b. $\left(\frac{2}{15} \cdot \frac{3}{4}\right)^2 = \left(\frac{\overset{1}{\cancel{2}}}{\cancel{3} \cdot 5} \cdot \frac{\overset{1}{\cancel{3}}}{2 \cdot 2}\right)^2$ Perform the multiplication within the parentheses. Simplify.

$\phantom{\left(\frac{2}{15} \cdot \frac{3}{4}\right)^2} = \left(\frac{1}{10}\right)^2$ Multiply fractions within parentheses.

$\phantom{\left(\frac{2}{15} \cdot \frac{3}{4}\right)^2} = \frac{1}{100}$

Answers

10. $10^2 = 100$; $\left(\frac{1}{10}\right)^2 = \frac{1}{100}$

11. $10^5 = 100,000$; $\left(\frac{1}{10}\right)^5 = \frac{1}{100,000}$

12. $\frac{16}{9}$ **13.** $\frac{1}{1000}$

3. Area of a Triangle

Recall that the area of a rectangle with length l and width w is given by

$$A = l \times w$$

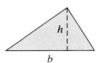

Area of a Triangle

The formula for the area of a triangle is given by $A = \frac{1}{2}bh$, read "one-half base times height."

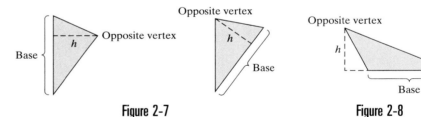

Concept Connections

Identify the measure of the base and height.

14.

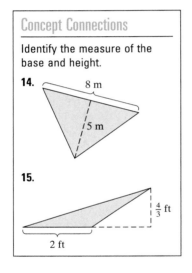
8 m
5 m

15.

$\frac{4}{3}$ ft

2 ft

The value of b is the measure of the base of the triangle. The value of h is the measure of the height of the triangle. The base b may be chosen as the length of any of the sides of the triangle. However, once you have chosen the base, the height must be measured as the shortest distance from the base to the opposite vertex (or point) of the triangle. For example, Figure 2-7 shows the same triangle with different choices for the base. Figure 2-8 shows a situation in which the height must be drawn "outside" the triangle. In such a case, notice that the height is drawn down to an imaginary extension of the base line.

Opposite vertex

Opposite vertex

Base

h

Opposite vertex

h

Base

Opposite vertex

h

Base

Figure 2-7   **Figure 2-8**

In Example 6 we demonstrate how to apply the formula to find the area of a triangle.

---

example 6   Finding the Area of a Triangle

Find the area of the triangle.

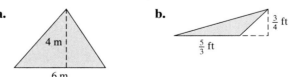

a.

4 m

6 m

b.

$\frac{3}{4}$ ft

$\frac{5}{3}$ ft

Answers

14. Base is 8 m; height is 5 m
15. Base is 2 ft; height is $\frac{4}{3}$ ft

Solution:

a. $b = 6$ m   and $h = 4$ m   Identify the measure of the base and the height.

$A = \dfrac{1}{2}bh$

$= \dfrac{1}{2}(6\,\text{m})(4\,\text{m})$ Apply the formula for the area of a triangle.

$= \dfrac{1}{2}\left(\dfrac{6}{1}\text{m}\right)\left(\dfrac{4}{1}\text{m}\right)$ Write the whole numbers as fractions.

$= \dfrac{1}{2}\left(\dfrac{\overset{3}{6}}{1}\text{m}\right)\left(\dfrac{4}{1}\text{m}\right)$ Simplify.

$= \dfrac{12}{1}\,\text{m}^2$ Multiply numerators. Multiply denominators.

$= 12\,\text{m}^2$ The area of the triangle is 12 square meters (m²).

b. $b = \dfrac{5}{3}$ ft   and $h = \dfrac{3}{4}$ ft   Identify the measure of the base and the height.

$A = \dfrac{1}{2}bh$

$= \dfrac{1}{2}\left(\dfrac{5}{3}\text{ft}\right)\left(\dfrac{3}{4}\text{ft}\right)$ Apply the formula for the area of a triangle.

$= \dfrac{1}{2}\left(\dfrac{5}{\overset{}{\underset{1}{3}}}\text{ft}\right)\left(\dfrac{\overset{1}{3}}{4}\text{ft}\right)$ Simplify.

$= \dfrac{5}{8}\,\text{ft}^2$ The area of the triangle is $\dfrac{5}{8}$ square feet (ft²).

 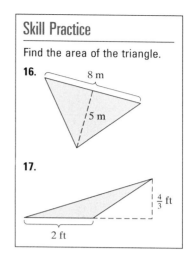
---

example 7   Find the Area of a Composite Geometric Figure

Find the area.

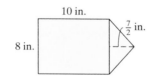
10 in.
8 in.
$\frac{7}{2}$ in.

Solution:

The total area is the sum of the areas of the rectangular region and the triangular region. That is,

Total area =
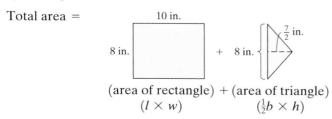
10 in.
8 in.
$+$ 8 in.
$\frac{7}{2}$ in.

(area of rectangle) + (area of triangle)
($l \times w$)   ($\frac{1}{2}b \times h$)

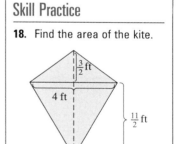

$$= (10 \text{ in.})(8 \text{ in.}) + \frac{1}{2}(8 \text{ in.})\left(\frac{7}{2} \text{ in.}\right)$$

$$= 80 \text{ in.}^2 + \frac{1}{2}\left(\frac{8}{1} \text{ in.}\right)\left(\frac{7}{2} \text{ in.}\right)$$

$$= 80 \text{ in.}^2 + \frac{1}{2}\left(\frac{\overset{1}{2} \cdot 2 \cdot \overset{1}{2}}{1} \text{ in.}\right)\left(\frac{7}{\underset{1}{2}} \text{ in.}\right)$$

$$= 80 \text{ in.}^2 + 14 \text{ in.}^2$$

$$= 94 \text{ in.}^2$$

The total area of the region is 94 square inches (in.^{2}).

4. Applications of Multiplying Fractions

Multiplying Fractions in an Application

In 2005, the population of Texas comprised roughly $\frac{3}{40}$ of the population of the United States. If the U.S. population was approximately 292,000,000, approximate the population of Texas.

Solution:

We must find $\frac{3}{40}$ of the U.S. population. This translates to

$$\frac{3}{40} \cdot 292,000,000 = \frac{3}{40} \cdot \frac{292,000,000}{1}$$

$$= \frac{876,000,000}{40} \quad \longleftarrow \quad \text{Multiply the numerators.}$$
$$\qquad\qquad\qquad\; \longleftarrow \quad \text{Multiply the denominators.}$$

$$= \frac{876,000,0\cancel{00}}{4\cancel{0}} \qquad\quad \text{Simplify by a factor of 10.}$$

$$= \frac{87,600,000}{4} \qquad\quad \begin{array}{l}\text{Simplify by first writing the fraction}\\ \text{as a division problem: } 87,600,000 \div 4.\end{array}$$

$$\begin{array}{r} 21,900,000 \\ 4\overline{)87,600,000} \\ \underline{-8} \\ 7 \\ \underline{-4} \\ 36 \\ \underline{-36} \\ 0 \end{array}$$

$$= 21,900,000 \qquad\qquad \text{Divide.}$$

The population of Texas was approximately 21,900,000.

---

19. Find the population of Tennessee if it is approximately $\frac{1}{50}$ of the U.S. population. Assume that the U.S. population was approximately 292,000,000.

Answer

19. 5,840,000

 **Practice Exercises**

Study Skills Exercises

1. Write down the page number(s) for the Chapter Summary for this chapter. _____
Describe one way in which you can use the summary found at the end of each chapter.

2. Define the key term **power of one-tenth**.

Review Exercises

For Exercises 3–7, identify the numerator and the denominator. Then simplify the fraction to lowest terms.

3. $\dfrac{10}{14}$ **4.** $\dfrac{32}{36}$ **5.** $\dfrac{25}{15}$ **6.** $\dfrac{2100}{7000}$ **7.** $\dfrac{7200}{90{,}000}$

Objective 1: Multiplication of Fractions

8. Shade the portion of the figure that represents $\frac{1}{3}$ of $\frac{1}{2}$.

9. Shade the portion of the figure that represents $\frac{1}{4}$ of $\frac{1}{4}$.

10. Shade the portion of the figure that represents $\frac{1}{3}$ of $\frac{1}{4}$.

11. Find $\frac{1}{2}$ of $\frac{1}{4}$.   **12.** Find $\frac{2}{3}$ of $\frac{1}{5}$.   **13.** Find $\frac{3}{4}$ of 8.   **14.** Find $\frac{2}{5}$ of 20.

For Exercises 15–26, multiply the fractions. Write the answer as a fraction. **(See Example 1.)**

15. $\dfrac{1}{2} \times \dfrac{3}{8}$ **16.** $\dfrac{2}{3} \times \dfrac{1}{3}$ **17.** $\dfrac{14}{9} \cdot \dfrac{1}{9}$ **18.** $\dfrac{1}{8} \cdot \dfrac{9}{8}$

19. $\left(\dfrac{12}{7}\right)\left(\dfrac{2}{5}\right)$

20. $\left(\dfrac{9}{10}\right)\left(\dfrac{7}{4}\right)$

21. $8 \cdot \left(\dfrac{1}{11}\right)$

22. $3 \cdot \left(\dfrac{2}{7}\right)$

23. $\dfrac{4}{5} \cdot 6$

24. $\dfrac{5}{8} \cdot 5$

25. $\dfrac{13}{9} \times \dfrac{5}{4}$

26. $\dfrac{6}{5} \times \dfrac{7}{5}$

For Exercises 27–50, multiply the fractions and simplify to lowest terms. Write the answer as a fraction or whole number. **(See Examples 2–4.)**

27. $\dfrac{2}{9} \times \dfrac{3}{5}$

28. $\dfrac{1}{8} \times \dfrac{4}{7}$

29. $\dfrac{5}{6} \times \dfrac{3}{4}$

30. $\dfrac{7}{12} \times \dfrac{18}{5}$

31. $\dfrac{21}{5} \cdot \dfrac{25}{12}$

32. $\dfrac{16}{25} \cdot \dfrac{15}{32}$

33. $\dfrac{24}{15} \cdot \dfrac{5}{3}$

34. $\dfrac{49}{24} \cdot \dfrac{6}{7}$

35. $\left(\dfrac{6}{11}\right)\left(\dfrac{22}{15}\right)$

36. $\left(\dfrac{12}{45}\right)\left(\dfrac{5}{4}\right)$

37. $\left(\dfrac{17}{9}\right)\left(\dfrac{72}{17}\right)$

38. $\left(\dfrac{39}{11}\right)\left(\dfrac{11}{13}\right)$

39. $\dfrac{21}{4} \cdot \dfrac{16}{7}$

40. $\dfrac{85}{6} \cdot \dfrac{12}{10}$

41. $12 \times \dfrac{15}{42}$

42. $4 \times \dfrac{8}{92}$

43. $\dfrac{9}{15} \times \dfrac{16}{3} \times \dfrac{25}{8}$

44. $\dfrac{49}{8} \times \dfrac{4}{5} \times \dfrac{20}{7}$

45. $\dfrac{5}{2} \times \dfrac{10}{21} \times \dfrac{7}{5}$

46. $\dfrac{55}{9} \times \dfrac{18}{32} \times \dfrac{24}{11}$

47. $\dfrac{7}{10} \cdot \dfrac{3}{28} \cdot 5$

48. $\dfrac{11}{18} \cdot \dfrac{2}{20} \cdot 15$

49. $\dfrac{100}{49} \times 21 \times \dfrac{14}{25}$

50. $\dfrac{38}{22} \times 11 \times \dfrac{5}{19}$

Objective 2: Fractions and the Order of Operations

51. Find the powers of $\dfrac{1}{10}$.

    a. $\left(\dfrac{1}{10}\right)^3$ **b.** $\left(\dfrac{1}{10}\right)^4$ **c.** $\left(\dfrac{1}{10}\right)^6$

For Exercises 52–63, simplify. Write the answer as a fraction or whole number. **(See Example 5.)**

52. $\left(\dfrac{1}{9}\right)^2$

53. $\left(\dfrac{1}{4}\right)^2$

54. $\left(\dfrac{3}{2}\right)^3$

55. $\left(\dfrac{4}{3}\right)^3$

56. $\left(4 \cdot \dfrac{3}{4}\right)^3$

57. $\left(5 \cdot \dfrac{2}{5}\right)^3$

58. $\left(\dfrac{1}{9} \cdot \dfrac{3}{5}\right)^2$

59. $\left(\dfrac{10}{3} \cdot \dfrac{1}{100}\right)^2$

60. $\dfrac{1}{3} \cdot \left(\dfrac{21}{4} \cdot \dfrac{8}{7}\right)$

61. $\dfrac{1}{6} \cdot \left(\dfrac{24}{5} \cdot \dfrac{30}{8}\right)$

62. $\dfrac{16}{9} \cdot \left(\dfrac{1}{2}\right)^3$

63. $\dfrac{28}{6} \cdot \left(\dfrac{3}{2}\right)^2$

Objective 3: Area of a Triangle

For Exercises 64–67, label the height with h and the base with b as shown in the figure.

64.   **65.**   **66.**   **67.**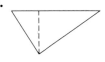

For Exercises 68–77, find the area of the figure. **(See Example 6.)**

68.
8 cm
11 cm

69.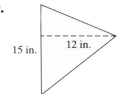
15 in.   12 in.

70.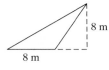
8 m
8 m

71.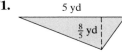
5 yd
$\frac{8}{5}$ yd

72.
1 ft
$\frac{7}{4}$ ft

73.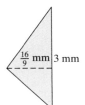
$\frac{16}{9}$ mm   3 mm

74.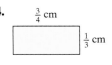
$\frac{3}{4}$ cm
$\frac{1}{3}$ cm

75.
3 m
$\frac{8}{3}$ m

76.
$\frac{15}{16}$ in.
$\frac{13}{16}$ in.

77.
$\frac{3}{4}$ ft
$\frac{23}{24}$ ft

For Exercises 78–81, find the area of the shaded region. **(See Example 7.)**

78.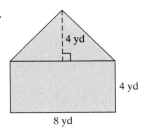
4 yd
4 yd
8 yd

79.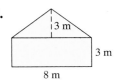
3 m
3 m
8 m

80.

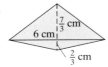

$\frac{7}{3}$ cm

6 cm

$\frac{2}{3}$ cm

81.

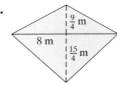

$\frac{9}{4}$ m

8 m

$\frac{15}{4}$ m

Objective 4: Applications of Multiplying Fractions

82. Ms. Robbins' car holds 16 gallons (gal) of gas. If the fuel gauge indicates that there is $\frac{5}{8}$ of a tank left, how many gallons of gas are left in the tank? **(See Example 8.)**

83. Land in a rural part of Bowie County, Texas, sells for $11,000 per acre. If Mr. Stembridge purchased $\frac{5}{4}$ acre, how much did it cost?

84. Jim has half a pizza left over from dinner. If he eats $\frac{1}{4}$ of what's left for breakfast, how much pizza did he eat for breakfast?

85. In a certain sample of individuals, $\frac{2}{5}$ are known to have blood type O. Of the individuals with blood type O, $\frac{1}{4}$ are Rh-negative. What fraction of the individuals in the sample have O negative blood?

86. A recipe for chocolate chip cookies calls for $\frac{2}{3}$ cup of sugar. If Erin triples the recipe, how many cups of sugar should she use?

87. Nancy spends $\frac{3}{4}$ hour 3 times a day walking and playing with her dog. What is the total time she spends walking and playing with the dog each day?

88. The Bishop Gaming Center hosts a football pool. There is $1200 in prize money. The first-place winner receives $\frac{2}{3}$ of the prize money. The second-place winner receives $\frac{1}{4}$ of the prize money, and the third-place winner receives $\frac{1}{12}$ of the prize money. How much money does each person get?

89. Frankie's lawn measures 40 yd by 36 yd. In the morning he mowed $\frac{2}{3}$ of the lawn. How many square yards of lawn did he already mow? How much is left to be mowed?

Expanding Your Skills

90. Evaluate.

    a. $\left(\dfrac{1}{6}\right)^2$ **b.** $\sqrt{\dfrac{1}{36}}$

91. Evaluate.

    a. $\left(\dfrac{2}{7}\right)^2$ **b.** $\sqrt{\dfrac{4}{49}}$

For Exercises 92–95, evaluate the square roots.

92. $\sqrt{\dfrac{1}{25}}$ **93.** $\sqrt{\dfrac{1}{100}}$ **94.** $\sqrt{\dfrac{64}{81}}$ **95.** $\sqrt{\dfrac{9}{4}}$

96. Find the next number in the sequence: $\frac{1}{2}, \frac{1}{4}, \frac{1}{8}, \frac{1}{16},$ _____

97. Find the next number in the sequence: $\frac{2}{3}, \frac{2}{9}, \frac{2}{27},$ _____

98. Which is greater, $\frac{1}{2}$ of $\frac{1}{8}$ or $\frac{1}{8}$ of $\frac{1}{2}$?

99. Which is greater, $\frac{2}{3}$ of $\frac{1}{4}$ or $\frac{1}{4}$ of $\frac{2}{3}$?

Calculator Connections

Topic: Multiplying fractions

Calculator Exercises

For Exercises 100–103, use a calculator to multiply the numerators and to multiply the denominators. Do not simplify the result to lowest terms.

100. $\dfrac{341}{415} \times \dfrac{71}{91}$

101. $\dfrac{37}{52} \cdot \dfrac{15}{103}$

102. $461 \cdot \dfrac{28}{313}$

103. $\dfrac{593}{225} \cdot 464$

---

section 2.5   Division of Fractions and Applications

1. Reciprocal of a Fraction

Two numbers whose product is 1 are said to be *reciprocals* of each other. For example, consider the product of $\frac{3}{8}$ and $\frac{8}{3}$.

$$\frac{3}{8} \cdot \frac{8}{3} = \frac{\overset{1}{3}}{8} \cdot \frac{\overset{1}{8}}{\underset{1}{3}} = 1$$

Because the product equals 1, we say that $\frac{3}{8}$ is the reciprocal of $\frac{8}{3}$ and vice versa.

To divide fractions, first we need to learn how to find the reciprocal of a fraction.

Finding the Reciprocal of a Fraction

To find the **reciprocal** of a nonzero fraction, interchange the numerator and denominator of the fraction. Thus, the reciprocal of $\frac{a}{b}$ is $\frac{b}{a}$ (provided $a \neq 0$ and $b \neq 0$).

Objectives

1. Reciprocal of a Fraction
2. Division of Fractions
3. Order of Operations
4. Applications of Multiplication and Division of Fractions

Concept Connections

Fill in the blank.

1. The product of a number and its reciprocal is _____.

Answer

1. 1

---

example 1   **Finding Reciprocals**

Find the reciprocal.

a. $\frac{2}{5}$ **b.** $\frac{1}{9}$ **c.** 5   **d.** 0

Solution:

a. The reciprocal of $\frac{2}{5}$ is $\frac{5}{2}$.

b. The reciprocal of $\frac{1}{9}$ is $\frac{9}{1}$, or 9.

c. First write the whole number 5 as the improper fraction $\frac{5}{1}$. The reciprocal of $\frac{5}{1}$ is $\frac{1}{5}$.

d. The number 0 has no reciprocal because $\frac{1}{0}$ is undefined.

---

2. Division of Fractions

To understand the division of fractions, we draw an analogy to the division of whole numbers. The statement $6 \div 2$ asks, How many groups of 2 can be found among 6 wholes? The answer is 3.

$$6 \div 2 = 3$$

In fractional form, the statement $6 \div 2 = 3$ can be written as $\frac{6}{2} = 3$. This result can also be found by multiplying.

$$6 \cdot \frac{1}{2} = \frac{6}{1} \cdot \frac{1}{2} = \frac{6}{2} = 3$$

That is, to divide by 2 is equivalent to multiplying by the reciprocal $\frac{1}{2}$.

Now let's look at an analogy where we have a fraction divided by a fraction. The statement $\frac{2}{3} \div \frac{1}{6}$ asks, How many increments of $\frac{1}{6}$ can be found within $\frac{2}{3}$ of a whole? The answer is 4.

From the figure,

$$\frac{2}{3} \div \frac{1}{6} = 4$$

4 increments of $\frac{1}{6}$

$\frac{2}{3}$ of a whole

As with the division of whole numbers, the quotient $\frac{2}{3} \div \frac{1}{6}$ can be found by multiplying: $\frac{2}{3} \cdot \frac{6}{1} = \frac{12}{3} = 4$. That is, division by $\frac{1}{6}$ is equivalent to multiplication by $\frac{6}{1}$.

Dividing Fractions

To divide two fractions, multiply the dividend (the "first" fraction) by the reciprocal of the divisor (the "second" fraction).

The process to divide fractions can be written symbolically as

Change division to
multiplication.

$$\frac{a}{b} \div \frac{c}{d} = \frac{a}{b} \cdot \frac{d}{c} \quad \text{provided } b, c, \text{ and } d \text{ are not } 0$$

Take the reciprocal
of the divisor.

example 2   Dividing Fractions

Divide and simplify, if possible. Write the answer as a fraction.

a. $\dfrac{2}{5} \div \dfrac{7}{4}$ **b.** $\dfrac{2}{27} \div \dfrac{8}{15}$ **c.** $\dfrac{35}{14} \div 7$ **d.** $12 \div \dfrac{8}{3}$

Solution:

a. $\dfrac{2}{5} \div \dfrac{7}{4} = \dfrac{2}{5} \cdot \dfrac{4}{7}$ Multiply by the reciprocal of the divisor ("second" fraction).

$= \dfrac{2 \cdot 4}{5 \cdot 7}$ Multiply numerators. Multiply denominators.

$= \dfrac{8}{35}$

b. $\dfrac{2}{27} \div \dfrac{8}{15} = \dfrac{2}{27} \cdot \dfrac{15}{8}$ Multiply by the reciprocal of the divisor ("second" fraction).

$= \dfrac{\overset{1}{2}}{\underset{1}{3 \cdot 3 \cdot 3}} \cdot \dfrac{3 \cdot \overset{1}{5}}{\underset{1}{2 \cdot 2 \cdot 2}}$ Factor and simplify.

$= \dfrac{5}{36}$ Multiply.

c. $\dfrac{35}{14} \div 7 = \dfrac{35}{14} \div \dfrac{7}{1}$ Write the whole number 7 as an improper fraction *before* multiplying by the reciprocal.

$= \dfrac{35}{14} \cdot \dfrac{1}{7}$ Multiply by the reciprocal of the divisor.

$= \dfrac{5 \cdot \overset{1}{7}}{2 \cdot \underset{1}{7}} \cdot \dfrac{1}{7}$ Factor and simplify.

$= \dfrac{5}{14}$ Multiply.

Avoiding Mistakes: Do not try to simplify until after taking the reciprocal of the divisor. In Example 2(a) it would be incorrect to "cancel" the 2 and the 4 in the expression $\frac{2}{5} \div \frac{7}{4}$.

d. $12 \div \dfrac{8}{3} = \dfrac{12}{1} \div \dfrac{8}{3}$ Write the whole number 12 as an improper fraction.

$= \dfrac{12}{1} \cdot \dfrac{3}{8}$ Multiply by the reciprocal of the divisor.

$= \dfrac{\overset{1}{2} \cdot \overset{1}{2} \cdot 3}{1} \cdot \dfrac{3}{\underset{1}{2} \cdot \underset{1}{2} \cdot 2}$ Factor and simplify.

$= \dfrac{9}{2}$ Multiply.

3. Order of Operations

When simplifying fractional expressions with more than one operation, be sure to follow the order of operations. Simplify within parentheses first. Then simplify expressions with exponents, followed by multiplication or division in the order of appearance from left to right.

example 3   Applying the Order of Operations

Simplify. Write the answer as a fraction.

a. $\dfrac{2}{3} \div \dfrac{4}{9} \div 6$ **b.** $\left(\dfrac{3}{5} \div \dfrac{2}{15}\right)^2$

Solution:

a. $\dfrac{2}{3} \div \dfrac{4}{9} \div 6$

$= \left(\dfrac{2}{3} \div \dfrac{4}{9}\right) \div 6$ We will divide from left to right. To emphasize this order, we can insert parentheses around the first two fractions.

$= \left(\dfrac{\overset{1}{2}}{3} \cdot \dfrac{\overset{3}{9}}{\underset{2}{4}}\right) \div 6$ Simplify within parentheses.

$= \left(\dfrac{3}{2}\right) \div \dfrac{6}{1}$ Simplify within parentheses and write the whole number as an improper fraction.

$= \left(\dfrac{3}{2}\right) \cdot \dfrac{1}{6}$ Multiply by the reciprocal of the divisor.

$= \dfrac{\overset{1}{3}}{2} \cdot \dfrac{1}{\underset{2}{6}}$ Simplify.

$= \dfrac{1}{4}$ Multiply.

Tip: In Example 3(a) we could also have written each division as multiplication of the reciprocal right from the start.

$\dfrac{2}{3} \div \dfrac{4}{9} \div 6 = \left(\dfrac{2}{3} \cdot \dfrac{9}{4}\right) \cdot \dfrac{1}{6}$

$= \dfrac{\overset{1}{2}}{3} \cdot \dfrac{\overset{1}{3} \cdot \overset{1}{3}}{\underset{1}{2} \cdot \underset{1}{2}} \cdot \dfrac{1}{2 \cdot \underset{1}{3}}$

Factor and simplify.

$= \dfrac{1}{4}$ Multiply.

Answers

12. 2   13. $\dfrac{9}{196}$

b. $\left(\dfrac{3}{5} \div \dfrac{2}{15}\right)^2$ Perform operations within parentheses first.

$= \left(\dfrac{3}{5} \cdot \dfrac{15}{2}\right)^2$ Multiply by the reciprocal of the divisor.

$= \left(\dfrac{3}{\cancel{5}_1} \cdot \dfrac{\overset{3}{\cancel{15}}}{2}\right)^2$ Simplify.

$= \left(\dfrac{9}{2}\right)^2$ Multiply within parentheses.

$= \dfrac{9}{2} \cdot \dfrac{9}{2}$ With an exponent of 2, multiply 2 factors of the base.

$= \dfrac{81}{4}$ Multiply.

4. Applications of Multiplication and Division of Fractions

Sometimes it is difficult to determine whether multiplication or division is appropriate to solve an application problem. Division is generally used for a problem that requires you to separate or "split up" a quantity into pieces. Multiplication is generally used if it is necessary to take a fractional part of a quantity.

example 4   Using Division in an Application

A road crew must mow the grassy median along a stretch of highway I-95. If they can mow $\frac{5}{8}$ mile (mi) in 1 hr, how long will it take them to mow a 15-mi stretch?

Solution:

Read and familiarize.

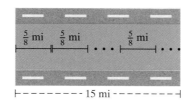

Strategy/operation: From the figure, we must separate or "split up" a 15-mi stretch of highway into pieces that are $\frac{5}{8}$ mi in length. Therefore, we must divide 15 by $\frac{5}{8}$.

$15 \div \dfrac{5}{8} = \dfrac{15}{1} \cdot \dfrac{8}{5}$ Write the whole number as a fraction. Multiply by the reciprocal of the divisor.

$= \dfrac{\overset{3}{\cancel{15}}}{1} \cdot \dfrac{8}{\cancel{5}_1}$

$= 24$

The 15-mi stretch of highway will take 24 hr to mow.

Skill Practice

14. A cookie recipe requires $\frac{2}{5}$ package of chocolate chips for each batch of cookies. If a restaurant has 20 packages of chocolate chips, how many batches of cookies can it make?

Answer

14. 50 batches

Skill Practice

15. A $\frac{25}{2}$-yd ditch will be dug to put in a new water line. If piping comes in segments of $\frac{5}{4}$ yd, how many segments are needed to line the ditch?

---

example 5   Using Division in an Application

A $\frac{9}{4}$-ft length of wire must be cut into pieces of equal length that are $\frac{3}{8}$ ft long. How many pieces can be cut?

Solution:

Read and familiarize.
Operation: Here we divide the total length of wire into pieces of equal length.

$$\frac{9}{4} \div \frac{3}{8} = \frac{9}{4} \cdot \frac{8}{3}$$ Multiply by the reciprocal of the divisor.

$$= \frac{\overset{3}{9}}{\underset{1}{4}} \cdot \frac{\overset{2}{8}}{\underset{1}{3}}$$ Simplify.

$$= 6$$

Six pieces of wire can be cut.

Skill Practice

16. A new school will cost $20,000,000 to build, and the state will pay $\frac{3}{5}$ of the cost.

   a. How much will the state pay?

   b. How much will the state not pay?

   c. The county school district issues bonds to pay $\frac{4}{5}$ of the money not covered by the state. How much money will be covered by bonds?

---

example 6   Using Multiplication in an Application

Carson estimates that his total cost for college for 1 year is $12,600. He has scholarship money to pay $\frac{2}{3}$ of the cost.

   a. How much money is the scholarship worth?

   b. How much money will Carson have to pay?

   c. If Carson's parents help him by paying $\frac{1}{3}$ of the amount not paid by the scholarship, how much money will be paid by Carson's parents?

Solution:

   a. Carson's scholarship will pay $\frac{2}{3}$ of $12,600. Because we are looking for a fraction of a quantity, we multiply.

$$\frac{2}{3} \cdot 12{,}600 = \frac{2}{3} \cdot \frac{12{,}600}{1}$$

$$= \frac{2}{\underset{1}{3}} \cdot \frac{\overset{4200}{12{,}600}}{1}$$

$$= 8400$$

   The scholarship will pay $8400.

   b. Carson will have to pay the remaining portion of the cost. This can be found by subtraction.

$$\$12{,}600 - \$8400 = \$4200$$

   Carson will have to pay $4200.

Answers

15. 10 segments of piping
16. a. $12,000,000
    b. $8,000,000
    c. $6,400,000

Tip: The answer to Example 6(b) could also have been found by noting that the scholarship paid $\frac{2}{3}$ of the cost. This means that Carson must pay $\frac{1}{3}$ of the cost, or

$$\frac{1}{\underset{1}{3}} \cdot \frac{\$12,\!600}{1} = \frac{1}{3} \cdot \frac{\overset{4200}{\cancel{\$12,\!600}}}{1}$$

$$= \$4200$$

c. Carson's parents will pay $\frac{1}{3}$ of $4200.

$$\frac{1}{\underset{1}{3}} \cdot \frac{\overset{1400}{\cancel{4200}}}{1}$$

Carson's parents will pay $1400.

---

example 7	Using Multiplication in an Application

Three-fifths of the students in the freshman class are female. Of these students, $\frac{5}{9}$ are over the age of 25. What fraction of the freshman class is female over the age of 25?

Solution:

Read and familiarize.
Strategy/operation: We must find $\frac{5}{9}$ of $\frac{3}{5}$ of one whole freshman class. This implies multiplication.

$$\frac{5}{9} \times \frac{3}{5} = \frac{\overset{1}{\cancel{5}}}{\underset{3}{\cancel{9}}} \times \frac{\overset{1}{\cancel{3}}}{\underset{1}{\cancel{5}}} = \frac{1}{3}$$

One-third of the freshman class consists of female students over the age of 25.

Skill Practice

17. In a certain police department, $\frac{1}{3}$ of the department is female. Of the female officers, $\frac{2}{7}$ were promoted within the last year. What fraction of the police department consists of females who were promoted within the last year?

Answer

17. $\dfrac{2}{21}$

---

section 2.5	**Practice Exercises**

Boost *your* **GRADE** at
mathzone.com!

MathZone

- Practice Problems
- Self-Tests
- NetTutor
- e-Professors
- Videos

Study Skills Exercises

1. Find the page numbers for the Chapter Review Exercises, the Chapter Test, and the Cumulative Review Exercises for this chapter.

Chapter Review Exercises, page(s) _____.

Chapter Test, page(s) _____.

Cumulative Review Exercise(s) _____.

Compare these features and state the advantages of each.

2. Define the key term **reciprocal**.

Review Exercises

For Exercises 3–11, multiply and simplify to lowest terms. Write the answer as a fraction or whole number.

3. $\dfrac{9}{11} \times \dfrac{22}{5}$

4. $\dfrac{24}{7} \cdot \dfrac{7}{8}$

5. $\dfrac{34}{5} \cdot \dfrac{5}{17}$

6. $3 \cdot \left(\dfrac{7}{6} \right)$

7. $8 \cdot \left(\dfrac{5}{24} \right)$

8. $\left(\dfrac{2}{7} \right)\left(\dfrac{7}{2} \right)$

9. $\left(\dfrac{9}{5} \right)\left(\dfrac{5}{9} \right)$

10. $\dfrac{1}{10} \times 10$

11. $\dfrac{1}{3} \times 3$

Objective 1: Reciprocal of a Fraction

12. For each number, determine whether the number has a reciprocal.

    a. $\dfrac{1}{2}$ **b.** $\dfrac{5}{3}$ **c.** 6 **d.** 0

For Exercises 13–20, find the reciprocal of the number, if it exists. **(See Example 1.)**

13. $\dfrac{7}{8}$

14. $\dfrac{5}{6}$

15. $\dfrac{10}{9}$

16. $\dfrac{14}{5}$

17. 4

18. 9

19. 0

20. $\dfrac{0}{4}$

Objective 2: Division of Fractions

For Exercises 21–24, fill in the blank.

21. Dividing by 3 is the same as multiplying by _____.

22. Dividing by 5 is the same as multiplying by _____.

23. Dividing by 8 is the same as _____ by $\dfrac{1}{8}$.

24. Dividing by 12 is the same as _____ by $\dfrac{1}{12}$.

For Exercises 25–50, divide and simplify the answer to lowest terms. Write the answer as a fraction or whole number. **(See Example 2.)**

25. $\dfrac{2}{15} \div \dfrac{5}{12}$

26. $\dfrac{11}{3} \div \dfrac{6}{5}$

27. $\dfrac{7}{13} \div \dfrac{2}{5}$

28. $\dfrac{8}{7} \div \dfrac{3}{10}$

29. $\dfrac{14}{3} \div \dfrac{6}{5}$

30. $\dfrac{11}{2} \div \dfrac{3}{4}$

31. $\dfrac{15}{2} \div \dfrac{3}{2}$

32. $\dfrac{9}{10} \div \dfrac{9}{2}$

33. $\dfrac{3}{4} \div \dfrac{3}{4}$

34. $\dfrac{6}{5} \div \dfrac{6}{5}$

35. $7 \div \dfrac{2}{3}$

36. $4 \div \dfrac{3}{5}$

37. $\dfrac{10}{9} \div \dfrac{1}{18}$

38. $\dfrac{4}{3} \div \dfrac{1}{3}$

39. $12 \div \dfrac{3}{4}$

40. $24 \div \dfrac{8}{5}$

41. $\dfrac{12}{5} \div 4$

42. $\dfrac{20}{6} \div 5$

43. $\dfrac{9}{50} \div \dfrac{18}{25}$

44. $\dfrac{30}{40} \div \dfrac{15}{8}$

45. $\dfrac{9}{100} \div \dfrac{13}{1000}$

46. $\dfrac{1000}{17} \div \dfrac{10}{3}$

47. $\dfrac{36}{5} \div \dfrac{9}{25}$

48. $\dfrac{13}{5} \div \dfrac{17}{10}$

49. $\dfrac{44}{3} \div \dfrac{2}{7}$

50. $\dfrac{31}{15} \div \dfrac{10}{3}$

Mixed Exercises

For Exercises 51–66, multiply or divide as indicated. Write the answer as a fraction or whole number.

51. $\dfrac{7}{8} \div \dfrac{1}{4}$

52. $\dfrac{7}{12} \div \dfrac{5}{3}$

53. $\dfrac{5}{8} \cdot \dfrac{2}{9}$

54. $\dfrac{1}{16} \cdot \dfrac{4}{3}$

55. $6 \cdot \dfrac{4}{3}$

56. $12 \cdot \dfrac{5}{6}$

57. $\dfrac{16}{5} \div 8$

58. $\dfrac{42}{11} \div 7$

59. $\dfrac{16}{3} \div \dfrac{2}{5}$

60. $\dfrac{17}{8} \div \dfrac{1}{4}$

61. $\dfrac{1}{8} \cdot 16$

62. $\dfrac{2}{3} \cdot 9$

63. $\dfrac{22}{7} \cdot \dfrac{5}{16}$

64. $\dfrac{40}{21} \cdot \dfrac{18}{25}$

65. $8 \div \dfrac{16}{3}$

66. $5 \div \dfrac{15}{4}$

Objective 3: Order of Operations

67. Explain the difference in the process to evaluate $\frac{2}{3} \cdot 6$ versus $\frac{2}{3} \div 6$. Then evaluate each expression.

68. Explain the difference in the process to evaluate $8 \cdot \frac{2}{3}$ versus $8 \div \frac{2}{3}$. Then evaluate each expression.

For Exercises 69–78, simplify by using the order of operations. Write the answer as a fraction or whole number.
(See Example 3.)

69. $\dfrac{54}{21} \div \dfrac{2}{3} \div \dfrac{9}{7}$

70. $\dfrac{48}{56} \div \dfrac{3}{8} \div \dfrac{8}{7}$

71. $\left(\dfrac{3}{5} \div \dfrac{6}{7} \right) \cdot \dfrac{5}{3}$

72. $\left(\dfrac{5}{8} \div \dfrac{35}{16} \right) \cdot \dfrac{1}{4}$

73. $\left(\dfrac{3}{8}\right)^2 \div \dfrac{9}{14}$　　　**74.** $\dfrac{7}{8} \div \left(\dfrac{1}{2}\right)^2$　　　**75.** $\left(\dfrac{2}{5} \div \dfrac{8}{3}\right)^2$　　　**76.** $\left(\dfrac{5}{12} \div \dfrac{2}{3}\right)^2$

77. $\left(\dfrac{63}{8} \div \dfrac{9}{4}\right)^2 \cdot 4$　　　**78.** $\left(\dfrac{25}{3} \div \dfrac{50}{9}\right)^2 \cdot 8$

Objective 4: Applications of Multiplication and Division of Fractions

79. How many eighths are in $\frac{9}{4}$?

80. How many sixths are in $\frac{4}{3}$?

81. If one cup is $\frac{1}{16}$ gal, how many cups of orange juice can be filled from $\frac{3}{2}$ gal? **(See Example 5.)**

82. If 1 centimeter (cm) is $\frac{1}{100}$ meter (m), how many centimeters are a $\frac{5}{4}$-m piece of rope?

83. During the month of December, a department store wraps packages free of charge. Each package requires $\frac{2}{3}$ yd of ribbon. If Li used up a 36-yd roll of ribbon, how many packages were wrapped? **(See Example 4.)**

84. A developer sells lots of land in increments of $\frac{3}{4}$ acre. If the developer has 60 acres, how many lots can be sold?

85. Dorci buys 16 sheets of plywood, each $\frac{3}{4}$ in. thick, to cover her windows in the event of a hurricane. She stacks the wood in the garage. How high will the stack be?

86. Davey built a bookshelf 36 in. long. Can the shelf hold a set of encyclopedias if there are 24 books and each book averages $\frac{5}{4}$ in. thick? Explain your answer.

87. A radio station allows 18 minutes (min) of advertising each hour. How many 40-second ($\frac{2}{3}$-min) commercials can be run in

    a. 1 hr　　　**b.** 1 day

88. A television station has 20 min of advertising each hour. How many 30-sec ($\frac{1}{2}$-min) commercials can be run in

    a. 1 hr　　　**b.** 1 day

89. A landowner has $\frac{9}{4}$ acres of land. She plans to sell $\frac{1}{3}$ of the land. **(See Example 7.)**

    a. How much land will she sell?　　　**b.** How much land will she retain?

90. Josh must read 24 pages for his English class and 18 pages for psychology. He has read $\frac{1}{6}$ of the pages.

    a. How many pages has he read?　　　**b.** How many pages does he still have to read?

91. A lab technician has $\frac{7}{4}$ liters (L) of alcohol. If she needs samples of $\frac{1}{8}$ L, how many samples can she prepare?

92. Troy has a $\frac{7}{8}$-in. nail that he must hammer into a board. Each strike of the hammer moves the nail $\frac{1}{16}$ in. into the board. How many strikes of the hammer must he make?

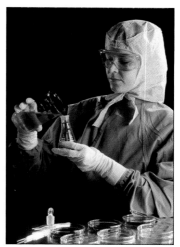

93. Ricardo wants to buy a new house for $240,000. The bank requires $\frac{1}{10}$ of the cost of the house as a down payment. As a gift, Ricardo's mother will pay $\frac{2}{3}$ of the down payment. **(See Example 6.)**

   a. How much money will Ricardo's mother pay toward the down payment?

   b. How much money will Ricardo have to pay toward the down payment?

   c. How much is left over for Ricardo to finance?

94. Althea wants to buy a Toyota Camry for a total cost of $18,000. The dealer requires $\frac{1}{12}$ of the money as a down payment. Althea's parents have agreed to pay one-half of the down payment for her.

   a. How much money will Althea's parents pay toward the down payment?

   b. How much will Althea pay toward the down payment?   **c.** How much will Althea have to finance?

Expanding Your Skills

95. The rectangle shown here has an area of 30 ft². Find the length.

$\frac{5}{2}$ ft

?

96. The rectangle shown here has an area of 8 m². Find the width.

?

14 m

Calculator Connections

Topic: Dividing fractions

Calculator Exercises

For Exercises 97–100, use a calculator to divide. Do not simplify the result to lowest terms.

97. $\dfrac{177}{37} \div \dfrac{23}{226}$ **98.** $\dfrac{412}{337} \div \dfrac{77}{53}$

99. $\dfrac{681}{214} \div 112$ **100.** $584 \div \dfrac{3317}{25}$

Objectives

1. Multiplication of Mixed Numbers
2. Division of Mixed Numbers
3. Applications of Multiplication and Division of Mixed Numbers

1. Multiplication of Mixed Numbers

To multiply mixed numbers, we follow these steps.

Steps to Multiply Mixed Numbers

1. Change each mixed number to an improper fraction.
2. Multiply the improper fractions and simplify to lowest terms, if possible (see Section 2.4).

Answers greater than 1 may be written as an improper fraction or as a mixed number depending on the directions of the problem.

Example 1 demonstrates this process.

Skill Practice

Multiply and write the answer as a mixed number or whole number.

1. $\left(4\frac{3}{5}\right)\left(5\frac{5}{6}\right)$

2. $16\frac{1}{2} \cdot 3\frac{7}{11}$

3. $7\frac{1}{6} \cdot 10$

example 1   Multiplying Mixed Numbers

Multiply and write the answer as a mixed number or whole number.

  a. $\left(3\frac{1}{5}\right)\left(4\frac{3}{4}\right)$ **b.** $25\frac{1}{2} \cdot 4\frac{2}{3}$ **c.** $12 \cdot 8\frac{7}{9}$

Solution:

a. $\left(3\frac{1}{5}\right)\left(4\frac{3}{4}\right) = \dfrac{16}{5} \cdot \dfrac{19}{4}$ Write each mixed number as an improper fraction.

$= \dfrac{\overset{4}{\cancel{16}}}{5} \cdot \dfrac{19}{\underset{1}{\cancel{4}}}$ Simplify.

$= \dfrac{76}{5}$ Multiply.

$= 15\dfrac{1}{5}$ Write the improper fraction as a mixed number.

$$\begin{array}{r} 15 \\ 5\overline{)76} \\ -5 \\ \hline 26 \\ -25 \\ \hline 1 \end{array}$$

Tip: To check whether the answer from Example 1(a) is reasonable, we can round each factor and estimate the product.

$3\frac{1}{5}$ rounds to 3.

$4\frac{3}{4}$ rounds to 5.

Thus, $\left(3\frac{1}{5}\right)\left(4\frac{3}{4}\right) \approx (3)(5) = 15$, which is close to $15\frac{1}{5}$.

Answers

1. $26\frac{5}{6}$
2. 60
3. $71\frac{2}{3}$

b. $25\dfrac{1}{2} \cdot 4\dfrac{2}{3} = \dfrac{51}{2} \cdot \dfrac{14}{3}$ Write each mixed number as an improper fraction.

$= \dfrac{\overset{17}{\cancel{51}}}{\underset{1}{2}} \cdot \dfrac{\overset{7}{\cancel{14}}}{\underset{1}{3}}$ Simplify.

$= \dfrac{119}{1}$ Multiply.

$= 119$

Avoiding Mistakes: Do not try to multiply mixed numbers by multiplying the whole-number parts and multiplying the fractional parts. You will not get the correct answer.

For the expression $25\frac{1}{2} \cdot 4\frac{2}{3}$, it would be incorrect to multiply (25)(4) and $\frac{1}{2} \cdot \frac{2}{3}$. Notice that these values do not equal 119.

c. $12 \cdot 8\dfrac{7}{9} = \dfrac{12}{1} \cdot \dfrac{79}{9}$ Write the whole number and mixed number as improper fractions.

$= \dfrac{\overset{4}{\cancel{12}}}{1} \cdot \dfrac{79}{\underset{3}{\cancel{9}}}$ Simplify.

$= \dfrac{316}{3}$ Multiply.

$= 105\dfrac{1}{3}$ Write the improper fraction as a mixed number.

$$\begin{array}{r} 105 \\ 3\overline{)316} \\ \underline{-3} \\ 16 \\ \underline{-15} \\ 1 \end{array}$$

Concept Connections

4. Explain how you could check whether the answer to Example 1(c) is reasonable.

2. Division of Mixed Numbers

To divide mixed numbers, we use the following steps.

> **Steps to Divide Mixed Numbers**
>
> **1.** Change each mixed number to an improper fraction.
> **2.** Divide the improper fractions and simplify to lowest terms, if possible. Recall that to divide fractions, we multiply the dividend by the reciprocal of the divisor (see Section 2.5).
>
> Answers greater than 1 may be written as an improper fraction or as a mixed number depending on the directions of the problem.

Skill Practice

Divide and write the answer as a mixed number or whole number.

5. $10\dfrac{1}{3} \div 2\dfrac{5}{6}$ **6.** $8 \div 4\dfrac{4}{5}$

7. $12\dfrac{4}{9} \div 8$

example 2   Dividing Mixed Numbers

Divide and write the answer as a mixed number or whole number.

a. $7\dfrac{1}{2} \div 4\dfrac{2}{3}$ **b.** $6 \div 5\dfrac{1}{7}$ **c.** $13\dfrac{5}{6} \div 7$

Answers

4. Round $8\frac{7}{9}$ to 9. Then
   $12 \cdot 8\frac{7}{9} \approx 12 \cdot 9 = 108$ which is close to $105\frac{1}{3}$.

5. $3\dfrac{11}{17}$ 6. $1\dfrac{2}{3}$ 7. $1\dfrac{5}{9}$

Solution:

a. $7\frac{1}{2} \div 4\frac{2}{3} = \frac{15}{2} \div \frac{14}{3}$ Write the mixed numbers as improper fractions.

$= \frac{15}{2} \cdot \frac{3}{14}$ Multiply by the reciprocal of the divisor.

$= \frac{45}{28}$ Multiply.

$= 1\frac{17}{28}$ Write the improper fraction as a mixed number.

b. $6 \div 5\frac{1}{7} = \frac{6}{1} \div \frac{36}{7}$ Write the whole number and mixed number as improper fractions.

$= \frac{6}{1} \cdot \frac{7}{36}$ Multiply by the reciprocal of the divisor.

$= \frac{\overset{1}{6}}{1} \cdot \frac{7}{\underset{6}{36}}$ Simplify.

$= \frac{7}{6}$ Multiply.

$= 1\frac{1}{6}$ Write the improper fraction as a mixed number.

c. $13\frac{5}{6} \div 7 = \frac{83}{6} \div \frac{7}{1}$ Write the whole number and mixed number as improper fractions.

$= \frac{83}{6} \cdot \frac{1}{7}$ Multiply by the reciprocal of the divisor.

$= \frac{83}{42}$ Multiply.

$= 1\frac{41}{42}$ Write the improper fraction as a mixed number.

3. Applications of Multiplication and Division of Mixed Numbers

Examples 3 and 4 demonstrate multiplication and division of mixed numbers in day-to-day applications.

Skill Practice

8. A recipe calls for $2\frac{3}{4}$ cups of flour. How much flour is required for $2\frac{1}{2}$ times the recipe?

example 3   Applying Multiplication of Mixed Numbers

Sometimes homeowners tape the windows in their homes before a hurricane to prevent glass from shattering. Antonio must buy masking tape to tape the windows of his house as shown (Figure 2-9). Each diagonal of the window is $5\frac{2}{5}$ ft. If Antonio has 8 windows of this size, how much masking tape does he need?

Figure 2-9

Answer

8. $6\frac{7}{8}$ cups of flour

Solution:

One approach is to find the amount of tape required for each window. Then multiply by the number of windows.

The length of tape needed for each window:

$$2 \cdot \left(5\frac{2}{5} \text{ ft}\right) = \frac{2}{1} \cdot \frac{27}{5} \text{ ft} \qquad \text{Write the whole number and mixed number as improper fractions.}$$

$$= \frac{54}{5} \text{ ft} \qquad \text{Multiply.}$$

Amount needed for 8 windows:

$$8 \cdot \left(\frac{54}{5} \text{ ft}\right) = \frac{8}{1} \cdot \frac{54}{5} \text{ ft}$$

$$= \frac{432}{5} \text{ ft}$$

$$= 86\frac{2}{5} \text{ ft}$$

The homeowner requires $86\frac{2}{5}$ ft of tape.

Tip: The solution to Example 3 can be computed quickly by estimating. In this case, the homeowner might want to *overestimate* the answer, to be sure that he doesn't run short of tape before the hurricane. We can round the value $5\frac{2}{5}$ ft to 6 ft.

Two strips are needed for each window, and there are 8 windows. Therefore our estimate is

$$2(6 \text{ ft})(8) = 96 \text{ ft}$$

---

example 4   Applying Division of Mixed Numbers

A construction site brings in $6\frac{2}{3}$ tons of soil. Each truck holds $\frac{2}{3}$ ton. How many truckloads are necessary?

Solution:

The $6\frac{2}{3}$ tons of soil must be distributed in $\frac{2}{3}$-ton increments. This will require division.

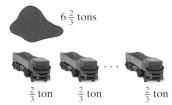

$$6\frac{2}{3} \div \frac{2}{3} = \frac{20}{3} \div \frac{2}{3} \qquad \text{Write the mixed number as an improper fraction.}$$

$$= \frac{20}{3} \cdot \frac{3}{2} \qquad \text{Multiply by the reciprocal of the divisor.}$$

$$= \frac{\overset{10}{\cancel{20}}}{\underset{1}{\cancel{3}}} \cdot \frac{\overset{1}{\cancel{3}}}{\underset{1}{\cancel{2}}} \qquad \text{Simplify.}$$

$$= \frac{10}{1} \qquad \text{Multiply.}$$

$$= 10$$

A total of 10 truckloads of soil will be required.

Skill Practice

9. A department store wraps packages for \$2 each. Ribbon $2\frac{5}{8}$ ft long is used to wrap each package. How many packages can be wrapped from a roll of ribbon 168 ft long?

Answer

9. 64 packages

section 2.6   Practice Exercises

Study Skills Exercise

1. Sometimes, test anxiety can be eliminated by adequate preparation and practice. List some places where you can find extra problems for practice.

Review Exercises

For Exercises 2–7, multiply or divide the fractions. Write your answers as fractions.

2. $\dfrac{5}{6} \cdot \dfrac{2}{9}$

3. $\dfrac{13}{5} \cdot \dfrac{10}{9}$

4. $\dfrac{20}{9} \div \dfrac{10}{3}$

5. $\dfrac{42}{11} \div \dfrac{7}{2}$

6. $\dfrac{32}{15} \div 4$

7. $\dfrac{52}{18} \div 13$

8. Explain the process to change a mixed number to an improper fraction.

For Exercises 9–16, write the mixed number as an improper fraction.

9. $3\dfrac{2}{5}$

10. $2\dfrac{7}{10}$

11. $1\dfrac{4}{7}$

12. $4\dfrac{1}{8}$

13. $12\dfrac{5}{6}$

14. $5\dfrac{2}{11}$

15. $9\dfrac{3}{4}$

16. $15\dfrac{1}{2}$

Objective 1: Multiplication of Mixed Numbers

For Exercises 17–36, multiply the mixed numbers. Write the answer as a mixed number or whole number.
(See Example 1.)

17. $\left(2\dfrac{2}{5}\right)\left(3\dfrac{1}{12}\right)$

18. $\left(5\dfrac{1}{5}\right)\left(3\dfrac{3}{4}\right)$

19. $2\dfrac{1}{3} \cdot \dfrac{5}{7}$

20. $6\dfrac{1}{8} \cdot \dfrac{4}{7}$

21. $4\dfrac{2}{9} \cdot 9$

22. $3\dfrac{1}{3} \cdot 6$

23. $\left(5\dfrac{3}{16}\right)\left(5\dfrac{1}{3}\right)$

24. $\left(8\dfrac{2}{3}\right)\left(2\dfrac{1}{13}\right)$

25. $\dfrac{2}{3} \cdot 2\dfrac{7}{10}$

26. $\dfrac{4}{3} \cdot 5\dfrac{1}{8}$

27. $\left(7\dfrac{1}{4}\right) \cdot 10$

28. $\left(2\dfrac{2}{3}\right) \cdot 3$

29. $4\dfrac{5}{8} \cdot 0$

30. $0 \cdot 6\dfrac{1}{10}$

31. $\left(3\dfrac{1}{2}\right)\left(2\dfrac{1}{7}\right)$

32. $\left(1\dfrac{3}{10}\right)\left(1\dfrac{1}{4}\right)$

33. $\left(5\dfrac{2}{5}\right)\left(\dfrac{2}{9}\right)\left(1\dfrac{4}{5}\right)$

34. $\left(6\dfrac{1}{8}\right)\left(2\dfrac{3}{4}\right)\left(\dfrac{8}{7}\right)$

35. $\left(3\dfrac{2}{5}\right)\left(\dfrac{7}{34}\right)\left(3\dfrac{3}{4}\right)$

36. $\left(5\dfrac{1}{6}\right)\left(1\dfrac{4}{7}\right)\left(\dfrac{14}{33}\right)$

Objective 2: Division of Mixed Numbers

For Exercises 37–54, divide the mixed numbers. Write the answer as a mixed number, proper fraction, or whole number. **(See Example 2.)**

37. $1\dfrac{7}{10} \div 2\dfrac{3}{4}$

38. $5\dfrac{1}{10} \div \dfrac{3}{4}$

39. $5\dfrac{8}{9} \div 1\dfrac{1}{3}$

40. $12\dfrac{4}{5} \div 2\dfrac{3}{5}$

41. $2\dfrac{1}{2} \div 1\dfrac{1}{16}$

42. $7\dfrac{3}{5} \div 1\dfrac{7}{12}$

43. $4\dfrac{1}{2} \div 2\dfrac{1}{4}$

44. $5\dfrac{5}{6} \div 2\dfrac{1}{3}$

45. $0 \div 6\dfrac{7}{12}$

46. $0 \div 1\dfrac{9}{11}$

47. $2\dfrac{5}{6} \div \dfrac{1}{6}$

48. $6\dfrac{1}{2} \div \dfrac{1}{2}$

49. $1\dfrac{1}{3} \div \dfrac{2}{7}$

50. $2\dfrac{1}{7} \div \dfrac{5}{13}$

51. $3\dfrac{1}{2} \div 2$

52. $4\dfrac{2}{3} \div 3$

53. $7\dfrac{1}{8} \div 1\dfrac{1}{3} \div 2\dfrac{1}{4}$

54. $3\dfrac{1}{8} \div 5\dfrac{5}{7} \div 1\dfrac{5}{16}$

Objective 3: Applications of Multiplication and Division of Mixed Numbers

55. Tabitha charges \$8 per hour for baby sitting. If she works for $4\dfrac{3}{4}$ hr, how much should she be paid? **(See Example 3.)**

56. Kurt bought $2\dfrac{2}{3}$ acres of land. If land costs \$10,500 per acre, how much will the land cost him?

57. The age of a small kitten can be approximated by the following rule. The kitten's age is given as 1 week for every quarter pound of weight. **(See Example 4.)**

   a. Approximately how old is a $1\dfrac{3}{4}$-lb kitten?

   b. Approximately how old is a $2\dfrac{1}{8}$-lb kitten?

58. Richard's estate is to be split equally among his three children. If his estate is worth \$$1\dfrac{3}{4}$ million, how much will each child inherit?

59. Lucy earns \$14 per hour and Ricky earns \$10 per hour. Suppose Lucy worked $35\frac{1}{2}$ hr last week and Ricky worked $42\frac{1}{2}$ hr.

   a. Who earned more money and by how much?

   b. How much did they earn altogether?

60. A roll of wallpaper covers an area of 28 ft². If the roll is $1\frac{17}{24}$ ft wide, how long is the roll?

Mixed Exercises

For Exercises 61–74, perform the indicated operation. Write the answer as a mixed number, proper fraction, or whole number.

61. $2\frac{1}{5} \div 1\frac{1}{10}$

62. $3\frac{3}{4} \cdot 1\frac{5}{6}$

63. $6 \div 1\frac{1}{8}$

64. $8 \div 2\frac{1}{3}$

65. $4\frac{1}{12} \cdot 0$

66. $5\frac{1}{3} \cdot 6$

67. $10\frac{1}{2} \div 9$

68. $\frac{2}{7} \cdot 1\frac{8}{9}$

69. $0 \div 9\frac{2}{3}$

70. $\frac{3}{8} \div 2\frac{1}{2}$

71. $12 \cdot \frac{1}{8}$

72. $20 \cdot \frac{2}{15}$

73. $6\frac{8}{9} \div 0$

74. $0 \cdot 2\frac{1}{8}$

Expanding Your Skills

75. A landscaper will use a decorative concrete border around the cactus garden shown. Each concrete brick is $1\frac{1}{4}$ ft long and costs \$3. What is the total cost of the border?

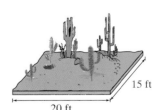

15 ft

20 ft

76. Sara drives a total of $64\frac{1}{2}$ mi to work and back. Her car gets $21\frac{1}{2}$ miles per gallon (mpg) of gas. If gas is \$2 per gallon, how much does it cost her to commute to and from work each day?

chapter 2 | summary

section 2.1 Introduction to Fractions and Mixed Numbers

Key Concepts	Examples

Key Concepts

A **fraction** represents a part of a whole unit. For example, $\frac{1}{3}$ represents one part of a whole unit that is divided into 3 equal pieces.

In the fraction $\frac{1}{3}$, the "top" number, 1, is the **numerator**, and the "bottom" number, 3, is the **denominator**.

A fraction in which the numerator is less than the denominator is called a **proper fraction**. A fraction in which the numerator is greater than or equal to the denominator is called an **improper fraction**.

An improper fraction can be written as a **mixed number** by dividing the numerator by the denominator. Write the quotient as a whole number, and write the remainder over the divisor.

A mixed number can be written as an improper fraction by multiplying the denominator by the whole number and adding the numerator. Then write that total over the denominator.

Fractions can be represented on a number line. For example,

represents $\frac{3}{10}$.

Examples

Example 1

$\frac{1}{3}$ of the pie is shaded.

Example 2

For the fraction $\frac{7}{9}$, the numerator is 7 and the denominator is 9.

Example 3

$\frac{5}{3}$ is an improper fraction, $\frac{3}{5}$ is a proper fraction, and $\frac{3}{3}$ is an improper fraction.

Example 4

$\frac{10}{3}$ can be written as $3\frac{1}{3}$ because $3\overline{)10}$

$$\begin{array}{r} 3 \\ 3\overline{)10} \\ -9 \\ \hline 1 \end{array}$$

Example 5

$2\frac{4}{5}$ can be written as $\frac{14}{5}$ because $\dfrac{5 \cdot 2 + 4}{5} = \dfrac{14}{5}$.

Example 6

section 2.2 Prime Numbers and Factorization

Key Concepts

A **factorization** of a number is a product of factors that equal the number.

A number is divisible by another if their quotient leaves no remainder.

Divisibility Rules for 2, 3, 5, and 10

- A whole number is divisible by 2 if the ones-place digit is 0, 2, 4, 6, or 8.
- A whole number is divisible by 3 if the sum of the digits of the number is divisible by 3.
- A whole number is divisible by 5 if the ones-place digit is 0 or 5.
- A whole number is divisible by 10 if the ones-place digit is 0.

A **prime number** is a whole number that has exactly two factors, 1 and itself.

Composite numbers are whole numbers that have more than two factors. The numbers 0 and 1 are neither prime nor composite.

Prime Factorization

The **prime factorization** of a number is the factorization in which every factor is a prime number.

A factor of a number n is any number that divides evenly into n. For example, the factors of 80 are 1, 2, 4, 5, 8, 10, 16, 20, 40, and 80.

Examples

Example 1

$4 \cdot 4$ and $8 \cdot 2$ are two factorizations of 16.

Example 2

15 is divisible by 5 because 5 divides into 15 evenly (3 times).

Example 3

382 is divisible by 2.
640 is divisible by 2, 5, and 10.
735 is divisible by 3 and 5.

Example 4

9 is a composite number.
2 is a prime number.
1 is neither prime nor composite.

Example 5

$$
\begin{array}{r}
7 \\
3\overline{)21} \\
3\overline{)63} \\
3\overline{)189} \\
2\overline{)378} \\
2\overline{)756}
\end{array}
$$

The prime factorization of 756 is

$$2 \cdot 2 \cdot 3 \cdot 3 \cdot 3 \cdot 7 \quad \text{or} \quad 2^2 \cdot 3^3 \cdot 7$$

section 2.3 Simplifying Fractions to Lowest Terms

Key Concepts	Examples

Key Concepts

Equivalent fractions are fractions that represent the same portion of a whole unit.

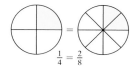

$$\frac{1}{4} = \frac{2}{8}$$

To determine if two fractions are equivalent, calculate the cross products. If the cross products are equal, then the fractions are equivalent.

To simplify fractions to **lowest terms**, use the fundamental principle of fractions:

Consider the fraction $\frac{a}{b}$ and the nonzero number c. Then

$$\frac{a}{b} = \frac{a \div c}{b \div c}$$

To simplify fractions with common powers of 10, "strike through" the common zeros first.

Examples

Example 1

a. Compare $\frac{5}{3}$ and $\frac{6}{4}$.

$$\frac{5}{3} \diagdown \diagup \frac{6}{4}$$

$20 \neq 18$

Fractions are not equivalent

b. Compare $\frac{4}{5}$ and $\frac{8}{10}$.

$$\frac{4}{5} \diagdown \diagup \frac{8}{10}$$

$40 = 40$

Fractions are equivalent

Example 2

$$\frac{25}{15} = \frac{5 \cdot \cancel{5}}{3 \cdot \cancel{5}} = \frac{5}{3}$$

$$\frac{27}{75} = \frac{\cancel{3} \cdot 3 \cdot 3}{\cancel{3} \cdot 5 \cdot 5} = \frac{9}{25}$$

Example 3

$$\frac{3,\cancel{000}}{12,\cancel{000}} = \frac{3}{12} = \frac{\cancel{3}}{2 \cdot 2 \cdot \cancel{3}} = \frac{1}{4}$$

section 2.4   Multiplication of Fractions and Applications

<table>
<tr>
<td>

Key Concepts

Multiplication of Fractions

To multiply fractions, write the product of the numerators over the product of the denominators. Then simplify the resulting fraction, if possible.

</td>
<td>

Examples

Example 1

$$\frac{4}{7} \times \frac{6}{5} = \frac{24}{35}$$

$$\left(\frac{15}{16}\right)\left(\frac{4}{5}\right) = \frac{3 \cdot \overset{1}{\cancel{5}} \cdot \overset{1}{\cancel{2}} \cdot \overset{1}{\cancel{2}}}{2 \cdot 2 \cdot 2 \cdot 2 \cdot \cancel{5}} = \frac{3}{4}$$

</td>
</tr>
<tr>
<td>

When multiplying a whole number and a fraction, first write the whole number as a fraction by writing the whole number over 1.

</td>
<td>

Example 2

$$8 \cdot \left(\frac{5}{6}\right) = \frac{8}{1} \cdot \frac{5}{6} = \frac{2 \cdot 2 \cdot 2 \cdot 5}{1 \cdot 2 \cdot \overset{1}{\cancel{3}}} = \frac{20}{3} \text{ or } 6\frac{2}{3}$$

</td>
</tr>
<tr>
<td>

For powers of fractions and multiple operations, use the order of operations to simplify.

</td>
<td>

Example 3

$$\left(\frac{3}{2}\right)^3 \cdot \left(\frac{2}{21} \times \frac{14}{5}\right) = \frac{27}{8} \cdot \left(\frac{2}{21} \times \frac{14}{5}\right)$$

$$= \frac{27}{8} \cdot \left(\frac{2}{\underset{3}{\cancel{21}}} \times \frac{\overset{2}{\cancel{14}}}{5}\right)$$

$$= \frac{\overset{9}{\cancel{27}}}{\underset{2}{8}} \cdot \frac{\overset{1}{\cancel{4}}}{\underset{5}{\cancel{15}}} = \frac{9}{10}$$

</td>
</tr>
<tr>
<td>

Powers of one-tenth can be expressed as a fraction with 1 in the numerator. The denominator has a 1 followed by the same number of zeros as the exponent of the base of $\frac{1}{10}$.

</td>
<td>

Example 4

$$\left(\frac{1}{10}\right)^4 = \frac{1}{10,000}$$

</td>
</tr>
<tr>
<td>

The formula for the area of a triangle is given by $A = \frac{1}{2}bh$.

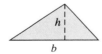

Area is expressed in square units such as ft², in.², yd², m², and cm².

</td>
<td>

Example 5

The area of the triangle is

$$A = \frac{1}{2}bh$$

$$= \frac{1}{2} \cdot \frac{7}{2} \cdot \frac{16}{7}$$

$$= 4$$

The area is 4 m².

</td>
</tr>
</table>

section 2.5 Division of Fractions and Applications

Key Concepts

The **reciprocal** of $\frac{a}{b}$ is $\frac{b}{a}$ for $a, b \neq 0$. The product of a fraction and its reciprocal is 1. For example, $\frac{6}{11} \cdot \frac{11}{6} = 1$.

Dividing Fractions

To divide two fractions, multiply the dividend (the "first" fraction) by the reciprocal of the divisor (the "second" fraction).

When dividing by a whole number, first write the whole number as a fraction by writing the whole number over 1. Then multiply by the reciprocal.

When simplifying expressions with more than one operation, follow the order of operations.

Examples

Example 1

The reciprocal of $\frac{5}{8}$ is $\frac{8}{5}$.
The reciprocal of 4 is $\frac{1}{4}$.

The number 0 does not have a reciprocal because $\frac{1}{0}$ is undefined.

Example 2

$$\frac{18}{25} \div \frac{30}{35} = \frac{\overset{3}{\cancel{18}}}{\underset{5}{\cancel{25}}} \cdot \frac{\overset{7}{\cancel{35}}}{\underset{5}{\cancel{30}}} = \frac{21}{25}$$

Example 3

$$\frac{9}{8} \div 4 = \frac{9}{8} \div \frac{4}{1} = \frac{9}{8} \cdot \frac{1}{4} = \frac{9}{32}$$

Example 4

$$\frac{5}{42} \div \left[\left(\frac{1}{6}\right)^2 \cdot 3\right] = \frac{5}{42} \div \left[\frac{1}{\underset{12}{\cancel{36}}} \cdot \frac{\overset{1}{\cancel{3}}}{1}\right]$$

$$= \frac{5}{42} \div \frac{1}{12}$$

$$= \frac{5}{\underset{7}{\cancel{42}}} \cdot \frac{\overset{2}{\cancel{12}}}{1} = \frac{10}{7}$$

Example 5

Tiffany was asked to provide folders for a meeting of 51 people. She had enough blue folders for $\frac{2}{3}$ of the participants. How many participants received blue folders?

Solution:

To find $\frac{2}{3}$ of 51, we multiply:

$$\frac{2}{\underset{1}{\cancel{3}}} \cdot \frac{\overset{17}{\cancel{51}}}{1} = 34$$

A total of 34 participants received blue folders.

section 2.6 Multiplication and Division of Mixed Numbers

Key Concepts

Multiply Mixed Numbers

1. Change each mixed number to an improper fraction.
2. Multiply the improper fractions and reduce to lowest terms, if possible.

Divide Mixed Numbers

1. Change each mixed number to an improper fraction.
2. Divide the improper fractions and reduce to lowest terms, if possible. Recall that to divide fractions, we multiply the dividend by the reciprocal of the divisor.

Examples

Example 1

$$4\frac{4}{5} \cdot 2\frac{1}{2} = \frac{\overset{12}{\cancel{24}}}{5} \cdot \frac{\overset{1}{\cancel{5}}}{2} = \frac{12}{1} = 12$$

Example 2

$$6\frac{2}{3} \div 2\frac{7}{9} = \frac{20}{3} \div \frac{25}{9} = \frac{20}{\underset{1}{\cancel{3}}} \cdot \frac{\overset{3}{\cancel{9}}}{\underset{5}{\cancel{25}}} = \frac{12}{5} = 2\frac{2}{5}$$

Example 3

A box of oatmeal contains $1\frac{1}{8}$ lb of oatmeal. If a single serving is $\frac{1}{8}$ lb, how many servings are there in a box?

$$1\frac{1}{8} \div \frac{1}{8} = \frac{9}{8} \div \frac{1}{8} = \frac{9}{\underset{1}{\cancel{8}}} \cdot \frac{\overset{1}{\cancel{8}}}{1} = \frac{9}{1} = 9$$

There are 9 servings in a box of oatmeal.

---

chapter 2 | review exercises

Section 2.1

For Exercises 1–2, write a fraction that represents the shaded area.

1.

2.

3. a. Write a fraction that has denominator 3 and numerator 5.

   b. Label the fraction as proper or improper.

4. a. Write a fraction that has numerator 1 and denominator 6.

   b. Label the fraction as proper or improper.

5. In an office supply store, 7 of the 15 computers displayed are laptops. Write a fraction representing the computers that are laptops.

For Exercises 6–7, write a fraction and a mixed number that represent the shaded area.

6.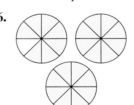

7.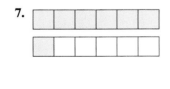

For Exercises 8–9, convert the mixed number to a fraction.

8. $6\frac{1}{7}$

9. $11\frac{2}{5}$

10. How many fourths are in $4\frac{1}{4}$?

For Exercises 11–12, convert the improper fraction to a mixed number.

11. $\frac{47}{9}$

12. $\frac{23}{21}$

For Exercises 13–15, locate the numbers on the number line.

13. $\dfrac{10}{5}$ **14.** $\dfrac{7}{8}$ **15.** $\dfrac{13}{8}$

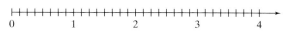

For Exercises 16–17, divide. Write the answer as a mixed number.

16. $7\overline{)941}$ **17.** $26\overline{)1582}$

Section 2.2

For Exercises 18–20, refer to this list of numbers: 21, 43, 51, 55, 58, 124, 140, 260, 1200.

18. List all the numbers that are divisible by 3.

19. List all the numbers that are divisible by 5.

20. List all the numbers that are divisible by 2.

For Exercises 21–24, determine if the number is prime, composite, or neither.

21. 61   **22.** 44

23. 1   **24.** 0

For Exercises 25–27, find the prime factorization.

25. 64   **26.** 330

27. 900

For Exercises 28–29, list all the factors of the number.

28. 48   **29.** 80

Section 2.3

For Exercises 30–31, determine if the fractions are equivalent. Fill in the blank with = or ≠.

30. $\dfrac{3}{6} \square \dfrac{5}{9}$ **31.** $\dfrac{15}{21} \square \dfrac{10}{14}$

For Exercises 32–39, simplify the fraction to lowest terms. Write the answer as a fraction.

32. $\dfrac{5}{20}$ **33.** $\dfrac{14}{49}$ **34.** $\dfrac{24}{16}$

35. $\dfrac{63}{27}$ **36.** $\dfrac{17}{17}$ **37.** $\dfrac{42}{21}$

38. $\dfrac{120}{150}$ **39.** $\dfrac{1400}{2000}$

40. On his final exam, Gareth got 42 out of 45 questions correct. What fraction of the test represents correct answers? What fraction represents incorrect answers?

41. Isaac proofread 6 pages of his 10-page term paper. Yulisa proofread 6 pages of her 15-page term paper.

   a. What fraction of his paper did Isaac proofread?

   b. What fraction of her paper did Yulisa proofread?

Section 2.4

For Exercises 42–47, multiply the fractions and simplify to lowest terms. Write the answer as a fraction or whole number.

42. $\dfrac{3}{5} \times \dfrac{2}{7}$ **43.** $\dfrac{4}{3} \times \dfrac{8}{3}$ **44.** $14 \cdot \dfrac{9}{2}$

45. $33 \cdot \dfrac{5}{11}$ **46.** $\dfrac{2}{9} \cdot \dfrac{5}{8} \cdot \dfrac{36}{25}$ **47.** $\dfrac{45}{7} \cdot \dfrac{6}{10} \cdot \dfrac{28}{63}$

For Exercises 48–51, evaluate by using the order of operations.

48. $\left(\dfrac{1}{10}\right)^{4}$ **49.** $\left(\dfrac{2}{5}\right)^{2} \cdot \left(\dfrac{1}{10}\right)^{2}$

50. $\left(\dfrac{3}{20} \cdot \dfrac{2}{3}\right)^{3}$ **51.** $\left(\dfrac{1}{10}\right)^{3}\left(\dfrac{1000}{17}\right)$

52. Write the formula for the area of a triangle.

53. Write the formula for the area of a rectangle.

For Exercises 54–56, find the area of the shaded region.

54.

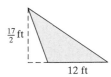

55.

56.

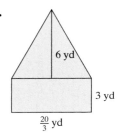

57. Maximus wants to build a workbench for his garage. He needs 4 boards, each cut into $\frac{7}{8}$-yd pieces. How many yards of lumber does he require?

For Exercises 58–61, refer to the graph. The graph represents the distribution of the students at a college by race/ethnicity.

Distribution of Student Body by Race/Ethnicity

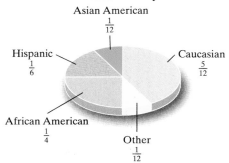

58. If the college has 3600 students, how many are African American?

59. If the college has 3600 students, how many are Asian American?

60. If the college has 3600 students, how many are Hispanic females (assume that one-half of the Hispanic student population is female)?

61. If the college has 3600 students, how many are Caucasian males (assume that one-half of the Caucasian student population is male)?

Section 2.5

For Exercises 62–63, multiply.

62. $\dfrac{3}{4} \cdot \dfrac{4}{3}$

63. $\dfrac{1}{12} \cdot 12$

For Exercises 64–67, find the reciprocal of the fraction, if it exists.

64. $\dfrac{7}{2}$

65. 7

66. 0

67. $\dfrac{1}{6}$

68. Dividing by 5 is the same as multiplying by _____.

69. Dividing by $\frac{2}{9}$ is the same as _____ by $\frac{9}{2}$.

For Exercises 70–75, divide and simplify the answer to lowest terms. Write the answer as a fraction or whole number.

70. $\dfrac{28}{15} \div \dfrac{21}{20}$

71. $\dfrac{7}{9} \div \dfrac{35}{63}$

72. $\dfrac{6}{7} \div 18$

73. $\dfrac{3}{10} \div \dfrac{9}{5}$

74. $\dfrac{200}{51} \div \dfrac{25}{17}$

75. $12 \div \dfrac{6}{7}$

For Exercises 76–79, simplify by using the order of operations. Write the answer as a fraction.

76. $\left(\dfrac{2}{19} \div \dfrac{8}{19}\right)^3$

77. $\left(\dfrac{12}{5}\right)^2 \div \dfrac{36}{5}$

78. $\dfrac{81}{55} \div \dfrac{3}{11} \div \dfrac{3}{2}$

79. $\dfrac{4}{13} \cdot \left(\dfrac{1}{2}\right)^3 \div 2$

For Exercises 80–81, translate to a mathematical statement. Then simplify.

80. How much is $\frac{4}{5}$ of 20?

81. How many $\frac{2}{3}$'s are in 18?

82. How many $\frac{2}{3}$-lb bags of candy can be filled from a 24-lb sack of candy?

83. Amelia worked only $\frac{4}{5}$ of her normal 40-hr workweek. If she makes \$18 per hour, how much money did she earn for the week?

84. A small patio floor will be made from square pieces of tile that are $\frac{4}{3}$ ft on a side. See the figure. Find the area of the patio (in square feet) if its dimensions are 10 tiles by 12 tiles.

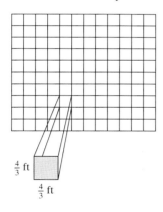

$\frac{4}{3}$ ft

$\frac{4}{3}$ ft

85. Chuck is an elementary school teacher and needs 22 pieces of wood, $\frac{3}{8}$ ft long, for a class project. If he has a 9-ft board from which to cut the pieces, will he have enough $\frac{3}{8}$-ft pieces for his class? Explain.

Section 2.6

For Exercises 86–96, multiply or divide as indicated.

86. $\left(3\dfrac{2}{3}\right)\left(6\dfrac{2}{5}\right)$

87. $\left(11\dfrac{1}{3}\right)\left(2\dfrac{3}{34}\right)$

88. $6\dfrac{1}{2} \cdot 1\dfrac{3}{13}$

89. $4 \cdot \left(5\dfrac{5}{8}\right)$

90. $45\dfrac{5}{13} \cdot 0$

91. $4\dfrac{5}{16} \div 2\dfrac{7}{8}$

92. $3\dfrac{5}{11} \div 3\dfrac{4}{5}$

93. $7 \div 1\dfrac{5}{9}$

94. $4\dfrac{6}{11} \div 2$

95. $10\dfrac{1}{5} \div 17$

96. $0 \div 3\dfrac{5}{12}$

97. It takes $1\frac{1}{4}$ gal of paint for Neva to paint her living room. If her great room (including the dining area) is $2\frac{1}{2}$ times larger than the living room, how many gallons will it take to paint that room?

98. A roll of ribbon contains $12\frac{1}{2}$ yd. How many pieces of length $1\frac{1}{4}$ yd can be cut from this roll?

chapter 2 | test

1. a. Write a fraction that represents the shaded portion of the figure.

  b. Is the fraction proper or improper?

2. a. Write a fraction that represents the total shaded portion of the 3 figures.

  b. Is the fraction proper or improper?

3. Write an improper fraction and a mixed number that represent the shaded region.

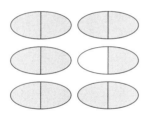

4. Is the fraction $\frac{7}{7}$ a proper or an improper fraction? Explain.

5. a. Write $\frac{44}{12}$ as a mixed number.

  b. Write $3\frac{7}{9}$ as an improper fraction.

For Exercises 6–9, plot the fraction on the number line.

6. $\frac{1}{2}$ ⊢+++++++++++⊢→
         0   1

7. $\frac{3}{4}$ ⊢+++++++++++⊢→
         0   1

8. $\frac{7}{12}$ ⊢+++++++++++⊢→
          0   1

9. $\frac{13}{5}$ ⊢+++++++++++++++++++++++→
           0   1   2   3   4

10. Label the following numbers as prime, composite, or neither.

  a. 15   **b.** 0

  c. 53   **d.** 1

  e. 29   **f.** 39

11. a. List all the factors of 80.

  b. Write the prime factorization of 80.

12. a. What is the divisibility rule for 3?

  b. Is 1,981,011 divisible by 3?

13. Determine whether 1155 is divisible by

  a. 2   **b.** 3

  c. 5   **d.** 10

For Exercises 14–15, determine if the fractions are equivalent. Then fill in the blank with either = or ≠ .

14. $\frac{15}{12}$ ☐ $\frac{5}{4}$ **15.** $\frac{2}{5}$ ☐ $\frac{4}{25}$

For Exercises 16–17, simplify the fractions to lowest terms.

16. $\frac{150}{105}$ **17.** $\frac{1,200,000}{1,400,000}$

18. Christine and Brad are putting their photographs in scrapbooks. Christine has placed 15 of her 25 photos and Brad has placed 16 of his 20 photos.

  a. What fractional part of the total photos has each person placed?

  b. Which person has a greater fractional part completed?

For Exercises 19–26, multiply or divide as indicated. Simplify the fraction to lowest terms.

19. $\dfrac{2}{9} \times \dfrac{57}{46}$

20. $\left(\dfrac{75}{24}\right) \cdot 4$

21. $\dfrac{28}{24} \div \dfrac{21}{8}$

22. $\dfrac{105}{42} \div 5$

23. $\dfrac{2}{18} \times \dfrac{9}{25} \times \dfrac{40}{6}$

24. $\dfrac{600}{1200} \div \dfrac{50}{65} \div \dfrac{13}{15}$

25. $\dfrac{10}{21} \div 4\dfrac{1}{6}$

26. $4\dfrac{4}{17} \cdot 2\dfrac{4}{15}$

27. Perform the order of operations. Simplify the fraction to lowest terms.

$$\dfrac{52}{72} \div \left[\left(\dfrac{1}{2}\right)^2 \cdot \dfrac{8}{3}\right]$$

28. Find the area of the triangle.

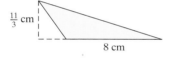

29. Which is greater, $20 \cdot \frac{1}{4}$ or $20 \div \frac{1}{4}$?

30. How many "quarter-pounders" can be made from 12 lb of ground beef?

31. The Humane Society has 120 dogs. Of the 120, $\frac{5}{8}$ are female. Among the female dogs, $\frac{1}{15}$ are pure breeds. How many of the dogs are female and pure breeds?

32. A zoning requirement indicates that a house built on less than 1 acre of land may take up no more than one-half of the land. If Liz and George purchased a $\frac{4}{5}$-acre lot of land, what is the maximum land area that they can use to build the house?

chapters 1 and 2 | cumulative review

1. Fill out the table with either the word name for the number or the number in standard form.

Mountain	Height (ft)	
	Standard Form	Words
Mt. Foraker (Alaska)		Seventeen thousand, four hundred
Mt. Kilimanjaro (Tanzania)	19,340	
El Libertador (Argentina)		Twenty-two thousand, forty-seven
Mont Blanc (France-Italy)	15,771	

For Exercises 2–13, perform the indicated operation.

2. $432 + 998$

3. $572 - 433$

4. 4122×52

5. $384 \div 16$

6. $23(81)$

7. $4\overline{)74}$

8. $\begin{array}{r} 3{,}000{,}000 \\ \times\, 40{,}000 \end{array}$

9. $\begin{array}{r} 1007 \\ -\, 823 \end{array}$

10. $\dfrac{48}{8}$

11. $6 + 2 \cdot 8$

12. $5^2 - 3^2$

13. $(5 - 3)^2$

For Exercises 14–18, match the algebraic expression with the property that it demonstrates.

14. $5 \cdot 8 = 8 \cdot 5$ **a.** Commutative property of addition

15. $4(3 + 2)$
$= 4 \cdot 3 + 4 \cdot 2$ **b.** Associative property of addition

16. $(12 + 3) + 5$
$= 12 + (3 + 5)$ **c.** Distributive property of multiplication over addition

17. $8 \cdot (7 \cdot 2)$
$= (8 \cdot 7) \cdot 2$ **d.** Commutative property of multiplication

18. $32 + 9$
$= 9 + 32$ **e.** Associative property of multiplication

19. Write a fraction that represents the shaded area.

a.

b.

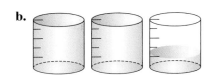

20. Identify each fraction as proper or improper.

a. $\dfrac{7}{8}$ **b.** $\dfrac{8}{7}$ **c.** $\dfrac{8}{8}$

21. a. List all the factors of 30.

b. Write the prime factorization of 30.

22. Simplify the fraction to lowest terms.

a. $\dfrac{144}{84}$ **b.** $\dfrac{60,000}{150,000}$

23. Multiply and simplify to lowest terms. $\dfrac{35}{27} \cdot \dfrac{51}{95}$

24. Divide and simplify to lowest terms. $5\frac{2}{3} \div 6\frac{4}{5}$

25. Is multiplication of fractions a commutative operation? Explain, using the fractions $\frac{8}{13}$ and $\frac{5}{16}$.

26. Is multiplication of fractions an associative operation? Explain, using the fractions $\frac{1}{2}, \frac{2}{9}$, and $\frac{5}{3}$.

27. Simplify, using the order of operations. Simplify the answer to lowest terms. $\left(\frac{5}{6} \cdot \frac{12}{25}\right)^2 \div \frac{2}{3}$

28. Find the area of the rectangle.

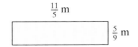

29. Find the area of the triangle.

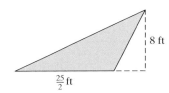

30. At one college $\frac{3}{4}$ of the students are male, and of the males, $\frac{1}{10}$ are from out of state. What fraction of the students are males who are from out of state?

Fractions and Mixed Numbers: Addition and Subtraction

3

In this chapter we learn about addition and subtraction of fractions as well as operations on mixed numbers. For example, suppose a student works 3 part-time jobs. She spends $8\frac{2}{3}$ hr per week delivering newspapers. She also spends $4\frac{1}{2}$ hr per week tutoring and $3\frac{3}{4}$ hr per week taking notes for a blind student. See Exercise 86 in Section 3.4 to determine the total number of hours per week that the student works.

chapter 3 | preview

The exercises in this chapter preview contain concepts that have not yet been presented. These exercises are provided for students who want to compare their levels of understanding before and after studying the chapter. Alternatively, you may prefer to work these exercises when the chapter is completed and before taking the exam.

Section 3.1

For Exercises 1–3, add or subtract the like fractions. Write the answer as a fraction or whole number simplified to lowest terms.

1. $\dfrac{53}{12} + \dfrac{25}{12}$ **2.** $\dfrac{28}{9} - \dfrac{10}{9}$ **3.** $\dfrac{19}{12} - \dfrac{5}{12} + \dfrac{7}{12}$

Section 3.2

4. a. List eight multiples of 9. Answers may vary.

   b. List eight multiples of 12. Answers may vary.

   c. From the lists from parts (a) and (b), list two common multiples of 9 and 12.

   d. Identify the least common multiple (LCM) of 9 and 12.

5. What is the LCM of the numbers 8, 12, and 18?

6. Rewrite each fraction with the indicated denominator.

   a. $\dfrac{8}{15} = \dfrac{}{60}$ **b.** $\dfrac{3}{5} = \dfrac{}{60}$

   c. $\dfrac{11}{20} = \dfrac{}{60}$ **d.** $\dfrac{7}{12} = \dfrac{}{60}$

7. Rank the fractions from Exercise 6 from least to greatest.

Section 3.3

For Exercises 8–11, add or subtract the fractions. Write the answer as a fraction or whole number, and simplify to lowest terms.

8. $\dfrac{23}{5} + \dfrac{13}{10}$ **9.** $\dfrac{31}{16} - \dfrac{5}{4}$

10. $6 - \dfrac{5}{4}$ **11.** $\dfrac{23}{18} - \dfrac{19}{36} + \dfrac{3}{4}$

Section 3.4

For Exercises 12–14, add and subtract the mixed numbers. Write the answer as a mixed number.

12. $14\dfrac{9}{10} + 2\dfrac{49}{100}$ **13.** $7 - 4\dfrac{13}{14}$

14. $15\dfrac{3}{8} - 12\dfrac{5}{8} + 1\dfrac{3}{4}$

Section 3.5

For Exercises 15–16, simplify the expressions by using the order of operations. Simplify the answer to lowest terms.

15. $\left(1\dfrac{3}{8} - \dfrac{15}{16}\right) \div \left(\dfrac{3}{4}\right)^2$ **16.** $1\dfrac{1}{9} + \left(\dfrac{10}{27} \cdot 1\dfrac{4}{5}\right)^3$

17. Lou wants to paint the front of his cottage. Find the area of the front of the house, given the figure.

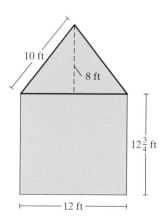

section 3.1   Addition and Subtraction of Like Fractions

1. Addition of Like Fractions

In Chapter 2 we learned how to multiply and divide fractions. The main focus of this chapter is to add and subtract fractions. The operation of addition can be thought of as combining like groups of objects. For example:

$$3 \text{ apples} + 1 \text{ apple} = 4 \text{ apples}$$

three-fifths + one-fifth = four-fifths

$$\frac{3}{5} + \frac{1}{5} = \frac{4}{5}$$

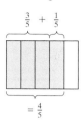

The fractions $\frac{3}{5}$ and $\frac{1}{5}$ are said to be **like fractions** because their denominators are the same. That is, their denominators represent the same part of a whole. In general, two or more like fractions may be added according to the following procedure.

> **Adding Like Fractions**
> 1. Add the numerators.
> 2. Write the sum over the common denominator.
> 3. Simplify the fraction to lowest terms, if possible.

example 1   Adding Like Fractions

Add. Write the answer as a fraction or whole number.

a. $\frac{1}{4} + \frac{5}{4}$ **b.** $\frac{3}{10} + \frac{7}{10}$ **c.** $\frac{7}{15} + \frac{2}{15} + \frac{1}{15}$

Solution:

a. $\frac{1}{4} + \frac{5}{4} = \frac{1+5}{4}$ Add the numerators.

$= \frac{6}{4}$ Write the sum over the common denominator.

$= \frac{\overset{3}{\cancel{6}}}{\underset{2}{\cancel{4}}}$ Simplify to lowest terms.

$= \frac{3}{2}$

> **Avoiding Mistakes:** Notice that when adding fractions, we do not add the denominators. We add *only* the numerators.

b. $\frac{3}{10} + \frac{7}{10} = \frac{3+7}{10}$ Add the numerators.

$= \frac{10}{10}$ Write the sum over the common denominator.

$= 1$ Simplify to lowest terms.

Objectives

1. Addition of Like Fractions
2. Subtraction of Like Fractions
3. Order of Operations
4. Applications of Addition and Subtraction of Fractions

Concept Connections

Determine whether the fractions are like or unlike.

1. $\frac{4}{7}$ and $\frac{7}{3}$ **2.** $\frac{9}{8}$ and $\frac{1}{8}$

Skill Practice

Add. Write the answer as a fraction or whole number.

3. $\frac{2}{9} + \frac{4}{9}$ **4.** $\frac{11}{5} + \frac{4}{5}$

5. $\frac{7}{12} + \frac{5}{12} + \frac{11}{12}$

Answers

1. Unlike   2. Like   3. $\frac{2}{3}$

4. 3   5. $\frac{23}{12}$

c. $\dfrac{7}{15} + \dfrac{2}{15} + \dfrac{1}{15} = \dfrac{7 + 2 + 1}{15}$ Add the numerators.

$= \dfrac{10}{15}$ Write the sum over the common denominator.

$= \dfrac{2}{3}$ Simplify to lowest terms.

Concept Connections

6. Which is the correct sum for $\frac{2}{3} + \frac{5}{3}$?

$\dfrac{7}{6}$ or $\dfrac{7}{3}$

7. Which is the correct product for $\frac{2}{3} \cdot \frac{5}{3}$?

$\dfrac{10}{9}$ or $\dfrac{10}{3}$

Avoiding Mistakes: Note that the process to add fractions is different from the process to multiply fractions.

$$\dfrac{2}{7} \times \dfrac{3}{7} = \dfrac{6}{49} \quad \text{but} \quad \dfrac{2}{7} + \dfrac{3}{7} = \dfrac{5}{7}$$

2. Subtraction of Like Fractions

Subtracting like fractions is performed in a manner similar to adding like fractions.

Subtracting Like Fractions

1. Subtract the numerators.

2. Write the difference over the common denominator.

3. Simplify the fraction to lowest terms, if possible.

Skill Practice

Subtract. Write the answer as a fraction or whole number.

8. $\dfrac{7}{2} - \dfrac{3}{2}$ **9.** $\dfrac{14}{11} - \dfrac{8}{11}$

10. $\dfrac{5}{6} + \dfrac{7}{6} - \dfrac{2}{6}$

---

example 2   Subtracting Like Fractions

Subtract. Write the answer as a fraction or whole number.

a. $\dfrac{13}{9} - \dfrac{2}{9}$ **b.** $\dfrac{7}{3} - \dfrac{1}{3}$

Solution:

a. $\dfrac{13}{9} - \dfrac{2}{9} = \dfrac{13 - 2}{9}$ Subtract the numerators.

$= \dfrac{11}{9}$ Write the difference over the common denominator. The fraction is already in lowest terms because 11 and 9 share no common factors.

b. $\dfrac{7}{3} - \dfrac{1}{3} = \dfrac{7 - 1}{3}$ Subtract the numerators.

$= \dfrac{6}{3}$ Write the difference over the common denominator.

$= 2$

Answers

6. $\dfrac{7}{3}$ **7.** $\dfrac{10}{9}$ **8.** 2

9. $\dfrac{6}{11}$ **10.** $\dfrac{5}{3}$

3. Order of Operations

Example 3 reviews the order of operations. Simplify within parentheses first. Then simplify expressions with exponents, followed by multiplication or division in the order they appear from left to right. Addition or subtraction is performed last in the order of appearance from left to right.

---

example 3   Applying the Order of Operations

Simplify.

a. $\left(\dfrac{2}{7} + \dfrac{1}{7}\right)^2$ **b.** $\dfrac{3}{5} \div \dfrac{9}{10} + \dfrac{1}{3}$ **c.** $\dfrac{7}{15} - \dfrac{3}{15} + \dfrac{2}{15}$

Solution:

a. $\left(\dfrac{2}{7} + \dfrac{1}{7}\right)^2 = \left(\dfrac{2+1}{7}\right)^2$ Add fractions within parentheses first.

$= \left(\dfrac{3}{7}\right)^2$

$= \dfrac{3}{7} \cdot \dfrac{3}{7}$ Square the fraction $\frac{3}{7}$.

$= \dfrac{9}{49}$ The fraction is in lowest terms.

b. $\dfrac{3}{5} \div \dfrac{9}{10} + \dfrac{1}{3} = \left(\dfrac{3}{5} \div \dfrac{9}{10}\right) + \dfrac{1}{3}$ We can insert parentheses to emphasize that division is performed before addition.

$= \left(\dfrac{3}{5} \cdot \dfrac{10}{9}\right) + \dfrac{1}{3}$ Multiply by the reciprocal of the divisor.

$= \left(\dfrac{\overset{1}{3}}{\underset{1}{5}} \cdot \dfrac{\overset{2}{10}}{\underset{3}{9}}\right) + \dfrac{1}{3}$ Simplify.

$= \left(\dfrac{2}{3}\right) + \dfrac{1}{3}$ Multiply the fractions.

$= \dfrac{3}{3}$ Add the fractions.

$= 1$ Simplify.

c. $\dfrac{7}{15} - \dfrac{3}{15} + \dfrac{2}{15} = \dfrac{7 - 3 + 2}{15}$ Subtract and add the numerators.

$= \dfrac{6}{15}$ Write the result over the common denominator.

$= \dfrac{\overset{2}{6}}{\underset{5}{15}}$ Simplify to lowest terms.

$= \dfrac{2}{5}$

Skill Practice

Simplify.

11. $\left(\dfrac{3}{10} + \dfrac{2}{10}\right)^2$

12. $\dfrac{2}{7} \div \dfrac{4}{5} + \dfrac{3}{14}$

13. $\dfrac{8}{11} - \dfrac{3}{11} + \dfrac{7}{11}$

Answers

11. $\dfrac{1}{4}$ **12.** $\dfrac{4}{7}$ **13.** $\dfrac{12}{11}$

4. Applications of Addition and Subtraction of Fractions

Recall that the perimeter of a polygon is found by adding the lengths of the sides.

Skill Practice

14. Find the perimeter.

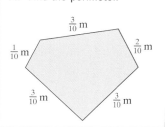

---

example 4   Finding Perimeter

Find the perimeter of the figure.

$\frac{10}{12}$ yd

$\frac{5}{12}$ yd $\frac{5}{12}$ yd

$\frac{15}{12}$ yd

Solution:

$$\text{Perimeter} = \frac{10}{12} + \frac{5}{12} + \frac{15}{12} + \frac{5}{12}$$

$$= \frac{10 + 5 + 15 + 5}{12}$$

$$= \frac{35}{12} \quad \text{or} \quad 2\frac{11}{12}$$

The perimeter is $2\frac{11}{12}$ yd.

Skill Practice

15. Jamie mixed $\frac{5}{8}$ gal of green paint with $\frac{7}{8}$ gal of white paint. Then she used $\frac{3}{8}$ gal of the mixture to paint a mural. How much paint is left over?

---

example 5   Applying Addition and Subtraction of Fractions

On Monday, a stock rose by $\frac{11}{16}$ point. On Tuesday it rose $\frac{3}{16}$ point, and on Wednesday it dropped $\frac{9}{16}$ point. What was the net gain for these 3 days?

Solution:

The net change in the stock is given by $\frac{11}{16} + \frac{3}{16} - \frac{9}{16}$.

$$\frac{11}{16} + \frac{3}{16} - \frac{9}{16} = \frac{11 + 3 - 9}{16}$$

$$= \frac{5}{16}$$

The stock rose by $\frac{5}{16}$ point.

Answers

14. $\frac{6}{5}$ or $1\frac{1}{5}$ m

15. $\frac{9}{8}$ or $1\frac{1}{8}$ gal

section 3.1   Practice Exercises

Study Skills Exercises

1. How can you utilize the margin exercises in the text?

2. Define the key term **like fractions**.

Objective 1: Addition of Like Fractions

For Exercises 3–7, add the like units.

3. 3 ft + 5 ft

4. 7 chairs + 2 chairs

5. 7 m + 13 m

6. 8 thirds + 2 thirds

7. 1 fourth + 6 fourths

For Exercises 8–9, shade in the portion of the third figure that represents the addition of the first two figures.

8.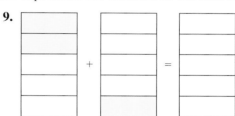

9.

10. Explain the difference between evaluating the two expressions $\frac{2}{5} \times \frac{7}{5}$ and $\frac{2}{5} + \frac{7}{5}$.

For Exercises 11–22, add the like fractions. Write the answer as a fraction or whole number. **(See Example 1.)**

11. $\dfrac{6}{11} + \dfrac{7}{11}$

12. $\dfrac{5}{3} + \dfrac{2}{3}$

13. $\dfrac{6}{5} + \dfrac{3}{5}$

14. $\dfrac{3}{10} + \dfrac{4}{10}$

15. $\dfrac{1}{4} + \dfrac{3}{4}$

16. $\dfrac{1}{8} + \dfrac{3}{8}$

17. $\dfrac{2}{9} + \dfrac{4}{9}$

18. $\dfrac{3}{2} + \dfrac{5}{2}$

19. $\dfrac{3}{20} + \dfrac{8}{20} + \dfrac{15}{20}$

20. $\dfrac{5}{8} + \dfrac{4}{8} + \dfrac{9}{8}$

21. $\dfrac{18}{14} + \dfrac{11}{14} + \dfrac{6}{14}$

22. $\dfrac{7}{18} + \dfrac{22}{18} + \dfrac{10}{18}$

23. Bethany pours $\frac{1}{4}$ cup of bleach into a container and then adds $\frac{9}{4}$ cups of water. How many cups of bleach and water mixture does she have?

24. Austin rode his bike $\frac{7}{6}$ mi before he got a flat tire. He then had to walk another $\frac{1}{6}$ mi. How far did Austin travel?

Objective 2: Subtraction of Like Fractions

For Exercises 25–28, subtract the like units.

25. 15 baskets − 4 baskets   **26.** 52 cards − 13 cards   **27.** 7 fifths − 1 fifth   **28.** 18 tenths − 11 tenths

For Exercises 29–30, shade in the portion of the third figure that represents the subtraction of the first two figures.

29. − =   **30.** 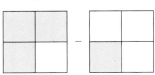 − =

For Exercises 31–42, subtract the like fractions. Write the answer as a fraction or whole number. **(See Example 2.)**

31. $\dfrac{9}{8} - \dfrac{6}{8}$

32. $\dfrac{7}{9} - \dfrac{6}{9}$

33. $\dfrac{9}{2} - \dfrac{6}{2}$

34. $\dfrac{10}{4} - \dfrac{5}{4}$

35. $\dfrac{13}{3} - \dfrac{7}{3}$

36. $\dfrac{13}{10} - \dfrac{3}{10}$

37. $\dfrac{23}{12} - \dfrac{15}{12}$

38. $\dfrac{13}{6} - \dfrac{5}{6}$

39. $\dfrac{28}{25} - \dfrac{14}{25} - \dfrac{4}{25}$

40. $\dfrac{34}{15} - \dfrac{6}{15} - \dfrac{3}{15}$

41. $\dfrac{10}{16} - \dfrac{1}{16} - \dfrac{5}{16}$

42. $\dfrac{31}{40} - \dfrac{14}{40} - \dfrac{12}{40}$

43. A chemist has $\frac{5}{8}$ grams (g) of NaCl (salt). If she uses $\frac{3}{8}$ g, how much is left?

44. Jason bought $\frac{11}{4}$ acres of land and then sold $\frac{3}{4}$ acre. How much land does he have left?

Mixed Exercises

For Exercises 45–54, add or subtract as indicated. Write the answer as a fraction or whole number.

45. $\dfrac{7}{8} + \dfrac{5}{8}$

46. $\dfrac{1}{21} + \dfrac{13}{21}$

47. $\dfrac{14}{5} - \dfrac{2}{5}$

48. $\dfrac{5}{3} - \dfrac{2}{3}$

49. $\dfrac{6}{13} + \dfrac{7}{13}$

50. $\dfrac{20}{35} + \dfrac{12}{35}$

51. $\dfrac{14}{15} + \dfrac{2}{15} - \dfrac{4}{15}$

52. $\dfrac{19}{6} - \dfrac{11}{6} + \dfrac{5}{6}$

53. $\dfrac{7}{2} - \dfrac{3}{2} + \dfrac{1}{2}$

54. $\dfrac{8}{3} + \dfrac{2}{3} - \dfrac{1}{3}$

Objective 3: Order of Operations

For Exercises 55–64, simplify the expression by using the order of operations. Write the answer as a fraction or whole number. (See Example 3.)

55. $\dfrac{6}{5} + \dfrac{7}{5} - \dfrac{4}{5}$

56. $\dfrac{10}{3} - \dfrac{2}{3} + \dfrac{5}{3}$

 57. $\dfrac{5}{4} \div \dfrac{3}{2} + \dfrac{5}{6}$

58. $\dfrac{1}{7} \div \dfrac{2}{21} + \dfrac{5}{2}$

59. $\dfrac{3}{7} + \dfrac{13}{14} \cdot 2$

60. $\dfrac{13}{6} - \dfrac{5}{18} \cdot 3$

61. $\left(\dfrac{7}{3} - \dfrac{5}{3} \right)^3$

62. $\left(\dfrac{11}{10} - \dfrac{2}{10} \right)^2$

63. $\left(\dfrac{2}{21} + \dfrac{11}{21} \right) \div \dfrac{1}{7}$

64. $\left(\dfrac{17}{30} - \dfrac{12}{30} \right) \div \dfrac{5}{6}$

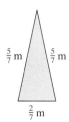 **65.** Gail mixed $\frac{1}{10}$ gal of red paint with $\frac{7}{10}$ gal of white paint. She used $\frac{3}{10}$ gal of the mixture to paint a room. How much mixture was left over? (See Example 5.)

66. Emeril mixed $\frac{3}{8}$ cup balsamic vinegar with $\frac{4}{8}$ cup oil to make a salad dressing. Then he used $\frac{1}{8}$ cup of the mixture for a large salad. How much oil and vinegar mixture was left over?

67. A chemist mixed $\frac{5}{8}$ liter (L) of water with $\frac{7}{8}$ L of alcohol. Then he used one-quarter of the mixture in an experiment. How much mixture did he use?

68. Malcom planted tomatoes in $\frac{2}{7}$ of his garden. He planted cucumbers in $\frac{3}{7}$ of the garden and cabbage in $\frac{1}{7}$. The remaining $\frac{1}{7}$ still has not yet been planted. A deer came in one night and ate $\frac{1}{3}$ of the plants in the planted area. What fraction of the garden did the deer eat?

Objective 4: Applications of Addition and Subtraction of Fractions

For Exercises 69–70, find the perimeter.

69.

$\frac{5}{7}$ m $\frac{5}{7}$ m

$\frac{2}{7}$ m

70.

$\frac{20}{9}$ ft $\frac{23}{9}$ ft

$\frac{11}{9}$ ft

71. Find the perimeter of the stamp. (See Example 4.)

$\frac{13}{16}$ in.

$\frac{15}{16}$ in.

72. Find the perimeter of the top of the table.

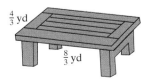

$\frac{4}{3}$ yd

$\frac{8}{3}$ yd

73. Thilan has taken up a new exercise program. He walks 6 days per week. One week he walked the distances given in the table.

Day	Distance
Monday	$\frac{4}{10}$ mi
Tuesday	$\frac{7}{10}$ mi
Wednesday	$\frac{9}{10}$ mi
Thursday	$\frac{5}{10}$ mi
Friday	$\frac{13}{10}$ mi
Saturday	$\frac{17}{10}$ mi

a. Find the total distance he walked for the week.

b. Find the average distance walked per day.

74. Denzel recorded the weekly rainfall for his town for 4 weeks of summer.

Week	Amount of rainfall
1	$\frac{2}{10}$ in.
2	$\frac{7}{10}$ in.
3	$\frac{9}{10}$ in.
4	$\frac{17}{10}$ in.

a. Find the total amount of rainfall for this 4-week period.

b. Find the average rainfall per week.

For Exercises 75–78, find the perimeter and the area.

75.

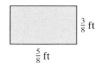

$\frac{3}{8}$ ft

$\frac{5}{8}$ ft

76.

$\frac{7}{8}$ m

$\frac{15}{8}$ m

77.

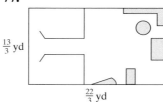

$\frac{13}{3}$ yd

$\frac{22}{3}$ yd

78.

May						
Sun	Mon	Tues	Wed	Thur	Fri	Sat

$\frac{25}{4}$ in.

$\frac{30}{4}$ in.

For Exercises 79–82, translate the phrase to a mathematical expression, then simplify.

79. The sum of three-fifths and two-fifths

80. Seven-ninths more than five-ninths

81. The difference of eleven-fifteenths and eight-fifteenths

82. Two-sevenths subtracted from five-sevenths

section 3.2   Least Common Multiple

1. Least Common Multiple
2. Finding the LCM by Using Prime Factors
3. Finding the LCM by Using Division by Primes (Optional)
4. Applications of the LCM
5. Ordering Fractions

1. Least Common Multiple

In Section 3.1 we learned how to add and subtract like fractions. To add or subtract fractions with different denominators, we must learn how to convert unlike fractions into like fractions. An essential concept in this process is the idea of a least common multiple of two or more numbers.

When we multiply a number by the whole numbers 1, 2, 3, and so on, we form the **multiples** of the number. For example, some of the multiples of 6 and 9 are shown below.

Concept Connections

1. Explain the difference between a multiple of a number and a factor of a number.

Multiples of 6	Multiples of 9
$6 \times 1 = 6$	$9 \times 1 = 9$
$6 \times 2 = 12$	$9 \times 2 = 18$
$6 \times 3 = 18$	$9 \times 3 = 27$
$6 \times 4 = 24$	$9 \times 4 = 36$
$6 \times 5 = 30$	$9 \times 5 = 45$
$6 \times 6 = 36$	$9 \times 6 = 54$
$6 \times 7 = 42$	$9 \times 7 = 63$
$6 \times 8 = 48$	$9 \times 8 = 72$
$6 \times 9 = 54$	$9 \times 9 = 81$

The **least common multiple (LCM)** of two given numbers is the smallest whole number that is a multiple of each given number. For example, the LCM of 6 and 9 is 18.

Multiples of 6:   6, 12, 18, 24, 30, 36, 42, . . .

Multiples of 9:   9, 18, 27, 36, 45, 54, 63, . . .

Tip: There are infinitely many numbers that are common multiples of both 6 and 9. These include 18, 36, 54, 72, and so on. However, 18 is the smallest, and is therefore the *least* common multiple.

If one number is a multiple of another number, then the LCM is the larger of the two numbers. For example, the LCM of 4 and 8 is 8. Similarly, the LCM of 3 and 6 is 6.

Multiples of 4:	4, 8, 12, 16, . . .	Multiples of 3:	3, 6, 9, 12, . . .
Multiples of 8:	8, 16, 24, 32, . . .	Multiples of 6:	6, 12, 18, 24, . . .

example 1   Finding the LCM by Listing Multiples

Find the LCM of the given numbers by listing several multiples of each number.

a. 15 and 12   **b.** 10, 15, and 8

Skill Practice

Find the LCM by listing several multiples of each number.

2. 15 and 25   3. 4, 6, and 10

Solution:

a. Multiples of 15:   15, 30, 45, 60
   Multiples of 12:   12, 24, 36, 48, 60

   The LCM of 15 and 12 is 60.

Answers

1. A multiple of a number is the product of the number and a whole number 1 or greater. A factor of a number is a value that divides evenly into the number.
2. 75   3. 60

b. Multiples of 10:   10, 20, 30, 40, 50, 60, 70, 80, 90, 100, 110, 120
Multiples of 15:   15, 30, 45, 60, 75, 90, 105, 120
Multiples of 8:   8, 16, 24, 32, 40, 48, 56, 64, 72, 80, 88, 96, 104, 112, 120

The LCM of 10, 15, and 8 is 120.

2. Finding the LCM by Using Prime Factors

In Example 1 we used the method of listing multiples to find the LCM of two or more numbers. As you can see, the solution to Example 1(b) required several long lists of multiples. Here we offer another method to find the LCM of two given numbers by using their prime factors.

Using Prime Factors to Find the LCM of Two Numbers

1. Write each number as a product of prime factors.
2. The LCM is the product of unique prime factors from both numbers. If a factor is repeated within the factorization of either number, use that factor the maximum number of times it appears in either factorization.

This process is demonstrated in Example 2.

Skill Practice

Find the LCM by using prime factors.

4. 9 and 24   **5.** 16 and 9

6. 36, 42, and 30

---

example 2   Finding the LCM by Using Prime Factors

Find the LCM.

a. 15 and 12   **b.** 18 and 8   **c.** 45, 54, and 10

Solution:

a. $15 = 3 \cdot 5$ Write each number as a product of
$12 = 2 \cdot 2 \cdot 3$ prime factors.

The factors 3 and 5 each occur a maximum of one time within a list of factors. Note that the factor 2 occurs twice in the second list of factors.

The LCM is $2 \cdot 2 \cdot 3 \cdot 5 = 60$.   The LCM is the product of the factors 2, 3, and 5, where 2 is repeated twice. Note that this is the same answer as in Example 1(a).

b. $18 = 2 \cdot 3 \cdot 3$ Write each number as a product of
$8 = 2 \cdot 2 \cdot 2$ prime factors.

Note that the factor 2 occurs most often (3 times) in the second list of factors. Note that the factor 3 occurs most often (2 times) in the first list of factors.

The LCM is $2 \cdot 2 \cdot 2 \cdot 3 \cdot 3 = 72$.   The LCM is the product of the factors 2 and 3, where 2 is repeated 3 times and 3 is repeated twice.

c. $45 = 3 \cdot 3 \cdot 5$ Write each number as a product of
$54 = 2 \cdot 3 \cdot 3 \cdot 3$ prime factors.
$10 = 2 \cdot 5$

Note that the factor 3 occurs most often (3 times) in the second list of factors. The factors 2 and 5 each occur a maximum of 1 time within a list of factors.

The LCM is $2 \cdot 3 \cdot 3 \cdot 3 \cdot 5 = 270$.

Answers

4. 72   5. 144   6. 1260

3. Finding the LCM by Using Division by Primes (Optional)

We present a third method for finding least common multiples. We systematically divide by prime numbers to determine which will be a factor of the LCM. This method is particularly helpful if three or more numbers are involved. Try each method, and then you and your instructor can decide which method works best for you.

---

example 3   Finding the LCM by Using Division by Primes

Find the LCM of 32, 48, and 30 by using division of prime factors.

Solution:

To begin this process, find any prime number that divides evenly into any of the numbers. Then divide and write the quotient as shown. We begin by dividing by the smallest prime number, 2.

$$2\overline{)32 \ \ 48 \ \ 30}$$
$$16 \ \ 24 \ \ 15$$

Repeat this process and bring down any number that is not divisible by the chosen prime.

$$2\overline{)32 \ \ 48 \ \ 30}$$
$$2\overline{)16 \ \ 24 \ \ 15}$$
$$8 \ \ 12 \ \ 15 \quad \longleftarrow \text{Bring down the 15.}$$

Continue until all quotients are 1. The LCM is the product of the prime factors at the left.

$$2\overline{)32 \ \ 48 \ \ 30}$$
$$2\overline{)16 \ \ 24 \ \ 15}$$
$$2\overline{)8 \ \ 12 \ \ 15}$$
$$2\overline{)4 \ \ 6 \ \ 15}$$
$$2\overline{)2 \ \ 3 \ \ 15}$$
$$3\overline{)1 \ \ 3 \ \ 15} \quad \longleftarrow$$
$$5\overline{)1 \ \ 1 \ \ 5}$$
$$1 \ \ 1 \ \ 1$$

At this point, the prime number 2 does not divide evenly into any of the quotients. We try the next-greater prime number, 3.

The LCM is $2 \cdot 2 \cdot 2 \cdot 2 \cdot 3 \cdot 5 = 480$.

---

4. Applications of the LCM

example 4   Using the LCM in an Application

A tile wall is to be made from 6-in., 8-in., and 12-in. square tiles. A design is made by alternating rows with different-size tiles. The first row uses only 6-in. tiles, the second row uses only 8-in. tiles, and the third row uses only 12-in. tiles. Neglecting the grout seams, what is the shortest length of wall space that can be covered using only whole tiles?

Skill Practice

Find the LCM by using division by prime factors.

7. 20, 36, and 15

8. 30, 22, and 45

Skill Practice

9. Three runners run on an oval track. One runner takes 60 sec to complete the loop. The second runner requires 75 sec, and the third runner requires 90 sec. Suppose the runners begin "lined up" at the same point on the track. Find the minimum amount of time required for all three runners to be lined up again.

Answers

7. 180   8. 990
9. After 900 sec (15 min) the runners will again be "lined up."

Solution:

The length of the first row must be a multiple of 6 in., the length of the second row must be a multiple of 8 in., and the length of the third row must be a multiple of 12 in. Therefore, the shortest-length wall that can be covered is given by the LCM of 6, 8, and 12.

$$6 = 2 \cdot 3$$
$$8 = 2 \cdot 2 \cdot 2$$
$$12 = 2 \cdot 2 \cdot 3$$

The LCM is $2 \cdot 2 \cdot 2 \cdot 3 = 24$. The shortest-length wall is 24 in.

This means that four 6-in. tiles can be placed on the first row, three 8-in. tiles can be placed on the second row, and two 12-in. tiles can be placed in the third row. See Figure 3-1.

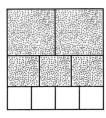

Figure 3-1

5. Ordering Fractions

The concept of the least common multiple is important when we compare the size of two fractions.

Comparing the fractions $\frac{3}{5}$ and $\frac{2}{5}$ is relatively easy because 3 parts out of 5 is clearly greater than 2 parts out of 5. Thus,

$$\frac{3}{5} > \frac{2}{5}$$

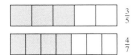

However, to compare the fractions $\frac{3}{5}$ and $\frac{4}{7}$ is more difficult. We have 3 pieces that are each one-fifth of a whole and 4 pieces that are each one-seventh of a whole.

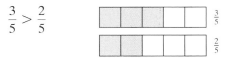

Because the pieces are of different sizes, it is difficult to make a comparison. For this reason, we want to express the fractions $\frac{3}{5}$ and $\frac{4}{7}$ as equivalent fractions with the same denominator, called a common denominator. The **least common denominator (LCD)** of two fractions is the LCM of the denominators of the fractions. Thus, the LCD of $\frac{3}{5}$ and $\frac{4}{7}$ is 35.

To convert $\frac{3}{5}$ and $\frac{4}{7}$ to equivalent fractions with a denominator of 35, we use the multiplication principle of fractions. This says that we can multiply the numerator and denominator of a fraction by the same nonzero number, and the value of the fraction remains the same.

Concept Connections

10. Shade the figures to determine which fraction represents a larger portion of a whole, $\frac{5}{6}$ or $\frac{7}{9}$.

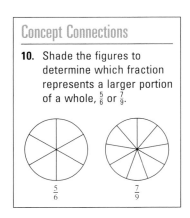

Answer

10.

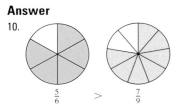

Tip: When we multiply the numerator and denominator of the fraction $\frac{3}{5}$, we are actually multiplying the fraction by 1. This is so because $\frac{7}{7} = 1$. Hence,

$$\frac{3}{5} = \frac{3}{5} \times 1 = \frac{3}{5} \times \frac{7}{7} = \frac{3 \times 7}{5 \times 7} = \frac{21}{35}$$

$$\frac{3}{5} = \frac{}{35}$$

What number must we multiply 5 by to get 35?

$$\frac{3 \times 7}{5 \times 7} = \frac{21}{35}$$

Multiply numerator and denominator by 7.

$$\frac{4}{7} = \frac{}{35}$$

What number must we multiply 7 by to get 35?

$$\frac{4 \times 5}{7 \times 5} = \frac{20}{35}$$

Multiply numerator and denominator by 5.

Comparing the fractions $\frac{3}{5}$ and $\frac{4}{7}$ is equivalent to comparing $\frac{21}{35}$ and $\frac{20}{35}$. Therefore, we have $\frac{21}{35} > \frac{20}{35}$, which means that $\frac{3}{5} > \frac{4}{7}$.

---

example 5   Comparing Two Fractions

Fill in the blank with $<$, $>$, or $=$.

$$\frac{9}{8} \ \square \ \frac{7}{6}$$

Solution:

The fractions have different denominators and cannot be compared by inspection. The LCD is 24. We need to convert each fraction to an equivalent fraction with a denominator of 24.

$$\frac{9}{8} = \frac{9 \times 3}{8 \times 3} = \frac{27}{24} \qquad \text{Multiply numerator and denominator by 3, because } 8 \times 3 = 24.$$

$$\frac{7}{6} = \frac{7 \times 4}{6 \times 4} = \frac{28}{24} \qquad \text{Multiply numerator and denominator by 4, because } 6 \times 4 = 24.$$

Because $\dfrac{27}{24} < \dfrac{28}{24}$, then $\dfrac{9}{8} \boxed{<} \dfrac{7}{6}$.

The relationship between $\frac{9}{8}$ and $\frac{7}{6}$ is shown on the number lines in Figure 3-2.

Skill Practice

Fill in the blank with $<$, $>$, or $=$.

11. $\dfrac{4}{7} \ \square \ \dfrac{5}{9}$

12. $\dfrac{1}{3} \ \square \ \dfrac{3}{8}$

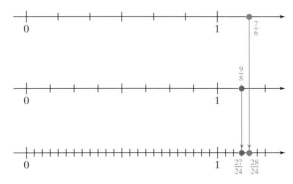

Figure 3-2

Answers

11. $>$
12. $<$

Skill Practice

13. Rank the fractions from least to greatest.

$$\frac{5}{9}, \frac{8}{15}, \text{ and } \frac{3}{5}$$

Concept Connections

14. Determine which fraction is greater by observation.

a. $\frac{3}{5}$ or $\frac{2}{5}$ **b.** $\frac{3}{5}$ or $\frac{3}{4}$

c. $\frac{5}{12}$ or $\frac{7}{12}$ **d.** $\frac{9}{14}$ or $\frac{9}{16}$

Answers

13. $\frac{8}{15}, \frac{5}{9}, \frac{3}{5}$

14. a. $\frac{3}{5}$ b. $\frac{3}{4}$ c. $\frac{7}{12}$ d. $\frac{9}{14}$

---

example 6   Ranking Fractions in Order from Least to Greatest

Rank the fractions from least to greatest.

$$\frac{9}{20}, \frac{7}{15}, \frac{4}{9}$$

Solution:

We want to convert each fraction to an equivalent fraction with a common denominator. The least common denominator is the LCM of 20, 15, and 9.

$$\left.\begin{array}{l} 20 = 2 \cdot 2 \cdot 5 \\ 15 = 3 \cdot 5 \\ 9 = 3 \cdot 3 \end{array}\right\}$$ The least common denominator is $2 \cdot 2 \cdot 3 \cdot 3 \cdot 5 = 180$.

Now convert each fraction to an equivalent fraction with a denominator of 180.

$$\frac{9}{20} = \frac{9 \times 9}{20 \times 9} = \frac{81}{180}$$ Multiply numerator and denominator by 9 because $20 \times 9 = 180$.

$$\frac{7}{15} = \frac{7 \times 12}{15 \times 12} = \frac{84}{180}$$ Multiply numerator and denominator by 12 because $15 \times 12 = 180$.

$$\frac{4}{9} = \frac{4 \times 20}{9 \times 20} = \frac{80}{180}$$ Multiply numerator and denominator by 20 because $9 \times 20 = 180$.

Ranking the fractions from least to greatest, we have $\frac{80}{180}, \frac{81}{180}, \frac{84}{180}$. This is equivalent to $\frac{4}{9}, \frac{9}{20}, \frac{7}{15}$.

---

section 3.2   Practice Exercises

Boost *your* GRADE at mathzone.com!

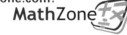

- Practice Problems
- Self-Tests
- NetTutor
- e-Professors
- Videos

Study Skills Exercises

1. Where do you usually do your homework? Is this the best place for you to concentrate? Explain.

2. Define the key terms.

    a. Multiple   **b. Least common multiple (LCM)**   **c. Least common denominator (LCD)**

Review Exercises

For Exercises 3–8, add and subtract as indicated. Write the answer as a whole number or fraction simplified to lowest terms.

3. $\dfrac{19}{6} - \dfrac{16}{6}$

4. $\dfrac{28}{4} - \dfrac{22}{4}$

5. $\dfrac{31}{15} + \dfrac{2}{15} - \dfrac{8}{15}$

6. $\dfrac{8}{5} + \dfrac{12}{5}$

7. $\dfrac{11}{3} + \dfrac{7}{3}$

8. $\dfrac{5}{19} - \dfrac{2}{19}$

Objective 1: Least Common Multiple

9. a. Circle the multiples of 24: 4, 8, 48, 72, 12, 240

   b. Circle the factors of 24: 4, 8, 48, 72, 12, 240

10. a. Circle the multiples of 30: 15, 90, 120, 3, 5, 60

   b. Circle the factors of 30: 15, 90, 120, 3, 5, 60

11. a. Circle the multiples of 36: 72, 6, 360, 12, 9, 108

   b. Circle the factors of 36: 72, 6, 360, 12, 9, 108

12. a. Circle the multiples of 28: 7, 4, 2, 56, 140, 280

   b. Circle the factors of 28: 7, 4, 2, 56, 140, 280

For Exercises 13–18, list five multiples of the given number. (Answers may vary.)

13. 5   **14.** 7   **15.** 14

16. 18   **17.** 16   **18.** 20

For Exercises 19–24, find the LCM by listing several multiples of each number. **(See Example 1.)**

19. 10 and 25   **20.** 21 and 14   **21.** 16 and 12

22. 20 and 12   **23.** 8, 10, and 12   **24.** 4, 6, and 14

Objective 2: Finding the LCM by Using Prime Factors

For Exercises 25–30, find the prime factorization.

25. 24   **26.** 42   **27.** 40

28. 80   **29.** 36   **30.** 64

For Exercises 31–40, find the LCM by using the prime factorization of each number. **(See Example 2.)**

31. 18 and 24   **32.** 9 and 30   **33.** 12 and 15   **34.** 27 and 45

35. 15 and 25   **36.** 16 and 24   **37.** 20, 18, and 27   **38.** 9, 15, and 42

39. 12, 15, and 20   **40.** 20, 30, and 40

Objective 3: Finding the LCM by Using Division by Primes

For Exercises 41–48, find the LCM by dividing by prime numbers. **(See Example 3.)**

41. 24 and 30

42. 14 and 35

43. 42 and 70

44. 6 and 21

45. 16, 24, and 30

46. 20, 42, and 35

47. 6, 12, 18, and 20

48. 21, 35, 50, and 75

Mixed Exercises

For Exercises 49–66, find the LCM by using any method.

49. 8 and 32

50. 7 and 14

51. 40 and 24

52. 21 and 28

53. 36 and 45

54. 48 and 36

55. 16 and 10

56. 6 and 15

57. 9 and 24

58. 32 and 40

59. 4, 6, and 14

60. 15, 30, and 45

61. 8, 18, and 20

62. 18, 15, and 20

63. 20, 30, and 40

64. 8, 10, and 12

65. 5, 15, 18, and 20

66. 28, 10, 21, and 35

Objective 4: Applications of the LCM

67. A tile floor is to be made from 10-in., 12-in., and 15-in. square tiles. A design is made by alternating rows with different-size tiles. The first row uses only 10-in. tiles, the second row uses only 12-in. tiles, and the third row uses only 15-in. tiles. Neglecting the grout seams, what is the shortest length of floor space that can be covered evenly by each row? **(See Example 4.)**

68. Four satellites revolve around the earth once every 6, 8, 10, and 15 hr, respectively. If the satellites are initially "lined up," how many hours must pass before they will again be lined up?

69. Mercury, Venus, and Earth revolve around the Sun approximately once every 3 months, 7 months, and 12 months, respectively (see the figure). If the planets begin "lined up," what is the minimum number of months required for them to be aligned again? (Assume that the planets lie roughly in the same plane.)

70. A bricklayer lays bricks of size 6, 8, and 10 in. She plans to lay one row of 6-in. bricks, a second row above of 8-in. bricks, and a third row of 10-in. bricks.

a. Neglecting the mortar seams, what is the shortest-length wall that can be covered?

b. Can a wall of length 12 ft (144 in.) be covered evenly, using this design?

Objective 5: Ordering Fractions

For Exercises 71–76, rewrite each fraction with the indicated denominators.

71. $\dfrac{2}{3} = \dfrac{}{21}$

72. $\dfrac{7}{4} = \dfrac{}{32}$

73. $\dfrac{5}{8} = \dfrac{}{16}$

74. $\dfrac{2}{9} = \dfrac{}{27}$

75. $\dfrac{3}{4} = \dfrac{}{16}$

76. $\dfrac{3}{10} = \dfrac{}{50}$

For Exercises 77–82, fill in the blanks with $<$, $>$, or $=$. **(See Example 5.)**

77. $\dfrac{7}{8} \square \dfrac{3}{4}$

78. $\dfrac{7}{15} \square \dfrac{11}{20}$

79. $\dfrac{13}{10} \square \dfrac{22}{15}$

80. $\dfrac{15}{4} \square \dfrac{21}{6}$

81. $\dfrac{3}{12} \square \dfrac{2}{8}$

82. $\dfrac{5}{20} \square \dfrac{4}{16}$

83. Which of the following fractions has the greatest value? $\dfrac{2}{3}, \dfrac{7}{8}, \dfrac{5}{6}, \dfrac{1}{2}$

84. Which of the following fractions has the least value? $\dfrac{1}{6}, \dfrac{1}{4}, \dfrac{2}{15}, \dfrac{2}{9}$

For Exercises 85–90, rank the fractions from least to greatest. **(See Example 6.)**

85. $\dfrac{7}{8}, \dfrac{2}{3}, \dfrac{3}{4}$

86. $\dfrac{5}{12}, \dfrac{3}{8}, \dfrac{2}{3}$

87. $\dfrac{5}{16}, \dfrac{3}{8}, \dfrac{1}{4}$

88. $\dfrac{2}{5}, \dfrac{3}{10}, \dfrac{5}{6}$

89. $\dfrac{4}{3}, \dfrac{13}{12}, \dfrac{17}{15}$

90. $\dfrac{5}{7}, \dfrac{11}{21}, \dfrac{18}{35}$

91. Susan buys $\frac{2}{3}$ lb of smoked turkey, $\frac{3}{5}$ lb of ham, and $\frac{5}{8}$ lb of roast beef. Which type of meat did she buy in the greatest amount? Which type did she buy in the least amount?

92. For a party, Aman had $\frac{3}{4}$ lb of cheddar cheese, $\frac{7}{8}$ lb of Swiss cheese, and $\frac{4}{5}$ lb of pepper jack cheese. Which type of cheese is in the least amount? Which type is in the greatest amount?

Expanding Your Skills

93. Which of the following fractions is between $\frac{1}{4}$ and $\frac{5}{6}$? Identify all that apply.

   a. $\frac{5}{12}$ **b.** $\frac{2}{3}$ **c.** $\frac{1}{8}$

94. Which of the following fractions is between $\frac{1}{3}$ and $\frac{11}{15}$? Identify all that apply.

   a. $\frac{2}{3}$ **b.** $\frac{4}{5}$ **c.** $\frac{2}{5}$

Objectives

1. Addition and Subtraction of Unlike Fractions
2. Order of Operations
3. Applications Involving Unlike Fractions

Skill Practice

Add.

1. $\frac{1}{10} + \frac{1}{15}$

2. $\frac{4}{7} + \frac{3}{14}$

Answers

1. $\frac{1}{6}$

2. $\frac{11}{14}$

section 3.3   Addition and Subtraction of Unlike Fractions

1. Addition and Subtraction of Unlike Fractions

In this section we use the concept of the LCD to help us add and subtract unlike fractions. The first step in adding or subtracting unlike fractions is to identify the LCD. Then we change the unlike fractions to like fractions having the LCD as the denominator.

example 1   Adding Unlike Fractions

Add.

$$\frac{1}{6} + \frac{3}{4}$$

Solution:

The LCD of $\frac{1}{6}$ and $\frac{3}{4}$ is 12. We can convert each individual fraction to an equivalent fraction with 12 as the denominator.

$$\frac{1}{6} = \frac{1 \cdot 2}{6 \cdot 2} = \frac{2}{12}$$ Multiply numerator and denominator by 2 because $6 \cdot 2 = 12$.

$$\frac{3}{4} = \frac{3 \cdot 3}{4 \cdot 3} = \frac{9}{12}$$ Multiply numerator and denominator by 3 because $4 \cdot 3 = 12$.

Thus, $\frac{1}{6} + \frac{3}{4}$ becomes $\frac{2}{12} + \frac{9}{12} = \frac{11}{12}$.

Tip: In Example 1, we multiplied the fraction $\frac{1}{6}$ by $\frac{2}{2}$. This is equivalent to multiplying $\frac{1}{6}$ by 1 and does not change the value.

$$\frac{1}{6} = \frac{1}{6} \cdot 1 = \frac{1}{6} \cdot \frac{2}{2} = \frac{2}{12}$$

The fraction $\frac{2}{12}$ is equivalent to $\frac{1}{6}$.

Adding fractions can be visualized by using a diagram. For example, the sum $\frac{1}{2} + \frac{1}{3}$ is illustrated in Figure 3-3.

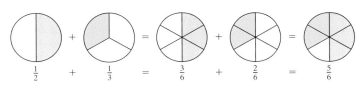

$$\frac{1}{2} \quad + \quad \frac{1}{3} \quad = \quad \frac{3}{6} \quad + \quad \frac{2}{6} \quad = \quad \frac{5}{6}$$

Figure 3-3

The general procedure to add or subtract unlike fractions is outlined as follows.

Steps to Add or Subtract Unlike Fractions

1. Identify the LCD.
2. Write each individual fraction as an equivalent fraction with the LCD.
3. Add or subtract the resulting fractions as indicated.
4. Simplify to lowest terms, if possible.

---

example 2 **Adding and Subtracting Unlike Fractions**

Add or subtract as indicated.

a. $\dfrac{3}{10} + \dfrac{1}{5}$ 　　　**b.** $\dfrac{17}{18} - \dfrac{1}{6}$ 　　　**c.** $\dfrac{2}{7} + \dfrac{4}{5} - \dfrac{1}{10}$

Solution:

a. $\dfrac{3}{10} + \dfrac{1}{5}$ 　　　The LCD is 10. We must convert $\frac{1}{5}$ to an equivalent fraction with 10 as the denominator

$= \dfrac{3}{10} + \dfrac{1 \cdot 2}{5 \cdot 2}$ 　　　Multiply numerator and denominator by 2 because $5 \cdot 2 = 10$.

$= \dfrac{3}{10} + \dfrac{2}{10}$ 　　　The fractions are now like.

$= \dfrac{3 + 2}{10}$ 　　　Add the like fractions.

$= \dfrac{5}{10}$

$= \dfrac{\overset{1}{\cancel{5}}}{\underset{2}{\cancel{10}}}$ 　　　Simplify to lowest terms.

$= \dfrac{1}{2}$

Avoiding Mistakes: Do not confuse addition of fractions with multiplication of fractions. In multiplication, we multiply denominators. In addition we do not add denominators. We get a common denominator and then add only the numerators.

Concept Connections

3. Use the figure to add the fractions.

$$\frac{1}{3} + \frac{2}{5}$$

Skill Practice

Add or subtract as indicated. Write the answer as a fraction.

4. $\dfrac{9}{5} + \dfrac{1}{10}$

5. $\dfrac{5}{21} - \dfrac{1}{7}$

6. $\dfrac{4}{15} - \dfrac{1}{10} + \dfrac{9}{20}$

Answers

3.

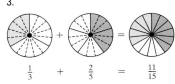

$$\frac{1}{3} \quad + \quad \frac{2}{5} \quad = \quad \frac{11}{15}$$

4. $\dfrac{19}{10}$ 　　5. $\dfrac{2}{21}$ 　　6. $\dfrac{37}{60}$

b. $\dfrac{17}{18} - \dfrac{1}{6}$ The LCD is 18.

$= \dfrac{17}{18} - \dfrac{1 \cdot 3}{6 \cdot 3}$ Multiply numerator and denominator by 3 because $6 \cdot 3 = 18$.

$= \dfrac{17}{18} - \dfrac{3}{18}$ The fractions are now like.

$= \dfrac{14}{18}$ Subtract like fractions.

$= \dfrac{\overset{7}{\cancel{14}}}{\underset{9}{\cancel{18}}}$ Simplify to lowest terms.

$= \dfrac{7}{9}$

c. $\dfrac{2}{7} + \dfrac{4}{5} - \dfrac{1}{10}$ First find the LCD.

$\left. \begin{array}{l} 7 = 7 \\ 5 = 5 \\ 10 = 2 \cdot 5 \end{array} \right\}$ The LCD is $2 \cdot 5 \cdot 7 = 70$.

$= \dfrac{2 \cdot 10}{7 \cdot 10} + \dfrac{4 \cdot 14}{5 \cdot 14} - \dfrac{1 \cdot 7}{10 \cdot 7}$ Convert each fraction to an equivalent fraction with the LCD.

$= \dfrac{20}{70} + \dfrac{56}{70} - \dfrac{7}{70}$ The fractions are like fractions.

$= \dfrac{20 + 56 - 7}{70}$ Add and subtract as indicated.

$= \dfrac{69}{70}$ The fraction is in lowest terms.

Sometimes when denominators are large, it is helpful to write the denominators as a product of prime factors. This is demonstrated in Example 3.

Skill Practice

Add or subtract as indicated.

7. $\dfrac{9}{8} - \dfrac{3}{32} - \dfrac{1}{20}$

8. $\dfrac{7}{18} + \dfrac{4}{15} - \dfrac{17}{30}$

example 3   Adding and Subtracting Unlike Fractions

Add or subtract as indicated.

$$\dfrac{7}{12} - \dfrac{2}{15} + \dfrac{5}{48}$$

Solution:

$\dfrac{7}{12} - \dfrac{2}{15} + \dfrac{5}{48}$ To find the LCD, factor each denominator.

$= \dfrac{7}{2 \cdot 2 \cdot 3} - \dfrac{2}{3 \cdot 5} + \dfrac{5}{2 \cdot 2 \cdot 2 \cdot 2 \cdot 3}$

$\left. \begin{array}{l} 12 = 2 \cdot 2 \cdot 3 \\ 15 = 3 \cdot 5 \\ 48 = 2 \cdot 2 \cdot 2 \cdot 2 \cdot 3 \end{array} \right\}$ The LCD is $2 \cdot 2 \cdot 2 \cdot 2 \cdot 3 \cdot 5 = 240$.

Answers

7. $\dfrac{157}{160}$ 8. $\dfrac{4}{45}$

We want to convert each fraction to an equivalent fraction having a denominator of $2 \cdot 2 \cdot 2 \cdot 2 \cdot 3 \cdot 5 = 240$. Multiply numerator and denominator of each original fraction by the factors missing from the denominator.

$$= \frac{7 \cdot (2 \cdot 2 \cdot 5)}{2 \cdot 2 \cdot 3 \cdot (2 \cdot 2 \cdot 5)} - \frac{2 \cdot (2 \cdot 2 \cdot 2 \cdot 2)}{3 \cdot 5 \cdot (2 \cdot 2 \cdot 2 \cdot 2)} + \frac{5 \cdot (5)}{2 \cdot 2 \cdot 2 \cdot 2 \cdot 3 \cdot (5)}$$

$$= \frac{140}{240} - \frac{32}{240} + \frac{25}{240} \qquad \text{The fractions are now like fractions.}$$

$$= \frac{140 - 32 + 25}{240} \qquad \text{Add and subtract as indicated.}$$

$$= \frac{133}{240} \qquad \text{The fraction is in lowest terms.}$$

2. Order of Operations

In Example 4 we must apply the order of operations to simplify the expressions.

example 4	Applying the Order of Operations

Simplify:

a. $\left(\frac{1}{4} + \frac{2}{3} \right)^2$ **b.** $\frac{5}{12} - \frac{1}{4} \div \frac{3}{2}$

Solution:

a. $\left(\frac{1}{4} + \frac{2}{3} \right)^2$ Perform the operation within parentheses first.

$$= \left(\frac{1 \cdot 3}{4 \cdot 3} + \frac{2 \cdot 4}{3 \cdot 4} \right)^2 \qquad \text{The common denominator is 12.}$$

$$= \left(\frac{3}{12} + \frac{8}{12} \right)^2 \qquad \text{The fractions within parentheses are now like fractions.}$$

$$= \left(\frac{11}{12} \right)^2 \qquad \text{Add fractions within parentheses.}$$

$$= \frac{11}{12} \cdot \frac{11}{12} \qquad \text{To square a number, multiply the number by itself.}$$

$$= \frac{121}{144} \qquad \text{The fraction is in lowest terms.}$$

b. $\frac{5}{12} - \frac{1}{4} \div \frac{3}{2}$ Perform the division before the subtraction.

$$= \frac{5}{12} - \frac{1}{\overset{}{\underset{2}{4}}} \cdot \frac{\overset{1}{2}}{3} \qquad \text{To divide fractions, multiply by the reciprocal of the divisor.}$$

Concept Connections

9. In what order would you perform the operations for this expression?

$$\left(\frac{2}{3} - \frac{1}{7} \right)^2$$

Skill Practice

Simplify.

10. $\left(\frac{2}{3} - \frac{1}{7} \right)^2$

11. $\frac{4}{15} \div \frac{2}{5} - \frac{1}{6}$

Answers

9. Subtract the fractions within parentheses. Then square the result.

10. $\dfrac{121}{441}$ 11. $\dfrac{1}{2}$

$$= \frac{5}{12} - \frac{1}{6} \qquad \text{The common denominator is 12.}$$

$$= \frac{5}{12} - \frac{1 \cdot 2}{6 \cdot 2} \qquad \text{Multiply numerator and denominator by 2 because } 6 \cdot 2 = 12.$$

$$= \frac{5}{12} - \frac{2}{12} \qquad \text{The fractions are now like fractions.}$$

$$= \frac{3}{12} \qquad \text{Subtract.}$$

$$= \frac{1}{4} \qquad \text{Simplify to lowest terms.}$$

3. Applications Involving Unlike Fractions

12. On Monday, $\frac{2}{3}$ in. of rain fell on a certain town. On Tuesday, $\frac{1}{5}$ in. of rain fell. How much rain fell during these 2 days?

example 5   Applying Operations on Unlike Fractions

A new Kelly Safari SUV tire has $\frac{7}{16}$-in. tread. After being driven 50,000 mi, the tread depth has worn down to $\frac{7}{32}$ in. By how much has the tread depth worn away?

Solution:

In this case we are looking for the difference in the tread depth.

$$\begin{pmatrix} \text{Difference in} \\ \text{tread depth} \end{pmatrix} = \begin{pmatrix} \text{original} \\ \text{tread depth} \end{pmatrix} - \begin{pmatrix} \text{final} \\ \text{tread depth} \end{pmatrix}$$

$$= \frac{7}{16} - \frac{7}{32} \qquad \text{The LCD is 32.}$$

$$= \frac{7 \cdot 2}{16 \cdot 2} - \frac{7}{32} \qquad \text{Multiply numerator and denominator by 2 because } 16 \cdot 2 = 32.$$

$$= \frac{14}{32} - \frac{7}{32} \qquad \text{The fractions are now like.}$$

$$= \frac{7}{32} \qquad \text{Subtract.}$$

The tire lost $\frac{7}{32}$ in. in tread depth after 50,000 mi of driving.

13. Maggie mixed $\frac{3}{4}$ gal of "winter wheat" paint with $\frac{1}{3}$ gal of "forest green." Then she used $\frac{3}{4}$ of the mixture to paint a wall.

    a. How much paint did she use?

    b. How much paint is left over?

example 6   Applying Operations on Unlike Fractions

An oil tank contains 2 liters (L) of oil. A slow leak has occurred, and oil leaks out at a rate of $\frac{1}{16}$ L per day. After 7 days a mechanic notices the leak and pours $\frac{3}{8}$ L of oil back into the tank. How much oil is now in the tank?

Solution:

To find the current amount in the tank, we must account for the amount lost and the amount added.

Answers

12. The total amount of rain was $\frac{13}{15}$ in.

13. **a.** $\frac{13}{16}$ gal was used.

    b. $\frac{13}{48}$ gal was left over.

The amount lost is given by the amount lost per day times 7 days.

$$\left(\frac{1}{16}\right)(7) = \frac{1}{16} \cdot \frac{7}{1} = \frac{7}{16}$$ The amount lost in 7 days is $\frac{7}{16}$ L.

Therefore, the current amount in the tank is given by

$$\begin{array}{l}\text{Current} \\ \text{amount}\end{array} = \begin{pmatrix}\text{original} \\ \text{amount}\end{pmatrix} - \begin{pmatrix}\text{amount} \\ \text{lost}\end{pmatrix} + \begin{pmatrix}\text{amount} \\ \text{replaced}\end{pmatrix}$$

$$= 2 - \frac{7}{16} + \frac{3}{8}$$

$$= \frac{2}{1} - \frac{7}{16} + \frac{3}{8} \qquad \text{Write the whole number as a fraction.}$$

$$= \frac{2 \cdot 16}{1 \cdot 16} - \frac{7}{16} + \frac{3 \cdot 2}{8 \cdot 2} \qquad \text{The LCD is 16.}$$

$$= \frac{32}{16} - \frac{7}{16} + \frac{6}{16} \qquad \text{The fractions are now like fractions.}$$

$$= \frac{32 - 7 + 6}{16} \qquad \text{Add and subtract as indicated.}$$

$$= \frac{31}{16} \qquad \text{The fraction is in lowest terms.}$$

The tank now contains $\frac{31}{16}$ L or equivalently $1\frac{15}{16}$ L.

---

example 7   **Finding Perimeter**

A parcel of land has the following dimensions. Find the perimeter.

Solution:

To find the perimeter, we add the lengths of the sides.

$$\frac{1}{8} + \frac{1}{4} + \frac{5}{12} + \frac{1}{3} \qquad \text{The LCD is 24.}$$

$$= \frac{1 \cdot 3}{8 \cdot 3} + \frac{1 \cdot 6}{4 \cdot 6} + \frac{5 \cdot 2}{12 \cdot 2} + \frac{1 \cdot 8}{3 \cdot 8} \qquad \begin{array}{l}\text{Convert the} \\ \text{fractions to like fractions.}\end{array}$$

$$= \frac{3}{24} + \frac{6}{24} + \frac{10}{24} + \frac{8}{24}$$

$$= \frac{27}{24} \qquad \text{Add the fractions.}$$

$$= \frac{9}{8} \qquad \text{Simplify to lowest terms.}$$

The perimeter is $\frac{9}{8}$ mi or equivalently $1\frac{1}{8}$ mi.

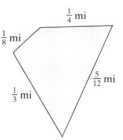

$\frac{1}{4}$ mi

$\frac{1}{8}$ mi

$\frac{5}{12}$ mi

$\frac{1}{3}$ mi

Skill Practice

14. Twelve members of a college hiking club hiked the perimeter of a canyon. How far did they hike?

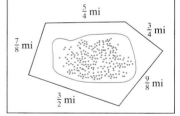

$\frac{5}{4}$ mi

$\frac{3}{4}$ mi

$\frac{7}{8}$ mi

$\frac{9}{8}$ mi

$\frac{3}{2}$ mi

Answer

14. They hiked $\frac{11}{2}$ mi. or equivalently $5\frac{1}{2}$ mi.

section 3.3   Practice Exercises

Study Skills Exercise

1. Do you need complete silence, or do you listen to music while you do your homework?

Try something different today so that you can compare and choose the best situation for you.

Review Exercises

For Exercises 2–13, rewrite the fraction with the given denominator.

2. $\dfrac{3}{5} = \dfrac{}{15}$

3. $\dfrac{6}{7} = \dfrac{}{14}$

4. $\dfrac{4}{9} = \dfrac{}{36}$

5. $\dfrac{2}{3} = \dfrac{}{21}$

6. $\dfrac{3}{1} = \dfrac{}{10}$

7. $\dfrac{5}{1} = \dfrac{}{5}$

8. $\dfrac{4}{1} = \dfrac{}{12}$

9. $\dfrac{2}{1} = \dfrac{}{4}$

10. $\dfrac{3}{4} = \dfrac{}{12}$

11. $\dfrac{4}{5} = \dfrac{}{100}$

12. $\dfrac{3}{2} = \dfrac{}{18}$

13. $\dfrac{1}{8} = \dfrac{}{40}$

Objective 1: Addition and Subtraction of Unlike Fractions

14. Explain the difference between the procedures to add fractions and to multiply fractions.

For Exercises 15–48, add or subtract. Write the answer as a fraction simplified to lowest terms. **(See Examples 1, 2, and 3.)**

15. $\dfrac{7}{8} + \dfrac{5}{16}$

16. $\dfrac{2}{9} + \dfrac{1}{18}$

17. $\dfrac{1}{10} + \dfrac{3}{20}$

18. $\dfrac{4}{15} + \dfrac{3}{5}$

19. $\dfrac{1}{4} + \dfrac{0}{3}$

20. $\dfrac{0}{5} + \dfrac{2}{3}$

21. $\dfrac{5}{6} + \dfrac{8}{7}$

22. $\dfrac{2}{11} + \dfrac{4}{5}$

23. $\dfrac{7}{8} - \dfrac{1}{2}$

24. $\dfrac{9}{10} - \dfrac{4}{5}$

25. $\dfrac{13}{12} - \dfrac{3}{4}$

26. $\dfrac{29}{30} - \dfrac{7}{10}$

27. $\dfrac{5}{2} - \dfrac{3}{5}$

28. $\dfrac{6}{5} - \dfrac{5}{6}$

29. $\dfrac{5}{8} - \dfrac{0}{11}$

30. $\dfrac{7}{12} - \dfrac{0}{5}$

31. $2 + \dfrac{9}{8}$

32. $3 + \dfrac{11}{9}$

33. $4 - \dfrac{4}{3}$

34. $2 - \dfrac{3}{8}$

35. $\dfrac{14}{3} + 1$

36. $\dfrac{12}{5} + 2$

37. $\dfrac{16}{7} - 2$

38. $\dfrac{15}{4} - 3$

39. $\dfrac{7}{10} + \dfrac{19}{100}$

40. $\dfrac{3}{10} + \dfrac{27}{100}$

41. $\dfrac{1}{10} - \dfrac{9}{100}$

42. $\dfrac{3}{100} - \dfrac{21}{1000}$

43. $\dfrac{3}{10} + \dfrac{9}{100} + \dfrac{1}{1000}$

44. $\dfrac{1}{10} + \dfrac{3}{100} + \dfrac{7}{1000}$

45. $\dfrac{5}{8} + \dfrac{3}{10} - \dfrac{1}{12}$

46. $\dfrac{7}{12} - \dfrac{2}{15} + \dfrac{5}{18}$

47. $\dfrac{1}{20} + \dfrac{5}{8} - \dfrac{7}{24}$

48. $\dfrac{5}{3} - \dfrac{7}{6} + \dfrac{5}{8}$

Objective 2: Order of Operations

For Exercises 49–62, simplify by applying the order of operations. Write the answer as a fraction. **(See Example 4.)**

49. $\dfrac{4}{5} + \dfrac{5}{8} \cdot \dfrac{16}{35}$

50. $\dfrac{1}{6} + \dfrac{3}{7} \cdot \dfrac{14}{15}$

51. $\dfrac{2}{3} \div \dfrac{1}{2} - \dfrac{3}{4}$

52. $\dfrac{3}{5} \div \dfrac{6}{7} - \dfrac{2}{5}$

53. $\left(\dfrac{11}{12} + \dfrac{1}{9} \right) \div \dfrac{7}{9}$

54. $\left(\dfrac{5}{6} - \dfrac{3}{8} \right) \div \dfrac{1}{4}$

55. $\left(\dfrac{7}{10} - \dfrac{1}{5} \right) \cdot \dfrac{8}{3}$

56. $\left(\dfrac{2}{5} + \dfrac{9}{10} \right) \cdot \dfrac{5}{6}$

57. $\left(\dfrac{1}{2} - \dfrac{1}{3} \right)^2$

58. $\left(\dfrac{2}{3} + \dfrac{1}{6} \right)^2$

59. $\left(\dfrac{2}{5} \right)^3 + \dfrac{1}{25}$

60. $\left(\dfrac{3}{2} \right)^3 - \dfrac{5}{4}$

61. $\left(\dfrac{1}{4} \right)^2 \div \left(\dfrac{5}{6} - \dfrac{2}{3} \right) + \dfrac{7}{12}$

62. $\left(\dfrac{1}{2} + \dfrac{1}{3} \right) \cdot \left(\dfrac{2}{5} \right)^2 + \dfrac{3}{10}$

Objective 3: Applications Involving Unlike Fractions

63. When doing her laundry, Inez added $\frac{3}{4}$ cup of bleach to $\frac{3}{8}$ cup of liquid detergent. How much total liquid is added to her wash?

64. What is the smallest possible length of screw needed to pass through two pieces of wood, one that is $\frac{7}{8}$ in. thick and one that is $\frac{1}{2}$ in. thick?

65. Mrs. Morgan has a bottle of eardrops that contains $\frac{2}{5}$ ounce (oz) of solution. If one dose is $\frac{1}{8}$ oz for each ear (that is, $\frac{1}{4}$ oz), is there enough solution in the bottle? If so, how much will be left over? **(See Example 5.)**

66. In one week it rained $\frac{5}{16}$ in. If a garden needs $\frac{9}{8}$ in. of water per week, how much more water does it need?

67. A contractor hired two electricians to do a job. One did $\frac{3}{5}$ of the job and the other did $\frac{3}{8}$ of the job. Did the job get completed? If not, what fraction of the job is left? **(See Example 6.)**

68. Mehule wants to take his laptop on a trip along with a manuscript that is $\frac{3}{8}$ in. thick. If the computer is $\frac{9}{4}$ in. deep, will the computer and manuscript fit into a case that is 3 in. deep? If so, what will be the clearance?

69. The information in the graph shows the distribution of a college student body by class.

    a. What fraction of the student body consists of upper classmen (juniors and seniors)?

    b. What fraction of the student body consists of freshmen and sophomores?

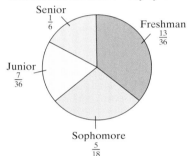

Distribution of Student Body by Class

70. A group of college students took part in a survey. One of the survey questions read:

"Do you think the government should spend more money on research to produce alternative forms of fuel?"

The results of the survey are shown in the figure.

    a. What fraction of the survey participants chose to strongly agree or agree?

    b. What fraction of the survey participants chose to strongly disagree or disagree?

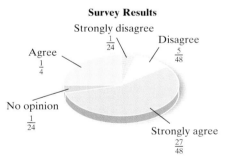

Survey Results

For Exercises 71–72, find the perimeter. **(See Example 7.)**

71.

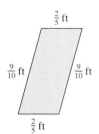

$\frac{2}{5}$ ft

$\frac{9}{10}$ ft $\frac{9}{10}$ ft

$\frac{2}{5}$ ft

72.

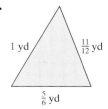

1 yd $\frac{11}{12}$ yd

$\frac{5}{6}$ yd

For Exercises 73–74, find the missing dimensions. Then calculate the perimeter.

73.

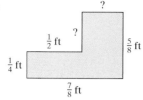

74.

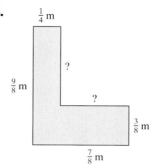

Expanding Your Skills

75. Which fraction is closest to $\dfrac{1}{2}$?

    a. $\dfrac{3}{4}$ **b.** $\dfrac{7}{10}$ **c.** $\dfrac{5}{6}$

76. Which fraction is closest to $\dfrac{3}{4}$?

    a. $\dfrac{5}{8}$ **b.** $\dfrac{7}{12}$ **c.** $\dfrac{5}{6}$

Calculator Connections

Topic: Adding and subtracting fractions

Calculator Exercises

For Exercises 77–80, the LCD for two fractions is given. Use a calculator to write each fraction as an equivalent fraction having the LCD as its denominator. Then add or subtract as indicated. Do not simplify the answer to lowest terms.

77. $\dfrac{37}{98} + \dfrac{59}{164}$; LCD = 8036

78. $\dfrac{11}{72} + \dfrac{33}{153}$; LCD = 1224

79. $\dfrac{89}{93} - \dfrac{19}{54}$; LCD = 1674

80. $\dfrac{21}{65} - \dfrac{3}{208}$; LCD = 1040

chapter 3 | midchapter review

1. Which operation on fractions requires that we obtain a common denominator? Select all that apply: addition, subtraction, multiplication, division.

2. Explain the procedure to subtract fractions with the same denominator.

3. Explain the procedure to add fractions with different denominators.

4. Explain the difference between the procedures to multiply and to divide fractions.

For Exercises 5–20, perform the indicated operation. Write the answer as a fraction simplified to lowest terms.

5. $\dfrac{2}{5} + \dfrac{7}{5}$ 6. $\dfrac{2}{5} \times \dfrac{7}{5}$ 7. $\dfrac{4}{3} \times \dfrac{1}{3}$

8. $\dfrac{4}{3} + \dfrac{1}{3}$ 9. $\dfrac{3}{4} - \dfrac{2}{7}$ 10. $\dfrac{3}{4} \times \dfrac{2}{7}$

11. $\dfrac{5}{3} \times \dfrac{10}{9}$ 12. $\dfrac{5}{3} - \dfrac{10}{9}$ 13. $\dfrac{5}{8} \div \dfrac{9}{8}$

14. $\dfrac{5}{8} + \dfrac{9}{8}$ 15. $\dfrac{5}{2} - \dfrac{1}{2}$ 16. $\dfrac{5}{2} \div \dfrac{1}{2}$

17. $\dfrac{13}{20} - \dfrac{9}{14}$ 18. $\dfrac{13}{20} \div \dfrac{9}{14}$ 19. $\dfrac{11}{12} \div \dfrac{5}{16}$

20. $\dfrac{11}{12} - \dfrac{5}{16}$

Objectives

1. Addition of Mixed Numbers
2. Subtraction of Mixed Numbers
3. Addition and Subtraction of Mixed Numbers by Using Improper Fractions
4. Applications of Mixed Numbers

Skill Practice

Add.

1. $1\dfrac{1}{5} + 9\dfrac{2}{5}$

2. $7\dfrac{2}{15} + 2\dfrac{1}{15}$

Answers

1. $10\dfrac{3}{5}$ 2. $9\dfrac{1}{5}$

---

section 3.4   Addition and Subtraction of Mixed Numbers

1. Addition of Mixed Numbers

In this section we learn to add and subtract mixed numbers. To find the sum of two or more mixed numbers, add the whole-number parts and add the fractional parts.

example 1   Adding Mixed Numbers

Add.

$$1\dfrac{5}{9} + 2\dfrac{1}{9}$$

Solution:

$$
\begin{array}{r}
1\dfrac{5}{9} \\
+\,2\dfrac{1}{9} \\
\hline
3\dfrac{6}{9}
\end{array}
$$

Add the whole numbers. ⟵⟶ Add the fractional parts.

The sum is $3\dfrac{6}{9}$ which simplifies to $3\dfrac{2}{3}$.

Tip: To understand why mixed numbers can be added in this way, recall that $1\dfrac{5}{9} = 1 + \dfrac{5}{9}$ and $2\dfrac{1}{9} = 2 + \dfrac{1}{9}$. Therefore,

$$1\dfrac{5}{9} + 2\dfrac{1}{9} = 1 + \dfrac{5}{9} + 2 + \dfrac{1}{9}$$

$$= 3 + \dfrac{6}{9}$$

$$= 3\dfrac{6}{9}$$

$$= 3\dfrac{2}{3}$$

When we perform operations on mixed numbers, it is often desirable to estimate the answer first. When rounding a mixed number, we offer the following convention.

1. If the fractional part of a mixed number is greater than or equal to $\frac{1}{2}$ (that is, if the numerator is one-half of the denominator or more), round to the next-greater whole number.

2. If the fractional part of the mixed number is less than $\frac{1}{2}$ (that is, if the numerator is less than one-half of the denominator), the mixed number rounds down to the whole number.

Concept Connections

Round to the nearest whole number.

3. $8\frac{6}{7}$ 4. $4\frac{2}{5}$

example 2   Adding Mixed Numbers

Estimate the sum and then find the actual sum.

$$42\frac{1}{12} + 17\frac{7}{8}$$

Solution:

To estimate the sum, we first round the addends.

$$42\frac{1}{12} \quad \text{rounds to} \quad 42$$

$$+ 17\frac{7}{8} \quad \text{rounds to} \quad \underline{+ 18}$$

$$\phantom{+ 17\frac{7}{8} \quad \text{rounds to} \quad} 60 \quad \text{The estimated value is 60.}$$

To find the actual sum, we must first write the fractional parts as like fractions. The LCD is 24.

$$42\frac{1}{12} = \quad 42\frac{1 \cdot 2}{12 \cdot 2} = \quad 42\frac{2}{24}$$

$$+ 17\frac{7}{8} = + 17\frac{7 \cdot 3}{8 \cdot 3} = + 17\frac{21}{24}$$

$$\phantom{+ 17\frac{7}{8} = + 17\frac{7 \cdot 3}{8 \cdot 3} = } 59\frac{23}{24}$$

The actual sum is $59\frac{23}{24}$. This is close to our estimate of 60.

Skill Practice

Estimate the sum and then find the actual sum.

5. $38\frac{2}{3} + 12\frac{1}{6}$

6. $6\frac{1}{11} + 3\frac{1}{2}$

example 3   Adding Mixed Numbers With Carrying

Estimate the sum and then find the actual sum.

$$7\frac{5}{6} + 3\frac{3}{5}$$

Solution:

$$7\frac{5}{6} \quad \text{rounds to} \quad 8$$

$$+ 3\frac{3}{5} \quad \text{rounds to} \quad \underline{+ 4}$$

$$\phantom{+ 3\frac{3}{5} \quad \text{rounds to} \quad} 12 \quad \text{The estimated value is 12.}$$

Concept Connections

7. Explain how you would rewrite $2\frac{9}{8}$ as a mixed number containing a proper fraction.

Answers

3. 9   4. 4
5. Estimate: 51; actual sum: $50\frac{5}{6}$
6. Estimate: 10; actual sum: $9\frac{13}{22}$
7. Write the improper fraction $\frac{9}{8}$ as $1\frac{1}{8}$, and add the result to 2. The result is $3\frac{1}{8}$.

To find the actual sum, we must first write the fractional parts as like fractions. The LCD is 30.

$$
\begin{aligned}
7\frac{5}{6} &= \ 7\frac{5 \cdot 5}{6 \cdot 5} = \ 7\frac{25}{30} \\
+\ 3\frac{3}{5} &= +\ 3\frac{3 \cdot 6}{5 \cdot 6} = +\ 3\frac{18}{30} \\
&\hphantom{=+3\frac{3\cdot6}{5\cdot6}==} 10\frac{43}{30}
\end{aligned}
$$

Notice that the number $\frac{43}{30}$ is an improper fraction. By convention, a mixed number is written as a whole number and a *proper* fraction. We have $\frac{43}{30} = 1\frac{13}{30}$. Therefore,

$$10\frac{43}{30} = 10 + 1\frac{13}{30} = 11\frac{13}{30}$$

The sum is $11\frac{13}{30}$. This is close to our estimate of 12.

---

2. Subtraction of Mixed Numbers

To subtract mixed numbers, we subtract the fractional parts and subtract the whole-number parts.

example 4   Subtracting Mixed Numbers

Subtract.

$$15\frac{2}{3} - 4\frac{1}{6}.$$

Solution:

To subtract the fractional parts, we need a common denominator. The LCD is 6.

$$
\begin{aligned}
15\frac{2}{3} &= \ 15\frac{2 \cdot 2}{3 \cdot 2} = \ 15\frac{4}{6} \\
-\ 4\frac{1}{6} &= -\ 4\frac{1}{6} \hphantom{\cdot 2} = -\ 4\frac{1}{6} \\
&\hphantom{=15\frac{2\cdot2}{3\cdot2}==} 11\frac{3}{6}
\end{aligned}
$$

Subtract the whole numbers. ⟶   ⟵ Subtract the fractional parts.

The difference is $11\frac{3}{6}$ which simplifies to $11\frac{1}{2}$.

Borrowing is sometimes necessary when subtracting mixed numbers. This occurs when the fractional part in the subtrahend is larger than the fractional part in the minuend.

example 5   Subtracting Mixed Numbers With Borrowing

Subtract.

$$4 - 2\frac{5}{8}$$

Solution:

$$\begin{array}{r} 4 \\ -\ 2\dfrac{5}{8} \\ \hline \end{array}$$ In this case, we have no fractional part from which to subtract.

$$\begin{array}{r} \overset{3}{4}\dfrac{8}{8} \\ -\ 2\dfrac{5}{8} \\ \hline 1\dfrac{3}{8} \end{array}$$ We can borrow 1 or equivalently $\frac{8}{8}$ from the whole number 4.

Tip: The borrowed 1 is written as $\frac{8}{8}$ because the common denominator is 8.

The difference is $1\frac{3}{8}$.

Skill Practice

12. Subtract.

$$10 - 3\dfrac{1}{6}$$

Tip: The subtraction problem $4 - 2\frac{5}{8} = 1\frac{3}{8}$ can be checked by adding:

$$1\dfrac{3}{8} + 2\dfrac{5}{8} = 3\dfrac{8}{8} = 3 + 1 = 4 \ \checkmark$$

---

example 6   Subtracting Mixed Numbers With Borrowing

Subtract.

a. $17\dfrac{2}{7} - 11\dfrac{5}{7}$ **b.** $14\dfrac{2}{9} - 9\dfrac{3}{5}$

Solution:

a. We cannot subtract $\frac{5}{7}$ from $\frac{2}{7}$. Therefore, borrow 1 from 17. The borrowed 1 is written as $\frac{7}{7}$ because the common denominator is 7.

$$\begin{array}{rcccc} 17\dfrac{2}{7} & = & \overset{16}{\cancel{17}}\dfrac{2}{7} + \dfrac{7}{7} = & 16\dfrac{9}{7} \\ -\ 11\dfrac{5}{7} & = & -\ 11\dfrac{5}{7} \qquad = & -\ 11\dfrac{5}{7} \\ \hline & & & 5\dfrac{4}{7} \end{array}$$

The difference is $5\frac{4}{7}$.

b. To subtract the fractional parts, we need a common denominator. The LCD is 45.

$$\left.\begin{array}{rcccc} 14\dfrac{2}{9} & = & 14\dfrac{2 \cdot 5}{9 \cdot 5} & = & 14\dfrac{10}{45} \\ -\ 9\dfrac{3}{5} & = & -\ 9\dfrac{3 \cdot 9}{5 \cdot 9} & = & -\ 9\dfrac{27}{45} \end{array}\right\}$$ We cannot subtract $\frac{27}{45}$ from $\frac{10}{45}$. Therefore, borrow 1 (or equivalently $\frac{45}{45}$) from 14.

$$\begin{array}{rcccc} & = & \overset{13}{\cancel{14}}\dfrac{10}{45} + \dfrac{45}{45} = & 13\dfrac{55}{45} \\ & = & -\ 9\dfrac{27}{45} \qquad = & -\ 9\dfrac{27}{45} \\ \hline & & & 4\dfrac{28}{45} \end{array}$$

The difference is $4\frac{28}{45}$.

Skill Practice

Subtract.

13. $24\dfrac{2}{7} - 8\dfrac{5}{6}$

14. $9\dfrac{2}{3} - 8\dfrac{3}{4}$

Answers

12. $6\dfrac{5}{6}$ **13.** $15\dfrac{19}{42}$ **14.** $\dfrac{11}{12}$

3. Addition and Subtraction of Mixed Numbers by Using Improper Fractions

We have shown how to add and subtract mixed numbers by writing the numbers in columns. Another approach to add or subtract mixed numbers is to write the numbers first as improper fractions. Then add or subtract the fractions, as you learned in Section 3.3. To demonstrate this process, we add the mixed numbers from Example 3.

example 7   Adding Mixed Numbers by Using Improper Fractions

Add.

$$7\frac{5}{6} + 3\frac{3}{5}$$

Solution:

$$7\frac{5}{6} + 3\frac{3}{5} = \frac{47}{6} + \frac{18}{5}$$ Write each mixed number as an improper fraction.

$$= \frac{47 \cdot 5}{6 \cdot 5} + \frac{18 \cdot 6}{5 \cdot 6}$$ Convert the fractions to like fractions. The LCD is 30.

$$= \frac{235}{30} + \frac{108}{30}$$ The fractions are now like fractions.

$$= \frac{343}{30}$$ Add the like fractions.

$$= 11\frac{13}{30}$$ Convert the improper fraction to a mixed number.

$$\begin{array}{r} 11 \\ 30\overline{)343} \qquad 11\frac{13}{30} \\ -30 \\ \hline 43 \\ -30 \\ \hline 13 \end{array}$$

The mixed number $11\frac{13}{30}$ is the same as the value obtained in Example 3.

As you can see from Example 7, when we convert mixed numbers to improper fractions, the numerators of the fractions become larger numbers. Thus, we must add (or subtract) larger numerators than if we had used the method involving columns. This is one drawback. However, an advantage to converting to improper fractions first is that there is no need for carrying or borrowing.

4. Applications of Mixed Numbers

Mixed numbers come up often in real-world applications.

example 8	Subtracting Mixed Numbers in an Application

The average height of a 3-year-old girl is $38\frac{1}{3}$ in. The average height of a 4-year-old girl is $41\frac{3}{4}$ in. On average, by how much does a girl grow between the ages of 3 and 4?

Solution:

We use subtraction to find the difference in heights.

$$41\frac{3}{4} = \quad 41\frac{3\cdot3}{4\cdot3} = \quad 41\frac{9}{12}$$
$$-\,38\frac{1}{3} = -\,38\frac{1\cdot4}{3\cdot4} = -\,38\frac{4}{12}$$
$$\overline{\qquad\qquad\qquad\qquad\qquad 3\frac{5}{12}}$$

The average amount of growth is $3\frac{5}{12}$ in.

Skill Practice

17. On December 1, the snow base at the Bear Mountain Ski Resort was $4\frac{1}{3}$ ft. By January 1, the base was $6\frac{1}{2}$ ft. By how much did the base amount of snow increase?

example 9	Adding Mixed Numbers in an Application

A cat "tree" is built from plywood and then covered in carpeting. What is the height of the model shown in Figure 3-4?

Solution:

The height is given by the sum of the heights of the individual parts. Also note that each mixed number (with the exception of $16\frac{1}{2}$) has a fractional part expressed in fourths. But $16\frac{1}{2} = 16\frac{2}{4}$. Therefore, the sum becomes

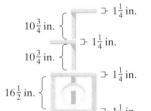

Figure 3-4

$$16\frac{2}{4}$$
$$10\frac{3}{4}$$
$$10\frac{3}{4}$$
$$1\frac{1}{4}$$
$$1\frac{1}{4}$$
$$1\frac{1}{4}$$
$$+\,1\frac{1}{4}$$
$$\overline{40\frac{12}{4}}$$ Because $\frac{12}{4} = 3$, then $40\frac{12}{4} = 40 + 3 = 43$.

The total height of the cat tree is 43 in.

Skill Practice

18. Mr. Barter runs 6 days a week. His daily mileage for one week is given in the table. How far did he run for the week?

Day	Mileage
Monday	$2\frac{1}{4}$
Tuesday	$4\frac{1}{2}$
Wednesday	0
Thursday	5
Friday	$6\frac{1}{3}$
Saturday	$8\frac{1}{3}$
Sunday	$3\frac{1}{2}$

Answers

17. $2\frac{1}{6}$ ft   18. $29\frac{11}{12}$ mi

section 3.4 Practice Exercises

Boost *your* GRADE at mathzone.com!

- Practice Problems
- Self-Tests
- NetTutor
- e-Professors
- Videos

Study Skills Exercise

1. a. Do you believe that you have math anxiety?   If yes, why do you think so?

b. Of the list below, circle the activities that you think can help someone with math anxiety.

Deep breathing   Reading a book about math anxiety

Scheduling extra study time   Keeping a positive attitude

Review Exercises

For Exercises 2–8, add or subtract as indicated. Write the answer as a fraction or whole number.

2. $\dfrac{3}{16} + \dfrac{7}{12}$

3. $\dfrac{25}{8} - \dfrac{23}{24}$

4. $\dfrac{9}{5} + 3$

5. $4 - \dfrac{15}{7}$

6. $\dfrac{23}{6} + \dfrac{5}{6} - \dfrac{2}{3}$

7. $\dfrac{125}{32} - \dfrac{51}{32} - \dfrac{58}{32}$

8. $\dfrac{17}{10} - \dfrac{23}{100} + \dfrac{321}{1000}$

Objective 1: Addition of Mixed Numbers

For Exercises 9–16, add the mixed numbers. **(See Example 1.)**

9. $\begin{aligned}&2\dfrac{1}{11}\\[4pt]+\;&5\dfrac{3}{11}\\\hline\end{aligned}$

10. $\begin{aligned}&5\dfrac{2}{7}\\[4pt]+\;&4\dfrac{3}{7}\\\hline\end{aligned}$

11. $\begin{aligned}&12\dfrac{1}{14}\\[4pt]+\;&3\dfrac{5}{14}\\\hline\end{aligned}$

12. $\begin{aligned}&1\dfrac{3}{20}\\[4pt]+\;&17\dfrac{7}{20}\\\hline\end{aligned}$

13. $\begin{aligned}&4\dfrac{5}{16}\\[4pt]+\;&11\dfrac{1}{4}\\\hline\end{aligned}$

14. $\begin{aligned}&21\dfrac{2}{9}\\[4pt]+\;&10\dfrac{1}{3}\\\hline\end{aligned}$

15. $\begin{aligned}&6\dfrac{2}{3}\\[4pt]+\;&4\dfrac{1}{5}\\\hline\end{aligned}$

16. $\begin{aligned}&7\dfrac{1}{6}\\[4pt]+\;&3\dfrac{5}{8}\\\hline\end{aligned}$

For Exercises 17–24, round the mixed number to the nearest whole number.

17. $5\dfrac{1}{3}$

18. $2\dfrac{7}{8}$

19. $1\dfrac{3}{5}$

20. $6\dfrac{3}{7}$

21. $14\dfrac{15}{16}$

22. $7\dfrac{5}{12}$

23. $21\dfrac{5}{9}$

24. $15\dfrac{8}{11}$

For Exercises 25–32, write the mixed number in proper form (that is, as a whole number along with a proper fraction that is simplified to lowest terms).

25. $2\dfrac{6}{5}$

26. $4\dfrac{8}{7}$

27. $7\dfrac{5}{3}$

28. $1\dfrac{9}{5}$

29. $10\dfrac{14}{12}$

30. $15\dfrac{12}{8}$

31. $5\dfrac{28}{21}$

32. $8\dfrac{20}{16}$

For Exercises 33–38, round the numbers to estimate the answer. Then find the exact sum. **(See Examples 2 and 3.)**

	Estimate	Exact
33.	7	$6\dfrac{3}{4}$
	$+\ 8$	$+\ 7\dfrac{3}{4}$
	$\overline{15}$	$\overline{}$

	Estimate	Exact
34.		$8\dfrac{3}{5}$
	$+$	$+\ 13\dfrac{4}{5}$
	$\overline{}$	$\overline{}$

	Estimate	Exact
35.		$14\dfrac{7}{8}$
	$+$	$+\ 8\dfrac{1}{4}$
	$\overline{}$	$\overline{}$

	Estimate	Exact
36.		$21\dfrac{3}{5}$
	$+$	$+\ 24\dfrac{9}{10}$
	$\overline{}$	$\overline{}$

	Estimate	Exact
37.		$3\dfrac{7}{16}$
	$+$	$+\ 15\dfrac{11}{12}$
	$\overline{}$	$\overline{}$

	Estimate	Exact
38.		$7\dfrac{7}{9}$
	$+$	$+\ 8\dfrac{5}{6}$
	$\overline{}$	$\overline{}$

For Exercises 39–42, add the mixed numbers to the whole numbers.

39. $3 + 6\dfrac{7}{8}$

40. $5 + 11\dfrac{1}{13}$

41. $32\dfrac{2}{7} + 10$

42. $2\dfrac{18}{37} + 16$

Objective 2: Subtraction of Mixed Numbers

For Exercises 43–50, subtract the mixed numbers. **(See Example 4.)**

43. $\begin{array}{r} 21\dfrac{9}{10} \\ -\ 10\dfrac{3}{10} \\ \hline \end{array}$

44. $\begin{array}{r} 19\dfrac{2}{3} \\ -\ 4\dfrac{1}{3} \\ \hline \end{array}$

45. $\begin{array}{r} 5\dfrac{9}{15} \\ -\ 3\dfrac{7}{15} \\ \hline \end{array}$

46. $\begin{array}{r} 33\dfrac{11}{12} \\ -\ 14\dfrac{5}{12} \\ \hline \end{array}$

47. $\begin{array}{r} 18\dfrac{5}{6} \\ -\ 6\dfrac{2}{3} \\ \hline \end{array}$

48. $\begin{array}{r} 21\dfrac{17}{20} \\ -\ 20\dfrac{1}{10} \\ \hline \end{array}$

49. $\begin{array}{r} 11\dfrac{5}{7} \\ -\ 9\dfrac{5}{14} \\ \hline \end{array}$

50. $\begin{array}{r} 5\dfrac{9}{11} \\ -\ 2\dfrac{13}{22} \\ \hline \end{array}$

51. Rewrite the number 1 as a fraction having the following denominators.

a. 3   **b.** 5   **c.** 12   **d.** 6

52. Rewrite the number 1 as a fraction having the following denominators.

 a. 20   **b.** 7   **c.** 10   **d.** 4

For Exercises 53–58, round the numbers to estimate the answer. Then find the exact difference. **(See Examples 5 and 6.)**

Estimate	Exact		Estimate	Exact		Estimate	Exact
53. 25	$25\frac{1}{4}$	**54.**		$36\frac{1}{5}$	▫ **55.**		$17\frac{1}{6}$
-14	$-13\frac{3}{4}$			$-12\frac{3}{5}$			$-15\frac{5}{12}$
$\overline{11}$	$\overline{}$		$\overline{}$	$\overline{}$		$\overline{}$	$\overline{}$

Estimate	Exact		Estimate	Exact		Estimate	Exact
56.	$22\frac{5}{18}$	**57.**		$46\frac{3}{7}$	**58.**		$23\frac{1}{2}$
	$-10\frac{7}{9}$			$-38\frac{1}{2}$			$-18\frac{10}{13}$
$\overline{}$	$\overline{}$		$\overline{}$	$\overline{}$		$\overline{}$	$\overline{}$

For Exercises 59–66, subtract the mixed numbers and whole numbers. Write the answers as mixed numbers or proper fractions.

▫ **59.** $6 - 2\frac{5}{6}$ **60.** $9 - 4\frac{1}{2}$ **61.** $12 - 9\frac{2}{9}$ **62.** $10 - 9\frac{1}{3}$

63. $5\frac{3}{17} - 3$ **64.** $16\frac{4}{11} - 5$ **65.** $23\frac{5}{14} - 17$ **66.** $21\frac{3}{4} - 10$

Objective 3: Addition and Subtraction of Mixed Numbers by Using Improper Fractions

For Exercises 67–82, add or subtract the mixed numbers by using improper fractions. Write the answer as mixed numbers, if possible. **(See Example 7.)**

67. $2\frac{2}{3} + 4\frac{5}{8}$ **68.** $5\frac{1}{4} - 3\frac{1}{2}$ **69.** $1\frac{11}{15} + 4\frac{2}{5}$ **70.** $2\frac{10}{11} + 2\frac{1}{2}$

71. $3\frac{7}{8} - 3\frac{3}{16}$ **72.** $3\frac{1}{6} - 1\frac{23}{24}$ **73.** $4\frac{1}{12} + 5\frac{1}{9}$ **74.** $10\frac{2}{25} - 7\frac{13}{20}$

75. $9\frac{5}{32} - 8\frac{1}{4}$ **76.** $4\frac{3}{40} - 2\frac{7}{8}$ **77.** $6\frac{11}{14} + 4\frac{1}{6}$ **78.** $8\frac{3}{22} + 4\frac{1}{4}$

79. $12\frac{1}{5} - 11\frac{2}{7}$ **80.** $5\frac{11}{30} + 5\frac{3}{4}$ **81.** $10\frac{1}{8} - 2\frac{17}{18}$ **82.** $3\frac{8}{21} + 6\frac{8}{9}$

Objective 4: Applications of Mixed Numbers

Bird		Length
Cuban Bee Hummingbird		$2\frac{1}{4}$ inches
Sedge Wren		$3\frac{1}{2}$ inches
Great Carolina Wren		$5\frac{1}{2}$ inches
Belted Kingfisher		$11\frac{1}{4}$ inches to 15 inches

For Exercises 83–85, use the chart to answer the questions.

83. If a Belted Kingfisher measures $11\frac{1}{4}$ inches in length, how much longer is the Belted Kingfisher than the Sedge Wren?

84. How much longer is the Great Carolina Wren than the Cuban Bee Hummingbird?

85. Estimate or measure the length of your index finger. Which is longer, your index finger or a Cuban Bee Hummingbird?

86. A student has 3 part-time jobs. She tutors, delivers newspapers, and takes notes for a blind student. During a typical week she works $8\frac{2}{3}$ hr delivering newspapers, $4\frac{1}{2}$ hr tutoring, and $3\frac{3}{4}$ hr note-taking. What is the total number of hours worked in a typical week?

87. A scouting group hikes around a lake. Find the total distance.

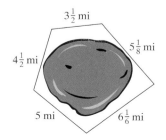

$3\frac{1}{2}$ mi

$5\frac{1}{8}$ mi

$4\frac{1}{2}$ mi

5 mi

$6\frac{1}{6}$ mi

88. Lara has a bike path on which she rides. One day she found a different path that she refers to as her shortcut. Find the perimeter of the original path and the perimeter of the shortcut. How much shorter is the shortcut?

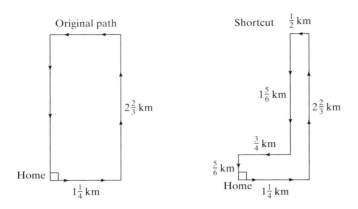

89. A water gauge in a pond measured $25\frac{7}{8}$ in. on Monday. After 2 days of rain and runoff, the gauge read $32\frac{1}{2}$ in. By how much did the water level rise? **(See Example 8.)**

90. A flight from Atlanta to San Diego takes $5\frac{1}{3}$ hr. After $2\frac{1}{2}$ hr, how much time remains?

91. In a triathlon, an athlete must swim $\frac{1}{4}$ mi, bike $10\frac{1}{2}$ mi, and run $3\frac{1}{5}$ mi. What is the total distance? **(See Example 9.)**

92. A contractor ordered three loads of gravel. The orders were for $2\frac{1}{2}$ tons, $3\frac{1}{8}$ tons, and $4\frac{1}{3}$ tons. What is the total amount of gravel ordered?

93. The number of hours worked per day for a plumber is given in the table. How many more hours did he work on Monday than on Saturday?

Monday	Tuesday	Wednesday	Thursday	Friday	Saturday	Sunday
$9\frac{1}{6}$ hr	$7\frac{3}{4}$ hr	$8\frac{1}{3}$ hr	$8\frac{1}{2}$ hr	$4\frac{1}{2}$ hr	$3\frac{3}{4}$ hr	0

94. Vertical blinds were purchased for a window that is $3\frac{5}{12}$ ft high. The blinds are $3\frac{3}{4}$ ft in length. Find the distance that the blinds will hang below the window.

Expanding Your Skills

For Exercises 95–98, fill in the blank to complete the pattern.

95. $1, 1\frac{1}{3}, 1\frac{2}{3}, 2, 2\frac{1}{3}, \square$

96. $\frac{1}{4}, 1, 1\frac{3}{4}, 2\frac{1}{2}, 3\frac{1}{4}, \square$

97. $\frac{5}{6}, 1\frac{1}{6}, 1\frac{1}{2}, 1\frac{5}{6}, \square$

98. $\frac{1}{2}, 1\frac{1}{4}, 2, 2\frac{3}{4}, 3\frac{1}{2}, \square$

Order of Operations and Applications of Fractions

1. Order of Operations

At this point in the text, we have learned how to add, subtract, multiply, and divide whole numbers, fractions, and mixed numbers. In this section we practice putting all these skills to use. We begin first by reviewing the order of operations.

Order of Operations

1. Perform all operations inside parentheses first.
2. Simplify expressions containing exponents or square roots.
3. Perform multiplication or division in the order that they appear from left to right.
4. Perform addition or subtraction in the order that they appear from left to right.

example 1   Applying the Order of Operations

Simplify.

a. $\left(3 - \dfrac{3}{4}\right)^2$ **b.** $1\dfrac{2}{3} \div 6 \cdot \left(\dfrac{3}{10}\right)$ **c.** $\left(\dfrac{2}{5}\right)^2 + \left(2\dfrac{5}{8}\right) \cdot \dfrac{3}{7}$

Solution:

a. $\left(3 - \dfrac{3}{4}\right)^2 = \left(\dfrac{3}{1} - \dfrac{3}{4}\right)^2$ We must first subtract the numbers within parentheses. Write the whole number as an improper fraction.

$= \left(\dfrac{3 \cdot 4}{1 \cdot 4} - \dfrac{3}{4}\right)^2$ Convert the fractions to like fractions. The LCD is 4.

$= \left(\dfrac{12}{4} - \dfrac{3}{4}\right)^2$ The fractions are now like.

$= \left(\dfrac{9}{4}\right)^2$ Subtract.

$= \dfrac{9}{4} \cdot \dfrac{9}{4}$ Square the quantity $\dfrac{9}{4}$.

$= \dfrac{81}{16}$ or $5\dfrac{1}{16}$

b. $1\dfrac{2}{3} \div 6 \cdot \left(\dfrac{3}{10}\right) = \dfrac{5}{3} \div \dfrac{6}{1} \cdot \dfrac{3}{10}$ Write the mixed number and whole number as improper fractions. In this expression, we must multiply and divide in order from left to right.

$= \dfrac{5}{3} \cdot \dfrac{1}{6} \cdot \dfrac{3}{10}$ Multiply by the reciprocal of the second fraction.

$= \dfrac{\overset{1}{\cancel{5}}}{\underset{1}{\cancel{3}}} \cdot \dfrac{1}{6} \cdot \dfrac{\overset{1}{\cancel{3}}}{\underset{2}{\cancel{10}}}$ Simplify.

$= \dfrac{1}{12}$ Multiply.

Objectives

1. Order of Operations
2. Applications of Fractions and Mixed Numbers
3. Applications to Geometry

Concept Connections

State the order in which you would perform the operations for the given expression.

1. $4 + \left(\dfrac{1}{3}\right)^2$

2. $2\dfrac{3}{7} + 6 \cdot \dfrac{1}{14}$

3. $\left(4 - 2\dfrac{2}{5}\right)^2 \cdot \dfrac{5}{7}$

Skill Practice

Simplify.

4. $4 + \left(\dfrac{1}{3}\right)^2$

5. $2\dfrac{3}{7} + 6 \cdot \dfrac{1}{14}$

6. $\left(4 - 2\dfrac{2}{5}\right)^2 \cdot \dfrac{5}{7}$

Answers

1. Square $\frac{1}{3}$ first. Then add the result to 4.
2. Multiply first. Then add the result to $2\frac{3}{7}$.
3. Subtract inside the parentheses first. Then square the result. Finally, multiply by $\frac{5}{7}$.
4. $\dfrac{37}{9}$ or $4\dfrac{1}{9}$ 5. $\dfrac{20}{7}$ or $2\dfrac{6}{7}$
6. $\dfrac{64}{35}$ or $1\dfrac{29}{35}$

c. $\left(\dfrac{2}{5}\right)^2 + \left(2\dfrac{5}{8}\right) \cdot \dfrac{3}{7}$ Perform the exponent operation first.

$= \dfrac{2}{5} \cdot \dfrac{2}{5} + \left(2\dfrac{5}{8}\right) \cdot \dfrac{3}{7}$ Square the quantity $\frac{2}{5}$.

$= \dfrac{4}{25} + \dfrac{21}{8} \cdot \dfrac{3}{7}$ Write the mixed number as an improper fraction.

$= \dfrac{4}{25} + \dfrac{\overset{3}{\cancel{21}}}{8} \cdot \dfrac{3}{\underset{1}{\cancel{7}}}$ Multiply before adding. Simplify common factors within the second two fractions. Do not try to "cancel" the 4 and the 8. The fraction $\frac{4}{25}$ is being *added*, not multiplied.

$= \dfrac{4}{25} + \dfrac{9}{8}$

$= \dfrac{4 \cdot 8}{25 \cdot 8} + \dfrac{9 \cdot 25}{8 \cdot 25}$ Add the fractions. The LCD is $25 \cdot 8 = 200$.

$= \dfrac{32}{200} + \dfrac{225}{200}$ The fractions are now like.

$= \dfrac{257}{200}$ or $1\dfrac{57}{200}$ The answer may be written as either an improper fraction or a mixed number.

Skill Practice

7. The graph gives the winning distance for men's and women's discus throw for selected Olympic games.

Winning Discus Throw Results for Selected Olympic Games

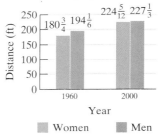

a. How much farther was the men's throw than the women's throw in 2000?
b. What was the average throw for women for these two years?

2. Applications of Fractions and Mixed Numbers

Examples 2 and 3 use operations on fractions and mixed numbers in real-world applications.

example 2   Using Mixed Numbers in a Sports Application

The graph in Figure 3-5 gives the winning height for the men's high jump for selected Olympic games.

a. What is the difference between the winning high jump in 1992 versus 1948?

b. What is the average height from these four Olympics?

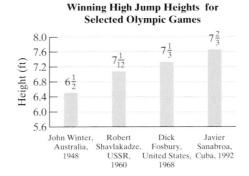

Figure 3-5

Answers

7. a. $2\dfrac{11}{12}$ ft   b. $202\dfrac{7}{12}$ ft

Solution:

a. The word "difference" implies subtraction. We subtract the 1948 height from the 1992 height.

$$7\frac{2}{3} = 7\frac{2 \cdot 2}{3 \cdot 2} = 7\frac{4}{6}$$
$$-6\frac{1}{2} = -6\frac{1 \cdot 3}{2 \cdot 3} = -6\frac{3}{6}$$
$$\overline{\hspace{4cm}1\frac{1}{6}}$$

The difference between the winning heights in 1992 and 1948 is $1\frac{1}{6}$ ft.

b. The average height is found by taking the sum of the four heights and dividing by 4. The sum of the heights is given by

$$\left.\begin{array}{rcl} 6\frac{1}{2} & = & 6\frac{6}{12} \\[2mm] 7\frac{1}{12} & = & 7\frac{1}{12} \\[2mm] 7\frac{1}{3} & = & 7\frac{4}{12} \\[2mm] +7\frac{2}{3} & = & +7\frac{8}{12} \end{array}\right\} \quad \text{The LCD of all four fractions is 12.}$$

$$27\frac{19}{12} = 27 + 1\frac{7}{12} = 28\frac{7}{12}$$

Now divide by 4 to find the average:

$$28\frac{7}{12} \div 4 = \frac{343}{12} \div \frac{4}{1} \qquad \text{Write the mixed number and whole number as improper fractions.}$$

$$= \frac{343}{12} \cdot \frac{1}{4} \qquad \text{Multiply by the reciprocal of the divisor.}$$

$$= \frac{343}{48} \qquad \text{Multiply.}$$

$$= 7\frac{7}{48} \qquad \text{Write the result as a mixed number.} \quad \begin{array}{r} 7 \\ 48\overline{)343} \\ -336 \\ \hline 7 \end{array}$$

The average winning height for the men's high jump for the selected years is $7\frac{7}{48}$ ft.

---

example 3   Using Multiplication of Fractions in a Finance Application

Sheila wants to buy a certain stock at $\$5\frac{3}{4}$ per share. How much money will it cost her to buy 300 shares?

Skill Practice

8. Sylvia paid $2820 for stock that cost $\$17\frac{5}{8}$ per share. How many shares did she buy?

Answer

8. Sylvia bought 160 shares.

Solution:

The value $5\frac{3}{4}$ gives the cost for 1 share of stock. For 300 shares we multiply.

$$\text{Total cost} = \left(5\frac{3}{4}\right)(300) = \left(\frac{23}{4}\right)\left(\frac{300}{1}\right)$$

$$= \left(\frac{23}{\underset{1}{4}}\right)\left(\frac{\overset{75}{300}}{1}\right)$$

$$= \frac{1725}{1}$$

The total cost for 300 shares of stock is $1725.

Tip: The answer can be estimated by rounding $5\frac{3}{4}$ to $6. Then ($6)(300) = $1800 which is close to $1725.

3. Applications to Geometry

In Examples 4 and 5, we review the concepts of perimeter and area in applications.

Skill Practice

9. Kent must put down mulch to cover the garden area shown in the figure. Each bag of mulch covers $8\frac{1}{2}$ yd^2 of ground. If Kent already has 2 bags, how many more bags does he need?

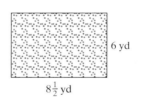

6 yd

$8\frac{1}{2}$ yd

example 4   Using Mixed Numbers in a Construction Application

Thom has $22\frac{1}{3}$ ft of molding left over from a previous carpentry job. He needs to apply molding around the base of the room shown in Figure 3-6. How much more molding does he need?

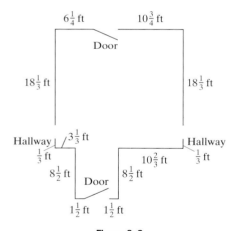

Figure 3-6

Solution:

To find the amount of molding still needed, we need to find the total required for the room and subtract $22\frac{1}{3}$ ft.

We find the total distance by adding the lengths of the segments of wall without including distances representing the hallways or doors.

Total length of wall
$$= 6\frac{1}{4} + 10\frac{3}{4} + 18\frac{1}{3} + \frac{1}{3} + 10\frac{2}{3} + 8\frac{1}{2} + 1\frac{1}{2} + 1\frac{1}{2} + 8\frac{1}{2} + 3\frac{1}{3} + \frac{1}{3} + 18\frac{1}{3}$$

Recall that mixed numbers represent addition of the whole-number part and the fractional part. Therefore, we can use the commutative and associative

Answer

9. Kent needs 4 more bags.

properties of addition to arrange the addends in an order convenient for computation.

$$= (6 + 10 + 18 + 10 + 8 + 1 + 1 + 8 + 3 + 18)$$

$$+ \left(\underbrace{\frac{1}{4} + \frac{3}{4}} + \underbrace{\frac{1}{3} + \frac{1}{3} + \frac{2}{3}} + \underbrace{\frac{1}{2} + \frac{1}{2} + \frac{1}{2} + \frac{1}{2}} + \underbrace{\frac{1}{3} + \frac{1}{3} + \frac{1}{3}} \right)$$

Add groups of like fractions.

$$= 83 + \left(\frac{4}{4} + \frac{4}{3} + \frac{4}{2} + \frac{3}{3} \right)$$

$$= 83 + \left(1 + 1\frac{1}{3} + 2 + 1 \right)$$

$$= 88\frac{1}{3}$$

There is $88\frac{1}{3}$ ft of molding required for this room. The carpenter already has $22\frac{1}{3}$ ft. The remaining amount is found by subtracting $22\frac{1}{3}$ ft from the total.

$$88\frac{1}{3} - 22\frac{1}{3} = 66$$

The carpenter still needs 66 ft of molding.

---

example 5   Using Fractions and Mixed Numbers in a Geometry Application

Jason and Sara plan to paint a side of their house (Figure 3-7).

  a. How much area will they have to paint?

  b. They want to string Christmas lights around the triangular portion of the house. What length is required for the string of lights?

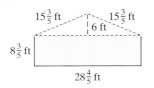

Figure 3-7

Solution:

  a. The area of the side of the house is given by the sum of the rectangular area and the triangular area.

$$\text{Area of the triangle} = \frac{1}{2}bh$$

$$= \frac{1}{2}\left(28\frac{4}{5} \right)(6)$$

$$= \frac{1}{2}\left(\frac{\overset{72}{\cancel{144}}}{5} \right)\left(\frac{6}{1} \right)$$

$$= \frac{432}{5} \text{ or } 86\frac{2}{5}$$

$$\text{Area of rectangle} = l \cdot w$$

$$= \left(28\frac{4}{5} \right)\left(8\frac{3}{5} \right)$$

$$= \frac{144}{5} \cdot \frac{43}{5}$$

$$= \frac{6192}{25} \text{ or } 247\frac{17}{25}$$

Skill Practice

10. A homeowner wants to sod the yard and fence the perimeter as shown in the figure.

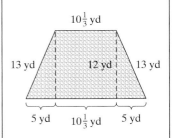

  a. How many square yards of sod are required?

  b. How much fencing is required?

Answers

10. **a.** 184 yd^2 of sod is needed.

   b. $56\frac{2}{3}$ yd of fencing is needed.

The total area is given by

$$86\frac{2}{5}\ \text{ft}^2 \longrightarrow 86\frac{10}{25}\ \text{ft}^2$$

$$+\ 247\frac{17}{25}\ \text{ft}^2 \longrightarrow +\ 247\frac{17}{25}\ \text{ft}^2$$

$$333\frac{27}{25}\ \text{ft}^2 \quad \text{or} \quad 334\frac{2}{25}\ \text{ft}^2$$

The total area is $334\frac{2}{25}$ ft².

b. The perimeter of the triangle is found by adding the lengths of the sides.

$15\frac{3}{5}$ ft $15\frac{3}{5}$ ft

$28\frac{4}{5}$ ft

$$15\frac{3}{5} + 15\frac{3}{5} + 28\frac{4}{5} = 58\frac{10}{5}$$

Because $\frac{10}{5} = 2$, we have $58\frac{10}{5} = 58 + \frac{10}{5} = 58 + 2 = 60$. Jason and Sara will require a string of lights 60 ft long.

section 3.5   Practice Exercises

Study Skills Exercise

1. When you take a test, go through the test, doing all the problems that you know first. Then go back and work on the problems that were more difficult. Give yourself a time limit for how much time you spend on each problem (maybe 3 to 5 minutes the first time through). Circle the importance of each statement.

	Not important	Somewhat important	Very important
a. Read through the entire test first.	1	2	3
b. If time allows, go back and check each problem.	1	2	3
c. Write out all steps instead of doing the work in your head.	1	2	3

Review Exercises

For Exercises 2–8, perform the indicated operation.

2. $7\dfrac{3}{10} + 2\dfrac{14}{15}$

3. $16 - 3\dfrac{7}{9}$

4. $5\dfrac{5}{8} \cdot 2\dfrac{1}{9}$

5. $7\dfrac{1}{9} \div 2\dfrac{2}{3}$

6. $24\dfrac{3}{5} - 14\dfrac{3}{4}$

7. $\left(1\dfrac{5}{6}\right)^2$

8. $13\dfrac{1}{14} + 4\dfrac{5}{7}$

For Exercises 9–12, convert the mixed number to an improper fraction.

9. $5\dfrac{2}{13}$

10. $2\dfrac{7}{11}$

11. $3\dfrac{9}{10}$

12. $1\dfrac{15}{16}$

For Exercises 13–16, convert the improper fraction to a mixed number.

13. $\dfrac{29}{5}$

14. $\dfrac{50}{7}$

15. $\dfrac{30}{19}$

16. $\dfrac{25}{8}$

Objective 1: Order of Operations

For Exercises 17–32, simplify, using the order of operations. Write the answer as a mixed number, if possible. **(See Example 1.)**

17. $1\dfrac{5}{6} \cdot 2\dfrac{1}{2} \div 1\dfrac{1}{4}$

18. $2\dfrac{1}{7} \div 1\dfrac{1}{3} \cdot \dfrac{7}{10}$

19. $6\dfrac{1}{6} + 2\dfrac{1}{3} \div 1\dfrac{3}{4}$

20. $8\dfrac{7}{9} + 2\dfrac{1}{6} \cdot 3\dfrac{1}{3}$

21. $6 - 5\dfrac{1}{7} \cdot \dfrac{1}{3}$

22. $11 - 6\dfrac{1}{3} \div 1\dfrac{1}{6}$

23. $\left(3\dfrac{1}{4} + 1\dfrac{5}{8}\right) \cdot 2\dfrac{2}{3}$

24. $\left(1\dfrac{3}{5} + 2\dfrac{4}{7}\right) \cdot 5\dfrac{5}{6}$

25. $\left(1\dfrac{1}{5}\right)^2 \cdot \left(1\dfrac{7}{9} - 1\dfrac{5}{12}\right)$

26. $\left(6\dfrac{3}{4} - 2\dfrac{1}{8}\right) \div \left(1\dfrac{1}{2}\right)^3$

27. $\left(1\dfrac{1}{3}\right)^3 \div \left(2\dfrac{7}{9} + 1\dfrac{2}{3}\right)$

28. $\left(2\dfrac{1}{2} + 1\dfrac{7}{8}\right) \cdot \left(1\dfrac{1}{7}\right)^2$

29. $\left(3\dfrac{1}{4} + \dfrac{3}{5}\right) \cdot \left(3\dfrac{1}{3} + \dfrac{5}{11}\right)$

30. $\left(4\dfrac{4}{15} - 3\dfrac{1}{5}\right) \cdot \left(1\dfrac{1}{3} + \dfrac{11}{12}\right)$

31. $\left(5 - 1\dfrac{7}{8}\right) \div \left(3 - \dfrac{13}{16}\right)$

32. $\left(4 + 2\dfrac{1}{9}\right) \div \left(2 - 1\dfrac{11}{36}\right)$

Objective 2: Applications of Fractions and Mixed Numbers

33. Acceleration on cars is often compared by the time it takes to go from 0 to 60 miles per hour (mph). The graph gives the times for four cars. **(See Example 2.)**

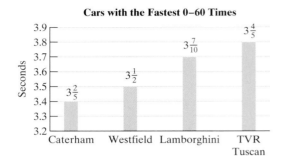

Cars with the Fastest 0–60 Times

a. What is the difference between the times for the Lamborghini and the Caterham?

b. Find the average time for these four cars.

34. Luis keeps a portfolio and tracks the number of hours he spends working on math each day.

a. How much time did Luis spend last week working on math?

b. Find the average amount of time spent per day working on math.

Time Spent Studying Math

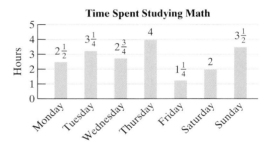

35. Six women go on a weight loss program for 4 weeks. The amount of weight lost for each woman is given in the graph.

a. Find the total weight loss.

b. Find the average weight loss per person.

Weight Loss

c. Find the difference between the maximum weight loss and the minimum weight loss.

36. Geoff ran $4\frac{1}{4}$ mi on Monday, $2\frac{1}{2}$ mi on Tuesday, 3 mi on Wednesday, and $8\frac{3}{4}$ mi on Thursday.

a. What is the total distance he ran?

b. What is the average distance he ran per day?

37. A financial analyst followed a certain stock. On Monday, the stock was at $15\frac{5}{8}$. By Friday, it had fallen to $11\frac{3}{4}$. By how much did the stock drop?

38. On Monday a stock closed at $12\frac{3}{4}$. On Tuesday, the stock rose $1\frac{1}{2}$. What was the closing price of the stock on Tuesday?

39. George will get $\frac{1}{3}$ of an $80,250 inheritance. How much money will he receive? **(See Example 3.)**

40. Aaron pays about $\frac{7}{25}$ of his annual salary in federal income tax. How much tax does he pay if his salary is $45,000 per year?

41. A $15\frac{1}{4}$-ft cable is cut into 4 pieces of equal length. How long is each piece?

42. Twenty-seven pounds of candy is distributed in $\frac{3}{4}$-lb bags. How many bags can be filled?

43. A cheese plate advertises a total of 3 lb of assorted cheeses. At the end of a party, there was $\frac{1}{4}$ lb of Swiss cheese, $\frac{1}{3}$ lb of cheddar, and $\frac{1}{6}$ lb of Jack cheese left over. How many pounds of cheese were eaten?

44. Meade gave $\frac{1}{4}$ of a candy bar to Max and then ate $\frac{1}{3}$ of the candy bar himself. What fraction of the candy bar is left?

45. A bread recipe calls for $3\frac{1}{4}$ cups of flour. If the Daily Bread bakery has large bags of flour containing 65 cups each, how many loaves of bread can be made?

46. A carpenter worked $37\frac{1}{4}$ hr last week and earned $894. What is his hourly rate?

47. If interest rates average $6\frac{1}{2}$ points and go up $\frac{3}{4}$ point, what is the new rate?

48. The annual consumption of tea in Hong Kong is $1\frac{9}{25}$ kg per capita. The per capita tea consumption in the United Kingdom is $\frac{21}{25}$ kg more than that in Hong Kong. What is the per capita amount of tea consumed in the United Kingdom?

49. Stephanie is planning to sew her bridesmaids' dresses. The pattern calls for $2\frac{1}{2}$ yd of material for one dress with an additional $1\frac{1}{4}$ yd for the matching jacket. If she has three bridesmaids, how many yards of material will she need to buy?

50. Grace travels $1\frac{1}{4}$ hr to work each day. If she works 5 days a week, how much time is spent traveling to and from work?

51. Wilma has $6\frac{1}{2}$ lb of mixed nuts. If she gives Fred one-half of the mixture and Barney one-third of the mixture, how many pounds of nuts does she have left?

52. Jeremy mowed $\frac{2}{3}$ of his front lawn in the morning and then $\frac{1}{4}$ of the lawn in the evening. What portion of the lawn still needs to be mowed?

53. Joan waters her plants each day with $22\frac{3}{4}$ gal of water. With a new irrigation system in place, she uses only $17\frac{2}{3}$ gal of water. How many gallons of water does she save in a 30-day period with the new irrigation system?

54. A school fund-raiser began a bake sale with $36\frac{1}{2}$ lb of cookies. At the end of the day $5\frac{3}{4}$ lb remained. How many pounds of cookies were sold?

Objective 3: Applications to Geometry

55. A decorator has $14\frac{2}{3}$ ft of wallpaper border. How much more does she need to place wallpaper border around the walls of the bathroom shown in the figure? (*Note:* No border is needed on the door.) **(See Example 4.)**

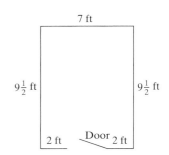

56. Leonie wants to put a border of shrubs around her garden, as seen in the figure. Find the distance that must be lined with shrubs.

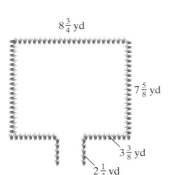

57. A stop sign is in the shape of an 8-sided polygon in which each of the sides is $12\frac{1}{2}$ in. Find the perimeter.

58. Matt needs to replace gutters on his townhouse. If the front and back of the townhouse are each $20\frac{1}{2}$ ft long and the side of the house is $35\frac{1}{3}$ ft long, how much gutter should he buy to go around the three sides of the house?

59. The shutters on the front of the house need to be painted. Determine the area of the four shutters.

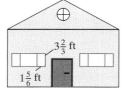

60. The cost of a new roof depends on the area of the roof. Find the area of the roof of the house in the figure.

$14\frac{1}{4}$ ft

$35\frac{7}{8}$ ft

61. Find the area of the triangular portion of the roof.

$8\frac{1}{2}$ ft

50 ft

62. Find the area of the calculator screen.

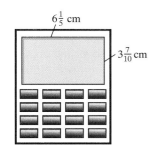

$6\frac{1}{5}$ cm

$3\frac{7}{10}$ cm

63. The Krajewskis want to improve the backyard by fertilizing the grass and putting up a decorative fence. To know how much fertilizer to use, they must know the area of the yard. **(See Example 5.)**

    a. Using the figure, determine the area of the yard.

    b. How many meters of fencing will they need?

16 m

$14\frac{3}{10}$ m

$13\frac{1}{5}$ m

$5\frac{1}{2}$ m

64. A rectangular patio is $20\frac{2}{3}$ ft by $10\frac{1}{2}$ ft. Find the area and perimeter.

Expanding Your Skills

65. Richard wants to paint his garage. Determine the area that needs to be painted. (He will paint the garage door, front, back, and sides of the garage, but not the roof.)

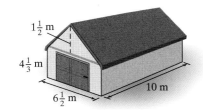

$1\frac{1}{2}$ m

$4\frac{1}{3}$ m

$6\frac{1}{2}$ m

10 m

chapter 3 | summary

section 3.1 Addition and Subtraction of Like Fractions

Key Concepts

Adding Like Fractions

1. Add the numerators.
2. Write the sum over the common denominator.
3. Simplify the fraction to lowest terms, if possible.

Subtracting Like Fractions

1. Subtract the numerators.
2. Write the difference over the common denominator.
3. Simplify the fraction to lowest terms, if possible.

When simplifying expressions with more than one operation, follow the order of operations.

Example 4 is an application involving subtraction of **like fractions**.

Examples

Example 1

$$\frac{5}{8} + \frac{7}{8} = \frac{12}{8} = \frac{\overset{1}{\cancel{2}} \cdot \overset{1}{\cancel{2}} \cdot 3}{\underset{1}{\cancel{2}} \cdot \underset{1}{\cancel{2}} \cdot 2} = \frac{3}{2}$$

Example 2

$$\frac{25}{10} - \frac{7}{10} = \frac{18}{10} = \frac{\overset{1}{\cancel{2}} \cdot 3 \cdot 3}{\underset{1}{\cancel{2}} \cdot 5} = \frac{9}{5}$$

Example 3

$$\left(\frac{5}{3}\right)^2 - \left(\frac{2}{9} + \frac{5}{9}\right) = \left(\frac{5}{3}\right)^2 - \left(\frac{7}{9}\right)$$

$$= \frac{25}{9} - \frac{7}{9}$$

$$= \frac{18}{9}$$

$$= 2$$

Example 4

A nail that is $\frac{13}{8}$ in. long is driven through a board that is $\frac{11}{8}$ in. thick. How much of the nail extends beyond the board?

$$\frac{13}{8} - \frac{11}{8} = \frac{2}{8} = \frac{1}{4}$$

The nail will extend $\frac{1}{4}$ in.

section 3.2 Least Common Multiple

Key Concepts	Examples

Key Concepts

The numbers obtained by multiplying a number by the whole numbers 1, 2, 3, and so on are called the **multiples** of the number.

The **least common multiple (LCM)** of two given numbers is the smallest whole number that is a multiple of each given number.

Using Prime Factors to Find the LCM of Two Numbers

1. Write each number as a product of prime factors.
2. The LCM is the product of unique prime factors from both numbers. If a factor is repeated within the factorization of either number, use that factor the maximum number of times it appears in either factorization.

To determine the order of two fractions with different denominators, rewrite each fraction as an equivalent fraction with a common denominator.

The **least common denominator (LCD)** of two fractions is the LCM of their denominators.

Examples

Example 1

The numbers 5, 10, 15, 20, 25, 30, 35, and 40 are several multiples of 5.

Example 2

Find the LCM of 8 and 10.
Some multiples of 8 are 8, 16, 24, 32, 40.
Some multiples of 10 are 10, 20, 30, 40.

40 is the least common multiple.

Example 3

Find the LCM for the numbers 24 and 16.

$24 = 2 \cdot 2 \cdot 2 \cdot 3$

$16 = 2 \cdot 2 \cdot 2 \cdot 2$

$\text{LCM} = 2 \cdot 2 \cdot 2 \cdot 2 \cdot 3 = 48$

Example 4

Given $\frac{5}{6} \,\square\, \frac{7}{8}$, fill in the blank with $<$, $>$, or $=$.

First write the fractions with a common denominator.

$6 = 2 \cdot 3$

$8 = 2 \cdot 2 \cdot 2$

LCD of $\frac{5}{6}$ and $\frac{7}{8}$ is $2 \cdot 2 \cdot 2 \cdot 3 = 24$

$$\frac{5}{6} = \frac{5 \cdot 4}{6 \cdot 4} = \frac{20}{24} \qquad \frac{7}{8} = \frac{7 \cdot 3}{8 \cdot 3} = \frac{21}{24}$$

Since $\frac{20}{24} < \frac{21}{24}$, then $\frac{5}{6} < \frac{7}{8}$.

section 3.3 Addition and Subtraction of Unlike Fractions

Key Concepts

To add or subtract unlike fractions, first we must write each fraction as an equivalent fraction with a common denominator.

<u>Steps to Add or Subtract Unlike Fractions</u>

1. Identify the LCD.
2. Write each individual fraction as an equivalent fraction with the LCD.
3. Add or subtract the resulting fractions as indicated.
4. Simplify to lowest terms, if possible.

Examples

Example 1

Add $\dfrac{3}{10} + \dfrac{4}{15}$.

LCD = 30: $\dfrac{3}{10} = \dfrac{3 \cdot 3}{10 \cdot 3} = \dfrac{9}{30}$

$\dfrac{4}{15} = \dfrac{4 \cdot 2}{15 \cdot 2} = \dfrac{8}{30}$

$\dfrac{3}{10} + \dfrac{4}{15} = \dfrac{9}{30} + \dfrac{8}{30} = \dfrac{17}{30}$

Example 2

Subtract $\dfrac{7}{15} - \dfrac{5}{21}$.

LCD = 105: $\dfrac{7}{15} = \dfrac{7 \cdot 7}{15 \cdot 7} = \dfrac{49}{105}$

$\dfrac{5}{21} = \dfrac{5 \cdot 5}{21 \cdot 5} = \dfrac{25}{105}$

$\dfrac{7}{15} - \dfrac{5}{21} = \dfrac{49}{105} - \dfrac{25}{105} = \dfrac{24}{105} = \dfrac{8}{35}$

When simplifying expressions with more than one operation, follow the order of operations.

Example 3

$$\frac{2}{5} \cdot \left(\frac{5}{2} - \frac{7}{3} \right) + \frac{13}{30} = \frac{2}{5} \cdot \left(\frac{15}{6} - \frac{14}{6} \right) + \frac{13}{30}$$

$$= \frac{2}{5} \cdot \left(\frac{1}{6} \right) + \frac{13}{30}$$

$$= \frac{2}{30} + \frac{13}{30}$$

$$= \frac{15}{30}$$

$$= \frac{1}{2}$$

section 3.4 Addition and Subtraction of Mixed Numbers

Key Concepts

Addition of Mixed Numbers

To find the sum of two or more mixed numbers, add the whole-number parts and add the fractional parts.

Subtraction of Mixed Numbers

To subtract mixed numbers, subtract the fractional parts and subtract the whole-number parts.

When the fractional part in the subtrahend is larger than the fractional part in the minuend, we borrow from the whole number.

We can also add or subtract mixed numbers by writing the numbers as improper fractions. Then add or subtract the fractions.

Example 6 illustrates an application involving addition and subtraction of mixed numbers.

Examples

Example 1

$$
\begin{aligned}
3\frac{5}{8} &= 3\frac{10}{16} \\
+\ 1\frac{1}{16} &= 1\frac{1}{16} \\
\hline
&\quad 4\frac{11}{16}
\end{aligned}
$$

Example 2

$$
\begin{aligned}
2\frac{9}{10} &= 2\frac{27}{30} \\
+\ 6\frac{5}{6} &= 6\frac{25}{30} \\
\hline
8\frac{52}{30} &= 8 + 1\frac{22}{30} \\
&= 9\frac{11}{15}
\end{aligned}
$$

Example 3

$$
\begin{aligned}
5\frac{3}{4} &= 5\frac{9}{12} \\
-\ 2\frac{2}{3} &= 2\frac{8}{12} \\
\hline
&\quad 3\frac{1}{12}
\end{aligned}
$$

Example 4

$$
\begin{aligned}
7\frac{1}{2} &= \overset{6}{7}\frac{\overset{5+10}{5}}{10} = 6\frac{15}{10} \\
-\ 3\frac{4}{5} &= 3\frac{8}{10} = 3\frac{8}{10} \\
\hline
&\qquad\qquad\qquad 3\frac{7}{10}
\end{aligned}
$$

Example 5

$$
4\frac{7}{8} + 2\frac{1}{16} - 3\frac{1}{4} = \frac{39}{8} + \frac{33}{16} - \frac{13}{4}
$$

$$
= \frac{78}{16} + \frac{33}{16} - \frac{52}{16} = \frac{59}{16} = 3\frac{11}{16}
$$

Example 6

A certain stock rose $5\frac{1}{2}$ points on Monday, dropped $2\frac{1}{4}$ points on Tuesday, and rose $3\frac{3}{4}$ points on Wednesday. What is the net change in stock for the 3 days?

$$
5\frac{1}{2} - 2\frac{1}{4} + 3\frac{3}{4} = \frac{11}{2} - \frac{9}{4} + \frac{15}{4}
$$

$$
= \frac{22}{4} - \frac{9}{4} + \frac{15}{4}
$$

$$
= \frac{28}{4}
$$

$$
= 7
$$

The stock rose 7 points.

section 3.5 Order of Operations and Applications of Fractions

Key Concepts

Order of Operations

1. Perform all operations inside parentheses first.
2. Simplify expressions containing exponents or square roots.
3. Perform multiplication or division in the order that they appear from left to right.
4. Perform addition or subtraction in the order that they appear from left to right.

Example 2 is an example of an application involving mixed numbers and fractions.

Examples

Example 1

$$4\frac{2}{9} + 2\frac{1}{12} \cdot 5\frac{1}{3} = 4\frac{2}{9} + \frac{25}{\overset{}{\underset{3}{12}}} \cdot \frac{\overset{4}{16}}{3}$$

$$= 4\frac{2}{9} + \frac{100}{9}$$

$$= \frac{38}{9} + \frac{100}{9}$$

$$= \frac{138}{9} = 15\frac{1}{3}$$

Example 2

Wallace has a budget of $750 for putting a curb around an area in his front yard. Curbing costs $3\frac{1}{2}$ per foot. To stay within his budget, how many feet of curbing can he afford?

Solution:

$$750 \div 3\frac{1}{2} = \frac{750}{1} \div \frac{7}{2}$$

$$= \frac{750}{1} \cdot \frac{2}{7}$$

$$= \frac{1500}{7} = 214\frac{2}{7}$$

Wallace can purchase up to $214\frac{2}{7}$ ft of curbing.

chapter 3 | review exercises

Section 3.1

For Exercises 1–4, add or subtract the like units.

1. 5 books + 3 books
2. 12 cm + 6 cm

3. 25 mi – 13 mi
4. 13 CDs – 2 CDs

5. Explain what is meant by the term *like fractions*.

6. Give an example of two like fractions and two unlike fractions. Answers may vary.

For Exercises 7–14, add or subtract the like fractions. Simplify the answer to lowest terms.

7. $\frac{5}{6} + \frac{4}{6}$

8. $\frac{4}{15} + \frac{6}{15}$

9. $\frac{5}{12} + \frac{1}{12}$

10. $\frac{2}{9} + \frac{7}{9}$

11. $\frac{15}{7} - \frac{6}{7}$

12. $\frac{14}{3} - \frac{10}{3}$

13. $\frac{21}{5} - \frac{6}{5}$

14. $\frac{12}{11} - \frac{9}{11}$

For Exercises 15–18, simplify the expression by using the order of operations.

15. $\dfrac{3}{8} \cdot \dfrac{3}{2} + \dfrac{3}{16}$

16. $\dfrac{4}{9} + \left(\dfrac{4}{3}\right)^2$

17. $\dfrac{21}{13} - \dfrac{5}{2} \div \dfrac{13}{4}$

18. $\left(\dfrac{7}{10} - \dfrac{2}{10}\right)^3 \cdot \dfrac{8}{7}$

19. Find the perimeter of the picture.

$\frac{13}{4}$ in.

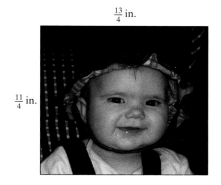

$\frac{11}{4}$ in.

20. Refer to the table that gives the average snowfall for selected cities.

	January	February	March
Spartanburg, SC	$\frac{25}{10}$ in.	$\frac{19}{10}$ in.	$\frac{12}{10}$ in.
Amarillo, TX	$\frac{39}{10}$ in.	$\frac{36}{10}$ in.	$\frac{28}{10}$ in.
Portland, OR	$\frac{33}{10}$ in.	$\frac{10}{10}$ in.	$\frac{4}{10}$ in.

a. What was the total snowfall for the months of January, February, and March for Spartanburg?

b. How much more snow did Amarillo get than Portland in the month of March?

Section 3.2

21. List four multiples for each number.

a. 7   **b.** 13   **c.** 22

22. Explain the difference between a common multiple and the *least* common multiple of a pair of numbers. Use the numbers 6 and 8 in your explanation.

23. List the factors for each number.

a. 100   **b.** 65   **c.** 70

24. Find the prime factorization.

a. 100   **b.** 65   **c.** 70

For Exercises 25–28, find the LCM by using any method.

25. 30 and 25

26. 22 and 144

27. 105 and 28

28. 16, 24, and 32

29. Sharon and Tonya signed up at a gym on the same day. Sharon will be able to go to the gym every third day and Tonya will go to the gym every fourth day. In how many days will they meet again at the gym?

For Exercises 30–31, rewrite each fraction with the indicated denominator.

30. $\dfrac{5}{16} = \dfrac{}{48}$

31. $\dfrac{9}{5} = \dfrac{}{35}$

For Exercises 32–34, fill in the blanks with $<$, $>$, or $=$.

32. $\dfrac{11}{24} \,\square\, \dfrac{7}{12}$

33. $\dfrac{5}{6} \,\square\, \dfrac{7}{9}$

34. $\dfrac{5}{6} \,\square\, \dfrac{15}{18}$

35. Rank the following numbers from least to greatest: $\dfrac{7}{10}, \dfrac{72}{105}, \dfrac{8}{15}, \dfrac{27}{35}$

Section 3.3

For Exercises 36–46, add or subtract. Write the answer as a fraction simplified to lowest terms.

36. $\dfrac{1}{8} + \dfrac{7}{12}$

37. $\dfrac{9}{10} - \dfrac{61}{100}$

38. $\dfrac{11}{25} - \dfrac{2}{5}$

39. $\dfrac{3}{26} + \dfrac{5}{13}$

40. $\dfrac{25}{11} + 2$

41. $4 - \dfrac{37}{20}$

42. $\dfrac{4}{15} - \dfrac{0}{3}$

43. $\dfrac{0}{17} + \dfrac{1}{34}$

44. $\dfrac{7}{100} - \dfrac{33}{1000}$

45. $\dfrac{2}{15} + \dfrac{5}{8} - \dfrac{1}{3}$

46. $\dfrac{11}{14} - \dfrac{4}{7} + \dfrac{3}{2}$

For Exercises 47–48, simplify by applying the order of operations.

47. $\left(\dfrac{2}{5} + \dfrac{1}{40}\right) \div \dfrac{15}{8} - \dfrac{4}{25}$

48. $\dfrac{20}{7} \cdot \left(\dfrac{11}{15} - \dfrac{1}{3}\right)^2 + \dfrac{1}{7}$

For Exercises 49–50, find (a) the perimeter and (b) the area.

49.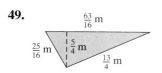

50.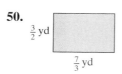

Section 3.4

For Exercises 51–62, add or subtract the mixed numbers.

51. $\begin{array}{r} 9\frac{8}{9} \\ + 1\frac{2}{7} \\ \hline \end{array}$

52. $\begin{array}{r} 10\frac{1}{2} \\ + 3\frac{15}{16} \\ \hline \end{array}$

53. $\begin{array}{r} 7\frac{5}{24} \\ - 4\frac{7}{12} \\ \hline \end{array}$

54. $\begin{array}{r} 5\frac{1}{6} \\ - 3\frac{1}{4} \\ \hline \end{array}$

55. $\begin{array}{r} 5\frac{3}{8} \\ - 2\frac{1}{3} \\ \hline \end{array}$

56. $\begin{array}{r} 3\frac{4}{5} \\ - 1\frac{4}{15} \\ \hline \end{array}$

57. $\begin{array}{r} 6\frac{4}{7} \\ + 5\frac{11}{14} \\ \hline \end{array}$

58. $\begin{array}{r} 3\frac{3}{8} \\ + 2\frac{13}{16} \\ \hline \end{array}$

59. $\begin{array}{r} 6 \\ - 2\frac{3}{5} \\ \hline \end{array}$

60. $\begin{array}{r} 8 \\ - 4\frac{11}{14} \\ \hline \end{array}$

61. $\begin{array}{r} 42\frac{1}{8} \\ + 21\frac{13}{16} \\ \hline \end{array}$

62. $\begin{array}{r} 38\frac{9}{10} \\ + 11\frac{3}{5} \\ \hline \end{array}$

For Exercises 63–66, round the numbers to estimate the answer. Then find the exact sum or difference.

63. $2\frac{1}{4} + 4\frac{2}{9} + 1\frac{29}{36}$

Estimate: _____

Exact: _____

64. $5\frac{2}{5} + 1\frac{9}{10} + 3\frac{19}{30}$

Estimate: _____

Exact: _____

65. $65\frac{1}{8} - 14\frac{9}{10}$

Estimate: _____

Exact: _____

66. $43\frac{13}{15} - 20\frac{23}{25}$

Estimate: _____

Exact: _____

67. Corry drove for $4\frac{1}{2}$ hr in the morning and $3\frac{2}{3}$ hr in the afternoon. Find the total number of hours he drove.

68. Denise owned $2\frac{1}{8}$ acres of land. If she sells $1\frac{1}{4}$ acres, how much will she have left?

Section 3.5

For Exercises 69–74, simplify by using the order of operations. Write the answer as a mixed number, if possible.

69. $1\frac{1}{5} + 4\frac{9}{10} \cdot 2\frac{2}{7}$

70. $5\frac{3}{4} - 23\frac{1}{2} \div 5\frac{2}{9}$

71. $\left(8\frac{1}{9} - 6\frac{2}{3}\right) \div 9\frac{3}{4}$

72. $\left(5\frac{1}{8} + 1\frac{1}{16}\right) \cdot 2\frac{10}{11}$

73. $\left(1\frac{1}{5}\right)^2 \cdot \left(4\frac{1}{2} + 3\frac{5}{6}\right)$

74. $\left(1\frac{5}{16}\right) \div \left(11\frac{1}{8} - 10\frac{3}{4}\right)^2$

75. In a certain region, the appraised value of a house is $\frac{9}{10}$ of its market value. If the market value of Owen's house is \$160,000, what is the appraised value?

76. Nuts 'N Things makes a nut mixture from $2\frac{1}{4}$ lb of cashews, $7\frac{3}{4}$ lb of peanuts, and $2\frac{1}{2}$ lb of pecans. The mixture is then divided into 10 bags. How many pounds are in each bag?

chapter 3 | test

For Exercises 1–2, add or subtract the like fractions.

1. $\dfrac{4}{5} + \dfrac{3}{5}$

2. $\dfrac{23}{16} - \dfrac{15}{16}$

3. Explain the difference between evaluating these two expressions:

$$\dfrac{5}{11} - \dfrac{3}{11} \quad \text{and} \quad \dfrac{5}{11} \times \dfrac{3}{11}$$

4. a. List four multiples of 24.

b. List all factors of 24.

c. Write the prime factorization of 24.

5. Find the LCM for the numbers 16, 24, and 30.

For Exercises 6–8, write each fraction with the indicated denominator.

6. $\dfrac{5}{9} = \dfrac{}{63}$

7. $\dfrac{11}{21} = \dfrac{}{63}$

8. $\dfrac{4}{7} = \dfrac{}{63}$

9. Rank the fractions in Exercises 6–8 from least to greatest.

For Exercises 10–13, add or subtract as indicated. Write the answer as a fraction.

10. $\dfrac{3}{8} + \dfrac{3}{16}$

11. $\dfrac{7}{3} - \dfrac{14}{27}$

12. $\dfrac{7}{12} - \dfrac{1}{4}$

13. $\dfrac{3}{5} + \dfrac{1}{15}$

For Exercises 14–17, add or subtract the mixed numbers. Write the answer as a mixed number.

14. $6\dfrac{3}{4} + 10\dfrac{5}{8}$

15. $12\dfrac{6}{11} - 9\dfrac{10}{11}$

16. $22\dfrac{1}{4} + 35\dfrac{1}{2} + 2\dfrac{2}{3}$

17. $15\dfrac{1}{6} - 12\dfrac{3}{8} - 1\dfrac{7}{24}$

For Exercises 18–21, perform the indicated operations.

18. $4\dfrac{5}{8} \cdot 2\dfrac{2}{3} - 8\dfrac{1}{6}$

19. $3\dfrac{1}{3} \div 2\dfrac{1}{2} + 5\dfrac{2}{3}$

20. $\left(\dfrac{2}{5}\right)^2 \div \left(1\dfrac{1}{10} + 2\dfrac{5}{6}\right)$ **21.** $\left(7\dfrac{1}{4} - 5\dfrac{1}{6}\right) \cdot 1\dfrac{3}{5}$

22. A fudge recipe calls for $1\dfrac{1}{2}$ lb of chocolate. How many pounds are required for $\dfrac{2}{3}$ of the recipe?

23. The towing capacity of the 2004 Ford Expedition is $4\dfrac{19}{40}$ times that of the 2004 Buick Rendezvous. If the Rendezvous can tow 1 ton (2000 lb), what is the towing capacity of the Expedition (in pounds)?

24. Find the area and perimeter of this parking area.

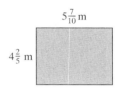

25. Justin has a budget of $14,000 to redecorate his kitchen. If he spends $\dfrac{3}{28}$ of the money on a stove and $\dfrac{1}{7}$ on a refrigerator, how much is left for new cabinets?

26. The figure gives the national records for long jump and high jump for men's and women's indoor track and field according to the International Association of Athletics Federation.

a. What is the difference for the record long jump between men and women?

b. What is the average height for high jump for men and women?

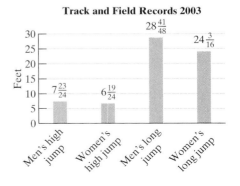

chapters 1–3 | cumulative review

1. Write the number in words: 23,400,806

2. Find the sum of 72 and 24.

3. Find the difference of 72 and 24.

4. Find the product of 72 and 24.

5. Find the quotient of 72 and 24.

6. Round the numbers to the ten-thousands place to estimate the product: $54{,}923 \times 28{,}543$.

7. Write the expression by using exponents:
   $4 \cdot 4 \cdot 5 \cdot 5 \cdot 5 \cdot 5 \cdot 8 \cdot 8$

8. $72 \div (4^2 - 10) \cdot 3$

9. List all the prime numbers between 15 and 35.

10. Write the prime factorization of 70.

11. Label the numerator and denominator of the fraction $\frac{21}{17}$.

12. What fraction is represented by the figure?

13. Kevin delivered 22 pizzas one evening. Of the 22 pizzas, 17 had pepperoni. What fraction of the pizzas had pepperoni? What fraction did not have pepperoni?

14. Label the fractions as proper or improper.

    a. $\frac{13}{5}$ b. $\frac{5}{13}$ c. $\frac{13}{13}$

15. Which of the numbers is divisible by 3 and 5?

    a. 2390   b. 1245   c. 9321

16. Label the numbers as prime, composite, or neither.

    a. 51   b. 52   c. 53

17. Find the prime factorization of 360.

18. Simplify the fraction to lowest terms: $\frac{180}{900}$

19. Multiply: $\frac{15}{16} \cdot \frac{2}{5}$

20. Divide: $\frac{20}{63} \div \frac{5}{9}$

21. Subtract: $\frac{13}{8} - \frac{7}{8}$

22. Add: $\frac{3}{16} + \frac{5}{16} + \frac{25}{16}$

23. Subtract: $4 - \frac{18}{5}$

24. Multiply: $6\frac{1}{10} \cdot 2\frac{3}{11}$

25. Divide: $2\frac{3}{5} \div 1\frac{7}{10}$

26. Simplify the expression.
    $$\left(8\frac{1}{4} \div 2\frac{3}{4}\right)^2 \cdot \frac{5}{18} + \frac{5}{6}$$

27. To approximate the distance around a circle, multiply the diameter by the fraction $\frac{22}{7}$. Find the distance around a circle with diameter 28 cm.

28. Find the perimeter of the triangle.

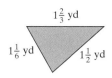

$1\frac{2}{3}$ yd

$1\frac{1}{6}$ yd $1\frac{1}{2}$ yd

29. Find the area of the triangle.

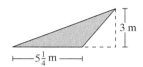

3 m

$5\frac{1}{4}$ m

30. The figure gives the earthquake intensity measured on the Richter scale for four major earthquakes in 2002 and 2003.

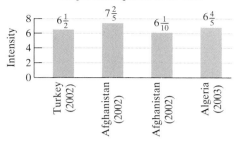

Major Earthquakes 2002–2003

Intensity

$6\frac{1}{2}$ $7\frac{2}{5}$ $6\frac{1}{10}$ $6\frac{4}{5}$

Turkey (2002)   Afghanistan (2002)   Afghanistan (2002)   Algeria (2003)

a. What was the difference in intensity between the two earthquakes that took place in Afghanistan?

b. What is the average intensity of these earthquakes?

Decimals

4

Chapter 4 is devoted to the study of decimal numbers. We begin with a discussion of place values and then perform addition, subtraction, multiplication, and division. The applications of decimal numbers are far-reaching. For example, in 1993, Becky and Keith Dilley were the proud parents of sextuplets (six children from a multiple birth). The children, now in their teen years, weighed just over 2 lb each at birth. See Exercise 56 of Section 4.2 to determine the total birth weight of the Dilley sextuplets.

chapter 4   preview

The exercises in this chapter preview contain concepts that have not yet been presented. These exercises are provided for students who want to compare their levels of understanding before and after studying the chapter. Alternatively, you may prefer to work these exercises when the chapter is completed and before taking the exam.

Section 4.1

1. Write the word name for the decimal 45.03.

2. Convert each decimal to a mixed number or fraction.

   a. 3.18   **b.** 0.005

3. Fill in the blanks with > or <.

   a. 4.05 ☐ 4.50   **b.** 6.731 ☐ 6.713

4. Round the numbers to the nearest hundredth.

   a. 190.371   **b.** 89.2967

Section 4.2

5. Add. $70.12 + 3.881$ **6.** Subtract. $5.4 - 3.281$

7. Find the ending balance of the checkbook register.

Check No.	Description	Credit	Debit	Balance
				$1166.29
3211	Water bill		$ 49.24	
3212	Mortgage		892.11	
	Transfer	$ 100.00		
	Paycheck	1021.33		
3213	Groceries		49.90	

Section 4.3

8. Multiply. 5.06×4.3

9. Multiply. 23.889×1000

10. Multiply. 72.91×0.01

11. Find the area of the pocket calculator shown.

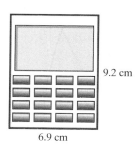

9.2 cm

6.9 cm

Section 4.4

12. Divide. $41.82 \div 6$ **13.** Divide. $3.74 \div 0.03$

14. Divide and round your answer to the nearest tenth. $2.1\overline{)5.413}$

Section 4.5

15. Complete the table.

	Decimal form	Fraction form
a.	0.55	
b.		$1\frac{8}{9}$
c.	$2.\overline{3}$	
d.		$4\frac{7}{20}$
e.		$\frac{3}{11}$

16. Determine what decimal number is represented by the point on the number line.

6.5   7.0

Section 4.6

17. Simplify. $0.5 \cdot (2.1)^2 + 1.105$

18. Simplify. $\left(1.2 \times \dfrac{1}{6}\right) + 3.5$

section 4.1   Decimal Notation and Rounding

Objectives

1. Decimal Notation
2. Writing Decimals as Mixed Numbers or Fractions
3. Ordering Decimal Numbers
4. Rounding Decimals

1. Decimal Notation

In Chapters 2 and 3, we studied fraction notation to denote equal parts of a whole. In this chapter we introduce decimal notation to denote equal parts of a whole. We first introduce the concept of a decimal fraction. A **decimal fraction** is a fraction whose denominator is a power of 10. The following are examples of decimal fractions.

$$\frac{3}{10} \text{ is read as "three-tenths"}$$

$$\frac{7}{100} \text{ is read as "seven-hundredths"}$$

$$\frac{9}{1000} \text{ is read as "nine-thousandths"}$$

Tip: Recall from Section 1.7 that powers of 10 are

$$10^1 = 10$$
$$10^2 = 100$$
$$10^3 = 1000$$

and so on.

We now want to write these fractions in **decimal notation**. This means that we will write the numbers by using place values, as we did with whole numbers. Therefore, we need to identify place positions for the tenths, hundredths, and thousandths place. The place value chart from Section 1.1 can be extended as shown in Figure 4-1.

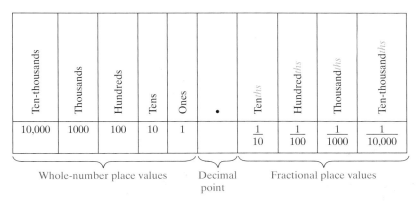

Figure 4-1

From Figure 4-1, we see that the decimal point separates the whole-number part from the fractional part. The place values for decimal fractions are located to the right of the decimal point. Their place value names are similar to those for whole numbers, but end in *ths*. Notice the correspondence between the tens place and the ten*ths* place. Similarly notice the hundreds place and the hundred*ths* place. Each place value on the left has a corresponding place value on the right, with the exception of the ones place. There is no "one*ths*" place.

example 1   Identify the Place Values

Identify the place value of each underlined digit.

   a. 30,804.0<u>9</u>   **b.** 0.8469<u>2</u>0   **c.** 2<u>9</u>3.604

Skill Practice

Identify the place value of each underlined digit.

   1. 24.38<u>6</u>   **2.** 1.0<u>5</u>724

   3. <u>1</u>316.42   **4.** 218.02168<u>4</u>

Answers

1. Thousandths   2. Hundredths
3. Hundreds   4. Millionths

Solution:

a. 30,804.0<u>9</u>   The digit 9 is in the hundredths place.

b. 0.8469<u>2</u>0   The digit 2 is in the hundred-thousandths place.

c. 2<u>9</u>3.604   The digit 9 is in the tens place.

For a whole number, the decimal point is understood to be after the ones place. However, it is usually not written, for example,

$$42. = 42$$

Using Figure 4-1, we can write the numbers $\frac{3}{10}$, $\frac{7}{100}$, and $\frac{9}{1000}$ in decimal notation.

Fraction	Word name	Decimal notation
$\frac{3}{10}$	Three-tenths	0.3
$\frac{7}{100}$	Seven-hundredths	0.07
$\frac{9}{1000}$	Nine-thousandths	0.009

tenths place

hundredths place

thousandths place

Tip: The 0 to the left of the decimal point is a placeholder so that the position of the decimal point can be easily identified. It does not contribute to the value of the number and can be omitted. Thus 0.3 and .3 are equal.

Now consider the number $15\frac{7}{10}$. This value represents 1 ten + 5 ones + 7 tenths. In decimal form we have 15.7.

15.7

1 ten ⎯⎯ ⎸⎸ ⎿⎯ 7 tenths

5 ones

The decimal point is interpreted as the word *and*. Thus, 15.7 is read as "fifteen *and* seven tenths." The number 356.29 can be represented as

$$356 + 2 \text{ tenths} + 9 \text{ hundredths} = 356 + \frac{2}{10} + \frac{9}{100}$$

$$= 356 + \frac{20}{100} + \frac{9}{100} \qquad \text{We can use the LCD of } 100 \text{ to add the fractions.}$$

$$= 356\frac{29}{100}$$

We can read the number 356.29 as "three hundred fifty-six and twenty-nine hundredths."

This discussion leads to a quicker method to read decimal numbers.

Concept Connections

5. Where is the decimal point located for the number 65?

6. Which is the correct decimal representation of $\frac{9}{10}$?

0.9   or   .9?

Reading a Decimal Number

1. The part of the number to the left of the decimal point is read as a whole number. *Note:* If there is no whole-number part, skip to step 3.

2. The decimal point is read *and*.

3. The part of the number to the right of the decimal point is read as a whole number but is followed by the name of the place position of the digit farthest to the right.

Answers

5. The decimal point is located after the ones-place digit, that is, 65.
6. Both are correct. The 0 in front of the decimal point is only for reference.

example 2   Reading Decimal Numbers

Write the word name for each number.

a. 1028.4   **b.** 2.0736   **c.** 0.478

Solution:

a. 1028.4   is written as "one thousand, twenty-eight and four-tenths."

b. 2.0736   is written as "two and seven hundred thirty-six ten-thousandths."

c. 0.478   is written as "four hundred seventy-eight thousandths."

2. Writing Decimals as Mixed Numbers or Fractions

A fractional part of a whole may be written as a fraction or as a decimal. To convert a decimal to an equivalent fraction, it is helpful to think of the decimal in words. For example:

Decimal	Word Name	Fraction	
0.3	Three tenths	$\frac{3}{10}$	
0.67	Sixty-seven hundredths	$\frac{67}{100}$	
0.048	Forty-eight thousandths	$\frac{48}{1000} = \frac{6}{125}$	(simplified)
6.8	Six and eight-tenths	$6\frac{8}{10} = 6\frac{4}{5}$	(simplified)

From the list, we notice several patterns that can be summarized as follows.

Converting a Decimal to a Mixed Number or Proper Fraction

1. The digits to the right of the decimal point are written as the numerator of the fraction.
2. The place value of the digit farthest to the right of the decimal point determines the denominator.
3. The whole-number part of the number is left unchanged.
4. Once the number is converted to a fraction or mixed number, simplify the fraction to lowest terms, if possible.

example 3   Writing Decimals as Proper Fractions or Mixed Numbers

Write the decimals as proper fractions or mixed numbers.

a. 0.847   **b.** 0.0025   **c.** 4.16

Solution:

a. $0.847 = \dfrac{847}{1000}$

thousandths place

b. $0.0025 = \dfrac{25}{10,000} = \dfrac{\overset{1}{\cancel{25}}}{\underset{400}{\cancel{10,000}}} = \dfrac{1}{400}$

ten-thousandths place

c. $4.16 = 4\dfrac{16}{100} = 4\dfrac{\overset{4}{\cancel{16}}}{\underset{25}{\cancel{100}}} = 4\dfrac{4}{25}$

hundredths place

Concept Connections

13. Which is a correct representation of 3.17?

$3\dfrac{17}{100}$ or $\dfrac{317}{100}$

A decimal number larger than 1 may be written as a mixed number or as an improper fraction. The number 4.16 from Example 3(c) may be expressed as follows.

$$4.16 = 4\dfrac{16}{100} = 4\dfrac{4}{25} \quad \text{or} \quad \dfrac{104}{25}$$

A quick way to obtain an improper fraction for a decimal number greater than 1 is outlined here.

Writing a Decimal Number Greater Than 1 as an Improper Fraction

1. The denominator is determined by the place position of the rightmost digit to the right of the decimal point.
2. The numerator is obtained by removing the decimal point of the original number. The resulting whole number is then written over the denominator.
3. Simplify the improper fraction to lowest terms, if possible.

For example,

Remove decimal point.

$$\overset{\frown}{4.16} = \dfrac{416}{100} = \dfrac{104}{25} \quad \text{(simplified)}$$

hundreds place

Skill Practice

Write the decimals as improper fractions.

14. 6.38

15. 15.1

example 4   Writing Decimals as Improper Fractions

Write the decimals as improper fractions.

a. 40.2   **b.** 2.113

Solution:

a. $40.2 = \dfrac{402}{10} = \dfrac{\overset{201}{\cancel{402}}}{\underset{5}{\cancel{10}}} = \dfrac{201}{5}$

b. $2.113 = \dfrac{2113}{1000}$ Note that the fraction is already in lowest terms.

Answers

13. They are both correct representations.

14. $\dfrac{319}{50}$ 15. $\dfrac{151}{10}$

3. Ordering Decimal Numbers

It is often necessary to compare the values of two decimal numbers. One way of doing this is to compare the numbers in fractional form. First note that adding 0 after the rightmost digit in a decimal number does not change its value. For example,

$$0.7 = 0.70 \quad \text{because} \quad \frac{7}{10} = \frac{70}{100}$$

---

example 5   **Comparing Decimal Numbers**

Write the numbers from least to greatest.

$$2.1, \quad 2.09, \quad 2.15$$

Solution:

First write each number with the same number of digits to the right of the decimal point.

2.10,   2.09,   2.15   We can now write each number as a decimal fraction with the same denominator. The value

$\dfrac{210}{100},$ $\dfrac{209}{100},$ $\dfrac{215}{100}$ 209 hundredths is smaller than 210 hundredths, which is smaller than 215 hundredths.

Writing the numbers from least to greatest, we have 2.09, 2.1, and 2.15.

A quicker way to compare two decimals is outlined next.

Comparing Two Positive Decimal Numbers

1. Starting at the left (and moving toward the right), compare the digits in each corresponding place position.
2. As we move from left to right, the first instance in which the digits differ determines the order of the numbers. The number having the greater digit is greater overall.

---

example 6   **Ordering Decimals**

Fill in the blank with $<$ or $>$.

a. 0.68 ☐ 0.7   **b.** 3.462 ☐ 3.4619

Solution:

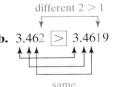

a. 0.68 $<$ 0.7   **b.** 3.462 $>$ 3.4619

Skill Practice

Write the numbers from least to greatest.

16. 3.7, 3.07, 3.69

17. 86.56, 86.559, 86.60

Skill Practice

Fill in the blank with $<$ or $>$.

18. 4.163 ☐ 4.159

19. 218.38 ☐ 218.41

Answers

16. 3.07, 3.69, 3.7
17. 86.559, 86.56, 86.60
18. $>$
19. $<$

4. Rounding Decimals

The process to round the decimal part of a number is nearly the same as rounding whole numbers (see Section 1.4). The main difference is that the digits to the right of the rounding place position are dropped instead of being replaced by zeros.

Rounding Decimals to a Place Value to the Right of the Decimal Point

1. Identify the digit one position to the right of the given place value.

2. If the digit in step 1 is 5 or greater, add 1 to the digit in the given place value. Then discard the digits to its right.

3. If the digit in step 1 is less than 5, discard it and any digits to its right.

Skill Practice

Round 187.26498 to the indicated place value.

20. Hundreds

21. Hundredths

22. Tenths

23. Thousandths

24. Ten-thousandths

example 7   Rounding Decimal Numbers

Round 14.795 to the indicated place value.

a. Tenths   **b.** Hundredths

Solution:

remaining digits discarded

a. $14.\overset{+1}{7}95 \approx 14.8$

tenths place — This digit is 5 or greater. Add 1 to the tenths place.

remaining digit discarded

b. $14.7\overset{+1}{9}5 \approx 14.80$

hundredths place — This digit is 5 or greater. Add 1 to the hundredths place.

In Example 7(b) the 0 in 14.80 indicates that the number was rounded to the hundredths place. It would be incorrect to drop the zero. Even though 14.8 has the same numerical value as 14.80, it implies a different level of accuracy. For example, when measurements are taken using some instrument such as a ruler or scale, the measured values are not exact. The place position to which a number is rounded reflects the accuracy of the measuring device. Thus, the value 14.8 lb indicates that the scale is accurate to the nearest tenth of a pound. The value 14.80 lb indicates that the scale is accurate to the nearest hundredth of a pound.

Concept Connections

25. Which number represents 46.4999 properly rounded to the nearest hundredth?

a. 46.5   **b.** 46.50

c. 46.500   **d.** 46.5000

e. All of these

example 8   Rounding Decimal Numbers

Round the number 4.02495 to the indicated place value.

a. Hundredths   **b.** Ten-thousandths

Solution:

remaining digits discarded

a. $4.02495 \approx 4.02$

hundredths place — This digit is less than 5. Discard it and all digits to the right.

Answers

20. 200   21. 187.26   22. 187.3
23. 187.265   24. 187.2650
25. **b**

remaining digit discarded

b. $4\ .\ 0\ 2\ 4\ \overset{+1}{9}\ \overset{\frown}{5} \approx 4.0250$

ten-thousandths place ——— This digit is 5 or greater. Add 1 to the ten-thousandths place.

■

One of the most common uses of decimal notation is to represent units of money. In the U.S. system of currency, one dollar ($1) equals one hundred cents (100¢). That is, 1 cent is equivalent to $\frac{1}{100}$ of a dollar. Therefore, when we are representing monetary amounts, the digits to the left of the decimal point represent whole dollars. The digits to the right represent the number of hundredths of a dollar (or the number of cents).

Thus, $36   represents 36 dollars

whereas $0.36   represents $\frac{36}{100}$ dollars or equivalently 36 cents (36¢)

Great Summer Savings!
Pima cotton trail shirt $24.50

example 9   Writing a Word Name for Monetary Values

Write a word name for $1094.45.

Solution:

The value $1094.45 represents one thousand, ninety-four and forty-five hundredths dollars.

This is equivalent to one thousand, ninety-four dollars and forty-five cents.

■

Answers

26. Two hundred sixteen and fifty-four hundredths dollars. This is equivalent to two hundred sixteen dollars and fifty-four cents.
27. One hundred six and seven hundredths dollars. This is equivalent to one hundred six dollars and seven cents.

section 4.1   **Practice Exercises**

Boost *your* **GRADE** at
mathzone.com!

MathZone

• Practice Problems   • e-Professors
• Self-Tests   • Videos
• NetTutor

Study Skills Exercises

1. After you get a test back, it is a good idea to correct the test so that you do not make the same errors again. One recommended approach is to use a clean sheet of paper and divide the paper down the middle vertically, as shown. For each problem that you missed on the test, rework the problem correctly on the left-hand side of the paper. Then write a written explanation on the right-hand side of the paper.

Take the time this week to make corrections from your last test.

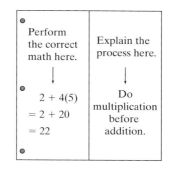

2. Define the key terms.

    a. Decimal fraction   **b. Decimal notation**

Objective 1: Decimal Notation

For Exercises 3–6, expand the powers of 10.

3. 10^2 **4.** 10^3 **5.** 10^4 **6.** 10^5

For Exercises 7–10, expand the powers of $\frac{1}{10}$.

7. $\left(\dfrac{1}{10}\right)^2$ **8.** $\left(\dfrac{1}{10}\right)^3$ **9.** $\left(\dfrac{1}{10}\right)^4$ **10.** $\left(\dfrac{1}{10}\right)^5$

For Exercises 11–21, identify the place value of each underlined digit. **(See Example 1.)**

11. 3.9̲83   **12.** 34.82̲   **13.** 440.39̲   **14.** 2̲48.94

15. 48̲9.02   **16.** 4.0928̲4   **17.** 9.283̲45   **18.** 0.321̲

19. 0.489̲   **20.** 58̲.211   **21.** 93̲.834

For Exercises 22–27, write the word name for each decimal fraction.

22. $\dfrac{9}{10}$ **23.** $\dfrac{7}{10}$ **24.** $\dfrac{23}{100}$ **25.** $\dfrac{19}{100}$

26. $\dfrac{33}{1000}$ **27.** $\dfrac{51}{1000}$

For Exercises 28–37, write the word name for the decimal. **(See Example 2.)**

28. 3.24   **29.** 4.26   **30.** 5.9   **31.** 3.4

32. 52.3   **33.** 21.5   **34.** 6.219   **35.** 7.338

36. 0.2419   **37.** 1.0201

Objective 2: Writing Decimals as Mixed Numbers or Fractions

For Exercises 38–47, write the decimal as a proper fraction or as a mixed number. **(See Example 3.)**

38. 3.7   **39.** 1.9   **40.** 2.8   **41.** 4.2

42. 0.25   **43.** 0.75   **44.** 0.55   **45.** 0.45

46. 20.812   **47.** 32.905

For Exercises 48–55, write the decimal as an improper fraction. **(See Example 4.)**

48. 8.4   **49.** 2.5   **50.** 3.14   **51.** 5.65

52. 23.5   **53.** 14.6   **54.** 11.91   **55.** 21.33

Objective 3: Ordering Decimal Numbers

For Exercises 56–63, fill in the blank with < or >. **(See Example 6.)**

56. 6.312 ☐ 6.321   **57.** 8.503 ☐ 8.530   **58.** 11.21 ☐ 11.2099   **59.** 10.51 ☐ 10.5098

60. 0.762 ☐ 0.76   **61.** 0.1291 ☐ 0.129   **62.** 51.72 ☐ 51.721   **63.** 49.06 ☐ 49.062

64. Which number is between 3.12 and 3.13? Circle all that apply.

     a. 3.127   **b.** 3.129   **c.** 3.134   **d.** 3.139

65. Which number is between 42.73 and 42.86? Circle all that apply.

     a. 42.81   **b.** 42.64   **c.** 42.79   **d.** 42.85

For Exercises 66–69, arrange the numbers from least to greatest. **(See Example 5.)**

66. 34.25, 34.2, 34.3, 34.29

67. 12.46, 12.4, 12.5, 12.49

68. 0.42, 0.043, $\frac{4}{10}$, 0.042, 0.43

69. 0.04999, 0.0499, $\frac{5}{10}$, 0.4999, 0.05001

70. The batting averages for four members of the Atlanta Braves as of May 2004 are given in the table. Rank the players' batting averages from lowest to highest.

Player	Average
J. Estrada	0.342
M. Giles	0.339
D. Hollis	0.364
N. Green	0.333

71. The average speed, in miles per hour (mph), of the Daytona 500 for selected years is given in the table. Rank the speeds from slowest to fastest.

Year	Driver	Speed (mph)
1986	Jeff Bodine	148.124
1989	Darrell Waltrip	148.466
1991	Ernie Irvan	148.148
1997	Jeff Gordon	148.295

Objective 4: Rounding Decimals

72. The numbers given all have equivalent value. However, suppose they represent measured values from a scale. Explain the difference in the interpretation of these numbers.

$$0.25, \quad 0.250, \quad 0.2500, \quad 0.25000$$

For Exercises 73–81, round the decimal to the indicated place value. **(See Examples 7–8.)**

73. 49.943; tenth

74. 12.7483; tenth

75. 33.416; hundredth

76. 4.359; hundredth

77. 9.0955; thousandth

78. 2.9592; thousandth

79. 21.0239; tenth

80. 15.9021; hundredth

 **81.** 16.804; hundredth

82. The average snail moves at a rate of about 0.00362005 miles per hour. Round the decimal value to the nearest ten thousandths place.

83. Which value is rounded to the nearest tenth, 7.1 or 7.10?

84. Which value is rounded to the nearest hundredth, 34.50 or 34.5?

85. Which number properly represents 3.499999 rounded to the thousandths place?

    a. 3.500   **b.** 3.5   **c.** 3.500000   **d.** 3.499

86. Which number properly represents 45.78999 rounded to the hundredths place?

    a. 45.78   **b.** 45.79   **c.** 45.79000   **d.** 45.790

For Exercises 87–90, round the number to the indicated place position.

	Number	Hundreds	Tens	Tenths	Hundredths	Thousandths
87.	349.2395					
88.	971.0948					
89.	79.0046					
90.	21.9754					

For Exercises 91–96, write the dollar amount in words. **(See Example 9.)**

91. $5.23

92. $8.11

93. $15.03

94. $82.44

95. $21.13

96. $32.45

Expanding Your Skills

97. What is the least number with three places to the right of the decimal that can be created with the digits 2, 9, and 7? Assume that the digits cannot be repeated.

98. What is the greatest number with three places to the right of the decimal that can be created from the digits 2, 9, and 7? Assume that the digits cannot be repeated.

---

section 4.2   Addition and Subtraction of Decimals

Objectives

1. Addition of Decimals
2. Subtraction of Decimals
3. Applications of Addition and Subtraction of Decimals

1. Addition of Decimals

In this section we learn to add and subtract decimals. To begin, consider the sum $5.67 + 3.12$.

$$5.67 = 5 + \frac{6}{10} + \frac{7}{100}$$
$$+ 3.12 = + 3 + \frac{1}{10} + \frac{2}{100}$$
$$8 + \frac{7}{10} + \frac{9}{100} = 8.79$$

Notice that the decimal points and place positions are lined up to add the numbers. In this way we can add digits with the same place values because we are effectively adding decimal fractions with like denominators. The intermediate step of using fraction notation is often skipped. We can get the same result more quickly by adding digits in like place positions.

Adding Decimals

1. Write the addends in a column with the decimal points and corresponding place values lined up. (You may insert additional zeros after the last digit to the right of the decimal point. These will act as placeholders so that each addend has the same number of digits to the right of the decimal point.)
2. Add the digits in columns from right to left, as you would whole numbers. The decimal point in the answer should be lined up with the decimal points from the addends.

Skill Practice

Add.

1. $64.13 + 8.26$

2. $184.218 + 14.12$

---

example 1   **Adding Decimals**

Add.

$$27.486 + 6.37$$

Solution:

$$\begin{array}{r} 27.486 \\ +\ 6.370 \\ \hline \end{array}$$ Line up decimal points.

           Insert an extra zero as a placeholder.

$$\begin{array}{r} {\scriptstyle 1\ \ 1} \\ 27.486 \\ +\ 6.370 \\ \hline 33.856 \end{array}$$ Add digits with common place values.

           Line up the decimal point in the answer.

With operations on decimals it is important to locate the correct position of the decimal point. A quick estimate can help you determine whether your answer is reasonable. From Example 1, we have

27.486	rounds to	27
6.370	rounds to	$+\ 6$
		33

The estimated value, 33, is close to the actual value of 33.856.

Concept Connections

3. Check your answer to problem 2 by estimation.

Skill Practice

Add.

4. $2.90741 + 15.13$

5. $36.2 + 4.106$

---

example 2   **Adding Decimals**

Add.

$$3.7026 + 43 + 816.3$$

Solution:

$$\begin{array}{r} 3.7026 \\ 43.0000 \\ +\ 816.3000 \\ \hline \end{array}$$ Line up decimal points.

           The decimal point and four zeros were inserted after the whole number.

           Insert three zeros.

$$\begin{array}{r} {\scriptstyle 1\ 1} \\ 3.7026 \\ 43.0000 \\ +\ 816.3000 \\ \hline 863.0026 \end{array}$$ Add digits with common place values.

           Line up decimal point in the answer.

The sum is 863.0026.

Concept Connections

6. Check your answer to problem 5 by estimation.

Tip: To check that the answer is reasonable, round each addend.

3.7026	rounds to	4
43	rounds to	$\overset{1}{43}$
816.3	rounds to	$+\ 816$
		863

which is close to the actual sum, 863.0026.

Answers

1. 72.39   2. 198.338
3. $184.218 + 14.12 \approx 184 + 14 = 198$, which is close to 198.338.
4. 18.03741   5. 40.306
6. $36.2 + 4.106 \approx 36 + 4 = 40$ which is close to 40.306.

2. Subtraction of Decimals

Subtraction of decimals is performed in much the same manner as addition.

> **Subtracting Decimals**
>
> 1. Write the numbers in a column with the decimal points and corresponding place values lined up. (You may insert additional zeros after the last digit to the right of the decimal point. These will act as placeholders so that each number has the same number of digits to the right of the decimal point.)
> 2. Subtract the digits in columns from right to left, as you would whole numbers. The decimal point in the answer should be lined up with the other decimal points.

---

example 3   **Subtracting Decimals**

Subtract.

a. $0.2868 - 0.056$　　　**b.** $139 - 28.63$　　　**c.** $192.4 - 89.387$

Solution:

a.
$$
\begin{array}{r}
0.2868 \\
-\ 0.0560 \\
\hline
0.2308
\end{array}
$$
Line up the decimal points.
Insert an extra zero as a placeholder.
Subtract digits with common place values.
Decimal point in the answer is lined up.

b.
$$
\begin{array}{r}
139.00 \\
-\ 28.63 \\
\end{array}
$$
Insert extra zeros as placeholders.
Line up the decimal points.

$$
\begin{array}{r}
\overset{8}{}\overset{\overset{9}{10}}{}\overset{10}{} \\
139.00 \\
-\ 28.63 \\
\hline
110.37
\end{array}
$$
Subtract digits with common place values.
Borrow where necessary.
Line up the decimal point in the answer.

c.
$$
\begin{array}{r}
192.400 \\
-\ 89.387 \\
\end{array}
$$
Insert extra zeros as placeholders.
Line up the decimal points.

$$
\begin{array}{r}
192.400 \\
-\ 89.387 \\
\hline
103.013
\end{array}
$$
Subtract digits with common place values.
Borrow where necessary.
Line up the decimal point in the answer.

---

example 4   **Adding and Subtracting Decimals**

Simplify.

$$27.819 - 13.78 + 9.6$$

Solution:

$\underbrace{27.819 - 13.78}\ + 9.6$ We must apply the order of operations.

First subtract: $27.819 - 13.78$

$$\begin{array}{r} {\scriptstyle 7\ \ 11} \\ 27.8\cancel{1}9 \\ -\ 13.780 \\ \hline 14.039 \end{array}$$

$= 14.039 + 9.6$ Now add 9.6 to the result.

$$\begin{array}{r} {\scriptstyle 1} \\ 14.039 \\ +\ 9.600 \\ \hline 23.639 \end{array}$$

$= 23.639$

■

3. Applications of Addition and Subtraction of Decimals

Decimals are used often in measurements and in day-to-day applications.

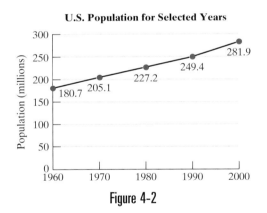
example 5   Subtracting Decimals in an Application

A graph of the U.S. population (in millions) is given for selected years (Figure 4-2).

  a. What is the difference in population between the years 1970 and 1960?

  b. What is the difference in population between the years 2000 and 1990?

U.S. Population for Selected Years

Figure 4-2

Solution:

  a. The difference in population between the years 1970 and 1960 is given by $205.1 - 180.7$.

$$\begin{array}{r} {\scriptstyle 1\ \ 10\ \ 4\ \ 11} \\ 2\,\cancel{0}\,\cancel{5}.\cancel{1} \\ -\,1\,8\,0.7 \\ \hline 2\,4.4 \end{array}$$ The difference in population is 24.4 million.

  b. The difference in population between the years 2000 and 1990 is given by $281.9 - 249.4$.

$$\begin{array}{r} {\scriptstyle 7\ \ 11} \\ 2\,8\,\cancel{1}.9 \\ -\,2\,4\,9.4 \\ \hline 3\,2.5 \end{array}$$ The difference in population is 32.5 million.

Comparing the values from parts (a) and (b), we see that the U.S. population increased during both 10-year periods. However, there was a greater increase between 1990 and 2000. This indicates that the rate of increase in population is increasing.

---

example 6   Applying Addition and Subtraction of Decimals in a Checkbook

Fill in the balance for each line in the checkbook register, shown in Figure 4-3. What is the ending balance?

Check No.	Description	Credit	Debit	Balance
				$684.60
2409	Doctor		$ 75.50	
2410	Mechanic		215.19	
2411	Home Depot		94.56	
	Paycheck	$981.46		
2412	Veterinarian		49.90	

Figure 4-3

Solution:

We begin with $684.60 in the checking account. For each debit, we subtract. For each credit, we add.

Check No.	Description	Credit	Debit	Balance	
				$ 684.60	
2409	Doctor		$ 75.50	609.10	= $684.60 − $75.50
2410	Mechanic		215.19	393.91	= $609.10 − $215.19
2411	Home Depot		94.56	299.35	= $393.91 − $94.56
	Paycheck	$981.46		1280.81	= $299.35 + $981.46
2412	Veterinarian		49.90	1230.91	= $1280.81 − $49.90

The ending balance is $1230.91.

---

example 7   Applying Decimals to Perimeter

a. Find the length of the side labeled x.

b. Find the length of the side labeled y.

c. Find the perimeter of the figure.

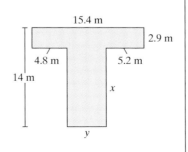

Solution:

a. If we extend the line segment labeled x with the dashed line as shown on page 248, we see that the sum of side x and the dashed line must equal 14 m. But the

as shown on page 248,

Skill Practice

12. Fill in the balance for each line in the checkbook register.

Check	Credit	Debit	Balance
			$437.80
1426		$82.50	
Pay	$514.02		
1427		26.04	

Skill Practice

13. Consider the figure.

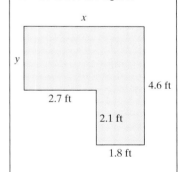

a. Find the length of side x.

b. Find the length of side y.

c. Find the perimeter.

Answers

12.

Check	Credit	Debit	Balance
			$ 437.80
1426		$82.50	355.30
Pay	$514.02		869.32
1427		26.04	843.28

13. a. Side x is 4.5 ft.
   b. Side y is 2.5 ft.
   c. The perimeter is 18.2 ft.

dashed line is the same length as the 2.9-m line segment on the right. Therefore, we can subtract $14 - 2.9$ to find the length of side x.

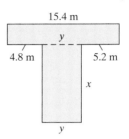

Length of side x:
$$\begin{array}{r} {\scriptstyle 3\ 10} \\ 1\cancel{4}.\cancel{0} \\ -\ 2.9 \\ \hline 11.1 \end{array}$$
Side x is 11.1 m long.

b. The dashed line in the figure has the same length as side y. We also know that $4.8 + 5.2 + y$ must equal 15.4. Because $4.8 + 5.2 = 10.0$, then

$$y = 15.4 - 10.0$$
$$= 5.4$$

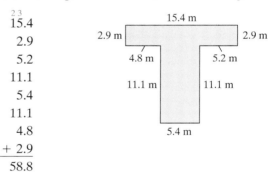

The length of side y is 5.4 m.

c. Now that we have the lengths of all sides, add them to get the perimeter.

$$\begin{array}{r} {\scriptstyle 2\ 3} \\ 15.4 \\ 2.9 \\ 5.2 \\ 11.1 \\ 5.4 \\ 11.1 \\ 4.8 \\ +\ 2.9 \\ \hline 58.8 \end{array}$$

The perimeter is 58.8 m.

Study Skills Exercise

1. Go to the online services that accompany this text (www.mhhe.com/moh). List two options that this online service offers that could help you in this course.

a. _____   **b.** _____

Review Exercises

2. Which number is equivalent to 5.03? Circle all that apply.

    a. 5.030   **b.** 5.30   **c.** 5.0300   **d.** 5.3

3. Which number is equivalent to $\frac{7}{100}$? Circle all that apply.

    a. 0.7   **b.** 0.07   **c.** 0.070   **d.** 0.007

4. Which number is equivalent to $\frac{9}{10}$? Circle all that apply.

    a. 0.09   **b.** 0.090   **c.** 0.90   **d.** 0.900

5. Which number is equivalent to 10.2? Circle all that apply.

    a. 10.200   **b.** 10.020   **c.** 10.02   **d.** 10.20

For Exercises 6–10, round the decimal to the indicated place value.

6. 23.489; tenth   **7.** 42.314; hundredth   **8.** 8.6025; thousandth

9. 0.981; tenth   **10.** 2.82998; ten-thousandth

Objective 1: Addition of Decimals

For Exercises 11–16, add the decimal numbers. Then round the numbers and find the sum to determine if your answer is reasonable.

Expression	Estimate		Expression	Estimate
11. 44.6 + 18.6	45 + 19 = 64		**12.** 28.2 + 23.2	
13. 5.306 + 3.645			**14.** 3.451 + 7.339	
15. 12.9 + 3.091			**16.** 4.125 + 5.9	

For Exercises 17–28, add the decimals. **(See Examples 1 and 2.)**

17. 78.9 + 0.9005   **18.** 44.2 + 0.7802   **19.** 23 + 8.0148   **20.** 7.9302 + 34

21. 34 + 23.0032 + 5.6   **22.** 23 + 8.01 + 1.0067   **23.** 68.394 + 32.02   **24.** 2.904 + 34.229

25. 103.94 + 24.5   **26.** 93.2 + 43.336   **27.** 54.2 + 23.993 + 3.87   **28.** 13.9001 + 72.4 + 34.13

Objective 2: Subtraction of Decimals

For Exercises 29–34, subtract the decimal numbers. Then round the numbers and find the difference to determine if your answer is reasonable.

Expression   Estimate   Expression   Estimate

29. $35.36 - 21.12$ $35 - 21 = 14$ **30.** $53.9 - 22.4$

31. $7.24 - 3.56$ **32.** $23.3 - 20.8$

33. $45.02 - 32.7$ **34.** $66.15 - 42.9$

For Exercises 35–46, subtract the decimal numbers. **(See Example 3.)**

35. $14.5 - 8.823$ **36.** $33.2 - 21.932$ **37.** $2 - 0.123$

38. $4 - 0.42$ **39.** $103.4 - 45.05 - 0.982$ **40.** $98.5 - 23.21 - 0.144$

41. $55.9 - 34.2354$ **42.** $49.1 - 24.481$ **43.** $18.003 - 3.238$

44. $21.03 - 16.446$ **45.** $183.01 - 23.452$ **46.** $164.23 - 44.3893$

For Exercises 47–54, add and subtract as indicated. **(See Example 4.)**

47. $6.007 + 12.74 - 3.4$ **48.** $3.005 + 25.127 - 13.7$ **49.** $23.37 - 21.9 + 5.111$

50. $0.78 - 0.028 + 6.1$ **51.** $8.962 + 51 - 40.05$ **52.** $11.957 + 45 - 3.55$

53. $5.3 + 5.03 + 5.003 - 5.0003$ **54.** $2.6 + 2.06 + 2.006 - 2.0006$

Objective 3: Applications of Addition and Subtraction of Decimals

55. The amount of time that it takes Mercury, Venus, Earth, and Mars to revolve about the Sun is given in the graph. **(See Example 5.)**

    a. How much longer does it take Mars to complete a revolution around the Sun than the Earth?

    b. How much longer does it take Venus than Mercury to revolve around the Sun?

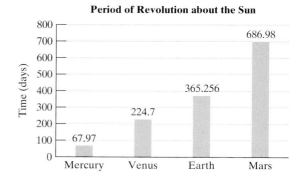

Period of Revolution about the Sun

56. The birth weights of the Dilley sextuplets are given in the graph.

    a. What is the difference between the weights of Julian and Quinn?

    b. What is the total weight of all six babies?

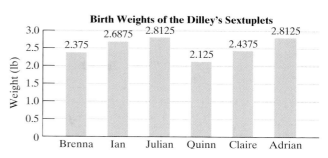

Birth Weights of the Dilley's Sextuplets

57. Water flows into a pool at a constant rate. The water level is recorded at several 1-hr intervals.

Time	Water Level
9:00 A.M.	4.2 in.
10:00 A.M.	5.9 in.
11:00 A.M.	7.6 in.
12:00 P.M.	9.3 in.

    a. From the table, how many inches is the water level rising each hour?

    b. At this rate, what will the water level be at 1:00 P.M.?

    c. At this rate, what will the water level be at 3:00 P.M.?

58. The gross earnings for the weekend of May 21–23, 2004, for three movies are given in the table.

    a. What was the difference between the earnings for *Shrek 2* and *Troy*?

    b. What was the difference between the earnings for *Troy* and *Van Helsing*?

    c. What was the total earnings for these three movies?

Movie	Gross Earnings ($ millions)
Shrek 2	104.3
Troy	10.1
Van Helsing	6.9

59. Fill in the balance for each line in the checkbook register shown in the figure. What was the ending balance? **(See Example 6.)**

Check No.	Description	Credit	Debit	Balance
				$ 245.62
2409	Electric bill		$ 52.48	
2410	Groceries		72.44	
2411	Department store		108.34	
	Payroll	$1084.90		
2412	Restaurant		23.87	
	Transfer from savings	200		

60. A section of a bank statement is shown in the figure. Find the mistake that was made by the bank.

Date	Action	Debit	Credit	Balance
				$1124.35
5/2/05	Check #4214	$749.32		375.03
5/3/05	Check #4215	37.29		337.74
5/4/05	Transfer		$ 400.00	737.74
5/5/05	Payroll		1451.21	2188.95
5/6/05	Cash withdrawal	150.00		688.95

61. The price of gasoline was $2.149 per gallon and rose another nine-tenths of a cent. What is the new price per gallon?

62. A laptop computer was originally priced at $1299.99 and was discounted to $998.95. By how much was it marked down?

63. The following table shows the widths of four U.S. coins. If you stacked three quarters and a dime in one pile and two nickels and two pennies in another pile, which pile would be higher?

Coin	Width
Quarter	1.75 mm
Dime	1.35 mm
Nickel	1.95 mm
Penny	1.55 mm

64. One service station charges $2.309 per gallon of gas while the station across the street charges $2.219. What is the difference between the price of gas at the two stations?

For Exercises 65–68, find the length of the sides labeled x and y. Then find the perimeter. **(See Example 7.)**

65.

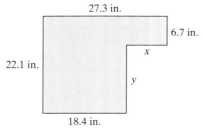

66.

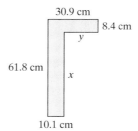

67.

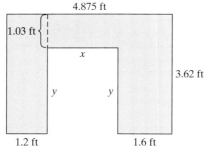

68.

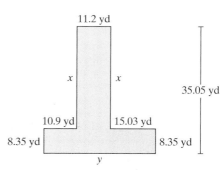

69. A city bus follows the route shown in the map. How far does it travel in one circuit?

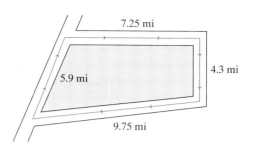

70. Santos built a new deck and needs to put a railing around the sides. He does not need railing where the deck is against the house. How much railing should he purchase?

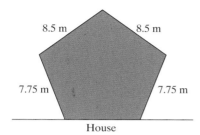

House

Expanding Your Skills

In a circle, the length of a line segment connecting two points on the circle and passing through the center is called a *diameter*.

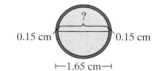

Use the definition of a diameter for Exercises 71–72.

71. The wire in a cable has a diameter of 6 mm. The insulation is 0.5 mm. What is the diameter of the cable with the insulation included?

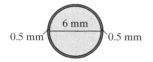

72. Find the inner diameter of the cable if the total diameter is 1.65 cm and the insulation is 0.15 cm thick.

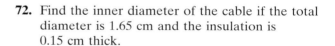

Calculator Connections

Topic: Entering decimals on a calculator

To enter decimals on a calculator, use the ☐·☐ key.

Expression	Keystrokes	Result
984.126 + 37.11	984 · 126 + 37 · 11 Enter	1021.236

Calculator Exercises

For Exercises 73–76, refer to the table showing the population for five countries for a recent year.

Country	Population (Millions)
United States	298.9
China	1310.6
Mexico	108.4
Germany	81.8
Canada	31.9

73. Find the difference between the populations of Mexico and Germany.

74. Find the difference between the population of the country with the greatest population and that of the country with the least population.

75. What is the sum of the populations of the five countries listed?

76. Is the sum of the populations of the United States, Mexico, Germany, and Canada greater than or less than the population of China?

Objectives

1. Multiplication of Decimals
2. Multiplication by a Power of 10 and by a Power of 0.1
3. Naming Large Numbers
4. Converting Dollars and Cents
5. Applications Involving Multiplication of Decimals
6. Finding Area

section 4.3   Multiplication of Decimals

1. Multiplication of Decimals

Multiplication of decimals is much like multiplication of whole numbers. However, we need to know where to place the decimal point in the product. Consider the product 0.3×0.41. One way to multiply these numbers is to write them first as decimal fractions.

$$0.3 \times 0.41 = \frac{3}{10} \times \frac{41}{100} = \frac{123}{1000} \text{ or } 0.123$$

Another method multiplies the factors vertically. First we multiply the numbers as though they were whole numbers. We temporarily disregard the decimal point in the product because it will be placed later.

$$\begin{array}{r} 0.41 \\ \times\ 0.3 \\ \hline 123 \end{array} \quad \longleftarrow \text{ decimal point not yet placed}$$

From the first method, we know that the correct answer to this problem is 0.123. Notice that 0.123 contains the same number of decimal places as the two factors combined. That is,

$$\begin{array}{rll} 0.41 & \longleftarrow & \text{2 decimal places} \\ \times\ 0.3 & \longleftarrow & \text{1 decimal place} \\ \hline .123 & \longleftarrow & \text{3 decimal places} \end{array}$$

In this example, notice that it was not necessary to line up the decimal point for the factors as we did with addition and subtraction. Instead, we write the factors "right-justified."

The process to multiply decimals is summarized as follows.

Concept Connections

1. How many decimal places will be in the product 2.72×1.4?

2. Explain the difference between the process to multiply 123×51 and the process to multiply 1.23×5.1.

Answers

1. 3
2. The actual process of vertical multiplication is the same for both cases. However, for the product 1.23×5.1, the decimal point must be placed so that the product has the same number of decimal places as both factors combined (in this case, 3).

Multiplying Two Decimals

1. Ignore the decimal point and multiply as you would whole numbers.

2. Place the decimal point in the product so that the number of decimal places equals the combined number of decimal places of both factors.

Note: You may need to insert zeros to the left of the whole-number product to get the correct number of decimal places in the answer.

In Example 1, we multiply decimals by using this process. To verify the correct position of the decimal point, we can check the answer by using estimation.

example 1   Multiplying Decimals

Multiply. Then use an estimate to check the location of the decimal point.

$$\begin{array}{r} 11.914 \\ \times\ 0.8 \end{array}$$

Solution:

$$\begin{array}{r} \overset{1\,7\ \ 1\,3}{11.914} \\ \times\ 0.8 \\ \hline 9.5312 \end{array} \quad \begin{array}{l} 3\ \text{decimal places} \\ +\ 1\ \text{decimal place} \\ \hline 4\ \text{decimal places} \end{array}$$

The product is 9.5312. To check the answer, we can round the factors and estimate the product. The purpose of the estimate is primarily to determine whether we have placed the decimal point correctly. Therefore, it is usually sufficient to round each factor so that there is only one nonzero digit. This is called **front-end rounding**. Thus,

$$\begin{array}{lll} 11.914 & \text{rounds to} & 10\ \rbrace \\ 0.8 & \text{rounds to} & 0.8\ \end{array} \quad \begin{array}{r} 10 \\ \times\ .8 \\ \hline 8 \end{array}$$

The first digit from both the exact value and the estimate is in the ones place. This suggests that the decimal point is placed correctly.

example 2   Multiplying Decimals

Multiply. Then use estimation to check the location of the decimal point.

$$29.3 \times 2.8$$

Solution:

Actual product: Estimate:

$$\begin{array}{r} \overset{1}{\underset{7\,2}{}} \\ 29.3 \\ \times\ 2.8 \\ \hline 2344 \\ 5860 \\ \hline 82.04 \end{array} \quad \begin{array}{l} 1\ \text{decimal place} \\ +\ 1\ \text{decimal place} \\ \\ \\ 2\ \text{decimal places} \end{array}$$

$$\begin{array}{lll} 29.3 & \text{rounds to} & 30 \\ 2.8 & \text{rounds to} & \times\ 3 \\ \hline & & 90 \end{array}$$

The first digit for both the actual product and the estimate is in the tens place. We are reasonably sure that we have located the decimal point correctly. The estimate 90 is close to 82.04.

The product is 82.04.

Example 3 illustrates the case where additional zeros must be placed in the product to obtain the correct number of decimal places.

example 3   Multiplying Decimals

Multiply. Then use estimation to check the location of the decimal point.

$$2.79 \times 0.0003$$

Solution:

Actual product:

$$\begin{array}{r} \overset{2\ 2}{2.79} \\ \times\ 0.0003 \\ \hline .000837 \end{array}$$ 2 decimal places
    + 4 decimal places
    6 decimal places
    (insert 3 zeros to the left)

Estimate:

0.0003   rounds to   0.0003
2.79   rounds to $\times$ 3
                    0.0009

The first digit for both the actual product and the estimate is in the ten-thousandths place. We are reasonably sure the decimal point is positioned correctly.

The product is 0.000837.

2. Multiplication by a Power of 10 and by a Power of 0.1

Consider the number 2.7 multiplied by the powers of ten, 10; that is, 10, 100, 1000 . . .

$$\begin{array}{r} 10 \\ \times\ 2.7 \\ \hline 70 \\ 200 \\ \hline 27.0 \end{array} \qquad \begin{array}{r} 100 \\ \times\ 2.7 \\ \hline 700 \\ 2000 \\ \hline 270.0 \end{array} \qquad \begin{array}{r} 1000 \\ \times\ 2.7 \\ \hline 7000 \\ 20000 \\ \hline 2700.0 \end{array}$$

Multiplying 2.7 by 10 has the effect of moving the decimal point 1 place to the right.
Multiplying 2.7 by 100 has the effect of moving the decimal point 2 places to the right.
Multiplying 2.7 by 1000 has the effect of moving the decimal point 3 places to the right.

This leads us to the following generalization.

Multiplying a Decimal by a Power of 10

Move the decimal point to the right the same number of decimal places as the number of zeros in the power of 10.

example 4   Multiplying Decimals by Powers of 10

Multiply.

   a. $14.78 \times 10,000$ **b.** 0.0064×100 **c.** $8.271 \times 1,000,000$

Solution:

   a. $14.78 \times 10,000 = 147,800$ Move the decimal point 4 places to the right.

   b. $0.0064 \times 100 = 0.64$ Move the decimal point 2 places to the right.

   c. $8.271 \times 1,000,000 = 8,271,000$ Move the decimal point 6 places to the right.

Skill Practice

Multiply.

9. 81.6×1000

10. $0.0000085 \times 10,000$

11. $2.396 \times 10,000,000$

Multiplying a decimal by 10, 100, 1000, and so on increases its value. Therefore, it makes sense to move the decimal point to the *right*. Now suppose we multiply a decimal by 0.1, 0.01, and 0.0001. These numbers represent the decimal fractions $\frac{1}{10}, \frac{1}{100}$, and $\frac{1}{1000}$, respectively, and are easily recognized as powers of 0.1 (see Section 2.4). Taking one-tenth of a number or one-hundredth of a number makes the number smaller. To multiply by 0.1, 0.01, 0.001, and so on (powers of 0.1), move the decimal point to the *left*.

$$
\begin{array}{r} 3.6 \\ \times\ 0.1 \\ \hline .36 \end{array}
\qquad
\begin{array}{r} 3.6 \\ \times\ 0.01 \\ \hline .036 \end{array}
\qquad
\begin{array}{r} 3.6 \\ \times\ 0.001 \\ \hline .0036 \end{array}
$$

Multiplying a Decimal by Powers of 0.1

Move the decimal point to the left the same number of places as there are decimal places in the power of 0.1.

Note: In this case, we move the decimal point the total number of decimal places in the power of 0.1, *not* the number of zeros. For example, when we multiply by 0.01, move the decimal point 2 places to the left.

Concept Connections

12. Explain the difference between multiplying a number by 100 versus 0.01.

example 5   Multiplying by Powers of 0.1

Multiply.

   a. 62.074×0.0001 **b.** 7965.3×0.1 **c.** 0.0057×0.00001

Solution:

   a. $62.074 \times 0.0001 = 0.0062047$ Move the decimal point 4 places to the left. Insert extra zeros.

   b. $7965.3 \times 0.1 = 796.53$ Move the decimal point 1 place to the left.

   c. $0.0057 \times 0.00001 = 0.000000057$ Move the decimal point 5 places to the left.

Skill Practice

Multiply.

13. 471.034×0.01

14. $9,437,214.5 \times 0.00001$

15. 0.0004×0.001

Answers
9. 81,600   **10.** 0.085   **11.** 23,960,000
12. Multiplying a number by 100 increases its value. Therefore, we move the decimal point to the right two places. Multiplying a number by 0.01 decreases its value. Therefore, move the decimal point to the left two places.
13. 4.71034   **14.** 94.372145
15. 0.0000004

3. Naming Large Numbers

Sometimes people prefer to use number names to express very large numbers. For example, we might say that the U.S. population in 2004 was approximately 280 million. To write this in decimal form, we note that 1 million = 1,000,000. In this case, we have 280 of this quantity. Thus,

$$280 \text{ million} = 280 \times 1,000,000 \text{ or } 280,000,000$$

---

example 6   Naming Large Numbers

Write the decimal number representing each word name.

a. The distance between the Earth and the Sun is approximately 92.9 million mi.

b. The number of deaths in the United States due to heart disease in 2000 was approximately 7 hundred thousand.

c. The number of barrels of crude oil and natural gas reserves in the United States in 2001 was estimated to be 22.4 billion.

Solution:

a. $92.9 \text{ million} = 92.9 \times 1,000,000 = 92,900,000$

b. $7 \text{ hundred thousand} = 7 \times 100,000 = 700,000$

c. $22.4 \text{ billion} = 22.4 \times 1,000,000,000 = 22,400,000,000$

---

4. Converting Dollars and Cents

The value $1.14 can be interpreted as $1 + 14¢. But since $1 = 100¢, then

$$\$1.14 = 100¢ + 14¢ = 114¢$$

From this discussion we have the following rule to convert dollars to cents.

Converting Dollars to Cents

To convert dollars to cents:

1. Multiply by 100. This has the effect of moving the decimal point to the *right* two places.

2. Drop the $ symbol and attach the ¢ symbol to the result.

To convert from cents to dollars, we reverse the process. That is, 114¢ = $1.14. The decimal point is moved to the *left*.

Converting Cents to Dollars

To convert cents to dollars:

1. Multiply by 0.01. This has the effect of moving the decimal point to the *left* two places.

2. Drop the ¢ symbol and place the $ symbol in front of the result.

example 7   Converting Dollars and Cents

a. Convert $6.82 to cents.   **b.** Convert 92¢ to dollars.

Solution:

a. $6.82 = 682¢   Move the decimal point 2 places to the *right*.

b. 92¢ = $0.92   Move the decimal point 2 places to the *left*.

Skill Practice

20. Convert $8.97 to cents.
21. Convert 4216¢ to dollars.

5. Applications Involving Multiplication of Decimals

example 8   Applying Decimal Multiplication

Jane Marie bought 8 cans of tennis balls for $1.98 each. She paid $1.03 in tax. What was the total bill?

Solution:

The cost of the tennis balls before tax is

$$8(\$1.98) = \$15.84$$

$$\begin{array}{r} {}^{7\ 6} \\ 1.98 \\ \times\ \ 8 \\ \hline 15.84 \end{array}$$

Adding the tax to this value, we have

$$\begin{pmatrix} \text{Total} \\ \text{cost} \end{pmatrix} = \begin{pmatrix} \text{Cost of} \\ \text{tennis balls} \end{pmatrix} + (\text{Tax})$$

$$\begin{array}{r} = \$15.84 \\ +\ 1.03 \\ \hline \$16.87 \end{array}$$ The total cost is $16.87.

Skill Practice

22. A book club ordered 12 books on www.amazon.com for $8.99 each. The shipping cost was $4.95. What was the total bill?

6. Finding Area

example 9   Finding the Area of a Rectangle

The *Mona Lisa* is perhaps the most famous painting in the world. It was painted by Leonardo da Vinci somewhere between 1503 and 1506 and now hangs in the Louvre in Paris, France. The dimensions of the painting are 30 in. by 20.875 in. (Figure 4-4). What is the total area?

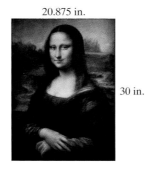

20.875 in.

30 in.

Figure 4-4

Skill Practice

23. The IMAX movie screen at the Museum of Science and Discovery in Ft. Lauderdale, Florida, is 18 m by 24.4 m. What is the area of the screen?

Answers
20. 897¢
21. $42.16
22. The total bill was $112.83.
23. The screen area is 439.2 m².

Solution:

Recall that the area of a rectangle is given by

$$A = \ell \cdot w$$

$$A = (30 \text{ in.})(20.875 \text{ in.})$$

$$= 626.25 \text{ in.}^2$$

$$\begin{array}{r} 20.875 \\ \times\ 30 \\ \hline 0 \\ 626250 \\ \hline 626.250 \end{array}$$

The area of the *Mona Lisa* is 626.25 in.²

---

section 4.3   Practice Exercises

Boost *your* GRADE at
mathzone.com!

- Practice Problems
- Self-Tests
- NetTutor

- e-Professors
- Videos

Study Skills Exercises

1. Look through pages 266 to 273 in your text. Write down a page number that contains

   a. Avoiding mistakes _____

   b. TIP box _____

   c. A key term (shown in bold) _____

2. Define the key term **front-end rounding**.

Review Exercises

For Exercises 3–6, expand the powers of 10 and 0.1.

3. 10^3 **4.** 0.1^3 **5.** 0.1^2 **6.** 10^2

Objecitve 1: Multiplication of Decimals

For Exercises 7–16, round each number by using front-end rounding.

7. 135   **8.** 481   **9.** 28   **10.** 52

11. 6721   **12.** 8207   **13.** 0.241   **14.** 0.339

15. 0.041   **16.** 0.056

For Exercises 17–26, multiply the decimals.

17.   0.8
       $\times$ 0.5

18.   0.6
       $\times$ 0.5

19. (0.9)(4)

20. (0.2)(9)

21.   0.4
       $\times$ 20

22.   0.9
       $\times$ 30

23. (60)(0.003)

24. (40)(0.005)

25.   22
       $\times$ 0.8

26.   31
       $\times$ 0.4

For Exercises 27–40, multiply the decimals. Then estimate the answer by using front-end rounding. **(See Examples 1–3.)**

Exact	Estimate		Exact	Estimate

27.   8.3
       $\times$ 4.5

Estimate: 8
       $\times$ 5
       40

28.   4.3
       $\times$ 9.2

29.   0.58
       $\times$ 7.2

30.   0.83
       $\times$ 6.5

31. 5.92×0.8

32. 9.14×0.6

33. (0.413)(7)

34. (0.321)(6)

35. 35.9×3.2

36. 41.7×6.1

37. 562×0.004

38. 984×0.009

39. 0.0004×3.6

40. 0.0008×6.5

Objective 2: Multiplication by a Power of 10 and by a Power of 0.1

41. If 417.43 is multiplied by 100, will the decimal point move to the left or to the right? By how many places?

42. If 2498.613 is multiplied by 10,000, will the decimal point move to the left or to the right? By how many places?

43. Multiply the numbers.

   a. 5.1×10 **b.** 5.1×100 **c.** 5.1×1000 **d.** $5.1 \times 10,000$

For Exercises 44–49, multiply the numbers by the powers of 10. **(See Example 4.)**

44. 34.9×100 **45.** 2.163×100 **46.** 96.59×1000 **47.** 18.22×1000

48. 2.001×10 **49.** 5.932×10

50. If 256.8 is multiplied by 0.001, will the decimal point move to the left or to the right? By how many places?

51. If 0.45 is multiplied by 0.1, will the decimal point move to the left or to the right? By how many places?

52. Multiply the numbers.

   a. 5.1×0.1 **b.** 5.1×0.01 **c.** 5.1×0.001 **d.** 5.1×0.0001

For Exercises 53–58, multiply the numbers by the powers of 0.1. **(See Example 5.)**

53. 93.3×0.01 **54.** 80.2×0.01 **55.** 54.03×0.001 **56.** 23.11×0.001

57. 0.5×0.0001 **58.** 0.8×0.0001

Objective 3: Naming Large Numbers

For Exercises 59–64, write the decimal number representing each word name. **(See Example 6.)**

59. The number of cattle in the United States is 96.7 million.

60. There are 42.515 million tons of yams produced in the United States.

61. About 16 thousand patents were granted in Spain during a recent year.

62. The musical *Miss Saigon* ran for about 4 thousand performances in a 10-year period.

63. The people in the United States have spent over $20.549 billion on DVDs.

64. The Bible is the highest-selling book of all time with over $6 billion in sales.

Objective 4: Converting Dollars and Cents

For Exercises 65–70, write the amount in terms of cents. **(See Example 7.)**

65. $3.24   **66.** $21.56   **67.** $61.34

68. $3.12   **69.** $0.37   **70.** $0.75

For Exercises 71–76, write the amount in terms of dollars. **(See Example 7.)**

71. 347¢

72. 512¢

73. 2041¢

74. 5712¢

75. 34¢

76. 12¢

77. a. Round $1.499 to the nearest dollar.

b. Round $1.499 to the nearest cent.

78. a. Round $20.599 to the nearest dollar.

b. Round $20.599 to the nearest cent.

Objective 5: Applications Involving Multiplication of Decimals

79. Corrugated boxes for shipping cost $2.27 each. How much will 10 boxes cost including tax that amounts to $1.59? **(See Example 8.)**

80. A bag of potato chips contains 11.5 oz. How many ounces are in 4 bags of chips?

81. The Athletic Department at Illinois Central College bought 20 pizzas for $10.95 each, 10 Greek salads for $3.95 each, and 60 soft drinks for $0.60 each. What was the total bill excluding tax?

82. A hotel gift shop ordered 40 T-shirts at $8.69 each, 10 hats at $3.95 each, and 20 beach towels at $4.99 each. What was the total cost of the merchandise?

83. Firestone tires cost $50.20 each. A set of four Lemans tires costs $197.99. How much can a person save by buying the set of four Lemans tires compared to four Firestone tires?

84. Certain DVDs titles are on sale for 2 for $32. If they regularly sell for $19.99, how much can a person save by buying 4 DVDs?

Objective 6: Finding Area

For Exercises 85–86, find the area. **(See Example 9.)**

85.

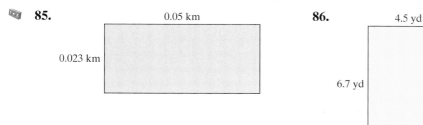

86.

87. Blake plans to build a rectangular patio that is 15 ft by 22.2 ft. What is the total area of the patio?

88. The front page of a newspaper is 56 cm by 31.5 cm. Find the area of the page.

89. Compare the quantities $(0.2)^2$ and 0.4. Are they equal?

90. Compare the quantities $(0.5)^2$ and 2.5. Are they equal?

For Exercises 91–96, simplify the expressions.

91. $(0.4)^2$ **92.** $(0.7)^2$ **93.** $(1.3)^2$ **94.** $(2.4)^2$

95. $(0.1)^3$ **96.** $(0.2)^3$

Expanding Your Skills

97. Evaluate.

    a. $(0.3)^2$ **b.** $\sqrt{0.09}$

98. Evaluate.

    a. $(0.5)^2$ **b.** $\sqrt{0.25}$

For Exercises 99–102, evaluate the square roots.

99. $\sqrt{0.01}$ **100.** $\sqrt{0.04}$ **101.** $\sqrt{0.36}$ **102.** $\sqrt{0.49}$

Calculator Connections

Topic: Multiplying decimals

Calculator Exercises

For Exercises 103–104, use a calculator to evaluate each expression.

103. $(43.75)^2$ **104.** $(9.3)^5$

105. A Hummer H2 SUV uses 1260 gal of gas to travel 12,000 mi per year. A Honda Accord uses 375 gal of gas to go the same distance. If gasoline costs $2.25 per gallon, how much is saved per year by driving a Honda Accord rather than a Hummer?

106. The actual time required for the Earth to revolve about the Sun is 365.256 days. Use this number to explain why we have a leap year every 4 years. (A leap year is a year in which February has an extra day, February 29.)

107. A homeowner pays the following average monthly expenses. What are the total expenses for the year?

Item	Monthly Charge
Mortgage payments	$678.75
Electricity	95.83
Water and sewer	47.02
House insurance	66.65
Property taxes	124.40
Internet connection	29.95
Phone	24.99
Cable TV	65.49
Association fees	50.00

chapter 4 | midchapter review

For Exercises 1–12, add, subtract, or multiply as indicated.

1. 123.04 + 100

2. 123.04 × 100

3. 123.04 – 100

4. 123.04 × 0.01

5. 123.04 – 0.01

6. 123.04 + 0.01

7. 10.82 – 0.1

8. 10.82 + 0.1

9. 10.82 × 0.1

10. 10.82 + 10

11. 10.82 × 10

12. 10.82 – 10

13. a. Add 4.8 + 2.391.

b. Add 2.391 + 4.8.

c. Are the answers to parts (a) and (b) the same?

14. What is the name of the property used in Exercise 13?

15. a. Multiply 4.8 × 2.391.

b. Multiply 2.391 × 4.8.

c. Are the answers to parts (a) and (b) the same?

16. What is the name of the property used in Exercise 15?

For Exercises 17–22, perform the indicated operation.

17. 0.004 × 6.21

18. 1.2 + 83.11

19. 4.6 – 0.0421

20. 5.1 × 0.0241

21. 65.02 + 2.012

22. 3.012 – 1.34

Objectives

1. Division of Decimals
2. Rounding a Quotient
3. Dividing by a Power of 10 and by a Power of 0.1
4. Applications of Decimal Division

Skill Practice

Divide. Check by using multiplication.

2. $153.6 \div 6$
3. $502.96 \div 8$

Answers

1. $8)\overline{116.32}$
2. 25.6
3. 62.87

---

section 4.4   Division of Decimals

1. Division of Decimals

Dividing decimals is much the same as dividing whole numbers. However, we must determine where to place the decimal point in the quotient. The first case we show involves dividing a decimal by a whole number.

First consider the quotient $3.5 \div 7$. We can write the numbers in fractional form and then divide.

$$3.5 \div 7 = \frac{35}{10} \div \frac{7}{1} = \frac{35}{10} \cdot \frac{1}{7} = \frac{35}{10} \cdot \frac{1}{7} = \frac{35}{70} = \frac{5}{10} = 0.5$$

Now consider the same problem by using the efficient method of long division: $7)\overline{3.5}$.

When the divisor is a whole number, we place the decimal point directly above the decimal point in the dividend. Then we divide as we would whole numbers.

decimal point placed above
the decimal point in the dividend.

$$\begin{array}{r} .5 \\ 7)\overline{3.5} \end{array}$$

Dividing a Decimal by a Whole Number

To divide by a whole number:

1. Place the decimal point in the quotient directly above the decimal point in the dividend.
2. Divide as you would whole numbers.

---

example 1   Dividing by a Whole Number

Divide and check the answer by multiplying.

$$30.55 \div 13$$

Solution:

Locate the decimal point in the quotient.

$$13)\overline{30.55}$$

$$\begin{array}{r} 2.35 \\ 13)\overline{30.55} \\ -26 \\ \hline 45 \\ -39 \\ \hline 65 \\ -65 \\ \hline 0 \end{array}$$

Divide as you would whole numbers.

Check by multiplying:

$$\begin{array}{r} 2.35 \\ \times\ 13 \\ \hline 705 \\ 2350 \\ \hline 30.55 \ \checkmark \end{array}$$

When dividing decimals, we do not use a remainder. Instead we insert zeros to the right of the dividend and continue dividing. This is demonstrated in Example 2.

---

example 2   Dividing by a Whole Number

Divide and check the answer by multiplying.

a. $3.5 \div 4$ **b.** $40\overline{)5}$

Solution:

Locate the decimal point in the quotient.

a. $4\overline{)3.5}$

$$
\begin{array}{r}
.8 \\
4\overline{)3.5} \\
-32 \\
\hline
3
\end{array}
$$

← Rather than using a remainder, we insert zeros in the dividend and continue dividing.

$$
\begin{array}{r}
.875 \\
4\overline{)3.500} \\
-32\downarrow \\
\hline
30 \\
-28\downarrow \\
\hline
20 \\
-20 \\
\hline
0
\end{array}
$$

Check by multiplying:

$$
\begin{array}{r}
{}^{3\ 2} \\
0.875 \\
\times\ 4 \\
\hline
3.500 \checkmark
\end{array}
$$

The quotient is 0.875.

b. $40\overline{)5.}$ ← The dividend is a whole number, and the decimal point is understood to be to its right. Insert the decimal point above it in the quotient.

$$
\begin{array}{r}
.125 \\
40\overline{)5.000} \\
-40\downarrow \\
\hline
100 \\
-80\downarrow \\
\hline
200 \\
-200 \\
\hline
0
\end{array}
$$

Since 40 is greater than 5, we need to insert zeros to the right of the dividend.

Check by multiplying.

$$
\begin{array}{r}
{}^{1\ 2} \\
.125 \\
\times\ 40 \\
\hline
000 \\
5000 \\
\hline
5.000 \checkmark
\end{array}
$$

The quotient is 0.125.

Sometimes when dividing decimals, the quotient follows a repeated pattern. The result is called a **repeating decimal**.

Skill Practice

Divide.

4. $6.8 \div 5$

5. $20\overline{)3}$

6. $87.5 \div 14$

Answers

4. 1.36
5. 0.15
6. 6.25

Divide.

7. $2.4 \div 9$

8. $57 \div 11$

example 3   Dividing Where the Quotient Is a Repeating Decimal

Divide.

a. $1.7 \div 30$ **b.** $11\overline{)68}$

Solution:

a.
$$\begin{array}{r} .05666\ldots \\ 30\overline{)1.70000} \\ -150 \\ \hline 200 \\ -180 \\ \hline 200 \\ -180 \\ \hline 200 \end{array}$$

Notice that as we continue to divide, we get the same values for each successive step. This causes a pattern of repeated digits in the quotient. Therefore, the quotient is a repeating decimal.

The quotient is 0.05666 . . . . To denote the repeated pattern, we often use a bar over the first occurrence of the repeat cycle to the right of the decimal point. That is,

$0.05666\ldots = 0.05\overline{6}$ ⟵ repeat bar

Avoiding Mistakes: In Example 3(a), notice that the repeat bar goes over only the 6. The 5 is not being repeated.

b.
$$\begin{array}{r} 6.1818\ldots \\ 11\overline{)68.0000} \\ -66 \\ \hline 20 \\ -11 \\ \hline 90 \\ -88 \\ \hline 20 \\ -11 \\ \hline 90 \\ -88 \\ \hline 20 \end{array}$$

Could have stopped here

Once again, we see a repeated pattern. The quotient is a repeating decimal. Notice that we could have stopped dividing when we obtained the second value of 20.

Avoiding Mistakes: Be sure to put the repeating bar over the entire block of numbers that is being repeated. In Example 3(b), the bar extends over both the 1 and the 8. We have $6.\overline{18}$.

The quotient is $6.\overline{18}$.

Identify the number as a terminating or repeating decimal.

9. 4.6666666   **10.** $4.\overline{6}$

11. $9.5\overline{37}$

12. 16.417417417

The numbers $0.05\overline{6}$ and $6.\overline{18}$ are examples of repeating decimals. A decimal that "stops" is called a **terminating decimal**. For example, 6.18 is a terminating decimal, whereas $6.\overline{18}$ is a repeating decimal.

In Examples 1–3, we performed division where the divisor was a whole number. Suppose now that we have a divisor that is *not* a whole number, for example, $0.56 \div 0.7$. Because division can also be expressed in fraction notation, we have

$$0.56 \div 0.7 = \frac{0.56}{0.7}$$

If we multiply numerator and denominator by 10, the denominator (divisor) becomes the whole number 7.

$$\frac{0.56}{0.7} = \frac{0.56 \times 10}{0.7 \times 10} = \frac{5.6}{7} \longrightarrow 7\overline{)5.6}$$

Answers

7. $0.2\overline{6}$ 8. $5.\overline{18}$
9. Terminating   10. Repeating
11. Repeating   12. Terminating

Recall that multiplying decimal numbers by 10 (or any power of 10, such as 100, 1000, etc.) has the effect of moving the decimal point to the right. We use this premise to divide decimal numbers when the divisor is not a whole number. This process is summarized as follows.

Dividing When the Divisor Is Not a Whole Number

1. Move the decimal point in the divisor to the right to make it a whole number.
2. Move the decimal point in the dividend to the right the same number of places as in step 1.
3. Place the decimal point in the quotient directly above the decimal point in the dividend.
4. Divide as you would whole numbers.

---

example 4   Dividing Decimals

Divide.

a. $0.56 \div 0.7$ **b.** $0.005\overline{)3.1}$ **c.** $50 \div 1.1$

Solution:

a. $.7\overline{).56}$ Move the decimal point in the divisor and dividend 1 place to the right.

$7\overline{)5.6}$ ⟵——— Line up the decimal point in the quotient.

$$\begin{array}{r} 0.8 \\ 7\overline{)5.6} \\ \underline{-5\ 6} \\ 0 \end{array}$$

The quotient is 0.8.

b. $.005\overline{)3.100}$ Move the decimal point in the divisor and dividend 3 places to the right. Insert additional zeros in the dividend if necessary. Line up the decimal point in the quotient.

$$\begin{array}{r} 620. \\ 5\overline{)3100.} \\ \underline{-30} \\ 10 \\ \underline{-10} \\ 00 \end{array}$$

The quotient is 620.

Answers
13. 1.6
14. 180
15. $116.\overline{6}$

c. $1.1\overline{)50.0}$ Move the decimal point in the divisor and dividend 1 place to the right. Insert an additional zero in the dividend. Line up the decimal point in the quotient.

$$
\begin{array}{r}
45.45\ldots \\
11\overline{)500.000} \\
-44 \\
\hline
60 \\
-55 \\
\hline
50 \\
-44 \\
\hline
60 \\
-55 \\
\hline
50 \\
\end{array}
$$

The quotient is a repeating decimal. Notice that the repeat cycle actually begins to the left of the decimal point. However, the repeat bar is placed on the first repeated block of digits to the *right* of the decimal point. Therefore, the quotient is $45.\overline{45}$.

> **Avoiding Mistakes:** The repeat bar is never written over the whole-number part of a quotient.

The quotient is $45.\overline{45}$.

2. Rounding a Quotient

In Example 4(c) we found that $50 \div 1.1 = 45.\overline{45}$. To check this result, we could multiply $45.\overline{45} \times 1.1$ and show that the product equals 50. However, at this point we do not have the tools to multiply repeating decimals. What we can do is round the quotient and then multiply to see if the product is *close* to 50.

example 5   Rounding a Repeating Decimal

Round $45.\overline{45}$ to the hundredths place. Then use the rounded value to estimate whether the product $45.\overline{45} \times 1.1$ is close to 50. (This will serve as a check to the division problem in Example 4(c).)

Solution:

To round the number $45.\overline{45}$, we must write out enough of the repeated pattern so that we can view the digit to the right of the rounding place. In this case, we must write out the number to the thousandths place.

$$45.\overline{45} = 45.454\cdots \approx 45.45$$

hundredths place   This digit is less than 5. Discard it and all others to its right.

Now multiply the rounded value by 1.1.

$$
\begin{array}{r}
45.45 \\
\times\ 1.1 \\
\hline
4545 \\
45450 \\
\hline
49.995 \\
\end{array}
$$

This value is close to 50. We are reasonably sure that we divided correctly in Example 4(c).

Sometimes we may want to round a quotient to a given place value. To do so, divide until you get a digit in the quotient one place value to the right of the rounding place. At this point, you may stop dividing and round the quotient.

example 6   Rounding a Quotient

Round the quotient to the tenths place.

$$47.3 \div 5.4$$

Solution:

5.4⟌47.3   Move the decimal point in the divisor and dividend 1 place to the right. Line up the decimal point in the quotient.

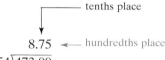

tenths place

```
       8.75  ←── hundredths place
    54)473.00
      −432
       410
      −378
       320
      −270
        50
```

To round the quotient to the tenths place, we must determine the hundredths-place digit and use it to base our decision on rounding. The hundredths-place digit is 5. Therefore, we round the quotient to the tenths place by increasing the tenths-place digit by 1 and discarding all digits to its right.

The quotient is approximately 8.8.

Skill Practice

Round the quotient to the indicated place value.

18. $42.68 \div 5.1$; hundredths

19. $84 \div 1.1$; tenths

Calculator Connections

Repeating decimals displayed on a calculator are rounded. This is so because the display cannot show an infinite number of digits.

On a scientific calculator, the repeating decimal $45.\overline{45}$ might appear as

| 45.45454545 |

---

3. Dividing by a Power of 10 and by a Power of 0.1

When we multiplied a number by 10, 100, 1000, and so on, we moved the decimal point to the right. However, dividing a number by 10, 100, or 1000 decreases its value. Therefore, we move the decimal point to the *left*.

For example, suppose 3.6 is divided by 10, 100, and 1000.

```
        .36              .036             .0036
    10)3.60         100)3.600        1000)3.6000
      −30             −300             −3000
       60              600              6000
      −60             −600             −6000
        0                0                0
```

Dividing by a Power of 10

To divide a number by a power of 10, move the decimal point to the *left* the same number of places as there are zeros in the power of 10.

Answers

18. 8.37
19. 76.4

example 7	Dividing by a Power of 10

Divide.

   a. $214.3 \div 10{,}000$ **b.** $0.03 \div 100$

Solution:

   a. $214.3 \div 10{,}000 = 0.02143$ Move the decimal point 4 places to the left. Insert an additional zero.

   b. $0.03 \div 100 = 0.0003$ Move the decimal point 2 places to the left. Insert two additional zeros.

To divide a number by 0.1, 0.01, 0.001, and so on, we are dividing by the fractions $\frac{1}{10}$, $\frac{1}{100}$, $\frac{1}{1000}$, etc. But dividing by these fractions is the same as multiplying by their reciprocals. Multiplying a number by 10, 100, and 1000 increases its value, and we must move the decimal point to the *right*.

$$4 \div 0.1 \;\; = 4 \div \frac{1}{10} = \frac{4}{1} \cdot \frac{10}{1} = 40$$

$$4 \div 0.01 \;\; = 4 \div \frac{1}{100} = \frac{4}{1} \cdot \frac{100}{1} = 400$$

$$4 \div 0.001 = 4 \div \frac{1}{1000} = \frac{4}{1} \cdot \frac{1000}{1} = 4000$$

From this discussion, we have the following rule.

> **Dividing by a Power of 0.1**
>
> To divide a number by a power of 0.1, move the decimal point to the *right* the same number of places as there are decimal places in the power of 0.1.

example 8	Dividing by a Power of 0.1

Divide.

   a. $316.24 \div 0.01$ **b.** $0.0057 \div 0.00001$

Solution:

   a. $316.24 \div 0.01 = 31{,}624$ Move the decimal point 2 places to the right.

   b. $0.00570 \div 0.00001 = 570$ Move the decimal point 5 places to the right. Insert an extra zero to the right of 0.0057.

4. Applications of Decimal Division

Examples 9 and 10 show how decimal division can be used in applications. Remember that division is used when we need to distribute a quantity into equal parts.

example 9   Applying Division of Decimals

A dinner costs $45.80, and the bill is to be split equally among 5 people. How much must each person pay?

Solution:

We want to distribute $45.80 equally among 5 people, so we must divide $45.80 ÷ 5.

```
      9.16
  5)45.80       Each person must pay $9.16.
   −45
    08
    −5
    30
   −30
     0
```

Skill Practice

24. A 46.5-ft cable is to be cut into 6 pieces of equal length. How long is each piece?

Division is also used in practical applications to express rates. In Example 10, we find the rate of speed in meters per second (m/sec) for the world record time in the men's 400-m run.

example 10   Using Division to Find a Rate of Speed

For a recent year, the world record time in the men's 400-m run was 43.2 sec. What is the speed in meters per second? Round to one decimal place.

Solution:

To find the rate of speed in meters per second, we must divide the distance in meters by the time in seconds.

```
43.2)400.0
```

```
                  tenths place
      9.25 ←— hundredths place
  432)4000.00
     −3888
      1120
      −864
      2560
     −2160
       400
```

To round the quotient to the tenths place, determine the hundredths-place digit and use it to make the decision on rounding. The hundredths-place digit is 5, which is 5 or greater. Therefore, add 1 to the tenths-place digit and discard all digits to its right.

The speed is approximately 9.3 m/sec.

Skill Practice

25. For a recent year Florence Griffith-Joyner ran a world-record time of 21.3 sec in the women's 200-m run. Find the speed in meters per second. Round to the nearest tenth.

Tip: In Example 10, we had to find speed in meters per second. The units of measurement required in the answer give a hint as to the order of the division. The word *per* implies division. So to obtain meters *per* second implies 400 m ÷ 43.2 sec.

Answers
24. Each piece is 7.75 ft.
25. The speed was 9.4 m/sec.

section 4.4   Practice Exercises

Study Skills Exercises

1. Meet some of the other students in your class. They can be good resources for asking questions and discussing the material that was covered in class. Write the names of two fellow students.

2. Define the key terms.

   a. Repeating decimal   **b. Terminating decimal**

Review Exercises

For Exercises 3–8, perform the indicated operation.

3. 5.28×1000 **4.** $8.003 - 2.2$ **5.** 11.8×0.32

6. $102.4 + 1.239$ **7.** $16.82 - 14.8$ **8.** 5.28×0.001

Objective 1: Division of Decimals

For Exercises 9–16, divide. Check the answer by using multiplication. **(See Example 1.)**

9. $8.1 \div 9$ Check: _____ $\times\ 9 = 8.1$ **10.** $4.8 \div 6$ Check: _____ $\times\ 6 = 4.8$

11. $6\overline{)1.08}$ Check: _____ $\times\ 6 = 1.08$ **12.** $4\overline{)2.08}$ Check: _____ $\times\ 4 = 2.08$

13. $4.24 \div 8$ **14.** $5.75 \div 25$ **15.** $5\overline{)105.5}$ **16.** $7\overline{)221.2}$

For Exercises 17–40, divide. Write the quotient in decimal form. **(See Examples 2–4.)**

17. $5\overline{)9.8}$ **18.** $30\overline{)2.07}$ **19.** $30.6 \div 12$ **20.** $16.5 \div 6$

21. $0.28 \div 8$ **22.** $0.35 \div 4$ **23.** $5\overline{)84.2}$ **24.** $2\overline{)89.1}$

25. $16 \div 3$ **26.** $52 \div 9$ **27.** $19 \div 6$ **28.** $9.1 \div 3$

29. $33\overline{)71}$ **30.** $11\overline{)42}$ **31.** $11\overline{)28}$ **32.** $33\overline{)202}$

33. $57.12 \div 1.02$ **34.** $95.89 \div 2.23$ **35.** $2.38 \div 0.8$ **36.** $5.51 \div 0.2$

37. $0.3\overline{)62.5}$ **38.** $1.05\overline{)22.4}$ **39.** $6.305 \div 0.13$ **40.** $42.9 \div 0.25$

Objective 2: Rounding a Quotient

41. Round $2.\overline{4}$ to the
  a. Tenths place
  b. Hundredths place
  c. Thousandths place
  (See Example 5.)

42. Round $5.\overline{2}$ to the
  a. Tenths place
  b. Hundredths place
  c. Thousandths place

43. Round $1.\overline{8}$ to the
  a. Tenths place
  b. Hundredths place
  c. Thousandths place

44. Round $4.\overline{7}$ to the
  a. Tenths place
  b. Hundredths place
  c. Thousandths place

45. Round $3.\overline{62}$ to the
  a. Tenths place
  b. Hundredths place
  c. Thousandths place

46. Round $9.\overline{38}$ to the
  a. Tenths place
  b. Hundredths place
  c. Thousandths place

For Exercises 47–54, divide. Round the answer to the indicated place value. Use the rounded quotient to check. **(See Example 6.)**

47. $7\overline{)1.8}$ hundredths **48.** $2.1\overline{)75.3}$ hundredths **49.** $54.9 \div 3.7$ tenths

50. $94.3 \div 21$ tenths **51.** $0.24\overline{)4.96}$ thousandths **52.** $2.46\overline{)27.88}$ thousandths

53. $0.9\overline{)32.1}$ hundredths **54.** $0.6\overline{)81.4}$ hundredths

Objective 3: Dividing by a Power of 10 and by a Power of 0.1

55. If 45.62 is divided by 100, will the decimal point move to the right or to the left? By how many places?

56. If 5689.233 is divided by 100,000, will the decimal point move to the right or to the left? By how many places?

For Exercises 57–64, divide by the powers of 10. **(See Example 7.)**

57. $3.923 \div 100$ **58.** $5.32 \div 100$ **59.** $98.02 \div 10$ **60.** $11.033 \div 10$

61. $0.027 \div 100$ **62.** $0.665 \div 100$ **63.** $1.02 \div 1000$ **64.** $8.1 \div 1000$

65. If 82.5 is divided by 0.1, will the decimal point move to the right or to the left? By how many places?

66. If 89.4201 is divided by 0.001, will the decimal point move to the right or to the left? By how many places?

For Exercises 67–74, divide by the powers of 0.1. **(See Example 8.)**

67. 5.03 ÷ 0.01

68. 4.44 ÷ 0.01

69. 0.992 ÷ 0.1

70. 0.31 ÷ 0.1

71. 3.2 ÷ 0.001

72. 1.1 ÷ 0.001

73. 123.4 ÷ 0.01

74. 420.6 ÷ 0.01

Objective 4: Applications of Decimal Division

When multiplying or dividing decimals, it is important to place the decimal point correctly. For Exercises 75–78, determine whether you think the number is reasonable or unreasonable. If the number is unreasonable, move the decimal point to a position that makes more sense.

75. Steve computed the gas mileage for his Honda Civic to be 3.2 miles per gallon.

76. The sale price of a new refrigerator is $96.0.

77. Mickey makes $8.50 per hour. He estimates his weekly paycheck to be $3400.

78. Jason works in a legal office. He computes the average annual income for the attorneys in his office to be $1400 per year.

For Exercises 79–84, solve the application. Check to see if your answers are reasonable.

79. A membership at a health club costs $560 per year. The club has a payment plan in which a member can pay $50 down and the rest in 12 equal payments. How much is each payment? **(See Example 9.)**

80. Brooke owes $39,628.68 on the mortgage for her house. If her monthly payment is $695.24, how many months does she still need to pay? How many years is this?

81. It is reported that on average 42,000 tennis balls are used and 650 matches are played at the Wimbledon tennis tournament each year. On average, how many tennis balls are used per match? Round to the nearest whole unit.

82. A package of dental floss contains 100 yd of floss. If Patty uses floss once a day and it lasts for 230 days, approximately how long is each piece that she uses? Round the answer to the nearest tenth of a yard.

83. In baseball, the batting average is found by dividing the number of hits by the number of times a batter was at bat. Babe Ruth was at bat 8399 times and had 2873 hits. What was his batting average? Round to the thousandths place. **(See Example 10.)**

84. Ty Cobb was at bat 11,434 times and had 4189 hits, giving him the all time best batting average. Find his average. Round to the thousandths place. (Refer to Exercise 83.)

Expanding Your Skills

85. What number is halfway between 47.26 and 47.27?

86. What number is halfway between 22.4 and 22.5?

87. Which numbers when divided by 8.6 will produce a quotient less than 12.4? Circle all that apply.

    a. 111.8   **b.** 103.2   **c.** 107.5   **d.** 105.78

88. Which numbers when divided by 5.3 will produce a quotient greater than 15.8? Circle all that apply.

    a. 84.8   **b.** 84.27   **c.** 83.21   **d.** 79.5

Calculator Connections

Topic: Dividing decimals

To divide numbers on a calculator, use the $\boxed{\div}$ key.

Expression	Keystrokes	Result
$.024\overline{)\,.0014064}$	.0014064 $\boxed{\div}$ .024 $\boxed{\text{Enter}}$	0.0586
$17 \div 3$	17 $\boxed{\div}$ 3 $\boxed{\text{Enter}}$	5.666666667

Note that the expression $17 \div 3$ results in the repeating decimal, $5.\overline{6}$. The calculator returns the value 5.666666667. This is not the exact value.

Calculator Exercises

89. A Chevy Blazer gets 16.5 mpg and a Toyota Corolla averages 36 mpg. Suppose a driver drives 12,000 mi/yr and the cost of gasoline is $2.25 per gallon. How much money would be saved by driving the Toyota Corolla rather than the Chevy Blazer? Round to the nearest dollar.

90. The estimated national debt for 2006 is $8,726,359 million (that is, roughly 8.7 trillion dollars). How much would each person have to pay if the national debt were divided evenly among the 298 million people living in the United States? Round to the nearest dollar.

91. Population *density* is defined to be the number of people per square mile of land area. If California has 41,373,000 people with a land area of 155,959 square miles (mi^2), what is the population density? Round to the nearest whole unit.

92. If Rhode Island has 1,158,000 people with a land area of 1045 mi^2, what is the population density? Round to the nearest whole unit.

Objectives

1. Writing Fractions as Decimals
2. Writing Decimals as Fractions
3. Decimals and the Number Line

Concept Connections

1. Write the fraction $\frac{2}{5}$ with a denominator of 10. Then write the equivalent decimal form.
2. Write the fraction $\frac{7}{20}$ with a denominator of 100. Then write the equivalent decimal form.

Skill Practice

Write each fraction or mixed number as a decimal.

3. $\frac{3}{8}$

4. $\frac{43}{20}$

5. $12\frac{5}{16}$

Answers

1. $\frac{4}{10}$; 0.4   2. $\frac{35}{100}$; 0.35

3. 0.375   4. 2.15   5. 12.3125

---

section 4.5   Fractions as Decimals

1. Writing Fractions as Decimals

Sometimes it is possible to convert a fraction to its equivalent decimal form by rewriting the fraction as a decimal fraction. That is, try to multiply the numerator and denominator by a number that will make the denominator a power of 10.

For example, the fraction $\frac{3}{5}$ can easily be written as an equivalent fraction with a denominator of 10.

$$\frac{3}{5} = \frac{3 \cdot 2}{5 \cdot 2} = \frac{6}{10} = 0.6$$

The fraction $\frac{3}{25}$ can easily be converted to a fraction with a denominator of 100.

$$\frac{3}{25} = \frac{3 \cdot 4}{25 \cdot 4} = \frac{12}{100} = 0.12$$

This technique is useful in some cases. However, some fractions such as $\frac{1}{3}$ cannot be converted to a fraction with a denominator that is a power of 10. This is so because 3 is not a factor of any power of 10. For this reason, we recommend the following alternative method.

Recall that a fraction bar implies division of the numerator by the denominator. If we carry out that division, we convert the fraction to its decimal form.

example 1   Writing Fractions as Decimals

Write each fraction or mixed number as a decimal.

a. $\frac{3}{5}$ **b.** $\frac{68}{25}$ **c.** $3\frac{5}{8}$

Solution:

a. $\frac{3}{5}$ means $3 \div 5$.

$$\begin{array}{r} .6 \\ 5\overline{)3.0} \\ -30 \\ \hline 0 \end{array}$$

Divide the numerator by the denominator.

$\frac{3}{5} = 0.6$

b. $\frac{68}{25}$ means $68 \div 25$.

$$\begin{array}{r} 2.72 \\ 25\overline{)68.00} \\ -50 \\ \hline 180 \\ -175 \\ \hline 50 \\ -50 \\ \hline 0 \end{array}$$

Divide the numerator by the denominator.

$\frac{68}{25} = 2.72$

c. $3\frac{5}{8} = 3 + 5 \div 8$

$$\begin{array}{r} .625 \\ 8\overline{)5.000} \\ -48 \\ \hline 20 \\ -16 \\ \hline 40 \\ -40 \\ \hline 0 \end{array}$$

Divide the numerator by the denominator.

$3\frac{5}{8} = 3 + 0.625$

$= 3.625$

The fractions in Example 1 are represented by terminating decimals. However, many fractions convert to repeating decimals.

| example 2 | Converting Fractions to Repeating Decimals |

Write each fraction as a decimal.

a. $\dfrac{4}{9}$ **b.** $\dfrac{5}{6}$ **c.** $\dfrac{4}{7}$

Solution:

a. $\dfrac{4}{9}$ means $4 \div 9$.

$$
\begin{array}{r}
.44\ldots \\
9\overline{)4.00} \\
-36 \\
\hline
40 \\
-36 \\
\hline
40
\end{array}
$$

The quotient is a repeating decimal.

$\dfrac{4}{9} = 0.\overline{4}$

b. $\dfrac{5}{6}$ means $5 \div 6$.

$$
\begin{array}{r}
.833\ldots \\
6\overline{)5.000} \\
-48 \\
\hline
20 \\
-18 \\
\hline
20 \\
-18 \\
\hline
20
\end{array}
$$

The quotient is a repeating decimal.

$\dfrac{5}{6} = 0.8\overline{3}$

c. $\dfrac{4}{7}$ means $4 \div 7$.

$$
\begin{array}{r}
.571428\ldots \\
7\overline{)4.000000} \\
-35 \\
\hline
50 \\
-49 \\
\hline
10 \\
-7 \\
\hline
30 \\
-28 \\
\hline
20 \\
-14 \\
\hline
60 \\
-56 \\
\hline
40
\end{array}
$$

The cycle will repeat.

$\dfrac{4}{7} = 0.571428571428571428571428\ldots = 0.\overline{571428}$

Skill Practice

Write each fraction as a decimal.

6. $\dfrac{8}{9}$

7. $\dfrac{1}{12}$

8. $\dfrac{3}{7}$

Calculator Connections

Repeating decimals displayed on a calculator must be rounded. For example, the repeating decimals from Example 2 might appear as follows.

$0.\overline{4}$ | 0.444444444 |

$0.8\overline{3}$ | 0.833333333 |

$0.\overline{571428}$ | 0.571428571 |

Several fractions are used quite often. Their decimal forms are worth memorizing and are presented in Table 4-1.

Answers

6. $0.\overline{8}$ 7. $0.08\overline{3}$ 8. $0.\overline{428571}$

9. Describe the pattern between the fractions $\frac{1}{9}$ and $\frac{2}{9}$ and their decimal forms.

table 4-1

$\frac{1}{4} = 0.25$	$\frac{2}{4} = \frac{1}{2} = 0.5$	$\frac{3}{4} = 0.75$	
$\frac{1}{9} = 0.\overline{1}$	$\frac{2}{9} = 0.\overline{2}$	$\frac{3}{9} = \frac{1}{3} = 0.\overline{3}$	$\frac{4}{9} = 0.\overline{4}$
$\frac{5}{9} = 0.\overline{5}$	$\frac{6}{9} = \frac{2}{3} = 0.\overline{6}$	$\frac{7}{9} = 0.\overline{7}$	$\frac{8}{9} = 0.\overline{8}$

Skill Practice

Convert the fraction to a decimal rounded to the indicated place value.

10. $\frac{9}{7}$; tenths

11. $\frac{17}{37}$; hundredths

---

example 3   Converting Fractions to Decimals with Rounding

Convert the fraction to a decimal rounded to the indicated place value.

a. $\frac{162}{7}$; tenths place   **b.** $\frac{21}{31}$; hundredths place

Solution:

a. $\frac{162}{7}$

$$
\begin{array}{r}
23.14 \\
7\overline{)162.00} \\
-14 \\
\hline
22 \\
-21 \\
\hline
10 \\
-7 \\
\hline
30 \\
-28 \\
\hline
2
\end{array}
$$

← tenths place
← hundredths place

To round to the tenths place, we must determine the hundredths-place digit and use it to base our decision on rounding. The hundredths-place digit is 4. Therefore, leave the tenths-place digit unchanged. The quotient rounds to 23.1.

The fraction $\frac{162}{7}$ is approximately 23.1.

b. $\frac{21}{31}$

$$
\begin{array}{r}
.677 \\
31\overline{)21.000} \\
-186 \\
\hline
240 \\
-217 \\
\hline
230 \\
-217 \\
\hline
13
\end{array}
$$

← hundredths place
← thousandths place

To round to the hundredths-place, we must determine the thousandths-place digit and use it to base our decision on rounding. The thousandths-place digit is 7. Therefore, increase the hundredths-place digit by 1. The rounded value is 0.68.

The fraction $\frac{21}{31}$ is approximately 0.68.

---

2. Writing Decimals as Fractions

In Section 4.1 we converted terminating decimals to fractions. We did this by writing the decimal as a decimal fraction and then reducing the fraction to lowest terms. For example,

$$0.46 = \frac{46}{100} = \frac{\overset{23}{\cancel{46}}}{\underset{50}{\cancel{100}}} = \frac{23}{50}$$

We do not yet have the tools to convert a repeating decimal to its equivalent fraction form. However, we can make use of our knowledge of the common fractions and their repeating decimal forms from Table 4-1.

---

example 4   Writing Decimals as Fractions and Fractions as Decimals

Complete the table.

	Decimal Form	Fractional Form
a.	0.475	
b.		$\frac{3}{16}$
c.		$2\frac{4}{5}$
d.	$0.\overline{6}$	
e.		$\frac{19}{11}$

Skill Practice

Complete the table.

	Decimal Form	Fractional Form
12.	0.875	
13.		$\frac{7}{20}$
14.		$2\frac{1}{3}$
15.	$0.\overline{7}$	

Solution:

a. $0.475 = \dfrac{475}{1000} = \dfrac{5 \cdot 5 \cdot 19}{2 \cdot 2 \cdot 2 \cdot 5 \cdot 5 \cdot 5} = \dfrac{19}{40}$

b. $\dfrac{3}{16} = 3 \div 16$

$$\begin{array}{r} .1875 \\ 16\overline{)3.0000} \\ -16 \\ \hline 140 \\ -128 \\ \hline 120 \\ -112 \\ \hline 80 \\ -80 \\ \hline 0 \end{array}$$

Therefore, $\dfrac{3}{16} = 0.1875$

c. To convert $2\frac{4}{5}$ to decimal form, we need to convert $\frac{4}{5}$ to decimal form. This may be done by dividing. Or we can easily convert $\frac{4}{5}$ to a decimal fraction with a denominator of 10.

$$\frac{4}{5} = \frac{4 \cdot 2}{5 \cdot 2} = \frac{8}{10} = 0.8$$

Therefore, $2\dfrac{4}{5} = 2.8$

d. From Table 4-1, the decimal $0.\overline{6} = \dfrac{2}{3}$.

e. $\dfrac{19}{11}$ means $19 \div 11$.

$$\begin{array}{r} 1.7272\ldots \\ 11\overline{)19.0000} \\ -11 \\ \hline 80 \\ -77 \\ \hline 30 \\ -22 \\ \hline 80 \end{array}$$

The cycle will repeat.

$\dfrac{19}{11} = 1.\overline{72}$

Answers

	Decimal Form	Fractional Form
12.	0.875	$\frac{7}{8}$
13.	0.35	$\frac{7}{20}$
14.	$2.\overline{3}$	$2\frac{1}{3}$
15.	$0.\overline{7}$	$\frac{7}{9}$

We can now complete the table.

	Decimal Form	Fractional Form
a.	0.475	$\frac{19}{40}$
b.	0.1875	$\frac{3}{16}$
c.	2.8	$2\frac{4}{5}$
d.	$0.\overline{6}$	$\frac{2}{3}$
e.	$1.\overline{72}$	$\frac{19}{11}$

3. Decimals and the Number Line

In Example 5, we rank the numbers from least to greatest and visualize the position of the numbers on the number line.

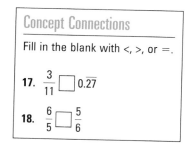
example 5   Ordering Decimals and Fractions

Rank the numbers from least to greatest. Then approximate the position of the points on the number line.

$$0.\overline{45},\ 0.45,\ \frac{1}{2}$$

Solution:

First note that $\frac{1}{2} = 0.5$ and that $0.\overline{45} = 0.454545\cdots$. By writing each number in decimal form, we can compare the decimals as we did in Section 4.1.

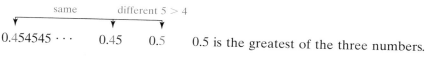

0.454545 ···   0.45   0.5   0.5 is the greatest of the three numbers.

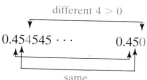

Notice that we inserted an extra zero to the right of 0.45 so that we could compare common place values. The number $0.\overline{45} > 0.45$.

Ranking the numbers from least to greatest, we have 0.45, $0.\overline{45}$, $\frac{1}{2}$.

The relative position of these numbers can be seen on the number line. First note that we have expanded the segment of the number line between 0.4 and 0.5 to see more place values to the right of the decimal point.

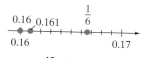
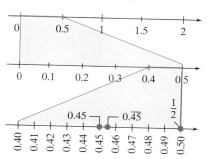

Recall that numbers that lie to the left on the number line have lesser value than numbers that lie to the right.

section 4.5	**Practice Exercises**

Boost *your* GRADE at mathzone.com!

- Practice Problems
- Self-Tests
- NetTutor
- e-Professors
- Videos

Study Skills Exercise

1. In a study group, check which activities you might try to help you learn and understand the material.

   ☐ Quiz one another by asking one another questions.

   ☐ Practice teaching one another.

   ☐ Share and compare class notes.

   ☐ Support and encourage one another.

   ☐ Work together on exercises and sample problems.

Review Exercises

For Exercises 2–5, write the decimal fraction in decimal form.

2. $\dfrac{9}{10}$ 3. $\dfrac{39}{100}$ 4. $\dfrac{141}{1000}$ 5. $\dfrac{71}{10,000}$

For Exercises 6–9, write the decimals as fractions.

6. 0.6   7. 0.0016   8. 0.35   9. 0.125

Objective 1: Writing Fraction as Decimals

For Exercises 10–16, write each fraction as a decimal fraction, that is, a fraction whose denominator is a power of 10. Then write the number in decimal form.

10. $\dfrac{7}{25}$ 11. $\dfrac{4}{25}$ 12. $\dfrac{316}{500}$ 13. $\dfrac{19}{500}$

14. $\dfrac{2}{5}$ 15. $\dfrac{4}{5}$ 16. $\dfrac{49}{50}$

For Exercises 17–32, write each fraction or mixed number as a decimal. **(See Example 1.)**

17. $\dfrac{7}{8}$ 18. $\dfrac{16}{64}$ 19. $\dfrac{5}{16}$ 20. $\dfrac{31}{50}$

21. $5\dfrac{3}{12}$ 22. $4\dfrac{1}{16}$ 23. $1\dfrac{1}{5}$ 24. $6\dfrac{5}{8}$

25. $\dfrac{18}{24}$ 26. $\dfrac{24}{40}$ 27. $2\dfrac{7}{20}$ 28. $3\dfrac{4}{25}$

29. $7\dfrac{9}{20}$ **30.** $3\dfrac{11}{25}$ **31.** $\dfrac{22}{25}$ **32.** $\dfrac{11}{20}$

For Exercises 33–48, write each fraction as a repeating decimal. **(See Example 2.)**

33. $3\dfrac{8}{9}$ **34.** $4\dfrac{7}{9}$ **35.** $\dfrac{7}{15}$ **36.** $\dfrac{5}{18}$

37. $\dfrac{19}{36}$ **38.** $\dfrac{7}{12}$ **39.** $\dfrac{6}{11}$ **40.** $\dfrac{8}{33}$

41. $\dfrac{14}{111}$ **42.** $\dfrac{58}{111}$ **43.** $\dfrac{41}{333}$ **44.** $\dfrac{68}{333}$

45. $\dfrac{1}{7}$ **46.** $\dfrac{2}{7}$ **47.** $\dfrac{1}{13}$ **48.** $\dfrac{9}{13}$

For Exercises 49–54, convert the fraction to a decimal and round to the indicated place value. **(See Example 3.)**

49. $\dfrac{15}{16}$; tenths   **50.** $\dfrac{3}{11}$; tenths   **51.** $\dfrac{5}{7}$; hundredths

52. $\dfrac{1}{8}$; hundredths   **53.** $\dfrac{25}{21}$; tenths   **54.** $\dfrac{18}{13}$; tenths

55. Show that $\frac{1}{4} = 0.25$, using two methods.

56. Show that $\frac{1}{2} = 0.5$, using two methods.

57. Write the fractions as decimals. Explain how to memorize the decimal form for these fractions with a denominator of 9.

     a. $\dfrac{1}{9}$ **b.** $\dfrac{2}{9}$ **c.** $\dfrac{4}{9}$ **d.** $\dfrac{5}{9}$

58. Write the fractions as decimals. Explain how to memorize the decimal forms for these fractions with a denominator of 3.

     a. $\dfrac{1}{3}$ **b.** $\dfrac{2}{3}$

59. Given that $\frac{1}{8} = 0.125$, use the method described in Exercises 57 and 58 to find the decimal form.

     a. $\dfrac{3}{8}$ **b.** $\dfrac{5}{8}$ **c.** $\dfrac{7}{8}$

60. Given that $\frac{1}{5}$ = 0.2, use the method described in Exercises 57 and 58 to find the decimal form.

 a. $\frac{2}{5}$ **b.** $\frac{3}{5}$ **c.** $\frac{4}{5}$

Objective 2: Writing Decimals as Fractions

For Exercises 61–64, complete the table. **(See Example 4.)**

61.

	Decimal Form	Fraction Form
a.	0.45	
b.		$1\frac{5}{8}$ or $\frac{13}{8}$
c.	$0.\overline{7}$	
d.		$\frac{5}{11}$

62.

	Decimal Form	Fraction Form
a.		$\frac{2}{3}$
b.	1.6	
c.		$\frac{152}{25}$
d.	$0.\overline{2}$	

63.

	Decimal Form	Fraction Form
a.	$0.\overline{3}$	
b.	2.125	
c.		$\frac{19}{22}$
d.		$\frac{42}{25}$

64.

	Decimal Form	Fraction Form
a.	0.75	
b.		$\frac{7}{11}$
c.	$1.\overline{8}$	
d.		$\frac{74}{25}$

Historically stock prices were given as fractions or mixed numbers, but are now given as decimals. For Exercises 65–66, complete the table that gives recent stock prices taken from the *Wall Street Journal*.

65.

Stock	Symbol	Closing Price ($) (Decimal)	Closing Price ($) (Fraction)
Corning	GLW	12.38	
Walgreen	WAG	34.95	
Brookstone	BKST		$19\frac{1}{2}$
SnapOn	SNA		$33\frac{11}{25}$

66.

Stock	Symbol	Closing Price ($) (Decimal)	Closing Price ($) (Fraction)
Dell	DELL	35.33	
StrideRite	SRR		$10\frac{18}{25}$
Intel	INTC	28.10	
Checkers	CHKR		$10\frac{3}{20}$

Objective 3: Decimals and the Number Line

For Exercises 67–78, insert the appropriate symbol. Choose from $<$, $>$, or $=$.

67. $0.2 \square \frac{1}{5}$

68. $1.5 \square \frac{3}{2}$

69. $0.2 \square 0.\overline{2}$

70. $\frac{3}{5} \square 0.\overline{6}$

71. $\frac{1}{3} \square 0.3$

72. $\frac{2}{3} \square 0.66$

73. $4\frac{1}{4} \square 4.\overline{25}$

74. $2.12 \square 2.\overline{12}$

75. $0.\overline{5} \square \frac{5}{9}$

76. $\frac{7}{4} \square 1.75$

77. $0.27 \square \frac{3}{11}$

78. $6.4\overline{3} \square 6.43$

For Exercises 79–82, rank the numbers from least to greatest. Then approximate the position of the points on the number line. **(See Example 5.)**

79. $0.\overline{1}, \frac{1}{10}, \frac{1}{5}$

80. $3\frac{1}{4}, 3\frac{1}{3}, 3.3$

81. $1.8, 1.75, 1.\overline{7}$

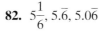

82. $5\frac{1}{6}, 5.\overline{6}, 5.0\overline{6}$

Expanding Your Skills

83. If $0.\overline{8} = \frac{8}{9}$, then what is the fraction form of $0.\overline{9}$?

For Exercises 84–86, simplify.

84. $1.\overline{9}$

85. $6.\overline{9}$

86. $15.\overline{9}$

section 4.6   Order of Operations and Applications of Decimals

Objectives

1. Order of Operations Involving Decimals
2. Calculations with Decimals and Fractions
3. Calculations with Round-off Error
4. Applications of Decimals and Fractions

1. Order of Operations Involving Decimals

In Example 1, we perform the order of operations with an expression involving decimal numbers.

example 1   Applying the Order of Operations by Using Decimal Numbers

Simplify.

$$16.4 - (6.7 - 3.5)^2$$

Solution:

$16.4 - (6.7 - 3.5)^2$

$= 16.4 - (3.2)^2$ Perform the subtraction within parentheses first.

$$\begin{array}{r} 6.7 \\ -\ 3.5 \\ \hline 3.2 \end{array}$$

$= 16.4 - 10.24$ Perform the operation involving the exponent.

$$\begin{array}{r} 3.2 \\ \times\ 3.2 \\ \hline 64 \\ 960 \\ \hline 10.24 \end{array}$$

$= 6.16$ Subtract. $\begin{array}{r} \overset{3\ 10}{1\,6.\,4\,\cancel{0}} \\ -\ 1\,0.\,2\,4 \\ \hline 6.\,1\,6 \end{array}$

Skill Practice

Simplify.

1. $34.1 - 3(1.6)^2$
2. $(5.8 - 4.3)^2 - 2$

2. Calculations with Decimals and Fractions

In Sections 4.1 and 4.5 we learned how to convert between fraction notation and decimal notation. In this section we apply the order of operations on fractions and decimals combined.

example 2   Dividing a Decimal and a Mixed Number

Divide.

$$1.52 \div 1\frac{3}{5}$$

Solution:

For this example we will show two approaches. In the first approach, we convert both numbers to fractional form and then divide. In the second approach, we convert both numbers to decimal form and then divide.

Skill Practice

Multiply $4.2\left(\frac{3}{4}\right)$ by the following approaches.

3. Approach 1: $\frac{42}{10} \cdot \frac{3}{4}$
4. Approach 1 (modified): $\frac{4.2}{1} \cdot \frac{3}{4}$
5. Approach 2: $(4.2)(0.75)$

Answers

1. 26.42   2. 0.25
3. $\frac{63}{20}$ or 3.15   4. $\frac{63}{20}$ or 3.15
5. 3.15

Approach 1

Convert both numbers to fractional form.

$$1.52 \div 1\frac{3}{5} = \frac{152}{100} \div \frac{8}{5}$$ Convert the decimal and mixed number to fractional form.

$$= \frac{152}{100} \cdot \frac{5}{8}$$ Multiply by the reciprocal of the divisor.

$$= \frac{\overset{19}{\cancel{152}}}{\underset{20}{\cancel{100}}} \cdot \frac{\overset{1}{\cancel{5}}}{\underset{1}{\cancel{8}}}$$ Simplify common factors.

$$= \frac{19}{20}$$ Multiply fractions.

We can write the quotient as $\frac{19}{20}$ or in its equivalent decimal form as 0.95.

Approach 1 (Modified)

A modification to approach 1 is to write the decimal 1.52 as $\frac{1.52}{1}$. (Recall that any number divided by 1 equals the number.)

$$1.52 \div 1\frac{3}{5} = \frac{1.52}{1} \div \frac{8}{5}$$

$$= \frac{1.52}{1} \cdot \frac{5}{8}$$ Multiply by the reciprocal of the divisor.

$$= \frac{7.6}{8}$$ Multiply fractions. Note that $1.52 \times 5 = 7.6$.

$$= 0.95$$ Divide. $\begin{array}{r} .95 \\ 8\overline{)7.60} \\ \underline{-72} \\ 40 \\ \underline{-40} \\ 0 \end{array}$

Approach 2

Convert the numbers to decimal form.

$$1.52 \div 1\frac{3}{5} = 1.52 \div 1.6$$ Convert the mixed number to a decimal.

$$= 0.95$$ Divide. $1.6\overline{)1.52}$ $\begin{array}{r} .95 \\ 16\overline{)15.20} \\ \underline{-144} \\ 80 \\ \underline{-80} \\ 0 \end{array}$

Skill Practice

6. Divide, using any approach.

$$21.7 \div \frac{7}{2}$$

Skill Practice

Simplify.

7. $5.7 \div \frac{1}{4} \cdot \frac{2}{5}$

8. $2.6 \cdot \frac{3}{10} \div 1\frac{1}{2}$

Answers

6. 6.2   7. 9.12   8. 0.52

---

example 3   **Applying the Order of Operations**

Simplify.

$$6.4 \times 2\frac{5}{8} \div \left(\frac{3}{5}\right)^2$$

Solution:

Approach 1
Convert all numbers to fractional form.

$$6.4 \times 2\frac{5}{8} \div \left(\frac{3}{5}\right)^2 = \frac{64}{10} \times \frac{21}{8} \div \left(\frac{3}{5}\right)^2 \qquad$$ Convert the decimal and mixed numbers to fractions.

$$= \frac{64}{10} \times \frac{21}{8} \div \frac{9}{25} \qquad$$ Square the quantity $\frac{3}{5}$.

$$= \frac{64}{10} \times \frac{21}{8} \times \frac{25}{9} \qquad$$ Multiply by the reciprocal of the divisor.

$$= \frac{\overset{4}{\cancel{\overset{8}{\cancel{64}}}}}{\underset{1}{\underset{2}{\cancel{10}}}} \times \frac{\overset{7}{\cancel{21}}}{\underset{1}{\cancel{8}}} \times \frac{\overset{5}{\cancel{25}}}{\underset{3}{\cancel{9}}} \qquad$$ Simplify common factors.

$$= \frac{140}{3} \text{ or } 46\frac{2}{3} \text{ or } 46.\overline{6} \qquad$$ Multiply.

Approach 2
Convert all numbers to decimal form.

$$6.4 \times 2\frac{5}{8} \div \left(\frac{3}{5}\right)^2 = 6.4 \times 2.625 \div (0.6)^2 \qquad$$ The fraction $\frac{5}{8} = 0.625$ and $\frac{3}{5} = 0.6$.

$$= 6.4 \times 2.625 \div 0.36 \qquad$$ Square the quantity 0.6. That is, $(0.6)(0.6) = 0.36$.

Multiply 6.4×2.625.

$$\begin{array}{r} 2.625 \\ \times\ 6.4 \\ \hline 10500 \\ 157500 \\ \hline 16.8000 \end{array}$$

$$= 16.8 \div 0.36 \qquad$$ Divide $16.8 \div 0.36$. $\quad.36\overline{)16.80}$

$$= 46.\overline{6}$$

$$\begin{array}{r} 46.6\ldots \\ 36\overline{)1680} \\ -144 \\ \hline 240 \\ -216 \\ \hline 240 \end{array}$$

3. Calculations with Round-off Error

example 4	Multiplying a Fraction and a Decimal

Multiply.

$$2.52 \cdot \left(\frac{5}{6}\right)$$

Solution:

Approach 1
Convert 2.52 to fractional form and then multiply fractions.

$$\frac{252}{100} \cdot \frac{5}{6} = \frac{\overset{126}{\cancel{252}}}{\underset{20}{\cancel{100}}} \cdot \frac{\overset{1}{\cancel{5}}}{\underset{3}{\cancel{6}}} = \frac{\overset{21}{\cancel{126}}}{\underset{10}{\cancel{60}}} = \frac{21}{10} \qquad \text{Simplify common factors and multiply fractions.}$$

$$= 2.1 \qquad \text{Divide: } 21 \div 10 = 2.1.$$

Approach 2
Convert $\frac{5}{6}$ to decimal form and then multiply the decimals. However, $\frac{5}{6} = 0.8\overline{3} = 0.8333\ldots$ and we do not know how to multiply repeating decimals. We can approximate the product by rounding the value $0.8\overline{3}$ to some desired level of accuracy. Suppose we round $0.8\overline{3}$ to the thousandths place. Then $0.8\overline{3} \approx 0.833$.

$$2.52 \cdot (0.8\overline{3}) \approx 2.52 \cdot (0.833) \qquad \text{Round } 0.8\overline{3} \approx 0.833.$$

$$= 2.09916 \qquad \text{Multiply decimals.}$$

$$\begin{array}{r} 2.52 \\ \times\ .833 \\ \hline 756 \\ 7560 \\ 201600 \\ \hline 2.09916 \end{array}$$

The approximated value 2.09916 is close to 2.1.

Notice that the second approach in Example 4 was not as accurate as the first method. This is so because we used "intermediate rounding." Intermediate rounding refers to rounding a number before it is used in a calculation. We rounded the number $0.8\overline{3}$ *before* multiplying. Keep in mind that a rounded number is not exact. Any calculation performed on a rounded number compounds the error.

To minimize the effects of round-off error, you can try to keep the fraction notation as long as possible in the expression. In this way, if you do choose to convert to decimal form, you perform the division in the last step. Any rounding is done at the end.

| example 5 | Dividing a Fraction and Decimal |

Divide $\frac{4}{7} \div 3.6$. Round the answer to the nearest hundredth.

Solution:

If we attempt to write $\frac{4}{7}$ as a decimal, we find that it is the repeating decimal $0.\overline{571428}$. Therefore, we choose to change 3.6 to fractional form: $3.6 = \frac{36}{10}$.

$$\frac{4}{7} \div 3.6 = \frac{4}{7} \div \frac{36}{10} \qquad \text{Write 3.6 as a fraction.}$$

$$= \frac{\overset{1}{4}}{7} \cdot \frac{10}{\underset{9}{36}} \qquad \text{Multiply by the reciprocal of the divisor.}$$

$$= \frac{10}{63} \qquad \text{Multiply and reduce to lowest terms.}$$

Concept Connections

10. Why is the product 18.9×0.4 not equal to the product $18.9 \times 0.\overline{4}$?

Skill Practice

11. Divide.

$$4.17 \div \frac{3}{2}$$

12. Divide. Round the answer to the nearest hundredth.

$$4.1 \div \frac{12}{5}$$

13. Multiply. Round the answer to the nearest hundredth.

$$\frac{3}{11} \times 2.4$$

Answers

10. The numbers 0.4 and $0.\overline{4}$ are not equal. Therefore, they will form a different product when multiplied by 18.9.
11. 2.78 or $\frac{139}{50}$ or $2\frac{39}{50}$ 12. 1.71
13. 0.65

We must write the answer in decimal form, rounded to the nearest hundredth.

$$
\begin{array}{r}
.158 \\
63\overline{)10.00} \\
-63 \\
\hline
370 \\
-315 \\
\hline
550 \\
-504 \\
\hline
46
\end{array}
$$

Divide. To round to the hundredths place, divide until we find the thousandths-place digit in the quotient. Use that digit to make a decision for rounding.

≈ 0.16 Round to the nearest hundredth.

4. Applications of Decimals and Fractions

example 6   Using Decimals and Fractions in a Consumer Application

Joanne filled the gas tank in her car and noted that the odometer read 22,341.9 mi. Ten days later she filled the tank again with $11\frac{1}{2}$ gal of gas. Her odometer reading at that time was 22,622.5 mi.

a. How many miles had she driven between fill-ups?

b. How many miles per gallon did she get?

Solution:

a. To find the number of miles driven, we need to subtract the initial odometer reading from the final reading.

$$
\begin{array}{r}
\overset{5\ \ 12\ \ 1\ \ 15}{2\,2\,,\,6\,2\,2.\,5} \\
-\ 2\,2\,,\,3\,4\,\ 1.\,9 \\
\hline
2\,8\ \ 0.\,6
\end{array}
$$

Recall that to add or subtract decimals, line up the decimal points.

Joanne had driven 280.6 mi between fill-ups.

b. To find the number of miles per gallon (mi/gal), we divide the number of miles driven by the number of gallons.

$280.6 \div 11\frac{1}{2} = 280.6 \div 11.5$ We convert to decimal form because the fraction $11\frac{1}{2}$ is recognized as 11.5.

$= 24.4$

$$
11.5\overline{)280.6}
$$

$$
\begin{array}{r}
24.4 \\
115\overline{)2806.0} \\
-230 \\
\hline
506 \\
-460 \\
\hline
460 \\
-460 \\
\hline
0
\end{array}
$$

Joanne got 24.4 mi/gal.

Answers

14. a. 821.5 mi   b. 62 mph

Skill Practice

15. A marathon is 26.2 mi. A runner stops $\frac{3}{5}$ of the way through the race.

    a. How many miles had he run?

    b. How many miles were left in the race?

---

example 7   Using Decimals and Fractions to Compute a Lawsuit Settlement

Althea won a legal settlement for $4105.20. Her lawyer received $\frac{1}{3}$ of the settlement.

    a. How much money did the lawyer get?

    b. How much money did Althea get?

Solution:

    a. The lawyer got $\frac{1}{3}$ of $4105.20. This implies multiplication.

$$\frac{1}{3} \cdot (4105.20)$$

          The fraction $\frac{1}{3}$ cannot be written as a terminating decimal.

$$= \frac{1}{3} \cdot \frac{4105.20}{1}$$

          We can write $4105.20 as $\frac{4105.20}{1}$ and multiply fractions.

$$= \frac{4105.20}{3}$$

          Multiply numerators. Multiply denominators.

$$= 1368.4$$

          Divide: $4105.20 \div 3 = 1368.4$.

The lawyer got $1368.40.

    b. Althea got the remaining portion of the money.

$$
\begin{array}{r}
\$4105.20 \\
-\ \ 1368.40 \\
\hline
\$2736.80
\end{array}
$$

Althea received $2736.80.

Skill Practice

16. A patchwork quilt is made up of quilted squares $1\frac{1}{4}$ ft on a side. Suppose a quilt measures 4 squares by 5 squares.

    a. Find the perimeter of the quilt.

    b. Find the area of the quilt.

---

example 8   Applying Decimals and Fractions to Geometry

A rectangular floor is 20 m by $17\frac{1}{2}$ m. If flooring costs $12.25 per square meter (m²), how much will it cost to put down flooring in the room?

Solution:

First draw a figure representing the rectangular floor (Figure 4-5). Because the cost of the flooring is given in terms of square meters, we need to know the area of the floor. (Recall that square units such as m², ft², and in.² are units of area.)

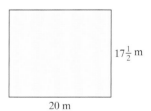

$17\frac{1}{2}$ m

20 m

Figure 4-5

For a rectangle,

$$A = l \cdot w$$

$$= (20)\left(17\frac{1}{2}\right)$$

$$= (20)(17.5)$$

    In this case, we convert each number to decimal form because later we will have to multiply by $12.25.

$$= 350$$

The floor is 350 m².

$$
\begin{array}{r}
{\scriptstyle 1\ 1} \\
17.5 \\
\times\ 20 \\
\hline
0 \\
3500 \\
\hline
350.0
\end{array}
$$

Now multiply the cost per square meter by the number of square meters.

Cost = ($12.25)(350)

$$= 4287.50$$

$$\begin{array}{r} 12.25 \\ \times\ 350 \\ \hline 0 \\ 61250 \\ 367500 \\ \hline 4287.50 \end{array}$$

The cost of the flooring is $4287.50.

---

example 9   Finding an Average

Table 4-2 represents the average snowfall for 6 winter months in Syracuse, New York. What is the mean (average) amount of snowfall per winter month? Round to the nearest tenth of an inch.

table 4-2

Month	Snowfall (in.)
November	9.3
December	26.8
January	29.6
February	26.2
March	17.3
April	4.0

Solution:

To find the average, we must add the values. Then divide by the number of values (in this case, 6).

To find the sum, line up the addends:

$$\begin{array}{r} {}^{1\ 4\ 2} \\ 9.3 \\ 26.8 \\ 29.6 \\ 26.2 \\ 17.3 \\ +\ \ 4.0 \\ \hline 113.2 \end{array}$$

The total snowfall for these 6 months is 113.2 in. To find the average, divide by 6.

Answer

17. The average is $484.41.

$$
\begin{array}{r}
18.86 \\
6\overline{)113.20} \\
\underline{-6} \\
53 \\
\underline{-48} \\
52 \\
\underline{-48} \\
40 \\
\underline{-36} \\
4
\end{array}
$$

To round to the tenths place, divide until we find the hundredths-place digit. Use that digit to make a decision for rounding.

The number 18.86 rounds to 18.9.

Syracuse averages about 18.9 in. of snow per month during these 6 months.

section 4.6 Practice Exercise

Boost *your* GRADE at mathzone.com!

MathZone

• Practice Problems   • e-Professors
• Self-Tests   • Videos
• NetTutor

Study Skills Exercise

1. In addition to studying the material for a test, here are some other activities that people use when preparing for a test. Circle the importance of each statement.

	Not important	Somewhat important	Very important
a. Get a good night's sleep the night before the test.	1	2	3
b. Eat a good breakfast on the day of the test.	1	2	3
c. Wear comfortable clothes on the day of the test.	1	2	3
d. Arrive early to class on the day of the test.	1	2	3

Review Exercises

For Exercises 2–9, perform the indicated operation.

2. $\left(\dfrac{24}{7}\right)\left(\dfrac{35}{36}\right)$

3. 34.1×9.2

4. $790.9 + 23.91$

5. $\dfrac{34}{9} + \dfrac{5}{27}$

6. $56.7 \div 1.2$

7. $\dfrac{55}{16} \div \dfrac{11}{4}$

8. $\dfrac{9}{4} - \dfrac{7}{8}$

9. $13 - 6.04$

Objective 1: Order of Operations Involving Decimals

10. List the order of operations.

For Exercises 11–20, simplify by using the order of operations. **(See Example 1.)**

11. $12.46 - 3.05 - 0.8^2$ **12.** $15.06 - 1.92 - 0.4^2$ **13.** $63.75 - 9.5(4)$ **14.** $6.84 + (3.6)(9)$

15. $(3.7 - 1.2)^2$ **16.** $(6.8 - 4.7)^2$ **17.** $6.8 \div 2 \div 1.7$ **18.** $8.4 \div 2 \div 2.1$

19. $2.2 + [9.34 + (1.2)^2]$ **20.** $(3.1)^2 - (4.2 \div 2.1)$

Objective 2: Calculations with Decimals and Fractions

For Exercises 21–26, simplify by using the order of operations. Express the answer in decimal form. **(See Examples 2–3.)**

21. $\dfrac{3}{4} \times 89.8$ **22.** $30.12 \times \dfrac{5}{8}$ **23.** $20.04 \div \dfrac{4}{5}$

24. $(78.2 - 60.2) \div \dfrac{9}{13}$ **25.** $14.4 \times \left(\dfrac{7}{4} - \dfrac{1}{8}\right)$ **26.** $6.5 + \dfrac{1}{8} \times 0.24$

Objective 3: Calculations with Round-off Error

For Exercises 27–32, perform the indicated operations. Round the answer to the nearest hundredth when necessary. **(See Examples 4–5.)**

27. $2.3 \times \dfrac{5}{9}$ **28.** $4.6 \times \dfrac{1}{6}$ **29.** $6.5 \div \dfrac{3}{5}$

30. $\dfrac{1}{12} \times 6.24 \div 2.1$ **31.** $(42.81 - 30.01) \div \dfrac{9}{2}$ **32.** $\dfrac{2}{7} \times 5.1 \times \dfrac{1}{10}$

For Exercises 33–36, perform the indicated operation. Write the answer as a repeating decimal.

33. $\dfrac{2}{9} \times 4.21$ **34.** $6.02 \div \dfrac{22}{23}$ **35.** $5.32 \div \dfrac{6}{5}$ **36.** $\dfrac{34}{11} \times 2.5$

Objective 4: Applications of Decimals and Fractions

37. Professor McGonagal earns $14.50 per hour for the first 40 hr worked each week. For hours worked over 40 hr, she earns time and a half. **(See Example 6.)**

    a. What is Professor McGonagal's hourly overtime wage?

    b. How much does she earn in a week in which she works 50 hr?

38. Jennifer earns $18.00 per hour. Alex's hourly wage is $\frac{3}{4}$ of what Jennifer makes per hour.

    a. How much does Alex make per hour?

    b. If they both work 40 hr, how much do they each make?

39. A cell phone plan calls for a $39.95 monthly fee and includes 450 min. For time on the phone over 450 min, the charge is $0.40 per minute. How much is Jorge charged for a month in which he talks for 597 min?

40. A night at a hotel in Dallas costs $129.95 with a nightly room tax of $20.75. The hotel also charges $1.10 per phone call made from the phone in the room. If Radcliff stays for 5 nights and makes 3 phone calls, how much is his total bill?

41. Susan's diet allows her 60 grams (g) of fat per day. If she has $\frac{1}{4}$ of her total fat grams for breakfast and a McDonald's Quarter Pounder (20.7 g of fat) for lunch, how many grams does she have left for dinner? **(See Example 7.)**

42. Todd is establishing his beneficiaries for his life insurance policy. The policy is for $150,000.00 and $\frac{1}{2}$ will go to his daughter, $\frac{3}{8}$ will go to his stepson, and the rest will go to his grandson. What dollar amount will go to the grandson?

43. Caren bought three packages of printer paper for $4.79 each. The sales tax for the merchandise was $0.86. If Caren paid with a $20 bill, how much change should she receive?

44. Mr. Timpson bought dinner for $28.42. He left a tip of $6.00. He paid the bill with two $20 bills. How much change should he receive?

45. Duncan earned test grades of 92, 84, 77, and 62. What is his test average?

46. Owen earned quiz grades of 19, 14, 16, and 20. What is his quiz average?

47. The average snowfall amounts for 5 winter months in Burlington, Vermont, are given in the table. Find the average snowfall per month. **(See Example 9.)**

Month	Snowfall (in.)
November	6.6
December	18.1
January	18.8
February	16.8
March	12.4

48. The average rainfall for 4 summer months for Houston, Texas, is given in the table. Find the average rainfall per month.

Month	Rainfall (in.)
June	5.6
July	5.2
August	3.3
September	3.7

49. The price of one share of stock of Microsoft for a 3-month period is given in the graph. Melanie bought stock on March 15 and sold it on April 30.

   a. By how much had the stock increased in value per share?

   b. If Melanie purchased 200 shares, how much money did she make?

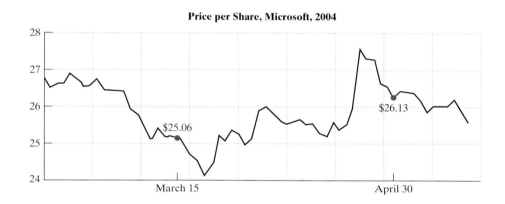

50. The price of one share of stock of Hershey Foods for a 3-month period is given in the graph. Taylor bought stock on March 1 and sold it on May 3.

   a. By how much had the stock increased in value per share?

   b. If Taylor purchased 250 shares, how much money did he make?

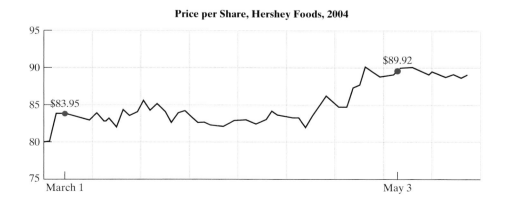

Use the following formula and table for Exercises 51–54. Round values to the tenths place if necessary.

In the spring of 1998, the National Institutes of Health (NIH) issued new guidelines for determining whether an individual is overweight. The standard of measure is called the *body mass index* (BMI). Body mass index is a measure of an individual's weight in relation to the person's height. The formula for body mass index is

$$\text{BMI} = \frac{703W}{h^2}$$

where W is weight in pounds and h is height in inches. The NIH categorizes body mass indices as shown in the table.

BMI	Weight Status
18.5–24.9	Considered ideal
25.0–29.9	Considered overweight
30.0 or above	Considered obese

51. Find your own body mass index.

52. a. Find the body mass index for a 150-lb person who is 5 ft 10 in. (70 in.) tall.

   b. Is this person considered ideal, overweight, or obese?

53. a. Find the body mass index for a 220-lb person who is 6 ft 0 in. (72 in.) tall.

   b. Is this person considered ideal, overweight, or obese?

54. a. Find the body mass index for a 141-lb person who is 5 ft 3 in. (63 in.) tall.

   b. Is this person considered ideal, overweight, or obese?

Expanding Your Skills

For Exercises 55–58, perform the indicated operation.

55. $0.\overline{3} \times 0.3 + 3.375$ **56.** $0.\overline{5} \div 0.\overline{2} - 0.75$ **57.** $(0.\overline{8} + 0.\overline{4}) \times 0.39$ **58.** $(0.\overline{7} - 0.\overline{6}) \times 5.4$

| Calculator Connections |

Topic: Applications using the order of operations with decimals

Calculator Exercises

59. Marty bought a home for $145,000. He paid $25,000 as a down payment and then financed the rest with a 30-yr mortgage. His monthly payments are $798.36 each and go toward paying off the loan and interest on the loan.

    a. How much money does Marty have to finance?

    b. How many months are in a 30-yr period?

    c. How much money will Marty pay over a 30-yr period to pay off the loan?

    d. How much money did Marty pay in interest over the 30-yr period?

60. Gwen bought a home for $109,000. She paid $15,000 as a down payment and then financed the rest with a 15-yr mortgage. Her monthly payments are $849.00 each and go toward paying off the loan and toward interest on the loan.

    a. How much money does Gwen have to finance?

    b. How many months are in a 15-yr period?

    c. How much money will Gwen pay over a 15-yr period to pay off the loan?

    d. How much money did Gwen pay in interest over the 15-yr period?

61. An inheritance for $80,460.60 is to be divided equally among four heirs. However, before the money can be distributed, approximately one-third of the money must go to the government for taxes. How much does each person get after the taxes have been subtracted?

62. For the fall semester, Sylvia bought 4 textbooks for $80.25, $42.50, $77.05, and $32.20. What is the average price per textbook?

chapter 4 | summary

section 4.1 Decimal Notation and Rounding

Key Concepts

A **decimal fraction** is a fraction whose denominator is a power of 10.

Identify the place values of a decimal number.

1 2 3 4 . 5 6 7 8

thousands
hundreds
tens
ones
decimal point
tenths
hundredths
thousandths
ten-thousandths

Reading a Decimal Number

1. The part of the number to the left of the decimal point is read as a whole number. *Note*: If there is not a whole-number part, skip to step 3.
2. The decimal point is read "and."
3. The part of the number to the right of the decimal point is read as a whole number but is followed by the name of the place position of the digit farthest to the right.

Converting a Decimal to a Mixed Number or Proper Fraction

1. The digits to the right of the decimal point are written as the numerator of the fraction.
2. The place value of the digit farthest to the right of the decimal point determines the denominator.
3. The whole-number part of the number is left unchanged.
4. Once the number is converted to a fraction or mixed number, simplify the fraction to lowest terms, if possible.

Writing a Decimal Number Greater Than 1 as an Improper Fraction

1. The denominator is determined by the place position of the rightmost digit to the right of the decimal point.
2. The numerator is obtained by removing the decimal point of the original number. The resulting whole number is then written over the denominator.
3. Simplify the improper fraction to lowest terms, if possible.

Examples

Example 1

$\frac{7}{10}$, $\frac{31}{100}$, and $\frac{191}{1000}$ are decimal fractions.

Example 2

In the number 34.914, the 1 is in the hundredths place.

Example 3

23.089 reads "twenty-three and eighty-nine thousandths."

Example 4

$$4.2 = 4\frac{\overset{1}{2}}{\underset{5}{10}} = 4\frac{1}{5}$$

Example 5

$$5.24 = \frac{\overset{131}{524}}{\underset{25}{100}} = \frac{131}{25}$$

Comparing Two Decimal Numbers

1. Starting at the left (and moving toward the right), compare the digits in each corresponding place position.
2. As we move from left to right, the first instance in which the digits differ determines the order of the numbers. The number having the greater digit is greater overall.

Rounding Decimals to a Place Value to the Right of the Decimal Point

1. Identify the digit one position to the right of the given place value.
2. If the digit in step 1 is 5 or greater, add 1 to the digit in the given place value. Then discard the digits to its right.
3. If the digit in step 1 is less than 5, discard it and any digits to its right.

In the U.S. system of currency, one dollar ($1) equals one hundred cents (100¢). That is, 1 cent is equivalent to $\frac{1}{100}$ of a dollar.

Example 6

$3.024 > 3.019$ because

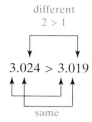

Example 7

Round 4.8935 to the nearest hundredth.

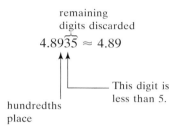

Example 8

$50.14 represents fifty and fourteen hundredths dollars, which is equivalent to fifty dollars and fourteen cents.

section 4.2   Addition and Subtraction of Decimals

<table>
<tr><td>

Key Concepts

Adding Decimals

1. Write the addends in a column with the decimal point and corresponding place values lined up. (You may insert additional zeros after the last digit to the right of the decimal point. These will act as placeholders so that each addend has the same number of digits to the right of the decimal point.)
2. Add the digits in columns from right to left as you would whole numbers. The decimal point in the answer should be lined up with the decimal points from the addends.

Subtracting Decimals

1. Write the numbers in a column with the decimal point and corresponding place values lined up. (You may insert additional zeros after the last digit to the right of the decimal point. These will act as placeholders so that each number has the same number of digits to the right of the decimal point.)
2. Subtract the digits in columns from right to left as you would whole numbers. The decimal point in the answer should be lined up with the other decimal points.

Example 3 illustrates an application involving addition and subtraction of decimals.

</td><td>

Examples

Example 1

Add $6.92 + 12 + 0.001$.

$$
\begin{array}{r}
6.920 \\
12.000 \\
+\ 0.001 \\
\hline
18.921
\end{array}
\left. \right\} \begin{array}{l} \text{Add zeros to the} \\ \text{right of the decimal} \\ \text{point as placeholders.} \end{array}
$$

Check by rounding:

6.92 rounds to 7 and 0.001 rounds to 0.

$7 + 12 + 0 = 19$, which is close to 18.921.

Example 2

Subtract $41.03 - 32.4$.

$$
\begin{array}{r}
\overset{3}{\cancel{4}}\ \overset{\overset{10}{\cancel{0}}}{\cancel{1}}.\overset{10}{\cancel{0}}\ 3 \\
-\ 3\ 2\ .\ 4\ 0 \\
\hline
8\ .\ 6\ 3
\end{array}
$$

Check by rounding:

41.03 rounds to 41 and 32.40 rounds to 32.

$41 - 32 = 9$, which is close to 8.63.

Example 3

Toni has $181.50 in her bank account. If she deposits $250 and then writes a check for $389.99, will she have enough left to withdraw $80 for concert tickets?

$$181.50 + 250 - 389.99 = 431.50 - 389.99$$
$$= 41.51$$

Toni has only $41.51 in her account which is not enough for the tickets.

</td></tr>
</table>

section 4.3 Multiplication of Decimals

Key Concepts

Multiplying Two Decimals

1. Ignore the decimal point and multiply as you would whole numbers.
2. Place the decimal point in the product so that the number of decimal places equals the combined number of decimal places of both factors.

Multiplying a Decimal by Powers of 10

Move the decimal point to the right the same number of decimal places as the number of zeros in the power of 10.

Multiplying a Decimal by Powers of 0.1

Move the decimal point to the left the same number of places as there are decimal places in the power of 0.1.

Large numbers are often expressed using number names, for example, 34.2 million.

Converting Dollars to Cents

1. Multiply by 100. This has the effect of moving the decimal point to the *right* two places.
2. Drop the $ symbol and attach the ¢ symbol to the result.

Converting Cents to Dollars

1. Multiply by 0.01. This has the effect of moving the decimal point to the *left* 2 places.
2. Drop the ¢ symbol and place the $ symbol in front of the result.

Examples 7 and 8 illustrate multiplication of decimals.

Example 7

A group of 5 college students attend a movie. The cost is discounted to $6.25 for students. What is the total cost of the tickets?

$6.25 \times 5 = 31.25$

The total cost is $31.25.

Examples

Example 1

Multiply 5.02×2.8.

$$
\begin{array}{r}
\overset{1}{5}.02 \\
\times\ 2.8 \\
\hline
4016 \\
10040 \\
\hline
14.056
\end{array}
$$

Example 2

$83.251 \times 100 = 8325.1$

Move 2 places
to the right.

Example 3

$149.02 \times 0.001 = 0.14902$

Move 3 places
to the left.

Example 4

$34.2 \text{ million} = 34.2 \times 1{,}000{,}000$

$= 34{,}200{,}000$

Example 5

Convert $5.12 to cents.

$5.12 = 512¢$

Example 6

Convert 2399¢ to dollars.

$2399¢ = \$23.99$

Example 8

Find the area of the TV remote control.

$16.7 \times 5.8 = 96.86$

The area is 96.86 cm².

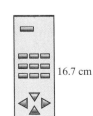

16.7 cm

5.8 cm

section 4.4 Division of Decimals

Key Concepts

Dividing a Decimal by a Whole Number

1. Place the decimal point in the quotient directly above the decimal point in the dividend.
2. Divide as you would whole numbers.

Dividing When the Divisor Is Not a Whole Number

1. Move the decimal point in the divisor to the right to make it a whole number.
2. Move the decimal point in the dividend to the right the same number of places as in step 1.
3. Place the decimal point in the quotient directly above the decimal point in the dividend.
4. Divide as you would whole numbers.

To round a repeating decimal, be sure to expand the repeating digits to one digit beyond the indicated rounding place.

Dividing by a Power of 10

To divide a number by a power of 10, move the decimal point to the *left* the same number of places as there are zeros in the power of 10.

Dividing by a Power of 0.1

To divide a number by a power of 0.1, move the decimal point to the *right* the same number of places as there are decimal places in the power of 0.1.

Examples

Example 1

$$62.6 \div 4$$

$$
\begin{array}{r}
15.65 \\
4\overline{)62.60} \\
-4 \\
\hline
22 \\
-20 \\
\hline
26 \\
-24 \\
\hline
20 \\
-20 \\
\hline
0
\end{array}
$$

Example 2

$$81.1 \div 0.9 \qquad .9\overline{)81.1}$$

$$
\begin{array}{r}
90.11\ldots \\
9\,\overline{)811.00} \\
-81 \\
\hline
01 \\
00 \\
\hline
10 \\
-9 \\
\hline
10
\end{array}
$$

The pattern repeats.

The answer is the repeating decimal $90.\overline{1}$.

Example 3

Round $6.\overline{56}$ to the thousandths place.

6.5656

thousandths place

The digit $6 > 5$ so increase the thousandths-place digit by 1.

$6.\overline{56} \approx 6.566$

Example 4

$$302.9 \div 100 = 3.029$$

Move 2 places to the left.

Example 5

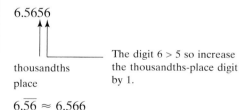

$$78.114 \div 0.001 = 78{,}114$$

Move 3 places to the right.

▪ section 4.5 **Fractions as Decimals**

Key Concepts

To write a fraction as a decimal, divide the numerator by the denominator. See Examples 1 and 2.

These are some common fractions represented by decimals.

$\dfrac{1}{4} = 0.25 \qquad \dfrac{1}{2} = 0.5 \qquad \dfrac{3}{4} = 0.75$

$\dfrac{1}{9} = 0.\overline{1} \qquad \dfrac{2}{9} = 0.\overline{2} \qquad \dfrac{1}{3} = 0.\overline{3}$

$\dfrac{4}{9} = 0.\overline{4} \qquad \dfrac{5}{9} = 0.\overline{5} \qquad \dfrac{2}{3} = 0.\overline{6}$

$\dfrac{7}{9} = 0.\overline{7} \qquad \dfrac{8}{9} = 0.\overline{8}$

To write a decimal as a fraction, first write the number as a decimal fraction and reduce. See Examples 3 and 4.

To rank decimals from least to greatest, compare corresponding digits from left to right.

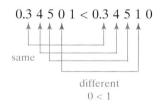

Examples

Example 1

$\dfrac{17}{20} = 0.85$

$$
\begin{array}{r}
.85 \\
20\overline{)17.00} \\
-160 \\
\hline
100 \\
-100 \\
\hline
0
\end{array}
$$

Example 2

$\dfrac{14}{3} = 4.\overline{6}$

$$
\begin{array}{r}
4.66\ldots \\
3\,\overline{)14.00} \\
-12 \\
\hline
20 \\
-18 \\
\hline
20
\end{array}
$$

The pattern repeats.

Example 3

$6.84 = \dfrac{\overset{171}{684}}{\underset{25}{100}} = \dfrac{171}{25} \quad \text{or} \quad 6\dfrac{21}{25}$

Example 4

$4.\overline{8} = 4\dfrac{8}{9} \quad \text{or} \quad \dfrac{44}{9}$

Example 5

Plot the decimals 2.6, $2.\overline{6}$, and 2.58 on a number line.

section 4.6 Order of Operations and Applications of Decimals

Key Concepts

Examples 1 and 2 apply the order of operations with fractions and decimals. There are two approaches to simplify: Write the expression in terms of all decimals, or write the expression in terms of all fractions.

Examples

Example 1

$$\left(1.6 - \frac{13}{25}\right) \div 8 = (1.6 - 0.52) \div 8$$
$$= (1.08) \div 8$$
$$= 0.135$$

Example 2

$$\frac{2}{3}\left(2.2 + \frac{7}{5}\right) = \frac{2}{3}\left(\frac{22}{10} + \frac{7}{5}\right)$$
$$= \frac{2}{3}\left(\frac{11}{5} + \frac{7}{5}\right)$$
$$= \frac{2}{\underset{1}{3}}\left(\frac{\overset{6}{18}}{5}\right) = \frac{12}{5} = 2.4$$

Example 3 illustrates an application of the order of operations involving decimals.

Example 3

For one month, Dan brought home paychecks worth $824.25, $840.05, $915.21, and $880.89. What was his average weekly pay?

$$(824.25 + 840.05 + 915.21 + 880.89) \div 4$$
$$= (3460.40) \div 4 = 865.1$$

Dan brings home an average of $865.10 a week.

---

chapter 4 | review exercises

Section 4.1

1. Identify the place values for each of the digits in the number 32.16.

2. Identify the place values for each of the digits in the number 2.079.

For Exercises 3–6, write the word name for the decimal.

3. 5.7

4. 10.21

5. 51.008

6. 109.01

For Exercises 7–8, write the decimal as a proper fraction or mixed number.

7. 4.8

8. 0.025

For Exercises 9–10, write the decimal as an improper fraction.

9. 1.3

10. 6.75

For Exercises 11–12, fill in the blank with either < or >.

11. 15.032 ☐ 15.03

12. 7.209 ☐ 7.22

13. The midseason earned run average (ERA) for five members of the National Baseball League for a recent year is given in the table. Rank the averages from least to greatest.

Player	ERA
G. Maddux	4.41
B. Laurence	4.40
B. Webb	4.48
V. Padilla	4.07
B. Meyers	4.24

For Exercises 14–15, round the decimal to the indicated place value.

14. 89.9245;   hundredths

15. 34.8895;   thousandths

16. A quality control manager tests the amount of cereal in several brands of breakfast cereal against the amount advertised on the box. She selects one box at random. She measures the contents of one 12.5-oz box and finds that the box has 12.46 oz.

  a. Is the amount in the box less than or greater than the advertised amount?

  b. If the quality control manager rounds the measured value to the tenths place, what is the value?

17. Which number is equivalent to 571.24? Circle all that apply.

  a. 571.240   **b.** 571.2400

  c. 571.024   **d.** 571.0024

18. Which number is equivalent to 3.709? Circle all that apply.

  a. 3.7   **b.** 3.7090

  c. 3.709000   **d.** 3.907

Section 4.2

For Exercises 19–26, add or subtract as indicated.

19. $45.03 + 4.713$ **20.** $239.3 + 33.92$

21. $34.89 - 29.44$ **22.** $5.002 - 3.1$

23. $221 - 23.04$ **24.** $34 + 4.993$

25. $17.3 + 3.109 - 12.6$

26. $189.22 + 13.1 - 120.055$

27. Find the values of x and y. Then find the perimeter of the figure.

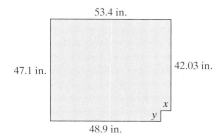

28. Gas prices for one month are shown in the table.

  a. Determine the difference in price between the consecutive weeks.

  b. Between which two weeks did the price increase the most?

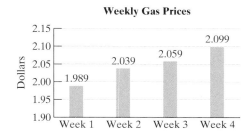

Section 4.3

For Exercises 29–36, multiply the decimals.

29. 3.9×2.1 **30.** 57.01×1.3

31. 60.1×4.4 **32.** 7.7×45

33. 85.49×1000

34. 1.0034×100

35. 92.01×0.01

36. 104.22×0.01

For Exercises 37–38, write the decimal number representing each word name.

37. The album with the most sales in the United States is *The Eagles Greatest Hits* with 28 million albums sold.

38. The population of Guadeloupe is approximately 4.32 hundred-thousand.

39. Write the amount in terms of dollars.

    a. 234¢

    b. 55¢

40. Write the amount in terms of cents.

    a. $5.25

    b. $0.12

41. A store advertises a package of two 9-volt batteries for sale at $1.99.

    a. What is the cost of buying 8 batteries?

    b. If another store has an 8-pack for the regular price of $9.99, how much can one save by buying batteries at the sale price?

42. If long-distance phone calls cost $0.07 per minute, how much will a 23-min long-distance call cost?

43. Find the area and perimeter of the rectangle.

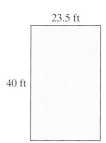

23.5 ft

40 ft

44. Population density gives the approximate number of people per square mile. The population density for Texas is given in the graph for selected years.

    a. Approximately how many people would have been located in a 200-mi² area in 1960?

    b. Approximately how many people would have been located in a 200-mi² area in 2000?

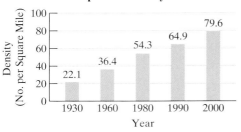

Figure for Exercise 44

Section 4.4

For Exercises 45–50, divide. Write the answer in decimal form.

45. $8.55 \div 0.5$

46. $64.2 \div 1.5$

47. $0.06\overline{)0.248}$

48. $0.3\overline{)2.63}$

49. $18.9 \div 0.7$

50. $0.036 \div 1.2$

51. For each number, round to the indicated place.

	8.$\overline{6}$	52.$\overline{52}$	0.$\overline{409}$
Tenths			
Hundredths			
Thousandths			
Ten-thousandths			

For Exercises 52–53, divide and round the answer to the nearest hundredth.

52. $104.6 \div 9$

53. $71.8 \div 6$

For Exercises 54–57, divide by powers of 10 and 0.1.

54. $493.93 \div 100$

55. $90.234 \div 10$

56. $553.8 \div 0.001$

57. $2.6 \div 0.01$

58. a. A package of 5 rolls of 35-mm film costs $9.99. What is the cost per roll? (Round the answer to the nearest cent, that is, the nearest hundredth of a dollar.)

    b. Another package of 4 rolls of 35-mm film costs $8.99. What is the cost per roll?

    c. Which of the two packages offers the better buy?

Section 4.5

For Exercises 59–61, write the fraction as a decimal fraction. Then write the number in decimal form.

59. $\dfrac{3}{5}$ **60.** $\dfrac{7}{20}$ **61.** $\dfrac{27}{500}$

For Exercises 62–65, write the fraction or mixed number as a decimal.

62. $2\dfrac{2}{5}$ **63.** $3\dfrac{13}{25}$

64. $\dfrac{24}{125}$ **65.** $\dfrac{7}{16}$

For Exercises 66–69, write the fraction as a repeating decimal.

66. $\dfrac{7}{12}$ **67.** $\dfrac{55}{36}$

68. $4\dfrac{7}{22}$ **69.** $\dfrac{2}{13}$

For Exercises 70–73, write the fraction as a decimal rounded to the nearest hundredth.

70. $\dfrac{5}{17}$ **71.** $\dfrac{20}{23}$

72. $\dfrac{11}{3}$ **73.** $\dfrac{17}{6}$

For Exercises 74–77, write the fraction or mixed number for the repeating decimal.

74. $0.\overline{2}$ **75.** $1.\overline{6}$

76. $3.\overline{3}$ **77.** $5.\overline{7}$

78. Complete the table, giving the closing value of small-cap stocks as reported in the *Wall Street Journal*.

Stock	Symbol	Closing Price ($) (Decimal)	Closing Price ($) (Fraction)
Isonics	ISON	1.66	
EnPointe	ENPT		$2\frac{1}{20}$
LucilleFarm	LUCY	1.75	
Workstream	WSTM		$2\frac{4}{5}$

For Exercises 79–81, insert the appropriate symbol. Choose from $<$, $>$, or $=$.

79. $1\dfrac{1}{3}$ ☐ 1.33 **80.** 2.25 ☐ $\dfrac{9}{4}$

81. 0.14 ☐ $\dfrac{1}{7}$

For Exercises 82–83, determine what decimal number is represented by the point on the number line.

82.

$$\underset{0.25 \qquad\qquad 0.30}{\vdash\!+\!+\!+\!\bullet\!+\!+\!\rightarrow}$$

83.

$$\underset{0.71 \qquad\qquad 0.72}{\vdash\!+\!\bullet\!+\!+\!+\!+\!+\!+\!\rightarrow}$$

Section 4.6

For Exercises 84–87, perform the indicated operations. Write the answer in decimal form.

84. $7.5 \div \dfrac{3}{2}$ **85.** $2(3.14)(20)$

86. $3.14(5)^2$ **87.** $\dfrac{1}{3}(3.14)(2)^2(6)$

88. The Pimsleur Spanish course is available online at one website for the following prices. How much money is saved by buying the combo package versus the three levels individually?

Level	Price
Spanish I	$189.95
Spanish II	199.95
Spanish III	219.95
Combo (Spanish I, II, III combined)	519.95

89. Marvin drives the route shown in the figure each day, making deliveries. He completes one-third of the route before lunch. How many more miles does he still have to drive after lunch?

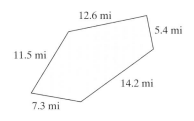

chapter 4 | test

1. Identify the place value of the underlined digit.

  a. 2<u>3</u>4.17   **b.** 234.1<u>7</u>

2. Write the word name for 509.024.

3. Write the decimal 1.26 as a mixed number and as a fraction.

4. The field goal percentages for four members of the San Antonio Spurs in the 2004 playoffs are given in the table. Rank the averages from least to greatest.

Player	Average
Tim Duncan	0.522
Jason Hart	0.550
Robert Horry	0.465
Devin Brown	0.486

5. Which statement is correct?

  a. $0.043 > 0.430$ **b.** $0.692 < 0.926$

  c. $0.078 < 0.0780$

For Exercises 6–13, perform the indicated operation.

6. $49.002 + 3.83$ **7.** $34.09 - 12.8$

8. 28.1×4.5 **9.** $25.4 \div 5$

10. $4 - 2.78$ **11.** $12.03 + 0.1943$

12. $39.82 \div 0.33$ **13.** 42.7×10.3

14. The temperature of a cake is recorded in 10-min intervals after it comes out of the oven. See the graph.

  a. What was the difference in temperature between 10 and 20 min?

b. What was the difference in temperature between 40 and 50 min?

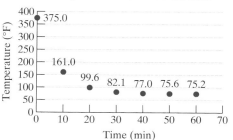

15. In the 2000 U.S. Presidential election, George Bush received approximately 50.5 million votes. In the same election, Al Gore received approximately 51.0 million votes.

  a. Write a decimal number representing the number of votes received by George Bush.

  b. Write a decimal number representing the number of votes received by Al Gore.

  c. What is the approximate difference between the number of votes received by Al Gore and the number of votes received by George Bush?

16. Explain the difference between dividing a number by 10 and multiplying a number by 10.

17. Explain the difference between dividing a number by 0.01 and multiplying a number by 0.01.

For Exercises 18–21, multiply or divide by the powers of 10 and 0.1.

18. 45.92×0.1 **19.** 579.23×100

20. $80.12 \div 0.01$ **21.** $2.931 \div 1000$

22. A picture is framed and matted as shown in the figure.

    a. Find the area of the picture itself.

    b. Find the area of the matting *only*.

    c. Find the area of the frame *only*.

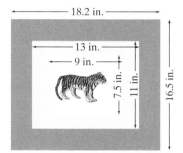

23. What is mathematically incorrect about the advertisement shown in the figure?

Note Cards 75 cents

.75¢

24. The table shows the winning times in seconds for women's speed skating for several years. Complete the table.

Year	Decimal	Fraction
1984		$41\frac{1}{50}$ sec
1988	39.10	
1992	40.33	
1994		$39\frac{1}{4}$

25. Rank the numbers and plot them on a number line.

$$3\frac{1}{2}, \ 3.\overline{5}, \ 3.2$$

For Exercises 26–27, simplify by using the order of operations.

26. $(8.7)\left(1.6 - \dfrac{1}{2}\right)$ **27.** $\dfrac{7}{3}\left(5.1 + \dfrac{3}{4}\right)$

28. Faulkner walked every day one week, and the distances are recorded in the table.

    a. Find the total distance walked.

    b. Find the average distance walked per day. Round to the nearest tenth of a mile.

Day	Distance
Monday	2.4 mi
Tuesday	1.6 mi
Wednesday	3.6 mi
Thursday	2.2 mi
Friday	2.8 mi
Saturday	2.9 mi
Sunday	4.3 mi

---

chapters 1–4 | cumulative review

1. Simplify: $(17 + 12) - (8 - 3) \cdot 3$

2. Convert the number to standard form:
4 thousands + 3 tens + 9 ones

3. Add: $3902 + 34 + 904$

4. Subtract: $4990 - 1118$

5. Multiply and round the answer to the thousands place: $23,444 \times 103$

6. Divide 4530 by 225. Then identify the dividend, divisor, whole-number part of the quotient, and remainder.

7. Explain how to check the division problem in Exercise 6.

8. The chart shows the retail sales for several companies. Find the difference between the highest and lowest sales in the chart.

Retail Sales

For Exercises 9–14, multiply or divide as indicated. Write the answer as a fraction.

9. $\dfrac{1}{5} \cdot \dfrac{6}{11}$

10. $\left(\dfrac{6}{15}\right)\left(\dfrac{10}{7}\right)$

11. $\left(\dfrac{7}{10}\right)^2$

12. $\left(\dfrac{32}{22}\right) \div \left(\dfrac{8}{11}\right)$

13. $\dfrac{8}{3} \div 4$

14. $\left(\dfrac{0}{5}\right) \div \left(\dfrac{3}{5}\right)$

15. A settlement for a lawsuit is made for \$15,000. The attorney gets $\frac{2}{5}$ of the settlement. How much is left?

16. Simplify by using the order of operations:
$$\frac{8}{25} + \frac{1}{5} \div \frac{5}{6} - \left(\frac{2}{5}\right)^2$$

For Exercises 17–20, add or subtract as indicated. Write the answer as a fraction.

17. $\dfrac{7}{10} + \dfrac{27}{100}$

18. $\dfrac{5}{11} + 3$

19. $5 - \dfrac{2}{7}$

20. $\dfrac{12}{5} - \dfrac{9}{10}$

21. Find the area and perimeter.

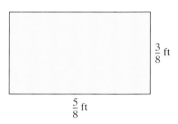

$\frac{3}{8}$ ft

$\frac{5}{8}$ ft

22. Cliff walks in the mornings for exercise. For the last 4 days he has walked for $\frac{9}{10}$, $1\frac{2}{5}$, $1\frac{1}{4}$, and $1\frac{1}{5}$ km. What is the average distance?

For Exercises 23–27, perform the indicated operations.

23. $50.9 + 123.23$

24. $700.8 - 32.01$

25. 301.1×0.25

26. $51.2 \div 3.2$

27. $\dfrac{4}{3}(3.14)(9)^2$

28. Divide $79.02 \div 1.7$ and round the answer to the nearest hundredth.

29. a. Multiply 0.004×938.12.

   b. Multiply 938.12×0.004.

   c. Identify the property that has been demonstrated in parts (a) and (b).

30. The table gives the average length of several bones in the human body. Complete the table by writing the mixed numbers as decimals and the decimals as mixed numbers.

Bone	Length (in.) (Decimal)	Length (in.) (Mixed Number)
Femur	19.875	
Fibula		$15\frac{15}{16}$
Humerus		$14\frac{3}{8}$
Innominate bone (hip)	7.5	

Ratio and Proportion

5

Chapter 5 is devoted to the study of ratios, rates, and proportions. Unit rates are extremely important when comparing two or more items. The equality of two rates or ratios, called a proportion, is used to solve numerous application problems. For example, proportions can be used to estimate the number of animals within a wilderness area. See Exercise 31 in Section 5.4 to estimate the number of bison in Yellowstone National Park in Wyoming.

chapter 5 | preview

The exercises in this chapter preview contain concepts that have not yet been presented. These exercises are provided for students who want to compare their levels of understanding before and after studying the chapter. Alternatively, you may prefer to work these exercises when the chapter is completed and before taking the exam.

Section 5.1

1. Write the ratio 3 to 11 in two other ways.

2. Terri bought 5 books for the fall semester, 3 of which were English literature books.

   a. Write a ratio of English literature books to total books.

   b. Write a ratio of English literature books to the books that are not English literature.

For Exercises 3–4, write the ratio as a fraction with a whole-number numerator and denominator. Simplify to lowest terms.

3. $6\frac{1}{2}$ to 10   4. 4.2 to 1.6

5. For a recent year at Sam Houston State University, there were approximately 13,000 students and 500 faculty. Write the student-to-faculty ratio in lowest terms.

Section 5.2

6. Write the rate in lowest terms: Ken caught 8 fish in 6 hr.

7. Write the unit rate: An L-1011 aircraft travels 2955 mi in 6 hr.

8. Write the unit cost for a box of tissue that cost $1.05 and contains 64 tissues. Round to 3 decimal places.

9. Iams dog food comes in different sizes: an 8-lb bag for $8.39, a 4-lb bag for $4.99, and a 40-lb bag for $25.99. Find the unit prices for each to determine the best buy. Round to 2 decimal places.

10. The average annual rainfall for Miami, Florida, is 58.5 in. What is the rate per month?

11. The federal minimum hourly wage in 1980 was $3.10. In 2000 it was $5.15.

    a. Compute the difference in minimum hourly wage between the years 2000 and 1980.

    b. Compute the rate representing the increase in minimum hourly wage per year. (Source: U.S. Bureau of Labor Statistics.)

Section 5.3

For Exercises 12–13, write a proportion for each statement.

12. 3 is to 15 as 6 is to 30.

13. The numbers 22 and 14 are proportional to the numbers 33 and 21.

14. Determine if the rates form a proportion:
    $$\frac{2\frac{1}{2}}{2} \stackrel{?}{=} \frac{25}{20}$$

15. Solve the proportion: $\dfrac{8.1}{0.9} = \dfrac{4.5}{x}$

Section 5.4

16. On a road map, 1 in. represents 7 mi. If the distance between two towns measures $3\frac{1}{4}$ in. on the map, what is the actual distance between the two towns?

17. In Hungary, the birthrate is 9.5 per 1000 people. If the population of Hungary is approximately 10,200,000, how many births are there in a given year?

18. Given that the triangles are similar, solve for x and y.

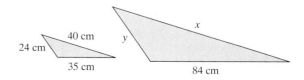

section 5.1   Ratios

1. Writing a Ratio

Thus far we have seen two interpretations of fractions.

- The fraction $\frac{5}{8}$ represents 5 parts of a whole that has been divided evenly into 8 pieces.
- The fraction $\frac{5}{8}$ represents $5 \div 8$.

Now we consider a third interpretation.

- The fraction $\frac{5}{8}$ represents the ratio of 5 to 8.

A **ratio** is a comparison of two quantities with the same units. There are three different ways to write a ratio.

Writing a Ratio

The ratio of a to b can be written as follows, provided $b \neq 0$.

1. a to b **2.** $a : b$ **3.** $\dfrac{a}{b}$

The colon means "to."   The fraction bar means "to."

Although there are three ways to write a ratio, we primarily use the fraction form.

---

example 1   Writing a Ratio

In an algebra class there are 15 women and 17 men.

a. Write the ratio of women to men.

b. Write the ratio of men to women.

c. Write the ratio of women to the total number of people in the class.

Solution:

It is important to observe the *order* of the quantities mentioned in a ratio. The first quantity mentioned is the numerator. The second quantity is the denominator.

a. The ratio of women to men is

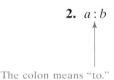

$$\frac{15}{17}$$

b. The ratio of men to women is

$$\frac{17}{15}$$

c. We must first find the total number of people in the class.

Total = number of women + number of men

$$= 15 + 17$$

$$= 32$$

Therefore the ratio of women to the total number of people in the class is

$$\frac{15}{32}$$

2. Simplifying a Ratio to Lowest Terms

It is often desirable to write a ratio in lowest terms. The process is similar to simplifying fractions to lowest terms.

example 2   Writing Ratios in Lowest Terms

Write each ratio in lowest terms.

a. 15 ft to 10 ft   **b.** $20 to $10

Solution:

Recall that in a ratio, the two numbers must have like units. In part (a) we are comparing feet to feet. In part (b) we are comparing dollars to dollars. We can "cancel" the like units in the numerator and denominator as we would common factors.

a. $\dfrac{15 \text{ ft}}{10 \text{ ft}} = \dfrac{3 \cdot 5 \text{ ft}}{2 \cdot 5 \text{ ft}}$

$\qquad = \dfrac{3 \cdot \overset{1}{\cancel{5}} \, \cancel{\text{ft}}}{2 \cdot \underset{1}{\cancel{5}} \, \cancel{\text{ft}}}$ Simplify common factors. "Cancel" common units.

$\qquad = \dfrac{3}{2}$

Even though the number $\frac{3}{2}$ is equivalent to $1\frac{1}{2}$, we do not write the ratio as a mixed number. Remember that a ratio is a comparison of *two* quantities. If you did convert $\frac{3}{2}$ to the mixed number $1\frac{1}{2}$, you would write the ratio as $\dfrac{1\frac{1}{2}}{1}$. This would imply that the numerator is one and one-half times as large as the denominator.

b. $\dfrac{\$20}{\$10} = \dfrac{\overset{2}{\cancel{\$20}}}{\underset{1}{\cancel{\$10}}}$ Simplify common factors. "Cancel" common units.

$\qquad = \dfrac{2}{1}$

Although the fraction $\frac{2}{1}$ is equivalent to 2, we do not generally write ratios as whole numbers. Again, a ratio compares *two* quantities. In this case, we say that there is a 2-to-1 ratio between the original dollar amounts.

3. Writing Ratios of Mixed Numbers and Decimals

It is often desirable to express a ratio in lowest terms by using whole numbers in the numerator and denominator. This is demonstrated in Examples 3 and 4.

---

example 3   Rewriting a Ratio as a Ratio of Whole Numbers

The length of a rectangular picture frame is 10.8 in., and the width is 8.64 in. Express the ratio of the length to the width. Then rewrite the ratio as a ratio of whole numbers reduced to lowest terms.

8.64 in.

10.8 in.

Solution:

The ratio of length to width is $\frac{10.8}{8.64}$. We now want to rewrite the ratio, using whole numbers in the numerator and denominator. Notice that if we multiply 8.64 by 100, the decimal point will move to the right 2 units. The result will be the whole number 864. Multiplying the numerator by the same number, 100, will produce a whole number in the numerator: $10.8 \times 100 = 1080$.

$$\frac{10.8}{8.64} = \frac{10.8 \times 100}{8.64 \times 100}$$ Multiply numerator and denominator by 100.

$$= \frac{1080}{864}$$

$$= \frac{\cancel{2} \cdot \cancel{2} \cdot \cancel{2} \cdot \cancel{3} \cdot \cancel{3} \cdot \cancel{3} \cdot 5}{\cancel{2} \cdot \cancel{2} \cdot \cancel{2} \cdot 2 \cdot 2 \cdot \cancel{3} \cdot \cancel{3} \cdot \cancel{3}}$$ Simplify common factors to lowest terms.

$$= \frac{5}{4}$$

The ratio of length to width is $\frac{5}{4}$.

In Example 3, we multiplied by 100 to move the decimal point *two* places to the right. Multiplying by 10 would not have been sufficient, because $8.64 \times 10 = 86.4$, which is not a whole number.

---

example 4   Rewriting a Ratio as a Ratio of Whole Numbers

Ling walked $2\frac{1}{4}$ mi on Monday and $3\frac{1}{2}$ mi on Tuesday. Write the ratio of miles walked Monday to miles walked Tuesday. Then rewrite the ratio as a ratio of whole numbers reduced to lowest terms.

Solution:

The ratio of miles walked on Monday to miles walked on Tuesday is $\dfrac{2\frac{1}{4}}{3\frac{1}{2}}$.

To convert this to a ratio of whole numbers, first we rewrite each mixed number as an improper fraction. Then we can divide the fractions and simplify.

$$\frac{2\frac{1}{4}}{3\frac{1}{2}} = \frac{\frac{9}{4}}{\frac{7}{2}}$$

Write the mixed numbers as improper fractions.
Recall that a fraction bar also implies division.

$$= \frac{9}{4} \div \frac{7}{2}$$

$$= \frac{9}{4} \cdot \frac{2}{7}$$ Multiply by the reciprocal of the divisor.

$$= \frac{9}{\overset{}{\underset{2}{4}}} \cdot \frac{\overset{1}{2}}{7}$$ Simplify common factors to lowest terms.

$$= \frac{9}{14}$$ This is a ratio of whole numbers in lowest terms.

4. Applications of Ratios

Ratios are used in a number of applications.

example 5   Using Ratios to Express Population Increase

The town of Roxbury, Connecticut, had 1825 people in the year 1990 and 2136 people in the year 2000. (Source: U.S. Bureau of the Census.) Write a ratio depicting the increase in population to the number of people in 1990.

Solution:

To write this ratio, we need to know the increase in population.

Increase in population = 2136 − 1825

= 311

The ratio of the increase in population to the number of people in 1990 is

Increase in population ⟶ $\dfrac{311}{1825}$ ⟵ number of people in 1990

When computing a ratio, it is important to use the same units of measurement. In Example 6, we compare two units of distance, but must convert to the same units of measurement.

example 6   Applying Ratios to Unit Conversion

A fence is 12 yd long and $1\frac{1}{2}$ ft high.

a. Find the ratio of the length to the height with all units measured in yards.

b. Find the ratio of length to height with all units measured in feet.

Solution:

a.

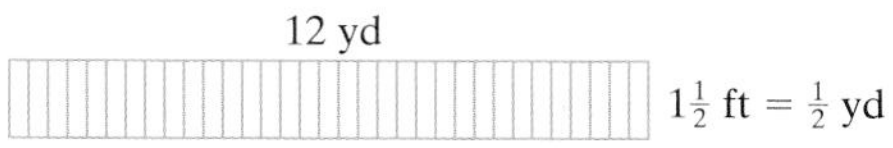

12 yd

$1\frac{1}{2}$ ft = $\frac{1}{2}$ yd

Figure 5-1

First recall that 3 ft = 1 yd. Therefore, $1\frac{1}{2}$ ft = $\frac{1}{2}$ yd. (Figure 5-1).

Measuring in yards, we see that the ratio of length to height is

$$\frac{12 \text{ yd}}{\frac{1}{2} \text{ yd}} = \frac{12}{1} \cdot \frac{2}{1} = \frac{24}{1}.$$

b. The length is 12 yd = 36 ft.

Measuring in feet, we see that the ratio of length to height is

$$\frac{36 \text{ ft}}{1\frac{1}{2} \text{ ft}} = \frac{36}{\frac{3}{2}} = \frac{\overset{12}{36}}{1} \cdot \frac{2}{\underset{1}{3}} = \frac{24}{1}.$$

Notice that regardless of the units used, the ratio is the same, 24 to 1. This means that the length is 24 times the height.

---

section 5.1   Practice Exercises

Boost *your* GRADE at
mathzone.com!

MathZone

- Practice Problems
- Self-Tests
- NetTutor

- e-Professors
- Videos

Study Skills Exercises

1. Does your school have a learning resource center or a tutoring center? If so, do you remember the location and hours of operation? Write them here.

Location of learning resource center or tutoring center:

Hours of operation:

2. Define the key term **ratio**.

Objective 1: Writing a Ratio

For Exercises 3–8, write the ratio in two other ways.

3. 5 to 6

4. 3 to 7

5. 11 : 4

6. 8 : 13

7. $\frac{1}{2}$

8. $\frac{1}{8}$

For Exercises 9–12, write the ratios in fraction form. **(See Example 1.)**

9. Nancy has 3 cats and 2 dogs.

    a. Write a ratio of cats to dogs.

    b. Write a ratio of dogs to cats.

    c. Write a ratio of cats to the total number of pets.

10. In a kindergarten classroom, there are 11 boys and 9 girls.

    a. Write a ratio of girls to boys.

    b. Write a ratio of boys to girls.

    c. Write a ratio of boys to the total number of children in class.

11. There are 52 cars in the parking lot, of which 21 are silver.

    a. Write a ratio of silver cars to the total number of cars.

    b. Write a ratio of silver cars to cars that are not silver.

12. On one city block, there are 21 houses and 10 of them have pools in the backyard.

    a. Write a ratio of the number of houses with a pool to the total number of houses.

    b. Write a ratio of the number of houses with a pool to the number of houses without a pool.

Objective 2: Simplifying a Ratio to Lowest Terms

For Exercises 13–24, write the ratio in lowest terms. **(See Example 2.)**

13. 4 yr to 6 yr	**14.** 10 lb to 14 lb	**15.** 5 mi to 25 mi	**16.** 20 ft to 12 ft
17. 8 m to 2 m	**18.** 14 oz to 7 oz	**19.** 33 cm to 15 cm	**20.** 21 days to 30 days
21. $60 to $50	**22.** 75¢ to 100¢	**23.** 18 in. to 36 in.	**24.** 3 cups to 9 cups

Objective 3: Writing Ratios of Mixed Numbers and Decimals

For Exercises 25–36, write the ratio in lowest terms with whole numbers in the numerator and denominator. **(See Examples 3 and 4.)**

25. 3.6 ft to 2.4 ft	**26.** 10.15 hr to 8.12 hr	**27.** 8 gal to $9\frac{1}{3}$ gal	**28.** 24 yd to $13\frac{1}{3}$ yd

29. $16\frac{4}{5}$ m to $18\frac{9}{10}$ m   **30.** $1\frac{1}{4}$ in. to $1\frac{3}{8}$ in.    **31.** $16.80 to $2.40   **32.** $18.50 to $3.70

33. $\frac{1}{2}$ day to 4 day   **34.** $\frac{1}{4}$ mi to $1\frac{1}{2}$ mi   **35.** 13.6 L to 10.2 L   **36.** 0.2 km to 0.08 km

Objective 4: Applications of Ratios

37. The population of Delaware was approximately 750 thousand in the year 2000 and in 2005 grew to approximately 800 thousand. Write a ratio depicting the increase in population to the number of people in 2000. **(See Example 5.)**

38. In the city of Goddard, Kansas, the population grew from 2100 in July 2000 to 2600 in July 2002. Write a ratio depicting the increase in population to the number of people in July 2000.

39. A name plaque is 6 in. wide and $1\frac{1}{3}$ ft long ($\frac{1}{3}$ ft is 4 in.). **(See Example 6.)**

    a. Find the ratio of width to length with all units in inches.

    b. Find the ratio of width to length with all units in feet.

40. A construction company needs 2 weeks to construct a family room and 3 days to add a porch.

    a. Find the ratio of the time it takes for constructing the porch to the time constructing the family room, with all units in weeks.

    b. Find the ratio of the time it takes for constructing the porch to the time constructing the family room, with all units in days.

For Exercises 41–44, refer to the table that shows the number of museums, historical sites, and similar institutions in the United States. Write each ratio in lowest terms. (Source: U.S. Census Bureau, Statistical Abstract of the United States.)

	Number of Establishments (Thousands)
Museums	41
Historical sites	9
Zoos and botanical gardens	5
Nature parks	5

41. Find the ratio of museums to nature parks.

42. Find the ratio of zoos and botanical gardens to nature parks.

43. Find the ratio of nature parks to the total number of establishments.

44. Find the ratio of museums to the total number of establishments.

For Exercises 45–48, refer to the table that shows the average spending per person for reading (books, newspapers, magazines, etc.) by age group. Write each ratio in lowest terms.

Age Group	Annual Average ($)
Under 25 years	60
25 to 34 years	111
35 to 44 years	136
45 to 54 years	172
55 to 64 years	183
65 to 74 years	159
75 years and over	128

45. Find the ratio of spending for the under-25 group to the spending for the group 75 years and over.

46. Find the ratio of spending for the group 25 to 34 years old to the spending for the group of 65 to 74 years old.

47. Find the ratio of spending for the group under 25 years old to the spending for the group of 55 to 64 years old.

48. Find the ratio of spending for the group 35 to 44 years old to the spending for the group 45 to 54 years old.

For Exercises 49–52, find the ratio of the shortest side to the longest side. Write each ratio in lowest terms with whole numbers in the numerator and denominator.

 49.

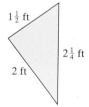

$1\frac{1}{2}$ ft $2\frac{1}{4}$ ft   2 ft

50.

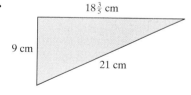

$18\frac{3}{5}$ cm   9 cm   21 cm

51.

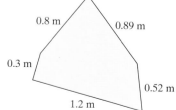

0.8 m   0.89 m   0.3 m   0.52 m   1.2 m

52.

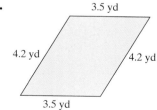

3.5 yd   4.2 yd   4.2 yd   3.5 yd

section 5.2 Rates

1. Definition of a Rate

A **rate** is a comparison of two quantities with *different* units, for example,

$$\frac{270 \text{ mi}}{13 \text{ gal}} \quad \text{and} \quad \frac{\$8.55}{1 \text{ hr}}$$

Several key words imply rates. These are given in Table 5-1.

table 5-1

Key Word	Example	Rate
Per	117 mi per 2 hr	$\dfrac{117 \text{ mi}}{2 \text{ hr}}$
For	$12 for 3 lb	$\dfrac{\$12}{3 \text{ lb}}$
In	400 m in 43.5 sec	$\dfrac{400 \text{ m}}{43.5 \text{ sec}}$
On	270 mi on 12 gal of gas	$\dfrac{270 \text{ mi}}{12 \text{ gal}}$

Because a rate is a comparison of two quantities with different units, it is important to include the units in both the numerator and the denominator. It is also desirable to write rates in lowest terms.

example 1 Writing Rates in Lowest Terms

Write each rate in lowest terms

a. In one region, there are approximately 640 trees on 12 acres.

b. Latonya drove 138 mi on 6 gal of gas.

c. Gail can type 625 words in 10 min.

Solution:

a. The rate of 640 trees on 12 acres can be expressed as $\dfrac{640 \text{ trees}}{12 \text{ acres}}$.

Now write this rate in lowest terms. $\dfrac{\overset{160}{\cancel{640}} \text{ trees}}{\underset{3}{\cancel{12}} \text{ acres}} = \dfrac{160 \text{ trees}}{3 \text{ acres}}$

b. The rate of 138 mi on 6 gal of gas can be expressed as $\dfrac{138 \text{ mi}}{6 \text{ gal}}$.

Now write this rate in lowest terms. $\dfrac{\overset{23}{\cancel{138}} \text{ mi}}{\underset{1}{\cancel{6}} \text{ gal}} = \dfrac{23 \text{ mi}}{1 \text{ gal}}$

When a rate has a denominator of 1 unit, such as $\frac{23 \text{ mi}}{1 \text{ gal}}$, the 1 is sometimes omitted. The division bar is read "per" and is written on the slant. Thus, $\frac{23 \text{ mi}}{1 \text{ gal}} = 23 \text{mi/gal}$ and is read "23 miles per gallon."

Objectives

1. Definition of a Rate
2. Unit Rates
3. Unit Costs
4. Applications of Rates

Concept Connections

1. Why is it important to include units when you are expressing a rate?

Skill Practice

Write each rate in lowest terms.

2. Maria reads 15 pages in 10 min.

3. A Chevrolet Corvette Z06 gets 163.4 mi on 8.6 gal of gas.

4. At Nash Community College in Rocky Mount, North Carolina, there are 2200 students and 120 instructors.

Answers

1. The units are different in the numerator and denominator and will not "cancel."
2. $\dfrac{3 \text{ pages}}{2 \text{ min}}$
3. $\dfrac{19 \text{ mi}}{1 \text{ gal}}$ or 19 mi/gal
4. $\dfrac{55 \text{ students}}{3 \text{ instructors}}$

c. The rate of 625 words in 10 min can be represented as $\dfrac{625 \text{ words}}{10 \text{ min}}$.

Now write this rate in lowest terms. $\dfrac{\overset{125}{\cancel{625}} \text{ words}}{\underset{2}{\cancel{10}} \text{ min}} = \dfrac{125 \text{ words}}{2 \text{ min}}$

This rate indicates that 625 words in 10 min is the same speed as 125 words in 2 min.

2. Unit Rates

A rate having a denominator of 1 unit is called a **unit rate**. Furthermore the number 1 is often omitted in the denominator, as in Example 1(b).

$\dfrac{23 \text{ mi}}{1 \text{ gal}} = 23 \text{ mi/gal}$ is read as "twenty-three miles per gallon."

$\dfrac{52 \text{ ft}}{1 \text{ sec}} = 52 \text{ ft/sec}$ is read as "fifty-two feet per second."

$\dfrac{\$15}{1 \text{ hr}} = \$15/\text{hr}$ is read as "fifteen dollars per hour."

To convert a rate to a unit rate, divide the numerator by the denominator and maintain the units of measurement.

example 2   Finding Unit Rates

Write each rate as a unit rate. Round to 3 decimal places if necessary.

a. A health club charges $125 for 20 visits. Find the unit rate in dollars per visit.

b. In 1960, Wilma Rudolph won the women's 200-m run in 24 sec. Find her speed in meters per second.

c. During one baseball season, Barry Bonds got 149 hits in 403 at bats. Find his batting average. (*Hint:* Batting average is defined as the number of hits per the number of at bats.)

Solution:

a. The rate of $125 for 20 visits can be expressed as $\dfrac{\$125}{20 \text{ visits}}$.

To convert this to a unit rate, divide $125 by 20 visits.

$\dfrac{\$125}{20 \text{ visits}} = \dfrac{\$6.25}{1 \text{ visit}}$ or $6.25/\text{visit}$

$$
\begin{array}{r}
6.25 \\
20\overline{)125.00} \\
-120 \\
\hline
50 \\
-40 \\
\hline
100 \\
-100 \\
\hline
0
\end{array}
$$

b. The rate of 200 m per 24 sec can be expressed as $\dfrac{200 \text{ m}}{24 \text{ sec}}$.

To convert this to a unit rate, divide 200 m by 24 sec.

$$\frac{200 \text{ m}}{24 \text{ sec}} \approx \frac{8.333 \text{ m}}{1 \text{ sec}} \quad \text{or approximately 8.333 m/sec}$$

$$\begin{array}{r} 8.\overline{3} \\ 24\overline{)200.00} \\ -192 \\ \hline 80 \\ -72 \\ \hline 80 \end{array}$$ The quotient repeats.

Wilma Rudolph's speed was approximately 8.333 m/sec.

c. The rate of 149 hits in 403 at bats can be expressed as $\dfrac{149 \text{ hits}}{403 \text{ at bats}}$.

To convert this to a unit rate, divide 149 hits by 403 at bats.

$$\frac{149 \text{ hits}}{403 \text{ at bats}} \approx \frac{0.370 \text{ hit}}{1 \text{ at bat}} \quad \text{or } 0.370 \text{ hit/at bat}$$

3. Unit Costs

A unit cost or unit price is the cost per 1 unit of something. At the grocery store, for example, you might purchase meat for $3.79/lb ($3.79 per 1 lb). Unit cost is useful in day-to-day life when we compare prices. Example 3 compares the prices of three different sizes of apple juice.

example 3   Finding Unit Costs

Apple juice comes in a variety of sizes and packaging options. Find the unit price per ounce and determine which is the best buy.

a. $1.69

Apple
Juice
48 oz

b. $2.39

Apple
Juice
64 oz

c. $2.99

Apple Juice
10-pack 6 oz each

Skill Practice

11. Gatorade comes in several size packages. Compute the unit price per ounce for each option (round to the nearest thousandth of a dollar). Then determine which is the best buy.

a. $2.99 for a 64-oz bottle

b. $3.99 for four 24-oz bottles

c. $3.79 for six 12-oz bottles

Solution:

When we compute a unit cost, the cost is always placed in the numerator of the rate. Furthermore, when we divide the cost by the amount, we need to obtain enough digits in the quotient to see the variation in unit price. In this example, we have rounded to the nearest thousandth of a dollar (nearest tenth of a cent). This means that we use the ten-thousandths-place digit in the quotient on which to base our decision on rounding.

Answers

11. a. $0.047/oz
    b. $0.042/oz; the 4-pack of 24-oz bottles is the best buy
    c. $0.053/oz

	Rate	Quotient	Unit Rate (Rounded)
a.	$\dfrac{\$1.69}{48\ \text{oz}}$	$\$1.69 \div 48\ \text{oz} \approx \$0.0352/\text{oz}$	$\$0.035/\text{oz}$ or $3.5¢/\text{oz}$
b.	$\dfrac{\$2.39}{64\ \text{oz}}$	$\$2.39 \div 64\ \text{oz} \approx \$0.0373/\text{oz}$	$\$0.037/\text{oz}$ or $3.7¢/\text{oz}$
c.	$\begin{array}{c} 6\ \text{oz} \times 10 = 60\ \text{oz} \\ \dfrac{\$2.99}{60\ \text{oz}} \end{array}$	$\$2.99 \div 60\ \text{oz} \approx \$0.0498/\text{oz}$	$\$0.050/\text{oz}$ or $5.0¢/\text{oz}$

From the chart, we see that the most economical buy is the 48-oz size because its unit rate is the least expensive.

Skill Practice

12. In Ecuador roughly 450 out of 500 adults can read. In Brazil, approximately 1700 out of 2000 adults can read.

   a. Compute the unit rate for Ecuador (this unit rate is called the *literacy rate*).

   b. Compute the unit rate for Brazil.

   c. Which country has the greater literacy rate?

Skill Practice

13. Pat can buy a 12-oz jar of peanut butter for $2.19. A 40-oz jar costs $4.39. Suppose she has a $0.75 off coupon for the larger jar and a 2-for-1 coupon for the smaller jar. Compute the unit price per ounce for each option (assume that she buys two of the smaller jars). Which size has the lower cost per ounce?

4. Applications of Rates

Examples 4 and 5 use unit rates for comparison in applications.

example 4   Computing Mortality Rates

Mortality rate is defined to be the total number of people who die due to some risk behavior by the total number of people who engage in the risk behavior. Based on the following statistics, compare the mortality rate for undergoing heart bypass surgery to the mortality rate of climbing Mt. Everest.

   a. Roughly 28 people will die for every 1000 who undergo heart bypass surgery. (Source: The Society of Thoracic Surgeons)

   b. As of June 2001, approximately 178 climbers have died on Mt. Everest out of approximately 2000 attempts at the summit.

Solution:

   a. Mortality rate for heart bypass surgery: $\frac{28}{1000} = 0.028$ death/surgery

   b. Mortality rate for attempting to climb Mt. Everest: $\frac{178}{2000} = 0.089$ death/climb

Comparing these rates shows that it is riskier to climb Mt. Everest than to have heart bypass surgery.

example 5   Comparing Rates to Determine the Better Buy

Joelle has a 50¢ coupon for a 32-oz jar of peanut butter that costs $3.69. She has a $0.75 coupon for a 12-oz jar that costs $1.99. Which is the better buy?

Solution:

First, we must subtract the amount of the coupon from the price for each jar.

   Cost of 32-oz jar: $\$3.69 - \$0.50 = \$3.19$

   Cost of 12-oz jar: $\$1.99 - \$0.75 = \$1.24$

Answers

12. a. 0.9 reader per adult
   b. 0.85 reader per adult
   c. Ecuador
13. Two 12-oz jars at a 2-for-1 discount: $0.09125/oz
   40-oz jar after the $0.75 coupon: $0.091
   The 40-oz jar with the coupon is cheaper.

Now we find the unit cost per ounce.

	Rate	Quotient	Unit rate (Rounded)
a.	$\dfrac{\$3.19}{32 \text{ oz}}$	$\$3.19 \div 32 \text{ oz} \approx \$0.0996875/\text{oz}$	$\$0.100/\text{oz}$ or $10.0¢/\text{oz}$
b.	$\dfrac{\$1.24}{12 \text{ oz}}$	$\$1.24 \div 12 \text{ oz} \approx \$0.10\overline{3}/\text{oz}$	$\$0.103/\text{oz}$ or $10.3¢/\text{oz}$

Despite having a greater coupon for the 12-oz jar, the 32-oz is the better buy.

section 5.2   Practice Exercises

Boost *your* GRADE at mathzone.com!

MathZone

- Practice Problems
- Self-Tests
- NetTutor
- e-Professors
- Videos

Study Skills Exercises

1. Budgeting enough time to do homework and to study for a class is one of the most important steps to success in a class. Use the weekly calendar to help you plan your time for your studies this week. Also write in other obligations such as the time required for your job, for your family, for sleeping, and for eating. Be realistic when you estimate the time for each activity.

Time	Mon.	Tues.	Wed.	Thurs.	Fri.	Sat.	Sun.
7–8							
8–9							
9–10							
10–11							
11–12							
12–1							
1–2							
2–3							
3–4							
4–5							
5–6							
6–7							
7–8							
8–9							
9–10							

2. Define the key terms.

    a. Rate   **b. Unit rate**

Review Exercises

3. Write the ratio 3 to 5 in two other ways.

4. Write the ratio 4:1 in two other ways.

For Exercises 5–9, write the ratio in lowest terms.

5. 36¢ to 27¢

6. $6\frac{3}{4}$ ft to $8\frac{1}{4}$ ft

7. 1.08 mi to 2.04 mi

8. $28.40 to $20.80

9. $7\frac{1}{3}$ m to $8\frac{2}{9}$ m

Objective 1: Definition of a Rate

For Exercises 10–21, write each rate in lowest terms. **(See Example 1.)**

10. A type of laminate flooring sells for $32 for 5 square feet ($ft^2$).

11. A remote control car can go up to 44 ft in 5 sec.

12. Elaine drives 234 mi in 4 hr.

13. Travis has 14 blooms on 6 of his plants.

14. Tyler earned $58 in 8 hr.

15. Neil can type only 336 words in 15 min.

16. A printer can print 13 pages in 26 sec.

17. During a bad storm there was 2 in. of rain in 6 hr.

18. There are 130 calories in 8 snack crackers.

19. The driveway is lined with 14 plants for a length of 22 ft.

20. An advertisement states that TV trays are selling for $30 for a set of 4 trays.

21. There are 50 students assigned to 4 advisers.

Objective 2: Unit Rates

22. Of the following rates, circle those that are unit rates.

    a. $\dfrac{3\text{ lb}}{\$1.00}$ **b.** $\dfrac{21\text{ ft}}{1\text{ sec}}$ **c.** 50 mi/hr   **d.** $\dfrac{232\text{ words}}{2\text{ min}}$

For Exercises 23–28, write each rate as a unit rate. **(See Example 2.)**

23. The Osborne family drove 452 mi in 4 days.

24. The book of poetry *The Prophet* by Kahlil Gibran has estimated sales of $6,000,000 over an 80-year period.

25. Philip drove 480 km in 5 hr.

26. Ian flew 1120 mi in 4 hr.

27. If Oscar bought an easy chair for $660 and plans to make 12 payments, what is the amount per payment?

28. The jockey David Gall had 7396 wins in 43 years of riding.

For Exercises 29–32, round the unit rates to the nearest hundredth when necessary.

29. At the market, bananas cost $1.50 for 4 lb.

30. Ceramic tiles sell for $13.08 for a box of 12 tiles. Find the price per tile.

31. Lottery prize money of $1,792,000 is for 7 people.

32. One WeightWatchers group lost 123 lb for its 11 members.

33. The record for men's speed skating was 34.32 sec for 500 m. Find the rate in seconds per meter. (Round to 3 decimal places.)

34. The record for women's speed skating was 37.22 sec for 500 m. Find the rate in seconds per meter. (Round to 3 decimal places.)

Objective 3: Unit Costs

For Exercises 35–42, find the unit costs (that is, dollars per unit). Round the answers to 3 decimal places when necessary. **(See Example 3.)**

35. Tide laundry detergent costs $4.99 for 100 oz.

36. Dove liquid body wash costs $3.49 for 12 oz.

37. Soda costs $1.99 for a 2-liter bottle.

38. A dinette set costs $221.00 for 4 chairs.

39. A set of 4 tires costs $210.

40. A package of 3 shirts costs $64.80.

41. A package of 6 newborn bodysuits costs $32.50.

42. A package of 8 AAA batteries costs $4.16.

Objective 4: Applications of Rates

43. The number of motor vehicles produced in the United States rose by 3,050,000 between 1990 and 2000. Compute the rate representing the increase in the number of vehicles produced per year during this time period. (Source: American Automobile Manufacturers Association.)

44. The number of motor vehicles produced in Japan decreased by 3,340,000 between 1990 and 2000. Compute the rate representing the decrease in the number of vehicles produced per year during this time period. (Source: American Automobile Manufacturers Association.)

45. a. The average yearly cost to attend a private four-year college rose $4150 between 1990 and 2000. Compute the rate representing the increase in cost per year.

    b. The average yearly cost to attend a public four-year college rose $1350 between 1990 and 2000. Compute the rate representing the increase in cost per year.

    c. Which type of institution had a greater rate of increase in cost? (Source: National Center for Education Statistics.) **(See Example 4.)**

46. a. Retail gasoline prices rose an average of $0.35 per gallon in the United States in the 10-year period between 1990 and 2000. Compute the rate representing the increase in the price of 1 gal of gas per year.

    b. In Germany gasoline prices rose $1.09 per gallon between 1990 and 2000. Compute the rate representing the increase in the price of 1 gal of gas per year.

    c. Which country had a greater rate of increase in gas prices? (Source: Energy Information Administration, U.S. Department of Energy.)

47. A cheetah can run 120 m in 4.1 sec. An antelope can run 50 m in 2.1 sec. Compare their speeds to determine the faster animal. Round to the nearest whole unit.

48. The United States has a rate of 8 out of 100,000 people under the age of 65 participating in hospice. There are 2491 hospice patients who are 65 years or older out of 1,000,000 people. Find the unit rates.

49. Corn comes in two size cans, 29 oz and 15 oz. The larger can costs $1.19, and the smaller can costs $0.77. Find the unit cost of each can, to choose the better buy. (Round to 3 decimal places.) **(See Exercise 5.)**

50. Napkins come in a variety of packages. Find the unit cost of each package to determine the better buy. (Round to 4 decimal places.)

    a. $2.99 for 400 napkins   **b.** $1.50 for 100 napkins

Calculator Connections

Topic: Applications of unit rates

Calculator Exercise

Don Shula coached football for 33 years. He had 328 wins and 156 losses. Tom Landry coached football for 29 years. He had 250 wins and 162 losses. Use this information to answer Exercises 51–52.

51. **a.** Compute a unit rate representing the number of wins per year for Don Shula. Round to 1 decimal place.

   b. Compute a unit rate representing the number of wins per year for Tom Landry. Round to 1 decimal place.

   c. Which coach had a better rate of wins per year?

52. **a.** Compute a unit rate representing the number of wins to the number of losses for Don Shula. Round to 1 decimal place.

   b. Compute a unit rate representing the number of wins to the number of losses for Tom Landry. Round to 1 decimal place.

   c. Which coach had a better win/loss rate?

53. Compare three brands of soap. Find the unit prices and determine the best buy. (Round to 2 decimal places.)

   a. Dove: $4.99 for a 6-bar pack of 4.5-oz bars

   b. Dial: $1.89 for a 3-bar pack of 4.5-oz bars

   c. Irish Spring: $2.99 for an 8-bar pack of 4.5-oz bars

54. Mayonnaise comes in 32-, 16-, and 8-oz jars. They are priced at $3.61, $2.19, and $1.19, respectively. Find the unit cost of each jar, to find the best buy. (Round to 3 decimal places.)

55. Albacore tuna comes in different-size cans. Find the unit cost of each package to find the best buy. (Round to 3 decimal places.)

   a. $3.59 for a 12-oz can   **b.** $4.99 for a 4-pack of 6-oz cans

   c. $2.99 for a 3-pack of 3-oz cans

56. Coke is sold in a variety of different packages. Find the unit cost of each package, to find the better buy. (Round to 3 decimal places.)

a. $4.99 for a case of 24 twelve-oz cans

b. $5.00 for a 12-pack of 8-oz cans

---

chapter 5 | midchapter review

1. What is the difference between a ratio and a rate?

2. Given a ratio and a rate, for which is it important to write the units?

For Exercises 3–10, write the reduced ratio or rate.

3. 28 mi in 6 hr

4. 42 ft to 20 ft

5. 234 cm to 432 cm

6. 36 bushels for 8 trees

7. 72 plants per 10 yd

8. $48 to $18

9. 12 quarts to 18 quarts

10. 6 gal per 200 mi

11. 20¢ to 80¢

12. 28 m per 21 sec

13. 50 tiles per 4 ft^2

14. 15 students to 3 teachers

Objectives

1. Definition of a Proportion
2. Determining Whether Two Ratios Form a Proportion
3. Solving Proportions

Skill Practice

Write a proportion for each statement.

1. 7 is to 28 as 13 is to 52.

2. $17 is to 2 hr as $102 is to 12 hr.

3. The numbers 5 and 11 are proportional to the numbers 15 and 33.

Answers

1. $\dfrac{7}{28} = \dfrac{13}{52}$ 2. $\dfrac{\$17}{2\,\text{hr}} = \dfrac{\$102}{12\,\text{hr}}$

3. $\dfrac{5}{11} = \dfrac{15}{33}$

section 5.3   Proportions

1. Definition of a Proportion

A statement indicating that two quantities are equal is called an **equation**. In this section we are interested in a special type of equation called a proportion. A **proportion** states that two ratios or rates are equal. For example,

$$\frac{1}{4} = \frac{10}{40} \text{ is a proportion.}$$

We know that the fractions $\frac{1}{4}$ and $\frac{10}{40}$ are equal because $\frac{10}{40}$ reduces to $\frac{1}{4}$.

We read the proportion $\frac{1}{4} = \frac{10}{40}$ as follows: "1 is to 4 as 10 is to 40."

We also say that the pair of numbers 1 and 4 is *proportional to* the pair of numbers 10 and 40.

example 1   Writing Proportions

Write a proportion for each statement.

a. 5 is to 12 as 30 is to 72.

b. 240 mi is to 4 hr as 300 mi is to 5 hr.

c. The numbers 3 and 7 are proportional to the numbers 12 and 28.

Solution:

a. $\dfrac{5}{12} = \dfrac{30}{72}$ 5 is to 12 as 30 is to 72.

b. $\dfrac{240 \text{ mi}}{4 \text{ hr}} = \dfrac{300 \text{ mi}}{5 \text{ hr}}$ 240 mi is to 4 hr as 300 mi is to 5 hr.

c. $\dfrac{3}{7} = \dfrac{12}{28}$ 3 and 7 are proportional to 12 and 28.

2. Determining Whether Two Ratios Form a Proportion

To determine whether two ratios form a proportion, we must determine whether the ratios are equal. Recall from Section 2.3 that two fractions are equal whenever their cross products are equal. That is,

$$\frac{a}{b} = \frac{c}{d} \quad \text{implies} \quad a \cdot d = b \cdot c$$

---

example 2   Determining Whether Two Ratios Form a Proportion

Determine whether the ratios form a proportion.

a. $\dfrac{3}{5} \overset{?}{=} \dfrac{9}{15}$ **b.** $\dfrac{2}{7} \overset{?}{=} \dfrac{18}{52}$

Solution:

a. $\dfrac{3}{5} \overset{?}{=} \dfrac{9}{15}$

$(3)(15) \overset{?}{=} (5)(9)$ Cross-multiply to form the cross products.

$45 = 45$ ✔

The cross products are equal. Therefore, the ratios form a proportion.

b. $\dfrac{2}{7} \overset{?}{=} \dfrac{18}{52}$

$(2)(52) \overset{?}{=} (7)(18)$ Cross-multiply to form the cross products.

$104 \neq 126$

The cross products are not equal. The ratios do not form a proportion.

Avoiding Mistakes: Cross multiplication can be performed only when there are two fractions separated by an equal sign.

---

example 3   Determining Whether Two Ratios Form a Proportion

Determine whether the ratios form a proportion.

a. $\dfrac{1\frac{2}{3}}{4} \overset{?}{=} \dfrac{10}{5\frac{1}{2}}$ **b.** $\dfrac{2.7}{5.3} \overset{?}{=} \dfrac{8.1}{15.9}$

Skill Practice

Determine whether the ratios form a proportion.

4. $\dfrac{4}{9} \overset{?}{=} \dfrac{12}{27}$

5. $\dfrac{5}{7} \overset{?}{=} \dfrac{3}{4}$

Skill Practice

Determine whether the ratios form a proportion.

6. $\dfrac{3\frac{1}{4}}{5} \overset{?}{=} \dfrac{8\frac{1}{8}}{12\frac{1}{2}}$ **7.** $\dfrac{4.2}{6.9} \overset{?}{=} \dfrac{6.3}{10}$

Answers

4. Yes   5. No   6. Yes   7. No

Solution:

a.

$$\frac{1\frac{2}{3}}{4} \overset{?}{=} \frac{10}{5\frac{1}{2}}$$

$$\left(1\frac{2}{3}\right)\left(5\frac{1}{2}\right) \overset{?}{=} (4)(10) \qquad \text{Cross-multiply to form the cross products.}$$

$$\frac{5}{3} \cdot \frac{11}{2} \overset{?}{=} 40 \qquad\qquad \text{Write the mixed numbers as improper fractions.}$$

$$\frac{55}{6} \neq 40 \qquad\qquad \text{Multiply fractions.}$$

The cross products are not equal. The ratios do not form a proportion.

b.

$$\frac{2.7}{5.3} \overset{?}{=} \frac{8.1}{15.9}$$

$$(2.7)(15.9) \overset{?}{=} (5.3)(8.1) \qquad \text{Cross-multiply to form the cross products.}$$

$$42.93 = 42.93 \; \checkmark \qquad\qquad \text{Multiply decimals.}$$

The cross products are equal. The ratios form a proportion.

Skill Practice

8. Determine whether the numbers $\frac{3}{5}$ and $\frac{1}{3}$ are proportional to the numbers 9 and 5.

9. Determine whether the numbers 1.2 and 2.5 are proportional to the numbers 2 and 5.

example 4   **Determining Whether Pairs of Numbers Are Proportional**

Determine whether the pair of numbers 8 and 14 is proportional to the pair of numbers $\frac{2}{3}$ and $\frac{7}{6}$.

Solution:

Two pairs of numbers are proportional if their ratios are equal.

$$\frac{8}{14} \overset{?}{=} \frac{\frac{2}{3}}{\frac{7}{6}}$$

$$(8)\left(\frac{7}{6}\right) \overset{?}{=} \left(\frac{2}{3}\right)(14) \qquad \text{Cross-multiply to form the cross products.}$$

$$\left(\frac{\overset{4}{8}}{1}\right)\left(\frac{7}{\underset{3}{6}}\right) \overset{?}{=} \left(\frac{2}{3}\right)\left(\frac{14}{1}\right) \qquad \begin{array}{l}\text{Write the whole numbers as improper}\\\text{fractions.}\end{array}$$

$$\frac{28}{3} = \frac{28}{3} \; \checkmark \qquad \text{Multiply the fractions.}$$

The cross products are equal. The numbers 8 and 14 are proportional to the numbers $\frac{2}{3}$ and $\frac{7}{6}$.

---

3. Solving Proportions

A proportion is made up of four values. If three of the four values are known, we can solve for the fourth.

Answers

8. Yes   9. No

Consider the proportion $\frac{x}{20} = \frac{3}{4}$. We let the variable x represent the unknown value in the proportion. To solve for x, we can equate the cross products to form an equivalent equation.

$$\frac{x}{20} = \frac{3}{4}$$

$4 \cdot x = 3 \cdot 20$ Cross-multiply to form the cross products.

$4 \cdot x = 60$ To determine the correct value of x, we ask, What number multiplied by 4 equals 60? The answer is 15 and can be found mentally by computing $60 \div 4 = 15$.

$\dfrac{4 \cdot x}{4} = \dfrac{60}{4}$ To show this step in the equation, divide both sides of the equation by 4.

$\dfrac{\overset{1}{\cancel{4}} \cdot x}{\underset{1}{\cancel{4}}} = \dfrac{\overset{15}{\cancel{60}}}{\underset{1}{\cancel{4}}}$ Simplify common factors within each fraction.

$1x = 15$

$x = 15$

We can check the value of x in the original proportion.

Check: $\dfrac{x}{20} = \dfrac{3}{4}$ $\xrightarrow{\text{substitute 15 for } x}$ $\dfrac{15}{20} \overset{?}{=} \dfrac{3}{4}$

$$(15)(4) \overset{?}{=} (3)(20)$$

$$60 = 60 \ ✔ \quad \text{The solution } x = 15 \text{ checks.}$$

This process uses the following important fact. *We may divide both sides of an equation by the same nonzero number.* By so doing, we produce an equivalent equation that has the same solution. In general, to solve an equation, the goal is to isolate the variable on one side of the equation. In the equation $4 \cdot x = 60$, we must eliminate the factor of 4 next to x. By dividing by 4 we form a ratio $\frac{4 \cdot x}{4}$. This reduces to $\frac{1 \cdot x}{1}$ which in turn simplifies to x.

The steps to solve a proportion are summarized next.

Steps to Solve a Proportion

1. Set the cross products equal to each other.
2. Divide both sides of the equation by the number being multiplied by the variable.
3. Check the solution in the original proportion.

---

example 5   Solving Proportions

Solve each proportion.

a. $\dfrac{x}{13} = \dfrac{6}{39}$ **b.** $\dfrac{4}{15} = \dfrac{9}{n}$

Concept Connections

10. Explain how to check the solution to a proportion.

Concept Connections

What number would you divide by on each side of the equation to solve for x?

11. $2x = 16$ 12. $1.5x = 6$

13. $14 = 7x$

Skill Practice

Solve the proportions. Be sure to check your answers.

14. $\dfrac{8}{7} = \dfrac{96}{x}$ 15. $\dfrac{2}{9} = \dfrac{n}{81}$

16. $\dfrac{3}{w} = \dfrac{21}{77}$

Answers

10. Substitute the solution back in for the variable in the original equation. Then see if the cross products are equal.
11. Divide by 2.   12. Divide by 1.5.
13. Divide by 7.   14. $x = 84$
15. $n = 18$ 16. $w = 11$

Solution:

a. $\dfrac{x}{13} = \dfrac{6}{39}$

$39 \cdot x = (6)(13)$ Set the cross products equal.

$39 \cdot x = 78$ Simplify.

$\dfrac{39 \cdot x}{39} = \dfrac{78}{39}$ Divide both sides by the number being multiplied by x (in this case, 39).

$\dfrac{\overset{1}{\cancel{39}} \cdot x}{\underset{1}{\cancel{39}}} = \dfrac{\overset{2}{\cancel{78}}}{\underset{1}{\cancel{39}}}$ Simplify common factors within each fraction.

$x = 2$ Check: $\dfrac{x}{13} = \dfrac{6}{39} \xrightarrow{\overset{\text{substitute}}{x\,=\,2}} \dfrac{2}{13} \overset{?}{=} \dfrac{6}{39}$

$(2)(39) \overset{?}{=} (6)(13)$

$78 = 78$ ✔

The solution $x = 2$ checks in the original proportion.

b. $\dfrac{4}{15} = \dfrac{9}{n}$ The variable can be represented by any letter.

$4 \cdot n = (9)(15)$ Set the cross products equal.

$4 \cdot n = 135$ Simplify.

$\dfrac{4 \cdot n}{4} = \dfrac{135}{4}$ Divide both sides by the number being multiplied by n (in this case, 4).

$\dfrac{\overset{1}{\cancel{4}} \cdot n}{\underset{1}{\cancel{4}}} = \dfrac{135}{4}$

$n = \dfrac{135}{4}$ The fraction $\frac{135}{4}$ is in lowest terms.

The solution may be written as $n = \frac{135}{4}$ or $n = 33\frac{3}{4}$ or $n = 33.75$.

To check the solution in the original proportion, we may use any of the three forms of the answer. We will use the decimal form.

Check: $\dfrac{4}{15} = \dfrac{9}{n} \xrightarrow{\text{substitute } n\,=\,33.75} \dfrac{4}{15} \overset{?}{=} \dfrac{9}{33.75}$

$(4)(33.75) \overset{?}{=} (9)(15)$

$135 = 135$ ✔   The solution checks.

Skill Practice

Solve the proportions. Be sure to check your answers.

17. $\dfrac{0.6}{x} = \dfrac{1.5}{2}$ **18.** $\dfrac{8}{3} = \dfrac{1\frac{3}{5}}{p}$

19. $\dfrac{\frac{1}{2}}{3.5} = \dfrac{x}{14}$

---

example 6   Solving Proportions

Solve each proportion.

a. $\dfrac{0.8}{3.1} = \dfrac{4}{p}$ **b.** $\dfrac{12}{8} = \dfrac{x}{\frac{2}{3}}$

Answers

17. $x = 0.8$ 18. $p = \dfrac{3}{5}$ or 0.6

19. $x = 2$

Solution:

a. $\dfrac{0.8}{3.1} = \dfrac{4}{p}$

$(0.8) \cdot p = (4)(3.1)$ Set the cross products equal.

$0.8p = 12.4$ Notice that in this case we dropped the $\cdot$ symbol for multiplication between 0.8 and p. If a variable and a constant are written adjacent to each other without an operator ($+, -, \times,$ or $\div$) between them, the operation is understood to be multiplication. That is, $0.8p = 0.8 \times p$.

$\dfrac{0.8p}{0.8} = \dfrac{12.4}{0.8}$ Divide both sides by the number being multiplied by p (in this case, 0.8).

$\dfrac{\overset{1}{\cancel{0.8}}p}{\underset{1}{\cancel{0.8}}} = \dfrac{12.4}{0.8}$

$p = 15.5$ Divide $12.4 \div 0.8 = 15.5$.

The check is left to the reader.

We chose to give the solution to Example 5(a) in decimal form because the values in the original proportion are decimal numbers. However, it would be correct to give the solution as a mixed number or fraction. The solution $p = 15.5$ is also equivalent to $p = 15\frac{1}{2}$ or $p = \frac{31}{2}$.

b. $\dfrac{12}{8} = \dfrac{x}{\frac{2}{3}}$

$(12)\left(\dfrac{2}{3}\right) = 8 \cdot x$ Set the cross products equal.

$\left(\dfrac{12}{1}\right)\left(\dfrac{2}{3}\right) = 8x$ Write the whole number as an improper fraction.

$\left(\dfrac{\overset{4}{\cancel{12}}}{1}\right)\left(\dfrac{2}{\underset{1}{\cancel{3}}}\right) = 8x$ Simplify common factors.

$\dfrac{8}{1} = 8x$ Multiply fractions.

$8 = 8x$ Simplify.

$\dfrac{8}{8} = \dfrac{8x}{8}$ Divide both sides by the number being multiplied by x (in this case, 8).

$\dfrac{\overset{1}{\cancel{8}}}{\underset{1}{\cancel{8}}} = \dfrac{\overset{1}{\cancel{8}}x}{\underset{1}{\cancel{8}}}$ Simplify common factors.

$1 = x$

The solution $x = 1$ checks in the original proportion.

section 5.3   Practice Exercises

Study Skills Exercises

1. You should not try to cram for tests. Instead, math is a subject that should be studied every day. This text gives you opportunities to review and practice as you work through the book.

   a. Find the page number for the Midchapter Review for this chapter.

   b. Find the page number for the Cumulative Review exercises for this chapter.

2. Define the key terms.

   a. Equation   **b. Proportion**

Review Exercises

For Exercises 3–8, write as a reduced ratio or rate.

3. 3 ft to 45 ft

4. 3 teachers for 45 students

5. 6 apples for 2 pies

6. 6 days to 2 days

7. 264 mi per 36 gal

8. $264 to $36

Objective 1: Definition of a Proportion

For Exercises 9–20, write a proportion for each statement. **(See Example 1.)**

9. 4 is to 16 as 5 is to 20.

10. 3 is to 18 as 4 is to 24.

11. 25 is to 15 as 10 is to 6.

12. 35 is to 14 as 20 is to 8.

13. The numbers 2 and 3 are proportional to the numbers 4 and 6.

14. The numbers 2 and 1 are proportional to the numbers 26 and 13.

15. The numbers 30 and 25 are proportional to the numbers 12 and 10.

16. The numbers 24 and 18 are proportional to the numbers 8 and 6.

17. $6.25 per hour is proportional to $187.50 per 30 hr.

18. $115 per week is proportional to $460 per 4 weeks.

19. 1 in. is to 7 mi as 5 in. is to 35 mi.

20. 16 flowers is to 5 plants as 32 flowers is to 10 plants.

Objective 2: Determining Whether Two Ratios Form a Proportion

For Exercises 21–28, determine whether the ratios form a proportion. **(See Examples 2–3.)**

21. $\dfrac{5}{18} \stackrel{?}{=} \dfrac{4}{16}$

22. $\dfrac{9}{10} \stackrel{?}{=} \dfrac{8}{9}$

23. $\dfrac{16}{24} \stackrel{?}{=} \dfrac{2}{3}$

24. $\dfrac{4}{5} \stackrel{?}{=} \dfrac{24}{30}$

25. $\dfrac{2\frac{1}{2}}{3\frac{2}{3}} \stackrel{?}{=} \dfrac{15}{22}$

26. $\dfrac{1\frac{3}{4}}{3} \stackrel{?}{=} \dfrac{7}{12}$

27. $\dfrac{2}{3.2} \stackrel{?}{=} \dfrac{10}{16}$

28. $\dfrac{4.7}{7} \stackrel{?}{=} \dfrac{23.5}{35}$

For Exercises 29–34, determine whether the pairs of numbers are proportional. **(See Example 4.)**

29. Are the numbers 48 and 18 proportional to the numbers 24 and 9?

30. Are the numbers 35 and 14 proportional to the numbers 5 and 2?

31. Are the numbers $2\frac{3}{8}$ and $1\frac{1}{2}$ proportional to the numbers $9\frac{1}{2}$ and 6?

32. Are the numbers $1\frac{2}{3}$ and $\frac{5}{6}$ proportional to the numbers 5 and $2\frac{1}{2}$?

33. Are the numbers 6.3 and 9 proportional to the numbers 12.6 and 16?

34. Are the numbers 7.1 and 2.4 proportional to the numbers 35.5 and 10?

Objective 3: Solving Proportions

For Exercises 35–42, what number would you divide by on each side of the equation to solve for the variable?

35. $2x = 8$

36. $3x = 27$

37. $5p = 30$

38. $7w = 49$

39. $32 = 8m$

40. $50 = 25y$

41. $0.15 = 0.6x$

42. $1.4 = 0.4z$

For Exercises 43–46, determine whether the given value is a solution to the proportion.

43. $\dfrac{x}{40} = \dfrac{1}{8}$; $x = 5$

44. $\dfrac{14}{x} = \dfrac{12}{18}$; $x = 21$

45. $\dfrac{12.4}{31} = \dfrac{8.2}{y}$; $y = 20$

46. $\dfrac{4.2}{9.8} = \dfrac{z}{36.4}$; $z = 15.2$

For Exercises 47–66, solve the proportion. Be sure to check your answers. **(See Examples 5–6.)**

47. $\dfrac{12}{16} = \dfrac{3}{x}$

48. $\dfrac{20}{28} = \dfrac{5}{x}$

49. $\dfrac{9}{21} = \dfrac{x}{7}$

50. $\dfrac{15}{10} = \dfrac{3}{x}$

51. $\dfrac{p}{12} = \dfrac{25}{4}$

52. $\dfrac{p}{8} = \dfrac{30}{24}$

53. $\dfrac{6}{n} = \dfrac{4}{8}$

54. $\dfrac{49}{n} = \dfrac{14}{18}$

55. $\dfrac{2}{3} = \dfrac{t}{18}$

56. $\dfrac{34}{51} = \dfrac{2}{t}$

57. $\dfrac{25}{100} = \dfrac{9}{y}$

58. $\dfrac{65}{15} = \dfrac{26}{y}$

59. $\dfrac{17}{x} = \dfrac{4\frac{1}{4}}{3}$

60. $\dfrac{26}{x} = \dfrac{5\frac{1}{5}}{6}$

61. $\dfrac{m}{12} = \dfrac{5}{8}$

62. $\dfrac{16}{12} = \dfrac{21}{a}$

63. $\dfrac{3\frac{1}{8}}{5} = \dfrac{18\frac{3}{4}}{k}$

64. $\dfrac{4\frac{3}{4}}{8} = \dfrac{9\frac{1}{2}}{k}$

65. $\dfrac{0.5}{h} = \dfrac{1.8}{9}$

66. $\dfrac{2.6}{h} = \dfrac{1.3}{0.5}$

Objectives

1. Applications of Proportions
2. Similar Triangles
3. Similar Polygons

section 5.4 Applications of Proportions and Similar Figures

1. Applications of Proportions

Proportions are used in a variety of applications. In Examples 1 through 4, we take information from the wording of a problem and form a proportion.

Skill Practice

1. Jacques bought 3 lb of tomatoes for $4.50. At this rate, how much would 7 lb cost?

example 1 Using a Proportion in a Consumer Application

Linda drove her Honda Accord 145 mi on 5 gal of gas. At this rate, how far can she drive on 12 gal?

Solution:

Let x represent the distance Linda can go on 12 gal.

Answer

1. The price for 7 lb would be $10.50.

This problem involves two rates. We can translate this to a proportion. Equate the two rates.

distance ⟶ $\dfrac{145 \text{ mi}}{5 \text{ gal}} = \dfrac{x \text{ mi}}{12 \text{ gal}}$ ⟵ distance

number of gallons ⟶ ⟵ number of gallons

Solve the proportion.

$(145)(12) = (5) \cdot x$ Cross-multiply to form the cross products.

$1740 = 5x$

$\dfrac{1740}{5} = \dfrac{\overset{1}{\cancel{5}}x}{\underset{1}{\cancel{5}}}$ Divide both sides by 5.

$348 = x$ Divide $1740 \div 5 = 348$.

Linda can drive 348 mi on 12 gal of gas.

> **Tip:** Notice that the two rates have the same units in the numerator (miles) and the same units in the denominator (gallons).

In Example 1 we could have set up a proportion in many different ways.

$\dfrac{145 \text{ mi}}{5 \text{ gal}} = \dfrac{x \text{ mi}}{12 \text{ gal}}$ or $\dfrac{5 \text{ gal}}{145 \text{ mi}} = \dfrac{12 \text{ gal}}{x \text{ mi}}$ or

$\dfrac{5 \text{ gal}}{12 \text{ gal}} = \dfrac{145 \text{ mi}}{x \text{ mi}}$ or $\dfrac{12 \text{ gal}}{5 \text{ gal}} = \dfrac{x \text{ mi}}{145 \text{ mi}}$

Notice that in each case, the cross products produce the same equation. We will generally set up the proportions so that the units in the numerators are the same and the units in the denominators are the same.

example 2   Using a Proportion in a Construction Application

If a cable 25 ft long weighs 1.2 lb, how much will a 120-ft cable weigh?

Solution:

Let w represent the weight of the 120-ft cable.   Label the unknown.

length ⟶ $\dfrac{25 \text{ ft}}{1.2 \text{ lb}} = \dfrac{120 \text{ ft}}{w \text{ lb}}$ ⟵ length   Translate to a proportion.

weight ⟶ ⟵ weight

$(25) \cdot w = (1.2)(120)$ Equate the cross products.

$25w = 144$

$\dfrac{25 w}{25} = \dfrac{144}{25}$ Divide both sides by 25.

$w = 5.76$ Divide $144 \div 25 = 5.76$.

The 120-ft cable weighs 5.76-lb.

Skill Practice

2. It takes 2.5 gal of paint to cover 900 ft² of wall. How much area could be painted with 4 gal of the same paint?

Answer

2. An area of 1440 ft² could be painted.

example 3   Using Proportions in a Geography Application

The distance between Phoenix and Los Angeles is 348 mi. On a certain map, this represents 6 in. On the same map, the distance between San Antonio and Little Rock is $8\frac{1}{2}$ in. What is the actual distance between San Antonio and Little Rock?

Solution:

Let d represent the distance between San Antonio and Little Rock.

actual distance $\longrightarrow$ $\dfrac{348 \text{ mi}}{6 \text{ in.}} = \dfrac{d \text{ mi}}{8\frac{1}{2} \text{ in.}}$ $\longleftarrow$ actual distance
distance on map $\longrightarrow$ $\longleftarrow$ distance on map

$$(348)(8\tfrac{1}{2}) = (6) \cdot d \qquad \text{Translate to a proportion.}$$

$\phantom{(348)(8\tfrac{1}{2}) = (6) \cdot d}$ Equate the cross products.

$$(348)(8.5) = 6d \qquad \text{Convert the values to decimal.}$$

$$2958 = 6d$$

$$\dfrac{2958}{6} = \dfrac{6d}{6} \qquad \text{Divide both sides by 6.}$$

$$493 = d \qquad \text{Divide } 2958 \div 6 = 493.$$

The distance between San Antonio and Little Rock is 493 mi.

example 4   Applying a Proportion to Environmental Science

A biologist wants to estimate the number of elk in a wildlife preserve. She sedates 25 elk and clips a small radio transmitter onto the ear of each animal. The elk return to the wild, and after 6 months, the biologist studies a sample of 120 elk in the preserve. Of the 120 elk sampled, 4 have radio transmitters. Approximately how many elk are in the whole preserve?

Solution:

Let n represent the number of elk in the whole preserve.

 Sample Population

number of elk in the sample
with radio transmitters $\longrightarrow$ $\dfrac{4}{120} = \dfrac{25}{n}$ $\longleftarrow$ number of elk in the population with radio transmitters
total elk in the sample $\longrightarrow$ $\longleftarrow$ total elk in the population

$$4 \cdot n = (120)(25) \qquad \text{Equate the cross products.}$$

$$4n = 3000$$

$$\dfrac{4n}{4} = \dfrac{3000}{4} \qquad \text{Divide both sides by 4.}$$

$$n = 750 \qquad \text{Divide } 3000 \div 4 = 750.$$

There are approximately 750 elk in the wildlife preserve.

2. Similar Triangles

Two triangles whose corresponding sides are proportional are called **similar triangles**. This means that the triangles have the same "shape" but may be different sizes. The following pairs of triangles are similar (Figure 5-2).

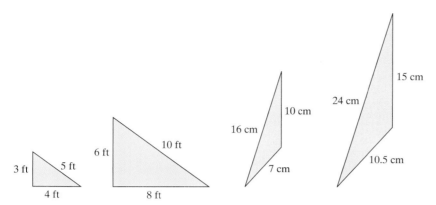

Figure 5-2

Consider the triangles:

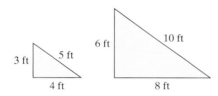

Notice that the ratio formed by the "left" sides of the triangles is $\frac{3}{6}$. This is the same as the ratio formed by the "bottom" sides, $\frac{4}{8}$, and is the same as the ratio formed by the "right" sides, $\frac{5}{10}$. Each ratio reduces to $\frac{1}{2}$. Because all ratios are equal, the corresponding sides are proportional.

| **example 5** | Finding the Unknown Sides in Similar Triangles |

The triangles in Figure 5-3 are similar.

a. Solve for x.   **b.** Solve for y.

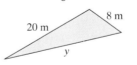

Figure 5-3

Solution:

a. Notice that the lengths of the "left" sides of both triangles are known. This forms a known ratio of $\frac{20}{5}$. Because the triangles are similar, the ratio of the other corresponding sides must be equal to $\frac{20}{5}$. To solve for x, we have

$$\frac{20}{5} = \frac{8}{x}$$ Translate to a proportion.

$$20 \cdot x = (5)(8)$$ Equate the cross products.

$$20x = 40$$

$$\frac{20x}{20} = \frac{40}{20}$$ Divide both sides of the equation by 20.

$$x = 2$$ Divide $40 \div 20 = 2$.

b. To solve for y, we have

$$\frac{20}{5} = \frac{y}{4} \qquad \text{Translate to a proportion.}$$

$$(20)(4) = 5 \cdot y \qquad \text{Equate the cross products.}$$

$$80 = 5y$$

$$\frac{80}{5} = \frac{5y}{5} \qquad \text{Divide both sides by 5.}$$

$$y = 16 \qquad \text{Divide } 80 \div 5 = 16.$$

The values for x and y are $x = 2$ and $y = 16$.

Skill Practice

7. Donna has drawn a pattern from which to make a scarf. Assuming that the triangles are similar (the pattern and scarf have the same shape), find the length of side x (in yards).

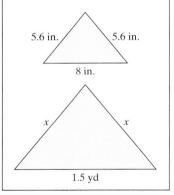

example 6   Using Similar Triangles in an Application

On a sunny day, a 6-ft man casts a 3.2 ft shadow on the ground. At the same time, a building casts an 80-ft shadow. How tall is the building?

Solution:

We illustrate the scenario in Figure 5-4. Note, however, that the distances are not drawn to scale.

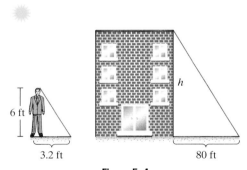

Figure 5-4

We will assume that the triangles are similar because the readings were taken at the same time of day (therefore, the triangles have the same "shape").

$$\frac{6}{h} = \frac{3.2}{80} \qquad \text{Translate to a proportion.}$$

$$(6)(80) = (3.2) \cdot h \qquad \text{Equate the cross products.}$$

$$480 = 3.2h$$

$$\frac{480}{3.2} = \frac{3.2h}{3.2} \qquad \text{Divide both sides by 3.2.}$$

$$150 = h \qquad \text{Divide } 480 \div 3.2 = 150.$$

The height of the building is 150 ft.

Answer

7. Side x is 1.05 yd long.

3. Similar Polygons

In addition to studying similar triangles, we present similar polygons. Recall from Section 1.2 that a polygon is a flat figure formed by line segments connected at their ends. If the corresponding sides of two polygons are proportional, then we have **similar polygons**. The polygons shown in Figure 5-5 are similar.

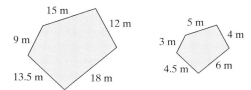

Figure 5-5

In Figure 5-5, notice that the ratios of corresponding sides are equal.

$$\frac{9}{3} = \frac{15}{5} = \frac{12}{4} = \frac{18}{6} = \frac{13.5}{4.5} \qquad \text{All simplify to } \frac{3}{1}.$$

That is, there is a 3 to 1 ratio between the lengths of the sides of the large polygon to those of the small polygon.

example 7	Using Similar Polygons in a Construction Application

A negative for a photograph is 3.5 cm by 2.5 cm. If the width of the resulting picture is 4 in., what is the length of the picture? See Figure 5-6.

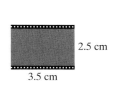

Figure 5-6

Solution:

The figures are similar because they have the same shape (they are both rectangles).

$$\frac{3.5 \text{ cm}}{x \text{ in.}} = \frac{2.5 \text{ cm}}{4 \text{ in.}} \qquad \text{Translate to a proportion.}$$

$$(3.5)(4) = (2.5) \cdot x$$

$$14 = 2.5x$$

$$\frac{14}{2.5} = \frac{2.5x}{2.5} \qquad \text{Divide both sides by 2.5.}$$

$$5.6 = x \qquad \text{Divide } 14 \div 2.5 = 5.6.$$

The picture is 5.6 in. long.

Skill Practice

8. The first- and second-place plaques for a softball tournament have the same shape but are different sizes. Find the length of side x.

10 in.

1st place

8 in.

x

2nd place

5 in.

Answer

8. Side x is 6.25 in.

section 5.4   Practice Exercises

Study Skills Exercise

1. Test yourself.

Yes _____ No _____ Did you have sufficient time to study for the test on this chapter? If not, what could you have done to create more time for studying?

Yes _____ No _____ Did you work all the assigned homework problems in this chapter?

Yes _____ No _____ If you encountered difficulty in this chapter, did you see your instructor or tutor for help?

Yes _____ No _____ Have you taken advantage of the textbook supplements such as the *Student Solutions Manual* and the Online Learning Center?

Review Exercises

For Exercises 2–5, use = or ≠ to make a true statement.

2. $\dfrac{4}{7} \ \square \ \dfrac{12}{21}$

3. $\dfrac{3}{13} \ \square \ \dfrac{15}{65}$

4. $\dfrac{2}{5} \ \square \ \dfrac{21}{55}$

5. $\dfrac{12}{7} \ \square \ \dfrac{35}{19}$

For Exercises 6–12, solve the proportion.

6. $\dfrac{2}{7} = \dfrac{3}{x}$

7. $\dfrac{4}{3} = \dfrac{n}{5}$

8. $\dfrac{p}{9} = \dfrac{1}{6}$

9. $\dfrac{3\frac{1}{2}}{k} = \dfrac{2\frac{1}{3}}{4}$

10. $\dfrac{2.4}{3} = \dfrac{m}{5}$

11. $\dfrac{3}{2.1} = \dfrac{7}{y}$

12. $\dfrac{1.2}{4} = \dfrac{3}{a}$

Objective 1: Applications of Proportions

13. Pam drives her Toyota Prius 244 mi in city driving on 4 gal of gas. At this rate how many miles can she drive on 10 gal of gas? **(See Example 1.)**

14. Didi takes her pulse for 10 sec and counts 13 beats. How many beats per minute is this?

15. To cement a garden path, it takes crushed rock and cement in a ratio of $\frac{13}{4}$. If a bag of 24 kg of cement is purchased, how much crushed rock will be needed? **(See Example 2.)**

16. Suppose two adults produce 62 lb of garbage in one week. At this rate, how many pounds will 48 adults produce in one week?

17. On a map, the distance from Sacramento, California, to San Francisco, California, is 8 cm. The legend gives the actual distance at 91 mi. On the same map, Faythe measured 7 cm from Sacramento to Modesto, California. What is the actual distance? (Round to the nearest mile.) **(See Example 3.)**

18. Property tax on a $150,000 house is $3000. At this rate how much tax would be paid on a $120,000 home?

19. In the U.S. Senate between 1999 and 2001, the ratio of Democrats to Republicans was 9 to 11. If there were 45 Democrats, how many Republicans were there?

20. Evelyn won an election by a ratio of 6 to 5. If she received 7230 votes, how many votes did her opponent receive?

21. If you flip a coin, the coin should come up heads about 1 time out of every 2 times it is flipped. If a coin is flipped 630 times, about how many heads do you expect to come up?

22. A die is a small cube used in games of chance. It has 6 sides, and each side has 1, 2, 3, 4, 5, or 6 dots painted on it. If you roll a die, the number 4 should come up about 1 time out of every 6 times the die is rolled. If you roll a die 366 times, about how many times do you expect the number 4 to come up?

23. A pitcher gave up 42 earned runs in 126 innings. Approximately how many earned runs will he give up in 1 game (9 innings)? This value is called the earned run average.

24. In one game Peyton Manning completed 34 passes for 357 yd. At this rate how many yards would be gained for 22 passes?

25. Pierre bought 41.3€ (Euros) with $50 American. At this rate, how many Euros can he buy with $900 American?

26. Erik bought $136 Canadian with $100 American. At this rate, how many Canadian dollars can he buy with $235 American?

27. At Daytona Beach Community College the ratio of women to men is 68 to 32.

   a. How many women would you expect among a group of 500 DBCC students?

   b. How many men would you expect among a group of 500 DBCC students?

28. Approximately 24 out of 100 Americans over the age of 12 smoke. How many smokers would you expect in a group of 850 Americans over the age of 12?

29. Park officials stocked a man-made lake with bass last year. To approximate the number of bass this year, a sample of 75 fish is taken out of the lake and tagged. Then later a different sample is taken, and it is found that 21 of 100 fish are tagged. Approximately how many fish are in the lake? Round to the nearest whole unit. **(See Example 4.)**

30. Laws have been instituted in Florida to help save the manatee. To establish the number in Florida, a sample of 150 manatees was marked and let free. A new sample was taken and found that there were 3 marked manatees out of 40 captured. What is the approximate population of manatees in Florida?

31. Yellowstone National Park in Wyoming has the largest population of free-roaming bison. To approximate the number of bison, 200 are captured and tagged and then let free to roam. Later, a sample of 120 bison is observed and 6 have tags. Approximate the population of bison in the park.

32. In Cass County, Michigan, there are about 20 white-tailed deer per square mile. If the county covers 492 mi^2, about how many deer are in the county?

Objective 2: Similar Triangles

For Exercises 33–36, the pairs of triangles are similar. Solve for x and y. **(See Example 5.)**

33.

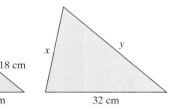

34.

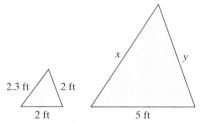

35.

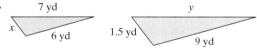

36.

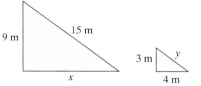

37. The height of a flagpole can be determined by comparing the shadow of the flagpole and the shadow of a yardstick. From the figure, determine the height of the flagpole. **(See Example 6.)**

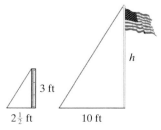

Figure for Exercise 37

38. A 15-ft flagpole casts a 4-ft shadow. How long will the shadow be for a 90-ft building?

39. A person 1.6 m tall stands next to a lifeguard observation platform. If the person casts a shadow of 1 m and the lifeguard platform casts a shadow of 1.5 m, how high is the platform?

40. A 32-ft tree casts a shadow of 18 ft. How long will the shadow be for a 22-ft tree?

Objective 3: Similar Polygons

For Exercises 41–44, the pairs of polygons are similar. Solve for the indicated variables.

41.

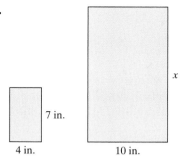

42.

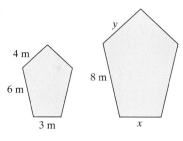

43.

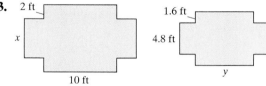

44.

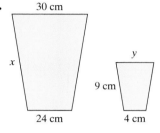

45. A carpenter makes a schematic drawing of a porch he plans to build. On the drawing, 2 in. represents 7 ft. Find the dimensions of the porch. **(See Example 7.)**

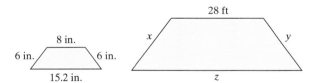

46. The Great Cookie Company has a sign on the front of its store as shown in the figure. The company would like to put the same shape sign in the back, but with dimensions $\frac{1}{3}$ as large. Find the lengths denoted by x and y.

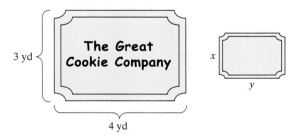

Calculator Connections

Topic: Solving proportions

Calculator Exercises

47. In a recent year the annual crime rate for Oklahoma was 4743 crimes per 100,000 people. If Oklahoma had approximately 3,500,000 people at that time, approximately how many crimes were committed? (Source: Oklahoma Department of Corrections.)

48. In a recent year the annual crime rate for Minnesota was 4465 crimes per 100,000 people. If Minnesota had approximately 5,000,000 people at that time, approximately how many crimes were committed? (Source: Minnesota Bureau of Criminal Apprehension.)

49. In a recent year the rate of breast cancer in women was 110 cases per 100,000 women. At that time the state of California had approximately 14,000,000 women. How many women in California would be expected to have breast cancer? (Source: Centers for Disease Control.)

50. In a recent year the rate of prostate disease in U.S. men was 118 cases per 1000 men. At that time the state of Massachusetts had approximately 2,500,000 men. How many men in Massachusetts would be expected to have prostate disease? (Source: National Center for Health Statistics.)

chapter 5 | summary

section 5.1 Ratios

Key Concepts

A **ratio** is a comparison of two quantities with the same units.

The ratio of a to b can be written as follows, provided $b \neq 0$.

1. a to b **2.** $a : b$ **3.** $\dfrac{a}{b}$

When we write a ratio in fraction form, we generally simplify it to lowest terms.

Ratios that contain mixed numbers, fractions, or decimals can be simplified to lowest terms with whole numbers in the numerator and denominator.

Ratios are used in a number of applications. See Example 5.

Examples

Example 1

Three forms of a ratio:

4 to 6   4 : 6 $\dfrac{4}{6}$

Example 2

A hockey team won 4 games out of 6. Write a ratio of games won to total games played and simplify to lowest terms.

$$\frac{4 \text{ games won}}{6 \text{ games played}} = \frac{2}{3}$$

Example 3

$$\frac{2\frac{1}{6}}{\frac{2}{3}} = 2\frac{1}{6} \div \frac{2}{3} = \frac{13}{\overset{}{\underset{2}{6}}} \cdot \frac{\overset{1}{3}}{2} = \frac{13}{4}$$

Example 4

$$\frac{2.1}{2.8} = \frac{2.1}{2.8} \cdot \frac{10}{10} = \frac{21}{28} = \frac{3 \cdot \overset{1}{7}}{2 \cdot 2 \cdot \underset{1}{7}} = \frac{3}{4}$$

Example 5

On a new car lot there were 28 foreign-made cars and 32 cars that were made in the United States. Write a ratio of the number of foreign-made cars to the total number of cars on the lot.

$28 + 32 = 60$ total cars

$$\frac{28}{60} = \frac{\overset{1}{2} \cdot \overset{1}{2} \cdot 7}{\underset{1}{2} \cdot \underset{1}{2} \cdot 3 \cdot 5} = \frac{7}{15}$$

section 5.2 Rates

Key Concepts

A **rate** compares two quantities with different units.

A rate having a denominator of 1 unit is called a **unit rate**. To find a unit rate, divide the numerator by the denominator.

A unit cost or unit price is the cost per 1 unit, for example, $1.21/lb or 43¢/oz. Comparing unit prices can help determine the best buy.

Examples

Example 1

New Jersey has 8,470,000 people living in 21 counties. Write a reduced ratio of people per county.

$$\frac{8{,}470{,}000 \text{ people}}{21 \text{ counties}} = \frac{1{,}210{,}000 \text{ people}}{3 \text{ counties}}$$

Example 2

If a race car traveled 1250 mi in 8 hr during a race, what is its speed in miles per hour?

$$\frac{1250 \text{ mi}}{8 \text{ hr}} = 156.25 \text{ mi/hr}$$

Example 3

Tide laundry detergent is offered in three sizes: $10.99 for 150 oz, $7.63 for 100 oz, and $4.99 for 50 oz. Find the unit prices to find the best buy.

$$\frac{\$10.99}{150 \text{ oz}} \approx \$0.0733/\text{oz}$$

$$\frac{\$7.63}{100 \text{ oz}} = \$0.0763/\text{oz}$$

$$\frac{\$4.99}{50 \text{ oz}} = \$0.0998/\text{oz}$$

The 150-oz package is the best buy because the unit cost is the least.

section 5.3 Proportions

Key Concepts

A statement indicating that two quantities are equal is called an **equation**.

A **proportion** states that two ratios or rates are equal.
$\dfrac{14}{21} = \dfrac{2}{3}$ is a proportion.

To determine if two ratios form a proportion, check to see if the cross products are equal, that is,

$\dfrac{a}{b} = \dfrac{c}{d}$ implies $a \cdot d = b \cdot c$

To solve a proportion, solve the equation formed by the cross products.

Examples

Example 1

Write each as a proportion.

a. 3 is to 8 as 9 is to 24.

$$\frac{3}{8} = \frac{9}{24}$$

b. 56 mi is to 2 gal as 84 mi is to 3 gal.

$$\frac{56 \text{ mi}}{2 \text{ gal}} = \frac{84 \text{ mi}}{3 \text{ gal}}$$

Example 2

$\dfrac{3}{8} \stackrel{?}{=} \dfrac{2\frac{1}{2}}{6\frac{2}{3}}$ $3 \cdot 6\dfrac{2}{3} \stackrel{?}{=} 8 \cdot 2\dfrac{1}{2}$

$\dfrac{3}{1} \cdot \dfrac{20}{3} \stackrel{?}{=} \dfrac{8}{1} \cdot \dfrac{5}{2}$

$20 = 20$ ✔

The ratios form a proportion.

Example 3

$\dfrac{4.1}{5.4} \stackrel{?}{=} \dfrac{1.25}{1.35}$ $(4.1)(1.35) \stackrel{?}{=} (5.4)(1.25)$

$5.535 \neq 6.75$

The ratios do not form a proportion.

Example 4

$\dfrac{5}{4} = \dfrac{18}{x}$ $\Rightarrow$ $5 \cdot x = 4 \cdot 18$

$5x = 72$

$\dfrac{\overset{1}{\cancel{5}}x}{\underset{1}{\cancel{5}}} = \dfrac{72}{5}$

$x = \dfrac{72}{5}$ or $14\dfrac{2}{5}$ or 14.4

section 5.4 Applications of Proportions and Similar Figures

Key Concepts

Example 1 demonstrates an application involving a proportion. Example 2 demonstrates the use of proportions involving similar triangles.

Example 1

According to the National Highway Traffic Safety Administration, 2 out of 5 traffic fatalities involve the use of alcohol. If there were 43,200 traffic fatalities in a recent year, how many involved the use of alcohol?

Let n represent the number of traffic fatalities involving alcohol.

Set up a proportion:

$$\frac{2 \text{ traffic fatalities w/alcohol}}{5 \text{ traffic fatalities}} = \frac{n}{43,200}$$

Solve the proportion:

$$2(43,200) = 5n$$

$$86,400 = 5n$$

$$\frac{86,400}{5} = \frac{\overset{1}{\cancel{5}}n}{\underset{1}{\cancel{5}}}$$

$$17,280 = n$$

17,280 traffic fatalities involved alcohol.

Examples

Example 2

On a sunny day, a tree 6-ft tall casts a 4-ft shadow. At the same time a telephone pole casts a 10-ft shadow. What is the height of the telephone pole?

A picture illustrates the situation. Let h represent the height of the telephone pole.

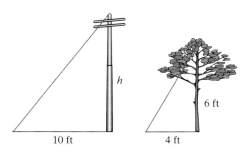

Set up a proportion:

$$\frac{10}{4} = \frac{h}{6}$$

Solve the proportion:

$$(10)(6) = 4h$$

$$60 = 4h$$

$$\frac{60}{4} = \frac{\overset{1}{\cancel{4}}h}{\underset{1}{\cancel{4}}}$$

$$15 = h \qquad \text{The pole is 15 ft high.}$$

chapter 5 | review exercises

Section 5.1

For Exercises 1–3, write the ratios in two other ways.

1. 5 : 4   **2.** 3 to 1   **3.** $\dfrac{8}{7}$

For Exercises 4–6, write the ratios in fraction form.

4. Saul had 3 daughters and 2 sons.

   a. Write a ratio of sons to daughters.

   b. Write a ratio of daughters to sons.

   c. Write a ratio of daughters to the total number of children.

5. In his refrigerator, Jonathan has 4 bottles of soda and 5 bottles of juice.

   a. Write a ratio of bottles of soda to bottles of juice.

   b. Write a ratio of bottles of juice to bottles of soda.

   c. Write a ratio of bottles of juice to the total number of bottles.

6. There are 12 face cards in a regular deck of 52 cards.

    a. Write a ratio of face cards to total cards.

    b. Write a ratio of face cards to cards that are not face cards.

For Exercises 7–10, write the ratio in lowest terms.

7. 52 cards to 13 cards   **8.** $21 to $15

9. 80 ft to 200 ft   **10.** 7 days to 28 days

For Exercises 11–14, write the ratio in lowest terms with whole numbers in the numerator and denominator.

11. $1\frac{1}{2}$ hr to $\frac{1}{3}$ hr   **12.** $\frac{2}{3}$ yd to $2\frac{1}{6}$ yd

13. $2.56 to $1.92   **14.** 42.5 mi to 3.25 mi

15. This year a high school had an increase of 320 students. The enrollment last year was 1200 students.

    a. How many students will be attending this year?

    b. Write a ratio of the increase in students to the total enrollment of students this year. Simplify to lowest terms.

16. A living room has dimensions of 3.8 m by 2.4 m. Find the ratio of length to width and reduce to lowest terms.

For Exercises 17–18, refer to the table that shows the number of personnel who smoke in a particular workplace.

	Smokers	**Nonsmokers**	**Totals**
Office personnel	12	20	32
Shop personnel	60	55	115

17. Find the ratio of office personnel who smoke to shop personnel who smoke.

18. Find the ratio of the total number of personnel who smoke to the total number of personnel.

Section 5.2

For Exercises 19–22, write each rate in lowest terms.

19. A concession stand sold 20 hot dogs in 45 min.

20. Mike can skate 4 mi in 34 min.

21. The CN Tower in Toronto, Canada, weighs 130,000 tons for a height of approximately 1800 ft.

22. Interpol reports that Denmark has a crime rate of 9460 per 100,000 people.

23. What is the difference between rates in lowest terms and unit rates?

24. Is it possible to compute a unit rate on any given rate?

For Exercises 25–28, write each rate as a unit rate.

25. A pheasant can fly 44 mi in $1\frac{1}{3}$ hr.

26. The temperature dropped 14° in 3.5 hr.

27. A hummingbird can flap its wings 2700 times in 30 sec.

28. It takes Perry's lawn company 66 min to cut 6 lawns.

For Exercises 29–32, find the unit costs. Round the answers to 3 decimal places when necessary.

29. Body lotion costs $5.99 for 10 oz.

30. Three towels cost $10.00.

31. A package of 40 trash bags costs $1.99.

32. Suntan lotion costs $5.99 for 8 oz. If Sherri has a coupon for $2.00 off, what will be the unit cost of the lotion after the coupon has been applied?

33. A 24-roll pack of bathroom tissue costs $6.99 without a discount card. The package is advertised at 17¢ per roll if the buyer uses the discount card. What is the difference in price per roll when the buyer uses the discount card? Round to the nearest cent.

34. The average annual rainfall for Los Angeles, California, is 13.2 in. What is the rate per month?

35. Retail gasoline prices rose an average of $0.57 per gallon in Japan between 1990 and 2000. Compute the increase in price per year for 1 gal of gas. (Source: Energy Information Administration, U.S. Department of Energy.)

36. The number of motor vehicles produced in Canada increased by 102,000 between 1990 and 2000. Compute the increase in the number of vehicles produced per year during this time period. (Source: American Automobile Manufacturers Association.)

Section 5.3

For Exercises 37–42, write a proportion for each statement.

37. 16 is to 14 as 12 is to $10\frac{1}{2}$.

38. 8 is to 20 as 6 is to 15.

39. The numbers 5 and 3 are proportional to the numbers 10 and 6.

40. The numbers 4 and 3 are proportional to the numbers 20 and 15.

41. $11 is to 1 hr as $88 is to 8 hr.

42. 2 in. is to 5 mi as 6 in. is to 15 mi.

For Exercises 43–46, determine whether the ratios form a proportion.

43. $\frac{64}{81} \stackrel{?}{=} \frac{8}{9}$

44. $\frac{3\frac{1}{2}}{7} \stackrel{?}{=} \frac{7}{14}$

45. $\frac{5.2}{3} \stackrel{?}{=} \frac{15.6}{9}$

46. $\frac{6}{10} \stackrel{?}{=} \frac{6.3}{10.3}$

For Exercises 47–50, determine whether the pairs of numbers are proportional.

47. Are the numbers $2\frac{1}{8}$ and $4\frac{3}{4}$ proportional to the numbers $3\frac{2}{5}$ and $7\frac{3}{5}$?

48. Are the numbers $5\frac{1}{2}$ and 6 proportional to the numbers $6\frac{1}{2}$ and 7?

49. Are the numbers 4.25 and 8 proportional to the numbers 5.25 and 10?

50. Are the numbers 12.4 and 9.2 proportional to the numbers 3.1 and 2.3?

For Exercises 51–56, solve the proportion.

51. $\frac{100}{16} = \frac{25}{x}$ 52. $\frac{y}{6} = \frac{45}{10}$ 53. $\frac{1\frac{6}{7}}{b} = \frac{13}{21}$

54. $\frac{p}{6\frac{1}{3}} = \frac{3}{9\frac{1}{2}}$ 55. $\frac{2.5}{6.8} = \frac{5}{h}$ 56. $\frac{0.3}{1.2} = \frac{k}{3.6}$

Section 5.4

57. One year of a dog's life is about the same as 7 years of a human life. If a dog is 12 years old in dog years, how does that equate to human years?

58. Lavu bought 11,000 Japanese yen with $100 American. At this rate, how many yen can he buy with $450 American?

59. The number of births in Alabama in 2002 was approximately 58,500. If the birthrate was about 13 per 1000, what was the approximate population of Alabama? (Round to the nearest person.)

60. If the tax on a $25.00 item is $1.20, what would be the tax on an item costing $145.00?

61. The triangles shown in the figure are similar. Solve for x and y.

62. The height of a building can be approximated by comparing the shadows of the building and of a meter stick. From the figure, find the height of the building.

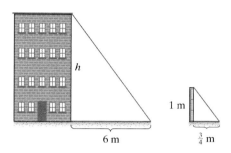

6 m $\frac{3}{4}$ m   1 m

63. The polygons shown in the figure are similar. Solve for x and y.

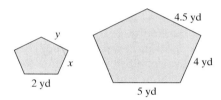

64. The figure shows two picture frames that are similar. Use this information to solve for x and y.

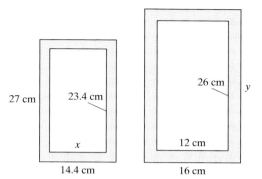

27 cm   23.4 cm   26 cm y
x 12 cm
14.4 cm   16 cm

---

chapter 5 | test

1. An elementary school has 25 teachers and 521 students. Write a ratio of teachers to students in three different ways.

2. In a marina, there were 17 sailboats out of a total of 23 boats.

a. Write a ratio of sailboats to total boats.

b. Write a ratio of sailboats to boats that are not sailboats.

For Exercises 3–4, write as a reduced ratio in fraction form.

3. For the 2002 WNBA season, the New York Liberty had a win-loss ratio of 18 to 14.

4. For the 2002 WNBA season, the Houston Comets had a win-loss ratio of 24 to 8.

5. Find the ratio of the shortest side to the longest side. Write the ratio in lowest terms.

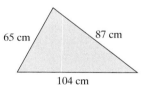

65 cm   87 cm
104 cm

6. a. In a recent year, the number of people in New Mexico whose income was below poverty level was 168 out of every 1000. Write this as a simplified ratio.

b. The poverty level in Iowa was 72 people to 1000 people. Write this as a simplified ratio.

c. Compare the ratios and comment.

7. Write as a simplified ratio in two ways: 30 sec to $1\frac{1}{2}$ min.

   a. By converting 30 sec to minutes.

   b. By converting $1\frac{1}{2}$ min to seconds.

For Exercises 8–10, write as a rate, simplified to lowest terms.

8. 255 mi per 6 hr

9. 20 lb in 6 weeks

10. There are 4 g of fat in 8 cookies.

For Exercises 11–12, write as a unit rate. Round to the nearest hundredth.

11. The element platinum had density of 2145 g per 100 cm^3.

12. There are approximately 104.8 oz of iron in 45.8 lb of rocks brought back from the moon. Round to 1 decimal place.

13. What is the unit cost for Raid Ant and Roach spray valued at $6.72 for 30 oz? Round to the nearest cent.

14. A package containing 3 toe rings is on sale for 2 packs for $3.00. What is the cost of 1 toe ring?

15. A generic pain reliever is on sale for 2 bottles for $3. Each bottle contains 30 tablets. Aleve pain reducer is sold in 24-capsule bottles for $1.92. Find the unit cost of each to determine the better buy.

16. What does it mean for two pairs of numbers to be proportional?

For Exercises 17–20, write a proportion for each statement.

17. 42 is to 15 as 28 is to 10.

18. 20 pages is to 12 min as 30 pages is to 18 min.

19. $15 an hour is proportional to $75 for 5 hr.

20. Are the numbers 105 and 55 proportional to the numbers 21 and 10?

For Exercises 21–24, solve the proportion.

21. $\dfrac{25}{p} = \dfrac{45}{63}$

22. $\dfrac{32}{20} = \dfrac{20}{x}$

23. $\dfrac{n}{9} = \dfrac{3\frac{1}{3}}{6}$

24. $\dfrac{y}{14} = \dfrac{7.2}{16.8}$

25. A computer can download 1.6 megabytes (MB) in 2.5 min. How long will it take to download a 4.8-MB file?

26. The number of representatives in the Kansas House of Representatives forms a ratio of 9 Democrats to 16 Republicans. If there are 45 Democrats, determine the number of Republicans.

27. Ms. Ehrlich wants to approximate the number of goldfish in her backyard pond. She scooped out 8 and marked them. Later she scooped out 10 and found that 3 were marked. Estimate the number of goldfish in her pond. Round to the nearest whole unit.

28. Given that the two triangles are similar, solve for x and y.

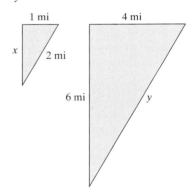

29. The plaques for the first-place and second-place teams in Little League are similar. Determine the length of the side labeled x.

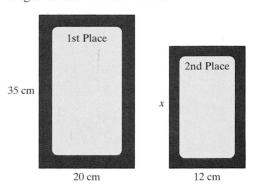

chapters 1–5 | cumulative review

1. Write the number 503,042 in words.

2. Estimate the sum by first rounding the numbers to the nearest hundred.

$$251 + 492 + 631$$

3. Multiply $226 \times 100,000$.

4. Divide and write the answer with a remainder: $355 \div 16$.

5. Explain how to check a division problem. Then check the problem in Exercise 4.

6. Simplify $2^2 \times (32 - 11) \div 14$.

7. Draw a figure that is shaded to represent the fraction $\frac{3}{7}$.

8. Simplify $\dfrac{245}{175}$.

9. Multiply $\dfrac{13}{2} \cdot \dfrac{3}{7}$.

10. Simplify $\left(\dfrac{3}{5}\right)^2$.

11. Bruce decides to share his 6-in. sub sandwich with his friend. If he gives $\frac{1}{4}$ of it to Dennis, how many inches of the sub does he have left?

12. Simplify $\left(\dfrac{7}{8} \div \dfrac{3}{4}\right) + \dfrac{5}{6}$.

13. Add $\dfrac{8}{9} + 3$.

14. Subtract $\dfrac{9}{13} - \dfrac{0}{3}$.

15. Emil wants to put a wallpaper border in his bathroom. The border will be on three walls (not in the shower) and not by the door. From the figure, determine how much wallpaper border will be needed.

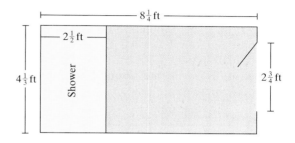

16. Tomoka Consolidated bought $82\frac{1}{4}$ acres of land. If it sold $\frac{3}{4}$ of the land, how much was sold? How much was left?

17. How many ninths are in $6\dfrac{5}{9}$?

18. Write the number 1004.701 in words.

19. Add and subtract: $23.88 + 11.3 - 7.123$.

20. Write 4.36 as an improper fraction.

21. Multiply 43.923×100.

22. Divide $237.9 \div 100$.

23. Find the perimeter of the figure. Write the answer in decimal form.

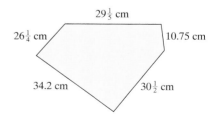

24. In the U.S. House of Representatives (2001–2003), the ratio of Democrats to Republicans was 212 to 221. Write this ratio in two other ways.

25. For a recent year at Southeastern Community College in North Carolina, there were 1950 students and 150 faculty. Write the student-to-faculty ratio in lowest terms.

26. The average annual rainfall for Mobile, Alabama, is 66.3 in. What is the rate per month?

27. Oregon has a land area of approximately 9600 mi^2. The population of Oregon is approximately 1,200,000. Compute the population density (recall that population density is a unit rate given by the number of people per square mile).

28. Determine whether the ratios form a proportion.

a. $\dfrac{7.5}{10} \stackrel{?}{=} \dfrac{9}{12}$

b. $\dfrac{31}{5} \stackrel{?}{=} \dfrac{33}{6}$

29. Solve the proportion $\dfrac{13}{11.7} = \dfrac{5}{x}$.

30. Jim can drive 150 mi on 6 gal. At this rate, how far can he travel on 4 gal?

Percents

6

In this chapter we present the concept of percent. Percents are used to measure the number of parts per 100 of some whole amount. For example, the earth is covered by approximately 360 million square kilometers of water. The surface of the earth covers 510 million square kilometers. See Exercise 58 of Section 6.4 to determine the percent of the earth's surface that is covered by water.

chapter 6   preview

The exercises in this chapter preview contain concepts that have not yet been presented. These exercises are provided for students who want to compare their levels of understanding before and after studying the chapter. Alternatively, you may prefer to work these exercises when the chapter is completed and before taking the exam.

Section 6.1

For Exercises 1–4, write a percent to express the shaded portion of the figure.

1.    **2.**

3. 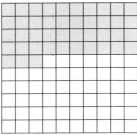   **4.**

For Exercises 5–7, write the percent as a decimal and as a fraction.

5. Water covers 70% of the earth's surface.

6. In 2000, the U.S. population was made up of 12.3% African Americans.

7. Thailand's population decreased by 0.4% in 1998.

Section 6.2

For Exercises 8–11, write the fraction as a percent.

8. $\dfrac{1}{4}$ **9.** $\dfrac{1}{3}$

10. $\dfrac{3}{8}$ **11.** $\dfrac{3}{2}$

12. The fraction of Americans who say they do not entertain at home is $\dfrac{11}{100}$. Write the fraction as a percent.

For Exercises 13–16, write the decimal as a percent.

13. 0.45   **14.** 2.1

15. 0.0025   **16.** 0.3

Sections 6.3–6.4

For Exercises 17–19, solve the percent equation.

17. The number 32 is 16% of what number?

18. What is 0.5% of 12,000?

19. The number 140 is what percent of 90? Round to the nearest tenth of a percent.

Section 6.5

20. Find the discount rate on the shirt that was reduced to $26.40 from its original price of $33.00.

21. If Rhoda buys a car for $25,400, what will her final price be if the sales tax rate is 9%?

Section 6.6

22. The price of oil in a recent year was $45.00 per barrel. Two years later, the price rose to $63.00 per barrel. What was the percent increase?

Section 6.7

23. How much interest is earned on a loan of $10,000 for 4 years with a simple interest rate of 7%?

section 6.1   Percents and Their Fraction and Decimal Forms

1. Definition of Percent

In this chapter we study the concept of percent. Literally, the word **percent** means *per one hundred.* To indicate percent, we use the percent symbol %. For example, 45% (read as "45 percent") of the land area in South America is rainforest (shaded in green). This means that if South America were divided into 100 squares of equal size, 45 of the 100 squares would cover rainforest. See Figures 6-1 and 6-2.

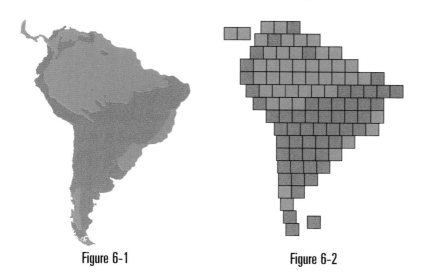

Figure 6-1   Figure 6-2

Consider another example. For a recent year, the population of Virginia could be described as follows.

21%	African American	21 out of 100 Virginians are African American
72%	Caucasian (non-Hispanic)	72 out of 100 Virginians are Caucasian (non-Hispanic)
3%	Asian American	3 out of 100 Virginians are Asian American
3%	Hispanic	3 out of 100 Virginians are Hispanic
1%	Other	1 out of 100 Virginians have other background

Figure 6-3 represents a sample of 100 residents of Virginia.

AA	AA	AA	C	C	C	C	C	C	C
AA	AA	C	C	C	C	C	C	C	C
AA	AA	C	C	C	C	C	C	C	C
AA	AA	C	C	C	C	C	C	C	H
AA	AA	C	C	C	C	C	C	C	H
AA	AA	C	C	C	C	C	C	C	H
AA	AA	C	C	C	C	C	C	C	A
AA	AA	C	C	C	C	C	C	C	A
AA	AA	C	C	C	C	C	C	C	A
AA	AA	C	C	C	C	C	C	C	O

AA African American
C   Caucasian (non-Hispanic)
A   Asian American
H   Hispanic
O   Other

Figure 6-3

Concept Connections

1. Shade the portion of the figure represented by 18%.

Answer

1.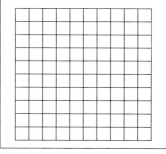

2. Converting Percents to Fractions

By definition, a percent represents a ratio of parts per 100. Therefore, we can write percents as fractions.

Percent		Fraction	Example/Interpretation
7%	$=$	$\dfrac{7}{100}$	A sales tax of 7% means that 7 cents in tax is charged for every 100 cents spent.
35%	$=$	$\dfrac{35}{100}$	To say that 35% of households own a cat means that 35 per every 100 households owns a cat.

Notice that $35\% = \dfrac{35}{100} = 35 \times \dfrac{1}{100} = 35 \div 100$.

From this discussion we have the following rule for converting percents to fractions.

Converting Percents to Fractions

1. Replace the symbol % by $\times \frac{1}{100}$ (or by $\div\ 100$).
2. Simplify the fraction to lowest terms, if possible.

---

example 1   **Converting Percents to Fractions**

Convert each percent to a fraction.

a. 56% **b.** 60% **c.** 125% **d.** 0.4%

Solution:

a. $56\% = 56 \times \dfrac{1}{100}$ Replace the % symbol by $\times \dfrac{1}{100}$.

$ = \dfrac{56}{100}$ Multiply.

$ = \dfrac{14}{25}$ Simplify to lowest terms.

b. $60\% = 60 \times \dfrac{1}{100}$ Replace the % symbol by $\times \dfrac{1}{100}$.

$ = \dfrac{60}{100}$ Multiply.

$ = \dfrac{3}{5}$ Simplify to lowest terms.

c. $125\% = 125 \times \dfrac{1}{100}$ Replace the % symbol by $\times \dfrac{1}{100}$.

$ = \dfrac{125}{100}$ Multiply.

$ = \dfrac{5}{4}$ or $1\dfrac{1}{4}$ Simplify to lowest terms.

d. $0.4\% = 0.4 \times \dfrac{1}{100}$ Replace the % symbol by $\times \dfrac{1}{100}$.

$\quad = \dfrac{4}{10} \times \dfrac{1}{100}$ Write 0.4 in fraction form.

$\quad = \dfrac{4}{1000}$ Multiply.

$\quad = \dfrac{1}{250}$ Simplify to lowest terms.

Note that $100\% = 100 \times \frac{1}{100} = 1$. That is, 100% represents 1 whole unit. In Example 1(c), $125\% = \frac{5}{4}$ or $1\frac{1}{4}$. This illustrates that any percent greater than 100% represents a quantity greater than 1 whole. Therefore, its fractional form may be expressed as an improper fraction or as a mixed number.

Note that $1\% = 1 \times \frac{1}{100} = \frac{1}{100}$. In Example 1(d), the value 0.4% represents a quantity less than 1%. Its fractional form is less than one-hundredth.

In Example 2 we convert some common percents to fraction form.

Concept Connections

Determine whether the percent represents a quantity greater than or less than 1 whole.

6. 1.92%

7. 19.2%

8. 192%

example 2   Converting Percents to Fractions

Convert the percents to fractions.

a. 25%   **b.** 10%   **c.** $33\frac{1}{3}\%$

Skill Practice

Convert the percents to fractions.

9. 75%

10. 150%

11. $66\frac{2}{3}\%$

Solution:

a. $25\% = 25 \times \dfrac{1}{100} = \dfrac{25}{100} = \dfrac{1}{4}$ Thus 25% represents one-quarter of a whole.

b. $10\% = 10 \times \dfrac{1}{100} = \dfrac{10}{100} = \dfrac{1}{10}$ Thus, 10% represents one-tenth of a whole.

c. $33\frac{1}{3}\% = 33\frac{1}{3} \times \dfrac{1}{100}$ Replace the % symbol by $\times \dfrac{1}{100}$.

$\quad = \dfrac{100}{3} \times \dfrac{1}{100}$ Convert the mixed number to an improper fraction.

$\quad = \dfrac{\overset{1}{\cancel{100}}}{3} \times \dfrac{1}{\underset{1}{\cancel{100}}}$ Simplify common factors.

$\quad = \dfrac{1}{3}$ Thus, $33\frac{1}{3}\%$ represents one-third of a whole.

3. Converting Percents to Decimals

To express part of a whole unit, we can use a percent, a fraction, or a decimal. We would like to be able to convert from one form to another. The procedure

Answers
6. Less than   7. Less than
8. Greater than   9. $\frac{3}{4}$
10. $\frac{3}{2}$ 11. $\frac{2}{3}$

to convert a percent to a decimal is the same as that to convert a percent to a fraction. We replace the % symbol by $\times \frac{1}{100}$. However, when converting to a decimal, it is usually more convenient to use the form $\times 0.01$.

Converting Percents to Decimals

Replace the % symbol by $\times 0.01$. (This is equivalent to $\times \frac{1}{100}$ and $\div 100$.)

Note: Multiplying a decimal by 0.01 is the same as moving the decimal point 2 places to the left.

---

example 3   Converting Percents to Decimals

Convert each percent to its decimal form.

a. 31%   **b.** 6.5%   **c.** 428%   **d.** $1\frac{3}{5}$%   **e.** 0.05%

Solution:

a. $31\% = 31 \times 0.01$ Replace the % symbol by $\times 0.01$.

$= 0.31$ Move the decimal point 2 places to the left.

b. $6.5\% = 6.5 \times 0.01$ Replace the % symbol by $\times 0.01$.

$= 0.065$ Move the decimal point 2 places to the left.

c. $428\% = 428 \times 0.01$ Because 428% is greater than 100% we expect the decimal form to be a number greater than 1.

$= 4.28$

d. $1\frac{3}{5}\% = 1.6 \times 0.01$ Convert the mixed number to decimal form.

$= 0.016$ Because the percent is just over 1%, we expect the decimal form to be just slightly greater than one-hundredth.

e. $0.05\% = 0.05 \times 0.01$ The value 0.05% is less than 1%. We expect the decimal form to be less than one-hundredth.

$= 0.0005$

Notice from Examples 1, 2, and 3 that we perform the same procedure to convert a percent to either a decimal or a fraction. In each case we multiply by $\frac{1}{100}$. When converting to a decimal, it is usually easier to use the form $\times 0.01$. When converting to a fraction, it is usually easier to use the form $\times \frac{1}{100}$. In both cases, this operation is also equivalent to dividing by 100.

4. Common Percents and Their Fraction and Decimal Forms

Table 6-1 shows some common percents and their equivalent fraction and decimal forms.

table 6-1

Percent	Fraction	Decimal	Example/Interpretation
100%	1	1.00	Of people who give birth, 100% are female.
50%	$\frac{1}{2}$	0.50	Of the population 50% is male. That is, one-half of the population is male.
25%	$\frac{1}{4}$	0.25	Approximately 25% of the U.S. population smokes. That is, one-quarter of the population smokes.
75%	$\frac{3}{4}$	0.75	Approximately 75% of homes have computers. That is, three-quarters of homes have computers.
10%	$\frac{1}{10}$	0.10	Of the population 10% is left-handed. That is, one-tenth of the population is left-handed.
1%	$\frac{1}{100}$	0.01	Approximately 1% of babies are born underweight. That is, about 1 in 100 babies is born underweight.
$33\frac{1}{3}\%$	$\frac{1}{3}$	$0.\overline{3}$	A basketball player made $33\frac{1}{3}\%$ of her shots. That is, she made about 1 basket for every 3 shots attempted.
$66\frac{2}{3}\%$	$\frac{2}{3}$	$0.\overline{6}$	Of the population $66\frac{2}{3}\%$ prefers chocolate ice cream to other flavors. That is, 2 out of 3 people prefer chocolate ice cream.

5. Applications of Percents

The next time you pick up a newspaper or magazine, notice that percents are used in abundance to convey information.

example 4   Converting Percent Notation to Decimal Notation

Find the decimal notation for the percent given in the sentence.

a. Fifty-one percent of applicants to U.S. medical schools in 2003–2004 were female. (Source: Association of American Medical Colleges.)

b. The price per gallon for regular unleaded gasoline in 2004 was 172% of what it was in 1984.

Solution:

a. $51\% = 51 \times 0.01 = 0.51$ Just over one-half of the applicants are female.

b. $172\% = 172 \times 0.01 = 1.72$ The cost of gas in 2004 was 1.72 times as great as in 1984.

Skill Practice

Find the decimal notation for the percent given in the sentence.

17. The U.S. unemployment rate in 2004 was 5.7%.

18. The public debt of the United States increased by 545% between 1982 and 2002.

Answers

17. 0.057   **18.** 5.45

Skill Practice

Find the fraction form for the percent given in the sentence.

19. In Pennsylvania, 15% of the residents are 65 or older.

20. One study found that teenage substance abuse rises by 40% during the summer months.

Answers

19. $\dfrac{3}{20}$ 20. $\dfrac{2}{5}$

---

example 5   Converting Percent Notation to Fraction Notation

Find the fraction form for the percent given in the sentence.

a. Forty-five percent of Americans use the Internet as a resource when planning vacations. (Source: *USA TODAY.*)

b. In 2001, Texas had the greatest percentage of residents not covered by health insurance. Approximately 23.5% of the residents were not covered.

Solution:

a. $45\% = 45 \times \dfrac{1}{100} = \dfrac{45}{100} = \dfrac{9}{20}$

Just under one-half of Americans planning a vacation use the Internet as a resource.

b. $23.5\% = 23.5 \times \dfrac{1}{100} = \dfrac{235}{10} \times \dfrac{1}{100}$

$= \dfrac{235}{1000} = \dfrac{47}{200}$

Almost one-quarter of Texans were not covered by health insurance in 2001.

---

section 6.1   Practice Exercises

Boost *your* GRADE at mathzone.com!

MathZone

• Practice Problems
• Self-Tests
• NetTutor

• e-Professors
• Videos

Study Skills Exercises

1. A test is a *grading* tool for your instructor. How can you turn it into a *learning* tool for you?

2. Define the key term **percent**.

Objective 1: Definition of Percent

For Exercises 3–8, use a percent to express the shaded portion of each drawing.

3.

4.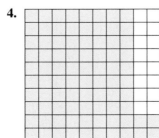

5.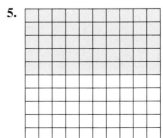

6. 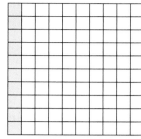   **7.** 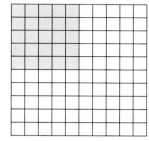   **8.**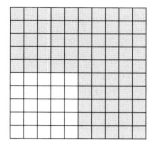

9. Shade the figure so that it represents 14%.

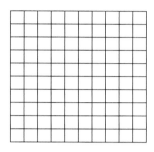

For Exercises 10–13, write a percent for each statement.

10. A bank pays $2 in interest for every $100 deposited.

 11. In South Dakota 5 out of every 100 people work in construction.

12. Out of 100 acres, 70 acres were planted with corn.

13. On TV, 26 out of every 100 minutes are filled with commercials.

Objective 2: Converting Percents to Fractions

14. Explain the procedure to change a percent to a fraction.

For Exercises 15–34, change the percent to a simplified fraction or mixed number. **(See Examples 1–2.)**

15. 13%	**16.** 41%	**17.** 84%	**18.** 32%
19. 25%	**20.** 20%	**21.** 35%	**22.** 75%
23. 115%	**24.** 150%	**25.** 175%	**26.** 120%
27. 0.5%	**28.** 0.2%	**29.** 0.25%	**30.** 0.75%
31. $66\frac{2}{3}\%$	**32.** $5\frac{1}{6}\%$	**33.** $24\frac{1}{2}\%$	**34.** $6\frac{1}{4}\%$

Objective 3: Converting Percents to Decimals

35. Explain the procedure to change a percent to a decimal.

For Exercises 36–51, change the percent to a decimal. **(See Example 3.)**

36. 58%   **37.** 72%   **38.** 15%   **39.** 66%

40. 8.5%   **41.** 12.9%   **42.** 72.31%   **43.** 41.05%

44. 142%   **45.** 201%   **46.** 110.1%   **47.** 126.5%

48. $26\frac{2}{5}\%$ **49.** $16\frac{1}{4}\%$ **50.** $55\frac{1}{20}\%$ **51.** $62\frac{1}{5}\%$

Objective 4: Common Percents and Their Fraction and Decimal Forms

For Exercises 52–57, use a percent to express the shaded portion of each drawing.

52.

53.

54.

55.

56.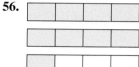

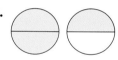 **57.**

For Exercises 58–63, match the percent with its fraction form.

58. $66\frac{2}{3}\%$ **a.** $\dfrac{3}{2}$

59. 10%   **b.** $\dfrac{3}{4}$

60. 90%   **c.** $\dfrac{2}{3}$

61. 75%   **d.** $\dfrac{1}{10}$

62. 25%   **e.** $\dfrac{9}{10}$

63. 150%   **f.** $\dfrac{1}{4}$

For Exercises 64–69, match the percent with its decimal form.

64. 30%   **a.** 0.01

65. $33\frac{1}{3}\%$ **b.** 0.50

66. 125%   **c.** 0.80

67. 50%   **d.** $0.\overline{3}$

68. 1%   **e.** 0.30

69. 80%   **f.** 1.25

70. In which direction do you move the decimal point when you convert a percent to a decimal? By how many places?

Objective 5: Applications of Percents

For Exercises 71–78, find the decimal and fraction equivalent of the percent given in the sentence. **(See Examples 4–5.)**

71. Between 1990 and 2000 the population in California grew by 13.8%.

72. Las Vegas is considered the fastest-growing city in the United States. Between 1990 and 2000 its population increased by 85.2%.

73. For a recent year, the unemployment rate in Kansas was 4.3%.

74. For a recent year, the unemployment rate in the United States was 5.8%.

75. In 2002, of the electricity used in the United States 20% was generated by nuclear power. (Source: U.S. Department of Energy.)

76. For a recent year the average U.S. income tax rate was 18.2%.

77. Thirty-five percent of Americans say they entertain at home once or twice a year. (Source: *USA TODAY.*)

78. Twenty-nine percent of Americans say they entertain at home once a month.

79. The graph represents the percent of dog owners who participate in certain activities to treat their dogs. Write the decimal and fraction forms of the percents given in the graph. (Source: American Animal Hospital Association.)

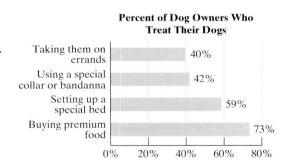

Percent of Dog Owners Who Treat Their Dogs

80. The graph represents the percent of people with at least a bachelor's degree for selected large cities. Write the decimal and fraction forms of the percents given in the graph. (Source: U.S. Census Bureau.)

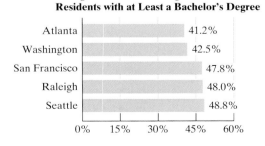

Large Cities with the Highest Percent of Residents with at Least a Bachelor's Degree

section 6.2   Fractions and Decimals and Their Percent Forms

1. Rule for Converting Fractions and Decimals to Percents

In Section 6.1 we converted percents to their equivalent fraction and decimal forms. This is done by replacing the % symbol by $\times \frac{1}{100}$. In this section we reverse the process. We convert fractions and decimals to percents by multiplying by 100 and applying the % symbol.

> **Converting Fractions and Decimals to Percent Form**
>
> Multiply the fraction or decimal by 100%.

2. Converting Decimals to Percents

In Example 1, we convert decimals to percents. As you read through Example 1, recall that multiplying a number by 100 moves the decimal point 2 places to the right.

Skill Practice

Convert each decimal to its percent form.

1. 0.46
2. 3.25
3. 2
4. 0.0006
5. 0.087

Tip: Multiplying a number by 100% is equivalent to multiplying the number by 1. Thus, the value of the number is not changed.

---

example 1   Converting Decimals to Percents

Convert each decimal to its equivalent percent form.

a. 0.62   **b.** 1.75   **c.** 1   **d.** 0.004   **e.** 8.9

Solution:

a. $0.62 = 0.62 \times 100\%$ Multiply by 100%.

     $= 62\%$ Multiplying by 100 moves the decimal point 2 places to the right.

b. $1.75 = 1.75 \times 100\%$ Multiply by 100%.

     $= 175\%$ The decimal number 1.75 is greater than 1. Therefore, we expect a percent greater than 100%.

c. $1 = 1 \times 100\%$ Multiply by 100%.

     $= 100\%$ Recall that 1 whole is equal to 100%.

d. $0.004 = 0.004 \times 100\%$ Multiply by 100%.

     $= 0.4\%$ Move the decimal point to the right 2 places.

e. $8.9 = 8.90 \times 100\%$ Multiply by 100%.

     $= 890\%$

---

3. Converting Fractions to Percents

In Example 2 we convert fractions to percent notation.

example 2   Converting Fractions to Percent Notation

Convert the fractions to percent notation.

a. $\dfrac{3}{5}$ **b.** $\dfrac{5}{8}$ **c.** $\dfrac{2}{3}$

Solution:

a. $\dfrac{3}{5} = \dfrac{3}{5} \times 100\%$ Multiply by 100%.

$= \dfrac{3}{5} \times \dfrac{100}{1}\%$ Convert the whole number to an improper fraction.

$= \dfrac{3}{\cancel{5}_{1}} \times \dfrac{\overset{20}{\cancel{100}}}{1}\%$ Multiply fractions and simplify to lowest terms.

$= 60\%$

Tip: In Example 2(a), we could also have converted $\frac{3}{5}$ to decimal form first (by dividing the numerator by the denominator) and then converted the decimal to a percent.

convert to decimal   convert to percent

$\frac{3}{5} = 0.60$ $=$ $0.60 \times 100\% = 60\%$

b. $\dfrac{5}{8} = \dfrac{5}{8} \times 100\%$ Multiply by 100%.

$= \dfrac{5}{8} \times \dfrac{100}{1}\%$ Convert the whole number to an improper fraction.

$= \dfrac{5}{\cancel{8}_{2}} \times \dfrac{\overset{25}{\cancel{100}}}{1}\%$ Multiply fractions and simplify to lowest terms.

$= \dfrac{125}{2}\%$

The number $\frac{125}{2}\%$ can be written as $62\frac{1}{2}\%$ or as 62.5%.

c. $\dfrac{2}{3} = \dfrac{2}{3} \times 100\%$ Multiply by 100%.

$= \dfrac{2}{3} \times \dfrac{100}{1}\%$ Convert the whole number to an improper fraction.

$= \dfrac{200}{3}\%$

The number $\frac{200}{3}\%$ can be written as $66\frac{2}{3}\%$ or as $66.\overline{6}\%$.

Skill Practice

Convert the fractions to percent notation.

6. $\dfrac{7}{10}$

7. $\dfrac{3}{16}$

8. $\dfrac{1}{9}$

Answers
6. 70%   7. 18.75%
8. $\dfrac{100}{9}\%$ or $11\frac{1}{9}\%$ or $11.\overline{1}\%$

Tip: In Example 2(c), first converting $\frac{2}{3}$ to a decimal before converting to percent notation would be cumbersome. This is so because $\frac{2}{3}$ is represented by a repeating decimal.

convert to decimal   convert to percent

$$\frac{2}{3} = 0.\overline{6} \qquad = \qquad 0.666\cdots \times 100\% = 66.\overline{6}\%$$

In Example 3, we convert an improper fraction and a mixed number to percent form.

Skill Practice

Convert to percent notation.

9. $1\dfrac{7}{10}$

10. $\dfrac{11}{4}$

---

example 3   Converting Improper Fractions and Mixed Numbers to Percents

Convert to percent notation.

a. $2\dfrac{1}{4}$ **b.** $\dfrac{13}{10}$

Solution:

a. $2\dfrac{1}{4} = 2\dfrac{1}{4} \times 100\%$ Multiply by 100%.

$\qquad = \dfrac{9}{4} \times \dfrac{100}{1}\%$ Convert the whole numbers to improper fractions.

$\qquad = \dfrac{9}{\underset{1}{4}} \times \dfrac{\overset{25}{\cancel{100}}}{1}\%$ Multiply and simplify to lowest terms.

$\qquad = 225\%$

b. $\dfrac{13}{10} = \dfrac{13}{10} \times 100\%$ Multiply by 100%.

$\qquad = \dfrac{13}{10} \times \dfrac{100}{1}\%$ Convert the whole number to an improper fraction.

$\qquad = \dfrac{13}{\underset{1}{\cancel{10}}} \times \dfrac{\overset{10}{\cancel{100}}}{1}\%$ Multiply and simplify to lowest terms.

$\qquad = 130\%$

Notice that both answers in Example 3 are greater than 100%. This is reasonable because any number greater than 1 whole unit represents a percent greater than 100%.

4. Approximating Percents

In Example 4 we approximate a percent from its fraction form.

Answers

9. 170%   **10.** 275%

example 4	Approximating a Percent

Write the fraction $\frac{5}{13}$ in percent notation to the nearest tenth of a percent.

Solution:

$$\frac{5}{13} = \frac{5}{13} \times 100\% \qquad \text{Multiply by 100\%.}$$

$$= \frac{5}{13} \times \frac{100}{1}\% \qquad \text{Write the whole number as an improper fraction.}$$

$$= \frac{500}{13}\%$$

To round to the nearest tenth of a percent, we must divide. We will obtain the hundredths-place digit in the quotient on which to base the decision on rounding.

Thus, $\dfrac{5}{13} \approx 38.5\%$.

$$\begin{array}{r} 38.46 \\ 13\overline{)500.00} \\ -39 \\ \hline 110 \\ -104 \\ \hline 60 \\ -52 \\ \hline 80 \\ -78 \\ \hline 2 \end{array}$$

Skill Practice

11. Write the fraction $\frac{3}{7}$ in percent notation to the nearest tenth of a percent.

12. Write the fraction $\frac{11}{3}$ in percent notation to the nearest hundredth of a percent.

Avoiding Mistakes: In Example 4, we converted a fraction to a percent where rounding was necessary. We converted the fraction to percent form *before* dividing and rounding. If you try to convert to decimal form first, you might round too soon.

5. Fractions, Decimals, Percents: A Summary

The diagram in Figure 6-4 illustrates the methods to convert fractions, decimals, and percents.

Concept Connections

13. To convert a decimal to a percent, in which direction do you move the decimal point?

14. To convert a percent to a decimal, in which direction do you move the decimal point?

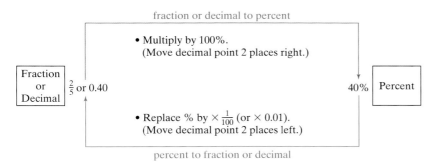

Figure 6-4

Answers

11. 42.9%   12. 366.67%
13. To the right 2 places
14. To the left 2 places

Skill Practice

Complete the table.

	Fraction	Decimal	Percent
15.		1.41	
16.	$\frac{1}{50}$		
17.			18%
18.		0.58	
19.			$33\frac{1}{3}\%$
20.	$\frac{7}{9}$		

---

example 5   Converting Fractions, Decimals, and Percents

Complete the table.

	Fraction	Decimal	Percent
a.		0.55	
b.	$\frac{1}{200}$		
c.			160%
d.		2.4	
e.			$66\frac{2}{3}\%$
f.	$\frac{2}{9}$		

Solution:

a. 0.55 to fraction: $0.55 = \dfrac{55}{100} = \dfrac{11}{20}$

0.55 to percent: $0.55 \times 100\% = 55\%$

b. $\dfrac{1}{200}$ to decimal: $1 \div 200 = 0.005$

$\dfrac{1}{200}$ to percent: $\dfrac{1}{200} \times 100\% = \dfrac{100}{200}\% = 0.5\%$

c. 160% to fraction: $160 \times \dfrac{1}{100} = \dfrac{160}{100} = \dfrac{8}{5}$ or $1\dfrac{3}{5}$

160% to decimal: $160 \times 0.01 = 1.6$

d. 2.4 to fraction: $\dfrac{24}{10} = \dfrac{12}{5}$ or $2\dfrac{2}{5}$

2.4 to percent: $2.4 \times 100\% = 240\%$

e. $66\dfrac{2}{3}\%$ to fraction: $66\dfrac{2}{3} \times \dfrac{1}{100} = \dfrac{\overset{2}{\cancel{200}}}{3} \times \dfrac{1}{\underset{1}{\cancel{100}}} = \dfrac{2}{3}$

$66\dfrac{2}{3}\%$ to decimal: $66\dfrac{2}{3} \times 0.01 = 66.\overline{6} \times 0.01 = 0.\overline{6}$

f. $\dfrac{2}{9}$ to decimal: $2 \div 9 = 0.\overline{2}$

$\dfrac{2}{9}$ to percent: $\dfrac{2}{9} \times 100\% = \dfrac{2}{9} \times \dfrac{100}{1}\% = \dfrac{200}{9}\% = 22\dfrac{2}{9}\%$ or $22.\overline{2}\%$

Answers

	Fraction	Decimal	Percent
15.	$\frac{141}{100}$ or $1\frac{41}{100}$	1.41	141%
16.	$\frac{1}{50}$	0.02	2%
17.	$\frac{9}{50}$	0.18	18%
18.	$\frac{29}{50}$	0.58	58%
19.	$\frac{1}{3}$	$0.\overline{3}$	$33\frac{1}{3}\%$
20.	$\frac{7}{9}$	$0.\overline{7}$	$77.\overline{7}\%$

The completed table is as follows.

	Fraction	Decimal	Percent
a.	$\frac{11}{20}$	0.55	55%
b.	$\frac{1}{200}$	0.005	0.5%
c.	$\frac{8}{5}$ or $1\frac{3}{5}$	1.6	160%
d.	$\frac{12}{5}$ or $2\frac{2}{5}$	2.4	240%
e.	$\frac{2}{3}$	$0.\overline{6}$	$66\frac{2}{3}\%$
f.	$\frac{2}{9}$	$0.\overline{2}$	$22\frac{2}{9}\%$ or $22.\overline{2}\%$

section 6.2   Practice Exercises

Study Skills Exercise

1. Do you remember your instructor's name, office hours, office location, and office phone? Write them here:

   Instructor's name: Instructor's office hours:

   Instructor's office location: Instructor's office phone:

Review Exercises

For Exercises 2–5, convert the percent to a fraction or mixed number.

2. 60%   **3.** 130%   **4.** $16\frac{1}{2}\%$ **5.** 0.5%

For Exercises 6–9, convert the percent to a decimal.

6. 80%   **7.** $6\frac{1}{3}\%$ **8.** 143%   **9.** 0.3%

Objective 1: Rule for Converting Fractions and Decimals to Percents

For Exercises 10–13, multiply.

10. $0.68 \times 100\%$ **11.** $1.62 \times 100\%$ **12.** $0.005 \times 100\%$ **13.** $0.26 \times 100\%$

14. Write the rule for multiplying a decimal by 100.

For Exercises 15–18, multiply.

15. $\dfrac{5}{4} \times 100\%$ **16.** $\dfrac{2}{5} \times 100\%$ **17.** $\dfrac{77}{100} \times 100\%$ **18.** $\dfrac{113}{100} \times 100\%$

19. Write the rule for multiplying a fraction and a whole number.

20. a. Convert 23% to a decimal.   **21. a.** Convert 17% to a decimal.
    b. Convert 0.23 to a percent.   **b.** Convert 0.17 to a percent.

22. a. Convert 81% to a fraction.   **23. a.** Convert 37% to a fraction.
    b. Convert $\frac{81}{100}$ to a percent.   **b.** Convert $\frac{37}{100}$ to a percent.

24. a. Convert 150% to a mixed number.
    b. Convert $1\frac{1}{2}$ to a percent.

Objective 2: Converting Decimals to Percents

For Exercises 25–36, convert the decimal to a percent. **(See Example 1.)**

25. 0.27   **26.** 0.51   **27.** 0.19   **28.** 0.33

29. 1.75   **30.** 2.8   **31.** 0.124   **32.** 0.277

33. 0.006   **34.** 0.0008   **35.** 1.014   **36.** 2.203

Objective 3: Converting Fractions to Percents

For Exercises 37–48, convert the fraction to a percent. **(See Example 2.)**

37. $\dfrac{71}{100}$ **38.** $\dfrac{89}{100}$ **39.** $\dfrac{19}{20}$ **40.** $\dfrac{7}{20}$

41. $\dfrac{7}{8}$ **42.** $\dfrac{5}{8}$ **43.** $\dfrac{13}{16}$ **44.** $\dfrac{11}{16}$

45. $\dfrac{5}{6}$ **46.** $\dfrac{5}{12}$ **47.** $\dfrac{4}{9}$ **48.** $\dfrac{1}{9}$

For Exercises 49–54, write the fraction as a percent.

49. One-quarter of Americans say they entertain at home 2 or more times a month. (Source: *USA TODAY*.)

50. According to the Centers for Disease Control (CDC), $\frac{37}{100}$ of U.S. teenage boys say they rarely or never wear their seatbelts.

51. According to the Centers for Disease Control, $\frac{1}{10}$ of teenage girls in the U.S. say they rarely or never wear their seatbelts.

52. In Italy, $\frac{3}{50}$ of the country's budget comes from tourism.

53. In a recent year, $\frac{2}{3}$ of the beds in U.S. hospitals were occupied.

54. In Georgia in 2001, approximately $\frac{1}{6}$ of the residents were not covered by health insurance. (Source: U.S. Bureau of the Census.)

For Exercises 55–62, convert the mixed number to percent notation. **(See Example 3.)**

55. $1\dfrac{3}{4}$ **56.** $\dfrac{7}{2}$ **57.** $\dfrac{27}{20}$ **58.** $2\dfrac{1}{8}$

59. $\dfrac{11}{9}$ **60.** $1\dfrac{5}{9}$ **61.** $1\dfrac{2}{3}$ **62.** $\dfrac{7}{6}$

Objective 4: Approximating Percents

For Exercises 63–70, write the fraction in percent notation to the nearest tenth of a percent. **(See Example 4.)**

63. $\dfrac{3}{7}$ **64.** $\dfrac{6}{7}$ **65.** $\dfrac{1}{13}$ **66.** $\dfrac{3}{13}$

67. $\dfrac{5}{11}$ **68.** $\dfrac{8}{11}$ **69.** $\dfrac{13}{15}$ **70.** $\dfrac{1}{15}$

Objective 5: Fractions, Decimals, Percents: A Summary

71. Explain the difference between $\frac{1}{2}$ and $\frac{1}{2}\%$.   **72.** Explain the difference between $\frac{3}{4}$ and $\frac{3}{4}\%$.

73. Explain the difference between 25% and 0.25%.   **74.** Explain the difference between 10% and 0.10%.

75. Which of the numbers represent 125%?

    a. 1.25   **b.** 0.125   **c.** $\dfrac{5}{4}$ **d.** $\dfrac{5}{4}\%$

76. Which of the numbers represent 60%?

    a. 6.0   **b.** 0.60%   **c.** 0.6   **d.** $\dfrac{3}{5}$

77. Which of the numbers represent 30%?

    a. $\dfrac{3}{10}$ **b.** $\dfrac{1}{3}$ **c.** 0.3   **d.** 0.03%

78. Which of the numbers represent 180%?

    a. 18   **b.** 1.8   **c.** $\dfrac{9}{5}$ **d.** $\dfrac{9}{5}\%$

For Exercises 79–82, complete the table. **(See Example 5.)**

79.

	Fraction	Decimal	Percent
a.	$\frac{1}{4}$		
b.		0.92	
c.			15%
d.		1.6	
e.	$\frac{1}{100}$		
f.			0.5%

80.

	Fraction	Decimal	Percent
a.			0.6%
b.	$\frac{2}{5}$		
c.		2	
d.	$\frac{1}{2}$		
e.		0.12	
f.			45%

81.

	Fraction	Decimal	Percent
a.			14%
b.		0.87	
c.		1	
d.	$\frac{1}{3}$		
e.			0.2%
f.	$\frac{19}{20}$		

82.

	Fraction	Decimal	Percent
a.		1.3	
b.			22%
c.	$\frac{3}{4}$		
d.		0.73	
e.			$22.\overline{2}\%$
f.	$\frac{1}{20}$		

Expanding Your Skills

83. Is the number 1.4 less than or greater than 100%?

84. Is the number 0.0087 less than or greater than 1%?

85. Is the number 0.052 less than or greater than 50%?

86. Is the number 25 less than or greater than 25%?

Objectives

1. Introduction to Percent Proportions
2. Identifying the Parts of a Percent Proportion
3. Solving Percent Proportions
4. Applications of Percent Proportions

section 6.3   Percent Proportions and Applications

1. Introduction to Percent Proportions

Recall that a percent is a ratio in parts per 100. For example, $50\% = \frac{50}{100}$. However, a percent can be represented by infinitely many equivalent fractions. Thus,

$$50\% = \frac{50}{100} = \frac{1}{2} = \frac{2}{4} = \frac{3}{6} \qquad \text{and infinitely many more}$$

Equating a percent to an equivalent ratio forms a proportion that we call a **percent proportion**. A percent proportion is a proportion in which one ratio is written with a denominator of 100. For example,

$$\frac{50}{100} = \frac{3}{6} \qquad \text{is a percent proportion}$$

2. Identifying the Parts of a Percent Proportion

We will be using percent proportions to solve a variety of application problems. But first we need to identify and label the parts of a percent proportion.

A percent proportion can be written in the form:

$$\frac{\text{Amount}}{\text{Base}} = p\% \qquad \text{or} \qquad \frac{\text{Amount}}{\text{Base}} = \frac{p}{100}$$

For example:

$$4 \text{ L out of } 8 \text{ L is } 50\% \qquad \frac{4}{8} = 50\% \qquad \text{or} \qquad \frac{4}{8} = \frac{50}{100}$$

amount   base   p

In this example, 8 L is some total (or base) quantity and 4 L is some part (or amount) of that whole. The ratio $\frac{4}{8}$ represents a fraction of the whole equal to 50%. In general, we offer the following guidelines for identifying the parts of a percent proportion.

Identifying the Parts of a Percent Proportion

A percent proportion can be written as

$$\frac{\text{Amount}}{\text{Base}} = p\% \qquad \text{or} \qquad \frac{\text{Amount}}{\text{Base}} = \frac{p}{100}.$$

- The **base** is the total or whole amount being considered. It often appears after the word *of* within a word problem.
- The **amount** is the part being compared to the base. It *sometimes* appears with the word *is* within a word problem.

---

example 1   Identifying Amount, Base, and *p* for a Percent Proportion

Identify the amount, base, and *p* value, and then set up a percent proportion.

a. 25% of 60 students is 15 students.   **b.** $32 is 50% of $64.

c. 5 of 1000 employees is 0.5%.

Solution:

For each problem, we recommend that you identify *p* first. It is the number in front of the symbol %. Then identify the base. In most cases it follows the word *of*. Then, by the process of elimination, find the amount.

a. 25% of 60 students is 15 students.

p	base	amount
(before % symbol)	(after the word *of*)	(the number left over)

$$\text{amount} \to \frac{15}{60} = \frac{25}{100} \leftarrow p \\ \text{base} \to \qquad\qquad \leftarrow 100$$

b. $32 is 50% of $64.

amount   p   base

$$\text{amount} \to \frac{32}{64} = \frac{50}{100} \leftarrow p \\ \text{base} \to \qquad\qquad \leftarrow 100$$

c. 5 of 1000 employees is 0.5%.

amount   base   p

$$\text{amount} \to \frac{5}{1000} = \frac{0.5}{100} \leftarrow p \\ \text{base} \to \qquad\qquad\quad \leftarrow 100$$

3. Solving Percent Proportions

In Example 1, we practiced identifying the parts of a percent proportion. Now we consider percent proportions in which one of these numbers is unknown. Furthermore, we will see that the examples come in three types:

- Amount is unknown.
- Base is unknown.
- Value p is unknown.

However, the process to solve each case is the same.

example 2   Solving Percent Proportions—Amount Unknown

a. What is 30% of 180?   b. 70% of 500 people is how many people?

Solution:

a. What is 30% of 180?   The base and value for p are known.

amount (x) p base   Let x represent the unknown amount.

$$\frac{x}{180} = \frac{30}{100}$$ Set up a percent proportion.

$$100 \cdot x = (30)(180)$$ Equate the cross products.

$$100x = 5400$$

$$\frac{\overset{}{100}x}{\underset{1}{100}} = \frac{\overset{54}{5400}}{\underset{1}{100}}$$ Divide both sides of the equation by 100.

$$x = 54$$ Simplify to lowest terms.

Therefore, 54 is 30% of 180.

Tip: We can check the answer to Example 2(a) as follows. Ten percent of a number is $\frac{1}{10}$ of the number. Furthermore, $\frac{1}{10}$ of 180 is 18. Thirty percent of 180 must be 3 times this amount.
$$3 \times 18 = 54 \quad ✔$$

b. 70% of 500 people is how many people?   The base and value for p are known.

p base   amount (x)   Let x represent the unknown amount.

$$\frac{x}{500} = \frac{70}{100}$$ Set up a percent proportion.

$$100 \cdot x = (70)(500)$$ Equate the cross products.

$$100x = 35,000$$

$$\frac{\overset{}{100}x}{\underset{1}{100}} = \frac{\overset{350}{35,000}}{\underset{1}{100}}$$ Divide both sides by 100.

$$x = 350$$ Simplify to lowest terms.

Therefore, 70% of 500 is 350.

Tip: To check the solution to Example 2(b), we can compute 10% of 500, which is 50. To find 70%, multiply this by 7. We have $7(50) = 350.$ ✔

example 3   Solving Percent Proportions—Base Unknown

a. 40% of what number is 25?   **b.** $13.50 is 150% of how many dollars?

Solution:

a. 40% of what number is 25?

$\underset{p}{|}$ $\underset{\text{base }(x)}{|}$ $\underset{\text{amount}}{|}$

The amount and value of p are known.

Let x represent the unknown base.

$$\frac{25}{x} = \frac{40}{100}$$ Set up a percent proportion.

$(25)(100) = 40 \cdot x$ Equate the cross products.

$2500 = 40x$

$$\frac{2500}{40} = \frac{\cancel{40}x}{\cancel{40}}$$ Divide both sides by 40.

$62.5 = x$

Therefore, 40% of 62.5 is 25.

b. $13.50 is 150% of how many dollars?

$\underset{\text{amount}}{|}$ $\underset{p}{|}$ $\underset{\text{base }(x)}{|}$

The amount and value of p are known.

Let x represent the unknown base.

$$\frac{13.50}{x} = \frac{150}{100}$$ Set up a percent proportion.

$150 \cdot x = (13.50)(100)$ Equate the cross products.

$150x = 1350$

$$\frac{\cancel{150}x}{\cancel{150}} = \frac{1350}{150}$$ Divide both sides by 150.

$x = 9$

Therefore, $13.50 is 150% of $9.

Tip: We can check the answer to Example 3(b) as follows: 150% of a quantity is the same as 1.5 times the quantity. Thus 1.5($9) = $13.50, as desired. ✔

example 4   Solving Percent Proportions—p Unknown

a. What percent of 80 mi is 12.4 mi?

b. 48 is what percent of 42? Round to the nearest percent.

Solution:

a. What percent of 80 mi is 12.4 mi?

$\underset{p}{|}$ $\underset{\text{base}}{|}$ $\underset{\text{amount}}{|}$

The amount and base are known.

The value of p is unknown.

$$\frac{12.4}{80} = \frac{p}{100}$$ Set up a percent equation.

$(12.4)(100) = 80 \cdot p$ Equate the cross products.

$1240 = 80p$

$$\frac{1240}{80} = \frac{\cancel{80}p}{\cancel{80}}$$ Divide both sides by 80.

$$15.5 = p$$

Therefore, 15.5% of 80 mi is 12.4 mi.

b. 48 is what percent of 42? Round to the nearest percent.

   amount p base   The value of p is unknown.

$$\frac{48}{42} = \frac{p}{100}$$ Set up a percent proportion.

$$(48)(100) = 42 \cdot p$$ Equate the cross products.

$$4800 = 42p$$

$$\frac{4800}{42} = \frac{\cancel{42}p}{\cancel{42}}$$ Divide both sides by 42.

$$114 \approx p$$ Note that $4800 \div 42 = 114.\overline{285714}$. Rounded to the nearest whole number, $p \approx 114$.

Therefore, 48 is approximately 114% of 42.

4. Applications of Percent Proportions

We now use percent proportions to solve application problems involving percents.

Skill Practice

10. In 2004, it was estimated that 24.7% of U.S. adults smoked tobacco products regularly. In a group of 2000 adults, how many would be expected to be smokers? Round to the nearest whole number.

example 5   Using Percents in Meteorology

Buffalo, New York, receives an average of 94 in. of snow each year. This year it had 120% of the normal annual snowfall. How much snow did Buffalo get this year?

Solution:

This situation can be translated to "the amount of snow Buffalo received is 120% of 94 in."

The percent is given as 120%. Therefore, $p = 120$.

94 in. *of* snow indicates the base.

Let x represent the amount of snow this year.

   Identify the known values and the unknown value within the percent proportion.

$$\frac{x}{94} = \frac{120}{100}$$ Set up a percent proportion.

$$100 \cdot x = (120)(94)$$ Equate the cross products.

$$100x = 11{,}280$$

$$\frac{\cancel{100}x}{\cancel{100}} = \frac{11{,}280}{100}$$ Divide both sides by 100.

$$x = 112.8$$

This year, Buffalo had 112.8 in. of snow.

Answer

10. 494 people

example 6   Using Percents in Statistics

Harvard University's freshman class for 2004 had 19% Asian American students. If this represented 385 students, how many students were admitted to the freshman class? Round to the nearest student.

Solution:

This situation can be translated to "385 is 19% of what number?"

The percent is given as 19%. Therefore, $p = 19$.

385 represents the amount taken out of an unknown total.

Let x represent the base (or total) number of incoming freshmen.

} Identify the known values and the unknown value within the percent proportion.

$$\frac{385}{x} = \frac{19}{100}$$ Set up a percent proportion.

$$(385)(100) = (19) \cdot x$$ Equate the cross products.

$$38{,}500 = 19x$$

$$\frac{38{,}500}{19} = \frac{19x}{19}$$ Divide both sides by 19.

$$2026 \approx x$$ Note that $38{,}500 \div 19 \approx 2026.3$. Rounded to the nearest whole unit (whole person), this is 2026.

The freshman class at Harvard in 2004 had approximately 2026 students.

Skill Practice

11. Eight students in a statistics class received a grade of A in the class. If this represents 19% of the class, how many students are in the class? Round to the nearest whole number.

Tip: We can check the answer to Example 6 by substituting $x = 2026$ back into the original proportion. The cross products will not be exactly the same because we had to round the value of x. However, the cross products should be *close*.

$$\frac{385}{2026} \overset{?}{\approx} \frac{19}{100}$$ Substitute $x = 2026$ into the proportion.

$$(385)(100) \overset{?}{\approx} (19)(2026)$$

$$38{,}500 \approx 38{,}494 \quad ✔ \quad \text{The values are close.}$$

---

example 7   Using Percents in Business

Suppose a tennis pro who is ranked 90th in the world on the men's professional tour earns $280,000 per year in tournament winnings and endorsements. A breakdown of his expenses is given in Figure 6-5. What percent of his salary goes toward his coach? Round to the nearest tenth of a percent.

Answer

11. About 42 students are in the class.

Major Expenses for Tennis Pro

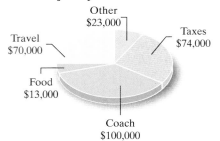

Figure 6-5

Skill Practice

12. The graph shows the number of donations made for various dollar amounts at an animal sanctuary. What percent of the donations was in the $100–$199 range? Round to the nearest tenth of a percent.

Number of Donations by Dollar Value

Solution:

This can be translated to "What *percent* of $280,000 is $100,000?"

The amount spent on the coach is $100,000.

The base (or total) earnings is $280,000.

The percent is not given. Therefore, *p* is unknown.

> Identify the known values and the unknown value within the percent proportion.

$$\frac{100{,}000}{280{,}000} = \frac{p}{100}$$ Set up a percent proportion.

$$\frac{100{,}\cancel{000}}{280{,}\cancel{000}} = \frac{p}{100}$$ The ratio on the left side of the equation can be simplified by a factor of 10,000. "Strike through" four zeros in the numerator and denominator.

$$\frac{10}{28} = \frac{p}{100}$$

$(10)(100) = (28) \cdot p$ Equate the cross products.

$1000 = 28p$

$$\frac{1000}{28} = \frac{28p}{28}$$ Divide both sides by 28.

$$\frac{1000}{28} = p$$

$35.7 \approx p$ Dividing $1000 \div 28$, we get approximately 35.7.

The tennis pro spends about 35.7% of his income on his coach.

$$
\begin{array}{r}
35.71 \\
28\overline{)1000.00} \\
-84 \\
\hline
160 \\
-140 \\
\hline
200 \\
-196 \\
\hline
40 \\
-28 \\
\hline
12
\end{array}
$$

Answer

12. Of the donations, approximately 14.1% are in the $100–$199 range.

section 6.3   Practice Exercises

Study Skills Exercises

1. Write down your instructor's policies for the following.

    a. Missing a test   **b.** Missing a class   **c.** Doing homework

2. Define the key terms.

    a. Percent proportion   **b. Base**   **c. Amount**

Review Exercises

For Exercises 3–5, convert the decimal to a percent.

3. 0.55   **4.** 1.30   **5.** 0.0006

For Exercises 6–8, convert the fraction to a percent.

6. $\dfrac{3}{8}$ **7.** $\dfrac{5}{2}$ **8.** $\dfrac{1}{100}$

For Exercises 9–11, convert the percent to a fraction.

9. $62\frac{1}{2}\%$ **10.** 2%   **11.** 77%

For Exercises 12–14, convert the percent to a decimal.

12. 82%   **13.** 0.3%   **14.** 100%

Objective 1: Introduction to Percent Proportions

For Example 15–19, determine if the proportion is a percent proportion.

15. $\dfrac{7}{100} = \dfrac{14}{200}$ **16.** $\dfrac{150}{300} = \dfrac{50}{100}$ **17.** $\dfrac{2}{3} = \dfrac{6}{9}$

18. $\dfrac{1\frac{1}{2}}{100} = \dfrac{3}{200}$ **19.** $\dfrac{\frac{3}{4}}{100} = \dfrac{3}{400}$

For Exercises 20–24, shade the figure to estimate the amount. The first exercise is given as an example.

Example: Find 60% of 80. (Answer: 48)

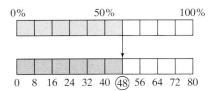

20. Find 40% of 60.

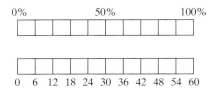

21. Find 75% of 60.

22. Find 15% of 240.

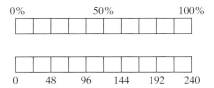

23. Find 120% of 40.

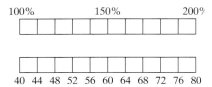

24. Find 170% of 30.

Objective 2: Identifying the Parts of a Percent Proportion

For Exercises 25–30, identify the amount, base, and *p* value. **(See Example 1.)**

25. 12 balloons is 60% of 20 balloons.

26. 25% of 400 cars is 100 cars.

27. $99 of $200 is 49.5%.

28. 45 of 50 children is 90%.

29. 50 hr is 125% of 40 hr.

30. 175% of 2 in. of rainfall is 3.5 in.

For Exercises 31–36, write the percent proportion.

31. 10% of 120 trees is 12 trees.

32. 15% of 20 pictures is 3 pictures.

33. 72 children is 80% of 90 children.

34. 21 dogs is 20% of 105 dogs.

35. 21,684 college students is 104% of 20,850 college students.

36. 103% of $40,000 is $41,200.

Objective 3: Solving Percent Proportions

For Exercises 37–46, solve the percent problems with an unknown amount. **(See Example 2.)**

37. What is $\frac{1}{2}$% of 40?

38. Find 35% of 412.

39. Compute 54% of 200 employees.

40. What is 1.8% of 900 grams?

41. Find 112% of 500.

42. Compute 106% of 1050.

43. Pedro pays 28% of his salary in income tax. If he makes $72,000 in taxable income, how much income tax does he pay?

44. A car dealer sets the sticker price of a car by taking 115% of the wholesale price. If a car sells wholesale at $17,000, what is the sticker price?

45. Jesse Ventura became the 38th governor of Minnesota by receiving 37% of the votes. If approximately 2,060,000 votes were cast, how many did Mr. Ventura get?

46. In a psychology class, 61.9% of the class consists of freshmen. If there are 42 students, how many are freshmen? Round to the nearest whole unit.

For Exercises 47–56, solve the percent problems with an unknown base. **(See Example 3.)**

47. 18 is 50% of what number?

48. 22% of what length is 44 ft?

49. 30% of what weight is 69 lb?

50. 70% of what number is 28?

51. 9 is $\frac{2}{3}$% of what number?

52. 9.5 is 200% of what number?

53. Albert saves $120 per month. If this is 7.5% of his monthly income, how much does he make per month?

54. Janie and Don left their house in South Bend, Indiana, to visit friends in Chicago. They drove 80% of the distance before stopping for lunch. If they had driven 56 mi before lunch, what would be the total distance from their house to their friends' house in Chicago?

55. Amiee read 14 e-mails which was only 40% of her total e-mails. What is her total number of e-mails?

56. A recent survey found that 5% of the population of the United States is unemployed. If Charlotte, North Carolina, has 32,000 unemployed, what is the population of Charlotte?

For Exercises 57–64, solve the percent problems with p unknown. **(See Example 4.)**

57. What percent of $120 is $42?

58. 112 is what percent of 400?

59. 84 is what percent of 70?

60. What percent of 12 letters is 4 letters?

61. What percent of 320 mi is 280 mi?

62. 54¢ is what percent of 48¢?

63. A student correctly answered 29 problems on a final exam of 40 problems. What percent of the questions did she answer correctly?

64. During his college basketball season, Jeff made 520 baskets out of 1280 attempts. What was his shooting percentage? Round to the nearest whole percent.

A metropolitan police force in the southeastern United States consists of 600 officers, 480 men and 120 women. Over the past 2 years, 160 officers on the police force have been promoted. The table shows the breakdown of promotions for male and female officers. Use the table for Exercises 65–68.

	Promoted	Not Promoted	Total
Male	140	340	480
Female	20	100	120
Total	160	440	600

65. What percent of the officers are female?

66. What percent of the officers are male?

67. What percent of the officers were promoted? Round to the nearest tenth of a percent.

68. What percent of the officers were not promoted? Round to the nearest tenth of a percent.

Objective 4: Applications of Percent Proportions

69. The rainfall at Birmingham Airport in the United Kingdom averages 56 mm per month. In August the amount of rain that fell was 125% of the average monthly rainfall. How much rain fell in August? **(See Example 5.)**

70. In a recent survey 38% of people in the United States say that gas prices have affected the type of vehicle they will buy. In a sample of 500 people who are in the market for a new vehicle, how many would you expect to be influenced by gas prices?

71. Harvard University reported that 209 African American students were admitted to the freshman class in a recent year. If this represents 11% of the total freshman class, how many freshmen were admitted? **(See Example 6.)**

72. Yellowstone National Park has 3366 mi² of undeveloped land. If this represents 99% of the total area, find the total area of the park.

73. During the 2001–2002 basketball season, Steve Smith of the San Antonio Spurs made 116 three-point shots out of 246 attempts. To the nearest tenth of a percent, find the percent of shots made. **(See Example 7.)**

74. As of the 2002–2003 football season, Peyton Manning had completed 1357 passes out of 2226 attempts. Find his completion percentage to the nearest tenth of a percent.

75. The graph in the figure shows the percent of households that own a dog according to the number of people residing in the household. (Source: American Veterinary Medical Association.)

   a. If 200 five-person households are surveyed, how many would you expect to own a dog?

   b. If 50 three-person households are surveyed, how many would you expect to own a dog?

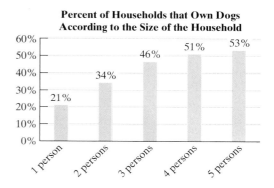

Percent of Households that Own Dogs According to the Size of the Household

76. President George W. Bush and his wife Laura paid approximately $228,000 in federal income tax in 2003 on an adjusted gross income of approximately $822,000. What was the President's tax rate? Round to the nearest percent. (Source: *Time*, April 26, 2004.)

A used car dealership sells several makes of vehicles. For Exercises 77–80, refer to the graph. Round the answers to the nearest whole unit.

77. If the dealership sold 215 vehicles in one month, how many were Chevys?

78. If the dealership sold 182 vehicles in one month, how many were Fords?

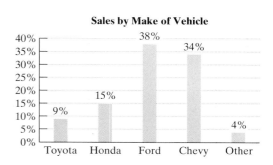

Sales by Make of Vehicle

79. If the dealership sold 27 Hondas in one month, how many total vehicles were sold?

80. If the dealership sold 10 cars in the "Other" category, how many total vehicles were sold?

Expanding Your Skills

81. Carson had $600 and spent 44% of it on clothes. Then he spent 20% of the remaining money on dinner. How much did he spend altogether?

82. Melissa took $52 to the mall and spent 24% on makeup. Then she spent one-half of the remaining money on lunch. How much did she spend altogether?

It is customary to leave a 15–20% tip for the server in a restaurant. However, when you are at a restaurant in a social setting, you probably do not want to take out a pencil and piece of paper to figure out the tip. It is more effective to compute the tip mentally. Try this method.

Step 1: First, if the bill is not a whole dollar amount, simplify the calculations by rounding the bill to the next-higher whole dollar.

Step 2: Take 10% of the bill. This is the same as taking one-tenth of the bill. Move the decimal point to the left 1 place.

Step 3: If you want to leave a 20% tip, double the value found in step 2.

Step 4: If you want to leave a 15% tip, first note that 15% is 5% + 10%. Therefore, add one-half of the value found in step 2 to the number in step 2.

83. Compute a 20% tip on a bill of $57.65. (*Hint:* Round up to $58 first.)

84. Compute a 20% tip on a bill of $18.79.

85. Compute a 15% tip on a dinner bill of $42.00.

86. Compute a 15% tip on a luncheon bill of $12.00.

Objectives

1. Solving Percent Equations—Amount Unknown
2. Solving Percent Equations—Base Unknown
3. Solving Percent Equations—Percent Unknown
4. Applications of Percent Equations

section 6.4   Percent Equations and Applications

1. Solving Percent Equations—Amount Unknown

In this section, we investigate an alternative method to solve applications involving percents. We use percent equations. A **percent equation** represents a percent proportion in an alternative form. For example, recall that we can write a percent proportion as follows:

$$p\% = \frac{\text{amount}}{\text{base}} \qquad \text{percent proportion}$$

This is equivalent to writing $\text{Amount} = (p\%) \cdot (\text{base})$ percent equation

To set up a percent equation, it is necessary to translate an English sentence into a mathematical equation. As you read through the examples in this section, you will notice several key words. In the phrase *percent of*, the word *of* implies multiplication. The verb *to be* (am, is, are, was, were, been) often implies =.

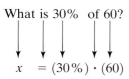 **Solving a Percent Equation—Amount Unknown**

What is 30% of 60?

Solution:

We translate the words to mathematical symbols.

What is 30% of 60?

$x = (30\%) \cdot (60)$

Translate.
Notice that in this context, the word *of* means to multiply.
Let x represent the unknown amount.

Notice that to find x, we must multiply 30% by 60. However, 30% means $\frac{30}{100}$ or 0.30. For the purpose of calculation, we *must* convert 30% to its equivalent decimal or fraction form. The equation becomes

$x = (0.30)(60)$

$= 18$

The value 18 is 30% of 60.

> **Tip:** The solution to Example 1 can be checked by noting that 10% of 60 is 6. Therefore, 30% is equal to $(3)(6) = 18$. ✓

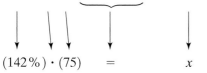 **Solving Percent Equations—Amount Unknown**

142% of 75 amounts to what number?

Solution:

142% of 75 amounts to what number?

$(142\%) \cdot (75) = x$

$(1.42)(75) = x$

$106.5 = x$

Translate. Let x represent the unknown amount.
The word *of* implies multiplication.

The phrase *amounts to* implies =.

Convert 142% to its decimal form (1.42).

Multiply.

Therefore, 142% of 75 amounts to 106.5.

Examples 1 and 2 illustrate that the percent equation gives us a quick way to find an unknown amount. For example, because $(p\%) \cdot (\text{base}) = \text{amount}$, we have

$$50\% \text{ of } 80 = 0.50(80) = 40$$
$$25\% \text{ of } 20 = 0.25(20) = 5$$
$$87\% \text{ of } 600 = 0.87(600) = 522$$
$$250\% \text{ of } 90 = 2.50(90) = 225$$

2. Solving Percent Equations—Base Unknown

Examples 3 and 4 illustrate the case in which the base is unknown.

example 3   Solving a Percent Equation—Base Unknown

225 is 40% of what number?

Solution:

225 is 40% of what number?   Translate. Let x represent the base number.

$225 = (0.40) \cdot \quad x$ Notice that we immediately converted 40% to its decimal form 0.40, so that we would not forget.

$225 = 0.40x$

$$\dfrac{225}{0.40} = \dfrac{0.40x}{0.40}$$ Divide both sides of the equation by the number multiplied by the variable x. In this case, divide by 0.40.

$562.5 = x$ Divide: $225 \div 0.40 = 562.5$.

The value 225 is 40% of 562.5.

example 4   Solving a Percent Equation—Base Unknown

0.19 is 0.2% of what number?

Solution:

0.19 is 0.2% of what number?   Translate. Let x represent the base number.

$0.19 = (0.002) \cdot x$ Convert 0.2% to its decimal form 0.002.

$0.19 = 0.002x$

$$\dfrac{0.19}{0.002} = \dfrac{0.002x}{0.002}$$ Divide both sides by the number multiplied by x. In this case, divide by 0.002.

$95 = x$ Divide: $0.19 \div 0.002 = 95$.

Therefore, 0.19 is 0.2% of 95.

3. Solving Percent Equations—Percent Unknown

Examples 5 and 6 demonstrate the process to find an unknown percent.

Answers

6. 117.5   7. 700

example 5   Solving a Percent Equation—Percent Unknown

75 is what percent of 250?

Solution:

$$\begin{array}{cccc} 75 & \text{is} & \text{what percent} & \text{of } 250? \\ \downarrow & \downarrow & \downarrow & \downarrow\;\downarrow \end{array}$$

$$75 = \quad x \quad\quad \cdot (250) \qquad \text{Let } x \text{ represent the unknown percent.}$$

$$75 = 250x$$

$$\frac{75}{250} = \frac{250x}{250} \qquad \text{Divide both sides by the number multiplied by the variable. In this case, divide by 250.}$$

$$0.3 = x \qquad\qquad\quad \text{Divide: } 75 \div 250 = 0.3.$$

At this point, we have $x = 0.3$. To write the value of x in percent form, multiply by 100%.

$$x = 0.3$$

$$= 0.3 \times 100\%$$

$$= 30\%$$

Thus, 75 is 30% of 250.

Avoiding Mistakes: When solving for an unknown percent using a percent equation, it is necessary to convert x to its percent form.

example 6   Solving a Percent Equation—Percent Unknown

What percent of $60 is $92? Round to the nearest tenth of a percent.

Solution:

$$\begin{array}{ccccc} \text{What percent} & \text{of} & \$60 & \text{is} & \$92? \\ \downarrow & & \downarrow & \downarrow & \downarrow \end{array}$$

$$x \quad \cdot (60) = 92 \qquad \text{Translate. Let } x \text{ represent the unknown percent.}$$

$$60x = 92$$

$$\frac{60x}{60} = \frac{92}{60} \qquad \text{Divide both sides by the number multiplied by the variable. In this case, divide by 60.}$$

$$x = 1.5\overline{3} \qquad \text{Divide: } 92 \div 60 = 1.5\overline{3}.$$

At this point, we have $x = 1.5\overline{3}$. To convert x to its percent form, multiply by 100%.

$$x = 1.5\overline{3}$$

$$= 1.5\overline{3} \times 100\% \qquad \text{Convert from decimal form to percent form.}$$

$$= (1.53333\ldots) \times 100\%$$

$$= 153.333\ldots\%$$

The hundredths-place digit is less than 5. Discard it and the digits to its right.

Round to the nearest tenth of a percent.

$$\approx 153.3\%$$

Therefore, $92 is approximately 153.3% of $60. (Notice that $92 is just over $1\frac{1}{2}$ times $60, so our answer seems reasonable.)

Tip: Notice that in Example 6 we converted the final answer to percent form first *before* rounding. With the number written in percent form, we are sure to round to the nearest tenth of a percent.

4. Applications of Percent Equations

In Examples 7, 8, and 9, we use percent equations in application problems. An important part of this process is to extract the base, amount, and percent from the wording of the problem.

example 7   Using a Percent Equation in Ecology

Forty-six panthers are thought to live in Florida's Big Cypress National Preserve. This represents 53% of the panthers living in Florida. How many panthers are there in Florida? Round to the nearest whole unit. (Source: U.S. Fish and Wildlife Services.)

Solution:

This problem translates to

"46 is 53% of the number of panthers living in Florida."

$$46 = (0.53) \cdot x$$ Let x represent the total number of panthers. Write 53% in decimal form as 0.53.

$$46 = 0.53x$$

$$\frac{46}{0.53} = \frac{0.53x}{0.53}$$ Divide both sides by the number multiplied by the variable. In this case, divide by 0.53.

$$87 \approx x$$ Divide: $46 \div 0.53 \approx 87$ (rounded to the nearest whole number).

There are approximately 87 panthers in Florida.

example 8   Using a Percent Equation in Sports Statistics

Steve Young of the San Francisco 49ers was ranked as the NFL's best passer (based on quarterback rating points). For one particular game he completed 23 of 30 passes. What percent of passes did he complete? Round to the nearest tenth of a percent.

Solution:

This problem translates to

"23 is what percent of 30?"

$$23 = \quad x \quad \cdot 30$$ Let x represent the unknown.

$$23 = 30x$$

$$\frac{23}{30} = \frac{30x}{30}$$ Divide both sides by 30.

$$0.767 \approx x$$ Divide $23 \div 30 \approx 0.767$.

The decimal value 0.767 has been rounded to 3 decimal places. We did this because the next step is to convert the decimal to a percent. Move the decimal point to the right 2 places and attach the % symbol. We have 76.7% which is rounded to the nearest tenth of a percent.

$$x \approx 0.767$$
$$= 0.767 \times 100\%$$
$$= 76.7\%$$

Steve Young completed approximately 76.7% of his passes.

example 9   Using a Percent Equation in Voting Statistics

Arnold Schwarzenegger received 50.5% of the votes in the 2003 California recall election for governor. If approximately 7.4 million votes were cast, how many votes did Schwarzenegger receive?

Solution:

This problem translates to

"What number is 50.5% of 7.4?"

$$x = 0.505 \cdot 7.4 \quad \text{Write 50.5\% in decimal form.}$$
$$x = (0.505)(7.4) \quad \text{Let } x \text{ represent the number of votes for Schwarzenegger.}$$
$$x = 3.737 \quad \text{Multiply.}$$

Arnold Schwarzenegger received approximately 3.737 million votes.

Skill Practice

12. In a science class, 85% of the students passed the class. If there were 40 people in the class, how many passed? Round to the nearest whole number.

Answer
12. 34 students passed the class.

section 6.4   Practice Exercises

Boost *your* GRADE at mathzone.com!

MathZone

- Practice Problems
- Self-Tests
- NetTutor
- e-Professors
- Videos

Study Skill Exercises

1. There's a saying, "Leave no stone unturned." In math, this means "leave no homework problem undone." Did you do all the assigned homework in Section 6.3? Do you understand the concepts well enough to move on to the homework in this section?

2. Define the key term **percent equation**.

Review Exercises

3. Explain how to solve the equation $26x = 65$.

For Exercises 4–8, solve the equation for the variable.

4. $3x = 27$

5. $12x = 48$

6. $0.15x = 45$

7. $0.32x = 60$

8. $1.02x = 841.5$

Objective 1: Solving Percent Equations—Amount Unknown

For Exercises 9–14, write the percent equation. Then solve for the unknown amount. **(See Examples 1–2.)**

9. What is 35% of 700?

10. Find 12% of 625.

11. 55% of 900 is what number?

12. What is 0.4% of 75?

13. Find 33% of 600.

14. 20% of 40.4 is what number?

15. What is a quick way to find 50% of a number?

16. What is a quick way to find 10% of a number?

17. Compute 200% of 14 mentally.

18. Compute 75% of 80 mentally.

19. Compute 50% of 40 mentally.

20. Compute 10% of 32 mentally.

21. Household bleach is 6% sodium hypochlorite (active ingredient). In a 64-oz bottle, how much is active ingredient?

22. One antifreeze solution is 40% alcohol. How much alcohol is in a 12.5-L mixture?

23. In football, Dan Marino completed 60% of his passes. If he attempted 8358 passes, how many did he complete? Round to the nearest whole unit.

24. To pass an exit exam, a student must pass a 60-question test with a score of 80% or better. What is the minimum number of questions she must answer correctly?

Objective 2: Solving Percent Equations—Base Unknown

For Exercises 25–30, write the percent equation. Then solve for the unknown base. **(See Examples 3–4.)**

25. 18 is 40% of what number?

26. 72 is 30% of what number?

27. 92% of what number is 41.4?

28. 84% of what number is 100.8?

29. 3.09 is 103% what number?

30. 189 is 105% of what number?

31. In tests of a new anti-inflammatory drug, it was found that 47 subjects experienced nausea. If this represents 4% of the sample, how many subjects were tested?

32. Ted typed 80% of his research paper before taking a break.

    a. If he typed 8 pages, how many total pages are in the paper?

    b. How many pages does he have left to type?

33. In a recent report, approximately 61.6 million Americans had some form of heart and blood vessel disease. If this represents 22% of the population, approximate the total population of the United States.

34. A city has a population of 245,300 which is 110% of the population last year. What was the population last year?

Objective 3: Solving Percent Equations—Percent Unknown

For Exercises 35–42, convert the decimal to a percent.

35. 0.13   **36.** 0.4   **37.** 1.08   **38.** 2.2

39. 0.005   **40.** 0.007   **41.** 0.17   **42.** 0.9

For Exercises 43–48, write the percent equation. Then solve for the unknown percent. **(See Examples 5–6.)**

43. What percent of 480 is 120?   **44.** 180 is what percent of 2000?   **45.** 666 is what percent of 740?

46. What percent of 60 is 2.88?   **47.** What percent of 300 is 375?   **48.** 34 is what percent of 400?

49. At a softball game, the concession stand had 120 hot dogs and sold 84 of them. What percent were sold?

50. The YMCA wants to raise $2500 for its summer program for disadvantaged children. If the YMCA has already raised $900, what percent of its goal has been achieved?

51. A one-year absentee record for a business is given in the table.

    a. From the table, determine the total number of employees.

    b. What percent missed exactly 3 days of work?

    c. What percent missed between 1 and 5 days, inclusive?

    d. What percent missed at least 4 days?

Number of Days Missed	Number of Employees
0	4
1	2
2	14
3	10
4	16
5	18
6	10
7	6

52. Of the 67 million acres of land in Colorado, approximately 3.7 million acres is rural forestland. What percent is rural forestland? (Source: U.S. Department of Agriculture.)

Objective 4: Applications of Percent Equations

53. In a recent year, children and adolescents comprised 6.3 million hospital stays. If this represents 18% of all hospital stays, what was the total number of hospital stays? **(See Example 7.)**

54. One fruit drink advertised that it contained "10% real fruit juice." In a 48-oz bottle,

    a. How much is fruit juice?   **b.** How much is something other than fruit juice?

55. Of the 87 panthers living in the wild in Florida, 11 are thought to live in Everglades National Park. To the nearest tenth of a percent, what percent is this? (Source: U.S. Fish and Wildlife Services.) **(See Example 8.)**

56. Forty-four percent of Americans use online travel sites to book hotel or airline reservations. If 400 people need to make airline or hotel reservations, how many would be expected to use online travel sites?

57. Fifty-two percent of American parents have started to put money away for their children's college educations. In a survey of 800 parents, how many would be expected to have started saving for their children's education? (Source: USA TODAY.) **(See Example 9.)**

58. The earth is covered by approximately 360 million km^2 of water. If the total surface area is 510 km^2, what percent is water? (Round to the nearest tenth of a percent.)

59. A television station plays commercials for 26% of its air time. In 60 min, how many minutes of commercials would be expected?

60. Sixty-five percent of the human body is water. For a 150-lb person, how many pounds are water?

For Exercises 61–64, use the graph.

61. If there were 10,000,000 people in the workforce in the 25–34 age group, how many made over $10 per hour?

62. If there were 6,600,000 people in the workforce in the 55–64 age group, how many made over $10 per hour?

63. If 4,000,000 people in the 16–24 age group made $10 or more per hour, how many total workers in this age group are there?

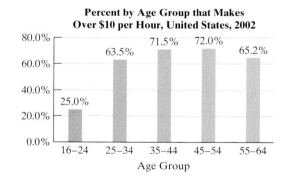

**Percent by Age Group that Makes
Over $10 per Hour, United States, 2002**

64. If 9,000,000 people in the 45–54 age group made $10 or more per hour, how many total workers in this age group are there?

Expanding Your Skills

The maximum recommended heart rate (in beats per minute) is given by 220 minus a person's age. For aerobic activity, it is recommended that individuals exercise at 60–85% of their maximum recommended heart rate. This is called the aerobic range. Use this information for Exercises 65–67.

65. a. Find the maximum recommended heart rate for a 20-year-old.

    b. Find the aerobic range for a 20-year-old.

66. a. Find the maximum recommended heart rate for a 42-year-old.

    b. Find the aerobic range for a 42-year-old.

67. a. Find the maximum recommended heart rate for yourself.

    b. Find the aerobic range for yourself.

---

chapter 6 | midchapter review

For Exercises 1–4, determine the percent represented by the shaded portion of the figure.

1.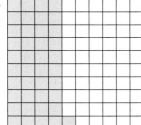

2.

3.

4.

a. $\dfrac{3}{100} = \dfrac{x}{12}$ **b.** $\dfrac{3}{12} = \dfrac{x}{100}$

c. $\dfrac{12}{3} = \dfrac{x}{100}$ **d.** $\dfrac{12}{52} = \dfrac{3}{x}$

7. Is 104% of 80 less than or greater than 80?

8. Is 8% of 50 less than or greater than 5?

9. Is 11% of 90 less than or greater than 9?

10. Is 52% of 200 less than or greater than 100?

5. Which proportion can be used to solve for the percent of a week that the weekend (Saturday and Sunday) represents?

a. $\dfrac{2}{7} = \dfrac{100}{x}$ **b.** $\dfrac{7}{52} = \dfrac{x}{100}$

c. $\dfrac{2}{7} = \dfrac{x}{100}$ **d.** $\dfrac{2}{100} = \dfrac{7}{x}$

6. Which proportion can be used to solve for the percent of a year represented by the months beginning with the letter J (January, June, July)?

For Exercises 11–16, solve the problem by using a percent proportion or a percent equation.

11. 6 is 0.2% of what number?

12. What percent of 500 is 120?

13. 12% of 40 is what number?

14. 27 is what percent of 180?

15. 150% of what number is 105?

16. What number is 30% of 120?

section 6.5 Applications Involving Sales Tax, Commission, Discount, and Markup

Percents are used in an abundance of applications in day-to-day life. In this section we investigate four common applications of percents:

- Sales tax
- Commission
- Discount
- Markup

1. Applications Involving Sales Tax

The first application involves computing sales tax. **Sales tax** is a tax based on a percent of the cost of merchandise.

Sales Tax Formula

$$\left(\begin{array}{c}\text{Amount of}\\\text{sales tax}\end{array}\right) = \left(\begin{array}{c}\text{Tax}\\\text{rate}\end{array}\right) \cdot \left(\begin{array}{c}\text{Cost of}\\\text{merchandise}\end{array}\right)$$

In this formula the tax rate is usually given by a percent. Also note that there are three parts to the formula, just as there are in the general percent equation (see Section 6.4). The sales tax formula is a special case of a percent equation.

Skill Practice

1. A graphing calculator costs $110. The sales tax rate is 4.5%.
   a. Compute the amount of tax.
   b. Compute the total cost.

Avoiding Mistakes: Notice that we must use the decimal form of the sales tax rate in the calculation.

Answers

1. a. The tax is $4.95.
   b. The total cost is $114.95.

example 1   Computing Sales Tax

Suppose a Toyota Camry sells for $20,000.

a. Compute the sales tax for a tax rate of 5.5%.
b. What is the total price of the car?

Solution:

a. Let x represent the amount of tax.   Label the unknown.

Tax rate $= 5.5\%$ Identify the parts of the formula.

Cost of merchandise $= \$20,000$

$$\left(\begin{array}{c}\text{Amount of}\\\text{sales tax}\end{array}\right) = \left(\begin{array}{c}\text{Tax}\\\text{rate}\end{array}\right) \cdot \left(\begin{array}{c}\text{Cost of}\\\text{merchandise}\end{array}\right)$$

$x = (5.5\%) \cdot (\$20,000)$ Substitute values into the sales tax formula.

$x = (0.055)(\$20,000)$ Convert the percent to its decimal form.

$x = \$1100$

The sales tax on the vehicle is $1100.

b. The total price is $20,000 + $1100 = $21,100.

example 2   Computing a Sales Tax Rate

Lindsay has just moved and must buy a new refrigerator for her home. The refrigerator costs $1200 and the sales tax is $48. Because she is new to the area, she does not know the sales tax rate. Use these figures to compute the tax rate.

Skill Practice

2. A DVD sells for $15. The sales tax is $0.90. What is the tax rate?

Solution:

Let x represent the sales tax rate.   Label the unknown.

Cost of merchandise = $1200   Identify the parts of the formula.

Amount of sales tax = $48

$$\left(\begin{array}{c}\text{Amount of}\\\text{sales tax}\end{array}\right) = \left(\begin{array}{c}\text{Tax}\\\text{rate}\end{array}\right) \cdot \left(\begin{array}{c}\text{Cost of}\\\text{merchandise}\end{array}\right)$$

$$48 \quad = \quad x \quad \cdot \quad (1200)$$ Substitute values into the sales tax formula.

$$48 = 1200x$$

$$\frac{48}{1200} = \frac{1200x}{1200}$$ Divide both sides by 1200.

$$0.04 = x$$

The question asks for the tax rate which is given in percent form.

$$x = 0.4$$

$$= 0.04 \times 100\%$$

$$= 4\%$$

The sales tax rate is 4%.

example 3   Finding Purchase Price

The tax on a new CD comes to $1.05. If the tax rate is 6%, find the cost of the CD.

Skill Practice

3. Sales tax on a new lawn mower is $21.50. If the tax rate is 5%, compute the price of the lawn mower before tax.

Solution:

Let x represent the cost of the CD.   Label the unknown.

Tax rate = 6%   Identify the parts of the formula.

Amount of tax = $1.05

$$\left(\begin{array}{c}\text{Amount of}\\\text{sales tax}\end{array}\right) = \left(\begin{array}{c}\text{Tax}\\\text{rate}\end{array}\right) \cdot \left(\begin{array}{c}\text{Cost of}\\\text{merchandise}\end{array}\right)$$

$$1.05 \quad = \quad (0.06) \quad \cdot \quad x$$ Notice that we immediately converted 6% to its percent form so that we would not forget.

$$1.05 = 0.06x$$

$$\frac{1.05}{0.06} = \frac{0.06x}{0.06}$$ Divide both sides by 0.06.

$$17.5 = x$$

The CD costs $17.50.

Answers

2. The tax rate is 6%.
3. The mower costs $430 before tax.

2. Applications Involving Commission

Salespeople often receive all or part of their salary in commission. **Commission** is a form of income based on a percent of sales.

Commission Formula

$$\begin{pmatrix} \text{Amount of} \\ \text{commission} \end{pmatrix} = \begin{pmatrix} \text{Commission} \\ \text{rate} \end{pmatrix} \cdot \begin{pmatrix} \text{Total} \\ \text{sales} \end{pmatrix}$$

In this formula, the commission rate is usually given as a percent.

Skill Practice

4. At a store that sells fine menswear, the sales people make 15% commission on sales. If one customer purchased $158 in merchandise, how much commission did the salesperson make?

example 4   Computing Commission

At a fine furniture store the salespeople make a 20% commission on the sale of merchandise. If a dining room set sells for $849, how much is the sales-person's commission?

Solution:

Let x represent the amount of commission.   Label the unknown.

Commission rate = 20%   Identify the parts of the
                                                     formula.
Total sales = $849

$$\begin{pmatrix} \text{Amount of} \\ \text{commission} \end{pmatrix} = \begin{pmatrix} \text{Commission} \\ \text{rate} \end{pmatrix} \cdot \begin{pmatrix} \text{Total} \\ \text{sales} \end{pmatrix}$$
$$\quad\quad x \quad\quad = \quad (0.20) \quad \cdot \quad (\$849)$$

Substitute values into the commission formula. Convert 20% to decimal form 0.20.

$$x = (0.20)(\$849)$$
$$\quad = \$169.80$$

The salesperson will make $169.80 in commission.

Skill Practice

5. Trevor sold a home for $160,000 and earned a $6400 commission. What is his commission rate?

example 5   Finding Commission Rate

Alexis works in real estate sales.

a. If she sells a $150,000 house and earns a commission of $10,500, what is her commission rate?

b. At this rate, how much will she earn by selling a $200,000 house?

Solution:

a. Let x represent the commission rate.   Label the unknown.

Total sales = $150,000   Identify the parts of the formula.

Amount of commission = $10,500

Answers

4. The salesperson made $23.70.
5. His commission rate is 4%.

$$\begin{pmatrix} \text{Amount of} \\ \text{commission} \end{pmatrix} = \begin{pmatrix} \text{Commission} \\ \text{rate} \end{pmatrix} \cdot \begin{pmatrix} \text{Total} \\ \text{sales} \end{pmatrix}$$

$$10,500 \quad = \quad x \quad \cdot (150,000)$$

Substitute values into the commission formula.

$$10,500 = 150,000x$$

$$\frac{10,500}{150,000} = \frac{\cancel{150,000}x}{\cancel{150,000}}$$

Divide both sides by 150,000.

$$0.07 = x$$

$$x = 0.07 \times 100\%$$

Convert to percent form.

$$= 7\%$$

The commission rate is 7%.

b. The commission on a $200,000 house is given by

Amount of commission = (0.07)($200,000) = $14,000

Alexis will earn $14,000 by selling a $200,000 house.

---

example 6   **Finding Sales Base**

Tonya is a real estate agent. She makes $10,000 as her annual base salary for the work she does in the office. In addition, she makes 8% commission on her total sales. If her salary for the year amounts to $106,000, what were her total sales?

Solution:

First note that her commission is her total salary minus the $10,000 for working in the office. Thus,

Amount of commission = $106,000 − $10,000 = $96,000

Let x represent Tonya's total sales.   Label the unknown.

Amount of commission = $96,000   Identify the parts of the formula.

Commission rate = 8%

$$\begin{pmatrix} \text{Amount of} \\ \text{commission} \end{pmatrix} = \begin{pmatrix} \text{Commission} \\ \text{rate} \end{pmatrix} \cdot \begin{pmatrix} \text{Total} \\ \text{sales} \end{pmatrix}$$

$$96,000 \quad = \quad (0.08) \quad \cdot \quad x$$

Substitute values into the commission formula.

$$96,000 = 0.08x$$

$$\frac{96,000}{0.08} = \frac{\cancel{0.08}x}{\cancel{0.08}}$$

Divide both sides by 0.08.

$$1,200,000 = x$$

Tonya's sales totaled $1,200,000 ($1.2 million).

Skill Practice

6. A sales rep for a pharmaceutical firm makes $50,000 as his base salary. In addition, he makes 6% commission on sales. If his salary for the year amounts to $98,000, what were his total sales?

Answer

6. He made $800,000 in sales.

3. Applications Involving Discount

When we go to the store, we often find items discounted or on sale. For example, a printer might be discounted 20%, or a blouse might be on sale for 30% off. We compute the amount of the **discount** (the savings) as follows.

> **Discount Formulas**
>
> $$\begin{pmatrix}\text{Amount of} \\ \text{discount}\end{pmatrix} = \begin{pmatrix}\text{Discount} \\ \text{rate}\end{pmatrix} \cdot \begin{pmatrix}\text{Original} \\ \text{price}\end{pmatrix}$$
>
> Sale price = Original price − Amount of discount

Skill Practice

7. Consider the advertisement for a new patio chair.

Patio Chair was $129

On Sale Now
30% OFF!

a. Find the amount of the discount.

b. Find the sale price.

example 7   Computing Discount

a. Find the amount of discount for the item shown in Figure 6-6.

b. Find the sale price.

Solution:

a. Let x represent the amount of discount.

Discount rate = 20%

Original price = $92.95

$$\begin{pmatrix}\text{Amount of} \\ \text{discount}\end{pmatrix} = \begin{pmatrix}\text{Discount} \\ \text{rate}\end{pmatrix} \cdot \begin{pmatrix}\text{Original} \\ \text{price}\end{pmatrix}$$
$$x \quad\;\; = \quad (0.20) \quad\cdot\quad (\$92.95)$$

$$= \$18.59$$

The discount is $18.59.

b. Sale price = Original price − Amount of discount

$$= \$92.95 - \$18.59$$

$$= \$74.36$$

The sale price is $74.36.

originally $92.95
ON SALE − 20% off!

Figure 6-6

Skill Practice

8. Find the discount rate.

Adventure Kayak, $600

On Sale Now, $480

example 8   Computing Discount Rate

A gold chain originally priced $500 is marked down to $375. What is the discount rate?

Solution:

First note that the amount of the discount is given by

Discount = Original price − Sale price

$$= \$500 - \$375$$

$$= \$125$$

Answers

7. a. The amount of the discount is $38.70.
   b. The sale price is $90.30.
8. The kayak is discounted 20%.

Let x represent the discount rate.   Label the unknown.

Original price = \$500   Identify the parts of the formula.

Amount of discount = \$125

$$\begin{pmatrix} \text{Amount of} \\ \text{discount} \end{pmatrix} = \begin{pmatrix} \text{Discount} \\ \text{rate} \end{pmatrix} \cdot \begin{pmatrix} \text{Original} \\ \text{price} \end{pmatrix}$$

$$125 \quad = \quad x \quad \cdot \quad 500 \qquad \text{Substitute values into discount formula.}$$

$$125 = 500x$$

$$\frac{125}{500} = \frac{500x}{500} \qquad \text{Divide both sides by 500.}$$

$$0.25 = x$$

Converting $x = 0.25$ to percent form, we have $x = 25\%$. The chain has been discounted 25%.

---

4. Applications Involving Markup

Retailers often buy goods from manufacturers or wholesalers. To make a profit, the retailer must increase the cost of the merchandise before reselling it. This is called **markup**.

> ### Markup Formulas
>
> $$\begin{pmatrix} \text{Amount of} \\ \text{markup} \end{pmatrix} = \begin{pmatrix} \text{Markup} \\ \text{rate} \end{pmatrix} \cdot \begin{pmatrix} \text{Original} \\ \text{price} \end{pmatrix}$$
>
> Retail price = Original price + Amount of markup

---

example 9   **Computing Markup**

A college bookstore marks up the price of books 40%.

  a. What is the markup for a math text that has a manufacturer price of \$66?

  b. What is the retail price of the book?

  c. If there is a 6% sales tax, how much will the book cost to take home?

Solution:

  a. Let x represent the amount of markup.   Label the unknown.

  Markup rate = 40%   Identify parts of the formula.

  Original price = \$66

  $$\begin{pmatrix} \text{Amount of} \\ \text{markup} \end{pmatrix} = \begin{pmatrix} \text{Markup} \\ \text{rate} \end{pmatrix} \cdot \begin{pmatrix} \text{Original} \\ \text{price} \end{pmatrix}$$

  $$x \quad = \quad (0.40) \quad \cdot \quad (\$66) \qquad \text{Use the decimal form of 40\%.}$$

  $$x = (0.40)(\$66)$$

  $$= \$26.40$$

  The amount of markup is \$26.40.

b. Retail price = Original price + Markup

$$= \$66 + \$26.40$$

$$= \$92.40$$

The retail price is $92.40.

c. Next we must find the amount of the sales tax. This value is added to the cost of the book. The sales tax rate is 6%.

$$\begin{pmatrix} \text{Amount of} \\ \text{sales tax} \end{pmatrix} = \begin{pmatrix} \text{Tax} \\ \text{rate} \end{pmatrix} \cdot \begin{pmatrix} \text{Cost of} \\ \text{merchandise} \end{pmatrix}$$

$$\text{Tax} \quad = (0.06) \cdot \quad (\$92.40)$$

$$\approx \$5.54 \qquad\qquad \text{Round the tax to the nearest cent.}$$

The total cost of the book is $92.40 + $5.54 = $97.94.

Skill Practice

10. A car dealership buys a Honda Accord from the manufacturer for $20,000 and sells the car for $22,400. What is the markup rate?

example 10   Computing Markup Rate

Suppose a car dealership buys a new car from the car manufacturer for $18,000 and sells the car for $19,800. What is the markup rate?

Solution:

The amount of markup is $19,800 − $18,000 = $1800.

Let x represent the markup rate.   Label the unknown.

Original price = $18,000   Identify parts of the formula.

Amount of markup = $1800

$$\begin{pmatrix} \text{Amount of} \\ \text{markup} \end{pmatrix} = \begin{pmatrix} \text{Markup} \\ \text{rate} \end{pmatrix} \cdot \begin{pmatrix} \text{Original} \\ \text{price} \end{pmatrix}$$

$$1800 \quad = \quad x \quad \cdot \ (18,000)$$

$$1800 = 18,000x$$

$$\frac{1800}{18,000} = \frac{18,000p}{18,000} \qquad\qquad \text{Divide both sides by 18,000.}$$

$$0.10 = x$$

Converting x to percent form, we have $x = 10\%$. The dealer markup is 10%.

It is important to note the similarities in the formulas presented in this section. To find the amount of sales tax, commission, discount, or markup, we multiply a rate (percent) by some original amount.

Answer

10. The markup rate is 12%.

Summary Formulas for Sales Tax, Commission, Discount, and Markup

$$\text{Amount} = \underline{\text{Rate} \times \text{Original amount}}$$

Sales tax: $\begin{pmatrix}\text{Amount of}\\\text{sales tax}\end{pmatrix} = \begin{pmatrix}\text{Tax}\\\text{rate}\end{pmatrix} \cdot \begin{pmatrix}\text{Cost of}\\\text{merchandise}\end{pmatrix}$

Commission: $\begin{pmatrix}\text{Amount of}\\\text{commission}\end{pmatrix} = \begin{pmatrix}\text{Commission}\\\text{rate}\end{pmatrix} \cdot \begin{pmatrix}\text{Total}\\\text{sales}\end{pmatrix}$

Discount: $\begin{pmatrix}\text{Amount of}\\\text{discount}\end{pmatrix} = \begin{pmatrix}\text{Discount}\\\text{rate}\end{pmatrix} \cdot \begin{pmatrix}\text{Original}\\\text{price}\end{pmatrix}$

Markup: $\begin{pmatrix}\text{Amount of}\\\text{markup}\end{pmatrix} = \begin{pmatrix}\text{Markup}\\\text{rate}\end{pmatrix} \cdot \begin{pmatrix}\text{Original}\\\text{price}\end{pmatrix}$

section 6.5   Practice Exercises

Boost *your* GRADE at
mathzone.com!

- Practice Problems
- Self-Tests
- NetTutor
- e-Professors
- Videos

Study Skills Exercises

1. Which of the following strategies can help you study for a test? Check all that apply.

☐ Read the Chapter Summary. The Chapter Summary for this chapter is on page ___.

☐ Do the Review Exercises at the end of the chapter. The Review Exercises for this chapter are on page ___.

☐ Do the Chapter Test at the end of the chapter. The Chapter Test for this chapter is on page ___.

2. Define the key terms.

   a. Sales tax   **b. Commission**   **c. Discount**   **d. Markup**

Review Exercises

For Exercises 3–5, find the answer mentally.

3. What is 15% of 80?   **4.** 20 is what percent of 60?

5. 14 is 50% of what number?

For Exercises 6–11, solve the percent problem by using either method from Sections 6.3 and 6.4.

6. 52 is 0.2% of what number?   **7.** What is 225% of 36?

8. 6 is what percent of 25?   **9.** 18 is 75% of what number?

10. What is 1.6% of 550?   **11.** 32.2 is what percent of 28?

Objective 1: Applications Involving Sales Tax

For Exercises 12–18, complete the table.

	Cost of Merchandise	Sales Tax Rate	Amount of Tax	Total Cost
12.	$ 20.00	5%		
13.	$ 56.00	6%		
14.	$ 12.50		$ 0.50	
15.	$212.00		$14.84	
16.		2%	$ 2.75	
17.	$ 55.00			$ 58.30
18.	$214.00			$220.42

19. A new coat costs $68.25. If the sales tax rate is 5%, what is the total bill? **(See Example 1.)**

20. Sales tax in Wisconsin is 4.5%. Compute the amount of tax on a new personal CD player that sells for $64.

21. The sales tax on a set of luggage is $16.80. If the luggage cost before tax is $240.00, what is the sales tax rate? **(See Example 2.)**

22. A new shirt is labeled at $42.00. Jon purchased the shirt and paid $44.10.

   a. How much was the sales tax?

   b. What is the sales tax rate? (Round to the nearest whole percent.)

23. The 6% sales tax on a fruit basket came to $2.67. What is the price of the fruit basket? **(See Example 3.)**

24. The sales tax on a bag of groceries came to $1.50. If the sales tax rate is 6%, what was the price of the groceries before tax?

Objective 2: Applications Involving Commission

For Exercises 25–30, complete the table.

	Total Sales	Commission Rate	Amount of Commission
25.	$ 20,000.00	5%	
26.	$ 540.00	16%	
27.	$125,000.00		$10,000.00
28.	$ 800.00		$ 24.00
29.		10%	$ 540.00
30.		15%	$ 159.00

31. Zach works in an insurance office. He receives a commission of 7% on new policies. How much did he make last month in commission if he sold $48,000 in new policies? **(See Example 4.)**

32. Marisa makes a commission of 15% on sales over $400. One day she sells $750 worth of merchandise.

    a. How much over $400 did Marisa sell?

    b. How much did she make in commission that day?

33. In one week, Rodney sold $2000.00 worth of sports equipment. He received $300.00 in commission. What is his commission rate? **(See Example 5.)**

34. A realtor sold a townhouse for $95,000. If he received a commission of $7600, what is his commission rate?

35. A realtor makes an annual salary of $25,000 plus a 3% commission on sales. If a realtor's salary is $67,000, what was the amount of her sales? **(See Example 6.)**

36. A salesperson receives a 5.5% commission on every car she sells. Her last commission was for $990.00. For how much did the car sell?

37. Jeff works as a pharmaceutical representative. He receives a 6% monthly commission on sales up to $60,000. He receives an 8.5% commission on all sales above $60,000. If he sold $86,000 worth of his product one month, how much did he receive in commission?

Objective 3: Applications Involving Discount

For Exercises 38–45, complete the table.

	Original Price	Discount Rate	Amount of Discount	Sale Price
38.	$ 56.00	20%		
39.	$175.00	15%		
40.	$900.00			$600.00
41.	$900.00			$630.00
42.			$ 8.50	$ 76.50
43.			$ 33.00	$ 77.00
44.		50%	$ 38.00	
45.		40%	$ 23.36	

46. Hospital employees get a 15% discount at the hospital cafeteria. If the lunch bill originally comes to $5.60, what is the price after the discount?

47. A health club membership costs $550 for one year. If a member pays up front in a lump sum, the member will receive a 10% discount. **(See Example 7.)**

    a. How much money is discounted?

    b. How much will the yearly membership cost with the discount?

48. A bathing suit is on sale for $45. If the regular price is $60, what is the discount rate?

49. A printer that sells for $229 is on sale for $183.20. What is the discount rate? (See Example 8.)

Explorer 4-person tent
On Sale 30% **OFF**

Was $269

50. Find the discount and the sale price of the tent in the given advertisement.

51. A set of dishes had an original price of $112. Then it was discounted 50%. A week later, the new sale price was discounted another 50%. At that time, was the set of dishes free? Explain why or why not.

52. Find the discount and the sale price of the bike in the given advertisement.

53. Find the discount and the discount rate of the chair from the given advertisement.

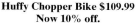

Huffy Chopper Bike $109.99
Now 10% off.

Accent Chair was $235.00
and is now $188.00

54. Find the discount and the discount rate of the watch from the given advertisement. Round the discount rate to the nearest tenth of a percent.

Prices Slashed!

$70.00
$50.00

Objective 4: Applications Involving Markup

For Exercises 55–62, complete the table.

	Original Price	Markup Rate	Amount of Markup	Retail Price
55.	$ 92.00	5%		
56.	$ 25.00	10%		
57.	$110.00			$118.80
58.	$ 50.00			$ 57.50
59.			$ 97.50	$422.50
60.			$175.00	$875.00
61.		20%	$ 9.00	
62.		18%	$ 31.50	

63. A business suit has a wholesale price of $150.00. A department store's markup rate is 18%. (See Example 9.)

    a. What is the markup for this suit?

    b. What is the retail price?

    c. If Antonio buys this suit including a 7% sales tax, how much will he pay?

64. An import/export business marks up imported merchandise by 110%. If a wicker chair imported from Singapore originally costs $84 from the manufacturer in Singapore, what is the markup price?

65. A table is purchased from the manufacturer for $300 and is sold retail at $375. What is the markup rate? **(See Example 10.)**

66. A $60 hairdryer is sold for $69. What is the markup rate?

67. A campus bookstore adds $43.20 to the cost of a science text. If the final cost is $123.20, what is the markup rate?

68. The retail price of a golf club is $420.00. If the golf store has marked up the price by $70, what is the markup rate?

---

section 6.6   Percent Increase and Decrease

1. Definition of Percent Increase and Decrease

Objectives

1. Definition of Percent Increase and Decrease
2. Computing Percent Increase
3. Computing Percent Decrease

Two important applications of percents are finding percent increase and percent decrease. For example,

- The price of gas increased 40% in 4 years.
- After taking a new drug for 3 months, a patient's cholesterol decreased by 35%.

When we compute **percent increase** or **percent decrease**, we are comparing the *change* between two given amounts to the *original amount*. The change (amount of increase or decrease) is found by subtraction. To compute the percent increase or decrease, we use the following formulas.

Computing Percent Increase or Percent Decrease

$$\left(\begin{array}{c}\text{Percent}\\\text{increase}\end{array}\right) = \left(\frac{\text{Amount of increase}}{\text{Original amount}}\right) \times 100\%$$

$$\left(\begin{array}{c}\text{Percent}\\\text{decrease}\end{array}\right) = \left(\frac{\text{Amount of decrease}}{\text{Original amount}}\right) \times 100\%$$

Concept Connections

1. Write the formula for percent increase.
2. Write the formula for percent decrease.

The formulas for percent increase and percent decrease are derived from the standard percent equation.

$$(\text{Amount of increase}) = (\text{Rate of increase})(\text{Original amount})$$

We can divide both sides of the equation by original amount:

$$\frac{\text{Amount of increase}}{\text{Original amount}} = \text{Rate of increase}$$

Answers

1. $\left(\begin{array}{c}\text{Percent}\\\text{increase}\end{array}\right) = \left(\frac{\text{Amount of increase}}{\text{Original amount}}\right) \times 100\%$

2. $\left(\begin{array}{c}\text{Percent}\\\text{decrease}\end{array}\right) = \left(\frac{\text{Amount of decrease}}{\text{Original amount}}\right) \times 100\%$

The rate of increase is a rate given in decimal form. To convert it to a percent, we multiply both sides by 100%.

$$\left(\frac{\text{Amount of increase}}{\text{Original amount}}\right) \times 100\% = \text{Percent increase}$$

2. Computing Percent Increase

In Example 1, we apply the percent increase formula.

3. After a raise, Denisha's salary increased from $42,500 to $44,200. What is the percent increase?

example 1   Computing Percent Increase

The price of gas climbed from an average of $1.50 per gallon to $2.10 per gallon over a 4-year period. Compute the percent increase.

Solution:

The original price was $1.50 per gallon.   Identify parts of the formula.

The final price after the increase is $2.10.

The amount of increase is given by subtraction.

Amount of increase = $2.10 − $1.50

                = $0.60   There was a $0.60 increase in price.

$$\left(\begin{array}{c}\text{Percent} \\ \text{increase}\end{array}\right) = \left(\frac{\text{Amount of increase}}{\text{Original amount}}\right) \times 100\%$$

$$= \frac{0.60}{1.50} \times 100\%$$ Apply the percent increase formula.

$$= 0.40 \times 100\%$$

$$= 40\%$$

There was a 40% increase in the price of gas.

Example 1 could also have been solved by using a percent proportion.

$$\frac{p}{100} = \frac{\text{amount of increase}}{\text{original amount}}$$

$$\frac{p}{100} = \frac{0.60}{1.50}$$ Substitute values into the proportion.

$$1.50p = (0.60)(100)$$ Equate the cross products.

$$1.50p = 60$$

$$\frac{1.50p}{1.50} = \frac{60}{1.50}$$ Divide both sides by 1.50

$$p = 40$$ There was a 40% increase in the price of gas.

Using a percent proportion applies knowledge you already have. However, it is not as efficient as simply applying the formula for percent increase.

Finding Percent Increase in an Application

The graph in Figure 6-7 depicts the number of sport utility vehicles sold (in millions) for selected years. Compute the percent increase between 1990 and 2000. Round to the nearest tenth of a percent.

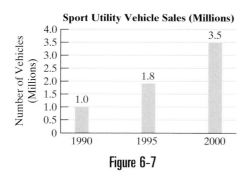

Figure 6-7

Solution:

Original amount = 1.0   Identify the parts of the

Final value after increase = 3.5   formula.

Amount of increase = 3.5 − 1.0 = 2.5

$$\left(\begin{array}{c}\text{Percent}\\\text{increase}\end{array}\right) = \left(\frac{\text{Amount of increase}}{\text{Original amount}}\right) \times 100\%$$

$$= \frac{2.5}{1.0} \times 100\% \qquad \text{Apply the formula.}$$

$$= 2.5 \times 100\%$$

$$= 250\%$$

The number of sport utility vehicles increased by 250% between 1990 and 2000.

3. Computing Percent Decrease

Example 3 demonstrates the use of the percent decrease formula.

Finding Percent Decrease in an Application

Amy's cholesterol dropped from 260 to 169 after she took a new experimental drug. Find the percent decrease.

Solution:

Original cholesterol level = 260   Identify parts of the formula.

Cholesterol level after decrease = 169

Amount of decrease = 260 − 169 = 91

Skill Practice

4. The number of cellular phone subscriptions in the United States for selected years is shown in the graph. Find the percent increase in subscriptions between 1996 and 2002. Round to the nearest percent.

Tip: The solution to Example 2 seems reasonable because the amount of increase was $2\frac{1}{2}$ times the original sales.

Skill Practice

5. After following a sensible diet and exercise program, Veronica's body weight dropped from 185 to 148 lb. What was her percent decrease in body weight?

Answers

4. There was a 225% increase in the number of subscriptions between 1996 and 2002.
5. Veronica lost 20% of her body weight.

$$\binom{\text{Percent}}{\text{decrease}} = \left(\frac{\text{Amount of decrease}}{\text{Original amount}} \right) \times 100\%$$

$$= \left(\frac{91}{260} \right) \times 100\%$$

$$= 0.35 \times 100\%$$

$$= 35\%$$

Amy's cholesterol decreased by 35%.

section 6.6   Practice Exercises

**Boost *your* GRADE at
mathzone.com!**

MathZone

• Practice Problems   • e-Professors
• Self-Tests   • Videos
• NetTutor

Study Skills Exercises

1. When you are taking a test, which of the following should you do? Check all that apply.

☐ Answer each question.   ☐ Show all work.

☐ Write neatly.   ☐ Include your scratch paper with your test.

☐ Check your work if time allows.   ☐ Correct your test after you get it back so that you do not make the same errors again.

2. Define the key terms.

a. Percent increase   **b. Percent decrease**

Review Exercises

3. The price of a sports coat is $65.

a. If Chris buys the sports coat today, what will the final price be after he adds a 5% sales tax?

b. Tomorrow the coat goes on sale for 20% off. How much would he spend tomorrow (including the 5% tax)?

c. How much will Chris save by waiting until tomorrow to buy the coat?

4. The price of a jogging suit is $32.50.

a. If Kira buys the jogging suit today, what will the final price be after a 6% sales tax is added?

b. Tomorrow the suit goes on sale for 25% off. How much would she spend tomorrow (including the 6% tax)?

c. How much will Kira save by waiting until tomorrow to buy the suit?

5. Katie earns a commission of 14% on all merchandise that she sells over $200. What is her commission if she sells a total of $425 of merchandise?

6. Sean earns a salary of $12 per hour and a commission of 3% on all sales. If Sean works a 30-hr week and sells $1290.00 of merchandise, what is his salary for that week?

7. Explain how to change a decimal into a percent.

For Exercises 8–11, change the decimal to a percent.

8. 0.23   **9.** 0.05   **10.** 0.88   **11.** 0.12

Objective 1: Definition of Percent Increase and Decrease

12. a. To find the amount of increase in the price of a product, should you subtract the original price from the increased price or the increased price from the original price?

b. To find the amount of decrease in the price of a product, should you subtract the original price from the decreased price or the decreased price from the original price?

For Exercises 13–20, (**a**) identify if there is an increase or decrease and (**b**) find the amount of increase or decrease.

13. 48 to 59   **14.** 78 to 123   **15.** 145 to 135   **16.** 190 to 109

17. 654 to 645   **18.** 24 to 42   **19.** 67 to 79   **20.** 205 to 105

Objective 2: Computing Percent Increase

21. Select the percent increase of a price that is double the original amount. For example, a book that originally cost $30 now costs $60.

    a. 200%   **b.** 2%   **c.** 100%   **d.** 150%

22. Select the percent increase of a price that is greater by $\frac{1}{2}$ of the original amount. For example, an employee made $20 per hour and now makes $30 per hour.

    a. 150%   **b.** 50%   **c.** $\frac{1}{2}$%   **d.** 200%

23. The U.S. government classified 8 million documents as secret in 2001. By 2003, this number had increased to 14 million. What is the percent increase? (Source: *Time*, April 12, 2004.) **(See Example 1.)**

24. The price of coal in 2002 was $27 per ton. In 2004, it was $57 per ton. What was the percent increase? Round to the nearest whole percent.

25. The price of steel in 2002 was $240 per ton. By 2004, the price had risen to $420 per ton. What was the percent increase?

26. The price of soybeans in 2002 was $4.50 per bushel. By 2004, the price had increased to $9.54 per bushel. What was the percent increase?

Objective 3: Computing Percent Decrease

27. Select the percent decrease of a price that is one-half of the original amount. For example, a bathing suit that originally cost $62 now costs $31.

a. 50%   **b.** $\frac{1}{2}$%   **c.** 100%   **d.** 5%

28. Select the percent decrease of a price that is one-quarter of the original amount. For example, a T-shirt that originally cost $40 now costs $10.

a. $\frac{1}{4}$%   **b.** 25%   **c.** $\frac{3}{4}$%   **d.** 75%

29. A stock closed at $12.60 per share on Monday. By Friday, the closing price was $11.97 per share. What was the percent decrease? **(See Example 3.)**

30. During a 5-year period, the number of participants collecting food stamps went from 27 million to 17 million. What is the percent decrease? Round to the nearest whole percent. (Source: U.S. Department of Agriculture.)

31. Gus, the cat, originally weighed 12 lb. He was diagnosed with a thyroid disorder, and Dr. Smith the veterinarian found that his weight had decreased to 10.2 lb. What percent of his body weight did Gus lose?

32. To lose weight, Kelly reduced her Calories from 3000 per day to 1800 per day. What is the percent decrease in Calories?

Calculator Connections

Topic: Applications of percent increase and percent decrease

Calculator Exercises

The figure represents the cost of tuition and fees for public four-year colleges in the United States for selected years. Use this information for Exercises 33–34. In each case, round to the nearest whole percent.

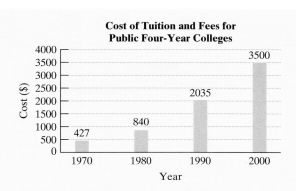

33. What was the percent increase in tuition and fees between 1970 and 1980? **(See Example 2.)**

34. What was the percent increase in tuition and fees between 1990 and 2000?

35. Stella's salary last year was $35,000. She got a raise and now makes $36,400 per year. What is the percent increase in her salary?

36. Julie's bill for cable television rose from $45.80 one year to $52.67 the next year. What is the percent increase?

The chart represents the population of Louisville, Kentucky, for certain years. Use this information for Exercises 37–38. In each case, round to the nearest whole percent.

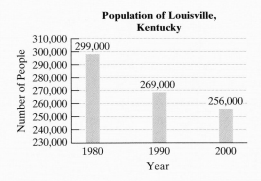

37. What is the percent decrease in population of Louisville, Kentucky, from 1980 to 1990?

38. What is the percent decrease in population of Louisville, Kentucky, from 1990 to 2000?

39. In 2001, Americans stayed in the hospital an average of 5 days. In 1970, the average length of stay was 8 days. Compute the percent decrease in the length of hospital stays from 1970 to 2001. (Source: National Center for Health Statistics.)

40. The number of savings banks went from 1533 in 2001 to 1467 in 2002. What is the percent decrease? Round to the nearest whole percent. (Source: Statistics on Banking.)

For Exercises 41–48, round the percent increase or decrease to the nearest tenth of a percent.

	Country	Population in 2000 (Millions)	Population in 2005 (Millions)	Change (Millions)	Percent Increase or Decrease
41.	Mexico	100.3	110.8		
42.	France	60.0	61.4		
43.	Bulgaria	8.15	8.11		
44.	Trinidad	1.09	1.075		

	Item	Value in 1995	Value in 2000	Change	Percent Increase or Decrease
45.	Number of unemployed people in the U.S.	7.4 million	5.6 million		
46.	Total pounds of fish caught commercially in the U.S.	10.5 billion lb	9.1 billion lb		
47.	U.S. federal debt	$4.9 trillion	$5.7 trillion		
48.	Number of motor vehicle deaths in the U.S.	43,400	42,500		

Objectives

1. Simple Interest
2. Compound Interest
3. Using the Compound Interest Formula (Optional)

section 6.7   Simple and Compound Interest

1. Simple Interest

In this section we use percents to compute simple and compound interest on an investment or a loan.

Banks hold large quantities of money for their customers. They keep some cash for day-to-day transactions, but invest the remaining portion of the money. As a result, banks often pay customers interest.

When making an investment, **simple interest** is the money that is earned on principal (**principal** is the original amount of money invested). That is, simple interest is a percent of the base amount (principal) that a person puts in the bank. When people take out a loan, the amount borrowed is the principal. The interest is a percent of the amount borrowed that you must pay back in addition to the principal.

The following formulas can be used to compute simple interest for an investment or a loan and to compute the total amount in the account.

Simple Interest Formulas

Simple interest $=$ Principal $\times$ Rate $\times$ Time $I = Prt$

Total amount $=$ Principal $+$ Interest $A = P + I$

where
$I =$ amount of interest
$P =$ amount of principal
$r =$ annual interest rate (in decimal form)
$t =$ time (years)
$A =$ total amount in an account

example 1   Computing Simple Interest

Suppose $2000 is invested in an account that earns 7% simple interest.

a. How much interest is earned after 3 years?

b. What is the total value of the account after 3 years?

Solution:

a. Principal: $P = \$2000$ Identify the parts of the formula.

Annual interest rate: $r = 7\%$

Time (in years): $t = 3$

Let I represent the amount of interest.   Label the unknown.

$I = Prt$

$= (\$2000)(7\%)(3)$ Substitute values into the formula.

$= \underbrace{(2000)(0.07)}(3)$ Convert 7% to decimal form.

$= \quad 140(3)$ Apply the order of operations.

$= 420$ Multiply from left to right.

The amount of interest earned is $420.

b. The total amount in the account is given by

$A = P + I$

$= \$2000 + \$420$

$= \$2420$

The total amount in the account is $2420.

> **Avoiding Mistakes:** It is important to use the decimal form of the interest rate when calculating interest.

When applying the simple interest formula, it is important that time be expressed in years. This is demonstrated in Example 2.

example 2   Computing Simple Interest

Clyde takes out a loan for $3500. He pays simple interest at a rate of 6% for 4 years 3 months.

a. How much money does he pay in interest?

b. How much total money must he pay to pay off the loan?

Solution:

a. $P = \$3500$ Identify parts of the formula.

    $r = 6\%$

    $t = 4\frac{1}{4}$ years or 4.25 years   3 months $= \frac{3}{12}$ year $= \frac{1}{4}$ year

  $I = Prt$

    $= (\$3500)(0.06)(4.25)$ Substitute values into the interest formula. Convert 6% to decimal form 0.06.

    $= \$892.50$ Multiply.

The interest earned is $892.50.

b. To find the total amount that must be paid, we have

  $A = P + I$

    $= \$3500 + \$892.50$

    $= \$4392.50$

The total amount that must be paid on the loan is $4392.50.

2. Compound Interest

Simple interest is based only on a percent of the original principal. However, many day-to-day applications involve compound interest. **Compound interest** is based on both the original principal and the interest earned.

To compare the difference between simple and compound interest, consider this scenario in Example 3.

example 3   Comparing Simple Interest and Compound Interest

Suppose $1000 is invested at 8% interest for 3 years.

a. Compute the total amount in the account after 3 years, using simple interest.

b. Compute the total amount in the account after 3 years of compounding interest annually.

Solution:

a. $P = \$1000$

   $r = 8\%$

   $t = 3$ years

  $I = Prt$

    $= (\$1000)(0.08)(3)$

    $= \$240$

The amount of simple interest earned is $240. The total amount in the account is $1000 + $240 = $1240.

b. To compute interest compounded annually over a period of 3 years, compute the interest earned in the first year. Then add the principal plus the interest earned in the first year. This value then becomes the principal on which to base the interest earned in the second year. We repeat this process, finding the interest for the second and third years based on the principal and interest earned in the preceding years. This process is outlined in a table.

Year	Interest Earned $I = Prt$	Total Amount in Account
First year	$I = (\$1000)(0.08)(1) = \$80$	$\$1000 + \$80 = \$1080$
Second year	$I = (\$1080)(0.08)(1) = \$86.40$	$\$1080 + \$86.40 = \$1166.40$
Third year	$I = (\$1166.40)(0.08)(1) = \$93.31$	$\$1166.40 + 93.31 = \mathbf{\$1259.71}$

The total amount in the account by compounding interest annually is $1259.71.

Skill Practice

7. Suppose $2000 is invested at 5% interest. Complete the table to compute the total amount in the account after 3 years if the interest is compounded annually.

Year	Interest	Total
1		
2		
3		

Notice that in Example 3 the final amount in the account is greater for the situation where interest is compounded. The difference is $1259.71 − $1240 = $19.71. By compounding interest we earn more money.

Interest may be compounded more than once per year.

Annually	1 time per year
Semiannually	2 times per year
Quarterly	4 times per year
Monthly	12 times per year
Daily	365 times per year

To compute compound interest, the calculations become very tedious. Banks use computers to perform the calculations quickly. You may want to use a calculator if calculators are allowed in your class.

Example 4 demonstrates interest compounded quarterly for $1\frac{1}{2}$ years.

example 4   Computing Compound Interest

Sadie invests $8000 in an account that earns 5% interest compounded quarterly. Compute the total amount in the account after $1\frac{1}{2}$ years.

Solution:

Complete the table. Note that each compound period is one-quarter of a year. Therefore, the value $t = 0.25$ years is used in the formula $I = Prt$. Also remember to use the decimal form of the interest rate, 0.05.

Answer

7.
Year	Interest	Total
1	$100	$2100
2	$105	$2205
3	$110.25	$2315.25

Skill Practice

8. Suppose Mason invests $10,000 in an account that earns 8% interest compounded semiannually. Complete the table to compute the total amount in the account after 2 years.

Period	Interest	Total
1		
2		
3		
4		

Compound Period	Interest Earned $I = Prt$	Total Amount in Account
1st quarter (first year)	$I = (\$8000)(0.05)(0.25) = \$100$	$\$8000 + \$100 = \$8100$
2nd quarter (first year)	$I = (\$8100)(0.05)(0.25) = \$101.25$	$\$8100 + \$101.25 = \$8201.25$
3rd quarter (first year)	$I = (\$8201.25)(0.05)(0.25) = \$102.52$	$\$8201.25 + \$102.52 = \$8303.77$
4th quarter (first year)	$I = (\$8303.77)(0.05)(0.25) = \$103.80$	$\$8303.77 + \$103.80 = \$8407.57$
1st quarter (second year)	$I = (\$8407.57)(0.05)(0.25) = \$105.09$	$\$8407.57 + \$105.09 = \$8512.66$
(2nd quarter (second year)	$I = (\$8512.66)(0.05)(0.25) = \$106.41$	$\$8512.66 + \$106.41 = \$8619.07$

The total amount of money in the account after $1\frac{1}{2}$ years is $8619.07.

3. Using the Compound Interest Formula (Optional)

As you can see from Examples 3 and 4, computing compound interest by hand is a cumbersome process. Can you imagine computing daily compound interest (365 times a year) by hand!

We now use a formula to compute compound interest. This formula requires the use of a scientific or graphing calculator. In particular, the calculator must have an exponent key $\boxed{y^x}$, $\boxed{x^y}$, or $\boxed{\char`\^}$.

Let A = total amount in an account

P = principal

r = annual interest rate

t = time, years

n = number of compounding periods per year

Then $A = P \cdot \left(1 + \dfrac{r}{n}\right)^{n \cdot t}$ computes the total amount in an account.

To use this formula, note the following guidelines:

- Rate r must be expressed in decimal form.
- Time t must be the total time of the investment in *years*.
- Number n is determined by the number of compounding periods:
  Annual $n = 1$
  Semiannual $n = 2$
  Quarterly $n = 4$
  Monthly $n = 12$
  Daily $n = 365$

Answer

8.

Period	Interest	Total
1	$400	$10,400
2	$416	$10,816
3	$432.64	$11,248.64
4	$449.95	$11,698.59

example 5 Computing Compound Interest by Using the Compound Interest Formula

Suppose $1000 is invested at 8% interest compounded annually for 3 years. Use the compound interest formula to find the total amount in the account after 3 years. Compare the result to the answer from Example 3(b).

Solution:

$P = \$1000$ .... Identify the parts of the formula.

$r = 8\%$ (0.08 in decimal form)

$t = 3$ years

$n = 1$ (annual compound interest is compounded 1 time per year)

$A = P \cdot \left(1 + \dfrac{r}{n}\right)^{n \cdot t}$

$\quad = 1000 \cdot \left(1 + \dfrac{0.08}{1}\right)^{1 \cdot 3}$ ...... Substitute values into the formula.

$\quad = 1000(1 + 0.08)^3$ ...... Apply the order of operations. Divide within parentheses. Simplify the exponent.

$\quad = 1000(1.08)^3$ .... Add within parentheses.

$\quad = 1000(1.259712)$ .... Evaluate $(1.08)^3 = (1.08)(1.08)(1.08) = 1.259712$.

$\quad = 1259.712$ .... Multiply.

$\quad \approx 1259.71$ .... Round to the nearest cent.

The total amount in the account after 3 years is $1259.71. This is the same value obtained in Example 3(b).

Skill Practice

9. Suppose $10,000 is invested at 8% interest compounded semiannually for 2 years. Use the formula.

$A = P \cdot \left(1 + \dfrac{r}{n}\right)^{n \cdot t}$

to find the total amount after 2 years. Compare the answer to margin exercise 8.

Calculator Connections

Topic: Using a calculator to compute compound interest

To enter the expression from Example 5 into a calculator, follow these keystrokes.

Expression	Keystrokes	Result
$1000 \cdot \left(1 + \dfrac{0.08}{1}\right)^{1 \cdot 3}$	1000 ⊠ ⟮ 1 ⊞ 0.08 ÷ 1 ⟯	
	∧ ⟮ 1 ⊠ 3 ⟯ Enter	1259.712

Some calculators may use the y^x key or x^y key.

Some calculators use the $=$ key.

Note: It is mandatory to insert parentheses ( ) around the product in the exponent.

---

example 6   Computing Compound Interest by Using the Compound Interest Formula

Suppose $8000 is invested in an account that earns 5% interest compounded quarterly for $1\frac{1}{2}$ years. Use the compound interest formula to compute the total amount in the account after $1\frac{1}{2}$ years. Compare the result to the answer from Example 4.

Answer

9. $11,698.59; this is the same as the result in margin exercise 8.

Skill Practice

10. Suppose $5000 is invested at 9% interest compounded monthly for 30 years. Use the formula for compound interest to find the total amount in the account after 30 years.

Solution:

$P = \$8000$ Identify the parts of the formula.

$r = 5\%$ (0.05 in decimal form)

$t = 1.5$ years

$n = 4$ (quarterly interest is compounded 4 times per year)

$$A = P \cdot \left(1 + \frac{r}{n}\right)^{n \cdot t}$$

$$= 8000 \cdot \left(1 + \frac{0.05}{4}\right)^{(4)(1.5)} \quad \text{Substitute values into the formula.}$$

$$= 8000(1 + 0.0125)^6 \quad \text{Apply the order of operations. Divide within parentheses. Simplify the exponent.}$$

$$= 8000(1.0125)^6 \quad \text{Add within parentheses.}$$

$$\approx 8000(1.077383181) \quad \text{Evaluate } (1.0125)^6. \text{ If your teacher allows the use of a calculator, consider using the exponent key.}$$

$$\approx 8619.07 \quad \text{Multiply and round to the nearest cent.}$$

The total amount in the account after $1\frac{1}{2}$ years is \$8619.07. This is the same value as in Example 4.

Calculator Connections

Topic: Using a calculator to compute compound interest

To enter the expression from Example 6 into a calculator, follow these keystrokes.

Expression	Keystrokes	Result
$8000 \cdot \left(1 + \dfrac{0.05}{4}\right)^{(4)(1.5)}$	8000 $\boxed{\times}$ $\boxed{(}$ 1 $\boxed{+}$ 0.05 $\boxed{\div}$ 4 $\boxed{)}$	
	$\boxed{\wedge}$ $\boxed{(}$ 4 $\boxed{\times}$ 1.5 $\boxed{)}$ $\boxed{\text{Enter}}$	8619.065444

Answer

10. The account is worth \$73,652.88 after 30 years.

---

section 6.7   Practice Exercises

Boost *your* GRADE at mathzone.com!

MathZone

- Practice Problems
- Self-Tests
- NetTutor
- e-Professors
- Videos

Study Skills Exercises

1. Instructors vary in what they emphasize on tests. For example, test material may come from the textbook, notes, handouts, homework, etc. What do you find that your instructor emphasizes?

2. Define the key terms.

    a. Simple interest   **b. Principal**   **c. Compound interest**

Review Exercises

For Exercises 3–7, find the percent increase or the percent decrease. Round to the nearest whole percent.

	U.S. National Parks	Visitors in 2000 (Thousands)	Visitors in 2002 (Thousands)	Change	Percent Increase or Decrease
3.	Bryce Canyon, UT	1099	886		
4.	Petrified Forest, AZ	605	572		
5.	Great Basin, NV	81	86		
6.	Kings Canyon, CA	529	545		
7.	Dry Tortugas, FL	84	80		

Objective 1: Simple Interest

For Exercises 8–14, find the simple interest and the total amount including interest. **(See Examples 1 and 2.)**

	Principal	Annual Interest Rate	Time, Years	Interest	Total Amount
8.	$6,000	5%	3	_____	_____
9.	$4,000	3%	2	_____	_____
10.	$5,050	6%	4	_____	_____
11.	$4,800	4%	3	_____	_____
12.	$12,000	4%	$4\frac{1}{2}$	_____	_____
13.	$6,230	7%	$6\frac{1}{3}$	_____	_____
14.	$9,220	8%	4	_____	_____

15. Dale deposited $2500 in an account that pays $3\frac{1}{2}$% simple interest for 4 years.

    a. How much interest will he earn in 4 years?

    b. What will be the total value of the account after 4 years?

16. Charlene invested $3400 at 4% simple interest for 5 years.

    a. How much interest will she earn in 5 years?

    b. What will be the total value of the account after 5 years?

17. Gloria borrowed $400 for 18 months at 8% simple interest.

    a. How much interest will Gloria have to pay?

    b. What will be the total amount that she has to pay back?

18. Floyd borrowed $1000 for 2 years 3 months at 8% simple interest.

    a. How much interest will Floyd have to pay?

    b. What will be the total amount that he has to pay back?

19. Jozef deposited $10,300 into an account paying 4% simple interest 5 years ago. If he withdraws the entire amount of money, how much will he have?

20. Heather invested $20,000 in an account that pays 6% simple interest. If she invests the money for 10 years, how much will she have?

21. Anne borrowed $4500 from a bank that charges 10% simple interest. If she repays the loan in $2\frac{1}{2}$ years, how much will she have to pay back?

22. Dan borrowed $750 from his brother who is charging 8% simple interest. If Dan pays his brother back in 6 months, how much does he have to pay back?

Objective 2: Compound Interest

23. Mary Ellen deposited $500 in a bank that has 4% interest compounded annually. Complete the table to determine the amount in her account after 3 years. **(See Example 3.)**

Year	Interest	Total
1		
2		
3		

24. Fatima invested $12,000 in an account that pays 5% interest compounded annually. Complete the table to determine the amount of her investment at the end of 3 years.

Year	Interest	Total
1		
2		
3		

25. If a bank compounds interest semiannually for 3 years, how many total compound periods are there?

26. If a bank compounds interest quarterly for 2 years, how many total compound periods are there?

27. If a bank compounds interest monthly for 2 years, how many total compound periods are there?

28. If a bank compounds interest monthly for $1\frac{1}{2}$ years, how many total compound periods are there?

29. Diana borrowed $6000 with interest compounded semiannually at 6%. If she pays off the loan in 1 year, find the total amount including interest. How much does she pay in interest? **(See Example 4.)**

30. A bank offers 4% interest compounded quarterly. If Jake invests $4000 in the account for 1 year, what will be his total amount at the end of that year? What is the amount of interest earned?

31. The amount of $8000 is invested at 4% for 3 years.

   a. Compute the ending balance if the bank calculates simple interest.

   b. Compute the ending balance if the bank calculates interest compounded annually.

   c. How much more interest is earned in the account with compound interest?

32. The amount of $12,000 is invested at 8% for 1 year.

   a. Compute the ending balance if the bank calculates simple interest.

   b. Compute the ending balance if the bank calculates interest compounded quarterly.

   c. How much more interest is earned in the account with compound interest?

Objective 3: Using the Compound Interest Formula (Optional)

33. For the formula $A = P \cdot \left(1 + \dfrac{r}{n}\right)^{n \cdot t}$, identify what each variable means.

34. If $1000 is deposited in an account paying 8% interest compounded monthly for 3 years, label the following variables: P, r, n, and t.

Calculator Connections

Topic: Computing compound interest

Calculator Exercises

For Exercises 35–42, find the total amount for the investment, using compound interest. **(See Examples 5–6.)**

	Principal	Annual Interest Rate	Time, Years	Compounded	Total Amount
35.	$5,000	4.5%	5	Annually	_____
36.	$12,000	5.25%	4	Annually	_____
37.	$6,000	5%	2	Semiannually	_____
38.	$4,000	3%	3	Semiannually	_____
39.	$10,000	6%	$1\frac{1}{2}$	Quarterly	_____
40.	$9,000	4%	2	Quarterly	_____
41.	$14,000	4.5%	3	Monthly	_____
42.	$9,000	8%	2	Monthly	_____

chapter 6 | summary

section 6.1 Percents and Their Fraction and Decimal Forms

Key Concepts

The word **percent** means *per one hundred.*

Converting Percents to Fractions

1. Replace the % symbol by $\times \frac{1}{100}$ (or by $\div 100$).
2. Simplify the fraction to lowest terms, if possible.

Converting Percents to Decimals

Replace the % symbol by $\times 0.01$.
(This is equivalent to $\times \frac{1}{100}$ and $\div 100$).

Note: Multiplying a decimal by 0.01 is the same as moving the decimal point 2 places to the left.

Examples

Example 1

40% means 40 per 100 or $\frac{40}{100}$.

Example 2

$84\% = 84 \times \frac{1}{100} = \frac{84}{100} = \frac{21}{25}$

Example 3

$33\frac{1}{3}\% = 33\frac{1}{3} \times \frac{1}{100} = \frac{100}{3} \times \frac{1}{100} = \frac{1}{3}$

Example 4

$24.5\% = 24.5 \times 0.01 = 0.245$

Example 5

$0.07\% = 0.07 \times 0.01 = 0.0007$
(Move the decimal point 2 places to the left.)

section 6.2 Fractions and Decimals and Their Percent Forms

Key Concepts

Converting Fractions and Decimals to Percent Form

Multiply the fraction or decimal by 100%.
(100% = 1.)

Examples

Example 1

$\frac{1}{5} = \frac{1}{5} \times 100\% = \frac{100}{5}\% = 20\%$

Example 2

$1.14 = 1.14 \times 100\% = 114\%$

Example 3

$\frac{2}{3} = 0.\overline{6} \times 100\% = 66.\overline{6}\%$ or $66\frac{2}{3}\%$

section 6.3 Percent Proportions and Applications

Key Concepts

A **percent proportion** is a proportion that equates a percent to an equivalent ratio.

A percent proportion can be written in the form

$$\frac{\text{Amount}}{\text{Base}} = p\% \quad \text{or} \quad \frac{\text{Amount}}{\text{Base}} = \frac{p}{100}$$

The **base** is the total or whole amount being considered. The **amount** is the part being compared to the base.

To solve a percent proportion, equate the cross products and divide by the factor with the variable. The variable can represent the amount, base, or p. Examples 4–6 demonstrate each type of percent problem.

Example 4

Of a sample of 400 people, 85% found relief using a particular pain reliever. How many people found relief?

Solve the proportion: $\dfrac{85}{100} = \dfrac{x}{400}$

$$85 \cdot 400 = 100x$$

$$\frac{34,000}{100} = \frac{100x}{100}$$

$$340 = x$$

340 people found relief.

Examples

Example 1

$\dfrac{36}{100} = \dfrac{9}{25}$ is a percent proportion.

Example 2

For the percent proportion $\dfrac{6}{100} = \dfrac{12}{200}$,
12 is the amount, 200 is the base, and $p = 6$.

Example 3

For the statement

75% of 84 employees is 63 employees,
63 is the amount, 84 is the base, and $p = 75$.

Example 5

44% of what number is 275?

Solve the proportion: $\dfrac{44}{100} = \dfrac{275}{x}$

$$44x = 275 \cdot 100$$

$$\frac{44x}{44} = \frac{27,500}{44}$$

$$x = 625$$

Example 6

There are approximately 750,000 career employees in the U.S. Postal Service. If 60,000 are mail handlers, what percent does this represent?

Solve the proportion: $\dfrac{p}{100} = \dfrac{60,000}{750,000}$

$$750,000p = 60,000 \cdot 100$$

$$\frac{750,000p}{750,000} = \frac{6,000,000}{750,000}$$

$$p = 8$$

Of postal employees 8% are mail handlers.

section 6.4 Percent Equations and Applications

Key Concepts

A **percent equation** represents a percent proportion in an alternative form:

Amount = $(p\%) \cdot$ (base)

Examples 1–3 demonstrate three types of percent problems.

Examples

Example 1

Of all breast cancer cases, 99% occur in women. Out of 2700 cases of breast cancer reported, how many are expected to occur in women?

Solve the equation: $x = (99\%)(2700)$

$$x = (0.99)(2700)$$

$$x = 2673$$

About 2673 cases are expected to occur in women.

Example 2

There are 599 endangered plants in the U.S. This represents 60.7% of the total number of endangered species. Find the total number of endangered species. Round to the nearest whole number.

Solve the equation: $599 = 0.607x$

$$\frac{599}{0.607} = \frac{0.607x}{0.607}$$

$$987 \approx x$$

There is a total of approximately 987 endangered species in the U.S.

Example 3

Of the car repairs performed on a certain day, 21 were repairs on transmissions. If 60 cars were repaired, what percent involved transmissions?

Solve the equation: $21 = 60x$

$$\frac{21}{60} = \frac{60x}{60}$$

$$0.35 = x$$

Because the problem asks for a percent, we have $x = 0.35$

$$= 0.35 \times 100\%$$

$$= 35\%$$

section 6.5 Applications Involving Sales Tax, Commission, Discount, and Markup

Key Concepts

To find **sales tax**, use the formula

$$\begin{pmatrix} \text{Amount of} \\ \text{sales tax} \end{pmatrix} = \begin{pmatrix} \text{Tax} \\ \text{rate} \end{pmatrix} \cdot \begin{pmatrix} \text{Cost of} \\ \text{merchandise} \end{pmatrix}$$

To find a **commission**, use the formula

$$\begin{pmatrix} \text{Amount of} \\ \text{commission} \end{pmatrix} = \begin{pmatrix} \text{Commission} \\ \text{rate} \end{pmatrix} \cdot \begin{pmatrix} \text{Total} \\ \text{sales} \end{pmatrix}$$

To find **discount** and sale price, use the formulas

$$\begin{pmatrix} \text{Amount of} \\ \text{discount} \end{pmatrix} = \begin{pmatrix} \text{Discount} \\ \text{rate} \end{pmatrix} \cdot \begin{pmatrix} \text{Original} \\ \text{price} \end{pmatrix}$$

Sale price = Original price − Amount of discount

To find **markup** and retail price, use the formulas

$$\begin{pmatrix} \text{Amount of} \\ \text{markup} \end{pmatrix} = \begin{pmatrix} \text{Markup} \\ \text{rate} \end{pmatrix} \cdot \begin{pmatrix} \text{Original} \\ \text{price} \end{pmatrix}$$

Retail price = Original price + Amount of markup

Examples

Example 1

A DVD is priced at $16.50, and the total amount paid is $17.82. To find the sales tax rate, first find the amount of tax.

$17.82 − $16.50 = $1.32

To compute the sales tax rate, solve:

$$1.32 = x \cdot 16.50$$

$$\frac{1.32}{16.50} = \frac{x \cdot \cancel{16.50}}{\cancel{16.50}}$$

$$0.08 = x$$

The sales tax rate is 8%.

Example 2

Fletcher makes 13% commission on the sale of all merchandise. If he sells $11,290 worth of merchandise, find how much Fletcher will earn.

$$x = (0.13)(11{,}290)$$

$$= 1467.7$$

Fletcher will earn $1467.70 in commission.

Example 3

Margaret found a ring that was originally $425 but is on sale for 30% off. To find the sale price, first find the amount of discount.

$$a = (0.30) \cdot (425)$$

$$= 127.5$$

The sale price is $425 − $127.50 = $297.50.

Example 4

A wholesale coat company marks up its coats 20% before selling them retail. To find the retail price of a $340 coat, first find the amount of markup.

$$a = (0.20) \cdot (340)$$

$$= 68$$

The retail price is $340 + $68 = $408.

section 6.6 Percent Increase and Decrease

Key Concepts

Percent increase or **percent decrease** compares the *change* between two given amounts to the *original amount.*

<u>Computing Percent Increase or Decrease</u>

$$\left(\begin{array}{l}\text{Percent}\\\text{increase}\end{array}\right) = \left(\frac{\text{Amount of increase}}{\text{Original amount}}\right) \times 100\%$$

$$\left(\begin{array}{l}\text{Percent}\\\text{decrease}\end{array}\right) = \left(\frac{\text{Amount of decrease}}{\text{Original amount}}\right) \times 100\%$$

Examples

Example 1

A child grows from 35 to 42 in. She increases $42 - 35 = 7$ in. The percent increase is

$$\frac{7}{35} \times 100\% = 0.20 \times 100\% = 20\%.$$

Example 2

Stephanie's weight went from 130 to 125 lb. Her weight decreased by 130 lb − 125 lb = 5 lb.

The percent decrease is

$$\frac{5}{130} \times 100\% \approx 0.038 \times 100\% = 3.8\%.$$

section 6.7 Simple and Compound Interest

Key Concepts

To find the **simple interest** made on a certain **principal**, use the formula $I = P \cdot r \cdot t$

where I = amount of interest

P = amount of principal

r = annual interest rate

t = time, years

The formula for the total amount in an account is $A = P + I$, where A = total amount in an account.

Many day-to-day applications involve compound interest. **Compound interest** is based on both the original principal and the interest earned.

Interest may be compounded more than once per year.

Annually   1 time per year
Semiannually   2 times per year
Quarterly   4 times per year
Monthly   12 times per year
Daily   365 times per year

The formula $A = P \cdot \left(1 + \dfrac{r}{n}\right)^{n \cdot t}$ computes the total amount in an account that uses compound interest

where A = total amount in an account

P = principal

r = annual interest rate

t = time (years)

n = number of compounding periods per year

Examples

Example 1

Betsey deposited $2200 in her account which pays 5.5% simple interest. To find the interest she will earn after 4 years, use the formula $I = Prt$ and solve for I.

$I = (2200)(0.055)(4)$

$= 484$

She will earn $484 interest.

To find the balance or total amount of her account, apply the formula $A = P + I$.

$A = 2200 + 484$

$= 2684$

Betsey's balance will be $2684.

Example 2

Betsey deposited $2200 in an account that paid 5.5% interest compounded annually for 3 years. Fill out the table to find the total amount in the account after 3 years.

Compound Interest	Interest Earned $I = Prt$	Total Amount $A = I + P$
Year 1	$121.00	$2321.00
Year 2	127.66	2448.66
Year 3	134.68	2583.34

Example 3

Gene borrows $1000 at 6% interest compounded semiannually. If he pays off the loan in 3 years, how much will he have to pay?

We are given $P = 1000$, $r = 0.06$, $n = 2$ (semiannually means twice a year), and $t = 3$.

$A = 1000\left(1 + \dfrac{0.06}{2}\right)^{2 \cdot 3}$

$= 1000(1.03)^6$

≈ 1194.05

Gene will have to pay $1194.05 to pay off the loan with interest.

chapter 6 | review exercises

Section 6.1

For Exercises 1–4, use a percent to express the shaded portion of each drawing.

1.   **2.**

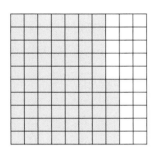

3.   **4.**

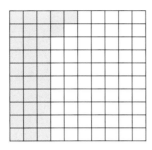

5. 68% can be expressed as which of the following forms? Circle all that apply.

  a. $\dfrac{68}{1000}$ **b.** $\dfrac{68}{100}$

  c. 0.68   **d.** 0.068

6. 0.4% can be expressed as which of the following forms? Circle all that apply.

  a. $\dfrac{4}{100}$ **b.** 0.04

  c. $\dfrac{0.4}{100}$ **d.** 0.004

For Exercises 7–12, match the percent with its fraction form.

7. 30%   **a.** $\dfrac{33}{100}$

8. $33\frac{1}{3}\%$ **b.** $\dfrac{1}{2}$

9. 33%   **c.** $\dfrac{2}{3}$

10. 50%   **d.** $\dfrac{1}{3}$

11. $66\frac{2}{3}\%$ **e.** $\dfrac{3}{5}$

12. 60%   **f.** $\dfrac{3}{10}$

For Exercises 13–18, match the percent with its decimal form.

13. 7.5%   **a.** 1

14. 75%   **b.** 0.25

15. 50%   **c.** 0.75

16. 100%   **d.** 0.0025

17. 0.25%   **e.** 0.075

18. 25%   **f.** 0.5

For Exercises 19–20, write the percent as a fraction and as a decimal.

19. Out of all the phone calls received per week, 42% were from solicitors.

20. In an hour-long TV show, 20% of the time is devoted to commercials.

The graph represents the average annual population growth rate for four of the fastest-growing cities. Use this graph for Exercises 21–24.

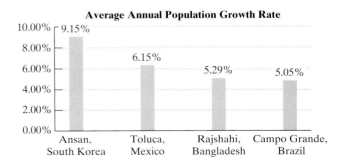

21. Write the rate for Toluca, Mexico, in decimal form.

22. Write the rate for Rajshahi, Bangladesh, in decimal form.

23. Write the rate for Ansan, South Korea, in fraction form.

24. Write the rate for Campo Grande, Brazil, in fraction form.

Section 6.2

For Exercises 25–28, convert the fraction to a percent.

25. $\dfrac{17}{100}$ **26.** $\dfrac{21}{50}$

27. $\dfrac{4}{5}$ **28.** $\dfrac{7}{4}$

For Exercises 29–32, convert the decimal to a percent.

29. 0.12   **30.** 1.1

31. 0.005   **32.** 0.4

For Exercises 33–36, write the fraction as a percent.

33. In a classroom the ratio of girls to boys is $\frac{14}{16}$.

34. In 2002, $\frac{11}{20}$ of the U.S. population were Internet users.

35. In 2000, $\frac{1}{5}$ of the families had one child under the age of 18. (Source: U.S. Bureau of the Census.)

36. In 2002, $\frac{1}{10}$ of the female population was widowed. (Source: U.S. Bureau of the Census.)

For Exercises 37–42, complete the table.

	Fraction	Decimal	Percent
37.	$\frac{9}{20}$		
38.		1	
39.			6%
40.		1.2	
41.	$\frac{9}{1000}$		
42.			75%

Section 6.3

For Exercises 43–46, identify the amount, base, and p values for the percent proportion $\dfrac{p}{100} = \dfrac{\text{Amount}}{\text{Base}}$.

43. 45% of \$150 is \$67.50.

44. 360 births is 12% of 3000 births.

45. 30.24 m^2 of 144 m^2 is 21%.

46. 106% of 30 gal is 31.8 gal.

For Exercises 47–50, write the percent proportion.

47. 6 books of 8 books is 75%

48. 15% of 180 lb is 27 lb.

49. 200% of \$420 is \$840.

50. 6 pines of 2000 trees is 0.3%.

For Exercises 51–56, solve the percent problems, using proportions.

51. What is 12% of 50?

52. $5\frac{3}{4}$% of 64 is what number?

53. 11 is what percent of 88?

54. 8 is what percent of 2500?

55. 13 is $33\frac{1}{3}$% of what number?

56. 24 is 120% of what number?

57. Based on recent statistics, one airline expects that 4.2% of its customers will be "no-shows." If the airline sold 260 seats, how many people would the airline expect as no-shows? Round to the nearest whole unit.

58. In a survey of college students, 58% said that they wore their seatbelts regularly. If this represents 493 people, how many people were surveyed?

59. Victoria spends $720 per month on rent. If her monthly take-home pay is $1800, what percent does she pay in rent? Round to the nearest whole percent.

60. Of the rental cars at the U-Rent-It company, 40% are compact cars. If this represents 26 cars, how many cars are on the lot?

Section 6.4

For Exercises 61–66, write as a percent equation and solve.

61. 18% of 900 is what number?

62. What number is 29% of 404?

63. 18.90 is what percent of 63?

64. What percent of 250 is 86?

65. 30 is 25% of what number?

66. 26 is 130% of what number?

67. A student buys a used book for $54.40. This is 80% of the original price. What was the original price? Round to the nearest cent.

68. Veronica has read 330 pages of a 600-page novel. What percent of the novel has she read?

69. Elaine tries to keep her fat intake to no more than 30% of her total calories. If she has a 2400-calorie diet, how many fat calories can she consume to stay within her goal?

70. In 1900, 4% of Americans were over 65 years old. By 1993, 13% of Americans were over 65. Suppose the U.S. population was 76,000,000 in 1900 and 258,000,000 in 1993.

  a. Find the number of Americans over 65 in the year 1900.

  b. Find the number of Americans over 65 in the year 1993.

Section 6.5

For Exercises 71–74, solve the problem involving sales tax.

71. A Sony TV costs $238. Find the sales tax if the rate is 6%.

72. The sales tax on sofa is $47.95. If the sofa costs $685.00 before tax, what is the sales tax rate?

73. To get a roll of film developed, it costs $6.75. The total bill came to $7.29.

  a. How much is the sales tax?

  b. What is the sale tax rate?

74. A resort hotel charges an 11% resort tax along with the 6% sales tax. If the hotel's one-night accommodation is $75.00, what will a tourist pay for 4 nights, including tax?

For Exercises 75–78, solve the problems involving commission.

75. At a recent auction, *Boy with a Pipe*, an early work by Pablo Picasso, sold for $104 million. The commission for the sale of the work was $11 million. What was the rate of commission? Round to the nearest tenth of a percent. (Source: *The New York Times.*)

76. Andre earns a commission of 12% on sales of restaurant supplies. If he sells $4075 in one week, how much commission will he earn?

77. Sela sells sportswear at a department store. She earns an hourly wage of $8, and she gets a 5% commission on all merchandise that she sells over $200. If Sela works an 8-hr day and sells $420 of merchandise, how much will she earn that day?

78. A house is sold for $160,000, and the real estate agent earned $9600 in commission. What is the commission rate?

For Exercises 79–82, solve the problems involving discount and markup.

79. Find the discount and the sale price of the movie if the regular price is $28.95.

80. This notebook computer cost $1747 originally. How much is the discount? After the $50 rebate, how much will a person pay for this computer?

81. A rug manufacturer sells a rug to a retail store for $160. The store then marks up the rug to $208. What is the markup rate?

82. Peg sold some homemade baskets to a store for $50 each. The store marks up all merchandise by 18%. What will be the retail price of the baskets after the markup?

Section 6.6

For Exercises 83–84, (a) identify if there is an increase or decrease and (b) find the percent of increase or decrease.

83. 86 to 107.5   84. 410 to 82

85. The number of species of animals on the endangered species list went from 263 in 1990 to 388 in 2003. Find the percent increase. Round to the nearest tenth of a percent. (Source: U.S. Fish and Wildlife Services.)

86. The number of corded phones went from 30 million in 2000 to 22 million in 2003. Find the percent decrease. Round to the nearest whole percent. (Source: Telecommunications Association.)

The number of subscribers to cellular telephones has increased from 1985. Use the information in the table to answer Exercises 87–88. In each case, round to the nearest whole percent.

Year	Number of Subscribers (1000s)
1985	300
1989	3,500
1993	16,000
1997	55,000
2001	128,000

87. What was the percent increase in cellular phone subscribers between 1993 and 2001?

88. What was the percent increase in cellular phone subscribers between 1985 and 1989?

Section 6.7

For Exercises 89–90, find the simple interest and the total amount including interest.

	Principal	Annual Interest Rate	Time, Years	Interest	Total Amount
89.	$10,200	3%	4	_____	_____
90.	$7000	4%	5	_____	_____

91. Jean-Luc borrowed $2500 at 5% simple interest. What is the total amount that he will pay back at the end of 18 months (1.5 years)?

92. Win loaned his brother $800 for 2 years. He is charging 2.5% simple interest. How much will his brother owe Win in 2 years?

93. Sydney deposited $6000 in a certificate of deposit that pays 4% interest compounded annually. Complete the table to determine her balance after 3 years.

Year	Interest	Total
1		
2		
3		

94. Nell deposited $10,000 in a money market account that pays 3% interest compounded semiannually. Complete the table to find her balance after 2 years.

Compound Periods	Interest	Total
Period 1 (end of first 6 months)		
Period 2 (end of year 1)		
Period 3 (end of 18 months)		
Period 4 (end of year 2)		

For Exercises 95–98, find the total amount for the investment, using compound interest. Use the formula $A = P \cdot \left(1 + \dfrac{r}{n}\right)^{n \cdot t}$.

	Principal	Annual Interest Rate	Time in Years	Compounded	Total Amount
95.	$850	8%	2	Quarterly	_____
96.	$2050	5%	5	Semiannually	_____
97.	$11,000	7.5%	6	Annually	_____
98.	$8200	4.5%	4	Monthly	_____

chapter 6 | test

1. Write a percent to express the shaded portion of the figure.

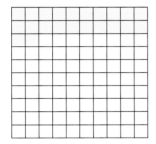

2. Shade the figure so that it represents 85%.

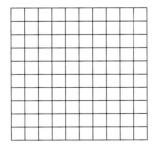

3. Write the percent in decimal form and in fraction form.

a. For a recent year, the unemployment rate of Illinois was 5.4%.

b. The incidence of breast cancer increased by 0.9% between 1992 and 2002.

c. For a certain city, gas prices increased by 170% in 10 years.

4. Write the following percents in fraction form.

a. 1%   **b.** 25%

c. $33\frac{1}{3}\%$ **d.** 50%

e. $66\frac{2}{3}\%$ **f.** 75%

g. 100%   **h.** 150%

For Exercises 5–6, write the percent as a decimal and as a fraction.

5. The incidence of colon cancer decreased by 0.6% between 1992 and 1999. (Source: National Cancer Institute.)

6. In 1950, 9.9% of the U.S. population was made up of African Americans.

7. Explain the process to write a fraction as a percent.

For Exercises 8–11, write the fraction as a percent. Round to the nearest tenth of a percent if necessary.

8. $\dfrac{3}{5}$ **9.** $\dfrac{1}{250}$

10. $\dfrac{7}{4}$ **11.** $\dfrac{5}{7}$

12. Explain the process to write a decimal as a percent.

For Exercises 13–16, write the decimal as a percent.

13. 0.32   **14.** 0.052

15. 1.3   **16.** 0.006

For Exercises 17–22, solve the percent problems.

17. What is 24% of 150?

18. What is 120% of 16?

19. 21 is 6% of what number?

20. 40% of what number is 80?

21. What percent of 220 is 198?

22. 75 is what percent of 150?

23. At McDonald's, a side salad without dressing has 10 mg of sodium. With a serving of Newman's Own Low-Fat Balsamic Dressing, the sodium content of the salad is 740 mg. (Source: www.mcdonalds.com)

    a. How much sodium is in the dressing itself?

    b. What percent of the sodium content in a side salad with dressing is from the dressing? Round to the nearest tenth of a percent.

The composition of the lower level of the earth's atmosphere is given in the figure (other gases are present in minute quantities). For Exercises 24–25, use the information in the graph.

Composition of the Earth's Atmosphere

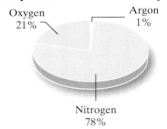

Oxygen 21%   Argon 1%   Nitrogen 78%

24. How much nitrogen would be expected in 500 m³ of atmosphere?

25. How much oxygen would be expected in 2000 m³ of atmosphere?

26. Brad bought a pair of blue jeans that cost $30.00. He wrote his check for $32.10.

    a. What is the amount of sales tax that he paid?

    b. What is the sales tax rate?

27. Charles earns a salary of $400 per week and gets a bonus of 6% commission on all merchandise that he sells. If Charles sells $3500 worth of merchandise, how much will he earn in that week?

28. Find the discount rate of the product in the advertisement.

Was $45.00
Now $18.00

End-of-Season Clearance! Quantities are limited so shop now!

29. The price of hogs in 2002 was $0.32 per pound. In 2004, the price rose to $0.44 per pound. What was the percent increase?

30. Maury borrowed $5000 at 8% simple interest. He plans to pay back the loan in 3 years.

    a. How much interest will he have to pay?

    b. What is the total amount that he has to pay back?

31. Dennis invested $20,000 in an account that pays 8% interest compounded semiannually. How much will he have at the end of 2 years?

32. Use the formula $A = P \cdot \left(1 + \dfrac{r}{n}\right)^{n \cdot t}$ to calculate the ending amount in an account that began with $25,000 invested at 4.5% compounded quarterly for 5 years.

---

chapters 1–6 | cumulative review

1. What is the name of the place value for the digit 6 in the number 26,009,235?

2. Fill in the table with either the word name for the number or the number in standard form.

Country		Area (mi²)	
	Standard Form	**Words**	
a. United States		Three million, five hundred thirty-nine thousand, two hundred forty-five	
b. Saudi Arabia	830,000		
c. Falkland Islands	4,700		
d. Colombia		Four hundred one thousand, forty-four	

3. Multiply: $\begin{array}{r} 34{,}882 \\ \times\ 100 \end{array}$

4. Divide: $9\overline{)783}$

5. Add: $234 + 44 + 6 + 2901$

6. Simplify: $\sqrt{16} - 6 \div 3 + 3^2$

7. Identify whether the fraction is proper or improper.

    a. $\dfrac{6}{6}$ **b.** $\dfrac{10}{7}$

    c. $\dfrac{7}{10}$ **d.** $\dfrac{1}{100}$

For Exercises 8–11, multiply or divide as indicated. Simplify the fraction to lowest terms.

8. $\dfrac{3}{8} \times \dfrac{32}{9}$

9. $\dfrac{42}{25} \div \dfrac{7}{100}$

10. $\dfrac{21}{2} \div 7$

11. $16 \times \dfrac{1}{24}$

12. Find the area of the figure.

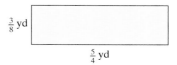

13. Find the perimeter of the figure.

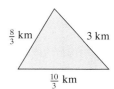

14. Add: $\dfrac{3}{10} + \dfrac{17}{100} + \dfrac{3}{1000}$

15. The value of the CBIN stock rose $\$\frac{7}{10}$ from $\$23\frac{3}{5}$. What is the current price?

16. A sheet of paper has the dimensions $13\frac{1}{2}$ in. by 17 in. What is the area of the paper?

17. a. List four multiples of 18.

b. List all factors of 18.

c. Write the prime factorization of 18.

18. Write a fraction that represents the shaded portion of each figure.

a.    **b.**

For Exercises 19–22, write the fraction as a decimal.

19. $\dfrac{3}{8}$

20. $\dfrac{4}{3}$

21. $\dfrac{7}{9}$

22. $\dfrac{3}{4}$

For Exercises 23–24, refer to the chart.

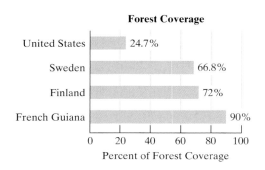

Forest Coverage

United States 24.7%
Sweden 66.8%
Finland 72%
French Guiana 90%

Percent of Forest Coverage

23. What is the difference in the percent of forest coverage in French Guiana and the United States?

24. What is the difference in the percent of forest coverage in Sweden and the United States?

For Exercises 25–28, multiply and divide by powers of 10.

25. 85×0.001

26. 85×100

27. $85 \div 10$

28. $85 \div 0.0001$

For Exercises 29–32, solve the proportions.

29. $\dfrac{3}{4} = \dfrac{15}{p}$

30. $\dfrac{2.5}{6} = \dfrac{p}{9}$

31. $\dfrac{4\frac{1}{3}}{p} = \dfrac{12}{18}$

32. $\dfrac{p}{100} = \dfrac{21.87}{81}$

33. If it takes $\frac{1}{2}$ hr to read 1 chapter of a book, how long will it take to read 5 chapters?

34. A 9.2-oz bag of candy costs $2.30. Find the unit cost, that is, the price per ounce.

35. A computer can download 1.6 megabytes (MB) in 2.5 min. How long will it take to download a 4.6 MB file? Round to the nearest tenth of a minute.

36. A DC-10 aircraft flew 1799 mi in 3.5 hr. Find the unit rate in miles per hour.

37. By 2008 it is projected that online grocery sales will reach $17.4 billion. In 2003, sales were only $3.7 billion. Find the percent increase. Round to the nearest whole percent. (Source: *USA TODAY.*)

38. New York City has the greatest population of any U.S. city. In 1900 it had 3.4 million people. By 2000, it had 8.0 million.

a. Compute the difference in population between the year 2000 and the year 1900.

b. Compute the rate representing the increase in population per year. (Source: U.S. Bureau of the Census.)

39. Kevin deposited $13,000 into a certificate of deposit that pays 3.2% simple interest. How much will Kevin have if he keeps the certificate for 5 years?

40. A mortgage charges 8% interest that is compounded monthly. Use the formula $A = P \cdot \left(1 + \dfrac{r}{n}\right)^{n \cdot t}$ to find the total amount of interest paid for a 10-year mortgage on $75,000.

Measurement

In this chapter we learn how to convert between different units of measurement. For example, what do each of these values have in common?

72 in.   2 yd   1.8288 m   188.88 cm   0.00114 mi

Each distance is approximately equal to 6 ft, the height of a tall man.

In this chapter, we present units of measurement in both the U.S. Customary System of measure and the metric system. In Exercise 64 in Section 7.3, we compare two Olympic speed skating races whose distances are 500 m and 5 km. See this exercise to determine which race is longer and by how much.

chapter 7   preview

The exercises in this chapter preview contain concepts that have not yet been presented. These exercises are provided for students who want to compare their levels of understanding before and after studying the chapter. Alternatively, you may prefer to work these exercises when the chapter is completed and before taking the exam.

For these exercises, refer to the table of conversions found in the back of the text.

Sections 7.1–7.2

For Exercises 1–6, convert the U.S. Customary units.

1. 6 yd = ____ ft

2. 0.5 yd = ____ in.

3. 200 sec = ____ min

4. 3 cups = ____ fl oz

5. 10 pt = ____ qt

6. 48 oz = ____ lb

7. Find the perimeter of the triangle. Give the answer in feet.

8. The Discovery Space Shuttle had a flight of 8 days 6 hr 30 min. What is the time in hours?

Sections 7.3–7.4

For Exercises 9–14, convert the metric units.

9. 5.903 kg = ____ g

10. 2260 mm = ____ m

11. 56 L = ____ cL

12. 3000 cc = ____ L

13. 2.5 dam = ____ dm

14. 0.45 kg = ____ cg

Section 7.5

For Exercises 15–20, convert the given units. Round to one decimal place.

15. 13 m ≈ ____ ft

16. 6 oz ≈ ____ g

17. 6 L ≈ ____ qt

18. 68 in. ≈ ____ cm

19. 10 kg ≈ ____ lb

20. 25 fl oz ≈ ____ mL

21. Convert 25°C to degrees Fahrenheit.

22. Convert 32°F to degrees Celsius.

Section 7.6 (Optional)

23. How much energy does it take to lift 1000 lb a distance of 5 ft?

For Exercises 24–25, convert the units of energy. Round to the nearest whole unit if necessary.

24. 190,000 ft·lb ≈ ____ Btu

25. 750 Btu ≈ ____ ft·lb

26. A swimmer burns 300 Cal/hr. How many Calories will she burn if she swims for 45 min 3 times this week?

27. A portable generator has 10 hp. Convert 10 hp to foot-pounds per second.

28. Find the power in raising 30 lb a distance of 4 ft in 15 sec.

section 7.1   Converting U.S. Customary Units of Length

1. U.S. Customary Units of Length

In many applications in day-to-day life, we need to measure things. To measure an object means to assign it a number and a **unit of measure**. We first present common units of measure for length which include inches (in.), feet (ft), yards (yd), and miles (mi). These units are part of the **U.S. Customary System of measurement** (sometimes called the English system of measurement).

The U.S. Customary units of length and some common equivalents are given in Table 7-1.

table 7-1

U.S. Customary Units of Length and Their Equivalents	
1 ft = 12 in.	1 in. = $\frac{1}{12}$ ft
1 yd = 3 ft	1 ft = $\frac{1}{3}$ yd
1 mi = 5280 ft	1 ft = $\frac{1}{5280}$ mi
1 mi = 1760 yd	1 yd = $\frac{1}{1760}$ mi

Note that sometimes units of feet are denoted with the ′ symbol. That is, 3 ft = 3′. Similarly, sometimes units of inches are denoted with the ″ symbol. That is, 4 in. = 4″.

2. Converting U.S. Customary Units of Length by Using Substitution

Examples 1 and 2 demonstrate how we can use substitution to convert between two units of length.

example 1   Converting Units of Length

a. 4 ft = _____ in.   **b.** 1.5 mi = _____ ft

Solution:

a. First note that 4 ft = 4 × 1 ft. Then we can substitute 1 ft = 12 in.

$$4 \text{ ft} = 4 \times 1 \text{ ft}$$
$$= 4 \times 12 \text{ in.} \qquad \text{Substitute } 1 \text{ ft} = 12 \text{ in.}$$
$$= 48 \text{ in.}$$

b. $1.5 \text{ mi} = 1.5 \times 1 \text{ mi}$
$$= 1.5 \times 5280 \text{ ft} \qquad \text{Substitute } 1 \text{ mi} = 5280 \text{ ft.}$$
$$= 7920 \text{ ft}$$

Skill Practice

Use substitution to convert units.

1. 6 ft = _____ in.
2. 3 mi = _____ yd

In Example 1 we converted feet to inches and miles to feet. In each case, we converted larger units of measure to smaller units of measure. In Example 2, we convert smaller units of measure to larger units of measure.

Answers
1. 72 in.
2. 5280 yd

Skill Practice

Use substitution to convert units.

3. 12.6 ft = _____ yd
4. 6160 yd = _____ mi

Tip: In Example 1(a) we converted feet to inches by multiplying by 12. In Example 2(a) we reversed this process to convert inches to feet. In this case we multiplied by $\frac{1}{12}$ which is the same as *dividing* by 12.

example 2   Converting Units of Length

a. 66 in. = _____ ft   **b.** 3960 ft = _____ yd

Solution:

a. 66 in. = 66 × 1 in.

$$= 66 \times \frac{1}{12} \text{ ft} \qquad \text{Substitute 1 in.} = \frac{1}{12} \text{ ft.}$$

$$= \frac{66}{12} \text{ ft} \qquad \text{Multiply.}$$

$$= \frac{11}{2} \text{ ft or } 5\frac{1}{2} \text{ ft or 5.5 ft}$$

b. 3960 ft = 3960 × 1 ft

$$= 3960 \times \frac{1}{3} \text{ yd} \qquad \text{Substitute 1 ft} = \frac{1}{3} \text{ yd.}$$

$$= \frac{3960}{3} \text{ yd} \qquad \text{Multiply.}$$

$$= 1320 \text{ yd} \qquad \text{Simplify.}$$

3. Converting U.S. Customary Units of Length by Using Unit Ratios

Examples 1 and 2 demonstrate how substitution can be used to convert between two units of measure. Another method is to multiply by a **conversion factor**. A conversion factor is a ratio of equivalent measures.

For example, note that 1 yd = 3 ft. Therefore, $\frac{1 \text{ yd}}{3 \text{ ft}} = 1$ and $\frac{3 \text{ ft}}{1 \text{ yd}} = 1$.

These ratios are called **unit ratios** (or **unit fractions**). In a unit ratio, the quotient is 1 because we are dividing measurements of equal length. To convert from one unit of measure to another, we can multiply by a unit ratio. We offer these guidelines to determine the proper unit ratio to use.

Concept Connections

Complete the unit ratio.

5. $\dfrac{1 \text{ ft}}{\text{in.}}$

6. $\dfrac{\text{yd}}{3 \text{ ft}}$

Choosing a Unit Ratio as a Conversion Factor

In a unit ratio,

- The unit of measure in the numerator should be the new unit you want to convert *to*.
- The unit of measure in the denominator should be the original unit you want to convert *from*.

example 3   Converting Units of Length by Using Unit Ratios

a. 1500 ft = _____ yd   **b.** 9240 yd = _____ mi   **c.** 8.2 mi = _____ ft

Answers

3. 4.2 yd   4. 3.5 mi
5. $\dfrac{1 \text{ ft}}{12 \text{ in.}}$ 6. $\dfrac{1 \text{ yd}}{3 \text{ ft}}$

Solution:

a. From Table 7-1, we have 1 yd = 3 ft.

$$1500 \text{ ft} = 1500 \text{ ft} \cdot \frac{1 \text{ yd}}{3 \text{ ft}} \quad \longleftarrow \text{ new unit to convert to} \\ \longleftarrow \text{ unit to convert from}$$

$$= \frac{1500 \text{ ft}}{1} \cdot \frac{1 \text{ yd}}{3 \text{ ft}}$$

Notice that the original units of ft reduce or "cancel" in much the same way as simplifying fractions. The unit yd remains in the final answer.

$$= \frac{1500}{3} \text{ yd}$$

$$= 500 \text{ yd}$$

b. From Table 7-1, we have 1 mi = 1760 yd.

$$9240 \text{ yd} = 9240 \text{ yd} \cdot \frac{1 \text{ mi}}{1760 \text{ yd}} \quad \longleftarrow \text{ new unit to convert to} \\ \longleftarrow \text{ unit to convert from}$$

$$= \frac{9240 \text{ yd}}{1} \cdot \frac{1 \text{ mi}}{1760 \text{ yd}}$$

The units yd cancel, leaving the answer in miles.

$$= \frac{9240}{1760} \text{ mi}$$

Multiply.

$$= 5.25 \text{ mi}$$

c. From Table 7-1 we have 1 mi = 5280 ft.

$$8.2 \text{ mi} = 8.2 \text{ mi} \cdot \frac{5280 \text{ ft}}{1 \text{ mi}} \quad \longleftarrow \text{ new unit to convert to} \\ \longleftarrow \text{ unit to convert from}$$

$$= \frac{8.2 \text{ mi}}{1} \cdot \frac{5280 \text{ ft}}{1 \text{ mi}}$$

The units mi cancel, leaving the answer in feet.

$$= 43{,}296 \text{ ft}$$

Skill Practice
Convert, using unit ratios.
7. 720 in. = _____ ft
8. 4224 ft = _____ mi
9. 8 mi = _____ yd

---

example 4   Making Multiple Conversions of Length

a. 0.25 mi = _____ in.   **b.** 22 in. = _____ yd

Solution:

a. To convert miles to inches, we use two conversion factors. The first unit ratio converts miles to feet. The second unit ratio converts feet to inches.

converts mi to ft   converts ft to in.

$$0.25 \text{ mi} = 0.25 \text{ mi} \cdot \frac{5280 \text{ ft}}{1 \text{ mi}} \cdot \frac{12 \text{ in.}}{1 \text{ ft}}$$

$$= \frac{0.25 \text{ mi}}{1} \cdot \frac{5280 \text{ ft}}{1 \text{ mi}} \cdot \frac{12 \text{ in.}}{1 \text{ ft}}$$

The units mi and ft reduce, leaving the answer in inches.

$$= 15{,}840 \text{ in.}$$

Skill Practice
Convert, using unit ratios.
10. 6.2 yd = _____ in.
11. 0.1 mi = _____ in.

Answers

7. 60 ft   8. 0.8 mi
9. 14,080 yd   10. 223.2 in.
11. 6336 in.

converts   converts
in. to ft   ft to yd

b. $22 \text{ in.} = 22 \text{ in.} \cdot \dfrac{1 \text{ ft}}{12 \text{ in.}} \cdot \dfrac{1 \text{ yd}}{3 \text{ ft}}$ Multiply by two conversion factors. The first converts inches to feet. The second converts feet to yards.

$= \dfrac{22 \text{ in.}}{1} \cdot \dfrac{1 \text{ ft}}{12 \text{ in.}} \cdot \dfrac{1 \text{ yd}}{3 \text{ ft}}$ The units in. and ft reduce, leaving the answer in yards.

$= \dfrac{22}{36} \text{ yd}$ Multiply fractions.

$= \dfrac{11}{18} \text{ yd}$ or $0.6\overline{1} \text{ yd}$ Simplify.

4. Adding and Subtracting Mixed Units

To add and subtract measurements, we must have like units. For example,

$$3 \text{ ft} + 8 \text{ ft} = 11 \text{ ft}$$

Sometimes, however, measurements have mixed units. For example, a drainpipe might be 4 ft 6 in. long or symbolically 4′6″. Measurements and calculations with mixed units act very much as mixed numbers.

Skill Practice

12. Add: 8′4″ + 4′10″

13. Subtract: 6′3″ − 4′9″

example 5   Adding and Subtracting Mixed Units of Measurement

a. Add 4′6″ + 2′9″.   **b.** Subtract 8′2″ − 3′6″.

Solution:

a. $4′6″ + 2′9″ =$ 4 ft +   6 in.
                      +2 ft +   9 in.
                      ‾‾‾‾‾‾‾‾‾‾‾‾‾‾‾‾
                       6 ft + 15 in.   Add like units.

= 6 ft + 1 ft + 3 in.   Note that 15 in. = 1 ft + 3 in.

= 7 ft 3 in.   or   7′3″

b. $8′2″ − 3′6″ =$ 8 ft + 2 in. = $\overset{7}{8}$ ft + $\overset{12 \text{ in.}}{2}$ in.   Borrow 1 ft = 12 in.
                    −(3 ft + 6 in.)   −(3 ft + 6 in.)

=   7 ft + 14 in.
   −(3 ft +   6 in.)
   ‾‾‾‾‾‾‾‾‾‾‾‾‾‾‾‾‾
    4 ft +   8 in.   or   4′8″

5. Applications

In Example 6 we convert to a common unit of measurement to find the perimeter of an object.

Answers

12. 13′2″   13. 1′6″

example 6   Finding Perimeter

Find the perimeter in feet.

2.2 mi

| | 1056 ft

Solution:

The perimeter is found by adding the lengths of the sides. Before we begin, the length and width must have the same units.

First note that $2.2 \text{ mi} = 2.2 \text{ mi} \cdot \dfrac{5280 \text{ ft}}{1 \text{ mi}}$

$$= 11{,}616 \text{ ft}$$

2.2 mi = 11,616 ft

| | 1056 ft

Perimeter $= 11{,}616 \text{ ft} + 1056 \text{ ft} + 11{,}616 \text{ ft} + 1056 \text{ ft}$

$$= 25{,}344 \text{ ft}$$

example 7   Applying U.S. Customary Units of Length

A plumber needs 5 pipes that are 8 ft 4 in. each. What is the total length of piping?

Solution:

First, note that 8 ft 4 in. = 8 ft + 4 in. Second, recall that multiplication represents repeated addition. Therefore, the total length is 5(8 ft + 4 in.). From Figure 7-1, we see that this product can be found by multiplying 5 by each measurement within parentheses. We can apply the distributive property introduced in Section 1.5. That is, $a(b + c) = ab + ac$.

$$5(8 \text{ ft} + 4 \text{ in.}) = 5(8 \text{ ft}) + 5(4 \text{ in.})$$

8 ft 4 in.

$5 \times (8 \text{ ft } 4 \text{ in.})$

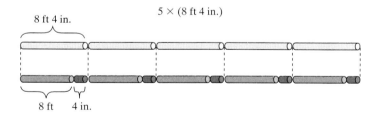

8 ft   4 in.

Figure 7-1

$$5(8 \text{ ft} + 4 \text{ in.}) = 5(8 \text{ ft}) + 5(4 \text{ in.})$$
$$= 40 \text{ ft} + 20 \text{ in.}$$
$$= 40 \text{ ft} + \overset{\frown}{1 \text{ ft} + 8 \text{ in.}} \qquad \text{Note that } 20 \text{ in.} = 1 \text{ ft} + 8 \text{ in.}$$
$$= 41 \text{ ft} + 8 \text{ in.}$$

The total length of piping is 41 ft 8 in.

Tip: Example 7 could have been solved by first converting 8 ft 4 in. to either inches or feet and then multiplying by 5. Converting to inches, we have

$$8 \text{ ft } 4 \text{ in.} = 96 \text{ in.} + 4 \text{ in.} = 100 \text{ in.}$$

Now multiply by 5. $5(100 \text{ in.}) = 500 \text{ in.}$

$$= 500 \text{ in.} \cdot \left(\frac{1 \text{ ft}}{12 \text{ in.}} \right) \longleftarrow \text{Convert back to feet.}$$

$$= \frac{500}{12} \text{ ft}$$

$$= 41 \text{ ft. } 8 \text{ in.}$$ Convert to mixed units by dividing.

$$\begin{array}{r} 41 \\ 12\overline{)500} \\ \underline{48} \\ 20 \\ \underline{12} \\ 8 \end{array}$$

Skill Practice

16. A 12-ft 6-in. length of ribbon is cut into 3 pieces of equal length. Find the length of each piece.

Answer

16. 4 ft 2 in.

---

example 8   Applying U.S. Customary Units of Length in Construction

A carpenter has a board 6′10″ in length. He must cut the board into two pieces of equal length. How long is each piece?

Solution:

The distance 6′10″ is equal to 6 ft 10 in. or 6 ft + 10 in. One method of dividing this total distance in half is to divide each individual unit in half. That is,

$$(6 \text{ ft} + 10 \text{ in.}) \div 2 = \frac{6 \text{ ft}}{2} + \frac{10 \text{ in.}}{2} = 3 \text{ ft} + 5 \text{ in.}   \text{ or }   3′5″$$

---

section 7.1   Practice Exercises

Boost *your* GRADE at mathzone.com!

MathZone

- Practice Problems
- Self-Tests
- NetTutor
- e-Professors
- Videos

Study Skills Exercises

1. Careless mistakes are usually caused by losing focus on what you are doing. What are some of the distractions that you encounter when doing homework?

List some ways you can avoid distractions while doing your homework.

2. Define the key terms.

    a. Unit of measurement

    b. U.S. Customary System of measurement

    c. Conversion factor

    d. Unit ratio (or unit fraction)

Objective 1: U.S. Customary Units of Length

For Exercises 3–8, fill in the blanks with the correct units. Refer to Table 7.1.

3. 5280 ft = 1 _____

4. $\frac{1}{12}$ ft = 1 _____

5. 1 yd = 3 _____

6. 1 ft = 12 _____

7. 1 ft = $\frac{1}{3}$ _____

8. 1 ft = $\frac{1}{5280}$ _____

Objective 2: Converting U.S. Customary Units of Length by Using Substitution

For Exercises 9–20, convert the units of length by using substitution. **(See Examples 1–2.)**

9. 2 yd = _____ ft

10. 2 mi = _____ yd

11. 6 ft = _____ in.

12. 1.25 mi = _____ ft

13. 2 mi = _____ ft

14. 5 ft = _____ in.

15. 24 ft = _____ yd

16. 36 in. = _____ ft

17. 9 in. = _____ ft

18. 10 ft = _____ yd

19. 1760 ft = _____ mi

20. 880 yd = _____ mi

Objective 3: Converting U.S. Customary Units of Length by Using Unit Ratios

21. Identify an appropriate ratio to convert feet to inches by using multiplication.

   a. $\frac{12 \text{ ft}}{1 \text{ in.}}$
   b. $\frac{12 \text{ in.}}{1 \text{ ft}}$
   c. $\frac{1 \text{ ft}}{12 \text{ in.}}$
   d. $\frac{1 \text{ in.}}{12 \text{ ft}}$

22. Identify an appropriate ratio to convert yards to miles by using multiplication.

   a. $\frac{1 \text{ mi}}{5280 \text{ ft}}$
   b. $\frac{1760 \text{ mi}}{1 \text{ yd}}$
   c. $\frac{1 \text{ mi}}{1760 \text{ yd}}$
   d. $\frac{1760 \text{ yd}}{1 \text{ mi}}$

23. Identify an appropriate ratio to convert yards to feet by using multiplication.

   a. $\frac{3 \text{ ft}}{1 \text{ yd}}$
   b. $\frac{1 \text{ yd}}{3 \text{ ft}}$
   c. $\frac{3 \text{ yd}}{1 \text{ ft}}$
   d. $\frac{1 \text{ ft}}{3 \text{ yd}}$

24. Identify an appropriate ratio to convert feet to miles by using multiplication.

   a. $\frac{5280 \text{ ft}}{1 \text{ mi}}$
   b. $\frac{1 \text{ ft}}{5280 \text{ mi}}$
   c. $\frac{5280 \text{ mi}}{1 \text{ ft}}$
   d. $\frac{1 \text{ mi}}{5280 \text{ ft}}$

For Exercises 25–36, convert the units of length by using unit ratios. **(See Example 3.)**

25. $4\frac{1}{2}$ in. = _____ ft

26. $2\frac{1}{3}$ yd = _____ ft

27. 3.5 ft = _____ in.

28. 9 ft = _____ yd

29. 11,880 ft = _____ mi

30. 0.75 mi = _____ ft

31. 6 yd = _____ ft

32. 5280 yd = _____ mi

33. 14 ft = _____ yd

34. 75 in. = _____ ft

35. 320 mi = _____ yd

36. $3\frac{1}{4}$ ft = _____ in.

For Exercises 37–44, convert the units of length, involving multiple conversions. **(See Example 4.)**

37. 171 in. = _____ yd

38. 0.3 mi = _____ in.

39. 2 yd = _____ in.

40. 12,672 in. = _____ mi

41. 0.8 mi = _____ in.

42. 900 in. = _____ yd

43. 6336 in. = _____ mi

44. 6 yd = _____ in.

Objective 4: Adding and Subtracting Mixed Units

45. a. Convert 6′4″ to inches.

    b. Convert 6′4″ to feet.

46. a. Convert 10 ft 8 in. to inches.

    b. Convert 10 ft 8 in. to feet.

47. a. Convert 2 yd 2 ft to feet.

    b. Convert 2 yd 2 ft to yards.

48. a. Convert 3′6″ to feet.

    b. Convert 3′6″ to inches.

For Exercises 49–58, add or subtract as indicated. **(See Example 5.)**

49. 1′3″ + 6′4″

50. 6′2″ + 4′6″

51. 2 ft 8 in. + 3 ft 4 in.

52. 5 ft 2 in. + 6 ft 10 in.

53. 4′10″ + 6′4″

54. 4′9″ + 3′9″

55. 8 ft 8 in. − 5 ft 4 in.

56. 3 ft 6 in. − 1 ft 5 in.

57. 9′2″ − 4′10″

58. 4′3″ − 2′9″

For Exercises 59–66, multiply or divide as indicated.

59. 2(4 ft 5 in.)

60. 4(5 ft 1 in.)

61. 6(4 ft 8 in.)

62. 8(2 ft 5 in.)

63. (6′4″) ÷ 2

64. (16′8″) ÷ 8

65. $\dfrac{18 \text{ ft } 3 \text{ in.}}{3}$

66. $\dfrac{10 \text{ ft } 10 \text{ in.}}{5}$

Objective 5: Applications

67. Find the perimeter in feet. **(See Example 6.)**

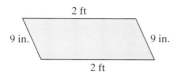

68. Find the perimeter in yards.

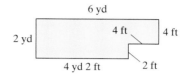

69. The garden pictured needs a decorative border. The border comes in pieces that are 1′6″ long. How many pieces of border are needed?

70. Monte fences a small field with panels of fencing that are 2 yd wide. How many panels of fencing does he need?

71. A plumber used two pieces of pipe for a job. One piece was 4′6″ and the other was 2′8″. How much pipe was used? **(See Example 7.)**

72. A carpenter needs to put wood molding around three sides of a room. Two sides are 6′8″ long, and the third side is 10′ long. How much molding should the carpenter purchase?

73. If you have 4 yd of rope and you use 5 ft, how much is left over? Express the answer in feet.

74. Refer to Exercise 72. If the carpenter buys 9 yd of molding, how much will he have left over?

75. A cable 6 ft 6 in. costs $3.25. What is the price per foot?

76. A piece of rope 8′6″ in length is cut in half. How long is each piece? **(See Example 8.)**

77. A picnic table requires 5 boards that are 6′ long, 4 boards that are 3′3″ long, and 2 boards that are 18″ long. Find the total length of lumber required.

78. A roll of ribbon is 60 yd. If you wrap 12 packages that each use 2.5 ft of ribbon, how much ribbon is left over?

Expanding Your Skills

In Section 1.5 we learned that area is measured in square units such as in.^{2}, ft^2, yd^2, and mi^2. Converting square units involves a different set of conversion factors. For example, 1 yd = 3 ft, but 1 yd^2 = 9 ft^2. To understand why, recall that the formula for the area of a rectangle is $A = l \times w$. In a square, the length and the width are the same distance s. Therefore the area of a square is given by the formula $A = s \times s$, where s is the length of a side. Thus,

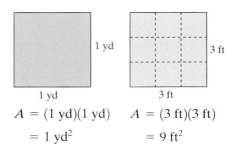

$$A = (1 \text{ yd})(1 \text{ yd}) \qquad A = (3 \text{ ft})(3 \text{ ft})$$
$$= 1 \text{ yd}^2 \qquad\qquad = 9 \text{ ft}^2$$

Instead of learning a new set of conversion factors, we can use multiples of the conversion factors that we mastered in Section 7.1.

Example: Converting Area

Convert 4 yd^2 to square feet.

Solution: We will use the unit ratio of $\dfrac{3 \text{ ft}}{1 \text{ yd}}$ twice.

$$\frac{4 \text{ yd}^2}{1} \cdot \underbrace{\frac{3 \text{ ft}}{1 \text{ yd}} \cdot \frac{3 \text{ ft}}{1 \text{ yd}}}_{\text{Multiply first.}} = \frac{4 \text{ yd}^2}{1} \cdot \frac{9 \text{ ft}^2}{1 \text{ yd}^2} = 36 \text{ ft}^2$$

For Exercises 79–82, convert the units of area by using multiple factors of the given unit ratio.

79. 54 ft^2 = _____ yd^2 (Use two factors of the ratio $\dfrac{1 \text{ yd}}{3 \text{ ft}}$.)

80. 432 in.^{2} = _____ ft^2 (Use two factors of the ratio $\dfrac{1 \text{ ft}}{12 \text{ in.}}$.)

81. 5 ft^2 = _____ in.^{2} (Use two factors of $\dfrac{12 \text{ in.}}{1 \text{ ft}}$.)

82. 3 yd^2 = _____ ft^2 (Use two factors of $\dfrac{3 \text{ ft}}{1 \text{ yd}}$.)

section 7.2   Converting U.S. Customary Units of Time, Weight, and Capacity

Objectives

1. U.S. Customary Units of Time, Weight, and Capacity
2. Converting Units of Time
3. Converting U.S. Customary Units of Weight
4. Converting U.S. Customary Units of Capacity

1. U.S. Customary Units of Time, Weight, and Capacity

In this section we convert U.S. Customary units of time, weight, and capacity. Table 7-2 summarizes several common units and their equivalent measures. These relationships should be memorized as common knowledge.

table 7-2

Summary of U.S. Customary Units of Length, Time, Weight, and Capacity	
Length	**Time**
1 foot (ft) = 12 inches (in.)	1 year (yr) = 365 days
1 yard (yd) = 3 feet (ft)	1 week (wk) = 7 days
1 mile (mi) = 5280 feet (ft)	1 day = 24 hours (hr)
1 mile (mi) = 1760 yards (yd)	1 hour (hr) = 60 minutes (min)
	1 minute (min) = 60 seconds (sec)
Capacity	**Weight**
3 teaspoons (tsp) = 1 tablespoon (T)	1 pound (lb) = 16 ounces (oz)
1 cup (c) = 8 fluid ounces (fl oz)	1 ton = 2000 pounds (lb)
1 pint (pt) = 2 cups (c)	
1 quart (qt) = 2 pints (pt)	
1 gallon (gal) = 4 quarts (qt)	

2. Converting Units of Time

In Example 1, we convert from one unit of time to another.

example 1   Converting Units of Time

a. 32 hr = ___ days

b. 36 hr = ___ sec

Solution:

a. $32 \text{ hr} = \frac{32 \text{ hr}}{1} \cdot \frac{1 \text{ day}}{24 \text{ hr}}$ ←—— new unit to convert to
←—— unit to convert from

Recall that
1 day = 24 hr.

$= \frac{32}{24} \text{ days}$ Multiply fractions.

$= \frac{4}{3} \text{ days or } 1\frac{1}{3} \text{ days}$ Simplify.

Answers

1. $\frac{2}{3}$ day   2. 86,400 sec

b. $36 \text{ hr} = \dfrac{36 \text{ hr}}{1} \cdot \overset{\text{converts}}{\underset{\text{hr to min}}{\dfrac{60 \text{ min}}{1 \text{ hr}}}} \cdot \overset{\text{converts}}{\underset{\text{min to sec}}{\dfrac{60 \text{ sec}}{1 \text{ min}}}}$ Multiply by two conversion factors.

$= 129{,}600 \text{ sec}$ Simplify.

Skill Practice

3. Ward Burton won the Daytona 500 in 2002 with a time of 2:22:45. Convert this time to minutes.

---

example 2   **Converting Units of Time**

After running a marathon, Dave crossed the finish line and noticed that the race clock read 2:20:30. Convert this time to minutes.

Solution:

The notation 2:20:30 means 2 hr 20 min 30 sec. We must convert 2 hr to minutes and 30 sec to minutes. Then we add the total number of minutes.

$$2 \text{ hr} = \frac{2 \text{ hr}}{1} \cdot \frac{60 \text{ min}}{1 \text{ hr}} = 120 \text{ min}$$

$$30 \text{ sec} = \frac{30 \text{ sec}}{1} \cdot \frac{1 \text{ min}}{60 \text{ sec}} = \frac{30}{60} \text{ min} = \frac{1}{2} \text{ min} \quad \text{or} \quad 0.5 \text{ min}$$

The total number of minutes is 120 min + 20 min + 0.5 min = 140.5 min
Dave finished the race in 140.5 min.

---

3. Converting U.S. Customary Units of Weight

Measurements of weight record the force of an object subject to gravity. In Example 3 we convert from one unit of weight to another.

Skill Practice

4. The blue whale is the largest animal on earth. It is so heavy that it would crush under its own weight if it were taken from the water. An average adult blue whale weighs 120 tons. Convert this to pounds.

5. A box of apples weighs 5 lb 12 oz. Convert this to ounces.

---

example 3   **Converting Units of Weight**

a. The average weight of an adult male African elephant is 12,400 lb. Convert this value to tons.

b. Convert the weight of a 7-lb 3-oz baby to ounces.

Solution:

a. Recall that 1 ton = 2000 lb.

$$12{,}400 \text{ lb} = \frac{12{,}400 \text{ lb}}{1} \cdot \frac{1 \text{ ton}}{2000 \text{ lb}}$$

$$= \frac{12{,}400}{2000} \text{ tons} \qquad \text{Multiply fractions.}$$

$$= \frac{31}{5} \text{ tons} \quad \text{or} \quad 6.2 \text{ tons}$$

An adult male African elephant weighs 6.2 tons.

Answers

3. 142.75 min   4. 240,000 lb
5. 92 oz

b. To convert 7 lb 3 oz to ounces, we must convert 7 lb to ounces.

$$7 \text{ lb} = \frac{7 \cancel{\text{ lb}}}{1} \cdot \frac{16 \text{ oz}}{1 \cancel{\text{ lb}}} \qquad \text{Recall that 1 lb} = 16 \text{ oz.}$$

$$= 112 \text{ oz}$$

The baby's total weight is 112 oz + 3 oz = 115 oz.

| example 4 | Applying U.S. Customary Units of Weight |

Jessica lifts four boxes of books. The boxes have the following weights: 16 lb 4 oz, 18 lb 8 oz, 12 lb 5 oz, and 22 lb 9 oz. How much weight did she lift altogether?

Solution:

$$
\begin{array}{ll}
\quad 16 \text{ lb} \quad 4 \text{ oz} & \text{Add like units in columns.} \\
\quad 18 \text{ lb} \quad 8 \text{ oz} & \\
\quad 12 \text{ lb} \quad 5 \text{ oz} & \\
\underline{+ \ 22 \text{ lb} \quad 9 \text{ oz}} & \\
\quad 68 \text{ lb} \quad 26 \text{ oz} = 68 \text{ lb} + 26 \text{ oz} & \\
\qquad\qquad\qquad = 68 \text{ lb} + \overset{\frown}{1 \text{ lb} + 10 \text{ oz}} & \text{Recall that 1 lb} = 16 \text{ oz.} \\
\qquad\qquad\qquad = 69 \text{ lb} \quad 10 \text{ oz} &
\end{array}
$$

Jessica lifted 69 lb 10 oz of books.

Skill Practice

6. A set of triplets weighed 4 lb 3 oz, 3 lb 9 oz, and 4 lb 5 oz. What is the total weight (in ounces) of all three babies?

4. Converting U.S. Customary Units of Capacity

A typical can of soda contains 12 fl oz. This is a measure of capacity. Capacity is the volume or amount that a container can hold. The U.S. Customary units of capacity are fluid ounces (fl oz), cup (c), pint (pt), quart (qt), and gallon (gal).

One fluid ounce is approximately the amount of liquid that two large spoonfuls will hold. One cup is the amount in an average-size cup of tea. While Table 7-2 summarizes the relationships among units of capacity, we also offer an illustration (Figure 7-2).

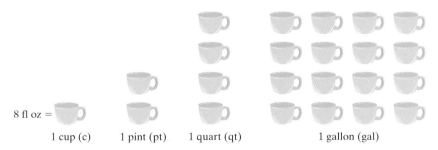

8 fl oz =

1 cup (c)   1 pint (pt)   1 quart (qt)   1 gallon (gal)

Figure 7-2

Concept Connections

7. From Figure 7-2, determine how many cups are in 1 gal.

8. From Figure 7-2, determine how many pints are in 1 gal.

Answers
6. 12 lb 1 oz   7. 16 c in 1 gal
8. 8 pt in 1 gal

Skill Practice

Convert.

9. 8.5 gal = _____ qt

10. 2.25 qt = _____ c

11. 40 fl oz = _____ qt

12. 6 pt = _____ gal

---

example 5   Converting Units of Capacity

a. 1.25 pt = _____ qt

b. 2 gal = _____ c

c. 48 fl oz = _____ gal

Solution:

a. $1.25 \text{ pt} = \dfrac{1.25 \text{ p\!t}}{1} \cdot \dfrac{1 \text{ qt}}{2 \text{ p\!t}}$ Recall that 1 qt = 2 pt.

$= \dfrac{1.25}{2} \text{ qt}$ Multiply fractions.

$= 0.625 \text{ qt}$ Simplify.

b. $2 \text{ gal} = 2 \text{ gal} \cdot \dfrac{4 \text{ qt}}{1 \text{ gal}} \cdot \dfrac{4 \text{ c}}{1 \text{ qt}}$ Use two conversion factors. The first unit ratio converts gallons to quarts. The second converts quarts to cups.

$= \dfrac{2 \text{ ga\!l}}{1} \cdot \dfrac{4 \text{ q\!t}}{1 \text{ ga\!l}} \cdot \dfrac{4 \text{ c}}{1 \text{ q\!t}}$

$= 32 \text{ c}$ Multiply.

c. $48 \text{ fl oz} = \dfrac{48 \text{ fl o\!z}}{1} \cdot \dfrac{1 \text{ c}}{8 \text{ fl o\!z}} \cdot \dfrac{1 \text{ q\!t}}{4 \text{ c}} \cdot \dfrac{1 \text{ gal}}{4 \text{ q\!t}}$ Convert from fluid ounces to cups, from cups to quarts, and from quarts to gallons.

$= \dfrac{48}{128} \text{ gal}$

$= \dfrac{3}{8} \text{ gal}$ or 0.375 gal

Avoiding Mistakes: It is important to note that ounces (oz) and fluid ounces (fl oz) are different quantities. An ounce (oz) is a measure of weight, and a fluid ounce (fl oz) is a measure of capacity. Furthermore,

16 oz = 1 lb

8 fl oz = 1 c

Skill Practice

13. A recipe calls for $3\frac{1}{2}$ c of tomato sauce. A jar of sauce holds 24 fl oz. Is there enough sauce in the jar for the recipe?

---

example 6   Applying Units of Capacity

A recipe calls for $1\frac{3}{4}$ c of chicken broth. A can of chicken broth holds 14.5 fl oz. Is there enough chicken broth in the can for the recipe?

Solution:

We need to convert each measurement to the same unit of measure for comparison. Converting $1\frac{3}{4}$ c to fluid ounces, we have

$1\dfrac{3}{4}\text{c} = \dfrac{7}{4} \text{c} \cdot \dfrac{8 \text{ fl oz}}{1 \text{ c}}$ Recall that 1 c = 8 fl oz.

$= \dfrac{56}{4} \text{ fl oz}$ Multiply fractions.

$= 14 \text{ fl oz}$ Simplify.

The recipe calls for $1\frac{3}{4}$ c or 14 fl oz of chicken broth. The can of chicken broth holds 14.5 fl oz which is enough.

Answers

9. 34 qt   10. 9 c

11. 1.25 qt   12. 0.75 gal

13. No, $3\frac{1}{2}$ c is equal to 28 oz. The jar only holds 24 fl oz.

section 7.2   Practice Exercises

Study Skills Exercise

1. Set goals for studying. Before you begin a homework assignment, estimate the time it will take for you to complete the assignment. This type of goal can make you more efficient in your work.

   Write down the time you expect to need to finish this assignment.

Review Exercises

For Exercises 2–8, complete the table.

	Object	in.	ft	yd	mi
2.	Distance to work				3 mi
3.	Length of a hallway		12 ft		
4.	Length of a car		14 ft		
5.	Height of a tree			6 yd	
6.	A basketball player's height	76 in.			
7.	Perimeter of a backyard			50 yd	
8.	Distance to the gas station				$1\frac{1}{2}$ mi

Objective 1: U.S. Customary Units of Time, Weight, and Capacity

For Exercises 9–18, fill in the blanks with the correct units.

9. 1 qt = 2 ___

10. 1 c = 8 ___

11. 1 lb = 16 ___

12. 1 pt = 2 ___

13. 1 yr = 365 ___

14. 1 ___ = 2000 lb

15. 1 gal = 4 ___

16. 1 ___ = 24 hr

17. 1 ___ = 60 min

18. 1 min = 60 ___

Objective 2: Converting Units of Time

For Exercises 19–28, convert the units of time. (**See Example 1.**)

19. 2 yr = ___ days

20. $1\frac{1}{2}$ days = ___ hr

21. 90 min = ___ hr

22. 3 wk = ___ days

23. 180 sec = ___ min

24. $3\frac{1}{2}$ hr = ___ min

25. 72 hr = ___ days

26. 28 days = ___ wk

27. 3600 sec = ___ hr

28. 168 hr = ___ wk

For Exercises 29–32, convert the time given as hr:min:sec to minutes. **(See Example 2.)**

29. 1:20:30   **30.** 3:10:45   **31.** 2:55:15   **32.** 1:40:30

33. Gil is a distance runner. The duration of his training runs for one week is given in the table. Find the total time that Gil ran that week, and express the answer by using mixed units.

Day	Time
Monday	1 hr 10 min
Tuesday	45 min
Wednesday	1 hr 20 min
Thursday	30 min
Friday	50 min
Saturday	Rest
Sunday	1 hr

34. A movie theater is holding a movie marathon in which it plans to show three movies. The lengths of the movies are 1 hr 45 min, 2 hr 10 min, and 2 hr 15 min. If Fernando stays for all three movies, what is his total viewing time?

35. In a team triathlon, Barb swims $\frac{1}{2}$ mi in 15 min 30 sec. Steve rides his bicycle 20 mi in 50 min 20 sec. Carlise runs 4 mi in 28 min 10 sec. Find the total time for the team.

36. Joe competes in a biathlon. He runs 5 mi in 32 min 8 sec. He rides his bike 25 mi in 1 hr 2 min 40 sec. Find the total time for his race.

Objective 3: Converting U.S. Customary Units of Weight

For Exercises 37–44, convert the units of weight. **(See Example 3.)**

37. 32 oz = ___ lb   **38.** 2500 lb = ___ tons   **39.** 2 tons = ___ lb   **40.** 8 oz = ___ lb

41. 4 lb = ___ oz   **42.** $3\frac{1}{4}$ tons = ___ lb   **43.** 3000 lb = ___ tons   **44.** 6 lb = ___ oz

For Exercises 45–50, add or subtract as indicated. **(See Example 4.)**

45. 6 lb 10 oz + 3 lb 14 oz   **46.** 12 lb 11 oz + 13 lb   **47.** 30 lb 10 oz − 22 lb 8 oz

48. 5 lb − 2 lb 5 oz   **49.** 10 lb − 3 lb 8 oz   **50.** 20 lb 3 oz + 15 oz

51. Byron lays sod in his backyard. Each piece of sod weighs 6 lb 4 oz. If he puts down 50 pieces, find the total weight.

52. A can of paint weighs 2.2 lb. How much would 6 cans weigh?

53. There is $2\frac{1}{2}$ tons of trash that needs to be moved. If a truck can handle 2500 lb in one trip, how many trips will the truck need to make to remove all the trash?

54. A box that contains 6 textbooks weighs 15 lb 12 oz. What is the weight of each textbook?

Objective 4: Converting U.S. Customary Units of Capacity

For Exercises 55–66, convert the units of capacity. (See Example 5.)

55. 16 fl oz = ____ c   **56.** 5 pt = ____ c   **57.** 6 gal = ____ qt   **58.** 8 pt = ____ qt

59. 1 gal = ____ c   **60.** 1 T = ____ tsp   **61.** 2 qt = ____ gal   **62.** 2 qt = ____ c

63. 1 pt = ____ fl oz   **64.** 32 fl oz = ____ qt   **65.** 2 T = ____ tsp   **66.** 2 gal = ____ pt

67. A recipe for minestrone soup calls for 3 c of spaghetti sauce. If a jar of sauce has 48 fl oz, is there enough for the recipe? (See Example 6.)

68. A recipe for punch calls for 6 c of apple juice. A bottle of juice has 2 qt. Is there enough juice in the bottle for the recipe?

69. A 24-fl-oz jar of spaghetti sauce sells for $2.69. Another jar that holds 1 qt of sauce sells for $3.29. Which is a better buy? Explain.

70. Tatiana went to purchase bottled water. She found the following options: a 12-pack of 1-pt bottles; a 1-gal jug; and a 6-pack of 24-fl-oz bottles. Which option should Tatiana choose to get the most water? Explain.

Calculator Connections

Topic: Converting units of capacity

For Exercises 71–79, complete the table.

	Object	fl oz	c	pt	qt	gal
71.	Bottle of canola oil	32 fl oz				
72.	Can of soda	12 fl oz				
73.	Laundry detergent					1 gal
74.	Container of gasoline					5 gal
75.	Bottle of Gatorade				0.5 qt	
76.	Container of orange juice				0.75 qt	
77.	Bottle of spring water		1 c			
78.	Milkshake			1 pt		
79.	Jug of maple syrup	64 fl oz				

Objectives

1. Introduction to the Metric System
2. Metric Units of Length
3. Converting Metric Units of Length
4. Converting Metric Units of Length by Using the Prefix Line
5. Applications

---

section 7.3 Metric Units of Length

1. Introduction to the Metric System

Throughout history the lack of standard units of measure led to much confusion in trade between countries. In 1790 the French Academy of Sciences adopted a simple, decimal-based system of units. This system is known today as the **metric system**. The metric system is the predominant system of measurement used in science.

The simplicity of the metric system is a result of having one basic unit of measure for each type of quantity (length, mass, and capacity). The base units are the **meter** for length, the **gram** for mass, and the **liter** for capacity. Other units of length, mass, and capacity in the metric system are multiples of 10 of the base unit.

2. Metric Units of Length

The **meter** (m) is the basic unit of length in the metric system. A meter is slightly longer than a yard.

| 1 meter | 1 m ≈ 39 in. |
| 1 yard | 1 yd = 36 in. |

The meter was defined in the late 1700s as one ten-millionth of the distance along the earth's surface from the North Pole to the Equator through Paris, France. Today the meter is defined as the distance traveled by light in a vacuum during $\frac{1}{299{,}792{,}458}$ sec.

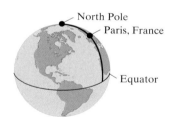

Six other common metric units of length are given in Table 7-3. Notice that each unit is related to the meter by a power of 10. This makes it particularly easy to convert from one unit to another.

table 7-3

Metric Units of Length and Their Equivalents

1 kilometer (km) = 1000 m	
1 hectometer (hm) = 100 m	
1 dekameter (dam) = 10 m	
1 meter (m) = 1 m	
1 decimeter (dm) = 0.1 m	$\left(\frac{1}{10}\,\text{m}\right)$
1 centimeter (cm) = 0.01 m	$\left(\frac{1}{100}\,\text{m}\right)$
1 millimeter (mm) = 0.001 m	$\left(\frac{1}{1000}\,\text{m}\right)$

Notice that each unit of length has a prefix followed by the word *meter* (kilometer, for example). You should memorize these prefixes along with their

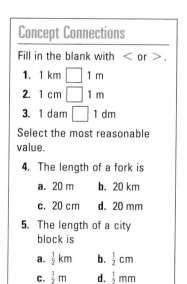

Answers

1. >   2. <   3. >   4. c
5. a

multiples of the basic unit, the meter. Furthermore, it is generally easiest to memorize the prefixes in order.

kilo-	hecto-	deka-	meter	deci-	centi-	milli-
× 1000	× 100	× 10	× 1	× 0.1	× 0.01	× 0.001

As you familiarize yourself with the metric units of length, it is helpful to have a sense of the distance represented by each unit (Figure 7-3).

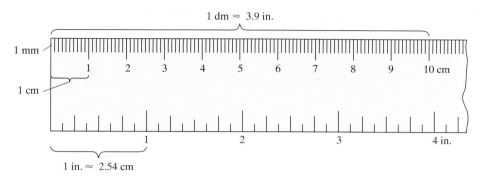

Figure 7-3

From Figure 7-3, you can see that 1 mm is about the thickness of 5 sheets of paper. One centimeter is about the width of a key on a calculator. One decimeter is about 4 in. In the U.S. Customary System, the mile is used to express longer distances. In the metric system, we use kilometers. For instance, the distance between Los Angeles and San Diego is about 120 mi or 193 km.

In Example 1 we practice measuring some familiar objects, using metric units of length.

example 1   Measuring Distances in Metric

Approximate the distance in centimeters and in millimeters.

a.

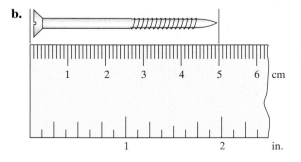

b.

Skill Practice

6. Approximate the length of the pin in centimeters and in millimeters.

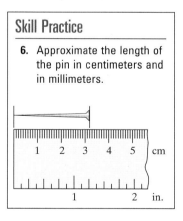

Answer

6. 3.2 cm or 32 mm

Solution:

a. The numbered lines on the ruler are units of centimeters. Each centimeter is divided into 10 mm. We see that the width of the penny is not quite 2 cm. We can approximate this distance as 1.8 cm or equivalently 18 mm.

b. The length of the screw appears to be approximately 5 cm or equivalently 50 mm.

3. Converting Metric Units of Length

In Example 2, we convert metric units of length by using unit ratios and substitution.

---

example 2   **Converting Metric Units of Length**

a. 10.4 km = _____ m

b. 88 mm = _____ m

Solution:

From Table 7-3, 1 km = 1000 m.

a. $10.4 \text{ km} = \dfrac{10.4 \text{ km}}{1} \cdot \dfrac{1000 \text{ m}}{1 \text{ km}}$ ←——— new unit to convert to
←——— unit to convert from

$= 10{,}400 \text{ m}$ Multiply.

b. From Table 7-3, 1 mm = 0.001 m. Using substitution, we have

$88 \text{ mm} = 88 \times 1 \text{ mm}$

$= 88 \times 0.001 \text{ m}$ Substitute 1 mm = 0.001 m.

$= 0.088 \text{ m}$ Multiply.

---

4. Converting Metric Units of Length by Using the Prefix Line

Recall that the place positions in our numbering system are based on powers of 10. For this reason, when we multiply a number by 10, 100, or 1000, we move the decimal point 1, 2, or 3 places, respectively, to the right. Similarly when we multiply by 0.1, 0.01, or 0.001, we move the decimal point to the left 1, 2, or 3 places, respectively.

Since the metric system is also based on powers of 10, we can convert between two metric units of length by moving the decimal point. The direction and number of place positions to move are based on the metric **prefix line**, shown in Figure 7-4.

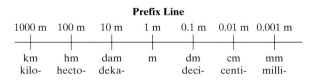

Figure 7-4

Tip: To use the prefix line effectively, you must know the order of the metric prefixes. Sometimes a mnemonic (memory device) can help. Consider the following sentence. The first letter of each word represents one of the metric prefixes.

kids	have	doughnuts	until	dad	calls	mom.
kilo-	hecto-	deka-	unit	deci-	centi-	milli-

represents the main
unit of measurement
(meter, liter, or gram)

Using the Prefix Line to Convert Metric Units

1. To use the prefix line, begin at the point on the line corresponding to the original unit you are given.

2. Then count the number of positions you need to move to reach the new unit of measurement.

3. Move the decimal point in the original measured value the same direction and same number of places as on the prefix line.

4. Replace the original unit with the new unit of measure.

---

example 3   Using the Prefix Line to Convert Metric Units of Length

Use the prefix line for each conversion.

a. 37.9 mm = _____ dm

b. 4700 cm = _____ km

c. 0.3 hm = _____ m

Solution:

a.

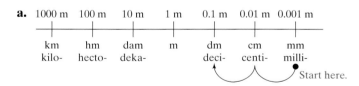

From the prefix line, we need to move the decimal point 2 places to the left.

$$37.9 \text{ mm} = 0.379 \text{ dm}$$

b.

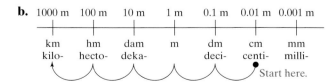

From the prefix line, we need to move the decimal point 5 places to the left. Add more zeros if necessary.

$$4700 \text{ cm} = 0.04700 \text{ km} = 0.047 \text{ km}$$

c.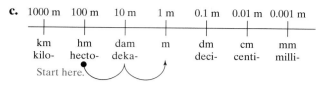

From the prefix line, we need to move the decimal point 2 places to the right.

$$0.3 \text{ hm} = 0\underset{\llcorner\!\!\lrcorner}{\,30.}\text{ m} = 30 \text{ m}$$

---

example 4   Converting Metric Units of Length

a. Mt. Everest, the highest mountain in the world, is 8850 m. How many kilometers is this?

b. Shaquille O'Neal is 2.159 m tall. How many centimeters is this?

Solution:

a. $8850 \text{ m} = 8.850 \text{ km}$

$= 8.85 \text{ km}$

b. $2.159 \text{ m} = 215.9 \text{ cm}$

$8850 \text{ m} = 8.85 \text{ km}$ $2.159 \text{ m} = 215.9 \text{ cm}$

---

5. Applications

example 5   Applying Metric Units of Length

On 1 tank of gas, Shena can travel 800 km. If her tank is $\frac{4}{5}$ full, can she travel from Springfield, Illinois, to Louisville, Kentucky, and back without having to buy gas (see Figure 7-5)?

Springfield

380 km

Louisville

Figure 7-5

Solution:

The distance Shena can go on $\frac{4}{5}$ tank of gas is given by $\frac{4}{5}(800 \text{ km}) = 640 \text{ km}$. The round trip between the cities is $2(380 \text{ km}) = 760 \text{ km}$. Therefore, Shena does not have enough gas to make the round trip.

section 7.3   Practice Exercises

Study Skills Exercises

1. Label the statements as True or False.

a. To do math, you must be born with a special skill. _____

b. There is only one way to solve a math problem. _____

c. Two answers can look different but are both correct. _____

2. Define the key terms.

a. Metric system   **b. Meter**   **c. Gram**   **d. Liter**   **e. Prefix line**

Review Exercises

For Exercises 3–9, convert the units of measurements.

3. 2200 yd = _____ mi   **4.** 8 c = _____ pt   **5.** 48 oz = _____ lb   **6.** 2 yd = _____ in.

7. 1 day = _____ min   **8.** 160 fl oz = _____ gal   **9.** 3.5 lb = _____ oz

Objective 1: Introduction to the Metric System

10. Identify the units that apply to length. Circle all that apply.

a. Yard   **b.** Ounce   **c.** Fluid ounce   **d.** Meter   **e.** Quart

f. Gram   **g.** Pound   **h.** Liter   **i.** Mile   **j.** Inch

11. Identify the units that apply to weight or mass. Circle all that apply.

a. Yard   **b.** Ounce   **c.** Fluid ounce   **d.** Meter   **e.** Quart

f. Gram   **g.** Pound   **h.** Liter   **i.** Mile   **j.** Inch

12. Identify the units that apply to capacity. Circle all that apply.

a. Yard   **b.** Ounce   **c.** Fluid ounce   **d.** Meter   **e.** Quart

f. Gram   **g.** Pound   **h.** Liter   **i.** Mile   **j.** Gram

Objective 2: Metric Units of Length

For Exercises 13–16, approximate each distance in centimeters and millimeters. **(See Example 1.)**

13.

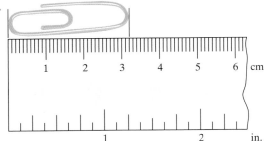

14.

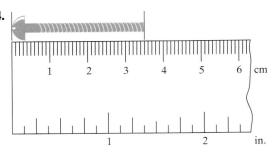

15.

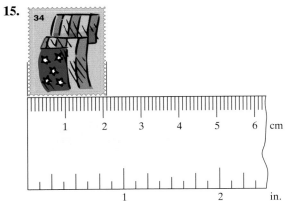

16.

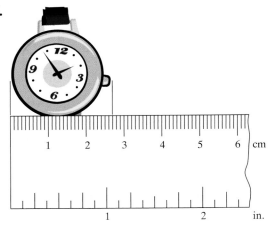

For Exercises 17–20, use a metric ruler to measure the dimensions of each figure in centimeters. Then find the perimeter and area.

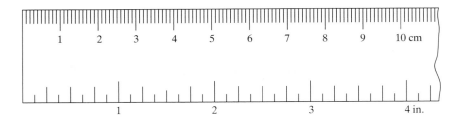

17. a. Length = _____   **b.** Width = _____

   c. Perimeter = _____   **d.** Area = _____

18. a. Length = _____   **b.** Width = _____

   c. Perimeter = _____   **d.** Area = _____

19. a. Length = _____   **b.** Width = _____

   c. Perimeter = _____   **d.** Area = _____

20. a. Length = _____   **b.** Width = _____

   c. Perimeter = _____   **d.** Area = _____

Figure for Exercise 19

For Exercises 21–26, select the most reasonable measurement.

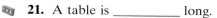 **21.** A table is _____ long.

   a. 2 m   **b.** 2 cm   **c.** 2 km   **d.** 2 hm

22. A picture frame is _____ wide.

   a. 12 dm   **b.** 12 mm   **c.** 12 m   **d.** 12 km

23. The distance between Albany, New York, and Buffalo, New York, is _____.

   a. 210 m   **b.** 2100 cm   **c.** 2.1 km   **d.** 210 km

Figure for Exercise 20

24. The distance between Denver and Colorado Springs is _____.

   a. 110 cm   **b.** 110 km   **c.** 11,000 km   **d.** 1100 mm

25. The height of a full-grown giraffe is approximately _____.

   a. 50 m   **b.** 0.05 m   **c.** 0.5 m   **d.** 5 m

26. The length of a canoe is _____.

   a. 5 m   **b.** 0.5 m   **c.** 0.05 m   **d.** 500 m

Objective 3: Converting Metric Units of Length

For Exercises 27–32, complete the unit ratios.

27. $\dfrac{1\text{ km}}{\underline{}\text{ m}}$
        28. $\dfrac{1\text{ hm}}{\underline{}\text{ m}}$
        29. $\dfrac{1\text{ m}}{\underline{}\text{ cm}}$

30. $\dfrac{1\text{ m}}{\underline{}\text{ mm}}$
        31. $\dfrac{1\text{ m}}{\underline{}\text{ dm}}$
        32. $\dfrac{1\text{ dam}}{\underline{}\text{ m}}$

For Exercises 33–40, convert metric units of length by using unit ratios or substitution. **(See Example 2.)**

33. 2430 m = _____ km   **34.** 52 hm = _____ m   **35.** 103 dm = _____ m   **36.** 1251 mm = _____ m

37. 50 m = _____ dam   **38.** 13 m = _____ mm   **39.** 4 km = _____ m   **40.** 5 m = _____ cm

Objective 4: Converting Metric Units of Length by Using the Prefix Line

For Exercises 41–52, convert metric units of length, using the prefix line. **(See Example 3.)**

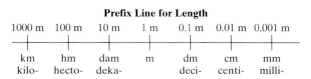

Prefix Line for Length

1000 m	100 m	10 m	1 m	0.1 m	0.01 m	0.001 m
km	hm	dam	m	dm	cm	mm
kilo-	hecto-	deka-		deci-	centi-	milli-

41. 4.31 km = _____ dam   **42.** 18 cm = _____ mm   **43.** 3328 dm = _____ km

44. 128 hm = _____ km   **45.** 345 dm = _____ dam   **46.** 450 mm = _____ dm

47. 0.25 km = _____ m   **48.** 3 hm = _____ m   **49.** 4003 cm = _____ dam

50. 6.8 m = _____ cm   **51.** 0.07 hm = _____ cm   **52.** 8 dam = _____ cm

53. The tallest sand sculpture was 20.91 m tall. How many centimeters is this? (Source: *Guinness Book of World Records*, 2003.) **(See Example 4.)**

54. The smallest steam engine is 16.24 mm long. How many centimeters is this? (Source: *Guinness Book of World Records*, 2003.)

55. The lowest point in Antarctica is 2538 m below sea level. How many kilometers is this?

56. The lowest point in South America is 40 m below sea level. How many kilometers is this?

57. The Trump World Tower in Atlanta is 269 m tall. How many kilometers is this?

58. The Trump Building in New York is 283 m tall. How many kilometers is this?

Objective 5: Applications

59. Veronique has a piece of molding 1 m long. Does she have enough to cut four pieces for the picture frame shown in the figure? **(See Example 5.)**

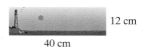
12 cm
40 cm

60. Rosanna has material 1.5 m long for a window curtain. If the window is 90 cm and she needs 10 cm for a hem at the bottom and 12 cm for finishing the top, does Rosanna have enough material?

61. A square tile is 110 mm in length. If they are placed side by side, how many tiles will it take to cover a length of wall 1.43 m long?

62. A parking area in an apartment complex is 0.108 km long. If parking spaces are 4.5 m wide, how many can fit along the length of the lot?

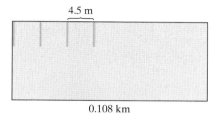

4.5 m

0.108 km

63. Find the missing length.

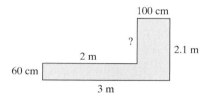

100 cm

?

2 m

2.1 m

60 cm

3 m

64. Two Olympic speed skating races for women are 500 m and 5 km. What is the difference (in meters) between the lengths of these races?

Expanding Your Skills

In the Expanding Your Skills of Section 7.1, we converted U.S. Customary units of area. We use the same procedure to convert metric units of area. This procedure involves multiplying by two unit ratios of length.

Example: Converting area

Convert 1000 mm^2 to square centimeters.

Solution:

$$\frac{1000 \text{ mm}^2}{1} \cdot \underbrace{\frac{1 \text{ cm}}{10 \text{ mm}} \cdot \frac{1 \text{ cm}}{10 \text{ mm}}}_{\text{Multiply first.}} = \frac{1000 \text{ mm}^2}{1} \cdot \frac{1 \text{ cm}^2}{100 \text{ mm}^2} = \frac{1000 \text{ cm}^2}{100} = 10 \text{ cm}^2$$

For Exercises 65–68, convert the units of area, using two factors of the given unit ratio.

65. 30,000 mm^2 = _____ dm^2 $\left(\text{Use } \dfrac{1 \text{ dm}}{100 \text{ mm}}.\right)$

66. 65,000,000 m^2 = _____ km^2 $\left(\text{Use } \dfrac{1 \text{ km}}{1000 \text{ m}}.\right)$

67. 4.1 m^2 = _____ cm^2 $\left(\text{Use } \dfrac{100 \text{ cm}}{1 \text{ m}}.\right)$

68. 0.56 hm^2 = _____ dm^2 $\left(\text{Use } \dfrac{1000 \text{ dm}}{1 \text{ hm}}.\right)$

Concept Connections

1. Which object could have a mass of 2 g?
   a. Rubber band
   b. Can of tuna fish
   c. Cell phone

Concept Connections

Fill in the blank with < or >.

2. 1 g ☐ 1 kg
3. 1 g ☐ 1 cg
4. 1 dg ☐ 1 dag

---

section 7.4   Metric Units of Mass and Capacity

1. Metric Units of Mass

In Section 7.2 we learned that the pound and ton are two measures of weight in the U.S. Customary System. Measurements of weight record the force of an object under the influence of gravity. The mass of an object is related to its weight. However, mass is not affected by gravity. Thus, the weight of an object will be different on earth than on the moon because the effect of gravity is different. However, the mass of the object will stay the same.

The fundamental unit of mass in the metric system is the **gram** (g). A penny is approximately 2.5 g (Figure 7-6). A paper clip is approximately 1 g (Figure 7-7).

≈ 2.5 g   ≈ 1 g

Figure 7-6   **Figure 7-7**

Other common metric units of mass are given in Table 7-4. Once again, notice that the metric units of mass are related to the gram by powers of 10.

table 7-4

Metric Units of Mass and Their Equivalents	
1 kilogram (kg) = 1000 g	
1 hectogram (hg) = 100 g	
1 dekagram (dag) = 10 g	
1 gram (g) = 1 g	
1 decigram (dg) = 0.1 g	$(\frac{1}{10}\,g)$
1 centigram (cg) = 0.01 g	$(\frac{1}{100}\,g)$
1 milligram (mg) = 0.001 g	$(\frac{1}{1000}\,g)$

On the surface of earth, 1 kg of mass is equivalent to approximately 2.2 lb of weight. Therefore, a 180-lb man has approximately 81.8 kg of mass.

180 lb ≈ 81.8 kg

Answers

1. a
2. <
3. >
4. <

2. Converting Metric Units of Mass

The metric prefix line for mass is shown in Figure 7-8. This can be used to convert from one unit of mass to another.

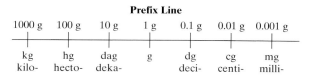

Figure 7-8

---

example 1   Converting Metric Units of Mass

a. 1.6 kg = _____ g

b. 1400 mg = _____ g

c. 214 hg = _____ cg

Solution:

a.

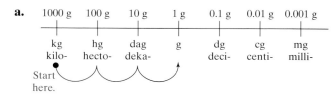

From the prefix line, move the decimal point 3 places to the right.

$$1.6 \text{ kg} = 1600 \text{ g}$$

b.

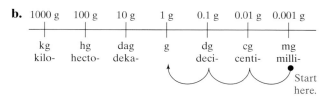

From the prefix line, move the decimal point 3 places to the left.

$$1400 \text{ mg} = 1.400 \text{ g} = 1.4 \text{ g}$$

c.

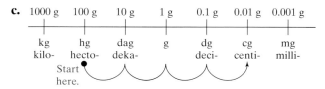

From the prefix line, move the decimal point 4 places to the right.

$$214 \text{ hg} = 2,140,000 \text{ cg}$$

Skill Practice

Convert.

5. 80 kg = _____ g

6. 49 cg = _____ g

7. 0.004 kg = _____ cg

8. 0.9 g = _____ mg

Answers

5. 80,000
6. 0.49 g
7. 400 cg
8. 900 mg

3. Metric Units of Capacity

The basic unit of capacity in the metric system is the **liter** (L). One liter is slightly more than 1 qt. Other common units of capacity are given in Table 7-5.

table 7-5

Metric Units of Capacity and Their Equivalents

1 kiloliter (kL) = 1000 L	
1 hectoliter (hL) = 100 L	
1 dekaliter (daL) = 10 L	
1 liter (L) = 1 L	
1 deciliter (dL) = 0.1 L	$(\frac{1}{10} L)$
1 centiliter (cL) = 0.01 L	$(\frac{1}{100} L)$
1 milliliter (mL) = 0.001 L	$(\frac{1}{1000} L)$

1 mL is also equivalent to a **cubic centimeter** (**cc** or **cm³**). The unit cc is often used to measure dosages of medicine. For example, after having an allergic reaction to a bee sting, a patient might be given 1 cc of adrenalin.

1 cc = 1 ml

4. Converting Metric Units of Capacity

The metric prefix line for capacity is similar to that of length and mass (Figure 7-9). It can be used to convert between metric units of capacity.

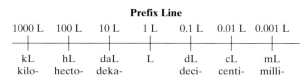

Figure 7-9

example 2   Converting Metric Units of Capacity

a. 5.5 L = _____ mL

b. 150 cL = _____ kL

c. 0.8 cL = _____ cc

Solution:

a.

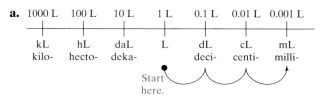

From the prefix line, move the decimal point 3 places to the right.

$$5.5 \text{ L} = 5500 \text{ mL}$$

b.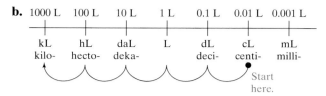

From the prefix line, move the decimal point 5 places to the left.

$$150 \text{ cL} = 0.00150 \text{ kL} = 0.0015 \text{ kL}$$

c. First recall that 1 cc = 1 mL. Therefore we must convert from centimeters to milliliters.

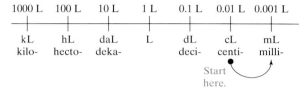

From the prefix line, move the decimal point 1 place to the right.

$$0.8 \text{ cL} = 8 \text{ mL} = 8 \text{ cc}$$

5. Summary of Metric Conversions

The prefix line in Figure 7-10 summarizes the relationships learned thus far.

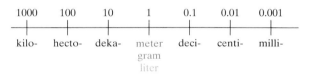

Figure 7-10

example 3   **Converting Metric Units**

a. The distance between San Jose and Santa Clara is 26 km. Convert this to meters.

b. A bottle of canola oil holds 946 mL. Convert this to liters.

c. The mass of a bag of rice is 90,700 cg. Convert this to grams.

d. A dose of an antiviral medicine is 0.5 cc. Convert this to milliliters.

Solution:

a. $26 \text{ km} = 26,000 \text{ m}$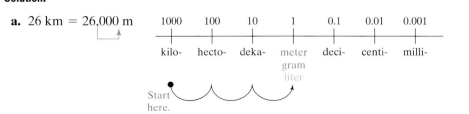

b. 946 mL = 0.946 L

1000	100	10	1	0.1	0.01	0.001
kilo-	hecto-	deka-	meter gram liter	deci-	centi-	milli-

Start here.

c. 90,700 cg = 907.00 g = 907 g

1000	100	10	1	0.1	0.01	0.001
kilo-	hecto-	deka-	meter gram liter	deci-	centi-	milli-

Start here.

d. Recall that 1 cc = 1 mL. Therefore, 0.5 cc = 0.5 mL.

6. Applications

Skill Practice

20. A chemist has 500 g of NaCl (table salt). He uses 500 cg, 1500 cg, and 1000 cg for three experiments. How much NaCl is left over?

example 4   Applying Metric Units of Measure

A chemist has a 1.5-L bottle of alcohol. She sets up four experiments and uses the following amounts of alcohol: 20 cL, 12 cL, 40 cL, 50 cL. How much alcohol will be left? Write the answer in centiliters.

Solution:

To determine the amount left over after the experiments, we must subtract the total amount used from the original 1.5-L bottle.

The total amount used is   20 cL + 12 cL + 40 cL + 50 cL = 122 cL.

To subtract this value from 1.5 L, we need to have like units of measurement. Note that 1.5 L = 150 cL. Thus, the amount left over is given by

$$150 \text{ cL} - 122 \text{ cL} = 28 \text{ cL}$$

The chemist has 28 cL of alcohol left.

Answer

20. 470 g

---

section 7.4   Practice Exercises

Boost *your* GRADE at mathzone.com!

MathZone

• Practice Problems
• Self-Tests
• NetTutor

• e-Professors
• Videos

Study Skills Exercises

1. To help you stay motivated for this class, list three reasons why you are taking the class.

2. Define the key terms.

   a. Gram   **b. Liter**   **c. Cubic centimeter**

Review Exercises

For Exercises 3–10, complete the table.

	Object	mm	cm	m	km
3.	Distance between Orlando and Miami				670
4.	Length of the Mississippi River				3766
5.	Length of a screw		2.5		
6.	Thickness of a pizza		3.2		
7.	Thickness of a dime	1.35			
8.	Diameter of a quarter	24.3			
9.	World record in men's long jump as of the year 2000			2.45	
10.	World record in women's pole vault as of the year 2000			4.6	

Objective 1: Metric Units of Mass

For Exercises 11–16, write the unit that corresponds to the abbreviation.

   11. cg   **12.** g   **13.** kg

   14. mg   **15.** dag   **16.** dg

For Exercises 17–22, complete the equivalent expressions.

   17. 1 dg = _____ g   **18.** 1 dag = _____ g   **19.** 1 cg = _____ g

   20. 1 kg = _____ g   **21.** 1 mg = _____ g   **22.** 1 hg = _____ g

Objective 2: Converting Metric Units of Mass

For Exercises 23–32, convert the units of mass. **(See Example 1.)**

   23. 539 g = _____ kg   **24.** 328 mg = _____ g

   25. 2.5 kg = _____ g   **26.** 2011 g = _____ kg

   27. 0.0334 g = _____ mg   **28.** 0.38 dag = _____ dg

   29. 90 dg = _____ hg   **30.** 0.003 kg = _____ dg

   31. 45 dag = _____ kg   **32.** 409 cg = _____ g

For Exercises 33–40, complete the table.

	Object	mg	cg	g	kg
33.	Bag of cat food				1.58
34.	Bag of flour				2.26
35.	Can of tuna			170	
36.	Bag of rice			907	
37.	Box of raisins		42,500		
38.	Hockey puck		17,000		
39.	Dose of acetaminophen	325			
40.	Olive	12			

Objective 3: Metric Units of Capacity

For Exercises 41–48, fill in the blank with >, <, or =.

41. 1 cL _____ 1 L

42. 1 kL _____ 1 daL

43. 1 L _____ 1 mL

44. 1 dL _____ 1 hL

45. 1 mL _____ 1 cc

46. 1 L _____ 1 cc

47. 1 cL _____ 1 kL

48. 1 mL _____ 1 cL

49. What does the abbreviation *cc* represent?

50. Which of the following are measures of capacity? Circle all that apply.

    a. cm   **b.** cc   **c.** cL   **d.** cg

Objective 4: Converting Metric Units of Capacity

For Exercises 51–60, convert the units of capacity. **(See Example 2.)**

51. 3200 mL = _____ L

52. 280 L = _____ kL

53. 7 L = _____ cL

54. 0.52 L = _____ mL

55. 42 mL = _____ dL

56. 0.88 L = _____ hL

57. 64 cc = _____ mL

58. 125 mL = _____ cc

59. 0.04 L = _____ cc

60. 38 cc = _____ L

For Exercises 61–68, complete the table.

	Object	mL	cL	L	kL
61.	1 Tablespoon	15			
62.	Bottle of vanilla extract	59			
63.	Bottle of vinegar		35.5		
64.	Bottle of soy sauce		29.6		
65.	Bottle of soda pop			2	
66.	Bottle of water			1	
67.	Capacity of a cooler				0.0377
68.	Capacity of a gasoline tank				0.0757

Objective 5: Summary of Metric Conversions

69. Identify the unit that applies to length.

    a. L   **b.** g   **c.** m

70. Identify the unit that applies to capacity.

    a. L   **b.** g   **c.** m

71. Identify the unit that applies to mass.

    a. L   **b.** g   **c.** m

72. Identify the units that apply to length.

    a. mL   **b.** mm   **c.** hg   **d.** cc   **e.** kg   **f.** hm   **g.** mL

73. Identify the units that apply to capacity.

    a. kg   **b.** km   **c.** cL   **d.** cc   **e.** hm   **f.** dag   **g.** mm

74. Identify the units that apply to mass.

    a. dg   **b.** hm   **c.** kL   **d.** cc   **e.** dm   **f.** kg   **g.** cL

For Exercises 75–80, convert the metric units as indicated. **(See Example 3.)**

75. The height of the tallest living tree is 112.014 m. Convert this to dekameters.

76. The Congo River is 4669 km long. Convert this to meters.

77. There is 600 mg of calcium in a multivitamin. Convert this to grams.

78. A soup can contains 305 g of soup. Convert this to milligrams.

79. A gasoline can has a capacity of 19 L. Convert this to kiloliters.

80. The capacity of a coffee cup is 0.25 L. Convert this to milliliters.

Objective 6: Applications

81. In one day, Stacy gets 600 mg of calcium in her daily vitamin, 500 mg in her calcium supplement, and 250 mg in the dairy products she ingests. How many grams of calcium will she ingest in one week? **(See Example 4.)**

82. Cliff drives his children to their sports activities outside of school. When he drives his son to baseball practice, it is a 6-km round trip. When he drives his daughter to basketball practice, it is a 1800-m round trip. If basketball practice is 3 times a week and baseball practice is twice a week, how many kilometers does Cliff drive?

83. A gas tank holds 58 L. If it costs $29 to fill up the tank, what is the price per liter?

84. A can of paint holds 120 L. How many kiloliters are contained in 8 cans?

85. A bottle of water holds 710 mL. How many liters are in a 6-pack?

86. A bottle of olive oil has 33 servings of 15 mL each. How many centiliters of oil does the bottle contain?

87. The drug amoxicillin is an antibiotic used to treat bacterial infections. A doctor orders 250 mg every 8 hr. How many grams of the drug would be given in 1 wk?

88. The drug acetaminophen is used as a pain reliever. A patient takes 325 mg every 6 hr. How many grams of the drug would be taken in a 3-day period?

89. A quart of milk has 130 mg of sodium per cup. How much sodium is in the whole bottle?

90. A ½-c serving of cereal has 180 mg of potassium. This is 5% of the recommended daily allowance of potassium. How many cups of cereal are needed to get 100% of the recommended daily allowance of potassium?

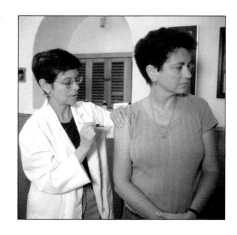

91. Dr. Boyd gives a patient 2 cc of Zantac. How many milliliters is this?

92. If a nurse mixed 11.5 mL of sterile water with 1.5 mL of oxacillin, how many cubic centimeters will this produce?

Expanding Your Skills

In the U.S. Customary System of measurement, 1 ton = 2000 lb. In the metric system, 1 metric ton = 1000 kg. Use this information to answer Exercises 93–96.

93. Convert 3300 kg to metric tons.

94. Convert 5780 kg to metric tons.

95. Convert 10.9 metric tons to kilograms.

96. Convert 8.5 metric tons to kilograms.

The metric system has other units of mass and capacity. For example, the microgram (μg) is 1 one-millionth of a gram (that is, 1 μg = 0.000001 g). Use this fact to answer Exercises 97–98.

97. Convert 0.02 mg to micrograms.

98. Convert 0.0004 cg to micrograms.

99. The drug Synthroid is used to treat thyroid disease. A patient is often started on a dose of 0.05 mg/day. How many micrograms is this?

---

chapter 7	midchapter review

For Exercises 1–20, convert the units of length, weight, mass, and capacity.

1. 36 c = _____ qt

2. 220 cm = _____ m

3. $\frac{3}{4}$ lb = _____ oz

4. 0.3 L = _____ mL

5. 12 ft = _____ yd

6. 6.03 kg = _____ g

7. 45 dm = _____ m

8. 9 in. = _____ ft

9. $\frac{1}{2}$ mi = _____ ft

10. 6000 lb = _____ tons

11. 8 pt = _____ qt

12. 1.5 tsp = _____ T

13. 21 m = _____ km

14. 68 mg = _____ cg

15. 36 mL = _____ cc

16. 64 oz = _____ lb

17. 4322 g = _____ kg

18. 5 m = _____ mm

19. 20 fl oz = _____ c

20. 510 sec = _____ min

Objectives

1. Summary of U.S. Customary and Metric Unit Equivalents
2. Converting U.S. Customary and Metric Units
3. Applications
4. Units of Temperature

 **section 7.5**   **Converting Between U.S. Customary and Metric Units**

1. Summary of U.S. Customary and Metric Unit Equivalents

In this section we learn how to convert between U.S. Customary and metric units of measure. Suppose, for example, that you take a trip to Europe. A street sign indicates that the distance to Paris is 45 km (Figure 7-11). This distance may be unfamiliar to you until you convert to miles.

| Paris | 45 km |
| Le Havre | 135 km |

Figure 7-11

Skill Practice

1. Use the fact that 1 mi ≈ 1.61 km to convert 184 km to miles. Round to the nearest mile.

example 1   Converting Metric Units to U.S. Customary Units

Use the fact that $1 \text{ mi} \approx 1.61 \text{ km}$ to convert 45 km to miles. Round to the nearest mile.

Solution:

$$45 \text{ km} \approx \frac{45 \text{ km}}{1} \cdot \frac{1 \text{ mi}}{1.61 \text{ km}}$$ Set up a unit ratio to convert kilometers to miles.

$$= \frac{45}{1.61} \text{ mi}$$ Multiply fractions.

$$\approx 28 \text{ mi}$$ Divide and round to the nearest mile.

The distance of 45 km to Paris is approximately 28 mi.

Table 7-6 summarizes some common metric and U.S. Customary equivalents.

table 7-6		
Length	**Weight/Mass (on Earth)**	**Capacity**
1 in. = 2.54 cm	1 lb ≈ 0.45 kg	1 qt ≈ 0.95 L
1 ft ≈ 0.305 m	1 oz ≈ 28 g	1 fl oz ≈ 30 mL = 30 cc
1 yd ≈ 0.914 m		
1 mi. ≈ 1.61 km		

2. Converting U.S. Customary and Metric Units

Using the U.S. Customary and metric equivalents given in Table 7-6, we can create unit ratios to convert between units.

Answer

1. 114 mi

example 2   Converting Units of Length

Fill in the blanks. Round to two decimal places, if necessary.

a. 18 cm = _____ in.   **b.** 15 yd ≈ _____ m   **c.** 82 m ≈ _____ ft

Solution:

a. $18 \text{ cm} = \dfrac{18 \text{ cm}}{1} \cdot \dfrac{1 \text{ in.}}{2.54 \text{ cm}}$ From Table 7-6, we know 1 in. = 2.54 cm.

$\quad\quad = \dfrac{18}{2.54} \text{ in.}$ Multiply fractions.

$\quad\quad \approx 7.09 \text{ in.}$ Divide and round to 2 decimal places.

b. $15 \text{ yd} \approx \dfrac{15 \text{ yd}}{1} \cdot \dfrac{0.914 \text{ m}}{1 \text{ yd}}$ From Table 7-6, we know 1 yd ≈ 0.914 m.

$\quad\quad = 13.71 \text{ m}$ Multiply.

c. $82 \text{ m} \approx \dfrac{8.2 \text{ m}}{1} \cdot \dfrac{1 \text{ ft}}{0.305 \text{ m}}$ From Table 7-6, we know 1 ft ≈ 0.305 m.

$\quad\quad = \dfrac{8.2}{0.305} \text{ ft}$ Multiply.

$\quad\quad = 26.89 \text{ ft}$ Divide and round to 2 decimal places.

Skill Practice

Convert. Round to 1 decimal place.

2. 4 m ≈ _____ ft

3. 3 in. ≈ _____ cm

4. 6500 yd ≈ _____ m

5. 6 km ≈ _____ mi

example 3   Converting Units of Weight and Mass

Fill in the blank. Round to 1 decimal place, if necessary.

a. 180 g ≈ _____ oz   **b.** 5.25 tons ≈ _____ kg

Solution:

a. $180 \text{ g} \approx \dfrac{180 \text{ g}}{1} \cdot \dfrac{1 \text{ oz}}{28 \text{ g}}$ From Table 7-6, we know 1 oz ≈ 28 g.

$\quad\quad = \dfrac{180}{28} \text{ oz}$

$\quad\quad \approx 6.4 \text{ oz}$ Divide and round to 1 decimal place.

b. We can first convert 5.25 tons to pounds. Then we can use the fact that 1 lb ≈ 0.45 kg.

$5.25 \text{ tons} = \dfrac{5.25 \text{ tons}}{1} \cdot \dfrac{2000 \text{ lb}}{1 \text{ ton}}$ Convert tons to pounds.

$\quad\quad = 10{,}500 \text{ lb}$

$\quad\quad \approx 10{,}500 \text{ lb} \cdot \dfrac{0.45 \text{ kg}}{1 \text{ lb}}$ Convert pounds to kilograms.

$\quad\quad = 4725 \text{ kg}$

Skill Practice

Convert. Round to 1 decimal place, if necessary.

6. 8 kg ≈ _____ lb

7. 4 tons ≈ _____ kg

8. 500 g ≈ _____ oz

Answers

2. 13.1 ft   3. 7.6 cm
4. 5941 m   5. 3.7 mi
6. 17.8 lb   7. 3600 kg
8. 17.9 oz

Skill Practice

Convert. Round to 1 decimal place, if necessary.

9. 120 mL $\approx$ _____ fl oz

10. 4 qt $\approx$ _____ L

---

example 4   Converting Units of Capacity

For each of the following conversions, round to 2 decimal places, if necessary.

a. 75 mL $\approx$ _____ fl oz   **b.** 3 qt $\approx$ _____ L

Solution:

a. $75 \text{ mL} \approx \dfrac{75 \text{ mL}}{1} \cdot \dfrac{1 \text{ fl oz}}{30 \text{ mL}}$ From Table 7-6, we know 1 fl oz $\approx$ 30 mL.

$\quad = \dfrac{75}{30} \text{ fl oz}$ Multiply fractions.

$\quad = 2.5 \text{ fl oz}$ Divide.

b. $3 \text{ qt} \approx \dfrac{3 \text{ qt}}{1} \cdot \dfrac{0.95 \text{ L}}{1 \text{ qt}}$ From Table 7-6, we know 1 qt $\approx$ 0.95 L.

$\quad = 2.85 \text{ L}$ Multiply.

---

3. Applications

Skill Practice

11. A 720-mL bottle of water sells for $0.79. A 32-oz bottle of water sells for $1.29. Compare the price per ounce to determine the better buy.

---

example 5   Converting Units in an Application

A 2-L bottle of soda sells for $2.19. A 32-oz bottle of soda sells for $1.59. Compare the price per quart of each bottle to determine the better buy.

Solution:

Note that 1 qt = 2 pt = 4 c = 32 fl oz. So a 32-oz bottle of soda costs $1.59 per quart. Next, if we can convert 2 L to quarts, we can compute the unit cost per quart and compare the results.

$2 \text{ L} \approx \dfrac{2 \text{ L}}{1} \cdot \dfrac{1 \text{ qt}}{0.95 \text{ L}}$ Recall that 1 qt = 0.95 L.

$\quad \approx \dfrac{2}{0.95} \text{ qt}$ Multiply fractions.

$\quad \approx 2.11 \text{ qt}$ Divide and round to 2 decimal places.

Now find the cost per quart. $\dfrac{\$2.19}{2.11 \text{ qt}} \approx \$1.04 \text{ per quart}$

The cost for the 2-L bottle is $1.04 per quart, whereas the cost for 32 oz is $1.59 per quart. Therefore, the 2-L bottle is a better buy.

---

example 6   Converting Units in an Application

In track and field, the 1500-m race is slightly less than 1 mi. How many yards less is it? Round to the nearest yard.

Answers

9. 4 fl oz   10. 3.8 L
11. 720 mL is 24 oz. The cost per ounce is $0.033. The unit price for the 32-oz bottle is $0.040 per ounce. The 720-mL bottle is the better buy.

Solution:

We know that 1 mi = 1760 yd. If we can convert 1500 m to yards, then we can subtract the results.

1 mi = 1760 yd

1500 m = ? yd

$$1500 \text{ m} \approx \frac{1500 \text{ m}}{1} \cdot \frac{1 \text{ yd}}{0.914 \text{ m}} \qquad \text{Recall that 1 yd} = 0.914 \text{ m.}$$

$$\approx \frac{1500}{0.914} \text{ yd} \qquad \text{Multiply fractions.}$$

$$\approx 1641 \text{ yd} \qquad \text{Divide and round to the nearest yard.}$$

Therefore, the difference between 1 mi and 1500 m is:

$$\underset{\downarrow}{(1 \text{ mi})} \quad - \quad \underset{\downarrow}{(1500 \text{ m})}$$

1760 yd − 1641 yd = 119 yd

Skill Practice

12. In track and field, the 800-m race is slightly shorter than a half-mile race. How many yards less is it? Round to the nearest yard.

4. Units of Temperature

In the United States, the **Fahrenheit** scale is used most often to measure temperature. On this scale, water freezes at 32°F and boils at 212°F. The symbol ° stands for "degrees," and °F means "degrees Fahrenheit."

Another scale used to measure temperature is the **Celsius** temperature scale. On this scale, water freezes at 0°C and boils at 100°C. The symbol °C stands for "degrees Celsius."

Figure 7-12 shows the relationship between the Celsius scale and the Fahrenheit scale.

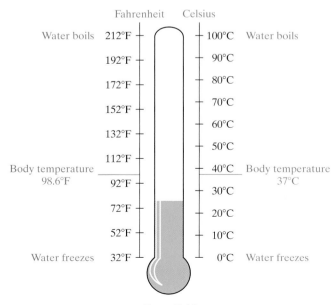

Figure 7-12

Fahrenheit   Celsius

Water boils   212°F   100°C   Water boils
               192°F   90°C
               172°F   80°C
               152°F   70°C
               132°F   60°C
                               50°C
               112°F
Body temperature   40°C   Body temperature
98.6°F   92°F   37°C
                                30°C
               72°F   20°C
               52°F   10°C
Water freezes   32°F   0°C   Water freezes

Concept Connections

Use Figure 7-12 to select the best choice.

13. The high temperature in Dallas for a day in July is

 a. 34°C   **b.** 34°F

14. The high temperature in Minneapolis for a day in March is

 a. 34°C   **b.** 34°F

Answers
12. 5 yd   13. a   14. b

To convert back and forth between the Fahrenheit and Celsius scales, we use the following formulas.

Conversions for Temperature Scale

To convert from °C to °F:

$$F = \frac{9}{5}C + 32$$

To convert from °F to °C:

$$C = \frac{5}{9}(F - 32)$$

---

example 7   Converting Units of Temperature

a. Convert a body temperature of 98.6°F to degrees Celsius.

b. Convert the temperature inside a refrigerator, 5°C, to degrees Fahrenheit.

Solution:

a. Because we want to convert degrees Fahrenheit to degrees Celsius, we use the formula $C = \frac{5}{9}(F - 32)$.

$$C = \frac{5}{9}(F - 32)$$

$$= \frac{5}{9}(98.6 - 32) \qquad \text{Substitute } F = 98.6.$$

$$= \frac{5}{9}(66.6) \qquad \text{Perform the operation inside parentheses first.}$$

$$= \frac{(5)(66.6)}{9}$$

$$= 37$$

Body temperature is 37°C.

b. Because we want to convert degrees Celsius to degrees Fahrenheit, we use the formula $F = \frac{9}{5}C + 32$

$$F = \frac{9}{5}C + 32$$

$$= \frac{9}{5} \cdot 5 + 32 \qquad \text{Substitute } C = 5.$$

$$= \frac{9}{5} \cdot \frac{5}{1} + 32$$

$$= 9 + 32$$

$$= 41$$

The temperature inside the refrigerator is 41°F.

section 7.5   Practice Exercises

Study Skills Exercises

1. Make a list of all the section titles in the chapter that you are studying. Write each section title on a separate sheet of paper or index card. Go back and fill in the list of concepts under each section title. When you are studying for the test, try to make up an exercise that corresponds to each concept and then work the exercise. To get started, write a problem for the objective of converting metric units of capacity from Section 7.4.

2. Define the key terms.

   a. Celsius   **b. Fahrenheit**

Review Exercises

For Exercises 3–6, select the equivalent amounts of mass. (*Hint:* There may be more than one answer for each exercise.)

3. 500 g
4. 500 mg
5. 500 cg
6. 500 kg

   a. 500,000 g
   b. 5 g
   c. 500,000,000 mg
   d. 0.5 kg
   e. 5000 mg
   f. 50,000 cg
   g. 0.5 g
   h. 50 cg

For Exercises 7–10, select the equivalent amounts of capacity.

7. 200 L
8. 200 kL
9. 200 mL
10. 200 cL

   a. 2000 mL
   b. 200 cc
   c. 0.2 kL
   d. 20,000,000 cL
   e. 200,000 L
   f. 200,000 mL
   g. 0.2 L
   h. 2 L

Objective 1: Summary of U.S. Customary and Metric Unit Equivalents

11. Fill in the blanks to complete the guidelines in choosing a unit ratio as a conversion factor.

    a. The unit of measure in the _____ of the fraction should be the new unit you want to convert *to*.

b. The unit of measure in the _____ of the fraction should be the original unit you want to convert *from*.

12. Identify an appropriate ratio to convert yards to meters by using multiplication.

a. $\dfrac{1 \text{ yd}}{0.914 \text{ m}}$ **b.** $\dfrac{0.914 \text{ m}}{1 \text{ yd}}$ **c.** $\dfrac{0.914 \text{ yd}}{1 \text{ m}}$ **d.** $\dfrac{1 \text{ m}}{0.914 \text{ yd}}$

13. Identify an appropriate ratio to convert pounds to kilograms by using multiplication.

a. $\dfrac{0.45 \text{ lb}}{1 \text{ kg}}$ **b.** $\dfrac{1 \text{ kg}}{0.45 \text{ lb}}$ **c.** $\dfrac{1 \text{ lb}}{0.45 \text{ kg}}$ **d.** $\dfrac{0.45 \text{ kg}}{1 \text{ lb}}$

14. Identify an appropriate ratio to convert quarts to liters by using multiplication.

a. $\dfrac{0.95 \text{ L}}{1 \text{ qt}}$ **b.** $\dfrac{1 \text{ qt}}{0.95 \text{ L}}$ **c.** $\dfrac{0.95 \text{ qt}}{1 \text{ L}}$ **d.** $\dfrac{1 \text{ L}}{0.95 \text{ qt}}$

15. Identify an appropriate ratio to convert miles to kilometers by using multiplication.

a. $\dfrac{1 \text{ mi}}{1.61 \text{ km}}$ **b.** $\dfrac{1 \text{ km}}{1.61 \text{ mi}}$ **c.** $\dfrac{1.61 \text{ km}}{1 \text{ mi}}$ **d.** $\dfrac{1.61 \text{ mi}}{1 \text{ km}}$

16. Identify an appropriate ratio to convert ounces to grams by using multiplication.

a. $\dfrac{28 \text{ oz}}{1 \text{ g}}$ **b.** $\dfrac{1 \text{ oz}}{28 \text{ g}}$ **c.** $\dfrac{28 \text{ g}}{1 \text{ oz}}$ **d.** $\dfrac{1 \text{ g}}{28 \text{ oz}}$

17. Identify an appropriate ratio to convert milliliters to fluid ounces by using multiplication.

a. $\dfrac{30 \text{ fl oz}}{1 \text{ mL}}$ **b.** $\dfrac{1 \text{ fl oz}}{30 \text{ mL}}$ **c.** $\dfrac{30 \text{ mL}}{1 \text{ fl oz}}$ **d.** $\dfrac{1 \text{ mL}}{30 \text{ fl oz}}$

18. Identify an appropriate ratio to convert inches to centimeters by using multiplication.

a. $\dfrac{1 \text{ in.}}{2.54 \text{ cm}}$ **b.** $\dfrac{2.54 \text{ in.}}{1 \text{ cm}}$ **c.** $\dfrac{1 \text{ cm}}{2.54 \text{ in.}}$ **d.** $\dfrac{2.54 \text{ cm}}{1 \text{ in.}}$

Objective 2: Converting U.S. Customary and Metric Units

For Exercises 19–26, convert the units of length. Round the answer to 1 decimal place, if necessary. **(See Examples 1–2.)**

19. 2 in. ≈ _____ cm   **20.** 4 ft ≈ _____ m   **21.** 8 m ≈ _____ yd

22. 120 km ≈ _____ mi   **23.** 400 ft ≈ _____ m   **24.** 0.75 m ≈ _____ yd

25. 2.5 km ≈ _____ mi   **26.** 1.5 cm ≈ _____ in.

For Exercises 27–34, convert the units of weight and mass. Round the answer to 1 decimal place, if necessary. **(See Example 3.)**

27. 6 oz ≈ _____ g

28. 6 lb ≈ _____ kg

29. 4 kg ≈ _____ lb

30. 10 g ≈ _____ oz

31. 14 g ≈ _____ oz

32. 0.54 kg ≈ _____ lb

33. 0.3 lb ≈ _____ kg

34. 1980 kg ≈ _____ tons

For Exercises 35–40, convert the units of capacity. Round the answer to 1 decimal place, if necessary. **(See Example 4.)**

35. 6 qt ≈ _____ L

36. 5 fl oz ≈ _____ mL

37. 120 mL ≈ _____ fl oz

38. 19 L ≈ _____ qt

39. 960 cc ≈ _____ fl oz

40. 0.5 fl oz ≈ _____ cc

Objective 3: Applications

41. A cross-country skiing race is 30 km long. Is this length more or less than 18 mi? **(See Example 6.)**

42. A can of cat food is 85 g. How many ounces is this? Round to the nearest ounce.

43. A 2-lb box of sugar costs $3.19. A box that contains single-serving packets contains 354 g and costs $1.49. Find the unit costs in dollars per ounce to determine the better buy. **(See Example 5.)**

44. A gas tank holds 12.5 gal of gas. How many liters is this? Round to the nearest tenth of a liter.

45. Carly Patterson of the U.S. Olympic gymnastic team weighs 97 lb. How many kilograms is this?

46. Mr. Kurry drives an average 100 km/hr while he is in Canada. Determine his speed in miles per hour by converting 100 km to miles. Round to the nearest mile per hour.

47. In a recent year, the price of gas in Germany was $1.60 per liter. What is the price per gallon?

48. A jar of spaghetti sauce is 2 lb 8 oz. How many kilograms is this? Round to 2 decimal places.

49. The thickness of a hockey puck is 2.54 cm. How many inches is this?

50. A bottle of grape juice contains 1.9 L of juice. Is there enough juice to fill 10 glasses that hold 6 oz each? If yes, how many ounces will be left over?

51. Hockey player Mario Lemieux weighs 99,790 g. How many pounds is this? Round to the nearest pound.

52. The distance between two lightposts is 6 m. How many feet is this? Round to the nearest foot.

53. A nurse gives a patient 45 cc of saline solution. How many fluid ounces does this represent?

54. Cough syrup comes in a bottle that contains 4 fl oz. How many milliliters is this?

Objective 4: Units of Temperature

For Exercises 55–60, convert the temperatures by using the appropriate formula: $F = \frac{9}{5}C + 32$ or $C = \frac{5}{9}(F - 32)$. **(See Example 7.)**

55. $25°C = $ _____ °F

56. $113°F = $ _____ °C

57. $68°F = $ _____ °C

58. $15°C = $ _____ °F

59. $30°C = $ _____ °F

60. $104°F = $ _____ °C

61. The boiling point of the element boron is $4000°C$. Find the boiling point in degrees Fahrenheit.

62. The melting point of the element copper is $1085°C$. Find the melting point in degrees Fahrenheit.

63. If the outdoor temperature is $35°C$, is it a hot day or a cold day?

64. The high temperature in London, England, on a typical September day was $18°C$, and the low was $13°C$. Convert these temperatures to degrees Fahrenheit.

65. Use the fact that water boils at $100°C$ to show that the boiling point is $212°F$.

66. Use the fact that water freezes at $0°C$ to show that the temperature at which water freezes is $32°F$.

Expanding Your Skills

In the U.S. Customary System of measurement, 1 ton = 2000 lb. In the metric system, 1 metric ton = 1000 kg. Use this information to answer Exercises 67–70.

67. A Lincoln Navigator weighs 5700 lb. How many metric tons is this?

68. An elevator has a maximum capacity of 1200 lb. How many metric tons is this?

69. The average mass of a blue whale (the largest mammal in the world) is approximately 108 metric tons. How many pounds is this?

70. The mass of a Mini-Cooper is 1.25 metric tons. How many pounds is this? Round to the nearest pound.

section 7.6   Energy and Power (Optional)

1. Units of Energy

In this section, we discuss common units of energy and power. **Energy** is defined as the amount of work that a physical system is capable of performing. Energy is stored in fossil fuels (coal, wood, oil, and natural gas), the food we eat, and even our own bodies.

Three common units of energy that we discuss here are the

Foot-pound (ft·lb)

British thermal unit (Btu)

Kilocalorie (or simply Calorie, Cal)

One foot-pound (1 ft·lb) is equal to the amount of energy required to lift 1 lb a distance of 1 ft (see Figure 7-13).

Figure 7-13

Notice that the foot-pound (ft·lb) is a product of feet and pounds. Thus, to compute the amount of energy necessary to lift a weight a certain distance, we have

$$\text{Energy} = \binom{\text{Distance}}{\text{in feet}} \cdot \binom{\text{Weight}}{\text{in pounds}}$$

example 1   Computing Energy in Foot-pounds

How many foot-pounds of energy is required to lift a steel beam weighing 475 lb to the top of a 64-ft roof?

Solution:

$$\text{Energy} = \binom{\text{Distance}}{\text{in feet}} \cdot \binom{\text{Weight}}{\text{in pounds}}$$

$$= (64 \text{ ft}) \cdot (475 \text{ lb})$$

$$= 30{,}400 \text{ ft·lb}$$

The British thermal unit (Btu) is often used to measure energy consumption for heating and air conditioning systems. For example, a portable heater might have a rating of 60,000 Btu. To convert between foot-pounds and British thermal units, we have

$$1 \text{ Btu} \approx 778 \text{ ft·lb}$$

example 2　Converting Units of Energy

a. 1250 Btu ≈ _____ ft·lb

b. 5057 ft·lb ≈ _____ Btu

Solution:

a. $1250 \text{ Btu} \approx \dfrac{1250 \text{ Btu}}{1} \cdot \dfrac{778 \text{ ft·lb}}{1 \text{ Btu}}$　　Recall that 1 Btu ≈ 778 ft·lb.

$\approx 972{,}500 \text{ ft·lb}$

b. $5057 \text{ ft·lb} \approx \dfrac{5057 \text{ ft·lb}}{1} \cdot \dfrac{1 \text{ Btu}}{778 \text{ ft·lb}}$

$\approx 6.5 \text{ Btu}$

Heat is also a type of energy. In the metric system, heat is measured in calories or kilocalories. One kilocalorie (1 kcal) is equal to 1000 cal. However, kilocalories are often called **Calories** (spelled with a capital C and abbreviated as Cal). It is these larger units of Calories that are used frequently in dietary science.

example 3　Computing Calories Used in Exercise

Walking at a moderate pace burns 480 Cal/hr. If Molly walks a half-marathon in 3 hr 15 min, how many calories does she burn?

Solution:

We will assume that the rate of 480 Cal/hr burned remains constant. Therefore, we can set up a proportion to solve this problem.

First note that 3 hr 15 min is equal to 3.25 hr. Thus,

$$\text{number of Calories} \to \dfrac{480}{1} = \dfrac{x}{3.25} \leftarrow \text{number of Calories}$$

number of hours →　　　　　← number of hours

$(480)(3.25) = (1) \cdot x$　　Cross-multiply to form the cross products.

$1560 = x$　　Simplify.

Molly will burn 1560 Cal walking a half-marathon.

2. Units of Power

Power is the rate at which energy is released. One way to express power is in foot-pounds per second $\left(\dfrac{\text{ft·lb}}{\text{sec}}\right)$.

example 4　Finding Power

a. Find the power if a 50-lb box is lifted 10 ft in 4 sec.

b. Find the power if a 50-lb box is lifted 10 ft in 2 sec.

Solution:

a. Because power is expressed in units of $\dfrac{\text{ft·lb}}{\text{sec}}$ we have

$$\text{Power} = \frac{(10 \text{ ft})(50 \text{ lb})}{4 \text{ sec}} = \frac{500 \text{ ft·lb}}{4 \text{ sec}} = 125 \frac{\text{ft·lb}}{\text{sec}}$$

b. To find the power used to lift the same box the same distance in one-half the time, we have

$$\text{Power} = \frac{(10 \text{ ft})(50 \text{ lb})}{2 \text{ sec}} = \frac{500 \text{ ft·lb}}{2 \text{ sec}} = 250 \frac{\text{ft·lb}}{\text{sec}}$$

Notice that more power is required to lift the box more quickly.

Skill Practice

5. Find the power if an 80-lb box is lifted 20 ft in 10 sec.

6. Find the power if an 80-lb box is lifted 20 ft in 5 sec.

A unit of power that is used in the U.S. Customary System is the horsepower.

$$1 \text{ horsepower (hp)} = 550 \frac{\text{ft·lb}}{\text{sec}}$$

A measure of 1 hp loosely indicates that a horse can move 550 lb a distance of 1 ft in 1 sec.

---

example 5   Convert Units of Power

A Mercury engine is rated 200 hp. Convert this to $\dfrac{\text{ft·lb}}{\text{sec}}$.

Solution:

$$200 \text{ hp} = \frac{200 \text{ hp}}{1} \cdot \frac{550\frac{\text{ft·lb}}{\text{sec}}}{1 \text{ hp}} \qquad \text{Recall that } 1 \text{ hp} = 500 \frac{\text{ft·lb}}{\text{sec}}.$$

$$= 110,000 \frac{\text{ft·lb}}{\text{sec}}$$

Skill Practice

7. A lawn mower is rated 6.5 hp. Convert this to $\dfrac{\text{ft·lb}}{\text{sec}}$.

Answer

5. $160 \dfrac{\text{ft·lb}}{\text{sec}}$ 6. $320 \dfrac{\text{ft·lb}}{\text{sec}}$

7. $3575 \dfrac{\text{ft·lb}}{\text{sec}}$

---

section 7.6   Practice Exercises

Boost *your* GRADE at mathzone.com!

MathZone

- Practice Problems
- Self-Tests
- NetTutor
- e-Professors
- Videos

Study Skills Exercises

1. Write down the number of hours devoted to study, work, classes, and entertainment for one week. Compare the amount of time devoted to studying for school to the time spent on entertainment such as watching TV and movies.

	Work	Classes	Studying	Entertainment
Monday				
Tuesday				
Wednesday				
Thursday				
Friday				
Saturday				
Sunday				

2. Define the key terms.

   a. Energy   **b. Calories**

Review Exercises

For Exercises 3–6, complete the table by converting the temperatures. Round to the nearest tenth if necessary.

	Description	°F	°C
3.	Heat of oven for baking cookies	350°F	
4.	Temperature of a typical spring day in Oklahoma City		25°C
5.	Temperature of a typical winter day in Modesto, California		5°C
6.	Body temperature	98.6°F	

Objective 1: Units of Energy

For Exercises 7–12, find the energy in foot-pounds.

  7. Find the energy required to lift a 3800-lb car 6 ft. **(See Example 1.)**

  8. Find the energy required to lift a 3000-lb car 5 ft.

  9. Find the energy required to lift 200 lb a distance of 2 yd.

 10. Find the energy required to lift 50 lb a distance of 1.5 yd.

 11. How much energy would be required to lift 2.5 tons a distance of 3 ft?

 12. How much energy would be required to lift 1.5 tons a distance of 4 ft?

For Exercises 13–16, find the foot-pound equivalent for the heater or air conditioner. **(See Example 2.)**

13. "Toasty" Patio Heater (40,000 Btu)

14. Portable battery powered Heatpro space heater (3000 Btu)

15. Air conditioner with remote (14,000 Btu)

16. Window air conditioner (8000 Btu)

For Exercises 17–20, convert the units of energy. Round to the nearest whole unit, if necessary.

17. 4000 ft·lb ≈ _____ Btu

18. 53,000 ft·lb ≈ _____ Btu

19. 32,000 ft·lb ≈ _____ Btu

20. 1,000,000 ft·lb ≈ _____ Btu

21. The amount of energy from 1 gal of gasoline is 124,000 Btu. Find the amount of energy in foot-pounds.

22. The amount of energy from 1 barrel (bbl) (42 gal) of crude oil is 5,800,000 Btu. Find the amount of energy in foot-pounds.

23. The amount of energy from 1 gal of propane is 90,000 Btu. Find the amount of energy in foot-pounds.

24. The amount of energy from 1 ft^3 of natural gas is 1026 Btu. Find the amount of energy in foot-pounds.

For Exercises 25–30, convert to hours. Write the answer in both fraction form and decimal form, rounded to the nearest hundredth.

25. 40 min = _____ hr

26. 45 min = _____ hr

27. 1 hr 15 min = _____ hr

28. 2 hr 30 min = _____ hr

29. 2 hr 24 min = _____ hr

30. 1 hr 6 min = _____ hr

For Exercises 31–38, determine the number of Calories burned. Use the information given in the table. Round to the nearest whole unit, if necessary. **(See Example 3.)**

Activity	Cal/hr
Basketball game	560
Bicycling, 14–16 mph	700
Mowing the lawn	380
Lacrosse	500
Running, 10-min mile pace	590
Walking, moderate pace	280

31. Bicycling at 14–16 mph for 1 hr 20 min.

32. Running a 10-min mile pace for 45 min.

33. Walking for 2 hr 45 min.

34. Playing lacrosse for 2 hr 30 min.

35. Mowing the lawn for 48 min.

36. Playing basketball for 1 hr 40 min.

37. Running a 10-min mile pace for 30 min.

38. Mowing the lawn for 1 hr 15 min.

Objective 2: Units of Power

39. Find the power in foot-pounds per second of lifting 25 lb a distance of 5 ft in 5 sec. **(See Example 4.)**

40. Find the power in foot-pounds per second of lifting 40 lb a distance of 3 ft in 2 sec.

41. A machine can raise 200 lb a distance of 1 yd in 6 sec. Find the power in foot-pounds per second.

42. An engine can raise 300 lb a distance of 1.5 yd in 10 sec. Find the power in foot-pounds per second.

43. Find the power in foot-pounds per second of raising 1 ton a distance of 3 ft in 15 sec.

44. Find the power in foot-pounds per second of raising 1 ton a distance of 3 ft in 30 sec.

For Exercises 45–50, convert foot-pounds per second to horsepower.

45. $550 \dfrac{\text{ft·lb}}{\text{sec}} = $ _____ hp

46. $1100 \dfrac{\text{ft·lb}}{\text{sec}} = $ _____ hp

47. $4950 \dfrac{\text{ft·lb}}{\text{sec}} = $ _____ hp

48. $6050 \dfrac{\text{ft·lb}}{\text{sec}} = $ _____ hp

49. $1540 \dfrac{\text{ft·lb}}{\text{sec}} = \underline{\hspace{2cm}} \text{ hp}$

50. $1375 \dfrac{\text{ft·lb}}{\text{sec}} = \underline{\hspace{2cm}} \text{ hp}$

51. A new LS2 6.0-L small-block V-8 is the standard engine in the 2005 Corvette C6. It delivers peak output levels of 400 hp. Convert this to foot-pounds per second. **(See Example 5.)**

52. The 2003 Porsche 911 is rated at 315 hp. Convert this to foot-pounds per second.

53. A six-speed car will produce up to 550 hp. Convert this to foot-pounds per second.

54. A Dodge Ram pickup has a 215-hp engine. Convert this to foot-pounds per second.

55. The Nissan Titan has a 305-hp engine. Convert this to foot-pounds per second.

56. The Toyota Tacoma has a 142-hp engine. Convert this to foot-pounds per second.

Expanding Your Skills

Electric energy is often measured by the watthour (Wh). For example, a 60-W lightbulb will emit 60 Wh of energy in 1 hr. For Exercises 57–59, use the fact that 1 kilowatthour (kWh) = 1000 Wh.

57. A television is rated at 250 W. Suppose the television is on for 3 hr/day.

    a. How many watthours are used over a 30-day period?

    b. How many kilowatthours are used over a 30-day period?

    c. If the power company charges $0.11 per kilowatthour, what is the cost to run the television for a 30-day period?

58. A hot water heater is rated at 3500 W. Suppose the water heater comes on for 6 hr/day.

    a. How many watthours are used over a 30-day period?

    b. How many kilowatthours are used over a 30-day period?

    c. If the power company charges $0.082 per kilowatthour, what is the cost to run the hot water heater for a 30-day period?

59. A refrigerator uses 3000 Wh/day of energy. What is the cost of energy to operate the refrigerator for 30 days if the power company charges $0.071 per kilowatthour?

chapter 7 | summary

section 7.1 Converting U.S. Customary Units of Length

Key Concepts

The U.S. Customary System for measuring length is commonly used in the United States. Conversion of length can be done by substitution using these equivalents.

1 ft = 12 in.	1 in. = $\frac{1}{12}$ ft
1 yd = 3 ft	1 ft = $\frac{1}{3}$ yd
1 mi = 5280 ft	1 ft = $\frac{1}{5280}$ mi
1 mi = 1760 yd	1 yd = $\frac{1}{1760}$ mi

Conversion of length can also be done by multiplying by an appropriate conversion factor, such as $\frac{12 \text{ in.}}{1 \text{ ft}}$ or $\frac{1 \text{ mi}}{5280 \text{ ft}}$.

To choose a **conversion factor**, follow these guidelines:

- The unit of measure in the numerator is the new unit you want to convert *to*.
- The unit of measure in the denominator is the original unit you want to convert *from*.

When adding and subtracting measurements, add or subtract like units.

Examples

Example 1
To convert 18 in. to feet, write

$$18 \text{ in.} = 18 \times 1 \text{ in.}$$

$$= \frac{18}{1} \times \frac{1}{12} \text{ ft}$$

$$= \frac{18}{12} = \frac{3}{2} \text{ ft or } 1\frac{1}{2} \text{ ft}$$

Example 2
To convert 8 yd to feet, multiply.

$$8 \text{ yd} \cdot \frac{3 \text{ ft}}{1 \text{ yd}} = \frac{8 \text{ yd}}{1} \cdot \frac{3 \text{ ft}}{1 \text{ yd}} = 24 \text{ ft}$$

Example 3
To convert 60 in. to yards, multiply.

$$60 \text{ in.} \cdot \frac{1 \text{ ft}}{12 \text{ in.}} \cdot \frac{1 \text{ yd}}{3 \text{ ft}}$$

$$= \frac{60 \text{ in.}}{1} \cdot \frac{1 \text{ ft}}{12 \text{ in.}} \cdot \frac{1 \text{ yd}}{3 \text{ ft}}$$

$$= \frac{60 \text{ yd}}{36}$$

$$= \frac{5}{3} \text{ yd or } 1\frac{2}{3} \text{ yd}$$

Example 4
To add 3 ft 9 in. + 2 ft 10 in., add like terms.

$$
\begin{array}{r}
3 \text{ ft} + 9 \text{ in.} \\
+ 2 \text{ ft} + 10 \text{ in.} \\
\hline
5 \text{ ft} + 19 \text{ in.} = 5 \text{ ft} + 1 \text{ ft} + 7 \text{ in.} \\
= 6 \text{ ft } 7 \text{ in.}
\end{array}
$$

Conversion of units of length can be used in many applications. See Example 5.

Example 5

Copper tubing is needed to connect a water source to a refrigerator and a dishwasher.

a. If it takes 4 ft 10 in. of tubing for the refrigerator and 3 ft 6 in. for the dishwasher, what is the total length of tubing needed?
b. Suppose a plumber has 10 ft of tubing in his truck. How much will he have left after the installation?

Solution:

a. 4 ft + 10 in.
  3 ft + 6 in.
  7 ft + 16 in. = 7 ft + 1 ft + 4 in.
        = 8 ft 4 in.

The total tubing needed is 8 ft 4 in.

b. 10 ft   =   9 ft + 12 in.
   −(8 ft 4 in.)   =   −(8 ft +   4 in.)
   ─────────   ─────────────
                            1 ft +   8 in.

There will be 1 ft 8 in. of tubing left over.

section 7.2 Converting U.S. Customary Units of Time, Weight, and Capacity

Key Concepts

Several common U.S. Customary units—time, weight, and capacity—are given.

Time

1 year = 365 days

1 week = 7 days

1 day = 24 hours (hr)

1 hour (hr) = 60 minutes (min)

1 minute (min) = 60 seconds (sec)

Weight

1 pound (lb) = 16 ounces (oz)

1 ton = 2000 pounds (lb)

Capacity

1 cup (c) = 8 fluid ounces (fl oz)

1 pint (pt) = 2 cups (c)

1 quart (qt) = 2 pints (pt)

1 gallon (gal) = 4 quarts (qt)

Examples

Example 1

To convert 200 min to hours, multiply.

$$200 \text{ min} \cdot \frac{1 \text{ hr}}{60 \text{ min}}$$

$$= \frac{200 \text{ min}}{1} \cdot \frac{1 \text{ hr}}{60 \text{ min}}$$

$$= \frac{200}{60} \text{ hr}$$

$$= \frac{10}{3} \text{ hr or } 3\frac{1}{3} \text{ hr}$$

Example 2

To convert 6 lb to ounces, multiply.

$$6 \text{ lb} \cdot \frac{16 \text{ oz}}{1 \text{ lb}}$$

$$= \frac{6 \text{ lb}}{1} \cdot \frac{16 \text{ oz}}{1 \text{ lb}}$$

$$= \frac{96 \text{ oz}}{1} \text{ or } 96 \text{ oz}$$

Example 3

To convert 40 cups to gallons, multiply.

$$40 \text{ c} \cdot \frac{1 \text{ pt}}{2 \text{ c}} \cdot \frac{1 \text{ qt}}{2 \text{ pt}} \cdot \frac{1 \text{ gal}}{4 \text{ qt}}$$

$$= \frac{40 \text{ c}}{1} \cdot \frac{1 \text{ pt}}{2 \text{ c}} \cdot \frac{1 \text{ qt}}{2 \text{ pt}} \cdot \frac{1 \text{ gal}}{4 \text{ qt}}$$

$$= \frac{40}{16} \text{ gal}$$

$$= \frac{5}{2} \text{ gal or } 2\frac{1}{2} \text{ gal}$$

■■■■ section 7.3 **Metric Units of Length**

Key Concepts

The **metric system** offers other units for measuring length, mass, and capacity. The base units are the **meter** for length, the **gram** for mass, and the **liter** for capacity. Other units of length, mass, and capacity in the metric system are powers of 10 of the base unit.

Metric units of length and their equivalents are given.

1 kilometer (km) = 1000 m

1 hectometer (hm) = 100 m

1 dekameter (dam) = 10 m

1 meter (m) = 1 m

1 decimeter (dm) = 0.1 m $\left(\frac{1}{10}\ m\right)$

1 centimeter (cm) = 0.01 m $\left(\frac{1}{100}\ m\right)$

1 millimeter (mm) = 0.001 m $\left(\frac{1}{1000}\ m\right)$

It is useful to use a **prefix line** to help in converting from one metric length to another.

Prefix Line for Length

1000 m	100 m	10 m	1 m	0.1 m	0.01 m	0.001 m
km	hm	dam	m	dm	cm	mm
kilo-	hecto-	deka-		deci-	centi-	milli-

1. To use the prefix line, begin at the point on the line corresponding to the original unit you are given.
2. Then count the number of positions you need to move to reach the new unit of measurement.
3. Move the decimal point in the original measured value in the same direction and same number of places as on the prefix line.
4. Replace the original unit with the new unit of measure.

Examples

Example 1

To convert 2 km to meters by using substitution, write

$2\ km = 2 \times 1\ km$

$= 2 \times 1000\ m$

$= 2000\ m$

Example 2

To convert 62 cm to meters by using substitution, write

$62\ cm = 62 \times 1\ cm$

$= 62 \times 0.01\ m$

$= 0.62\ m$

Example 3

Prefix Line for Length

1000 m	100 m	10 m	1 m	0.1 m	0.01 m	0.001 m
km	hm	dam	m	dm	cm	mm
kilo-	hecto-	deka-		deci-	centi-	milli-

4.2 dam = 4200 cm

Example 4

Javier has 5 books that he wants to stack on a shelf. The thicknesses of the books are 40, 38, 35, 30, and 33 mm. If the shelf is 18 cm high, will the stack of the books fit?

Add: 40 + 38 + 35 + 30 + 33 = 176 mm

Convert 176 mm = 17.6 cm. Because 17.6 cm is less than 18 cm, the stack of books will fit on the shelf.

section 7.4 Metric Units of Mass and Capacity

Key Concepts

The prefix line can be used to convert metric units for mass, capacity, and length.

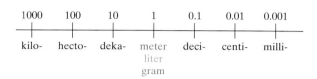

The metric unit conversions for mass are given in **grams**.

1 kilogram (kg) = 1000 g

1 hectogram (hg) = 100 g

1 dekagram (dag) = 10 g

1 gram (g) = 1 g

1 decigram (dg) = 0.1 g $\left(\frac{1}{10}\text{ g}\right)$

1 centigram (cg) = 0.01 g $\left(\frac{1}{100}\text{ g}\right)$

1 milligram (mg) = 0.001 g $\left(\frac{1}{1000}\text{ g}\right)$

The metric unit conversions for capacity are given in **liters**.

1 kiloliter (kL) = 1000 L

1 hectoliter (hL) = 100 L

1 dekaliter (daL) = 10 L

1 liter (L) = 1 L

1 deciliter (dL) = 0.1 L $\left(\frac{1}{10}\text{ L}\right)$

1 centiliter (cL) = 0.01 L $\left(\frac{1}{100}\text{ L}\right)$

1 milliliter (mL) = 0.001 L $\left(\frac{1}{1000}\text{ L}\right)$

Note that 1 mL = 1 cc.

Examples

Example 1

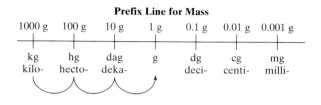

0.962 kg = 962 g

Example 2

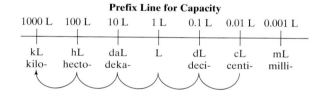

59,000 cL = 0.59000 kL

= 0.59 kL

section 7.5 Converting Between U.S. Customary and Metric Units

Key Concepts

The common conversions between U.S. Customary and metric systems are given.

Length

1 in. = 2.54 cm

1 ft ≈ 0.305 m

1 yd ≈ 0.914 m

1 mi ≈ 1.61 km

Weight/Mass (on Earth)

1 lb ≈ 0.45 kg

1 oz ≈ 28 g

Capacity

1 qt ≈ 0.95 L

1 fl oz ≈ 30 mL = 30 cc

To convert U.S. Customary units to metric or metric units to U.S. Customary, use unit ratios and the prefix line.

The U.S. Customary System uses the **Fahrenheit** scale (°F) to measure temperature. The metric system uses the **Celsius** scale (°C). The conversions are given.

To convert from °C to °F: $F = \dfrac{9}{5}C + 32$

To convert from °F to °C: $C = \dfrac{5}{9}(F - 32)$

Examples

Example 1

Convert 1200 yd to meters by using a unit ratio.

$$\frac{1200 \text{ yd}}{1} \cdot \frac{0.914 \text{ m}}{1 \text{ yd}} \approx 1096.8 \text{ m}$$

Example 2

To convert 900 cc to fluid ounces, recall that 900 cc = 900 mL. Then use a unit ratio to convert to fluid ounces.

$$\frac{900 \text{ mL}}{1} \cdot \frac{1 \text{ fl oz}}{30 \text{ mL}} = 30 \text{ fl oz}$$

Example 3

The average January temperature in Havana, Cuba, is 21°C. The average January temperature in Johannesburg, South Africa, is 69°F. Which temperature is warmer?

Convert 21°C to degrees Fahrenheit:

$$F = \frac{9}{5}C + 32$$

$$= \frac{9}{5}(21) + 32$$

$$= 37.8 + 32 = 69.8$$

The value 21°C = 69.8°F, which is 0.8 degree warmer than the temperature in Johannesburg.

section 7.6 Energy and Power (Optional)

Key Concepts

Energy is the amount of work that a physical system is capable of performing. Three units of energy are

Foot-pound (ft·lb)

British thermal unit (Btu)

Kilocalorie (Calorie, Cal)

One foot-pound (1 ft·lb) is equal to the amount of energy required to lift 1 lb a distance of 1 ft.

The British thermal unit (Btu) is often used to measure energy consumption for heating and aircondi-tioning systems.

To convert between foot-pounds and British thermal units, we use 1 Btu ≈ 778 ft·lb

The kilocalorie (**Calorie**) is used in exercise and dietary science.

Power is the rate at which energy is released. Power can be expressed in foot-pounds per second $\left(\frac{\text{ft·lb}}{\text{sec}}\right)$.

Another way that power can be expressed is in horse-power. To convert between foot-pounds per second and horsepower, we use

$1 \text{ horsepower (hp)} = 550 \dfrac{\text{ft·lb}}{\text{sec}}$

Examples

Example 1

The energy required to lift 6 lb a distance of 4 ft is 24 ft·lb. That is,

6 lb · 4 ft = 24 ft·lb

Example 2

The amount of energy from 1 gal of heating oil is 139,000 Btu. To convert this to foot-pounds, multiply.

$139{,}000 \text{ Btu} \approx 139{,}000 \text{ Btu} \cdot \dfrac{778 \text{ ft·lb}}{1 \text{ Btu}}$

$= 108{,}142{,}000 \text{ ft·lb}$

Example 3

When Julie plays tennis, she burns 580 Cal/hr. To find the number of Calories burned when she plays for 1 hr 15 min, solve the proportion.

$\dfrac{580}{1} = \dfrac{x}{1.25}$ (1 hr 15 min = 1.25 hr.)

$580(1.25) = (1)x$

$725 = x$ Julie will burn 725 Cal.

Example 4

The power to raise 120 lb a distance of 5 ft in 15 sec is given by

$\dfrac{120 \text{ lb} \cdot 5 \text{ ft}}{15 \text{ sec}} = 40 \dfrac{\text{ft·lb}}{\text{sec}}$

Example 5

Convert a 250-hp engine to $\left(\dfrac{\text{ft·lb}}{\text{sec}}\right)$.

$250 \text{ hp} \cdot \dfrac{550 \frac{\text{ft·lb}}{\text{sec}}}{1 \text{ hp}} = 137{,}500 \dfrac{\text{ft·lb}}{\text{sec}}$

chapter 7 | review exercises

Section 7.1

For Exercises 1–8, convert the units of length.

1. 48 in. = ___ ft

2. $3\frac{1}{4}$ ft = ___ in.

3. 2 mi = ___ yd

4. 2200 yd = ___ mi

5. 7040 ft = ___ mi

6. $\frac{1}{2}$ mi = ___ ft

7. 2 yd = ___ in.

8. 6336 in. = ___ mi

For Exercises 9–16, perform the indicated operations.

9. 3 ft 9 in. + 5 ft 6 in.

10. 4′11″ + 1′5″

11. 5′3″ − 2′5″

12. 12 ft 7 in. − 8 ft 10 in.

13. 4 × (5′3″)

14. 2 × (4 ft 8 in.)

15. 6 ft 3 in. ÷ 3

16. 6 yd 2 ft ÷ 2

17. Find the perimeter in feet.

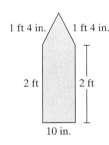

1 ft 4 in.   1 ft 4 in.

2 ft   2 ft

10 in.

18. A roll of wire contains 50 yd of wire. If Ivan uses 48 ft, how much wire is left?

Section 7.2

For Exercises 19–30, convert the units of time, weight, and capacity.

19. 72 hr = ___ days

20. 6 min = ___ sec

21. 5 lb = ___ oz

22. 1 wk = ___ hr

23. 12 fl oz = ___ c

24. 0.25 ton = ___ lb

25. 3500 lb = ___ tons

26. 150 min = ___ hr

27. 1800 sec = ___ hr

28. 2 gal = ___ pt

29. 12 oz = ___ lb

30. 16 qt = ___ gal

31. A runner finished a race with a time of 2:24:30. Convert the time to minutes.

32. Margaret Johansson gave birth to triplets who weighed 3 lb 10 oz, 4 lb 2 oz, and 4 lb 1 oz. What was the total weight of the triplets?

33. One and one-half tons of dirt are to be equally distributed to 8 locations. How many pounds will go to each location?

34. A case of 12-fl-oz cans of soda contains 24 cans. How many gallons of soda are in the case?

Section 7.3

For Exercises 35–36, use a metric ruler to measure the dimensions of each figure in centimeters.

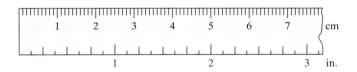

cm

1   2   3

in.

35. Post-it notes

Sticky Notes

36. Credit card

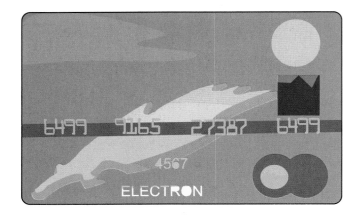

For Exercises 37–40, select the most reasonable measurement.

37. A pencil is _____ long.

    a. 16 mm   **b.** 16 cm

    c. 16 m   **d.** 16 km

38. A mosquito is _____ long.

    a. 12 mm   **b.** 12 cm

    c. 12 m   **d.** 12 km

39. A two-story house is _____ tall.

    a. 9 mm   **b.** 9 cm

    c. 9 m   **d.** 9 km

40. The distance between Houston and Dallas is _____ .

    a. 362 mm   **b.** 362 cm

    c. 362 m   **d.** 362 km

For Exercises 41–50, convert the metric units of length.

41. 52 cm = ___ mm   **42.** 91 mm = ___ cm

43. 2.338 km = ___ m   **44.** 93 m = ___ km

45. 34 dm = ___ m   **46.** 2.1 m = ___ dm

47. 4 cm = ___ dam   **48.** 3 dam = ___ cm

49. 1.2 hm = ___ dm   **50.** 4023 dm = ___ km

51. The highest point in Arizona is 3.851 km above sea level. The highest point in Louisiana is 163 m. What is the difference between these elevations in meters?

52. Determine the perimeter of the triangle in centimeters.

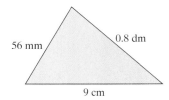

Section 7.4

For Exercises 53–58, convert the metric units of mass.

53. 6.1 g = ___ cg   **54.** 420 g = ___ kg

55. 3212 mg = ___ g   **56.** 0.7 hg = ___ g

57. 5 cg = ___ mg   **58.** 0.1 dag = ___ cg

For Exercises 59–64, convert the metric units of capacity.

59. 300 mL = ___ L   **60.** 2.4 hL = ___ L

61. 830 cL = ___ L   **62.** 124 mL = ___ cc

63. 225 cc = ___ cL   **64.** 0.49 kL = ___ L

65. The dimensions of a dining room table are 2 m by 125 cm. Convert the units to meters and find the perimeter and area of the tabletop.

66. A bottle of apple juice contains 1.2 L of juice. If a glass holds 24 cL, how many glasses can be filled from this bottle?

67. An adult has a mass of 68 kg. A baby has a mass of 3200 g. What is the difference in their masses, in kilograms?

68. From a wooden board 2 m long, Jesse needs to cut 3 pieces that are each 75 cm long. Is the 2-m length of board long enough for the 3 pieces?

69. A standard hypodermic syringe holds 3 cc of fluid. If a nurse uses 1.8 mL of the fluid, how much is left in the syringe?

70. A medication comes in 250-mg capsules. If Clayton took 3 capsules a day for 10 days, how many grams of the medication did he take?

Section 7.5

For Exercises 71–80, convert the units of length, capacity, mass, and weight. Round to the nearest hundredth, if necessary.

71. 6.2 in. ≈ _____ cm

72. 75 mL ≈ _____ fl oz

73. 140 g ≈ _____ oz

74. 5 L ≈ _____ qt

75. 3.4 ft ≈ _____ m

76. 100 lb ≈ _____ kg

77. 120 km ≈ _____ mi

78. 6 qt ≈ _____ L

79. 1.5 fl oz ≈ _____ cc

80. 12.5 tons ≈ _____ kg

81. The height of a computer desk is 30 in. The height of the chair is 38 cm. What is the difference in height between the desk and chair, in centimeters?

82. A bag of snack crackers contains 7.2 oz. If one serving is 30 g, approximately how many servings are in one bag?

83. A prescription for cough syrup indicates that 30 mL should be taken twice a day for 7 days. What is the total amount, in liters, of cough syrup to be taken?

84. The Boston Marathon is 42.195 km long. Convert this distance to miles. Round to the tenths place.

85. Write the formula to convert degrees Fahrenheit to degrees Celsius.

86. When roasting a turkey, the meat thermometer should register between 180°F and 185°F to indicate that the turkey is done. Convert these temperatures to degrees Celsius. Round to the nearest tenth, if necessary.

87. Write the formula to convert degrees Celsius to degrees Fahrenheit.

88. The average October temperature for Toronto, Ontario, Canada, is 8°C. Convert this temperature to degrees Fahrenheit.

Section 7.6 (Optional)

89. Find the energy in foot-pounds required to lift 2800 lb a distance of 5.5 ft.

90. How much energy in foot-pounds would be required to raise 350 lb a distance of 2 yd?

For Exercises 91–92, find the foot-pound equivalent for the heater or air conditioner.

91. Indigo Flame Heater 15,000 Btu to 30,000 Btu

92. Room air conditioner 20,000 Btu

For Exercises 93–94, convert foot-pounds to British thermal units. Round to the nearest whole unit, if necessary.

93. 12,448,000 ft·lb ≈ ___ Btu

94. 77,800 ft·lb ≈ ___ Btu

95. Kristi burns 170 Cal in 1 hr when she takes a casual walk. She burns 220 Cal/hr if she takes a brisk walk. If Kristi plans to walk for 4 hr this week, how many more Calories will she burn if she walks briskly?

96. Leon burns 440 Cal/hr playing baseball. If he has 3 baseball practices this week, each lasting about 1 hr 30 min, plus a 2-hr 15-min baseball game on Saturday, how many Calories will Leon burn from playing baseball?

97. Find the power in foot-pounds per second of lifting 240 lb a distance of 5 ft in 12 sec.

98. A machine can raise 500 lb a distance of 1.5 yd in 20 sec. Find the power in foot-pounds per second.

For Exercises 99–100, convert foot-pounds per second to horsepower.

99. $275 \dfrac{\text{ft·lb}}{\text{sec}} = $ ___ hp

100. $6600 \dfrac{\text{ft·lb}}{\text{sec}} = $ ___ hp

For Exercises 101–104, complete the table by converting the horsepower of the vehicles to foot-pounds per second.

		hp	ft·lb/sec
101.	Dodge Viper	450	
102.	Ferrari 355 F1	375	
103.	Chevrolet Corvette	345	
104.	Porsche Carrera	300	

chapter 7 | test

1. Identify the units that apply to measuring length. Circle all that apply.

 a. Pound   **b.** Ounce   **c.** Meter

 d. Mile   **e.** Gram   **f.** Pint

 g. Feet   **h.** Liter   **i.** Fluid ounce

 j. Kilometer

2. Identify the units that apply to measuring capacity. Circle all that apply.

 a. Pound   **b.** Ounce   **c.** Meter

 d. Mile   **e.** Gram   **f.** Pint

 g. Foot   **h.** Liter   **i.** Fluid ounce

 j. Kilometer

3. Identify the units that apply to measuring mass or weight. Circle all that apply.

 a. Pound   **b.** Ounce   **c.** Meter

 d. Mile   **e.** Gram   **f.** Pint

 g. Foot   **h.** Liter   **i.** Fluid ounce

 j. Kilometer

4. A backyard needs 25 ft of fencing. How many yards is this?

5. It's estimated that an adult *Tyrannosaurus Rex* weighed approximately 11,000 lb. How many tons is this?

6. Two exits on the highway are 52,800 ft apart. How many miles is this?

7. A recipe for brownies calls for $\frac{3}{4}$ c of milk and 4 oz of water. What is the total amount of liquid in ounces?

8. A television show has 1200 sec of commercials. How many minutes is this?

9. Find the perimeter of the rectangle in feet.

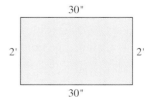

10. A decorator wraps a gift, using two pieces of ribbon. One is 1′10″ and the other is 2′4″. Find the total length of ribbon used.

11. When Stephen was born, he weighed 8 lb 1 oz. When he left the hospital, he weighed 7 lb 10 oz. How much weight did he lose after he was born?

12. A decorative pillow requires 3 ft 11 in. of fringe around the perimeter of the pillow. If 5 pillows are produced, how much fringe is required?

13. Josh ran a race and finished with the time of 1:15:15. Convert this time to minutes.

14. Approximate the width of the nut in centimeters and millimeters.

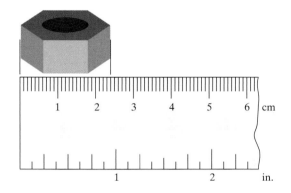

15. Select the most reasonable measurement for the length of a living room.

   a. 5 mm   **b.** 5 cm   **c.** 5 m   **d.** 5 km

16. The span of the Mackinac Bridge in Michigan is 1158 m. What is this length in kilometers?

17. A tablespoon (1 T) contains 0.015 L. How many milliliters is this?

18. a. What does the abbreviation *cc* stand for?

   b. Convert 235 mL to cubic centimeters.

   c. Convert 1 L to cubic centimeters.

19. A can of diced tomatoes is 411 g. Convert 411 g to centigrams.

20. A box of crackers is 210 g. If a serving of crackers is 30,000 mg, how many servings are in the box?

21. A bottle of Sprite contains 2 L. What is the capacity in quarts? Round to the nearest tenth.

22. Maurice Greene was one of the premier sprinters in U.S. track and field. His best race is the 100-m. How many yards is this? Round to the nearest yard.

23. The distance between two exits on a highway is 4.5 km. How far is this in miles? Round to the nearest tenth.

24. Breckenridge, Colorado, is 9603 ft above sea level. What is this height in meters? Round to the nearest meter.

25. A snowy egret stands about 20 in. tall and has a 38-in. wingspan. Convert both values to centimeters.

26. The mass of a laptop computer is 5000 g. What is the weight in pounds? Round to the nearest pound.

27. The oven temperature needed to bake cookies is 375°F. What is this temperature in degrees Celsius? Round to the nearest tenth.

28. The average January temperature in Albuquerque, New Mexico, is 2°C. Convert this temperature to degrees Fahrenheit.

29. During an average 45-min workout, Rich burns 275 Cal. How many Calories will he burn if he works out 4 times this week?

30. How much energy in foot-pounds is required to lift 45 lb a distance of 2 yd?

31. Find the foot-pound equivalent for an aboveground heater for pool and spa, rated at 100,000 Btu.

32. An electric blower is rated 4.5 hp. Convert this to foot-pounds per second.

chapters 1–7 | cumulative review

1. Round each number to the indicated place.

   a. 2499; thousands place   **b.** 42,099; tens place

For Exercises 2–3, refer to the figure.

2. Find the perimeter.

3. Find the area.

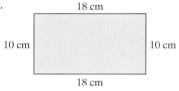

18 cm

10 cm   10 cm

18 cm

4. Simplify, using the order of operations.
$144 \div 9 \div (17 - 3 \cdot 5)^2$

5. The table gives the amount of money spent on research and development for five major U.S. companies. (Source: The Financial Times Ltd.)

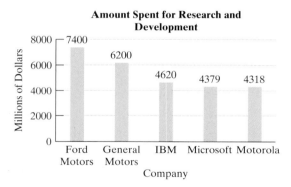

Amount Spent for Research and Development

a. Which company spends the most on research and development? What is that amount?

b. What is the difference between the amount spent at IBM and the amount spent at Motorola?

c. What is the total amount spent for research and development for these five companies?

6. Write an equivalent fraction with the indicated denominator. $\frac{2}{13} = \frac{}{39}$

7. Is 32,542 divisible by 3? Explain why or why not.

8. Write the prime factorization of 108.

9. Find the area of the triangle.

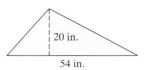

20 in.

54 in.

10. Keesha wants to bake oatmeal cookies, but she has only 1 c of oatmeal. The recipe calls for 3 c. Does Keesha have enough oatmeal to make $\frac{1}{4}$ of the recipe?

11. Find the LCD of $\frac{1}{2}$, $\frac{6}{5}$, and $\frac{3}{10}$.

12. Simplify. $\frac{1}{2} + \frac{6}{5} - \frac{3}{10}$

For Exercises 13–16, perform the indicated operations with mixed numbers.

13. $6\frac{2}{3} + 2\frac{5}{6}$ **14.** $6\frac{2}{3} \cdot 2\frac{5}{6}$

15. $6\frac{2}{3} \div 2\frac{5}{6}$ **16.** $6\frac{2}{3} - 2\frac{5}{6}$

For Exercises 17–22, complete the table.

	Fraction	Decimal
17.	$\frac{1}{3}$	
18.		0.45
19.		1.25
20.	$\frac{7}{2}$	
21.	$\frac{3}{8}$	
22.		0.04

23. A football team won 6 games and lost 5.

   a. Write a ratio of wins to losses.

   b. Write a ratio of wins to total games played.

24. In 1990, the men's median income in the United States (median is a type of average) was $20,300. By 2000 the median income had risen to $26,400. In 1990, the women's median income was $10,100. By 2000 the women's median income had risen to $14,800.

   a. Compute the difference in men's income between the years 2000 and 1990.

   b. Compute the unit rate representing the increase in income per year for men.

   c. Compute the difference in women's income between the years 2000 and 1990.

   d. Compute the unit rate representing the increase in income per year for women.

   e. For which group, men or women, is the rate of increase in salary greater? (Source: U.S. Bureau of Labor Statistics.)

25. If a car dealership sells 18 cars in a 5-day period, how many cars can the dealership expect to sell in 25 days?

26. A hospital ward has 40 beds and employs 6 nurses. What is the unit rate of beds per nurse? Round to the nearest tenth.

27. Are 6 and 8 proportional to 2 and 3? Explain why or why not.

28. Four-fifths of the drinks sold at a movie theater were soda. What percent is this?

29. Assume that these figures are similar. Solve for x.

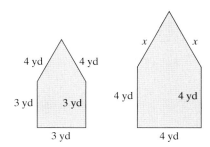

30. Of the trees in a forest, 62% were saved from a fire. If this represents 1420 trees, how many trees were originally in the forest? Round to the nearest whole tree.

31. Of 60 people, 45% had eaten at least one meal at McDonald's this week. How many people ate at McDonald's?

32. The sales tax on a $21 meal is $1.26. What is the sales tax rate?

33. The commission rate that Yvonne earns is 14% of sales. If she earned $2100 in commission, how much did she sell?

34. If $5000 is invested at 3.4% simple interest for 6 years, how much interest is earned?

For Exercises 35–40, convert the units of capacity, length, mass, and weight. Round to 1 decimal place, if necessary.

35. 5800 g = ___ kg   **36.** 5.8 kg ≈ ___ lb

37. 72 in. ≈ ___ cm   **38.** 72 in. = ___ ft

39. $3\frac{1}{2}$ qt = ___ pt   **40.** $3\frac{1}{2}$ qt ≈ ___ L

Geometry

8

In this chapter, we study geometry. We begin by first categorizing familiar figures such as squares, rectangles, triangles, and circles. Then we study the concepts of perimeter and area. We identify familiar three-dimensional solids such as cylinders, cones, spheres, and rectangular solids and learn how to find their volumes.

The applications of geometry are far-reaching. For example, in Exercise 55 of Section 8.4, we find the perimeter and area of a roller rink shaped as a rectangle with semicircles on each end. Using the values shown, we can determine how much railing and how much flooring must be purchased for the rink.

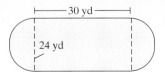

chapter 8   preview

The exercises in this chapter preview contain concepts that have not yet been presented. These exercises are provided for students who want to compare their levels of understanding before and after studying the chapter. Alternatively, you may prefer to work these exercises when the chapter is completed and before taking the exam.

Section 8.1

1. If $m(\angle X) = 22°$, find the measure of the complement and supplement.

For Exercises 2–5, answer true or false.

2. Intersecting lines have exactly two points of intersection.

3. If two lines are not parallel, then they must be perpendicular.

4. An obtuse angle is one with measure between 90° and 180°.

5. A 90° angle is a right angle.

Section 8.2

6. Evaluate the square roots. If necessary, use a calculator and round to 2 decimal places.

   a. $\sqrt{49}$ **b.** $\sqrt{50}$

7. Use the Pythagorean theorem to find the length of the missing side.

Section 8.3

For Exercises 8–9, find the area of each figure.

   8. Trapezoid   **9.** Triangle

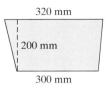 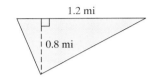

Section 8.4

10. Find the area A and the circumference C of the circle. Use 3.14 for π.

11. Find the area of the composite figure. Use $\frac{22}{7}$ for π.

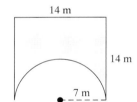

Section 8.5

For Exercises 12–15, match the volume formula with the description.

12. $V = s^3$ **a.** Right circular cylinder

13. $V = lwh$ **b.** Sphere

14. $V = \frac{4}{3}\pi r^3$ **c.** Rectangular solid

15. $V = \pi r^2 h$ **d.** Cube

16. Find the volume of a ball if the diameter of the ball is approximately 2 in. Use 3.14 for π and round the answer to the nearest whole unit.

section 8.1   Lines and Angles

1. Basic Definitions

In this chapter we will introduce some basic concepts of geometry.

A **point** is a specific location in space. We often symbolize a point by a dot and label it with a capital letter such as *P*.

A **line** consists of infinitely many points that follow a straight path. A line extends forever in both directions. This is illustrated by arrowheads at both ends. Figure 8-1 shows a line through points *A* and *B*. The line can be represented as $\overleftrightarrow{AB}$ or as $\overleftrightarrow{BA}$.

Figure 8-1

A **line segment** is a part of a line between two distinct endpoints. A line segment with endpoints *P* and *Q* can be denoted $\overline{PQ}$ or $\overline{QP}$. In Figure 8-2, point *S* is on the line segment $\overline{PQ}$.

Figure 8-2

Concept Connections

1. Are the rays $\overrightarrow{AB}$ and $\overrightarrow{BA}$ the same?
2. Are the lines $\overleftrightarrow{PQ}$ and $\overleftrightarrow{QP}$ the same?

A **ray** is the part of a line that includes an endpoint and all points on one side of the endpoint. In Figure 8-3, ray $\overrightarrow{PQ}$ is named by using the endpoint *P* and another point *Q* on the ray. Notice that the rays $\overrightarrow{PQ}$ and $\overrightarrow{QP}$ are different because they point in different directions. See Figure 8-3.

Ray $\overrightarrow{PQ}$ Ray $\overrightarrow{QP}$

Figure 8-3

Tip: A ray has only one endpoint, which is always written first.

---

example 1   Identifying Points, Lines, Line Segments, and Rays

Identify each as a point, line, line segment, or ray.

a. $\overleftrightarrow{MN}$ **b.** $\overrightarrow{NM}$ **c.** $\overline{MN}$ **d.** • *S*

Solution:

a. The double arrowheads indicate that $\overleftrightarrow{MN}$ is a line.

b. The single arrowhead indicates that $\overrightarrow{NM}$ is a ray with endpoint *N*.

c. The bar drawn above the letters $\overline{MN}$ indicates a line segment with endpoints *M* and *N*.

d. The dot represents a point.

Skill Practice

Identify each as a point, line, line segment, or ray.

3. $\overline{RS}$ **4.** • *Q*
5. $\overrightarrow{XY}$ **6.** $\overleftrightarrow{TV}$

Answers

1. No   2. Yes
3. Line segment
4. Point   5. Ray   6. Line

2. Naming and Measuring Angles

An **angle** is a geometric figure formed by two rays that share a common endpoint. The common endpoint is called the **vertex** of the angle. In Figure 8-4, the rays $\overrightarrow{PR}$ and $\overrightarrow{PQ}$ share the endpoint P. These rays form the sides of the angle denoted by $\angle QPR$ or $\angle RPQ$. Notice that when we name an angle, the vertex must be the middle letter. Sometimes a small arc $\rangle$ is drawn to illustrate the location of an angle. In such a case, the angle may be named by using the symbol $\angle$ along with the letter of the vertex.

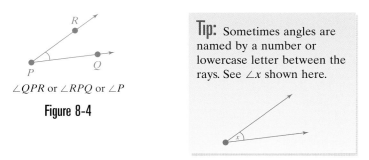

$\angle QPR$ or $\angle RPQ$ or $\angle P$

Figure 8-4

> **Tip:** Sometimes angles are named by a number or lowercase letter between the rays. See $\angle x$ shown here.

The most common unit to measure an angle is the degree, denoted by °. To become familiar with the measure of angles, consider the following benchmarks. Two rays that form a quarter turn of a circle make a 90° angle. A 90° angle is called a **right angle** and is often depicted with a □ symbol. Two rays that form a half turn of a circle make a 180° angle. A 180° angle is called a **straight angle** because it appears as a straight line. A full circle has 360°. For example, the second hand of a clock sweeps out an angle of 360° in 1 minute. See Figure 8-5.

> **Tip:** The ° symbol can be used for temperature or for measuring angles. The context of the problem tells us the units being used.

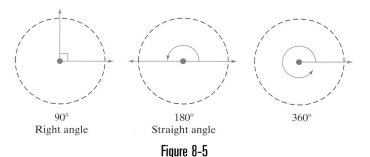

| 90° | 180° | 360° |
| Right angle | Straight angle | |

Figure 8-5

We can approximate the measure of an angle by using a tool called a *protractor*, shown in Figure 8-6. A protractor uses equally spaced tick marks around a semicircle to measure angles from 0° to 180°.

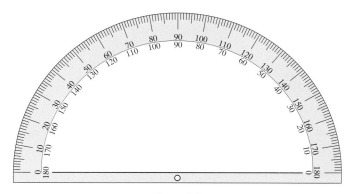

Figure 8-6

Example 2 shows how we can use a protractor to measure several angles. To denote the measure of an angle, we use the symbol m, written in front of the name of the angle. For example, if the measure of angle A is 30°, we write $m(\angle A) = 30°$.

example 2   Measuring Angles

Read the protractor to determine the measure of each angle.

a. $\angle AOB$ b. $\angle AOC$ c. $\angle AOD$ d. $\angle AOE$

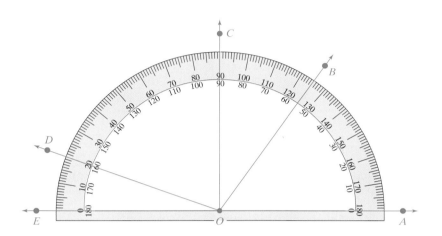

Solution:

We will use the inner scale on the protractor. This is done because we are measuring the angles in a counterclockwise direction, beginning at 0° along ray $\overrightarrow{OA}$.

a. $m(\angle AOB) = 55°$ On the inner scale, ray $\overrightarrow{OA}$ is aligned with 0° and ray $\overrightarrow{OB}$ passes through 55°. Therefore, $m(\angle AOB) = 55°$.

b. $m(\angle AOC) = 90°$ $\angle AOC$ is a right angle.

c. $m(\angle AOD) = 160°$

d. $m(\angle AOE) = 180°$ $\angle AOE$ is a straight angle.

An angle is said to be an **acute angle** if its measure is between 0° and 90°. An angle is said to be an **obtuse angle** if its measure is between 90° and 180°. See Figure 8-7.

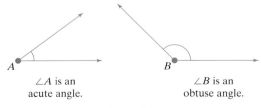

$\angle A$ is an acute angle. $\angle B$ is an obtuse angle.

Figure 8-7

Skill Practice

Read the protractor to determine the measure of each angle.

7. $\angle POQ$ 8. $\angle POR$
9. $\angle POS$ 10. $\angle POT$

Concept Connections

Answer true or false.

11. An angle whose measure is 102° is obtuse.

12. An angle whose measure is 98° is acute.

Answers
7. 20°   8. 45°   9. 125°
10. 180°   11. True   12. False

3. Complementary and Supplementary Angles

- Two angles are said to be equal or **congruent** if they have the same measure.
- Two angles are said to be **complementary** if the sum of their measures is 90°. In Figure 8-8, the complement of a 60° angle is a 30° angle, and vice versa.
- Two angles are said to be **supplementary** if the sum of their measures is 180°. In Figure 8-9, the supplement of a 60° angle is a 120° angle, and vice versa.

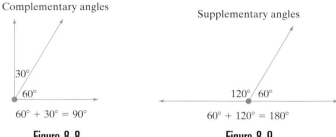

Complementary angles

30°
60°
60° + 30° = 90°

Figure 8-8

Supplementary angles

120° 60°
60° + 120° = 180°

Figure 8-9

Skill Practice

13. What is the supplement of a 35° angle?

14. What is the complement of a 52° angle?

---

example 3   Identifying Supplementary and Complementary Angles

a. What is the supplement of a 105° angle?

b. What is the complement of a 12° angle?

Solution:

a. The sum of a 105° angle and its supplement must equal 180°. Writing the related subtraction problem, we have

$$180° - 105° = 75°$$

The supplement is a 75° angle.

b. The sum of a 12° angle and its complement must equal 90°. Writing the related subtraction problem, we have

$$90° - 12° = 78°$$

The complement is a 78° angle.

4. Parallel and Perpendicular Lines

Two lines may intersect (cross) or may be parallel. **Parallel lines** lie on the same flat surface (called a plane), but never intersect. See Figure 8-10.

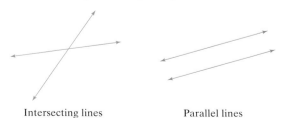

Tip: Sometimes we use the symbol ∥ to denote parallel lines.

Intersecting lines   Parallel lines

Figure 8-10

Answers

13. 145°   14. 38°

Notice that two intersecting lines form four angles. In Figure 8-11, $\angle a$ and $\angle c$ are said to be **vertical angles**. They appear on opposite sides of the vertex. Likewise, angles $\angle b$ and $\angle d$ are vertical angles. Vertical angles are equal in measure. Thus, $m(\angle a) = m(\angle c)$ and $m(\angle b) = m(\angle d)$.

If two lines intersect at a right angle, they are said to be **perpendicular lines** (Figure 8-12). With perpendicular lines, also notice that the vertical angles each measure 90°.

> **Tip:** Sometimes we use the symbol $\perp$ to denote perpendicular lines.

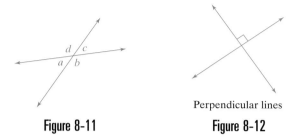

Figure 8-11

Perpendicular lines

Figure 8-12

5. Parallel Lines Cut by a Transversal

In Figure 8-13, lines L_1 and L_2 are parallel lines. If a third line m intersects the two parallel lines, that line is called a **transversal**. Suppose we label the eight angles formed by lines L_1, L_2, and m by the numbers 1–8.

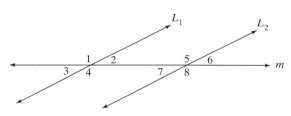

Figure 8-13

These angles have the following special properties.

Lines L_1 and L_2 Are Parallel; Line m Is a Transversal	Name of Angles	Property
	The following pairs of angles are called **alternate interior angles**: $\angle 2$ and $\angle 7$ $\angle 4$ and $\angle 5$	**Alternate interior angles are equal in measure.** $m(\angle 2) = m(\angle 7)$ $m(\angle 4) = m(\angle 5)$
	The following pairs of angles are called **alternate exterior angles**: $\angle 1$ and $\angle 8$ $\angle 3$ and $\angle 6$	**Alternate exterior angles are equal in measure.** $m(\angle 1) = m(\angle 8)$ $m(\angle 3) = m(\angle 6)$
	The following pairs of angles are called **corresponding angles**: $\angle 1$ and $\angle 5$ $\angle 2$ and $\angle 6$ $\angle 3$ and $\angle 7$ $\angle 4$ and $\angle 8$	**Corresponding angles are equal in measure.** $m(\angle 1) = m(\angle 5)$ $m(\angle 2) = m(\angle 6)$ $m(\angle 3) = m(\angle 7)$ $m(\angle 4) = m(\angle 8)$

Skill Practice

Assume that lines L_1 and L_2 are parallel. Find the measure of each angle. Explain how the angle is related to the given angle.

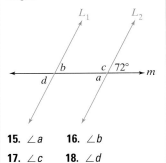

15. $\angle a$ **16.** $\angle b$

17. $\angle c$ **18.** $\angle d$

Answers

15. $72°$; vertical angles
16. $72°$; corresponding angles
17. $108°$; supplementary angles
18. $72°$; alternate exterior angles

example 4   Finding the Measure of Angles in a Diagram

Assume that lines L_1 and L_2 are parallel. Find the measure of each angle, and explain how the angle is related to the given angle of $65°$.

a. $\angle a$

b. $\angle b$

c. $\angle c$

d. $\angle d$

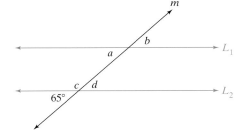

Solution:

a. $m(\angle a) = 65°$ $\angle a$ is a corresponding angle to the given angle.

b. $m(\angle b) = 65°$ $\angle b$ is an alternate exterior angle to the given angle.

c. $m(\angle c) = 115°$ $\angle c$ is the supplement to the given angle.

d. $m(\angle d) = 65°$ $\angle d$ and the given angle are vertical angles.

---

section 8.1   Practice Exercises

Study Skills Exercises

1. For your next test, make a memory sheet: On a 3×5 card (or several 3×5 cards), write all the formulas and rules that you need to know. Memorize all this information. Then when your instructor hands you the test, write down all the information that you can remember before you begin the test. Then you can take the test without worrying that you will forget something important. This process is referred to as a "memory dump." What important definitions and concepts have you learned in this section of the text?

2. Define the key terms.

a. Point   **b. Line**   **c. Line segment**   **d. Ray**

e. Angle   **f. Vertex**   **g. Right angle**   **h. Straight angle**

i. **Acute angle**   j. **Obtuse angle**   k. **Congruent**   l. **Complementary**

m. **Supplementary**   n. **Parallel lines**   o. **Vertical angles**   p. **Perpendicular lines**

q. **Transversal**   r. **Alternate exterior angles**   s. **Alternate interior angles**   t. **Corresponding angles**

Objective 1: Basic Definitions

3. Explain the difference between a line and a line segment.

4. Explain the difference between a line and a ray.

For Exercises 5–10, identify each figure as a line, line segment, ray, or point. **(See Example 1.)**

5. $K \quad H$

6. $K \quad H$

7. H

8. $H \quad K$

9. $H \quad K$

10. K

For Exercises 11–14, draw a figure that represents the expression.

11. $\overline{XY}$

12. A point named X

13. $\overleftrightarrow{YX}$

14. $\overrightarrow{XY}$

Objective 2: Naming and Measuring Angles

15. Sketch a right angle.

16. Sketch a straight angle.

For Exercises 17–22, use the protractor to determine the measure of each angle. **(See Example 2.)**

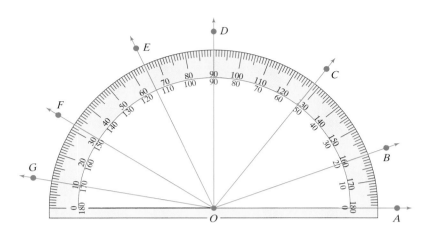

17. $m(\angle AOB)$ **18.** $m(\angle AOC)$ **19.** $m(\angle AOD)$

20. $m(\angle AOE)$ **21.** $m(\angle AOF)$ **22.** $m(\angle AOG)$

23. Sketch an acute angle.   **24.** Sketch an obtuse angle.

For Exercises 25–32, label each as an obtuse angle, acute angle, right angle, or straight angle.

25. $m(\angle A) = 90°$ **26.** $m(\angle E) = 91°$ **27.** $m(\angle B) = 98°$ **28.** $m(\angle F) = 30°$

29. $m(\angle C) = 2°$ **30.** $m(\angle G) = 130°$ **31.** $m(\angle D) = 180°$ **32.** $m(\angle H) = 45°$

Objective 3: Complementary and Supplementary Angles

For Exercises 33–40, the measure of an angle is given. Find the measure of the complement. **(See Example 3.)**

33. 80°   **34.** 5°   **35.** 27°   **36.** 64°

37. 29.5°   **38.** 13.2°   **39.** 89°   **40.** 1°

For Exercises 41–48, the measure of an angle is given. Find the measure of the supplement. **(See Example 3.)**

41. 80°   **42.** 5°   **43.** 127°   **44.** 124°

45. 37.4°   **46.** 173.9°   **47.** 179°   **48.** 1°

49. Can two supplementary angles both be obtuse?   **50.** Can two supplementary angles both be acute?

51. Can two complementary angles both be acute?

52. Can two complementary angles both be obtuse?

53. What angle is its own supplement?

54. What angle is its own complement?

Objective 4: Parallel and Perpendicular Lines

55. Sketch two lines that are parallel.

56. Sketch two lines that are *not* parallel.

57. Sketch two lines that are perpendicular.

58. Sketch two lines that are *not* perpendicular.

For Exercises 59–62, find the measure of angles *a*, *b*, *c*, and *d*.

59.

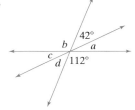

60.

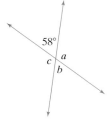

61.

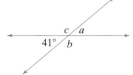

62.

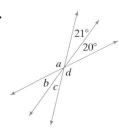

63. If two intersecting lines form vertical angles and each angle measures 90°, what can you say about the lines?

64. In the figure, describe the pair of angles, *a* and *b*, as complementary, vertical, or supplementary angles.

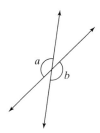

Objective 5: Parallel Lines Cut by a Transversal

For Exercises 65–68, refer to the figure.

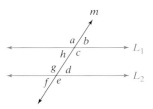

65. Name a pair of vertical angles.

66. Name a pair of alternate interior angles.

67. Name a pair of alternate exterior angles.

68. Name a pair of corresponding angles.

For Exercises 69–70, find the measure of angles a–g in the figure. Assume that L_1 and L_2 are parallel and that m is a transversal. **(See Example 4.)**

69.

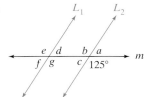

70.

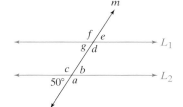

For Exercises 71–80, refer to the figure and answer true or false.

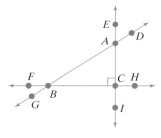

71. $\overleftrightarrow{AC}$ and $\overrightarrow{BC}$ are perpendicular lines.

72. $\overleftrightarrow{AB}$ and $\overrightarrow{AC}$ are perpendicular lines.

73. $\angle GBF$ is an acute angle.

74. $\angle EAD$ is an acute angle.

75. $\angle EAD$ and $\angle DAC$ are complementary angles.

76. $\angle GBF$ and $\angle FBA$ are complementary angles.

77. $\angle EAD$ and $\angle CAB$ are vertical angles.

78. $\angle ABC$ and $\angle FBG$ are vertical angles.

79. The point B is on $\overline{GA}$.

80. The point C is on $\overline{BH}$.

For Exercises 81–84, find the measure of ∠XYZ.

81.

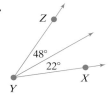

82.

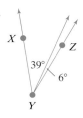

83.

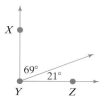

84.

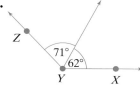

 85. Use the figure to find the measure of each angle.

   a. ∠AOB

   b. ∠EOD

   c. ∠AOE

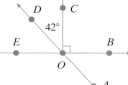

86. Use the figure to find the measure of each angle.

   a. ∠AOB

   b. ∠EOD

   c. ∠AOE

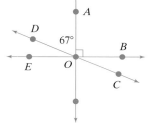

Expanding Your Skills

The second hand on a clock sweeps out a complete circle in 1 min. A circle forms a 360° arc. Use this information for Exercises 87–90.

  87. How many degrees does a second hand on a clock move in 30 sec?

  88. How many degrees does a second hand on a clock move in 15 sec?

  89. How many degrees does a second hand on a clock move in 20 sec?

  90. How many degrees does a second hand on a clock move in 45 sec?

Objectives

1. Categorizing Triangles
2. Square Roots
3. Pythagorean Theorem
4. Applications of the Pythagorean Theorem

section 8.2   Triangles and the Pythagorean Theorem

1. Categorizing Triangles

A triangle is a three-sided polygon. Furthermore, the sum of the measures of the angles within a triangle is 180°. Teachers often demonstrate this fact by tearing a triangular sheet of paper as shown in Figure 8-14. Then align the **vertices** (points) of the triangle to form a straight angle.

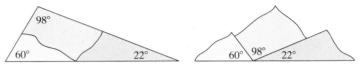

Figure 8-14

Skill Practice

Find the measure of angles *a* and b.

1.

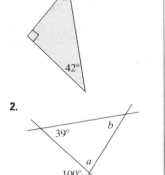

2.

example 1   Finding the Measure of Angles Within a Triangle

Find the measure of angles *a* and *b*.

a.

b.

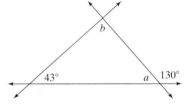

Solution:

a. Recall that the □ symbol represents a 90° angle.

$$38° + 90° + m(\angle a) = 180°$$ The sum of the angles within a triangle is 180°.

$$128° + m(\angle a) = 180°$$ Add the measures of the two known angles.

$$m(\angle a) = 180° - 128°$$ Write the related subtraction problem.

$$m(\angle a) = 52°$$

b. ∠*a* is the supplement of the 130° angle. Thus $m(\angle a) = 50°$.

$$43° + 50° + m(\angle b) = 180°$$ The sum of the angles within a triangle is 180°.

$$93° + m(\angle b) = 180°$$ Add the measures of the two known angles.

$$m(\angle b) = 180° - 93°$$ Write the related subtraction problem.

$$m(\angle b) = 87°$$

Answers

1. $m(\angle a) = 48°$
2. $m(\angle a) = 80°$
   $m(\angle b) = 61°$

Triangles may be categorized by the measures of their angles and by the number of equal sides or angles (Figures 8-15 and 8-16).

- An **acute triangle** is a triangle in which all three angles are acute.
- A **right triangle** is a triangle in which one angle is a right angle.
- An **obtuse triangle** is a triangle in which one angle is obtuse.

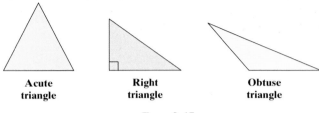

Acute   Right   Obtuse
triangle   triangle   triangle

Figure 8-15

- An **equilateral triangle** is a triangle in which all three sides (and all three angles) are equal in measure.
- An **isosceles triangle** is a triangle in which two sides are equal in length (the angles opposite the equal sides are also equal in measure).
- A **scalene triangle** is a triangle in which no sides (or angles) are equal in measure.

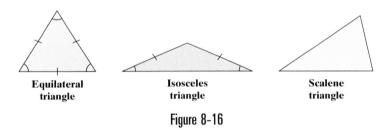

Equilateral   Isosceles   Scalene
triangle   triangle   triangle

Figure 8-16

Tip: Sometimes we use tick marks / to denote segments of equal length. Similarly, we sometimes use a small arc ⟩ to denote angles of equal measure.

2. Square Roots

In this section we present an important theorem called the Pythagorean theorem. To understand this theorem, we first need some background definitions.

Recall from Section 1.7 that to square a number means to find the product of the number and itself. Thus, $b^2 = b \cdot b$. For example,

$$6^2 = 6 \cdot 6 = 36$$

We now want to reverse this process by finding a square root of a number. Recall that this is denoted by the radical sign $\sqrt{}$. For example, $\sqrt{36}$ reads as "the positive square root of 36." Thus,

$$\sqrt{36} = 6 \qquad \text{because } 6^2 = 6 \cdot 6 = 36$$

Answers
3. b   4. a   5. c
6. b   7. a   8. c

example 2   Evaluating Square Roots

Simplify.

a. $\sqrt{64}$ b. $\sqrt{100}$ c. $\sqrt{1}$ d. $\sqrt{0}$

Solution:

a. $\sqrt{64} = 8$ because $8 \cdot 8 = 64$

b. $\sqrt{100} = 10$ because $10 \cdot 10 = 100$

c. $\sqrt{1} = 1$ because $1 \cdot 1 = 1$

d. $\sqrt{0} = 0$ because $0 \cdot 0 = 0$

Table 8-1 gives a list of several whole numbers, their squares, and the corresponding square roots.

table 8-1

$0^2 = 0 \rightarrow \sqrt{0} = 0$	$8^2 = 64 \rightarrow \sqrt{64} = 8$
$1^2 = 1 \rightarrow \sqrt{1} = 1$	$9^2 = 81 \rightarrow \sqrt{81} = 9$
$2^2 = 4 \rightarrow \sqrt{4} = 2$	$10^2 = 100 \rightarrow \sqrt{100} = 10$
$3^2 = 9 \rightarrow \sqrt{9} = 3$	$11^2 = 121 \rightarrow \sqrt{121} = 11$
$4^2 = 16 \rightarrow \sqrt{16} = 4$	$12^2 = 144 \rightarrow \sqrt{144} = 12$
$5^2 = 25 \rightarrow \sqrt{25} = 5$	$13^2 = 169 \rightarrow \sqrt{169} = 13$
$6^2 = 36 \rightarrow \sqrt{36} = 6$	$14^2 = 196 \rightarrow \sqrt{196} = 14$
$7^2 = 49 \rightarrow \sqrt{49} = 7$	$15^2 = 225 \rightarrow \sqrt{225} = 15$

Many square roots cannot be written as a whole number. For example, there is no whole number that when squared equals $\sqrt{26}$. However, we might speculate that $\sqrt{26}$ is a number slightly greater than 5 because $\sqrt{25} = 5$. A decimal approximation can be made by using a calculator.

$$\sqrt{26} \approx 5.099 \qquad \text{because } 5.099^2 = 25.999801 \approx 26$$

Calculator Connections

Topic: Entering square roots on a calculator

To enter a square root on a calculator, use the $\boxed{\sqrt{}}$ key. On some calculators, the $\boxed{\sqrt{}}$ function is associated with the x^2 key. In such a case, it is necessary to press $\boxed{2^{nd}}$ or $\boxed{\text{Shift}}$ first, followed by the $\boxed{x^2}$ key. Some calculators require the square root key to be entered first, before the number, while with others we enter the number first followed by $\boxed{\sqrt{}}$.

Expression	Keystrokes	Result
$\sqrt{26}$	26 $\boxed{\sqrt{}}$ $\boxed{=}$	5.099019514
	or $\boxed{\sqrt{}}$ 26 $\boxed{=}$	5.099019514
$\sqrt{9325}$	9325 $\boxed{\sqrt{}}$ $\boxed{=}$	96.56603958
	or $\boxed{\sqrt{}}$ 9325 $\boxed{=}$	96.56603958
$\sqrt{100}$	100 $\boxed{\sqrt{}}$ $\boxed{=}$	10
	or $\boxed{\sqrt{}}$ 100 $\boxed{=}$	10

3. Pythagorean Theorem

Recall that a right triangle is a triangle with a 90° angle. The two sides forming the right angle are called the **legs**. The side opposite the right angle is called the **hypotenuse**. Note that the hypotenuse is always the longest side. See Figure 8-17. We often use the letters a and b to represent the legs of a right triangle. The letter c is used to label the hypotenuse.

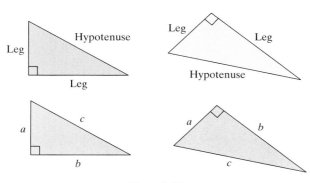

Figure 8-17

For any right triangle, the Pythagorean theorem gives us this important relationship among the lengths of the sides.

Pythagorean Theorem

For any right triangle,

$$(\text{Leg})^2 + (\text{Leg})^2 = (\text{Hypotenuse})^2$$

Using the letters a, b, and c to represent the legs and hypotenuse, respectively, we have

$$a^2 + b^2 = c^2$$

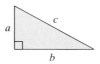

---

example 3   Finding the Length of the Hypotenuse of a Right Triangle

Find the length of the hypotenuse of the right triangle.

Solution:

The lengths of the legs are given.

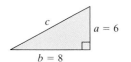

Label the triangle, using a, b, and c. It does not matter which leg is labeled a and which is labeled b.

Concept Connections

13. Label the triangle with the words *leg, leg,* and *hypotenuse.*

14. Label the triangle with the letters *a*, *b*, and *c*, where *c* represents the hypotenuse.

Answers
13.

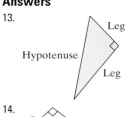

14.

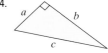

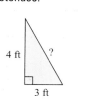
$$a^2 + b^2 = c^2 \qquad \text{Apply the Pythagorean theorem.}$$

$$(6)^2 + (8)^2 = c^2 \qquad \text{Substitute } a = 6 \text{ and } b = 8.$$

$$36 + 64 = c^2 \qquad \text{Simplify.}$$

$$100 = c^2 \qquad \text{The solution to this equation is the positive number, } c, \text{ that when squared equals 100.}$$

$$\sqrt{100} = c$$

$$10 = c \qquad \text{Simplify the square root of 100.}$$

The solution may be checked using the Pythagorean theorem.

$$a^2 + b^2 = c^2$$

$$(6)^2 + (8)^2 \stackrel{?}{=} (10)^2$$

$$36 + 64 = 100 \ \checkmark$$

In Example 4, we solve for one of the legs of a right triangle when the other leg and the hypotenuse are known.

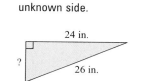
example 4   Finding the Length of a Leg in a Right Triangle

Find the length of the unknown side of the right triangle.

12

13   ?

Solution:

$a = 12$

$c = 13$ b

Label the triangle, using a, b, and c. One of the legs is unknown. It doesn't matter whether we call the unknown leg, a or b.

$$a^2 + b^2 = c^2 \qquad \text{Apply the Pythagorean theorem.}$$

$$(12)^2 + b^2 = (13)^2 \qquad \text{Substitute } a = 12 \text{ and } c = 13.$$

$$144 + b^2 = 169 \qquad \text{Simplify.}$$

Our goal, when solving an equation, is to isolate the variable. For this addition equation, we can write the related subtraction equation.

$$b^2 = 169 - 144 \qquad \text{Related subtraction equation.}$$

$$b^2 = 25 \qquad \text{The solution to this equation is the positive number } b \text{ that when squared equals 25.}$$

$$b = \sqrt{25}$$

$$b = 5 \qquad \text{Simplify the square root of 25.}$$

The solution may be checked by using the Pythagorean theorem.

$$a^2 + b^2 = c^2$$

$$(12)^2 + (5)^2 \stackrel{?}{=} (13)^2$$

$$144 + 25 = 169 \ \checkmark$$

Answers

15. 5 ft
16. 10 in.

4. Applications of the Pythagorean Theorem

In Example 5 we use the Pythagorean theorem in an application.

example 5   Using the Pythagorean Theorem in an Application

When Barb swam across a river, the current carried her 300 yd downstream from her starting point. If the river is 400 yd wide, how far did Barb swim?

Solution:

We first familiarize ourselves with the problem and draw a diagram (Figure 8-18). The distance Barb actually swims is the hypotenuse of the right triangle. Therefore, we label this distance c.

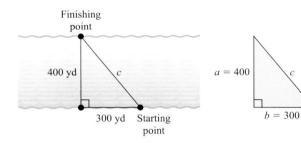

Figure 8-18

$$a^2 + b^2 = c^2 \qquad \text{Apply the Pythagorean theorem.}$$

$$(400)^2 + (300)^2 = c^2 \qquad \text{Substitute } a = 400 \text{ and } b = 300.$$

$$160{,}000 + 90{,}000 = c^2 \qquad \text{Simplify.}$$

$$250{,}000 = c^2 \qquad \text{Add. The solution to this equation is the positive number } c \text{ that when squared equals } 250{,}000.$$

$$\sqrt{250{,}000} = c$$

$$500 = c \qquad \text{Simplify.}$$

The distance that Barb swims is 500 yd.

Skill Practice

17. The bottom of a 17-ft ladder is placed 8 ft from the bottom of a building. How far up the building is the top of the ladder?

17 ft

8 ft

Answer
17. 15 ft

---

section 8.2   Practice Exercises

Study Skills Exercises

1. When you are solving an application involving geometry, be sure to draw a picture of the situation and label the known quantities with numbers and the unknown quantities with variables. This will help to solve the problem. After you read this section, what geometric figure do you think you will be drawing most often in this section?

2. Define the key terms.

   a. Vertices

   d. Obtuse triangle

   g. Scalene triangle

   j. Legs

   b. Acute triangle

   e. Equilateral triangle

   h. Pythagorean theorem

   c. Right triangle

   f. Isosceles triangle

   i. Hypotenuse

Review Exercises

3. Do $\angle ACB$ and $\angle BCA$ represent the same angle?

4. Is line segment $\overline{MN}$ the same as the line segment $\overline{NM}$?

5. Is ray $\overrightarrow{AB}$ the same as ray $\overrightarrow{BA}$?

6. Is the line $\overleftrightarrow{PQ}$ the same as the line $\overleftrightarrow{QP}$

7. Is a right angle an obtuse angle?

8. Can two acute angles be supplementary?

Objective 1: Categorizing Triangles

For Exercises 9–16, find the measures a and b. **(See Example 1.)**

9.

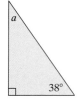

10.

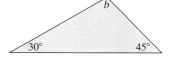

11.

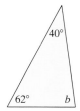

12.

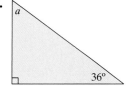

13.

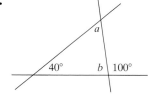

14.

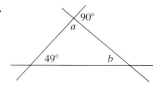

15.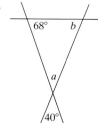

16.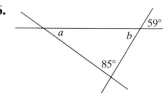

For Exercises 17–22, choose all figures that apply. The tick marks / denote segments of equal length, and small arcs ⟩ denote angles of equal measure.

17. Acute triangle

18. Obtuse triangle

19. Right triangle

20. Scalene triangle

21. Isosceles triangle

22. Equilateral triangle

a.

b.

c.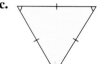

d.

e.

f.

g.

Objective 2: Square Roots

For Exercises 23–38, simplify the squares and the square roots. **(See Example 2.)**

23. $\sqrt{49}$ **24.** $\sqrt{64}$ **25.** 7^2 **26.** 8^2

27. 4^2 **28.** 5^2 **29.** $\sqrt{16}$ **30.** $\sqrt{25}$

31. $\sqrt{36}$ **32.** $\sqrt{100}$ **33.** 6^2 **34.** 10^2

35. 9^2 **36.** 3^2 **37.** $\sqrt{81}$ **38.** $\sqrt{9}$

For Exercises 39–44, complete the table. For the estimate, find two consecutive whole numbers between which the square root lies. The first row is done for you.

	Square Root	Estimate	Calculator Approximation (Round to 3 Decimal Places)
	$\sqrt{50}$	is between <u>7</u> and <u>8</u>	7.071
39.	$\sqrt{10}$	is between ____ and ____	
40.	$\sqrt{90}$	is between ____ and ____	
41.	$\sqrt{116}$	is between ____ and ____	
42.	$\sqrt{65}$	is between ____ and ____	
43.	$\sqrt{5}$	is between ____ and ____	
44.	$\sqrt{48}$	is between ____ and ____	

For Exercises 45–52, use a calculator to approximate the square roots to 3 decimal places.

45. 427.75   **46.** 3184.75   **47.** 1,246,000   **48.** 50,416,000

49. 0.49   **50.** 0.25   **51.** 0.56   **52.** 0.82

Objective 3: Pythagorean Theorem

For Exercises 53–56, find the length of the unknown side. (See Examples 3–4.)

53.

54.

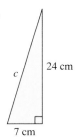

55.

56.

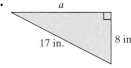

For Exercises 57–60, find the length of the unknown leg or hypotenuse.

57. Leg = 24 ft, hypotenuse = 26 ft   **58.** Leg = 9 km, hypotenuse = 41 km

59. Leg = 32 in., leg = 24 in.   **60.** Leg = 16 m, hypotenuse = 34 m

Objective 4: Applications of the Pythagorean Theorem

61. Find the length of the supporting brace. **(See Example 5.)**

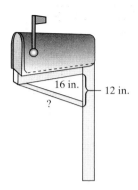

62. Find the length of the ramp.

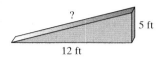

63. Find the height of the airplane above the ground.

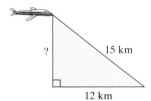

64. A 25-in. television measures 25 in. across the diagonal. If the length is 20 in., find the width.

65. A car travels east 24 mi and then south 7 mi. How far is the car from its starting point?

66. A 26-ft-long wire is to be tied from a stake in the ground to the top of a 24-ft pole. How far from the pole should the stake be placed?

For Exercises 67–70, find the perimeter.

67.

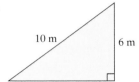

68.

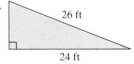

69.

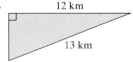

70.

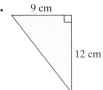

Expanding Your Skills

71. Find the length of side c by dividing this figure into a rectangle and a right triangle. Then find the perimeter of the figure.

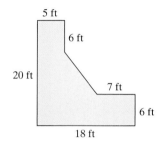

72. Find the perimeter of the figure.

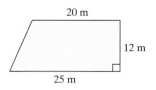

73. Find the perimeter of the figure.

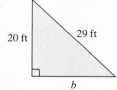

74. Tyler drives 20 mi north, 8 mi east, 4 mi south, and 4 mi east. How far is Tyler from his starting point?

Calculator Connections

Topic: Pythagorean theorem

For Exercises 75–80, find the length of the unknown side. Round to three decimal places if necessary.

75.

76.

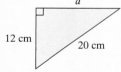

77. Leg = 5 mi, leg = 10 mi

78. Leg = 2 m, leg = 8 m

79. Leg = 12 in., hypotenuse = 22 in.

80. Leg = 15 ft, hypotenuse = 18 ft

81. A square tile is 1 ft on each side. What is the length of the diagonal? Round to the nearest hundredth of a foot.

82. A tennis court is 120 ft long and 60 ft wide. What is the length of the diagonal? Round to the nearest hundredth of a foot.

section 8.3   Quadrilaterals, Perimeter, and Area

1. Quadrilaterals

Recall that a **polygon** is a flat figure formed by line segments connected at their ends. A four-sided polygon is called a **quadrilateral**. Some quadrilaterals fall in the following categories.

A **parallelogram** is a quadrilateral with opposite sides parallel. It follows that opposite sides must be equal in length.

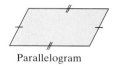
Parallelogram

A **rectangle** is a parallelogram with four right angles.

Rectangle

A **square** is a rectangle with sides of equal length.

Square

A **rhombus** is a parallelogram with sides of equal length. The angles are not necessarily equal, as with a square.

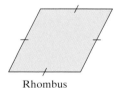
Rhombus

A **trapezoid** is a quadrilateral with one pair of parallel sides.

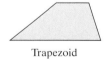
Trapezoid

Notice that some figures belong to more than one category. For example, a square is also a rectangle and a parallelogram.

2. Perimeter

Recall that the **perimeter** of a polygon is the distance around the figure. For example, we use perimeter to find the amount of fencing needed to enclose a yard. The perimeter of a polygon is found by adding the lengths of the sides. However, with some geometric figures we can shorten the process by using a formula.

Perimeter of a Square

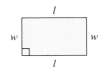

$$P = s + s + s + s$$
$$= 4s$$

Perimeter of a Rectangle

$$P = l + l + w + w$$
$$= 2l + 2w$$

We usually do not give perimeter formulas for other polygons. It is generally easier simply to add the lengths of the sides than to memorize numerous formulas.

Skill Practice

1. Find the perimeter.

4.3 in.

2. Find the perimeter in feet.

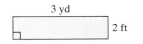
3 yd
2 ft

---

example 1   Finding Perimeter

Use an appropriate formula to find the perimeter of each figure.

a.

8.2 yd

b.   9 in.

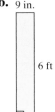
6 ft

Solution:

a. $P = 4s$ The figure is a square. Use $P = 4s$.

       $= 4(8.2 \text{ yd})$ Substitute $s = 8.2$ yd.

       $= 32.8 \text{ yd}$ Simplify.

b. First note that to add the lengths of the sides, we must have like units.

$$9 \text{ in.} = \frac{9 \text{ in.}}{1} \cdot \frac{1 \text{ ft}}{12 \text{ in.}} = \frac{9}{12} \text{ ft} = \frac{3}{4} \text{ ft} \text{ or } 0.75 \text{ ft}$$

   $P = 2l + 2w$ The figure is a rectangle. Use $P = 2l + 2w$.

       $= 2(6 \text{ ft}) + 2(0.75 \text{ ft})$ Substitute $l = 6$ ft and $w = 0.75$ ft.

       $= 12 \text{ ft} + 1.5 \text{ ft}$ Simplify.

       $= 13.5 \text{ ft}$

---

3. Area

Answers

1. 17.2 in.
2. 22 ft

The area of a region is the number of square units that can be enclosed within the region. The rectangle shown in Figure 8-19 encloses 6 square inches (in.^{2}). We would compute area, for example, if we wanted to determine how much sod was needed to cover a yard.

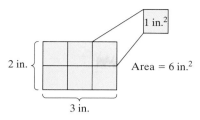

Figure 8-19

3. How many square centimeters are enclosed in the figure?

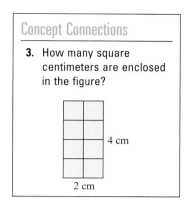

We offer the following convenient formulas to find the area enclosed within various figures. We begin with the familiar formulas for the area of a rectangle and square. These formulas were first introduced in Section 1.5.

Area of a Rectangle

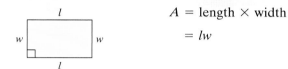

$A = \text{length} \times \text{width}$

$\quad = lw$

Area of a Square

A square is also a rectangle. Therefore, the area of a square is

$A = \text{length} \times \text{width}$

$\quad = s \cdot s$

$\quad = s^2$

To find the area of a parallelogram, consider cutting it into two pieces, as shown in Figure 8-20. Then realign the pieces to form a rectangle.

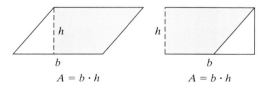

$A = b \cdot h$ $A = b \cdot h$

Figure 8-20

The area of a parallelogram is the corresponding area of the rectangle: $A = \text{base} \times \text{height}$, or simply $A = bh$. The height h is the distance between the base and its opposite side.

Tip: The height h of a parallelogram can also be drawn *outside* the parallelogram.

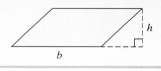

Area of a Parallelogram

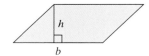

$A = \text{base} \times \text{height}$

$\quad = bh$

Answer

3. 8 cm^2

Skill Practice

4. Find the area.

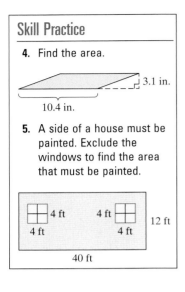

5. A side of a house must be painted. Exclude the windows to find the area that must be painted.

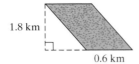

Finding Area

a. Find the area of the field.

b. Find the area of the matting.

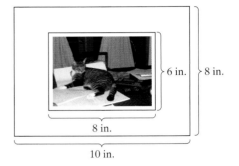

Solution:

a. The field is in the shape of a parallelogram. The base is 0.6 km and the height is 1.8 km.

$A = bh$ Area formula for a parallelogram.

$\quad = (0.6 \text{ km})(1.8 \text{ km})$ Substitute $b = 0.6$ km and $h = 1.8$ km.

$\quad = 1.08 \text{ km}^2$

The field is 1.08 km².

Tip: When two common units are multiplied, such as km · km, the resulting units are square units, such as km².

b. To find the area of the matting only, we can subtract the inner 6-in. × 8-in. area from the outer 8-in. × 10-in. area.

$$\text{outer area} \quad - \quad \text{inner area}$$

$$\text{Area of matting} = (10 \text{ in.})(8 \text{ in.}) - (8 \text{ in.})(6 \text{ in.})$$

$$= 80 \text{ in.}^2 - 48 \text{ in.}^2$$

$$= 32 \text{ in.}^2$$

The matting is 32 in.²

The formula for the area of a triangle was first introduced in Section 2.4. The formula is $A = \frac{1}{2}bh$, where b is the base of the triangle and h is the height. Notice that this formula gives one-half the value of the area of a parallelogram (Figure 8-21).

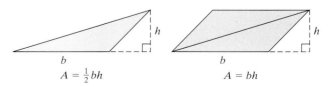

Figure 8-21

Answers

4. 32.24 in.²
5. 448 ft²

Area of a Triangle

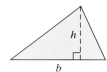

$$A = \tfrac{1}{2} \times \text{base} \times \text{height}$$
$$= \tfrac{1}{2}bh$$

The last formula we present here is the formula to find the area of a trapezoid. To develop the formula, we place two identical trapezoids next to each other (Figure 8-22). Together they form a parallelogram with base $a + b$. Notice that the height of a trapezoid is the distance between the two parallel sides.

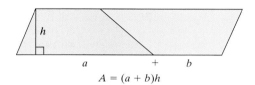

$$A = \tfrac{1}{2}(a + b)h \qquad A = (a + b)h$$

Figure 8-22

The area of the trapezoid is one-half the area of the parallelogram.

Area of a Trapezoid

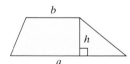

$$A = \tfrac{1}{2} \times (\text{sum of the parallel sides}) \times \text{height}$$
$$= \tfrac{1}{2}(a + b)h$$

example 3   Finding Area

Find the area of each region.

a.

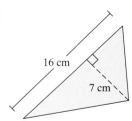

16 cm

7 cm

b.

5 ft

11 ft

9 ft

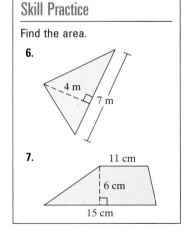
Solution:

a. $A = \tfrac{1}{2}bh$ Apply the formula for the area of a triangle.

$= \tfrac{1}{2}(16 \text{ cm})(7 \text{ cm})$ Substitute $b = 16$ cm and $h = 7$ cm.

$= \tfrac{1}{2}(\overset{8}{16} \text{ cm})(7 \text{ cm})$ Multiply fractions.

$= 56 \text{ cm}^2$

b. $A = \frac{1}{2}(a + b)h$ Apply the formula for the area of a trapezoid.

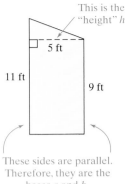

In this case, the two parallel sides are the left-hand side and the right-hand side. Therefore, these sides are the two bases, *a* and *b*.

The "height" is the distance between the two parallel sides.

$A = \frac{1}{2}(11 \text{ ft} + 9 \text{ ft})5 \text{ ft}$ Substitute $a = 11$ ft, $b = 9$ ft, and $h = 5$ ft.

$= \frac{1}{2}(20 \text{ ft})5 \text{ ft}$ Simplify within parentheses first.

$= \frac{1}{2}(\overset{10}{\cancel{10}} \text{ ft})5 \text{ ft}$ Multiply fractions.

$= 50 \text{ ft}^2$

Here is a summary of the area formulas presented thus far. These should be memorized as common knowledge.

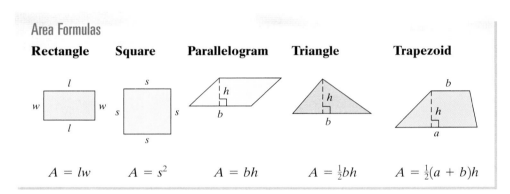

Area Formulas

Rectangle	Square	Parallelogram	Triangle	Trapezoid
$A = lw$	$A = s^2$	$A = bh$	$A = \frac{1}{2}bh$	$A = \frac{1}{2}(a + b)h$

4. Applications of Area

Skill Practice

8. Find the area of the shaded region.

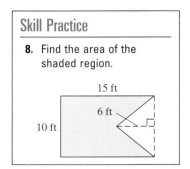

example 4   Finding the Area of a Composite Figure

Find the area of the shaded region.

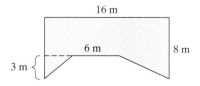

Solution:

The shaded region can be thought of as a large "outer" rectangle with a trapezoidal piece removed (Figure 8-23).

Answer

8. 120 ft²

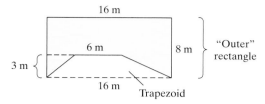

Figure 8-23

Therefore we can subtract the area of the trapezoid from the area of the outer rectangle.

$$\text{Area of rectangle: } A = lw$$
$$= (16 \text{ m})(8 \text{ m})$$
$$= 128 \text{ m}^2$$

$$\text{Area of trapezoid: } A = \tfrac{1}{2}(a + b)h$$
$$= \tfrac{1}{2}(16 \text{ m} + 6 \text{ m})(3 \text{ m})$$
$$= \tfrac{1}{2}(22 \text{ m})(3 \text{ m})$$
$$= 33 \text{ m}^2$$

The area of the shaded region is the area of the rectangle minus the area of the trapezoid.

$$\text{Area of shaded region} = 128 \text{ m}^2 - 33 \text{ m}^2$$
$$= 95 \text{ m}^2$$

---

example 5   Finding Area for a Landscaping Application

Sod can be purchased in palettes for \$225. If a palette contains 240 ft^2 of sod, how much will it cost to cover the area in Figure 8-24?

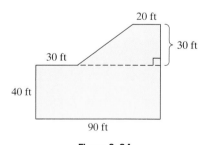

Figure 8-24

Skill Practice

9. A homeowner wants to apply water sealant to a pier that extends from the backyard to a lake. One gallon of sealant covers 160 ft^2 and sells for \$11.95. How much will it cost to cover the pier?

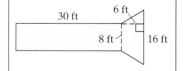

Solution:

To find the total cost, we need to know the total number of square feet. Then we can determine how many 240-ft^2 palettes are required.

Answer
9. \$23.90

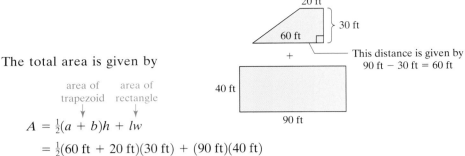

The total area is given by

area of   area of
trapezoid rectangle
   ↓   ↓
$A = \frac{1}{2}(a + b)h + lw$

$= \frac{1}{2}(60 \text{ ft} + 20 \text{ ft})(30 \text{ ft}) + (90 \text{ ft})(40 \text{ ft})$

$= \frac{1}{2}(80 \text{ ft})(30 \text{ ft}) + 3600 \text{ ft}^2$

$= 1200 \text{ ft}^2 + 3600 \text{ ft}^2$

$= 4800 \text{ ft}^2$ The total area is 4800 ft^2.

To determine how many 240-ft^2 palettes of sod are required, divide the total area by 240 ft^2.

Number of palettes: $4800 \text{ ft}^2 \div 240 \text{ ft}^2 = 20$

The total cost for 20 palettes is   ($225 per palette) $\times$ 20 palettes = $4500

The cost for the sod is $4500.

---

section 8.3   Practice Exercises

Study Skills Exercises

1. It may help to remember formulas if you understand how they were derived. For example, the perimeter of a square has the formula $P = 4s$. It was derived from the fact that perimeter measures the distance around a figure. Observe:

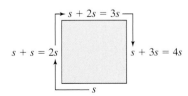

Explain how the formula for the perimeter of a rectangle ($P = 2l + 2w$) was derived.

2. Define the key terms.

   a. Polygon   **b. Quadrilateral**   **c. Parallelogram**   **d. Rectangle**

   e. Square   **f. Rhombus**   **g. Trapezoid**   **h. Perimeter**

Review Exercises

For Exercises 3–8, state the characteristics of each triangle.

3. Isosceles triangle   **4.** Right triangle

5. Acute triangle   **6.** Equilateral triangle

7. Obtuse triangle   **8.** Scalene triangle

Objective 1: Quadrilaterals

9. Write the definition of a quadrilateral.

For Exercises 10–14, state the characteristics of each quadrilateral.

10. Square   **11.** Trapezoid

12. Parallelogram   **13.** Rectangle

14. Rhombus

For Exercises 15–18, answer true or false.

15. All rectangles are parallelograms.   **16.** All trapezoids are parallelograms.

17. All rhombi (plural of rhombus) are squares.   **18.** All squares are rhombi.

Objective 2: Perimeter

For Exercises 19–24, find the perimeter. **(See Example 1.)**

19. Rectangle   **20.** Square

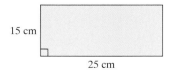

15 cm

25 cm

32 in.

21. Square

65 mm

22. Rectangle

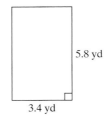

5.8 yd

3.4 yd

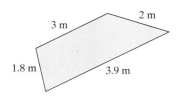 **23.** Trapezoid

3 m   2 m   1.8 m   3.9 m

24. Trapezoid

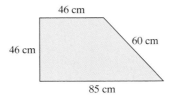

46 cm   60 cm   46 cm   85 cm

25. Find the perimeter of a triangle with sides 3 ft 8 in., 2 ft 10 in., and 4 ft.

26. Find the perimeter of a triangle with sides 4 ft 2 in., 3 ft, and 2 ft 9 in.

27. Find the perimeter of a rectangle with length 2 ft and width 6 in.

28. Find the perimeter of a rectangle with length 4 m and width 85 cm.

29. Find the lengths of the two missing sides labeled x and y. Then find the perimeter of the figure.

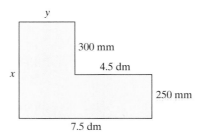

y 300 mm   4.5 dm x 250 mm   7.5 dm

30. Find the lengths of the two missing sides labeled a and b. Then find the perimeter of the figure.

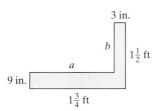

3 in. b $1\frac{1}{2}$ ft a 9 in. $1\frac{3}{4}$ ft

31. Rain gutters are going to be installed around the perimeter of the house. How many feet of rain gutters are needed?

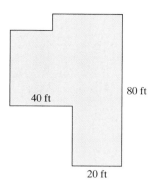

32. Wood molding needs to be installed around the perimeter of a living room floor. With no wood molding needed in the doorway, how much wood molding is needed?

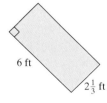

Objective 3: Area

For Exercises 33–42, find the area. **(See Examples 2–3.)**

33. Square

34. Rectangle

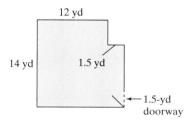

35. Triangle

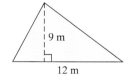

36. Parallelogram

37. Trapezoid

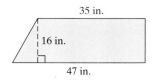

38. Triangle

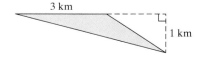

39. Parallelogram

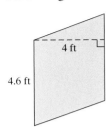

4 ft

4.6 ft

40. Square (write the answer in square feet)

3'6"

41. Rectangle (write the answer in square feet)

2 ft 3 in.

5 ft 6 in.

42. Trapezoid

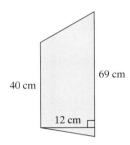

69 cm

40 cm

12 cm

Objective 4: Applications of Area

For Exercises 43–48, find the area of the shaded region. **(See Example 4.)**

43.

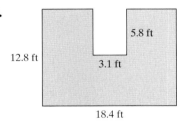

5.8 ft

12.8 ft

3.1 ft

18.4 ft

44.

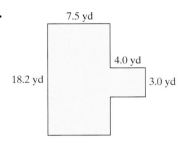

7.5 yd

4.0 yd

18.2 yd

3.0 yd

45.

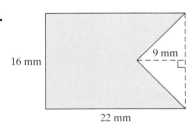

9 mm

16 mm

22 mm

46.

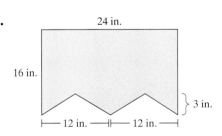

24 in.

16 in.

3 in.

12 in.   12 in.

47.

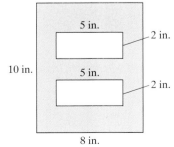

5 in.

2 in.

10 in.

5 in.

2 in.

8 in.

48.

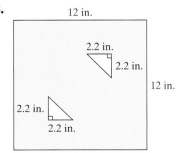

12 in.

2.2 in.

2.2 in.

12 in.

2.2 in.

2.2 in.

49. A rectangular living room is all to be carpeted except for the tiled portion in front of the fireplace. What is the area to be carpeted? What is the area to be tiled? **(See Example 5.)**

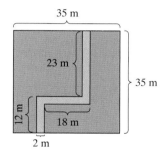

50. A square garden is developed with a 2-m-wide path as shown. How much area is left for flowers?

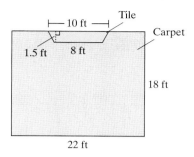

51. A patio area is to be covered with outdoor tile. What is the area of the patio?

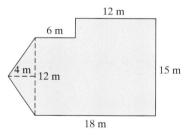

52. Find the area of the sign.

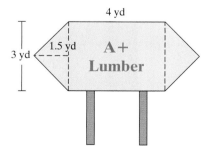

53. Find the area of the kite.

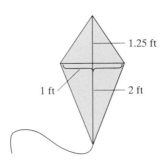

Expanding Your Skills

54. If the lengths of the sides of a square are doubled, by how many times is the area increased?

55. If the lengths of the sides of a square are tripled, by how many times is the area increased?

---

chapter 8 | midchapter review

For Exercises 1–4, state whether the unit can be used to measure perimeter or area.

1. ft

2. km

3. mi^2

4. cm^2

For Exercises 5–9, find both the area and the perimeter.

5. Square

5 ft

6. Rectangle

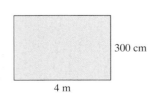

300 cm

4 m

7. Parallelogram

520 m   430 m

1.1 km

8. Triangle (use the Pythagorean theorem to find the third side)

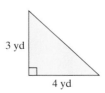

3 yd

4 yd

9. Trapezoid.

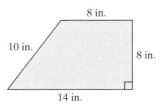

8 in.

10 in.   8 in.

14 in.

section 8.4   Circles, Circumference, and Area

Objectives

1. Basic Definitions
2. Circumference
3. Area
4. Applications

1. Basic Definitions

A **circle** is a figure consisting of all points located the same distance r from a fixed point C. The fixed point C is called the **center** of the circle. The distance r is the length of a **radius** of the circle. A radius of a circle is a line segment drawn from the center of a circle to a point on the circle.

The radius can be measured from the center to any point on a circle. A line segment connecting two points on a circle and passing through the center is called a **diameter** of the circle. In Figure 8-25, $\overline{AC}$ is a radius of the circle, and $\overline{AB}$ is a diameter.

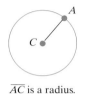

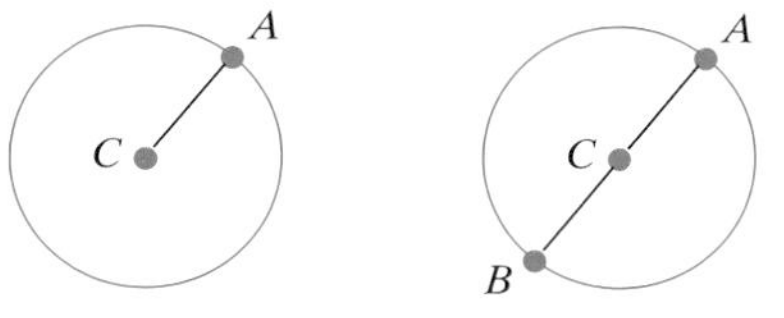

$\overline{AC}$ is a radius. $\overline{AB}$ is a diameter.

Figure 8-25

Notice that the length of a diameter is twice the radius. Therefore, we have

$$d = 2r \quad \text{or equivalently} \quad r = \frac{d}{2}$$

Concept Connections

1. Explain the difference between a radius of a circle and a diameter of a circle.

---

example 1   Finding Diameter and Radius

a. Find the length of a radius.   **b.** Find the length of a diameter.

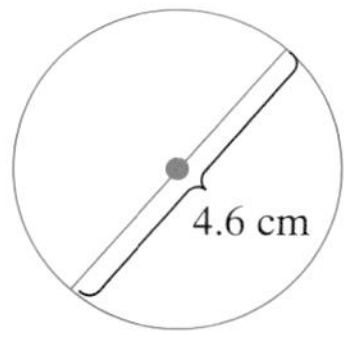

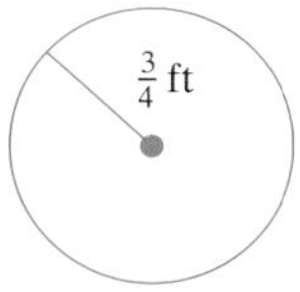

Solution:

a. $r = \dfrac{d}{2} = \dfrac{4.6 \text{ cm}}{2} = 2.3$ cm

b. $d = 2r = \overset{1}{2}\left(\dfrac{3}{\underset{2}{4}} \text{ ft}\right) = \dfrac{3}{2}$ ft or 1.5 ft

Skill Practice

2. Find the length of a radius.

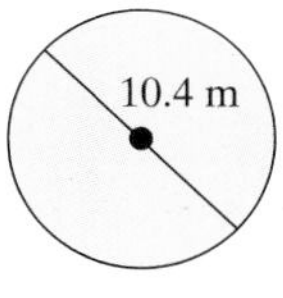

3. Find the length of a diameter.

2. Circumference

The distance around a circle is called the **circumference**. Early mathematicians discovered that the ratio of the circumference of a circle to its diameter $\frac{C}{d}$ is constant. This is true for any size circle. For example, the circumference of a can of beans is approximately 9.25 in., and its diameter is 3 in. The ratio $\frac{C}{d} = \frac{9.25 \text{ in.}}{3 \text{ in.}} \approx 3.1$. The same ratio is found for a can of paint. See Figure 8-26.

Answers

1. A radius of a circle is a line segment drawn from the center to any point on the circle. A diameter is a line segment through the center of a circle, with endpoints on the circle. The length of the diameter is 2 times the length of a radius.

2. 5.2 m   3. $\dfrac{5}{4}$ yd

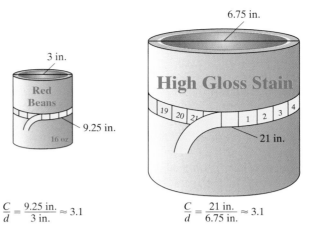

$$\frac{C}{d} = \frac{9.25 \text{ in.}}{3 \text{ in.}} \approx 3.1 \qquad\qquad \frac{C}{d} = \frac{21 \text{ in.}}{6.75 \text{ in.}} \approx 3.1$$

Figure 8-26

Concept Connections

4. What is the meaning of the circumference of a circle?

The constant value given by $\frac{C}{d}$ is called the number π (read "pi"). The number π is about 3.14 or about $\frac{22}{7}$. The relationship between the circumference and diameter of a circle gives us the following formulas for the circumference of a circle.

Circumference of a Circle

The circumference C of a circle is given by

$$C = \pi d \qquad \text{or} \qquad C = 2\pi r$$

where π is approximately 3.14 or $\frac{22}{7}$.

Tip: The circumference of a circle is similar to the perimeter of a polygon. It is the distance around the figure.

Skill Practice

Find the circumference. Use 3.14 for π.

5.

10 ft

6.

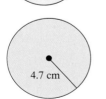

4.7 cm

example 2   **Finding Circumference**

Find the circumference. Use 3.14 for π.

a.

8 m

b.

5.1 ft

Solution:

a. The diameter is given, $d = 8$ m.

$$C = \pi d$$

$$= \pi(8 \text{ m}) \qquad \text{Substitute } d = 8 \text{ m.}$$

$$\approx (3.14)(8 \text{ m}) \qquad \text{Approximate } \pi \text{ by 3.14.}$$

$$= 25.12 \text{ m} \qquad \text{The circumference is approximately 25.12 m.}$$

Tip: In Example 2, the exact value of the circumference is 8π m. By using the approximation 3.14 for π, we approximate the circumference as 25.12 m.

Answers

4. The circumference of a circle is the distance around the circle.
5. 31.4 ft
6. 29.516 cm

b. The radius is given, $r = 5.1$ ft.

$C = 2\pi r$

$\quad = 2\pi(5.1 \text{ ft})$ Substitute $r = 5.1$ ft.

$\quad \approx 2(3.14)(5.1 \text{ ft})$ Approximate π by 3.14.

$\quad = (6.28)(5.1 \text{ ft})$

$\quad = 32.028 \text{ ft}$ The circumference is approximately 32.028 ft.

3. Area

The circumference of a circle is given by $C = 2\pi r$. The length of a **semicircle** (one-half of a circle) is one-half of this amount: $\frac{1}{2}2\pi r = \pi r$. To develop the formula for the area of a circle, consider the bottom half and top half of a circle cut into pie-shaped wedges. Unfold the figure as shown (Figure 8-27).

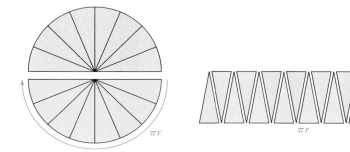

Figure 8-27

The resulting figure is nearly a parallelogram, with base $= \pi r$ and height approximately equal to the radius of the circle. The area is (base) $\times$ (height) $= (\pi r) \cdot r = \pi r^2$. This is the area formula for a circle.

Area of a Circle

The area A of a circle is given by

$$A = \pi r^2$$

Tip: To express the formula for the circumference of a circle, we can use either the radius ($C = 2\pi r$) or the diameter ($C = \pi d$).

     To find the area of a circle, we will always use the radius ($A = \pi r^2$).

example 3   **Finding the Area of a Circle**

a. Find the area. Use $\frac{22}{7}$ for π.

14 m

b. Find the area of a quarter. Use 3.14 for π. Round to the nearest whole unit.

24.4 mm

Skill Practice

7. Find the area of the circle. Use $\frac{22}{7}$ for π.

8. Find the area of the clock face. Use 3.14 for π. Round to the nearest whole unit.

Solution:

a. $A = \pi r^2$

$\approx \left(\dfrac{22}{7}\right) \cdot (14 \text{ m})^2$ Substitute $\pi \approx \frac{22}{7}$ and $r = 14$ m.

$= \left(\dfrac{22}{7}\right) \cdot (196 \text{ m}^2)$ Perform the order of operations.

$= \left(\dfrac{22}{7}\right) \cdot \left(\dfrac{\overset{28}{196}}{1} \text{ m}^2\right)$ Multiply fractions.

$= 616 \text{ m}^2$ The area is 616 m².

b. To compute the area, we first find the radius of the circle.

$r = \dfrac{d}{2} = \dfrac{24.4 \text{ mm}}{2} = 12.2 \text{ mm}$

$A = \pi r^2$

$\approx (3.14)(12.2 \text{ mm})^2$ Substitute 3.14 for π and $r = 12.2$ mm.

$= (3.14)(148.84 \text{ mm}^2)$ Perform the order of operations.

$= 467.3576 \text{ mm}^2$ Multiply.

$\approx 467 \text{ mm}^2$ Round to the nearest whole unit.

4. Applications

Skill Practice

9. The shaded region consists of a rectangle and three semicircles. Find the perimeter and area. Use 3.14 for π.

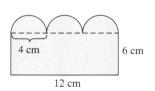

example 4   Finding Perimeter and Area of a Composite Figure

The region shown is formed by a rectangle and a semicircle. Find the perimeter and area. Use 3.14 for π.

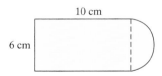

Solution:

The figure consists of a rectangle and a semicircle. We can label the figure further to identify the length of each side and the radius of the semicircle.

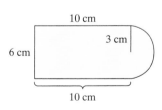

The perimeter is the sum of the three sides and the distance around the semicircle.

$$P = \underbrace{6 \text{ cm} + 10 \text{ cm} + 10 \text{ cm}}_{\text{sum of the 3 sides}} + \underbrace{\dfrac{1}{2} \cdot 2(3.14)(3 \text{ cm})}_{\text{distance around semicircle } (\frac{1}{2}2\pi r)}$$

$\approx 26 \text{ cm} + \dfrac{1}{2} \cdot 2(3.14)(3 \text{ cm})$ Multiply fractions.

$= 26 \text{ cm} + 9.42 \text{ cm}$ Add.

$= 35.42 \text{ cm}$ The perimeter is approximately 35.42 cm.

Answers

7. 1386 in.²   8. 69 in.²
9. Perimeter = 42.84 cm; area = 90.84 cm²

The area is the sum of the area of the rectangle and the area of the semicircle.

$$\underbrace{\text{area of rectangle}}_{} \quad \overbrace{\text{area of semicircle } (\tfrac{1}{2}\pi r^2)}^{}$$

$$A \approx (10 \text{ cm})(6 \text{ cm}) + \frac{1}{2}(3.14)(3 \text{ cm})^2$$

$$= 60 \text{ cm}^2 + \frac{1}{2}(3.14)(9 \text{ cm}^2) \qquad \text{Simplify exponents.}$$

$$= 60 \text{ cm}^2 + (1.57)(9 \text{ cm}^2) \qquad \text{Multiply.}$$

$$= 60 \text{ cm}^2 + 14.13 \text{ cm}^2 \qquad \text{Add like units.}$$

$$= 74.13 \text{ cm}^2 \qquad \text{The area is approximately } 74.13 \text{ cm}^2.$$

---

example 5   Finding Area in an Application

A rotating lawn sprinkler shoots water a distance of 20 ft. If the sprinkler rotates in a full circle, find the total area that can be watered.

Solution:

Draw a figure. The area that is watered is in the shape of a circle.

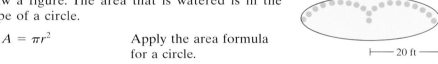

── 20 ft ──

$$A = \pi r^2 \qquad \text{Apply the area formula for a circle.}$$

$$\approx (3.14)(20 \text{ ft})^2 \qquad \text{Substitute 3.14 for } \pi \text{ and 20 ft for } r.$$

$$= (3.14)(400 \text{ ft}^2) \qquad \text{Simplify exponents.}$$

$$= 1256 \text{ ft}^2 \qquad \text{The total area to be watered is approximately } 1256 \text{ ft}^2.$$

Skill Practice

10. A revolving beacon shines light a distance of 200 m. If the beacon rotates in a full circle, find the area that can be lighted. Use 3.14 for π.

Answer

10. 125,600 m²

---

section 8.4   **Practice Exercises**

Boost *your* GRADE at
mathzone.com!

• Practice Problems   • e-Professors
• Self-Tests   • Videos
• NetTutor

Study Skills Exercises

1. To help remember the formulas for circles, list them together and observe the similarities and differences in the formulas.

Circumference = _____

Area = _____

Similarities: Differences:

2. Define the key terms.

   a. Circle   **b. Center**   **c. Radius**

   d. Diameter   **e. Circumference**   **f. Semicircle**

Review Exercises

3. Find the area of a rectangle with length 42 cm and width 30 cm.

4. Find the area of a parallelogram with base 42 cm and height 30 cm.

5. Find the area of a triangle with base 42 cm and height 30 cm.

6. How do the areas found in Exercises 3–5 compare to one another?

7. Could the formula $A = bh$ apply to finding the area of a rectangle? Explain.

Objective 1: Basic Definitions

8. How does the length of a radius of a circle compare to the length of a diameter?

For Exercises 9–14, find the length of a diameter. **(See Example 1.)**

9.   **10.**   **11.**

12.   **13.**   **14.**

For Exercises 15–20, find the length of a radius. **(See Example 1.)**

15.   **16.**   **17.**

18.   **19.**   **20.**

Objective 2: Circumference

21. Circumference is similar to which type of measure? (Circle the correct answer.)

    a. Area   **b.** Capacity   **c.** Perimeter   **d.** Weight

22. Indicate the type of units that could be associated with measuring circumference. Circle all that apply.

    a. ft   **b.** m²   **c.** Liters

    d. Meters   **e.** Grams   **f.** in.²

    g. Miles   **h.** Kilometers   **i.** Cubic centimeters

23. Define π in terms of the circumference and diameter of a circle.

24. Which of the following are *not* good approximations for π? Circle all that apply.

    a. 31.4   **b.** 3.14   **c.** $\dfrac{22}{7}$ **d.** $22\frac{1}{7}$

For Exercises 25–32, find the circumference of the circle. Use 3.14 for π. **(See Example 2.)**

25.

26.

27.

28.

29.

30.

31.

32.

For Exercises 33–38, use 3.14 for π.

33. Find the circumference of the can of soda.

34. Find the circumference of a can of tuna.

35. Find the circumference of a compact disk.

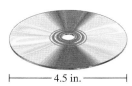

36. Find the outer circumference of a pipe with 3.5-in. diameter.

37. Find the outer circumference of a washer 2.2 cm in diameter.

38. Find the circumference of a pencil 5 mm in diameter.

Objective 3: Area

For Exercises 39–42, find the area of the circle. Use $\frac{22}{7}$ for π. (**See Example 3.**)

39.

7 m

40.

$\frac{7}{2}$ km

41.

21 cm

42.

42 in.

For Exercises 43–46, find the area. Use 3.14 for π. Round to the nearest whole unit. (**See Example 3.**)

43.

25 mm

44.

10 ft

45.

6.2 ft

46.

2.9 m

Objective 4: Applications

For Exercises 47–54, find the area of the shaded region. Use 3.14 for π. (**See Example 4.**)

47.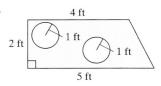

4 ft

2 ft   1 ft

1 ft

5 ft

48.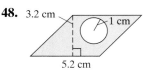

3.2 cm   1 cm

5.2 cm

49.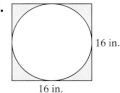

16 in.

16 in.

50.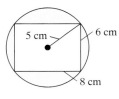

5 cm   6 cm

8 cm

51.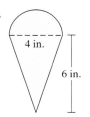

4 in.

6 in.

52.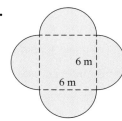

6 m

6 m

53.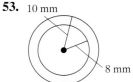

10 mm

8 mm

54.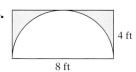

4 ft

8 ft

55. A roller hockey rink is a rectangle with a semicircle at each end.

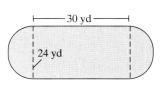

30 yd

24 yd

a. How much will it cost to put up a rail around the rink if railing costs $2.59 per *foot*?

b. How much will it cost to put down the floor if flooring costs $8.00 per square yard?

56. A ceiling fan blade rotates in a full circle. If the fan blades are 2 ft long, what is the area covered by the fan blades?

57. An outdoor torch lamp shines light a distance of 30 ft in all directions. What is the total ground area lighted? **(See Example 5.)**

58. How many times larger is the area of circle 1 than circle 2?

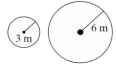

Circle 1   Circle 2

Expanding Your Skills

59. A bicycle wheel has a 26-in. diameter.

    a. Find the circumference. Use 3.14 for π.

    b. How many times will the wheel have to turn to go a distance of 1000 ft (12,000 in.)? Round to the nearest whole unit.

60. A hula hoop has a 20-in. diameter.

    a. Find the circumference. Use 3.14 for π.

    b. How many times will the hula hoop have to turn to roll down a 40-yd driveway? Round to the nearest whole unit.

61. Latasha has a bicycle, and the wheel has a 22-in. diameter. If the wheels of the bike turned 1000 times, how far did she travel? Use 3.14 for π. Give the answer to the nearest inch and to the nearest foot.

Calculator Connections

Topic: Applications of circles

62. Jeff uses surveyor tape to wrap around trees when making a hiking trail. The average diameter of the trees is 0.8 ft. Furthermore, 6 in. of tape is used to make each knot. How much tape is needed to mark 200 trees?

63. Find the area for each pizza. Then compute the unit cost per square inch. If the pizzas have the same thickness, which pizza is a better buy?

Diameter	Cost	Area	Cost per in.^{2}
8 in.	$ 6.50		
12 in.	12.40		

section 8.5 Volume

1. Introduction to Volume

In this section we learn how to compute volume. Volume is another word for capacity. We use volume, for example, to determine how much can be held in a moving van.

In addition to the units of capacity learned in Sections 7.2 and 7.4, volume can be measured in cubic units. For example, a cube that is 1 cm on a side has a volume of 1 cubic centimeter (1 cm^3 or cc). A cube that is 1 in. on a side has a volume of 1 cubic inch (1 in.^{3}). See Figure 8-28. Additional units of volume include cubic feet (ft^3), cubic yards (yd^3), cubic meters (m^3), and so on.

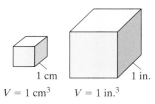

$V = 1$ cm^3 $V = 1$ in.^{3}

Figure 8-28

> **Tip:** Recall that 1 cubic centimeter can also be denoted as 1 cc. Furthermore, 1 cc = 1 mL.

2. Volume Formulas for Selected Solids

The formulas used to compute the volume of several common solids are given.

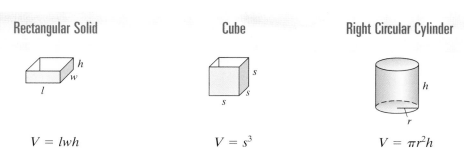

Rectangular Solid	Cube	Right Circular Cylinder
$V = lwh$	$V = s^3$	$V = \pi r^2 h$

Notice that the volume formulas for these three figures are given by the product of the area of the base and the height of the figure:

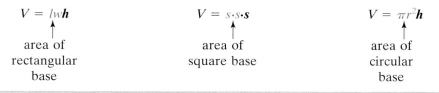

$V = lw\boldsymbol{h}$ $V = s \cdot s \cdot \boldsymbol{s}$ $V = \pi r^2 \boldsymbol{h}$

  ↑   ↑   ↑

area of rectangular base   area of square base   area of circular base

example 1   Finding Volume

Find the volume. Use 3.14 for π where applicable. Round to the nearest whole unit.

a.

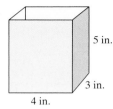

5 in.

3 in.

4 in.

b.

3.7 cm

11.2 cm

Frijoles

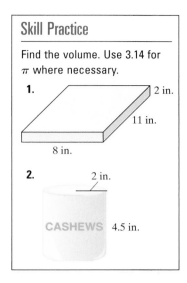

Solution:

a. $V = lwh$ Use the volume formula for a rectangular solid. Identify the length, width, and height.

$= (4 \text{ in.})(3 \text{ in.})(5 \text{ in.})$ $l = 4$ in., $w = 3$ in., and $h = 5$ in.

$= 60 \text{ in.}^3$

The volume of this solid is given by the number of cubic inches that can be enclosed. We can visualize the volume by "layering" cubes that are each 1 in. high (Figure 8-29). The number of cubes in each layer is equal to $4 \times 3 = 12$. Each layer has 12 cubes, and there are 5 layers. Thus, the total number of cubes is $12 \times 5 = 60$ for a volume of 60 in.^{3}

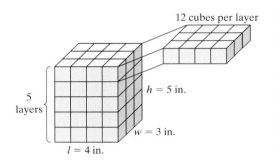

12 cubes per layer

5 layers

$h = 5$ in.

$w = 3$ in.

$l = 4$ in.

Figure 8-29

b. $V = \pi r^2 h$ Use the formula for the volume of a right circular cylinder.

$\approx (3.14)(3.7 \text{ cm})^2(11.2 \text{ cm})$ Substitute 3.14 for π, $r = 3.7$ cm, and $h = 11.2$ cm.

$= (3.14)(13.69 \text{ cm}^2)(11.2 \text{ cm})$ Simplify exponents first.

$= 481.44992 \text{ cm}^3$ Multiply from left to right.

$\approx 481 \text{ cm}^3$ Round to the nearest whole unit.

A right circular cone has the shape of a party hat. A sphere has the shape of a ball. To compute the volume of a cone and a sphere, we use the following formulas.

Tip: A **hemisphere** is one-half of a sphere. Therefore, the volume of a hemisphere is one-half that of a full sphere.

Hemisphere

$$V = \frac{1}{2} \cdot \left(\frac{4}{3}\pi r^3\right)$$

Right Circular Cone

$$V = \frac{1}{3}\pi r^2 h$$

Sphere

$$V = \frac{4}{3}\pi r^3$$

Tip: Notice that the formula for the volume of a right circular cone is $\frac{1}{3}$ that of a right circular cylinder.

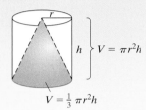

$$V = \pi r^2 h$$

$$V = \frac{1}{3}\pi r^2 h$$

Skill Practice

Find the volume. Use 3.14 for π.

3.
 $r = 3$ cm

4. $r = 3$ cm

5. |— 8 cm —|
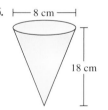
18 cm

Answers

3. 113.04 cm³
4. 56.52 cm³
5. 301.44 cm³

example 2   Finding Volume of a Cone and Sphere

Find the volume. Use 3.14 for π. Round to 1 decimal place.

a.
 $r = 6$ in.

b.

8 in.
|— 5 in. —|

Solution:

a. $V = \dfrac{4}{3}\pi r^3$ Use the formula for the volume of a sphere.

$\approx \dfrac{4}{3}(3.14)(6 \text{ in.})^3$ Substitute 3.14 for π and $r = 6$ in.

$= \dfrac{4}{3}(3.14)(216 \text{ in.}^3)$ Simplify exponents first.
$(6 \text{ in.})^3 = (6 \text{ in.})(6 \text{ in.})(6 \text{ in.}) = 216 \text{ in.}^3$

$= \dfrac{4}{3}\left(\dfrac{3.14}{1}\right)\left(\dfrac{216 \text{ in.}^3}{1}\right)$ Multiply fractions.

$= \dfrac{4}{\underset{1}{\cancel{3}}}\left(\dfrac{3.14}{1}\right)\left(\dfrac{\overset{72}{\cancel{216}} \text{ in.}^3}{1}\right)$ Simplify to lowest terms.

$= 904.32 \text{ in.}^3$ Multiply from left to right.

$\approx 904.3 \text{ in.}^3$ Round to 1 decimal place.

b. $V = \frac{1}{3}\pi r^2 h$

Use the formula for the volume of a right circular cone.

To find the radius we have

$r = \frac{d}{2} = \frac{5 \text{ in.}}{2} = 2.5 \text{ in.}$

$V \approx \frac{1}{3}(3.14)(2.5 \text{ in.})^2(8 \text{ in.})$

Substitute 3.14 for π, $r = 2.5$ in., and $h = 8$ in.

$= \frac{1}{3}\left(\frac{3.14}{1}\right)\left(\frac{6.25 \text{ in.}^2}{1}\right)\left(\frac{8 \text{ in.}}{1}\right)$

Simplify exponents first.

$= \frac{157}{3} \text{ in.}^3$

Multiply fractions.

$\approx 52.3 \text{ in.}^3$

Round to 1 decimal place.

3. Volumes of Composite Figures

example 3   Finding the Volume of a Composite Figure

Find the volume of the HEPA filter (Figure 8-30). Use 3.14 for π and round the answer to the nearest whole unit.

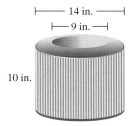

Figure 8-30

Solution:

The solid consists of an outer cylinder with a cylindrical core cut out of the center.

To find the volume, we can find the volume of the outer cylinder and subtract the volume of the inner cylinder.

The radius of the outer cylinder is

$r = \frac{d}{2} = \frac{14 \text{ in.}}{2} = 7 \text{ in.}$

The radius of the inner cylinder is

$r = \frac{d}{2} = \frac{9 \text{ in.}}{2} = 4.5 \text{ in.}$

The volume of the outer cylinder is

$V = \pi r^2 h$

$\approx (3.14)(7 \text{ in.})^2(10 \text{ in.})$

$= (3.14)(49 \text{ in.}^2)(10 \text{ in.})$

$= 1538.6 \text{ in.}^3$

The volume of the inner cylinder is

$V = \pi r^2 h$

$\approx (3.14)(4.5 \text{ in.})^2(10 \text{ in.})$

$= (3.14)(20.25 \text{ in.}^2)(10 \text{ in.})$

$= 635.85 \text{ in.}^3$

The volume of the Hepa filter is the difference of the outer cylinder and the inner cylinder.

volume of
outer cylinder
↓

volume of
inner cylinder
↓

Volume of filter $= 1538.6 \text{ in.}^3 - 635.85 \text{ in.}^3$

$= 902.75 \text{ in.}^3$

$\approx 903 \text{ in.}^3$

Round to the nearest whole unit.

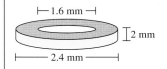

Study Skills Exercises

1. Apply what you have learned to real life situations. This can help you remember formulas and methods as well as give some meaning to math. Write down one real-life application to geometry.

2. Define the key terms.

   a. Rectangular solid   **b. Cube**   **c. Right circular cylinder**

   d. Right circular cone   **e. Sphere**   **f. Hemisphere**

Review Exercises

For Exercises 3–5, find the circumference C and area A of the circles. Use 3.14 for π.

3.
4 in.

4.
10 cm

5.
6 m

6. Find the area of the shaded region.

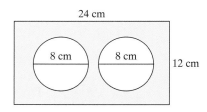

24 cm

8 cm   8 cm

12 cm

Objective 1: Introduction to Volume

7. Which of the units denote volume? Circle all that apply.

   a. ft^2 **b.** m^3 **c.** in.   **d.** cc   **e.** mi

8. Which of the units denote volume? Circle all that apply.

   a. yd   **b.** yd^2 **c.** yd^3 **d.** km   **e.** km^3

For Exercises 9–12, determine the area of the square and the volume of the cube.

9.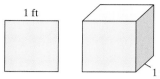
1 ft
1 ft

10.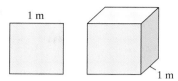
1 m
1 m

11.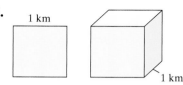
1 km
1 km

12.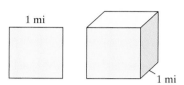
1 mi
1 mi

Objective 2: Volume Formulas for Selected Solids

For Exercises 13–24, find the volume. Use 3.14 for π where necessary. **(See Examples 1 and 2.)**

13.
1.4 cm
1.4 cm
1.4 cm

14.
4.5 m
4.5 m
4.5 m

15.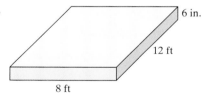
6 in.
12 ft
8 ft

16.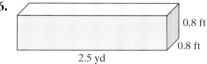
0.8 ft
0.8 ft
2.5 yd

17.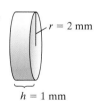
$r = 2$ mm
$h = 1$ mm

18.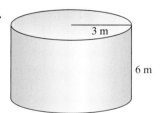
3 m
6 m

19.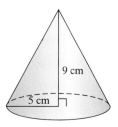
9 cm
5 cm

20.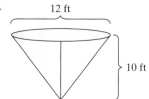
12 ft
10 ft

21. $r = 3$ yd

22. $d = 12$ in.

23. 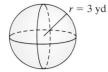 12 ft

24. 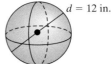 15 cm

For Exercises 25–30, use 3.14 for π. Round each value to the nearest whole unit.

25. The diameter of a volleyball is 8.2 in. Find the volume.

26. The diameter of a basketball is 9 in. Find the volume.

27. Find the volume of the sand pile.

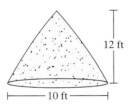

12 ft

10 ft

28. In decorating cakes, many people use an icing bag which has the shape of a cone. Find the volume of the icing bag.

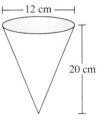

12 cm

20 cm

29. Find the volume of water (in cubic feet) that the pipe can hold.

50 ft

6 in.

30. Find the volume of the wastebasket that has the shape of a cylinder with the height of 15 in. and diameter of 10 in.

Objective 3: Volumes of Composite Figures

31. A gasoline storage tank is in the shape of a cylinder with hemispheres on each end. Find the volume. Round to the nearest whole unit. **(See Example 3.)**

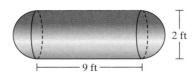

2 ft

9 ft

32. A silo is in the shape of a cylinder with a hemisphere on the top. Find the volume. Round to the nearest whole unit.

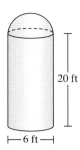

20 ft

6 ft

33. An ice cream cone is in the shape of a cone with a hemisphere on top. Assuming that ice cream is packed inside the cone, find the volume of the ice cream. Round to one decimal place.

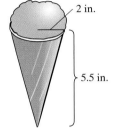

2 in.

5.5 in.

34. A birdbath is made from a hemisphere on a pedestal. Find the volume of water that the birdbath will hold. Round to the nearest whole unit.

2 cm

24 cm

For Exercises 35–40, find the volume of the shaded region. Use 3.14 for π if necessary.

35.

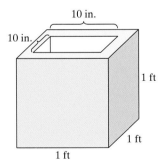

10 in.

10 in.

1 ft

1 ft

1 ft

1 ft

36.

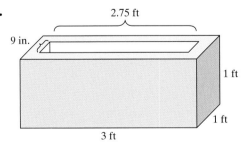

2.75 ft

9 in.

1 ft

1 ft

3 ft

37.

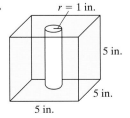

$r = 1$ in.

5 in.

5 in.

5 in.

38.

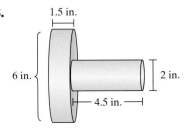

1.5 in.

6 in.

2 in.

4.5 in.

39.

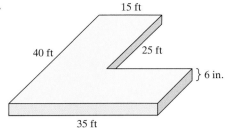

40.

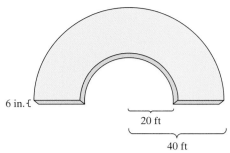

Calculator Connections

Topic: Applications of volume

41. The earth is approximately in the shape of a sphere. Find the volume of the earth if the radius is approximately 4000 mi. Use 3.14 for π.

42. The moon is approximately in the shape of a sphere. Find the volume of the moon if the radius is approximately 1080 mi. Use 3.14 for π.

The volume formulas for right circular cylinders and right circular cones are the same for slanted cylinders and cones. For Exercises 43–44, find the volume. Use 3.14 for π.

43.

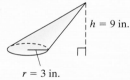

$h = 9$ in.

$r = 3$ in.

44.

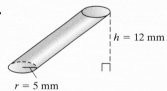

$h = 12$ mm

$r = 5$ mm

chapter 8 | summary

section 8.1 Lines and Angles

Key Concepts

Basic concepts of geometry:

A **point** is a specific location in space.

A **line** consists of infinitely many points that follow a straight path.

A **line segment** is a part of a line between two distinct endpoints.

A **ray** is the part of a line that includes an endpoint and all points on one side of the endpoint.

An **angle** is a geometric figure formed by two rays that share a common endpoint. The common endpoint is called the **vertex** of the angle.

Angles are generally measured in degrees.

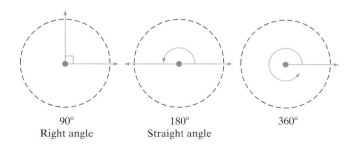

90°
Right angle

180°
Straight angle

360°

A **protractor** can be used to measure angles.

An angle is said to be **acute** if its measure is between 0° and 90°. An angle is said to be **obtuse** if its measure is between 90° and 180°.

Examples

Example 1

A point is symbolized by a dot named with a capital letter.

P

A line is symbolized by a line with arrows at the ends because the line extends forever in both directions. This line can be denoted $\overleftrightarrow{AB}$ or $\overleftrightarrow{BA}$.

This line segment can be denoted $\overline{PQ}$ or $\overline{QP}$.

A ray is denoted $\overrightarrow{RS}$. Note that the first letter is the endpoint.

Example 2

An angle is named by three letters, and the middle letter represents the vertex, such as $\angle DEF$. An angle can also be named by just the vertex, as in $\angle E$.

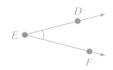

Example 3

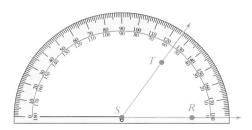

The measure of $\angle RST$ is 55°.

Two angles are said to be **complementary** if the sum of their measures is 90°. Two angles are said to be **supplementary** if the sum of their measures is 180°.

Parallel lines lie on the same flat surface (called a plane), but never intersect.

If two lines intersect at a right angle, the lines are said to be **perpendicular**.

Given two intersecting lines, **vertical angles** are angles that appear on opposite sides of the vertex.

When two parallel lines are crossed by a **transversal**, eight angles are formed.

Example 4

The complement of a 32° angle is a 58° angle. The supplement of a 32° angle is a 148° angle.

Example 5

Parallel lines:

Example 6

Perpendicular lines:

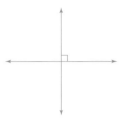

Example 7

Intersecting lines:

∠1 and ∠3 are vertical angles and are congruent. Also ∠2 and ∠4 are vertical angles and are congruent.

Example 8

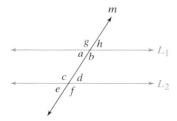

$m(\angle a) = m(\angle d)$ because they are **alternate interior angles**.
$m(\angle e) = m(\angle h)$ because they are **alternate exterior angles**.
$m(\angle c) = m(\angle g)$ because they are **corresponding angles**.

section 8.2 Triangles and the Pythagorean Theorem

Key Concepts

The sum of the measures of the angles of any triangle is 180°.

An **acute triangle** is a triangle in which all three angles are acute.

A **right triangle** is a triangle in which one angle is a right angle.

An **obtuse triangle** is a triangle in which one angle is obtuse.

An **equilateral triangle** is a triangle in which all three sides (and all three angles) are equal in measure.

An **isosceles triangle** is a triangle in which two sides are equal in length (the angles opposite the equal sides are also equal in measure).

A **scalene triangle** is a triangle in which no sides (or angles) are equal in measure.

Pythagorean Theorem

The sum of the squares of the **legs** of a right triangle equals the square of the **hypotenuse**.

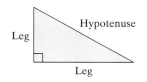
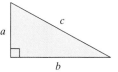

$$a^2 + b^2 = c^2$$

Examples

Example 1

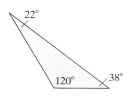

$$22° + 120° + 38° = 180°$$

Example 2

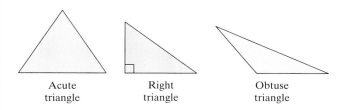

Acute   Right   Obtuse
triangle   triangle   triangle

Example 3

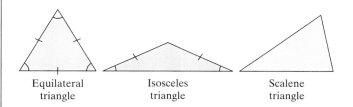

Equilateral   Isosceles   Scalene
triangle   triangle   triangle

Example 4

Given:

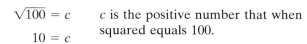

To find the length of the hypotenuse, solve for c.

$$6^2 + 8^2 = c^2$$

$$36 + 64 = c^2$$

$$100 = c^2$$

$$\sqrt{100} = c \qquad$$ c is the positive number that when squared equals 100.

$$10 = c$$

The length of the hypotenuse is 10 cm.

section 8.3 Quadrilaterals, Perimeter, and Area

Key Concepts

A four-sided **polygon** is called a **quadrilateral**.

A **parallelogram** is a quadrilateral with opposite sides parallel.

A **rectangle** is a parallelogram with four right angles.

A **square** is a rectangle with sides equal in length.

A **rhombus** is a parallelogram with sides equal in length.

A **trapezoid** is a quadrilateral with one pair of parallel sides.

Perimeter is the distance around a figure.

Perimeter of a square: $P = 4s$

Perimeter of a rectangle: $P = 2l + 2w$

Area is the number of square units that can be enclosed by a figure.

Area of a rectangle: $A = lw$
Area of a square: $A = s^2$
Area of a parallelogram: $A = bh$
Area of a triangle: $A = \frac{1}{2}bh$
Area of a trapezoid: $A = \frac{1}{2}(a + b)h$

Examples

Example 1

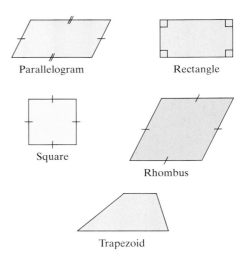

Parallelogram   Rectangle

Square

Rhombus

Trapezoid

Example 2

Given:

8 in.

22 in.

The perimeter of the rectangle is

$P = 2(22) + 2(8)$

$\quad = 44 + 16$

$\quad = 60$

The perimeter is 60 in.

Example 3

Given:

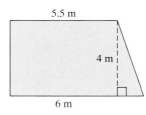

5.5 m

4 m

6 m

The area of the trapezoid is

$A = \frac{1}{2}(6 + 5.5)4$

$\quad = \frac{1}{2}(11.5)4$

$\quad = 23$

The area is 23 m².

section 8.4   Circles, Circumference, and Area

Key Concepts

A **circle** is a figure consisting of all points located the same distance r from a fixed point C. The fixed point C is called the **center** of the circle. The line segment from the center to any point on the circle is called a **radius** of the circle. A **diameter** of a circle is a line segment whose endpoints are on the circle and that passes through the center.

The length of a radius is one-half the length of a diameter. That is,

$$r = \frac{d}{2} \quad \text{or} \quad d = 2r$$

The **circumference of a circle** is the distance around the circle and can be found by using the formula $C = 2\pi r$ or $C = \pi d$.

The **area of a circle** is found by using the formula $A = \pi r^2$.

The number $\pi = \frac{\text{circumference}}{\text{diameter}}$. We use the approximation 3.14 or $\frac{22}{7}$.

Examples

Example 1

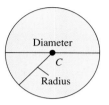

Example 2

If a circle has diameter 3.2 m, the radius measures 1.6 m.

Example 3

The circumference is $C = 2\pi r$

$$\approx 2(3.14)(24 \text{ in.})$$
$$\approx 150.72 \text{ in.}$$

The area is $A = \pi r^2$

$$\approx (3.14)(24 \text{ in.})^2$$
$$\approx 1808.64 \text{ in.}^2$$

Example 4

Find the area of the **semicircle**.

The diameter = 12 ft; therefore the radius = 6 ft. The area of a semicircle is one-half the area of a circle. $A \approx \frac{1}{2}(3.14)(6 \text{ ft})^2 = 56.52 \text{ ft}^2$

section 8.5 Volume

Key Concepts

Volume is another word for capacity.

Formulas for selected solids are given.

Rectangular solid

$V = lwh$

Cube

$V = s^3$

Right circular cylinder

$V = \pi r^2 h$

Right circular cone

$V = \dfrac{1}{3}\pi r^2 h$

Sphere

$V = \dfrac{4}{3}\pi r^3$

Examples

Example 1

Find the volume of a tissue box with dimensions 23.5 cm by 12 cm by 12 cm.

Volume of a rectangular solid: $V = lwh$

$V = (23.5\ \text{cm})(12\ \text{cm})(12\ \text{cm})$

$= 3384\ \text{cm}^3$

The volume is 3384 cm^3.

Example 2

Find the volume of the cone.

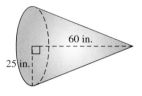

$V = \dfrac{1}{3}\pi r^2 h$

$\approx \dfrac{1}{3}(3.14)(25\ \text{in.})^2(60\ \text{in.})$

$= 39{,}250\ \text{in.}^3$

The volume is appoximately 39,250 in.^{3}.

chapter 8   review exercises

Section 8.1

For Exercises 1–4, match the symbol with a description.

1. $\overleftrightarrow{AB}$ **a.** Ray AB

2. $\overrightarrow{AB}$ **b.** Line segment AB

3. $\overrightarrow{BA}$ **c.** Ray BA

4. $\overline{AB}$ **d.** Line AB

For Exercises 5–8, refer to the figure.

5. Name a ray.

6. Name a line segment.

7. Name the vertex of the indicated angle.

8. Name the indicated angle.

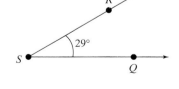

9. Describe the measure of an acute angle.

10. Describe the measure of an obtuse angle.

11. Describe the measure of a straight angle.

12. Describe the measure of a right angle.

13. Let $m(\angle X) = 32°$.

    a. Find the complement of $\angle X$.

    b. Find the supplement of $\angle X$.

14. Let $m(\angle T) = 20°$.

    a. Find the complement of $\angle T$.

    b. Find the supplement of $\angle T$.

For Exercises 15–18, refer to the figure to determine the measure of the indicated angle.

15. $m(\angle ABE)$

16. $m(\angle DBC)$

17. $m(\angle ABG)$

18. $m(\angle ABC)$

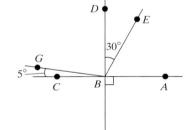

For Exercises 19–20, select the figure or figures that apply.

19. Two lines that are parallel

20. Two lines that are *not* perpendicular.

a.

b.   **c.**
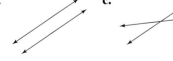

For Exercises 21–26, refer to the figure to determine if the statements are true or false.

L_1 is parallel to L_2

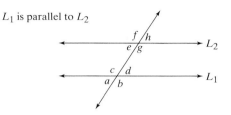

21. a and d are vertical angles.

22. g and d are corresponding angles.

23. e and d are alternate exterior angles.

24. a and h are alternate exterior angles.

25. g and c are alternate interior angles.

26. e and f are supplementary angles.

Section 8.2

For Exercises 27–28, find the measures of the angles x and y.

27.   **28.**

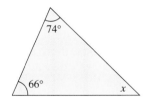   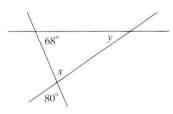

For Exercises 29–34, describe the characteristics of each type of triangle.

29. Obtuse triangle

30. Equilateral triangle

31. Right triangle

32. Acute triangle

33. Isosceles triangle

34. Scalene triangle

For Exercises 35–38, simplify the square roots.

35. $\sqrt{25}$ **36.** $\sqrt{49}$

37. $\sqrt{100}$ **38.** $\sqrt{64}$

For Exercises 39–42, use a calculator to approximate the square roots. Round to the nearest thousandth.

39. $\sqrt{14}$ **40.** $\sqrt{30}$

41. $\sqrt{5}$ **42.** $\sqrt{88}$

43. State the Pythagorean theorem in words.

For Exercises 44–45, find the length of the unknown side.

44.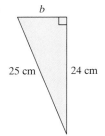

45.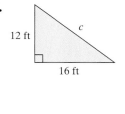

46. Kia is flying a kite. At one point the kite is 5 m from Kia horizontally and 12 m above her (see figure). How much string will be extended at this time? (Assume there is no slack.)

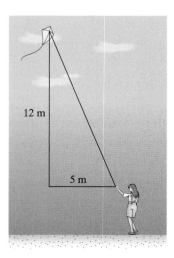

Section 8.3

For Exercises 47–50, indicate the similarities and differences of the quadrilaterals.

47. A rhombus and a square

48. A trapezoid and a parallelogram

49. A rectangle and a square

50. A rectangle and a parallelogram

51. Find the perimeter of the figure.

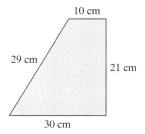

52. Find the perimeter of a triangle with sides of 4.2, 6.1, and 7.0 m.

53. Find the perimeter of a rectangle with length 16 mi and width 12 mi.

54. How much fencing is required to put up a chain link fence around a 120-yd by 80-yd playground?

55. The perimeter of a rectangle is 120 ft. The two shorter sides add up to 36 ft. What is the length of each of the longer sides?

56. The perimeter of a square is 62 ft. What is the length of each side?

57. Find the area of the triangle.

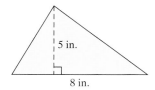

58. Fatima has a Persian rug 8.5 ft by 6 ft. What is the area?

59. A lot is 150 ft by 80 ft. Within the lot, there is a 12-ft easement along all edges. An easement is the portion of the lot on which nothing may be built. What is the area of the portion that may be used for building?

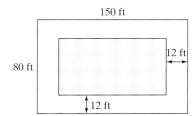

60. Find the area and the perimeter of the shaded triangle.

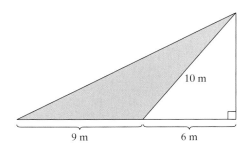

Section 8.4

61. Find the diameter of a circle whose radius is 45 mm.

62. Find the diameter of a circle whose radius is 3.2 ft.

63. Find the radius of a circle whose diameter is 45 mm.

64. Find the radius of a circle whose diameter is 3.2 ft.

For Exercises 65–68, find the circumference C and the area A of the circle.

65. Use 3.14 for π.

66. Use $\frac{22}{7}$ for π.

67. Use $\frac{22}{7}$ for π.

68. Use 3.14 for π.

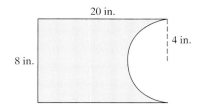

69. Find the area of the shaded region. Use 3.14 for π.

70. The diameter of Lupé's pocket watch is 6 cm. The diameter of his wristwatch is 3 cm.

  a. Find the area of the pocket watch. Use 3.14 for π.

  b. Find the area of the wristwatch.

  c. Is the area of the pocket watch twice the area of the wristwatch?

71. A sign is constructed from a square with a side length of 2 yd. The square has a semicircle on top with diameter the same length as the side of the square. What is the area of the sign?

Section 8.5

For Exercises 72–75, find the volume. Use 3.14 for π.

72.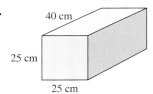

73.

74. 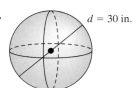 $d = 30$ in.

75.  4 km   3 km

76. Find the volume of a can of paint if the can is a cylinder with radius 6.5 in. and height 7.5 in. Round to the nearest whole unit.

77. Find the volume of a ball if the diameter of the ball is approximately 6 in. Round to the nearest whole unit.

For Exercises 78–79, find the volume of the shaded region. Use 3.14 for π if necessary. Round to the nearest whole unit.

78. 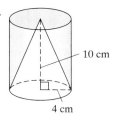 10 cm   4 cm

79. 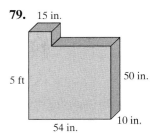 15 in.   5 ft   50 in.   54 in.   10 in.

80. A microwave oven is a rectangular solid with dimensions 1 ft by 1 ft 9 in. by 1 ft 4 in. Find the volume in cubic feet.

---

chapter 8 | test

For Exercises 1–10, decide which of the following is illustrated in the figure. Answer yes or no.

1. $\overline{AB}$

2. $\angle ABC$

3. $\overrightarrow{CD}$

4. $\overleftrightarrow{AB}$

5. $\angle DBA$

6. $\angle ACD$

7. $\overrightarrow{DC}$

8. $\overline{BD}$

9. $\overleftrightarrow{BD}$

10. $\overrightarrow{AB}$

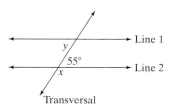

For Exercises 11–18, identify the angles as acute, obtuse, right, or straight.

11. $m(\angle A) = 100°$

12. $m(\angle B) = 89°$

13. $m(\angle C) = 73°$

14. $m(\angle D) = 99°$

15. $m(\angle E) = 90°$

16. $m(\angle F) = 1°$

17. $m(\angle G) = 180°$

18. $m(\angle H) = 45°$

19. Determine the measure of angles x and y. Assume that line 1 is parallel to line 2.

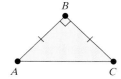

Line 1 y $55°$ Line 2 x Transversal

20. Given that the lengths of $\overline{AB}$ and $\overline{BC}$ are equal, what are the measures of $\angle A$ and $\angle C$?

B

A   C

21. From the figure, determine $m(\angle S)$.

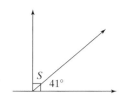

22. What is the sum of all the angles of a triangle?

23. What is the measure of $\angle A$?

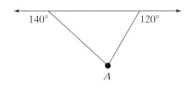

24. A firefighter places a 13-ft ladder against a wall of a burning building. If the bottom of the ladder is 5 ft from the base of the building, how far up the building will the ladder reach?

25. José is a landscaping artist and wants to make a walkway through a rectangular garden, as shown. What is the length of the walkway?

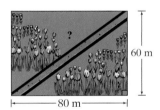

For Exercises 26–31, match the formula with the description.

26. Area of a trapezoid   **a.** $A = lw$

27. Area of a triangle   **b.** $P = 4s$

28. Perimeter of a rectangle   **c.** $A = \frac{1}{2}bh$

29. Perimeter of a square   **d.** $A = \frac{1}{2}(a + b) \cdot h$

30. Area of a rectangle   **e.** $A = bh$

31. Area of a parallelogram   **f.** $P = 2l + 2w$

32. An octagon is an eight-sided figure. A *regular* octagon has eight sides of equal length. A stop sign is in the shape of a regular octagon. Find the perimeter of the stop sign.

33. Jayne wants to put up wallpaper border for a 12-ft by 15-ft room. The border comes in 6-yd rolls. How many rolls would be needed?

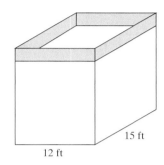

34. Find the area of the ceiling fan blade shown in the figure.

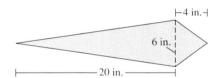

35. Which gives more pizza, a 12-in. by 8-in. rectangular pizza or a 12-in.-diameter round pizza? By how much? Assume that the pizzas have equal thicknesses. Use 3.14 for π.

36. Find the volume of the child's wading pool shown in the figure. Use 3.14 for π and round the answer to the nearest whole unit.

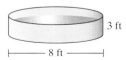

37. Find the volume of the briefcase.

14 in.

18 in.

5 in.

38. Find the volume of the figure. Use $\frac{22}{7}$ for π.

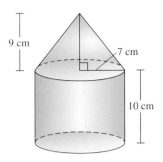

9 cm

7 cm

10 cm

---

chapters 1–8 | cumulative review

For Exercises 1–3, divide.

1. $80{,}535 \div 21$

2. $0 \div 21$

3. $21 \div 0$

For Exercises 4–5, refer to the table.

State	Population
Maine	1,275,000
New Hampshire	1,236,000
Vermont	609,000

4. Find the difference in the populations of Maine and Vermont.

5. Find the sum of the populations of Maine and New Hampshire.

6. Place the fractions on the number line: $\frac{1}{3}, \frac{5}{6},$ and $\frac{3}{5}$.

0   1

7. There is 14 oz of ketchup in a bottle. If $\frac{1}{4}$ is used, how many ounces are left?

For Exercises 8–10, simplify the expressions.

8. $6 \div \frac{1}{3}$

9. $\frac{1}{3} \div 6$

10. $\frac{2}{7} \div \frac{3}{7} \cdot \frac{9}{5}$

11. Find the LCM of 6, 4, and 10.

12. Add: $\frac{1}{6} + \frac{1}{4} + \frac{7}{10}$

13. Subtract: $\frac{13}{6} - \frac{3}{4} - \frac{3}{10}$

14. Write $\frac{132}{8}$ as a mixed number.

15. Write $5\frac{1}{9}$ as an improper fraction.

16. The price of a collectible Three Stooges glass is $11.99. How much will a set of four cost?

17. A sale advertises "Buy 2 get 1 free." If Geraldo buys three shirts and spends $26.98, how much money is he saving?

For Exercises 18–20, complete the table.

	Fraction	Decimal
18.	$\frac{3}{8}$	
19.		$0.\overline{2}$
20.		0.02

21. Simplify the ratio: $\dfrac{2\frac{1}{2}}{3\frac{3}{4}}$

22. Solve the proportion: $\dfrac{2}{9} = \dfrac{8.3}{n}$

23. A party consisting of 25 people requires about 7 pizzas. How many pizzas should be purchased if 60 people are expected? (Round to the nearest whole pizza.)

24. In the 2001–2002 basketball season, Shaquille O'Neal of the Los Angeles Lakers made 712 baskets out of 1229 attempted. Donyell Marshall of the Utah Jazz made 343 baskets out of 661 attempted.

   a. Compute the unit rate of baskets made per attempt for O'Neal. Round to 2 decimal places.

   b. Compute the unit rate of baskets per attempt for Marshall. Round to 2 decimal places.

   c. Compare the unit rates. Which player had a better shooting rate?

25. The operating cost for a Boeing 727 aircraft is approximately $8590 for 2.5 hr of operation. Find the unit rate in dollars per hour.

26. What is 22% of 240?

27. 65% of what number is 46.8?

28. 65 is what percent of 50?

29. A park bench costs $150 and is marked up to sell for $180. What is the percent markup?

30. A jacket normally sells for $85. If it is on sale for $71.40, what is the percent discount?

31. Find the perimeter in feet.

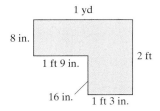

32. A piece of material is 60 in. wide. A sewing pattern requires a width of $4\frac{1}{2}$ ft. Is the material wide enough?

33. A recipe requires $\frac{1}{2}$ cup (c) of milk and 6 fl oz of pineapple juice. How many cups of total liquid are required for this recipe?

34. In Canada, just outside of London, Ontario, the speed limit is posted as 100 kilometers per hour (kph). Convert 100 km to miles to find the equivalent speed in miles per hour. Round to the nearest whole unit.

35. An LT1 Corvette has 300 hp. Convert this to $\dfrac{\text{ft·lb}}{\text{sec}}$.

For Exercises 36–37, find the perimeter.

36.

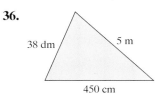

37.

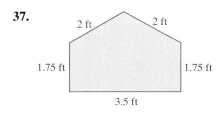

For Exercises 38–39, find the area. Use 3.14 for π if necessary.

38.   **39.**

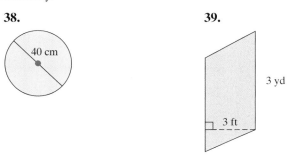

40. Find the volume of the hemisphere. Use 3.14 for π and round to the nearest whole unit.

$r = 6$ in.

Introduction to Statistics

9

9.1 Tables, Bar Graphs, Pictographs, and Line Graphs

9.2 Frequency Distributions and Histograms

9.3 Circle Graphs

9.4 Mean, Median, and Mode

9.5 Introduction to Probability

This chapter introduces the study of statistics. This includes interpreting and constructing a variety of statistical graphs such as bar graphs, line graphs, circle graphs, pictographs, and histograms. We also learn how to compute the mean, median, and mode to measure the "center" of a set of values. We conclude with an introduction to probability, which measures the likelihood of an event to occur.

We encounter statistics daily. For example, in Exercises 30–35, in Section 9.1, we see that the number of sport utility vehicles (SUVs) sold in the United States grew rapidly in the 1990s, but has since leveled off. Vehicle manufacturers may use this type of information to plan the number of vehicles to produce.

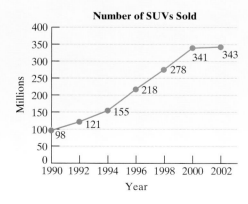

Number of SUVs Sold

chapter 9 | preview

The exercises in this chapter preview contain concepts that have not yet been presented. These exercises are provided for students who want to compare their levels of understanding before and after studying the chapter. Alternatively, you may prefer to work these exercises when the chapter is completed and before taking the exam.

Section 9.1

1. The table gives the number of teachers teaching pre-kindergarten and the total student/teacher ratio for all grades for 5 states.

State	Student/Teacher Ratio	Number of Pre-kindergarten Teachers
California	20.5	11,578
Florida	18.6	877
Illinois	16.0	1,017
New York	13.7	2,223
Texas	14.7	5,550

a. Which state has the lowest student/teacher ratio?

b. Which state has the greatest number of pre-kindergarten teachers?

Section 9.2

2. A list of the number of times people make roundtrip flights in one year is given. Complete the frequency distribution.

3	2	10	20	5	1
3	6	2	12	2	4
21	4	15	12	15	

Class Intervals (number of round-trip flights)	Tally	Frequency
1–5		
6–10		
11–15		
16–20		
21–25		

Section 9.3

3. The percent of public elementary and secondary students categorized by race/ethnicity is displayed in the pie graph. (Source: U. S. Department of Education, National Center for Education Statistics.)

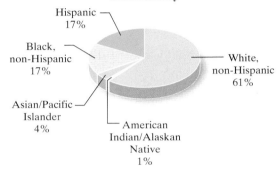

Percent of Public Elementary and Secondary Students, by Race/Ethnicity

Hispanic 17%
Black, non-Hispanic 17%
Asian/Pacific Islander 4%
American Indian/Alaskan Native 1%
White, non-Hispanic 61%

a. If a county has 2340 elementary and secondary students, how many can be expected to be Hispanic? Round to the nearest whole number.

b. Of 2340 students, how many can be expected to be Asian/Pacific Islander?

Section 9.4

4. An advertisement for a weight loss program gives a list of the weight (in pounds) lost by 8 customers. Find the mean, median, and mode if possible.

| 45 | 17 | 24 | 29.5 |
| 14 | 18 | 12 | 15.5 |

Section 9.5

5. A class has 13 freshmen and 7 sophomores. If one person is selected at random from the class, what is the probability that the person is a freshman?

6. In a game of cards, each player is dealt a hand that consists of 5 cards. A full house is a hand that has 3 of one kind and a pair. The probability of getting a full house from a deck of regular playing cards is approximately 0.0173. What is the probability of not getting a full house?

section 9.1   Tables, Bar Graphs, Pictographs, and Line Graphs

1. Introduction to Data and Tables

Statistics is the branch of mathematics that involves collecting, organizing, and analyzing **data** (information). One method to organize data is by using tables. A **table** uses rows and columns to reference information. The individual entries within a table are called **cells**.

example 1   Interpreting Data in a Table

Table 9-1 summarizes the maximum wind speed, number of reported deaths, and estimated cost for recent hurricanes that made landfall in the United States. (Source: National Oceanic and Atmospheric Administration)

Skill Practice

For Exercises 1–4, use the information in Table 9-1.

1. Which hurricane had the highest sustained wind speed at landfall?
2. Which hurricane was the most recent?
3. What was the difference in the death toll for Fran and Alicia?
4. How many times greater was the cost for Hugo than for Opal?

table 9-1

Hurricane	Date	Landfall	Maximum Sustained Winds at Landfall (mph)	Number of Reported Deaths	Estimated Cost ($ Billions)
Fran	1996	North Carolina, Virginia	115	37	5.8
Opal	1995	Florida	120	27	3.6
Andrew	1992	Florida, Louisiana	145	61	35.6
Hugo	1989	South Carolina	130	57	10.8
Alicia	1983	Texas	115	19	5.9

a. Which hurricane caused the greatest number of deaths?

b. Which hurricane was the most costly?

c. What was the difference in the maximum sustained winds for hurricane Opal and hurricane Fran?

d. How many times greater was the death toll for Hugo than for Alicia?

Solution:

a. The death toll is reported in the 5th column. The death toll for hurricane Andrew, 61, is the greatest value.

b. The estimated cost is reported in the 6th column. Hurricane Andrew was also the costliest hurricane at $35.6 billion.

c. The wind speeds are given in the 4th column. The difference in the wind speed for hurricane Opal and hurricane Fran is 120 mph − 115 mph = 5 mph.

d. There were 57 deaths from Hugo and 19 from Alicia. The ratio of deaths from Hugo to deaths from Alicia is given by

$$\frac{57}{19} = 3$$

There were 3 times as many deaths from Hugo as from Alicia.

Answers
1. Andrew
2. Fran
3. 18 deaths
4. 3 times greater

Skill Practice

5. A political poll was taken to determine the political party and gender of several registered voters. The following codes were used.

M = male
F = female
dem = Democrat
rep = Republican
ind = Independent

Complete the table given the following results.

F–dem	F–rep	M–dem
M–rep	M–ind	M–rep
F–dem	M–rep	F–dem
F–dem	M–dem	F–ind
M–dem	M–ind	M–rep
M–rep	F–rep	F–dem

	Male	Female
Democrat		
Republican		
Independent		

example 2   Constructing a Table from Observed Data

The following data were taken by a student conducting a study for a statistics class. The student observed the type of vehicle and gender of the driver for 18 vehicles in the school parking lot. Complete the table.

Male–car	Female–truck	Male–truck
Male–truck	Female–car	Male–truck
Female–car	Male–truck	Male–motorcycle
Female–car	Male–car	Female–car
Male–motorcycle	Female–car	Female–car
Female–motorcycle	Male–car	Male–car

Driver \ Vehicle	Car	Truck	Motorcycle
Male			
Female			

Solution:

We need to count the number of data values that fall in each of the six cells. One method is to go through the list of data one by one. For each value place a tally mark | in the appropriate cell. For example, the first data value male–car would go in the cell in the first row, first column.

Driver \ Vehicle	Car	Truck	Motorcycle
Male	IIII	IIII	II
Female	IIIII I	I	I

To form the completed table, count the number of tally marks in each cell. See Table 9-2.

table 9-2

Driver \ Vehicle	Car	Truck	Motorcycle
Male	4	4	2
Female	6	1	1

2. Bar Graphs

In Table 9-1 we saw that hurricane Hugo caused 3 times as many deaths as hurricane Alicia. This same information can be displayed in a graph. Figure 9-1 shows a bar graph of the number of deaths for each hurricane listed in Table 9-1. Notice that the height of the bar showing the deaths for Hugo is 3 times as high as the bar for Alicia.

Answer

5.

	Male	Female
Democrat	3	5
Republican	5	2
Independent	2	1

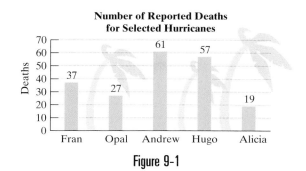

Figure 9-1

Notice that the bar graph compares the data values through the height of each bar. The bars in a bar graph may also be presented horizontally. For example, the double bar graph in Figure 9-2 illustrates the data from Example 2.

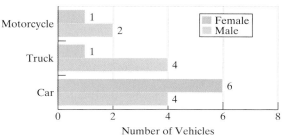

Figure 9-2

| example 3 | Constructing a Bar Graph |

The number of fat grams for five different ice cream brands and flavors is given in Table 9-3. Each value is based on a $\frac{1}{2}$-c serving. Construct a bar graph with vertical bars to depict this information.

table 9-3

Brand/Flavor	Number of Fat Grams (g) per $\frac{1}{2}$-c Serving
Breyers Strawberry	6
Edy's Grand Light Mint Chocolate Chip	4.5
Healthy Choice Chocolate Fudge Brownie	2
Ben and Jerry's Chocolate Chip Cookie Dough	15
Häagen-Dazs Vanilla Swiss Almond	20

Solution:

First draw a horizontal line and label the different food categories. Then draw a vertical line on the left-hand side of the graph as in Figure 9-3. The vertical line represents the number of grams of fat. The vertical scale must extend to at least 20 to accommodate the largest value in the table. In Figure 9-3, the vertical scale ranges from 0 to 24 in multiples (or steps) of 4.

Concept Connections

6. Complete the double bar graph for the data given in the table found in margin Exercise 5.

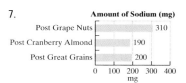

Skill Practice

7. The amount of sodium in milligrams (mg) per $\frac{1}{2}$-c serving for three different brands of cereal is given in the table. Construct a bar graph with horizontal bars.

Brand/Flavor	Amount of Sodium (mg)
Post Grape Nuts	310
Post Cranberry Almond	190
Post Great Grains	200

Answers

6.

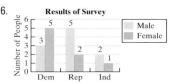

7.

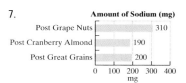

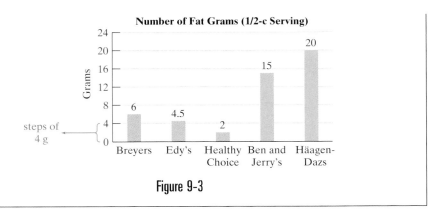

Figure 9-3

3. Pictographs

Sometimes a bar graph might use an icon or small image to convey a unit of measurement. This type of bar graph is called a **pictograph**.

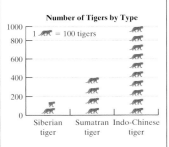
example 4   Interpreting a Pictograph

The number of music CDs sold in the United States is given in Figure 9-4 for selected years.

a. What is the value of each CD icon in the graph?

b. From the graph, estimate the number of CDs sold in 1997.

c. For which year were there approximately 900 million CDs sold?

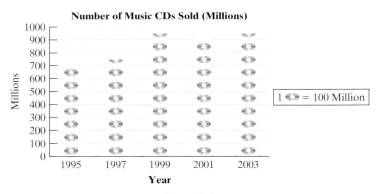

Figure 9-4

Solution:

a. The legend indicates that 1 = 100 million CDs sold.

b. The height of the "bar" in 1997 is given by $7\frac{1}{2}$ CDs. Therefore, there were approximately 750 million CDs sold in 1997.

c. The "bar" containing 9 CDs (900 million sold) corresponds to the year 2001.

4. Line Graphs

Line graphs are often used to track how one variable changes with respect to a second variable. For example, a line graph may illustrate a pattern or trend of a variable over time and allow us to make predictions.

---

example 5   Interpreting a Line Graph

Figure 9-5 shows the number of bachelor's degrees earned by men and women for selected years.

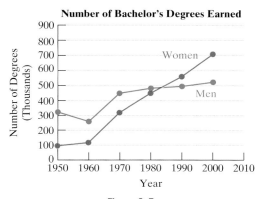

Figure 9-5

a. In 1950, who earned more bachelor's degrees, men or women?

b. In 2000, who earned more bachelor's degrees, men or women?

c. Use the trends in the graph to predict the number of bachelor's degrees earned by men and by women for the year 2010.

Solution:

a. The blue graph represents men, and the red graph represents women. In 1950, men earned approximately 320 thousand bachelor's degrees. Women earned approximately 100 thousand. Men earned more bachelor's degrees in 1950.

b. In 2000, women earned more bachelor's degrees.

c. To predict the number of bachelor's degrees earned by men and women in the year 2010, we need to extend both line graphs. See the dashed lines in Figure 9-6. The number of bachelor's degrees earned by women is predicted to be approximately 850 thousand. The number of degrees earned by men will be approximately 550 thousand.

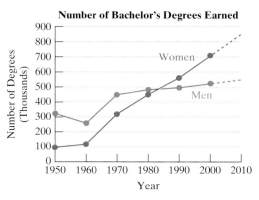

Figure 9-6

Skill Practice

14. The table gives the number of cars registered in the United States (in millions) for selected years (Source: U.S. Department of Transportation). Create a line graph for this information.

Year	Number of Cars (Millions)
1960	62
1970	89
1980	122
1990	134
2000	134

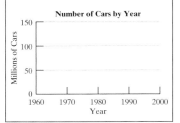

Answer

14.

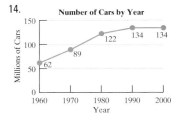

---

example 6   Creating a Line Graph

Table 9-4 gives the number of worldwide airline fatalities (excludes deaths caused by terrorism) for selected years. Use the data given in the table to create a line graph.

table 9-4	
Year	**Number of Deaths**
1986	641
1990	544
1994	1170
1998	904
2002	577

Solution:

First draw a horizontal line and label the year. Then draw a vertical line on the left-hand side of the graph, as in Figure 9-7. The vertical line represents the number of fatalities. In Figure 9-7, the vertical scale ranges from 0 to 1200 in multiples (or steps) of 200. For each year, draw a dot corresponding to the number of deaths for that year. Then connect the dots.

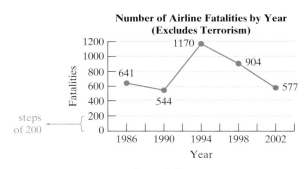

Figure 9-7

In Figure 9-7, we labeled the value at each data point because the exact values are difficult to read from the graph.

---

section 9.1   Practice Exercises

Boost *your* GRADE at
mathzone.com!

MathZone+/

- Practice Problems
- Self-Tests
- NetTutor
- e-Professors
- Videos

Study Skills Exercises

1. List three benefits of successfully completing this course.

2. Define the key terms.

    a. Statistics   **b. Data**   **c. Table**

    d. Cells   **e. Pictograph**   **f. Line graph**

Objective 1: Introduction to Data and Tables

For Exercises 3–7 refer to the table. The table represents the Seven Summits (the highest peaks from each continent). **(See Example 1.)**

Mountain	Continent	Height (ft)
Mt. Kilimanjaro	Africa	19,340
Elbrus	Europe	18,510
Aconcagua	South America	22,834
Denali	North America	20,320
Vinson Massif	Antarctica	16,864
Mt. Kosciusko	Australia	7,310
Mt. Everest	Asia	29,035

3. Which mountain is the highest?

4. In which continent does the highest mountain lie?

5. Which mountain among those listed is the lowest? In which continent does it lie?

6. How much higher is Mt. Aconcagua than Denali?

7. What is the difference between the heights of the highest mountain in Europe and the highest mountain in Australia?

For Exercises 8–13 refer to the table. The table gives the average ages (in years) for U.S. women and men married for the first time for selected years. (Source: U.S. Census Bureau.)

	Men	Women
1940	24.3	21.5
1960	22.8	20.3
1980	24.7	22.0
2000	26.8	25.1

8. By how much has the average age for women increased between 1940 and 2000?

9. By how much has the average age for men increased between 1940 and 2000?

10. What is the difference between the men's and women's average age at first marriage in 1940?

11. What is the difference between the men's and women's average age at first marriage in 2000?

12. Which group, men or women, had the consistently higher age at first marriage?

13. Which group, men or women, had a greater increase in age between 1940 and 2000?

14. The following data were taken from a survey of a 3rd-grade classroom. The survey denotes the gender of a student and whether the student owned a dog, a cat, or neither. Complete the table. Be sure to label the rows and columns.

Boy–dog	Boy–dog	Boy–cat	Boy–neither
Girl–dog	Girl–neither	Boy–dog	Girl–cat
Girl–neither	Girl–neither	Girl–dog	Girl–cat
Boy–dog	Girl–cat	Boy–neither	Girl–dog
Boy–neither	Girl–neither	Girl–cat	

	Dog	Cat	Neither
Boy			
Girl			

15. A survey was made of households to determine whether a family attends a church regularly and if there are children living at home. Complete the table. Be sure to label the rows and columns. **(See Example 2.)**

With children–attends church	Without children–attends church
With children–does not attend	Without children–does not attend
With children–attends church	With children–attends church
Without children–attends church	Without children–does not attend
With children–attends church	With children–attends church
With children–does not attend	With children–attends church
With children–attends church	With children–does not attend
Without children–does not attend	Without children–attends church
Without children–attends church	With children–does not attend

	Attends Church	Does Not Attend Church
With children		
Without children		

Objective 2: Bar Graphs

16. The table represents the world's major consumers of primary energy in 2000. All measurements are in quadrillions of Btu. *Note:* 1 quadrillion = 1,000,000,000,000,000. (Source: Energy Information Administration, U.S. Department of Energy.)

Country	Amount of Energy Consumed (Quadrillions of Btu)
India	11
Canada	13
Germany	14
Japan	22
Russia	28
China	37
United States	99

Construct a bar graph using horizontal bars by following these steps.

a. On the horizontal scale, make tick marks representing Btu's starting from 0 to 100 in increments (steps) of 10.

b. On the vertical scale, write the country names. Begin at the bottom with India.

c. Construct the bar graph. The length of each bar corresponds to the amount of energy consumed for each country.

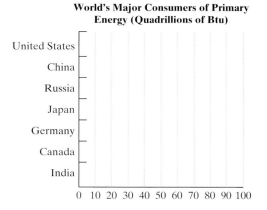

World's Major Consumers of Primary Energy (Quadrillions of Btu)

17. The following table shows the number of cellular telephone subscriptions by year in the United States. **(See Example 3.)**

Year	Number of Cellular Phone Subscriptions
1992	11,000
1994	24,100
1996	44,000
1998	69,200
2000	109,400
2002	142,600

Construct a bar graph with horizontal bars by following these steps.

a. On the horizontal scale, make tick marks starting from 0 to 150,000 in increments (steps) of 25,000.

b. On the vertical scale, write the year. Begin at the bottom with the year 1992.

c. Construct the bar graph. The length of each bar corresponds to the number of cellular phone subscriptions for each year.

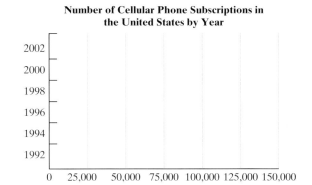

Number of Cellular Phone Subscriptions in the United States by Year

18. The following table shows the number of students per computer in the U.S. public schools, for selected years. (Source: National Center for Education Statistics.)

a. Which school year had the greatest number of students per computer? What was the value of the greatest number of students per computer?

School Year	Students per Computer
1983–1984	125
1986–1987	37
1989–1990	22
1992–1993	16
1995–1996	10
1998–1999	5.7
2001–2002	4.9

b. Draw a bar graph with vertical bars to illustrate these data. [*Hint:* Recall that the highest value on the vertical scale must be higher than the value from part (a). Also, select a suitable increment for the tick marks on the vertical scale.]

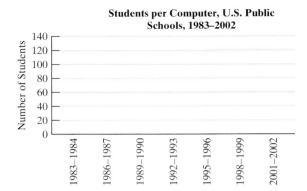

19. The number of new jobs for selected industries from March to April 2004 are given in the table. (Source: Bureau of Labor Statistics.)

Industry	Number of New Jobs
Health care	219,400
Temporary help	212,000
Construction	173,000
Food service	167,600
Retail	78,600

a. Which category has the greatest number of new jobs? How many new jobs is this?

b. Draw a bar graph with vertical bars to illustrate these data. [*Hint:* Recall that the highest value on the vertical scale must be higher than the value from part (a). Also, select a suitable increment for the tick marks on the vertical scale.]

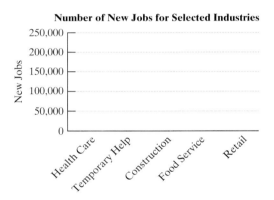

Objective 3: Pictographs

20. A local ice cream stand kept track of its ice cream sales for one weekend, as shown in the figure.

a. What does each ice cream icon represent?

b. From the graph, estimate the number of servings of ice cream sold on Saturday.

c. Which day had approximately 275 servings of ice cream sold?

21. Adults access the Internet to see weather updates and check on current news. The pictograph displays the percent of adult Internet users who access these topics. (See Example 4.)

a. What does each computer icon represent?

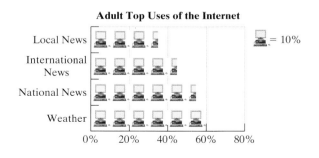

b. From the graph, estimate the percent of adult users that access the Internet for weather.

c. Which type of news is accessed about 45% of the time?

22. The figure displays the annual sales of books for three major companies.

   a. Estimate the sales for the company with the greatest annual sales.

   b. Estimate the total sales for all three companies.

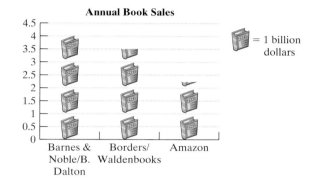

23. The largest populations of senior citizens in 2003 were in California, Florida, New York, Texas, and Pennsylvania, as shown in the figure. (U.S. Bureau of the Census.)

   a. Estimate the number of senior citizens living in Texas.

   b. How many more senior citizens are living in California than in Pennsylvania?

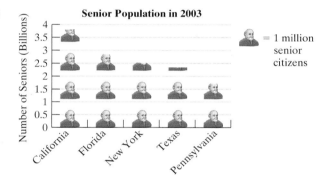

Objective 4: Line Graphs

For Exercises 24–29, use the graph provided. The graph shows the trend depicted by the percent of men and women 65 years or older in the labor force. (Source: Bureau of the Census.) **(See Example 5.)**

24. What was the difference in the percent of men and the percent of women over 65 in the labor force in the year 1920?

25. What was the difference in the percent of men and the percent of women over 65 in the labor force in the year 2000?

26. What was the overall trend in the percent of women over 65 in the labor force for the years shown in the graph?

27. What was the overall trend in the percent of men over 65 in the labor force for the years shown in the graph?

28. Use the graph to predict the number of men over 65 in the labor force in the year 2020. Answers will vary.

29. Use the graph to predict the number of women over 65 in the labor force in the year 2020. Answers will vary.

For Exercises 30–35, refer to the graph representing the number of sport utility vehicles (SUVs) sold in the United States (in millions) for selected years.

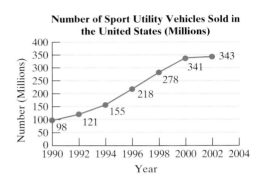

Number of Sport Utility Vehicles Sold in the United States (Millions)

30. In which year were the most sport utility vehicles sold? How many were sold?

31. In which year was the least number of sport utility vehicles sold? How many were sold?

32. What is the difference between the sales in 1996 and 1994?

33. What is the difference between the sales in 2002 and 2000?

34. In which 2-year period was the increase in sales the greatest?

35. In which 2-year period was the increase in sales the least?

36. The data shown here give the average height for boys based on age. (Source: National Parenting Council.) Make a line graph to illustrate these data by following these steps.

Age	Height (in.)
2	36
3	39
4	42
5	44
6	46.75
7	49
8	51
9	53.5

    a. On the horizontal scale, make tick marks representing age, starting from 0 to 10 in increments of 1.

    b. On the vertical scale, make tick marks representing height, starting from 0 to 60 in increments of 10.

    c. For each age value, plot a point for the corresponding height.

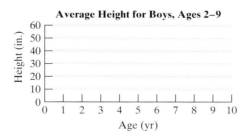

Average Height for Boys, Ages 2–9

37. The data shown here give the average height for girls based on age. (Source: National Parenting Council.) Make a line graph to illustrate these data by following these steps. **(See Example 6.)**

Age	Height (in.)
2	35
3	38.5
4	41.5
5	44
6	46
7	48
8	50.5
9	53

   a. On the horizontal scale, make tick marks representing age, starting from 0 to 10 in increments of 1.

   b. On the vertical scale, make tick marks representing height, starting from 0 to 60 in increments of 10.

   c. For each age value, plot a point for the corresponding height.

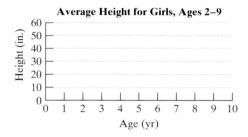

Expanding Your Skills

All packaged food items have to display nutritional facts so that the consumer can make informed choices. For Exercises 38–41, refer to the nutritional chart for Breyers French Vanilla ice cream.

38. How many servings are there per container? How much total fat is in one container of this ice cream?

39. How much total sodium is in one container of this ice cream?

40. If 8 g of fat is 13% of the daily value, what is the daily value? Round to 1 decimal place.

41. If 50 mg of cholesterol is 17% of the daily value, what is the daily value? Round to the nearest whole unit.

Nutrition Facts
Serving Size $\frac{1}{2}$ cup (68 g)
Servings per Container 14

Amount per Serving		
Calories 150		**Calories from Fat** 80

		% Daily Value
Total Fat	8 g	13%
Saturated fat	5 g	25%
Cholesterol	50 mg	17%
Sodium	45 mg	2%
Total Carbohydrate		
Dietary fiber	0 g	
Sugars	15 g	
Protein	3 g	

section 9.2 **Frequency Distributions and Histograms**

1. Frequency Distributions

The ages at the time of inauguration for 42 Presidents of the United States are given.

57	57	49	52	51	51	51	56	46
61	61	64	56	47	56	60	61	54
57	54	50	46	55	55	62	52	
57	68	48	54	54	51	43	69	
58	51	65	49	49	54	55	64	

The youngest President to take office to date was John F. Kennedy at 43 years old. The oldest was Ronald Reagan at 69 years old. Suppose we wanted to organize this information further by age groups. One way is to create a frequency distribution. A **frequency distribution** is a table displaying the number of values that fall within categories called **class intervals**. This is demonstrated in Example 1.

Skill Practice

1. The ages (in years) of individuals arrested on a certain day in Galveston, Texas, are listed.

18	20	35	46
19	26	24	32
28	25	30	34
22	29	39	19
18	19	26	40

Complete the table to form a frequency distribution.

Class (Age)	Tally	Frequency
18–23		
24–29		
30–35		
36–41		
42–47		

example 1 Creating a Frequency Distribution

Complete the table to form a frequency distribution for the ages of U.S. Presidents.

Class Intervals, Age (yr)	Tally	Frequency (Number of Presidents)
40–44		
45–49		
50–54		
55–59		
60–64		
65–69		

Solution:

The classes represent different age groups. Go through the list of ages, and use tally marks to track the number of Presidents that fall within each class. Tally marks are shown in red for the first five data values: 57, 61, 57, 57, and 58.

table 9-5

Class Intervals, Age (yr)	Tally	Frequency (Number of Presidents)
40–44	I	1
45–49	JHt II	7
50–54	JHt JHt III	13
55–59	JHt JHt I	11
60–64	JHt II	7
65–69	III	3

After completing the table, the frequency is a count of the tally marks within each class. See Table 9-5.

Answer

1.

Class (Age)	Tally	Frequency
18–23	JHt II	7
24–29	JHt I	6
30–35	IIII	4
36–41	II	2
42–47	I	1

| example 2 | Interpreting a Frequency Distribution |

Consider the frequency distribution in Table 9-5.

a. Which class had the most values?

b. How many values are represented in the table?

c. What percent of the Presidents were 60 years old or over at the time of inauguration?

Solution:

a. The 50–54 class had 13 data values, which is the highest frequency.

b. The number of data values is given by the sum of the frequencies.

$$\text{Total number of values} = 1 + 7 + 13 + 11 + 7 + 3$$
$$= 42$$

c. The number of Presidents 60–64 years old is 7. The number between 65 and 69 is 3. This means that there are 10 presidents who were 60 years or older. The percent is

$$\frac{10}{42} \approx 0.238 \quad \text{or} \quad 23.8\%$$

The percent of Presidents who were 60 or older is approximately 23.8%.

When creating a frequency distribution, keep these important guidelines in mind.

- The classes should be equally spaced. For instance, in Example 1, we would not want one class to represent 5 years of age and another to represent 20 years of age.
- The classes should not overlap. That is, a value should belong to one and only one class.
- In general, we usually create a frequency distribution with between 5 and 15 classes.

2. Histograms

A **histogram** is a special bar graph that illustrates data given in a frequency distribution. The class intervals are given on the horizontal scale. The height of each bar in a histogram represents the frequency for each class.

| example 3 | Constructing a Histogram |

Construct a histogram for the frequency distribution given in Example 1.

Solution:

Class Intervals, Age (yr)	Frequency (Number of Presidents)
40–44	1
45–49	7
50–54	13
55–59	11
60–64	7
65–69	3

Number Arrested by Age Group

To create a histogram of these data, we list the classes (ages of Presidents) on the horizontal scale. On the vertical scale we measure the frequency (Figure 9-8).

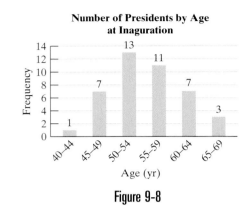

Figure 9-8

section 9.2   Practice Exercises

Study Skills Exercises

1. It is always helpful to read the material in a section and make some notes before it is presented in class. Writing notes ahead of time will free you to listen more in class and to pay special attention to the concepts that need clarification. Refer to your class syllabus and list the next two sections that will be covered in class and a time that you can read them beforehand.

2. Define the key terms.

   a. Frequency distribution   **b. Class intervals**   **c. Histogram**

Objective 1: Frequency Distributions

3. From the frequency distribution, determine the total number of data.

	Frequency
1–4	14
5–8	18
9–12	24
13–16	10
17–20	6

4. From the frequency distribution, determine the total number of data.

	Frequency
1–50	29
51–100	12
101–150	6
151–200	22
201–250	56
251–300	60

5. For the table in Exercise 3, which category contains the most data?

6. For the table in Exercise 4, which category contains the most data?

7. The retirement age (in years) for 20 college professors is given. Complete the frequency distribution.
(See Examples 1 and 2.)

| 67 | 56 | 68 | 70 | 60 | 65 | 73 | 72 | 56 | 65 |
| 71 | 66 | 72 | 69 | 65 | 65 | 63 | 65 | 68 | 70 |

Class Intervals (Age Group)	Tally	Frequency (Number of Professors)
56–58		
59–61		
62–64		
65–67		
68–70		
71–73		

a. Which class has the most values?

b. How many data values are represented in the table?

c. What percent of the professors retire when they are 68 to 70 years old?

8. The number of miles run in one day by 16 selected runners is given. Complete the frequency distribution.

| 2 | 4 | 7 | 3 | 8 | 4 | 5 | 7 |
| 4 | 6 | 4 | 3 | 4 | 2 | 4 | 10 |

Class Intervals (Number of Miles)	Tally	Frequency (Number of Runners)
1–2		
3–4		
5–6		
7–8		
9–10		

a. Which class has the most values?

b. How many data values are represented in the table?

c. What percent of the runners run 3 to 4 mi/day?

9. The number of gallons of gas purchased by 16 customers at a certain gas station is given. Complete the frequency distribution.

12.7	13.1	9.8	12.0	10.4	9.8	14.2	8.6
19.2	8.1	14.0	15.4	12.8	18.2	15.1	13.0

Class Intervals (Amount Purchased)	Tally	Frequency (Number of Customers)
8.0–9.9		
10.0–11.9		
12.0–13.9		
14.0–15.9		
16.0–17.9		
18.0–19.9		

  a. Which class has the most values?

  b. How many data values are represented in the table?

  c. What percent of the customers purchased 18 to 19.9 gal of gas?

10. The hourly salaries in dollars of 14 student employees at Miami-Dade College are given. Complete the frequency distribution.

5.95	6.00	7.20	6.15	5.85	5.95	8.00
6.50	7.00	7.25	6.95	7.25	7.50	8.05

Class Intervals (Hourly Salary, $)	Tally	Frequency (Number of Employees)
5.50–5.99		
6.00–6.49		
6.50–6.99		
7.00–7.49		
7.50–7.99		
8.00–8.49		

  a. Which class has the most values?

  b. How many data values are represented in the table?

  c. What percent of the employees earn $7.00 or more?

11. List three guidelines in setting up the class intervals for a frequency distribution.

12. Explain what is wrong with the following class intervals.

Class	Tally	Frequency
0–4		
5–10		
11–17		
18–25		
26–34		

13. Explain what is wrong with the following class intervals.

Class	Tally	Frequency
1–6		
7–11		
12–17		
18–23		
24–28		

14. Explain what is wrong with the following class intervals.

Class	Tally	Frequency
1–20		
21–40		

15. Explain what is wrong with the following class intervals.

Class	Tally	Frequency
1–33		
34–66		
67–99		

16. Explain what is wrong with the following class intervals.

Class	Tally	Frequency
10–12		
12–14		
14–16		
16–18		
18–20		

17. Explain what is wrong with the following class intervals.

Class	Tally	Frequency
1–5		
5–10		
10–15		
15–20		
20–25		
25–30		

18. The heights of 20 students at Valencia Community College are given. Complete the frequency distribution.

70	71	73	62	65	70	69	70	64	66
73	63	68	67	69	72	64	66	67	69

Class Interval (Height, in.)	Frequency (Number of Students)
62–63	
64–65	
66–67	
68–69	
70–71	
72–73	

19. The weights of 20 females who are each 5 ft 5 in. tall are given. Complete the frequency distribution.

138	120	115	162	145	118	152	123	141	155
176	144	125	137	130	121	148	116	139	161

Class Interval (Weight, lb)	Frequency (Number of Females)
115–124	
125–134	
135–144	
145–154	
155–164	
165–174	
175–184	

20. The weekly earnings in dollars of 21 high school seniors are given. Construct a frequency distribution.

142	110	56	120	216	310	32
176	108	136	100	84	188	154
96	184	74	98	204	68	54

Class Interval (Weekly Earnings, $)	Frequency (Number of Seniors)
0–49	
50–99	
100–149	
150–199	
200–249	
250–299	
300–349	

21. The amount withdrawn in dollars from a certain ATM is given for 20 customers. Construct a frequency distribution.

40	50	200	200	100	120	200	50	100	60
100	100	30	40	100	100	50	200	150	200

Class Interval (Amount, $)	Frequency (Number of Customers)
0–49	
50–99	
100–149	
150–199	
200–249	

Objective 2: Histograms

22. Construct a histogram for the frequency table in Exercise 18.

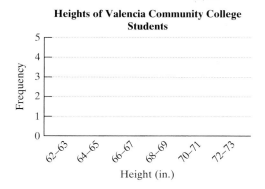

23. Construct a histogram for the frequency table in Exercise 19. **(See Example 3.)**

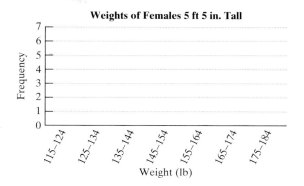

24. Construct a histogram for the frequency table in Exercise 20.

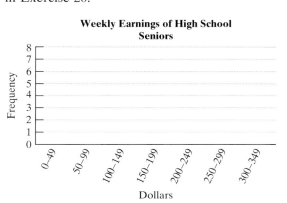

25. Construct a histogram for the frequency table in Exercise 21.

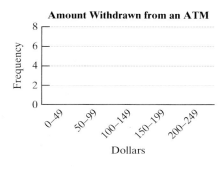

26. Construct a histogram, using the given data. Each number represents the number of Calories in a 100-g serving for selected fruits.

59	65	48	49	161	47	92	43	52	59
56	49	35	55	72	30	67	44	32	32
61	29	30							

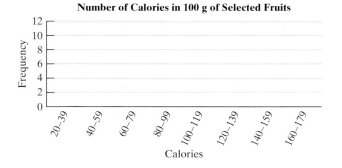

27. Construct a histogram, using the given data. Each number represents the number of Calories in a single serving of selected meats.

| 110 | 170 | 445 | 140 | 135 | 240 | 120 | 185 | 245 | 205 |

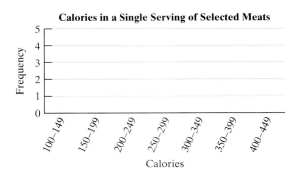

28. The lengths of twenty of the longest tunnels are given in kilometers. Construct a histogram.

| 20 | 17 | 15 | 14 | 20 | 16 | 15 | 13 | 19 | 15 |
| 14 | 19 | 15 | 14 | 19 | 15 | 14 | 19 | 15 | 14 |

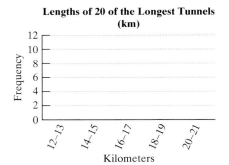

29. The list of data gives the number of children of the Presidents of the United States. Construct a histogram.

0	4	3	3	2	2	5	10	0	5
2	4	5	7	4	3	6	4	0	7
5	6	1	4	3	0	4	3	2	6
4	6	8	3	2	1	0	2	6	0
2	2

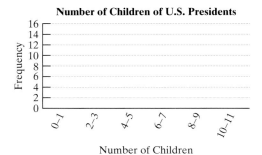

section 9.3   Circle Graphs

Objectives

1. Interpreting Circle Graphs
2. Circle Graphs and Percents
3. Constructing Circle Graphs

1. Interpreting Circle Graphs

Thus far we have used bar graphs, line graphs, and histograms to visualize data. A **circle graph** (or pie graph) is another type of graph used to show how a whole amount is divided into parts. Each part of the circle, called a **sector**, is like a slice of pie. The size of each piece corresponds to the fraction of the whole it represents.

example 1   Interpreting a Circle Graph

The grade distribution for a math test is shown in the circle graph (Figure 9-9).

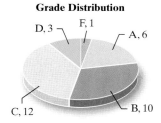

Grade Distribution

Figure 9-9

a. How many total grades are represented?

b. How many grades are B's?

c. How many times more C's are there than D's?

d. What percent of the grades were A's?

Solution:

a. The total number of grades is equal to the sum of the number of grades from each category.

$$\text{Total number of grades} = 6 + 10 + 12 + 3 + 1$$
$$= 32$$

b. The number of B's is represented by the red portion of the graph. There are 10 B's.

c. There are 12 C's and 3 D's. The ratio of C's to D's is $\frac{12}{3} = 4$. Therefore, there are 4 times as many C's as D's.

d. There are 6 A's. The percent of A's is given by

$$\frac{6}{32} = 0.1875$$

Therefore, the percent of A's is 18.75%.

Skill Practice

A used car dealership sells cars and trucks. Use the circle graph to answer Exercises 1–4.

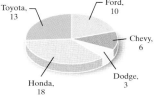

Vehicle Distribution by Make

1. How many total vehicles are represented?
2. How many are Toyotas?
3. How many more Hondas are there than Dodges?
4. What percent are Fords?

Answers

1. 50   2. 13
3. There are 15 more Hondas than Dodges.
4. 20% are Fords.

2. Circle Graphs and Percents

Sometimes circle graphs show data in percent form. This is illustrated in Example 2.

example 2   Calculating Amounts by Using a Circle Graph

A certain video rental store carries 2000 different videos. It groups its video collection by the categories shown in the graph (Figure 9-10).

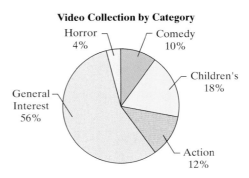

Video Collection by Category

Horror 4%   Comedy 10%   Children's 18%   Action 12%   General Interest 56%

Figure 9-10

a. How many videos are comedy?

b. How many videos are action or horror?

Solution:

a. First note that the store carries 2000 different videos. From the graph we know that 10% are comedies. Therefore, this question can be interpreted as

What is 10% of 2000?
$$x = (0.10) \cdot (2000)$$
$$= 200 \qquad \text{There are 200 comedies.}$$

b. From the graph we know that 12% of the videos are action and 4% are horror. This accounts for 16% of the total video collection. Therefore, this question asks

What is 16% of 2000?
$$x = (0.16) \cdot (2000)$$
$$= 320 \qquad \text{There are 320 videos that are either action or horror.}$$

3. Constructing Circle Graphs

Recall that a full circle is a 360° arc. To draw a circle graph, we must compute the number of degrees of arc for each sector. In Example 2, of the videos 10% are comedies. To draw the sector for this category, we must determine 10% of 360°.

$$10\% \text{ of } 360° = 0.10(360°) = 36°$$

The sector representing comedies should be drawn with a 36° angle. To do this, we can use a protractor (Figure 9-11). Recall from Section 8.1 that a protractor uses equally spaced tick marks around a semicircle to measure angles from 0° to 180°.

To draw a sector with a 36° arc, first draw a circle. Place the hole in the protractor over the center of the circle. Using the inner scale on the protractor, place a tick mark at 0° and at 36°. Use a straightedge to draw two line segments from the center of the circle to each tick mark. See Figure 9-11.

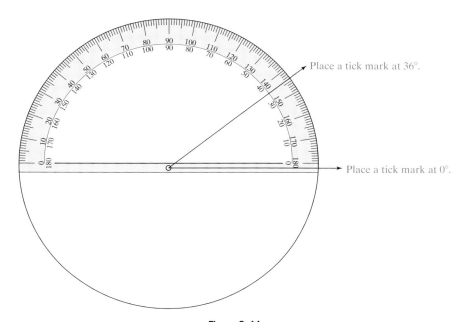

Place a tick mark at 36°.

Place a tick mark at 0°.

Figure 9-11

In Example 3, we use this technique to construct a circle graph.

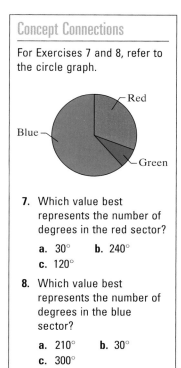

---

example 3   Constructing a Circle Graph

A teacher earns a monthly salary of $2400 after taxes. Her monthly budget is broken down in Table 9-6.

table 9-6

Budget Item	Monthly Value ($)
Rent	840
Utilities	210
Car (payment and insurance)	510
Groceries	360
Savings	300
Other	180

Construct a circle graph illustrating the information in this table. Label each sector of the graph with the percent that it represents.

Answers

7. c
8. a

Skill Practice

9. Several voters in Oregon were asked to identify the political party to which they belonged. Construct a circle graph. Label each sector of the graph with the percent that it represents.

Political Affiliation	Number
Democrat	900
Republican	720
Libertarian	36
Green Party	144

Solution:

This problem calls for two types of calculations: (1) For each budget item, we must compute the percent of the whole that it represents. (2) We must determine the number of degrees for each category. We can use a table to help organize our calculations.

Budget Item	Monthly Value ($)	Percent	Number of Degrees
Rent	840	$p\% = \dfrac{840}{2400} = 0.35$ or 35%	35% of 360° $= 0.35(360°)$ $= 126°$
Utilities	210	$p\% = \dfrac{210}{2400} = 0.0875$ or 8.75%	8.75% of 360° $= 0.0875(360°)$ $= 31.5°$
Car	510	$p\% = \dfrac{510}{2400} = 0.2125$ or 21.25%	21.25% of 360° $= 0.2125(360°)$ $= 76.5°$
Groceries	360	$p\% = \dfrac{360}{2400} = 0.15$ or 15%	15% of 360° $= 0.15(360°)$ $= 54°$
Savings	300	$p\% = \dfrac{300}{2400} = 0.125$ or 12.5%	12.5% of 360° $= 0.125(360°)$ $= 45°$
Other	180	$p\% = \dfrac{180}{2400} = 0.075$ or 7.5%	7.5% of 360° $= 0.075(360°)$ $= 27°$

Tip: To verify your calculations in Example 3, you can add the percent values to be sure they total to 100%. Similarly, you can add the degree values to be sure they total to 360°.

Now construct the circle graph. Use the degree measures found in the table for each sector. Label the graph with the percent for each sector (Figure 9-12).

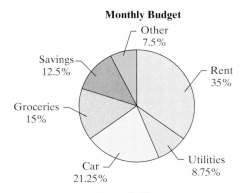

Figure 9-12

Answer

9.
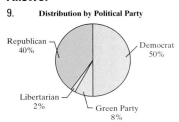

section 9.3   Practice Exercises

Study Skills Exercises

1. Some instructors are available to answer questions during evening hours via e-mail. Find out if you can contact your instructor by e-mail during evening hours or weekends, and write down the e-mail address.

2. Define the key terms.

   a. Circle graph   **b. Sector**

Objective 1: Interpreting Circle Graphs

For Exercises 3–10, refer to the figure. The figure represents five countries outside the United States according to the number of troops they had deployed in Iraq in March 2004. (Source: *Newsweek*, March 29, 2004.) **(See Example 1.)**

Number of Troops Deployed, March 2004

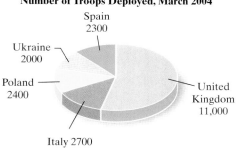

3. What is the total number of troops represented in this graph?

4. Which of these countries had the most troops deployed?

5. How many more troops were deployed from Italy than from Poland?

6. How many more troops were deployed from the United Kingdom than from Ukraine?

7. What percent of these troops came from Spain? Round to the nearest tenth of a percent.

8. What percent of these troops came from Poland? Round to the nearest tenth of a percent.

9. How many *times* more troops came from the United Kingdom than from Ukraine?

10. How many *times* more troops came from Poland than from Ukraine?

For Exercises 11–16, refer to the figure. The figure represents the average number of viewers for five daytime dramas. (Source: Nielsen Media Research.)

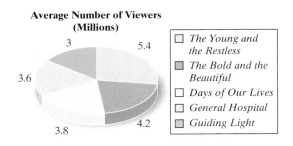

Average Number of Viewers (Millions)

☐ *The Young and the Restless*
▨ *The Bold and the Beautiful*
☐ *Days of Our Lives*
☐ *General Hospital*
▨ *Guiding Light*

11. How many viewers are represented?

12. How many viewers does the most popular daytime drama have?

13. How many times more viewers does *The Young and the Restless* have than *Guiding Light*?

14. How many times more viewers does *The Young and the Restless* have than *General Hospital*?

15. What percent of the viewers watch *General Hospital*?

16. What percent of the viewers watch *The Bold and the Beautiful*?

Objective 2: Circle Graphs and Percents

For Exercises 17–20, use the graph representing the type of music CDs found in a store containing approximately 8000 CDs. **(See Example 2.)**

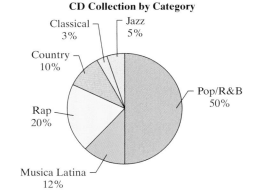

CD Collection by Category

17. How many CDs are musica Latina?

18. How many CDs are rap?

19. How many CDs are jazz or classical?

20. How many CDs are *not* Pop/R&B?

For Exercises 21–24, use the graph representing the states that hosted Super Bowl I through Super Bowl XXXVI (a total of 36 Super Bowls).

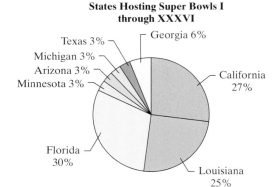

States Hosting Super Bowls I through XXXVI

21. How many Super Bowls were played in Louisiana?

22. How many Super Bowls were played in Florida? Round to the nearest whole number.

23. How many Super Bowls were played in Georgia? Round to the nearest whole number.

24. How many Super Bowls were played in Michigan? Round to the nearest whole number.

Objective 3: Constructing Circle Graphs

For Exercises 25–32, use a protractor to construct an angle of the given measure.

25. 20°

26. 70°

27. 125°

28. 270°

29. 195°

30. 5°

31. 300°

32. 90°

33. Draw a circle and divide it into sectors of 30°, 60°, 100°, and 170°.

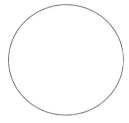

34. Draw a circle and divide it into sectors of 125°, 180°, and 55°.

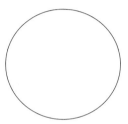

35. The table provided gives the expenses for one semester at college. **(See Example 3.)**

   a. Complete the table.

	Expenses	Percent	Number of Degrees
Tuition	$9000		
Books	600		
Housing	2400		

   b. Construct a circle graph to display the college expenses. Label the graph with percents.

College Expenses for a Semester

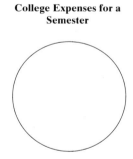

36. The table provided gives the number of establishments of the three largest pizza chains.

   a. Complete the table.

	Number of Stores	Percent	Number of Degrees
Pizza Hut	8100		
Domino's	7200		
Papa Johns	2700		

   b. Construct a circle graph. Label the graph with percents.

Number of Pizza Establishments

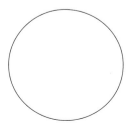

37. The Sunshine Nursery sells flowering plants, shrubs, ground cover, trees, and assorted flower pots. Construct a pie graph to show the distribution of the types of purchases.

Sunshine Nursery Distribution of Sales

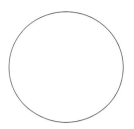

Types of Purchases	Percent of Distribution
Flowering plants	45%
Shrubs	13%
Ground cover	18%
Trees	20%
Flower pots	4%

38. The party affiliation of registered Latino voters for a recent year is as follows:

45% Democrat   20% Republican

13% Other   22% Independent

Construct a circle graph from this information.

Party Affiliation of Latino Voters

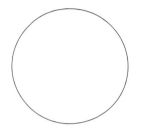

chapter 9 | midchapter review

One hundred people were surveyed and asked their blood type. The results are as follows:

20 were type A; 26 were type B; 15 were type AB; and 39 were type O.

For Exercises 1–2, use this information to construct the type of graph indicated.

1. Bar graph

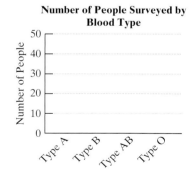

2. Circle graph (label the graph with percents)

Percent by Blood Type

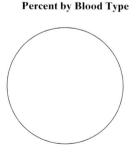

The mortgage rates were recorded for one year as follows:

July: 6.8%, September: 5.8%, November: 6%, January: 5.9%, March: 5.6%, May: 5.3%.

For Exercises 3–4, use this information to construct the type of graph indicated.

3. Line graph

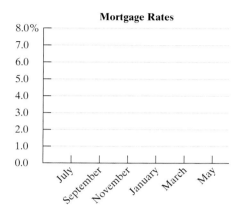

4. Bar graph

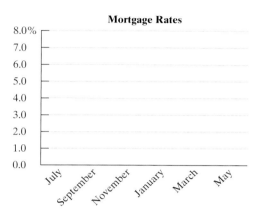

A survey was taken from 23 graduate students asking them their grade point averages. The results are as follows:

3.52	3.42	3.01	3.40	2.99	3.12	3.33	2.98	3.80	3.75
2.75	2.60	3.04	3.42	3.10	2.80	3.85	3.24	3.27	3.00
2.85	3.00	3.31							

For Exercises 5–6, complete the frequency table and construct a histogram.

5. Frequency table

Grade Point Average	Frequency
2.50–2.74	
2.75–2.99	
3.00–3.24	
3.25–3.49	
3.50–3.74	
3.75–3.99	

6. Histogram (use the information from the table in Exercise 5)

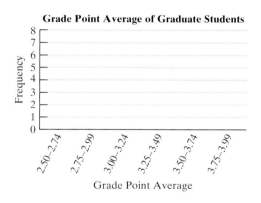

section 9.4 **Mean, Median, and Mode**

1. Mean

When given a list of numerical data, it is often desirable to obtain a single number that represents the central value of the data. In this section, we introduce three such values called the mean, median, and mode. The first calculation we present is the mean (or average) of a list of data values.

Mean

The **mean** (or average) of a set of numbers is the sum of the values divided by the number of values. We can write this as a formula.

$$\text{Mean} = \frac{\text{sum of the values}}{\text{number of values}}$$

Skill Practice

1. Marcelo bought six new and used CDs at a music store. The prices are given below.

   $11.99 $7.99 $14.99
   $ 4.99 $9.99 $ 6.99

   Find the mean price.

example 1   Finding the Mean of a Data Set

Ms. Guinn bought five textbooks for her classes for fall semester. The prices are given in Table 9-7.

table 9-7

Class	Price ($)
Art Appreciation	129.90
First Aid	46.99
Intermediate Algebra	94.95
English I	59.00
English I Workbook	36.99

Find the mean price of her textbooks.

Solution:

$$\text{Mean} = \frac{129.90 + 46.99 + 94.95 + 59.00 + 36.99}{5}$$

Divide the sum of the data by the number of values.

$$= \frac{367.83}{5}$$

When computing a mean, notice that the data are added first before dividing.

$$= 73.566$$

Divide.

The mean (or average) price per textbook is $73.57.

Answer

1. $9.49

example 2   Finding the Mean of a Data Set

A small business employs five workers. Their yearly salaries are

$42,000 $36,000 $45,000 $35,000 $38,000

a. Find the mean yearly salary for the five employees.

b. Suppose the owner of the business makes $218,000 per year. Find the mean salary for all six individuals (that is, include the owner's salary).

Solution:

a. Mean salary of five employees

$$= \frac{42,000 + 36,000 + 45,000 + 35,000 + 38,000}{5}$$

$$= \frac{196,000}{5} \quad \text{Add the data values.}$$

$$= 39,200 \quad \text{Divide.}$$

The mean salary for employees is $39,200.

b. Mean of all six individuals

$$= \frac{42,000 + 36,000 + 45,000 + 35,000 + 38,000 + 218,000}{6}$$

$$= \frac{414,000}{6}$$

$$= 69,000$$

The mean salary with the owner's salary included is $69,000.

2. Median

In Example 2, you may have noticed that the mean salary was greatly affected by the unusually high value of $218,000. For this reason, you may want to use a different measure of "center" called the median. The **median** is the "middle" number in an ordered list of numbers.

Median

To compute the median of a list of numbers, first arrange the numbers in order from least to greatest.
- If the number of data values in the list is *odd*, then the median is the middle number in the list.
- If the number of data values is *even*, there is no single middle number. Therefore, the median is the mean of the two middle numbers in the list.

To understand how to compute the median, consider the salaries of the five employees from Example 2, arranged in order.

Answers

2. $136,600   3. $330,500

Concept Connections

4. Find the median of the five housing prices given in margin Exercise 2.

$108,000	$149,000
$164,000	$118,000
$144,000	

5. Find the median of the six housing prices given in margin Exercise 3.

$108,000	$149,000
$164,000	$118,000
$144,000	$1,300,000

6. Was the mean or the median affected more by the presence of the $1.3 million house? See margin Exercises 2 and 3.

First five salaries:

35,000   36,000   38,000   42,000   45,000   Arrange the data in order.

Because there are five data values (an *odd* number), the median is the middle number.

The median is $38,000.

Now consider the scores of all six individuals (including the owner). Arrange the data in order.

35,000   36,000   38,000   42,000   45,000   218,000

$$\frac{38,000 + 42,000}{2}$$

There are six data values (an *even* number). The median is the mean of the two middle numbers.

$$= \frac{80,000}{2}$$

Add the two middle numbers.

$$= 40,000$$

Divide.

The median of all six salaries is $40,000.

The mean of all six salaries is $69,000, whereas the median is $40,000. This example shows that the median is a good choice for a central value when the data list has an unusually high (or low) value.

Skill Practice

7. The monthly rainfall for Houston, Texas, is given in the table. Find the median rainfall amount.

Month	Rainfall (in.)
Jan.	4.5
Feb.	3.0
March	3.2
April	3.5
May	5.1
June	6.8
July	4.3
Aug.	4.5
Sept.	5.6
Oct.	5.3
Nov.	4.5
Dec.	3.8

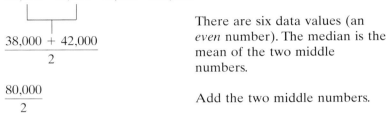 **example 3** Finding the Median of a Data Set

For a recent year, the student-to-teacher ratio for elementary schools is shown in Table 9-8. Find the median student-to-teacher ratio.

table 9-8

State	Student-to-Teacher Ratio
California	20.6
Illinois	16.1
Indiana	16.1
Maine	12.5
Mississippi	16.1
New Hampshire	14.5
North Dakota	13.4
Rhode Island	14.8
Utah	21.9
Wisconsin	14.1

Source: National Center for Education Statistics.

Answers

4. $144,000   5. $146,500
6. The mean was affected more than the median.
7. 4.5 in.

Solution:

First arrange the numbers in order from least to greatest:

ME	ND	WI	NH	RI	IL	IN	MS	CA	UT
12.5	13.4	14.1	14.5	14.8	16.1	16.1	16.1	20.6	21.9

$$\text{Median} = \frac{14.8 + 16.1}{2} = 15.45$$

There are 10 data values (an *even* number). Therefore, the median is the mean of the middle two numbers. The median student-to-teacher ratio is 15.45. This indicates that there are approximately 15 or 16 students per teacher.

3. Mode

A third representative value for a list of data is called the mode.

> **Mode**
>
> The **mode** of a set of data is the value or values that occur most often. If each value occurs the same number of times, then there is no mode.
>
> *Note:* If two values occur most often, then we say that the data are **bimodal**.

---

example 4   **Finding the Mode of a Data Set**

Find the mode of the student-to-teacher ratios from Example 3.

Solution:

12.5   13.4   14.1   14.5   14.8   16.1   16.1   16.1   20.6   21.9

The data value 16.1 appears the most often. Therefore, the mode is 16.1.

---

example 5   **Finding the Mode of a Data Set**

Find the mode of the list of average monthly temperatures for Albany, New York.

Jan.	Feb.	March	April	May	June	July	Aug.	Sept.	Oct.	Nov.	Dec.
22	25	35	47	58	66	71	69	61	49	39	26

Solution:

No data value occurs most often. There is no mode for this set of data.

Skill Practice

8. Find the mode of the rainfall amounts from margin Exercise 7.

4.5	3.0	3.2
3.5	5.1	6.8
4.3	4.5	5.6
5.3	4.5	3.8

Skill Practice

9. Find the mode of the weights in pounds of babies born one day at Brackenridge Hospital in Austin, Texas.

7.2	8.1	6.9
9.3	8.3	7.7
7.9	6.4	7.5

Answers

8. 4.5 in.   9. No mode

Skill Practice

10. The ages of children participating in an after-school sports program are given. Find the mode(s).

13	15	17	15
14	15	16	16
15	16	12	13
15	14	16	15
15	16	16	13
16	13	14	18

---

example 6   Finding the Mode of a Data Set

The grades for a quiz in college algebra are as follows. The scores are out of a possible 10 points.

9	4	6	9	9	8	2	1	4	9
5	10	10	5	7	7	9	8	7	3
9	7	10	7	10	1	7	4	5	6

Solution:

Sometimes arranging the data in order makes it easier to find the repeated values.

1	1	2	3	4	4	4	5	5	5
6	6	7	7	7	7	7	7	8	8
9	9	9	9	9	9	10	10	10	10

The score of 9 occurs 6 times. The score of 7 occurs 6 times. There are two modes, 9 and 7, because these scores both occur more than any other score. We say that these data are *bimodal*.

4. Weighted Mean

Sometimes data values in a list appear multiple times. In such a case, we can compute a **weighted mean**. In Example 7, each data value is "weighted" by the number of times it appears in the list.

Skill Practice

11. People throw coins in wishing wells, and often the money is used for charity. The number of coins collected from one wishing well is shown in the table. Compute the mean dollar amount per coin. Round to 3 decimal places.

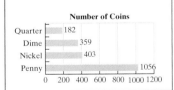

Number of Coins

---

example 7   Computing a Weighted Mean

Donations are made to a certain charitable organization in increments of $25, $50, $75, and $100, as shown in Figure 9-13. Find the mean amount donated.

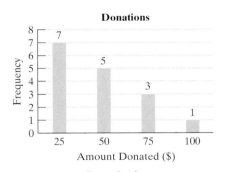

Figure 9-13

Solution:

Notice that there are 16 data values represented in the graph:

7 of these $\begin{cases} \$25 \\ \$25 \\ \$25 \\ \$25 \\ \$25 \\ \$25 \\ \$25 \end{cases}$ 5 of these $\begin{cases} \$50 \\ \$50 \\ \$50 \\ \$50 \\ \$50 \end{cases}$ 3 of these $\begin{cases} \$75 \\ \$75 \\ \$75 \end{cases}$ 1 of these $\{\$100$

The data value $25 occurs 7 times. Rather than adding $25 seven times, we can find the sum by multiplying $25(7) = $175. Similarly, the value $50 occurs 5 times for a sum of $50(5) = $250, and so on. To find the sum of all the values, multiply each data value by the number of times it occurs (its frequency). Then add the results. This process can be organized easily in a table.

Amount Donated ($)	Frequency	Product ($)
25	7	(25)(7) = 175
50	5	(50)(5) = 250
75	3	(75)(3) = 225
100	1	(100)(1) = 100
Total:	**16**	**750** ← sum of all data values

total number of donations made

The mean is the sum of all data values divided by the total number of donations made (total frequency).

$$\text{Mean} = \frac{750}{16} = 46.875$$

The mean amount donated is $46.88.

section 9.4   Practice Exercises

Boost your GRADE at mathzone.com!

MathZone

- Practice Problems
- Self-Tests
- NetTutor
- e-Professors
- Videos

Study Skills Exercises

1. Most people cannot concentrate on studying for more than 1 hr without taking a break. To make the most of your time, write down a schedule for your next study session. Include breaks where you can eat a meal, walk the dog, or perform other simple tasks that need to be completed.

2. Define the key terms.

   a. Mean   **b. Median**   **c. Mode**

   d. Bimodal   **e. Weighted mean**

Objective 1: Mean

For Exercises 3–8, find the mean of each set of numbers. **(See Example 1.)**

3. 4, 6, 5, 10, 4, 5, 8   **4.** 3, 8, 5, 7, 4, 2, 7, 4   **5.** 0, 5, 7, 4, 7, 2, 4, 3

6. 7, 6, 5, 10, 8, 4, 8, 6, 9   **7.** 10, 13, 18, 20, 15   **8.** 22, 14, 12, 16, 15

9. The wingspan of five butterflies is given in the table. Find the mean wingspan.

Butterfly	Wingspan (in.)
Queen Alexandra's birdwing	11.0
African giant swallowtail	9.1
Goliath birdwing	8.3
Buru opalescent birdwing	7.9
Chimaera birdwing	7.5

10. The number of wins in the American Baseball League, Central Division, for 2003 is given in the table. Find the mean number of wins.

Team	Number of Wins
Minnesota Twins	90
Chicago White Sox	86
Kansas City Royals	83
Cleveland Indians	68
Detroit Tigers	43

11. The flight times in hours for six flights between New York and Los Angeles are given. Find the mean flight time. Round to the nearest tenth of an hour.

5.5   6.0   5.8   5.8   6.0   5.6

12. The prices of four comparable pizzas are given. Find the mean price. Round to the nearest cent.

$9.99 $10.50 $9.59 $10.75

13. The number of Calories for six different chicken sandwiches and chicken salads is given in the table.

a. What is the mean number of Calories for a chicken sandwich? Round to the nearest whole unit.

b. What is the mean number of Calories for a salad with chicken? Round to the nearest whole unit.

c. What is the difference in the means?

Chicken Sandwiches	Salads with Chicken
360	310
370	325
380	350
400	390
400	440
470	500

14. The heights of the players from two NBA teams are given in the table. All heights are in inches.

a. Find the mean height for the players on the Philadelphia 76ers. Round to the nearest tenth of an inch.

b. Find the mean height for the players on the Milwaukee Bucks. Round to the nearest tenth of an inch.

c. What is the difference in the mean heights?

Philadelphia 76ers' Height (in.)	Milwaukee Bucks' Height (in.)
83	70
83	83
72	82
79	72
77	82
84	85
75	75
76	75
82	78
79	77
76	82
83	83
82	81

15. Zach received the following scores for his first four tests: 98%, 80%, 78%, 90%. **(See Example 2.)**

    a. Find Zach's mean test score.

    b. Zach got a 59% on his fifth test. Find the mean of all five tests.

    c. How did the low score of 59% affect the overall mean of five tests?

16. The prices of four steam irons are $50, $30, $25, and $45.

    a. Find the mean of these prices.

    b. An iron that costs $140 is added to the list. What is the mean of all five irons?

    c. How does the expensive iron affect the mean?

Objective 2: Median

For Exercises 17–22, find the median for each set of numbers. **(See Example 3.)**

17. 16, 14, 22, 13, 20, 19, 17

18. 32, 35, 22, 36, 30, 31, 38

19. 109, 118, 111, 110, 123, 100

20. 134, 132, 120, 135, 140, 118

21. 58, 55, 50, 40, 40, 55

22. 82, 90, 99, 82, 88, 87

23. The infant mortality rates for five countries are given in the table. Find the median.

Country	Infant Mortality Rate (Deaths per 1000)
Sweden	3.93
Japan	4.10
Finland	3.82
Andorra	4.09
Singapore	3.87

24. The inflation rates for five countries are given in the table. Find the median.

25. The ages (in years) of the last 10 Presidents at the time of their inauguration are given. Find the median age.

46, 64, 69, 52, 61, 56, 55, 43, 62, 60

Country	Inflation Rate (%)
Angola	1700
Sudan	133
Turkey	80
Venezuela	103
Bulgaria	311

26. A list of the number of commuter rail stations from eight systems is given. Find the median number of stations.

124, 227, 108, 167, 121, 177, 49, 18

27. The number of albums sold (in millions) as of 2002 is listed for the 10 best sellers. Find the median number of albums sold.

2.7, 3.0, 4.8, 7.4, 3.4, 2.6, 3.0, 3.0, 3.9, 3.2

28. The following list gives the number of international arrivals (in millions) in the year 2001 for nine countries around the world. Find the median number of arrivals.

33.2, 19.8, 23.4, 39.0, 15.3, 45.5, 13.7, 15.0, 76.5

Objective 3: Mode

For Exercises 29–34, find the mode(s) for each set of numbers. **(See Examples 4–6.)**

29. 4, 5, 3, 8, 4, 9, 4, 2, 1, 4

30. 12, 14, 13, 17, 19, 18, 19, 17, 17

31. 90%, 89%, 91%, 77%, 88%

32. 132, 253, 553, 255, 552, 234

33. 28, 21, 24, 23, 24, 30, 21

34. 45, 42, 40, 41, 49, 49, 42

35. The table gives the number of hazardous waste sites for selected states in 2003. Find the mode.

State	Number of Sites
Florida	51
New Jersey	112
Michigan	67
Washington	47
Texas	41
Wisconsin	39
California	96
Pennsylvania	94
Illinois	39
New York	90

36. The table gives the price of seven "smart" cell phones. Find the mode.

Brand and Model	Price ($)
Samsung	600
Kyocera	400
Sony Ericsson	800
PalmOne	450
Motorola	300
Siemens	600

37. The unemployment rates in percent for nine countries are given for the year 2002. Find the mode.

6.3%, 7.0%, 5.8%, 9.1%, 5.2%, 8.8%, 8.4%, 5.4%, 5.2%

38. A list of the number of children of the last 10 Presidents is given. Find the mode.

1, 6, 2, 2, 4, 4, 2, 2, 3, 2

39. The price per gallon for regular unleaded gas for five gas stations is given. Find the mode.

$2.49, $2.39, $2.51, $2.49, $2.51

40. The length of time (in minutes) of eight TV commercials is given. Find the mode.

1.00,   0.50,   1.00,   1.25,   2.00,   0.50,   1.00,   0.50

Objective 4: Weighted Mean

41. There are 20 students enrolled in a 12th-grade math class. The graph displays the number of students by age. First complete the table, and then find the mean. **(See Example 7.)**

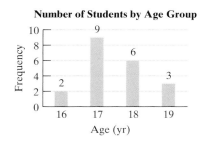

Number of Students by Age Group

Age (yr)	Number of Students	Product
16		
17		
18		
19		
Total:		

42. A survey was made in a neighborhood of 37 houses. The graph represents the number of residents who live in each house. Complete the table and determine the mean number of residents per house.

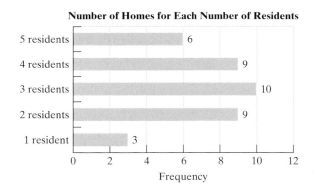

Number of Homes for Each Number of Residents

Number of Residents in Each House	Number of Houses	Product
1		
2		
3		
4		
5		
Total:		

43. Several instructors were asked the number of students who were initially enrolled in their classes. The results are represented in the graph. Find the weighted mean of the number of students initially enrolled in class. Round to the nearest whole unit.

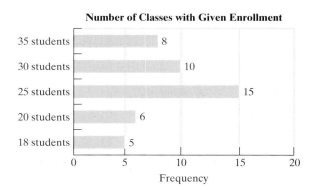

Number of Classes with Given Enrollment

44. During two months, Jake bought several hamburgers from three different fast-food restaurants. Find the mean price that he paid per hamburger. Round to the nearest cent.

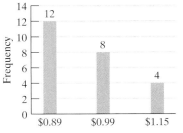

At most colleges and universities, weighted means are used to compute students' grade point averages (GPAs). At one college, the grades A–F are assigned numerical values as follows:

A	= 4.0	C	= 2.0
B+	= 3.5	D+	= 1.5
B	= 3.0	D	= 1.0
C+	= 2.5	F	= 0.0

Grade point average is a weighted mean where the "weights" for each grade are the number of credit-hours for that class. Use this information to answer Exercises 45–48.

45. Compute the GPA for the following grades. Round to the nearest hundredth.

Course	Grade	Number of Credit-Hours (Weights)
Intermediate Algebra	B	4
Theater	C	1
Music Appreciation	A	3
World History	D	5

46. Compute the GPA for the following grades. Round to the nearest hundredth.

Course	Grade	Number of Credit-Hours (Weights)
General Psychology	B+	3
Beginning Algebra	A	4
Student Success	A	1
Freshman English	B	3

47. Compute the GPA for the following grades. Round to the nearest hundredth.

Course	Grade	Number of Credit-Hours (Weights)
Business Calculus	B+	3
Biology	C	4
Library Research	F	1
American Literature	A	3

48. Compute the GPA for the following grades. Round to the nearest hundredth.

Course	Grade	Number of Credit-Hours (Weights)
University Physics	C+	5
Calculus I	A	4
Computer Programming	D	3
Swimming	A	1

1. Basic Definitions

The probability of an event measures the likelihood of the event to occur. It is of particular interest because of its application to everyday life.

- The probability of picking the winning six-number combination for the New York lotto grand prize is $\frac{1}{45,057,474}$.
- Genetic DNA analysis can be used to determine the risk that a child will be born with cystic fibrosis. If both parents test positive, the probability is 25% that a child will be born with cystic fibrosis.

To begin our discussion, we must first understand some basic definitions.

An activity with observable outcomes is called an **experiment**. The collection (or set) of all possible outcomes of an experiment is called the **sample space** of the experiment.

example 1   Determining the Sample Space of an Experiment

a. Suppose a single die is rolled. Determine the sample space of the experiment.

b. Suppose a coin is flipped. Determine the sample space of the experiment.

Solution:

a. A die is a single six-sided cube in which each side has between 1 and 6 dots painted on it. When the die is rolled, any of the six sides may come up.

The sample space is {1, 2, 3, 4, 5, 6}.   Notice that the symbols { } (called *set braces*) are used to enclose the elements of the sample space.

b. The coin may land as a head H or as a tail T. The sample space is {H, T}.

2. Probability of an Event

Any part of a sample space is called an **event**. For example, if we roll a die, the event of rolling number 5 or a greater number consists of the outcomes 5 and 6. In mathematics, we measure the likelihood of an event to occur by its probability.

Probability of an Event

$$\textbf{Probability of an event} = \frac{\text{number of elements in event}}{\text{number of elements in sample space}}$$

Note: From the definition, a probability value can never be negative or greater than 1.

Skill Practice

1. Suppose one marble is selected from the box shown, and the color is recorded. What is the sample space of the experiment?

2. For an individual birth, the gender of the baby is recorded. Determine the sample space for this experiment.

Tip: The word "die" is the singular form of the word "dice."

Answers

1. {red, green, blue, yellow}
2. {male, female}

Skill Practice

3. What is the probability of selecting a yellow ball from the box in margin Exercise 1?

---

example 2   Computing the Probability of an Event

a. Compute the probability of rolling a 5 or greater on a die.

b. Compute the probability of flipping a coin and having it land as a head.

Solution:

a. The event can occur in 2 ways: The die lands as a 5 or 6.
The sample space has 6 elements: 1, 2, 3, 4, 5, and 6.

The probability of rolling a 5 or greater: $\dfrac{2}{6}$ ← number of ways to roll a 5 or greater
       ← number of elements in the sample space

$\qquad = \dfrac{1}{3}$ Simplify to lowest terms.

b. The event can occur in 1 way (the coin lands head side up).
The sample space has 2 outcomes: heads or tails.

The probability of flipping a head on a coin: $\dfrac{1}{2}$ ← number of ways to get a head on a coin
       ← number of elements in the sample space

The value of a probability can be written as a fraction, as a decimal, or as a percent. For example, the probability of a coin landing as heads is $\frac{1}{2}$ or 0.5 or 50%. In words, this means that if we flip a coin many times, theoretically we expect one-half (50%) of the outcomes to land as a head.

Skill Practice

A group of registered voters has 9 Republicans, 8 Democrats, and 3 Independents. Suppose one person from the group is selected at random.

4. What is the probability that the person is a Democrat?

5. What is the probability that the person is an Independent?

6. What is the probability that the person is registered with the Libertarian Party?

---

example 3   Computing Probabilities

A class has 4 freshmen, 12 sophomores, and 6 juniors. If one individual is selected at random from the class, find the probability of selecting

a. A sophomore

b. A junior

c. A senior

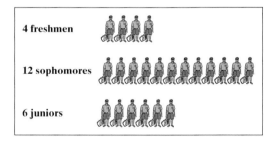

Solution:

In this case, there are 22 members of the class (4 freshmen + 12 sophomores + 6 juniors). This means that the sample space has 22 elements.

a. There are 12 sophomores in the class. The probability of selecting a sophomore is

$\dfrac{12}{22}$ There are 12 sophomores out of 22 people in the sample space.

$= \dfrac{6}{11}$ Simplify to lowest terms.

Answers

3. $\dfrac{1}{4}$ or 0.25

4. $\dfrac{8}{20} = \dfrac{2}{5}$ or 0.4

5. $\dfrac{3}{20}$

6. 0

b. There are 6 juniors out of 22 people in the sample space. The probability of selecting a junior is

$$\frac{6}{22} \quad \text{or} \quad \frac{3}{11}$$

c. There are no seniors in the class. The probability of selecting a senior is

$$\frac{0}{22} \quad \text{or} \quad 0$$

A probability of 0 indicates that the event is impossible. It is impossible to select a senior from a class that has no seniors.

---

3. Estimating Probabilities from Observed Data

We were able to compute the probabilities in Examples 2 and 3 because the sample space was known. Sometimes we need to collect information to help us estimate probabilities.

| example 4 | Estimating Probabilities from Observed Data |

One yogurt company has an instant win game. Some of the containers have a coupon for a free yogurt printed on the inside lid. After purchasing 40 yogurts over a period of several months, Chad records 5 wins.

a. Based on this observation, what is the probability of winning a prize on a given trial?

b. What is the probability of losing on a given trial?

Solution:

a. In this situation each of the 40 containers of yogurt may be thought of as a win or a loss. In this case, 5 trials out of 40 came out as winners. The probability of winning on a given trial is

$$\frac{5}{40} \quad \text{or} \quad \frac{1}{8}$$

b. If 5 trials came out as winners, then 35 were losers. The probability of losing on a given trial is

$$\frac{35}{40} \quad \text{or} \quad \frac{7}{8}$$

Skill Practice

In a carnival game, Erin will win a prize if she can toss a ring around the neck of a bowling pin. After observing 200 players who had gone before her, she learns that 15 players won a prize.

7. Based on this observation, what is the probability of winning a prize?

8. What is the probability of losing?

9. What percent of games came out as wins?

---

4. Complementary Events

The events in Example 4(a) and 4(b) are called complementary events. The **complement of an event** is the set of all elements in the sample space that are not in the event. In this case, 5 yogurts were winners and the remaining 35 were losers. Together these two events make up the entire sample space, yet they do not overlap. For this reason, the probability of an event plus the probability of its complement is 1.

Answers

7. $\dfrac{15}{200} = \dfrac{3}{40}$ or 0.075

8. $\dfrac{185}{200} = \dfrac{37}{40}$ or 0.925

9. 7.5%

Skill Practice

10. For one particular medicine, the probability that a patient will experience side effects is $\frac{1}{20}$. What is the probability that a patient will *not* experience side effects?

11. The probability that a flight arrives on time is 0.18. What is the probability that a flight will *not* arrive on time?

Answers

10. $\frac{19}{20}$

11. 0.82

---

example 5   Finding the Probability of Complementary Events

Find the indicated probability.

a. The probability of getting a winter cold is $\frac{3}{10}$. What is the probability of *not* getting a winter cold?

b. If the probability that a washing machine will break before the end of the warranty period is 0.0042, what is the probability that a washing machine will *not* break before the end of the warranty period?

Solution:

a. The probability of an event plus the probability of its complement must add up to 1. Therefore, we have an addition problem with a missing addend. This may also be expressed as subtraction.

$$\frac{3}{10} + ? = 1 \qquad \text{or equivalently} \qquad 1 - \frac{3}{10} = ?$$

$$\frac{10}{10} - \frac{3}{10} = \frac{7}{10} \qquad \text{Find a common denominator and subtract.}$$

There is a $\frac{7}{10}$ chance (70% chance) of *not* getting a winter cold.

b. The probability that a washing machine will break before the end of the warranty period is 0.0042. Then the probability that a machine will *not* break before the end of the warranty period is given by

$$1 - 0.0042 = 0.9958 \text{ or equivalently } 99.58\%$$

---

section 9.5   Practice Exercises

Boost *your* GRADE at mathzone.com!

MathZone

- Practice Problems
- Self-Tests
- NetTutor
- e-Professors
- Videos

Study Skills Exercises

1. A good way to determine what will be on a test is to look at both your notes and the exercises assigned by your instructor. List five kinds of problems that you think will be on the test for this chapter.

2. Define the key terms.

   a. Experiment   **b. Sample space**   **c. Event**   **d. Probability of an event**

   e. Complement of an event

Review Exercises

For Exercises 3–8, find the mean, median, and mode (if one exists).

3. 13, 16, 22, 25, 10

4. 62, 64, 62, 67, 40

5. 8, 9, 10, 7, 8, 8, 11, 10

6. 96%, 88%, 89%, 90%, 88%, 50%

7. 96%, 88%, 89%, 90%, 88%, 41%

8. 96%, 88%, 89%, 90%, 88%, 29%

Objectives 1: Basic Definitions

9. A card is chosen from a deck consisting of 10 cards numbered 1–10. Determine the sample space of this experiment. **(See Example 1.)**

10. A marble is chosen from a jar containing a yellow marble, a red marble, a blue marble, a green marble, and a white marble. Determine the sample space of this experiment.

11. Two dice are thrown, and the sum of the top sides is observed. Determine the sample space of this experiment.

12. A coin is tossed twice. Determine the sample space of this experiment.

13. Describe an event that could happen when rolling a die. Answers may vary.

14. Describe an event that could happen when tossing a coin twice. Answers may vary.

Objective 2: Probability of an Event

15. Which of the values can represent the probability of an event?

a. 1.62   **b.** $-\dfrac{7}{5}$ **c.** 0   **d.** 1

e. 200%   **f.** 4.5   **g.** 4.5%   **h.** 0.87

16. Which of the values can represent the probability of an event?

a. 1.5   **b.** 0   **c.** $\dfrac{2}{3}$ **d.** 1

e. 150%   **f.** 3.7   **g.** 3.7%   **h.** 0.92

17. If a single die is rolled, what is the probability that it will come up as a number less than 3? **(See Example 2.)**

18. If a single die is rolled, what is the probability that it will come up as a number greater than 5?

19. If a single die is rolled, what is the probability that it will come up with an even number?

A jar contains 8 marbles with 2 white, 5 green, and 1 blue. Use this information to answer Exercises 20–23.

20. What is the probability of choosing a green marble from the jar?

21. What is the probability of choosing a white marble from the jar?

22. What is the probability of choosing a blue marble from the jar?

23. What is the probability of choosing a purple marble from the jar?

24. If two dice are tossed, what is the probability of getting a sum of 1?

25. What is the definition of an impossible event?

26. If a die is tossed, what is the probability that a number from 1–6 will come up?

27. If one coin is tossed, what is the probability of getting a head or a tail?

28. What is the definition of a certain event?

29. In a deck of cards there are 12 face cards and 40 cards with numbers. What is the probability of selecting a face card from the deck? **(See Example 3.)**

30. In a deck of cards, 13 are diamonds, 13 are spades, 13 are clubs, and 13 are hearts. Find the probability of selecting a diamond from the deck.

31. A jar contains 7 yellow marbles, 5 red marbles, and 4 green marbles. What is the probability of selecting a red marble or a yellow marble?

32. A jar contains 10 black marbles, 12 white marbles, and 4 blue marbles. What is the probability of selecting a blue marble or a black marble?

Objective 3: Estimating Probabilities from Observed Data

33. The table displays the length of stay for vacationers at a small motel. **(See Example 4.)**

Number of Days Stayed	Frequency
2	14
3	13
4	18
5	28
6	11
7	30
8	6

  a. What is the probability that a vacationer will stay for 4 days?

  b. What is the probability that a vacationer will stay for less than 4 days?

  c. Based on the information from the table, what percent of vacationers stay for more than 6 days?

34. A number of students at a large university were asked if they owned a car. The table shows the results.

	Number of Car Owners	Number Who Do Not Own a Car
Dorm resident	32	88
Lives off campus	59	26

  a. What is the probability that a dorm resident selected at random owns a car?

  b. What is the probability that a student selected at random does not own a car?

35. A survey was made of 60 participants, asking if they drive an American-made car, a Japanese car, or a car manufactured in another foreign country. The table displays the results.

	Frequency
American	21
Japanese	30
Other	9

  a. What is the probability that a randomly selected car is manufactured in America?

  b. What percent of cars is manufactured in some country other than Japan?

36. The number of customer complaints for service representatives is given in the table. If one representative is picked at random, find the probability that the representative received

Number of Complaints	Number of Representatives
0	4
1	2
2	14
3	10
4	16
5	18
6	10
7	6

  a. Exactly 3 complaints

  b. Between 1 and 5 complaints, inclusive

  c. At least 4 complaints

  d. More than 5 complaints or less than 2 complaints

37. Mr. Gutierrez noted the times in which his students entered his classroom and constructed the following chart.

   a. What is the probability that a student will be early to class?

   b. What is the probability that a student will be late to class?

   c. What percent of students arrive on time or early? Round to the nearest whole percent.

Time	Number of Students
About 10 min early	1
About 5 min early	6
On time	11
About 5 min late	7
About 10 min late	3
About 15 min late	1

38. Each person at an office party purchased a raffle ticket. The graph shows the results of the raffle.

   a. How many people bought raffle tickets?

   b. What is the probability of winning the TV set?

   c. What percent of people won some type of prize?

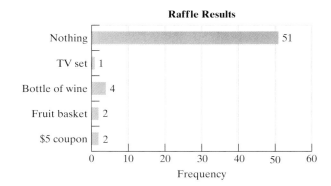

Objective 4: Complementary Events

39. If the probability of the horse Sugar 'N Spice to win is $\frac{2}{7}$, what is the probability that he will lose? **(See Example 5.)**

40. If the probability of being hit by lighting is $\frac{1}{1,000,000}$, what is the probability of not getting hit by lighting?

41. If the probability of having twins is 1.2%, what is the probability of not having twins?

42. The probability of a woman's surviving breast cancer is 88%. What is the probability that a woman would not survive?

Expanding Your Skills

43. A firm performing political polls is interested in determining attitudes regarding fuel consumption in the United States. One survey asked a sample of 530 randomly selected voters the following question.

   "Would you consider buying a hybrid vehicle that uses both electric and gasoline power if the vehicle were affordable?"

The following table shows the results of the survey by gender and response.

	Yes	No	Don't Know	Total
Male	184	14	8	206
Female	72	204	48	324
Total	256	218	56	530

Based on these results, if a registered voter is randomly selected from the population, find the probability that

a. The person will answer "Yes" *or* "Don't know"

b. A male will answer "no"

44. A survey was taken of a college class to identify the age of students who are registered to vote. The table shows the results of the survey.

Ages	Registered	Not Registered	Total
18–20	9	15	24
21–23	6	2	8
24–26	8	0	8
Total	23	17	40

a. What is the probability that a student selected at random will be registered?

b. What is the probability that a student of age 18–20 will be registered?

chapter 9 | summary

section 9.1 Tables, Bar Graphs, Pictographs, and Line Graphs

Key Concepts

Statistics is the branch of mathematics that involves collecting, organizing, and analyzing **data** (information). Information can often be organized in tables and graphs. The individual entries within a table are called **cells**.

Example 1

Construct a bar graph for the data in Example 1.

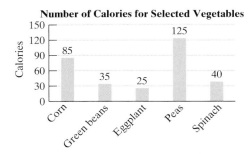

Number of Calories for Selected Vegetables

A **pictograph** uses an icon or small image to convey a unit of measurement.

Example 3

What is the value of each icon in the graph?

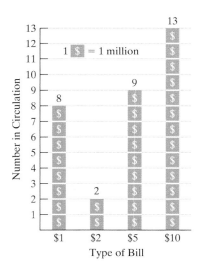

Each icon is worth 1,000,000 bills in circulation.

Examples

Example 2

The data in the table give the number of Calories for a 1-c serving of selected vegetables.

Vegetable (1 c)	Number of Calories
Corn	85
Green beans	35
Eggplant	25
Peas	125
Spinach	40

How many more Calories does 1 c of peas have than 1 c of corn?

125 Cal − 85 Cal = 40 Cal

Line graphs are often used to track how one variable changes with respect to the change in a second variable.

Example 4

In what year were there 52.3 million married-couple households?

Number of Married-Couple Households

From the graph, the year 1990 corresponds to 52.3 million married-couple households.

section 9.2 Frequency Distributions and Histograms

Key Concepts

A **frequency distribution** is a table displaying the number of data values that fall within specified intervals called **class intervals**.

When constructing a frequency distribution, keep these important guidelines in mind.
- The classes should be equally spaced.
- The classes should not overlap.
- In general, use between 5 and 15 classes.

A **histogram** is a special bar graph that illustrates data given in a frequency distribution. The class intervals are given on the horizontal scale. The height of each bar in a histogram measures the frequency for each class.

Examples

Example 1
Create a frequency distribution for the following data.

50	53	53	36	32
54	51	44	34	40
50	40	52	32	33
50	47	42	30	38

Class Intervals	Tally	Frequency
30–34	ⅢⅠ	5
35–39	‖	2
40–44	‖‖	4
45–49	Ⅰ	1
50–54	ⅢⅠ ‖‖	8

Example 2
Create a histogram for the data in Example 1.

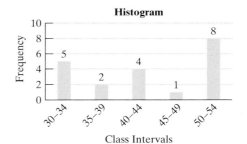

section 9.3 Circle Graphs

Key Concepts

A **circle graph** (or pie graph) is a type of graph used to show how a whole amount is divided into parts. Each part of the circle, called a **sector**, is like a slice of pie.

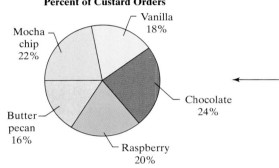

Percent of Custard Orders

Vanilla 18%
Mocha chip 22%
Butter pecan 16%
Raspberry 20%
Chocolate 24%

Examples

Example 1

At Ritter's Frozen Custard, the flavors for the day are given with the number of orders for each flavor.

Flavor	Number of Orders
Vanilla	180
Chocolate	240
Raspberry	200
Butter pecan	160
Mocha chip	220

Construct a circle graph for the data given. Label the sectors with percents.

section 9.4 Mean, Median, and Mode

Key Concepts

The **mean** (or average) of a set of numbers is the sum of the values divided by the number of values.

$$\text{Mean} = \frac{\text{sum of the values}}{\text{number of values}}$$

The **median** is the "middle" number in an ordered list of numbers. For an ordered list of numbers:
- If the number of data values is *odd*, then the median is the middle number in the list.
- If the number of data values is *even*, the median is the mean of the two middle numbers in the list.

The **mode** of a set of data is the value or values that occur most often. If each value occurs the same number of times, then there is no mode.

When data values in a list appear multiple times, we can compute a **weighted mean**.

Examples

Example 1

Find the mean test score: 92, 100, 86, 60, 90

$$\text{Mean} = \frac{92 + 100 + 86 + 60 + 90}{5}$$

$$= \frac{428}{5} = 85.6$$

Example 2

Find the median: 12   18   6   10   5

First order the list: 5   6   10   12   18
The median is the middle number, 10.

Example 3

Find the median: 15   20   20   32   40   45

The median is the average of 20 and 32:

$$\frac{20 + 32}{2} = \frac{52}{2} = 26 \qquad \text{The median is 26.}$$

Example 4

Find the mode: 7   2   5   7   7   4   6   10

The value 7 is the mode because it occurs most often.

Example 5

The ages of children in a day-care center are given in the table. Find the mean age.

Age (yr)	Frequency	Product
3	5	$3 \cdot 5 = 15$
4	10	$4 \cdot 10 = 40$
5	3	$5 \cdot 3 = 15$
6	2	$6 \cdot 2 = 12$
Total	**20**	**82**

$$\text{Mean} = \frac{82}{20} = 4.1$$

The mean age is 4.1 yr.

section 9.5 **Introduction to Probability**

Key Concepts

The collection (or set) of all possible outcomes of an experiment is called the **sample space**.

The **probability of event** is given by:

$$\frac{\text{number of elements in event}}{\text{number of elements in sample space}}$$

The **complement of an event** is the set of all elements in the sample space that are not in the event.

Examples

Example 1

Define the sample space for selecting a colored ball from the box.

Sample space = {red, blue, yellow, green}

Example 2

What is the probability of selecting a yellow ball from the box in Example 1?

Let A represent the event of picking a yellow ball. Then A = {yellow}.

$$P(A) = \frac{1}{4} \quad \begin{matrix} \longleftarrow \text{number of yellow balls} \\ \longleftarrow \text{number of balls in box} \end{matrix}$$

Example 3

49 CDs are in a shopping cart.

10 Rap
24 Rock
12 Latina
3 Classical

If one CD is selected at random, find the probability that

a. A rock CD is selected. $\dfrac{24}{49}$

b. A rock CD is *not* selected. This is the complementary event to part (a).

$$1 - \frac{24}{49} = \frac{25}{49}$$

chapter 9 | review exercises

Section 9.1

For Exercises 1–4, refer to the table. The table gives the number of Calories and the amount of fat, cholesterol, sodium, and total carbohydrate for one $\frac{1}{2}$-c serving of chocolate ice cream.

Ice Cream	Calories	Fat(g)	Cholesterol (mg)	Sodium (mg)	Carbohydrate (g)
Breyers	150	8	20	35	17
Häagen-Dazs	270	18	115	60	22
Edy's Grand	150	8	25	35	17
Blue Bell	160	8	35	70	18
Godiva	290	18	65	50	28

1. Which ice cream has the most calories?

2. Which ice cream has the least amount of cholesterol?

3. How many more times the sodium does Blue Bell have per serving than Edy's Grand?

4. What is the difference in the amount of carbohydrate for Godiva and Blue Bell?

From 1940 to the present, the number of U.S. farms has decreased. However, the average size of the farms has increased. The graph shows the average size of U.S. farms for selected years. Refer to the graph for Exercises 5–8. (Source: U.S. Department of Agriculture.)

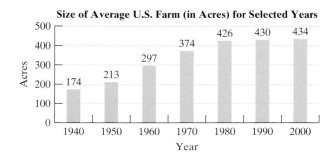

Size of Average U.S. Farm (in Acres) for Selected Years

5. What was the average size of the farms in 1970?

6. What is the difference between the average size farm in the year 2000 compared to 1940?

7. What is the difference between the average size farm in the year 1990 compared to 1980?

8. Between which 10-year interval was the increase the greatest?

The number of detached single-family homes in a certain county for selected years is shown in the pictograph. For Exercises 9–12, refer to the graph.

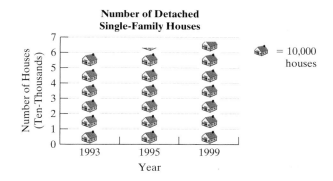

Number of Detached Single-Family Houses

= 10,000 houses

9. What does each icon represent?

10. From the graph, estimate the number of houses in 1995.

11. From the graph, estimate the number of houses in 1993.

12. Estimate the difference in the number of houses in 1999 and the number in 1993.

For Exercises 13–16, use the graph representing the total production of wheat over several years.

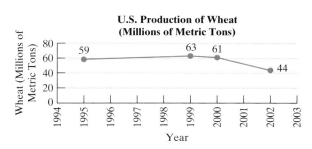

U.S. Production of Wheat (Millions of Metric Tons)

13. In which year was the wheat production the greatest?

14. What was the wheat production in 1995?

15. Does the recent trend appear to be increasing or decreasing?

16. Extend the line to estimate the wheat production in the year 2003.

Section 9.2

The ages of students in a Spanish class are given.

18 22 19 26 31 20 40 24 43 22
29 28 35 42 29 30 24 31 23 21

Use these data for Exercises 17–18.

17. Complete the frequency table.

Class Intervals (Age)	Frequency
18–21	
22–25	
26–29	
30–33	
34–37	
38–41	
42–45	

18. Construct a histogram of the data in Exercise 17.

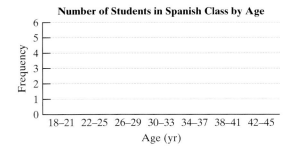

Number of Students in Spanish Class by Age

Section 9.3

The pie graph describes the types of hot subs offered at Larry's Sub Shop. Use the information in the graph for Exercises 19–21.

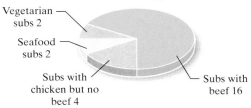

Number of Certain Types of Subs

Vegetarian subs 2
Seafood subs 2
Subs with chicken but no beef 4
Subs with beef 16

19. How many types of subs are offered at Larry's?

20. What fraction of the subs at Larry's is made with beef?

21. What fraction of the subs at Larry's is not made with beef?

22. A survey was conducted with 200 people, and they were asked their education level. The results of the survey are given in the table.

   a. Complete the table.

Education Level	Number of People	Percent	Number of Degrees
Grade school	10		
High school	50		
Some college	60		
Four-year degree	40		
Post graduate work	40		

   b. Construct a circle graph from the information in the table.

Percent by Education Level

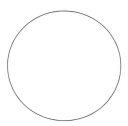

Section 9.4

23. For the list of quiz scores, find the mean, median, and mode(s).

20, 20, 18, 16, 18, 17, 16, 10, 20, 20, 15, 20

24. Juanita kept track of how many milligrams of calcium she took each day through vitamins and dairy products. Determine the mean number of milligrams of calcium per day. Round to the nearest 10 milligrams.

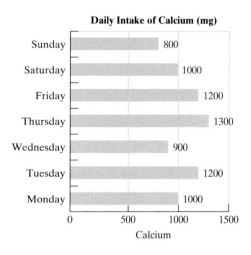

Daily Intake of Calcium (mg)

25. The seating capacity for five arenas used by the NBA is given in the table. Find the median number of seats.

Arena	Number of Seats
Phelps Arena, Atlanta	20,000
Fleet Center, Boston	18,624
Chevrolette Coliseum, Charlotte	23,799
United Center, Chicago	21,500
Gund Arena, Cleveland	20,562

26. The manager of a restaurant had his customers fill out evaluations on the service that they received. A scale of 1 to 5 was used, where 1 represents very poor service and 5 represents excellent service. Given the list of responses, determine the mode(s).

4 5 3 4 4 3 2 5 5 1 4 3 4 4 5
2 5 4 4 3 2 5 5 1 4

Section 9.5

27. Roberto has six pairs of socks, each a different color: blue, green, brown, black, gray, and white. If Roberto randomly chooses a pair of socks, write the sample space for this event.

28. Refer to Exercise 27. What is the probability that Roberto will select a pair of gray socks?

29. Which of the following numbers could represent a probability?

a. $\dfrac{1}{2}$ **b.** $\dfrac{5}{4}$ **c.** 0 **d.** 1

e. 25% **f.** 2.5 **g.** 6% **h.** 6

30. A bicycle shop sells a child's tricycle in three colors: red, blue, and pink. In the warehouse there are 8 red tricycles, 6 blue tricycles, and 2 pink tricycles. If Kevin selects a tricycle at random, what is the probability that he will pick a red tricycle?

31. Refer to Exercise 30. What is the probability of Kevin's selecting a green tricycle?

chapter 9 | test

1. The table represents the world's major producers of primary energy for a recent year. All measurements are in quadrillions of Btu.

Note: 1 quadrillion = 1,000,000,000,000,000. (Source: Energy Information Administration, U.S. Dept. of Energy.)

Country	Amount of Energy Produced (quadrillions of BTUs)
United States	72
Russia	43
China	35
Saudi Arabia	43
Canada	18
United Kingdom	11
Iran	10

Construct a bar graph using horizontal bars by following these steps.

a. On the horizontal scale, make tick marks representing Btu's, starting from 0 to 80 in increments (steps) of 10.

b. On the vertical scale, write the country names. Begin at the bottom with Iran.

c. Construct the bar graph. The length of each bar corresponds to the amount of energy produced for each country.

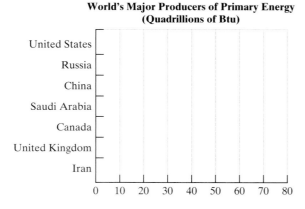

World's Major Producers of Primary Energy (Quadrillions of Btu)

2. Of the approximately 2.9 million workers in 1820 in the United States, 71.8% were employed in farm occupations. Since then, the percent of U.S. workers in farm occupations has declined drastically. The table shows the percent of total U.S. workers who worked in farm-related occupations for selected years. (Source: U.S. Department of Agriculture.)

Year	Percent of U.S. Workers in Farm Occupations
1820	72%
1860	59%
1900	38%
1940	17%
1980	3%

a. Which year had the greatest percent of U.S. workers employed in farm occupations? What is the value of the greatest percent?

b. Make a line graph with the year on the horizontal scale and the percent on the vertical scale. [*Hint:* Recall that the highest value on the vertical scale must be higher than the value from part (a). Also, select a suitable increment for the tick marks on the vertical scale.]

Percent of U.S. Workers in Farm Occupations

80%
70
60
50
40
30
20
10
0

1820   1860   1900   1940   1980

Year

c. Based on the graph, estimate the percent of U.S. workers employed in farm occupations for the year 1960.

3. The pictograph shows the flower sales for the first 5 months of the year for a flower shop.

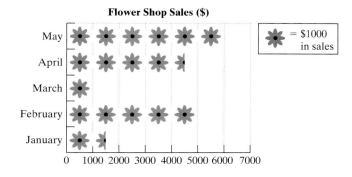

Flower Shop Sales ($)

= $1000 in sales

a. What is the value of each flower icon?

b. From the graph, estimate the sales for the month of April.

c. Which month brought in sales of $5000?

4. The tardiness record for a group of 30 employees at a retail store is given. The values denote the number of days that each employee was late to work. Complete the table.

3	0	5	4	7	4	0	3	2	8
7	4	8	3	2	5	3	2	1	0
4	5	6	7	2	3	4	2	1	3

Number of Days Late	Tally	Number of Employees
0		
1		
2		
3		
4		
5		
6		
7		
8		

5. A cellular phone company questioned 20 people at a mall, to determine approximately how many minutes each individual spent on the cell phone each month. Using the list of results, complete the frequency distribution and construct a histogram.

100	120	250	180	300	200	250	175
110	280	330	280	300	325	60	75
100	350	60	90

Number of Minutes Used Monthly	Tally	Frequency
51–100		
101–150		
151–200		
201–250		
251–300		
301–350		

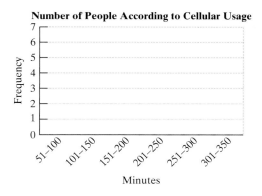

Number of People According to Cellular Usage

6. The circle graph shows the percent of different types of floor covering that people have on their living room floors.

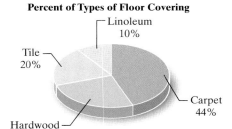

Percent of Types of Floor Covering

Linoleum 10%

Tile 20%

Carpet 44%

Hardwood 26%

a. If 150 people were questioned, how many would be expected to have carpet on their living room floor?

b. If 200 people were questioned, how many would be expected to have tile on their living room floor?

c. If 300 people were questioned, how many would be expected not to have linoleum on their living room floor?

7. The table represents the heights of the Seven Summits (the highest peaks from each continent).

Mountain	Continent	Height (ft)
Mt. Kilimanjaro	Africa	19,340
Elbrus	Europe	18,510
Aconcagua	South America	22,834
Denali	North America	20,320
Vinson Massif	Antarctica	16,864
Mt. Kosciusko	Australia	7,310
Mt. Everest	Asia	29,035

    a. What is the mean height of the Seven Summits? Round to the nearest whole unit.

    b. What is the median height?

    c. Is there a mode?

8. Mike and Darcy listed the amount of money paid for going to the movies for the past 3 months. This list contains the amount for 2 tickets. Find the mean and the median.

    $11 $14 $11 $16 $15 $16 $12 $16 $15 $20

9. An ad in the newspaper gave prices for different kinds of house paint. Each price is for 1 gal of paint. Determine the modal price.

    $14.97 $14.97 $16.97 $18.97 $9.97

    $11.97 $13.97 $14.97 $18.97 $17.97

    $15.97 $13.97 $16.97 $14.97

10. A board game has a die with eight sides with the numbers 1–8 printed on each side.

    a. What is the sample space for rolling the die one time?

    b. What is the probability of rolling a 6?

    c. What is the probability of rolling an even number?

    d. What is the probability of rolling a number less than 3?

11. At a party there is a cooler filled with ice and soft drinks: 6 cans of diet cola, 4 cans of ginger ale, 2 cans of root beer, and 2 cans of cream soda. A person takes a can of soda at random.

    a. What is the probability that the person selects a can of ginger ale?

    b. What is the probability of the person not selecting a can of ginger ale?

12. Compute the GPA for the following grades. Round to the nearest hundredth. Use this scale:

    A   = 4.0   C   = 2.0
    B+ = 3.5   D+ = 1.5
    B   = 3.0   D   = 1.0
    C+ = 2.5   F   = 0.0

Course	Grade	Number of Credit-Hours (Weights)
Art Appreciation	B	4
College Algebra	A	3
English II	C	3
Physical Fitness	A	1

---

chapters 1–9 | cumulative review

1. Identify the place value of the underlined digit.

    a. 23,990,192   **b.** 5,981,902   **c.** 3,019,226

2. Add.   2087 + 53 + 10,499 + 6

3. Estimate the product by first rounding each number to the nearest hundred.

    687 × 1243

4. Divide 651 by 23. Identify the divisor, dividend, whole part of the quotient, and remainder.

5. What fraction of this circle is shaded blue?

For Exercises 6–8, multiply or divide as indicated. Reduce the answers to lowest terms.

6. $\dfrac{12}{7} \cdot \dfrac{14}{36}$
   7. $\dfrac{105}{96} \div \dfrac{7}{16}$
   8. $\dfrac{5}{8} \div \dfrac{6}{15} \cdot \dfrac{24}{25}$

For Exercises 9–11, add or subtract as indicated. Reduce to lowest terms.

9. $\dfrac{97}{102} - \dfrac{63}{102}$
   10. $\dfrac{3}{10} + \dfrac{7}{100}$
   11. $\dfrac{1}{2} + \dfrac{5}{3} - \dfrac{1}{6}$

12. Simplify, using the order of operations.
$$2\dfrac{1}{8} \div 17 + \dfrac{7}{12} \cdot \dfrac{9}{14} - \dfrac{1}{3}$$

13. The table gives the prices of certain stocks and their increase or decrease from one day to the next. Complete the table.

Stock	Yesterday's Closing Price ($)	Increase/ Decrease	Today's Closing Price ($)
RylGold	13.28	0.27	
NetSolve	9.51	−0.17	
Metals USA	14.35	0.10	
PAM Transpt	18.09	0.09	
Steel Tch	21.63	−0.37	

For Exercises 14–16, multiply or divide by powers of 10.

14. 68.412×100
15. 68.412×0.1

16. $68.412 \div 0.001$

17. The number of unemployed Americans in 1980 was 7,600,000. By 2000, the number of unemployed was 5,700,000.

   a. By how much had the number of unemployed Americans decreased between 1980 and 2000?

   b. Compute the unit rate representing the decrease in unemployment per year. (Source: Bureau of Labor Statistics.)

18. Quick Cut Lawn Company can service 5 customers in $2\frac{3}{4}$ hr. Speedy Lawn Company can service 6 customers in 3 hr. Find the unit rate in time per customer for both lawn companies and decide which company is faster.

19. If Rosa can type a 4-page English paper in 50 min, how long will it take her to type a 10-page term paper?

20. Find the values of x and y, assuming that the two triangles are similar.

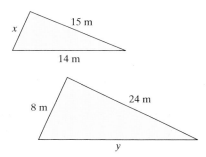

21. Out of a group of people, 95 said that they brushed their teeth twice a day. If this number represents 78% of the people surveyed, how many were surveyed? Round to the nearest whole unit.

22. The number of children accompanying their parents on business trips has jumped 230% in the last 10 years. If the number of children 10 years ago was 7.4 million, determine the present number.

23. Out of 120 people, 78 wear glasses. What percent does this represent?

24. A savings account pays 3.4% simple interest. If $1200 is invested for 5 years, what will be the balance?

25. Convert 2 ft 5 in. to inches.

26. Convert $4\frac{1}{2}$ gal to quarts.

27. Add. 3 yd 2 ft + 5 yd 2 ft

28. Subtract. 12 km − 2360 m

29. Divide 10 lb 12 oz by 4.

For Exercises 30–32, identify the type of angle. Choose from acute, obtuse, right, or straight.

30.

31.

32.

33. Find the area of the parallelogram.

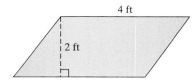

4 ft

2 ft

34. Find the volume of the cone. Let $\pi = \frac{22}{7}$.

7 m

3 m

35. The data shown in the table give the average weight for boys based on age. (Source: National Parenting Council.) Make a line graph to illustrate these data by following these steps. Age is represented on the horizontal scale. Weight is on the vertical scale.

Age	Weight (lb)
5	44.5
6	48.5
7	54.5
8	61.25
9	69
10	74.5
11	85
12	89

a. On the horizontal scale, make tick marks representing years, starting from 0 to 12 in increments of 1.

b. On the vertical scale, make tick marks representing pounds, starting from 0 to 100 in increments of 20.

c. For each age value, plot a point for the corresponding weight.

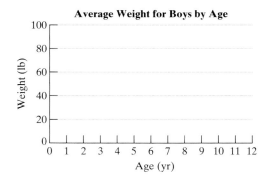

36. Use the circle graph to answer the questions. Round the answers to the nearest tenth of a percent.

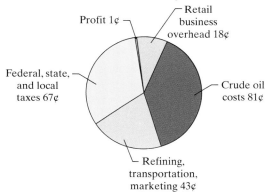

a. Approximately what percent is profit?

b. Approximately what percent is paid in taxes?

c. Approximately what percent is allocated to crude oil costs, refining, transportation, and marketing?

37. The number of videos rented per customer at a video store is given for 30 customers. Complete the table.

0	4	0	1	2	6	1	3	1	1
1	2	1	3	2	4	3	2	0	2
4	2	1	0	2	1	1	3	1	0

Number of Videos	Tally	Number of Customers
0		
1		
2		
3		
4		
5		
6		

A game has a spinner with four equal sections shaded in four different colors. Refer to the spinner for Exercises 38–40.

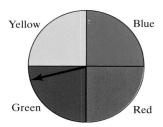

38. If a person spins the pointer once, determine the sample space.

39. What is the probability of the pointer landing on green?

40. What is the probability of the pointer *not* landing on green?

Real Numbers

10

10.1 Real Numbers and the Real Number Line

10.2 Addition of Real Numbers

10.3 Subtraction of Real Numbers

10.4 Multiplication and Division of Real Numbers

10.5 Order of Operations and Scientific Notation

In this chapter, we begin our study of algebra by learning how to add, subtract, multiply, and divide positive and negative numbers. Negative numbers are applied daily when discussing the balance on a bank account, temperatures below 0°, elevations below sea level, scoring in golf, and many other contexts. In Exercise 85 of Section 10.3 you will use subtraction of real numbers to find the range between the daytime and nighttime temperatures on the moon.

chapter 10   preview

The exercises in this chapter preview contain concepts that have not yet been presented. These exercises are provided for students who want to compare their levels of understanding before and after studying the chapter. Alternatively, you may prefer to work these exercises when the chapter is completed and before taking the exam.

Section 10.1

1. Write an integer that represents the numerical value.

   a. The number of people 65 and older living in the United States increased by approximately 3,740,000.

   b. The number of people living in poverty has decreased by 2,446,000.

2. Identify the numbers as rational or irrational numbers.

   a. -6 b. 0 c. $-\sqrt{5}$

   d. 1.3 e. $-\dfrac{2}{7}$

3. Determine the absolute value.

   a. $\left|-\dfrac{4}{3}\right|$ b. $|4.2|$

4. Find the opposite.

   a. -3.1 b. 15

Section 10.2

For Exercises 5–8, add.

5. $14 + (-22)$ 6. $-21 + (-34)$

7. $-18 + 18$ 8. $-\dfrac{2}{11} + \left(-\dfrac{6}{11}\right)$

9. Translate the phrase to a mathematical expression. Then simplify the expression. -2.3 increased by 8

Section 10.3

For Exercises 10–11, rewrite the subtraction problem as an equivalent addition problem. Then simplify.

10. $5 - 12 = $ _____ $= $ _____

11. $-34 - (-6) = $ _____ $= $ _____

For Exercises 12–15, subtract.

12. $4 - (-11)$ 13. $-4.3 - 11.9$

14. $-39 - (-39)$ 15. $0 - (-23)$

16. Simplify: $2 + (-6) - (-9) - 1 + 10$

17. Translate the phrase to a mathematical expression. Then simplify the expression. 16 subtracted from -2

Section 10.4

For Exercises 18–24, simplify.

18. $-8(-12)$ 19. $\dfrac{-57}{-3}$

20. $\dfrac{5}{12} \cdot \left(-\dfrac{16}{3}\right)$ 21. $\dfrac{0}{-10}$

22. $\dfrac{-3}{0}$ 23. $(-3)^4$

24. -3^4

Section 10.5

For Exercises 25–26, simplify by using the order of operations.

25. $-15 + (-3)^2 - 2(-4)$ 26. $5(3 - 5)^3 - 2(1 - 3)^3$

27. Convert the numbers to scientific notation.

    a. $4,502,000,000$ b. 0.00022301

28. Convert the numbers to decimal notation.

    a. 5.9×10^{-2} b. 1.08×10^5

section 10.1 Real Numbers and the Real Number Line

1. Integers

Thus far in the text we have worked with the number zero and numbers greater than zero. Numbers greater than zero are called **positive numbers**. Positive numbers lie to the right of zero on a number line (Figure 10-1).

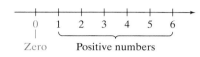

Figure 10-1

In some applications of mathematics we need to use *negative* numbers. For example,

- On a winter day in Detroit, the low temperature was 5 degrees below zero: $-5°$
- Tiger Woods' golf score in the U.S. Open was 3 below par: -3
- Maria is $128 overdrawn on her checking account. Her balance is: $-\$128$

The values $-5°$, -3, and $-\$128$ are negative numbers. **Negative numbers** lie to the left of zero on a number line (Figure 10-2).

Figure 10-2

The numbers . . . $-3, -2, -1, 0, 1, 2, 3, \ldots$ and so on are called **integers**.

---

example 1   Writing Integers

Write an integer that denotes each numerical value.

a. Liquid nitrogen freezes at 346°F below zero.

b. The shoreline of the Dead Sea on the border of Israel and Jordan is the lowest land area on earth. The "altitude" is 1300 ft below sea level.

Solution:

a. $-346°F$ **b.** -1300 ft

Skill Practice

Write an integer that denotes each number.

1. The average temperature at the South Pole in July is 65°C below zero.

2. The balance on Sylvia's checking account is overdrawn by $156.

2. Rational and Irrational Numbers

A number that can be written as a ratio of two integers is called a **rational number** (division by zero is excluded). For example, the following numbers are rational numbers.

$\dfrac{2}{3}$ because it is a ratio of 2 and 3.

$\dfrac{-5}{7}$ because it is a ratio of -5 and 7.

Answers
1. $-65°C$ 2. $-\$156$

8   because it is a ratio of 8 and 1. That is, $8 = \frac{8}{1}$.
This shows that an integer is also a rational number.

0.25   because it is a ratio of 25 and 100. That is, $0.25 = \frac{25}{100}$.
This shows that a terminating decimal is a rational number.

$0.\overline{3}$ because it is a ratio of 1 and 3. That is, $0.\overline{3} = \frac{1}{3}$.
This shows that a repeating decimal is a rational number.

> **Tip:** Rational numbers consist of all numbers that can be expressed as a terminating decimal or as a repeating decimal.

Every rational number can be matched with a point on a number line.

example 2   Locating Numbers on a Number Line

Plot each number on a number line.

a. -4.2 **b.** $\dfrac{15}{4}$ **c.** $-\dfrac{1}{3}$

Solution:

a. Because -4.2 is negative, it is located to the left of zero on the number line. Draw a dot two-tenths of a unit beyond -4.

b. Write $\frac{15}{4}$ as a mixed number or decimal.

$$\frac{15}{4} = 3\frac{3}{4} \text{ or } 3.75$$

Draw a dot three-fourths of a unit beyond 3 on the number line.

c. $-\dfrac{1}{3}$

Draw a dot one-third of a unit to the left of zero.

 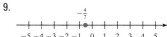

Some numbers are not rational numbers. That is, some numbers *cannot* be written as a ratio of two integers. These numbers are called **irrational numbers**. An irrational number expressed in decimal form is a nonrepeating and nonterminating decimal. Some examples of irrational numbers are π and the square roots of non-perfect squares such as $\sqrt{2}$, $\sqrt{3}$, and $\sqrt{5}$.

3. Real Numbers and the Real Number Line

All numbers we have studied thus far taken together form the **real numbers**. Furthermore, every real number corresponds to a point on a number line. For this reason, we sometimes call the number line the **real number line**.

The order between two real numbers can be determined by using the number line.

- A number a is less than b (denoted $a < b$) if a lies to the left of b on the number line. See Figure 10-3.
- A number a is greater than b (denoted $a > b$) if a lies to the right of b on the number line. See Figure 10-4.

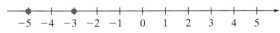

a is less than b
$a < b$

Figure 10-3

a is greater than b
$a > b$

Figure 10-4

---

example 3	**Determining Order**

Fill in the blank with $<$ or $>$.

a. $-3 \,\square\, -5$ **b.** $-2.7 \,\square\, 4.1$ **c.** $-\dfrac{4}{7} \,\square\, -\dfrac{3}{5}$

Solution:

a. $-3 \,\boxed{>}\, -5$ because -3 lies to the right of -5

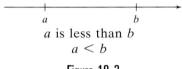

b. $-2.7 \,\boxed{<}\, 4.1$ because -2.7 lies to the left of 4.1

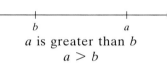

c. To determine order between two fractions, write the fractions with a common denominator.

$$-\dfrac{4 \cdot 5}{7 \cdot 5} \,\square\, -\dfrac{3 \cdot 7}{5 \cdot 7} \qquad \text{The common denominator is 35.}$$

$$-\dfrac{20}{35} \,\boxed{>}\, -\dfrac{21}{35}$$

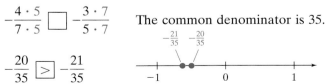

Notice that on the negative side of zero, -20 lies to the right of -21. Therefore, $-\frac{20}{35}$ is greater than $-\frac{21}{35}$.

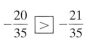

Skill Practice

Fill in the blank with $<$ or $>$.

10. $-4 \,\square\, -6$

11. $-\dfrac{2}{3} \,\square\, -\dfrac{3}{5}$

12. $5.1 \,\square\, -6.1$

Answers

10. $>$ 11. $<$ 12. $>$

4. Absolute Value

Notice on the number line that pairs of numbers such as 4 and −4 are the same distance from 0 (Figure 10-5). The distance between a number and zero on the number line is called its **absolute value**.

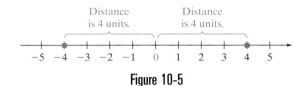

Figure 10-5

> ### Absolute Value
>
> The absolute value of a number a is denoted $|a|$. The value of $|a|$ is the distance between a and 0 on the number line.

example 4	Finding Absolute Value

Determine the absolute value.

a. $|-5|$ **b.** $|2.1|$ **c.** $\left|-\dfrac{3}{2}\right|$ **d.** $|0|$

Solution:

a. $|-5| = 5$ The number −5 is 5 units from 0 on the number line.

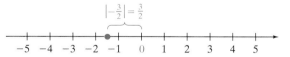

b. $|2.1| = 2.1$ The number 2.1 is 2.1 units from 0 on the number line.

c. $\left|-\dfrac{3}{2}\right| = \dfrac{3}{2}$ The number $-\dfrac{3}{2}$ is $\dfrac{3}{2}$ units (or 1.5 units) from 0.

d. $|0| = 0$ The number 0 is 0 units from 0 on the number line.

Tip: The absolute value of a nonzero number is always positive. The absolute value of 0 is 0.

5. Opposite

Two numbers that are the same distance from zero on the number line, but on opposite sides of zero are called **opposites**. For example, the numbers −2 and 2 are opposites (see Figure 10-6).

Answers

13. 6   **14.** 3.8

15. $\dfrac{5}{4}$ **16.** 0

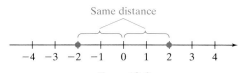

Figure 10-6

The opposite of a number a is denoted $-(a)$.

Original Number a	Opposite $-(a)$	Simplified Form	
5	$-(5)$	-5	The opposite of a positive number is a negative number.
2.4	$-(2.4)$	-2.4	
-7	$-(-7)$	7	The opposite of a negative number is a positive number.
$-\frac{1}{2}$	$-(-\frac{1}{2})$	$\frac{1}{2}$	

The opposite of a negative number is a positive number. Thus, for $a > 0$,

$-(-a) = a$ This is sometimes called the *double-negative property*.

---

example 5   Finding the Opposite of a Real Number

Find the opposite of each number.

a. 4   **b.** 6.3   **c.** -11 **d.** $-\dfrac{5}{2}$

Solution:

a. The opposite of 4 is -4.

b. The opposite of 6.3 is -6.3.

c. The opposite of -11 is 11.

d. The opposite of $-\frac{5}{2}$ is $\frac{5}{2}$.

Tip: To find the opposite of a number, we can simply change the sign.

Skill Practice

Find the opposite of each number.

17. 8   **18.** -4.8

19. -2 **20.** $\dfrac{5}{3}$

Answers

17. -8 18. 4.8
19. 2   20. $-\dfrac{5}{3}$

---

section 10.1   Practice Exercises

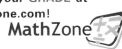
Study Skills Exercises

1. When working with signed numbers, keep a simple example in your mind, such as temperature. We understand that 10 degrees below zero is colder than 2 degrees below zero so the inequality $-10 < -2$ makes sense. Write down another example involving signed numbers that you can easily remember.

2. Define the key terms.

    a. Absolute value   **b. Integers**   **c. Irrational numbers**

    d. Negative numbers   **e. Opposite**   **f. Positive numbers**

    g. Rational numbers   **h. Real numbers**   **i. Real number line**

Objective 1: Integers

For Exercises 3–12, write an integer that represents each numerical value. **(See Example 1.)**

3. Death Valley, California, has elevation of 86 m below sea level.

4. In a card game, Jack lost $45.

5. Playing *Wheel of Fortune,* Sally won $3800.

6. Jim's golf score is 5 over par.

7. Rena lost $500 in the stock market in one month.

8. LaTonya earned $23 in interest on her saving account.

9. Patrick lost 14 lb on a diet.

10. Emily's score on a video game was 5000 points higher than that of the past record holder.

11. The number of Internet users rose by about 140,000.

12. A small business experienced a loss of $20,000 last year.

Objective 2: Rational and Irrational Numbers

For Exercises 13–24, plot the numbers on the number line. **(See Example 2.)**

13. $-2, 4$

14. $3, -1$

15. $2, -3$

16. $-1, 5$

17. $-\frac{7}{2}, \frac{17}{4}$

18. $\frac{4}{3}, -\frac{2}{3}$

19. $\frac{11}{4}, -\frac{7}{8}$

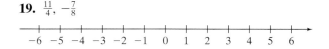

20. $-\frac{1}{4}, \frac{27}{5}$

21. $4.1, -0.8$

22. $0, -3.1$

23. −2.5, 1.6

24. −1.9, 4.2

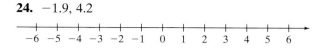

For Exercises 25–36, identify the number as rational or irrational.

25. $-\dfrac{2}{5}$

26. $-\dfrac{1}{9}$

27. 5

28. 3

29. 3.5

30. 1.1

31. $\sqrt{7}$

32. $\sqrt{11}$

33. π

34. 2π

35. $-\sqrt{4}$

36. $-\sqrt{9}$

Objective 3: Real Numbers and the Real Number Line

For Exercises 37–52, place the correct symbol, $>$ or $<$, between the two numbers. **(See Example 3.)**

37. 0 $\square$ −3

38. −1 $\square$ 0

39. −8 $\square$ −9

40. −5 $\square$ −2

41. −9.1 $\square$ 2.2

42. −1.5 $\square$ 1.5

43. −3.35 $\square$ −3.3

44. 0.9 $\square$ −0.5

45. $-\dfrac{2}{3}$ $\square$ $-\dfrac{5}{6}$

46. $-\dfrac{1}{5}$ $\square$ $-\dfrac{1}{4}$

47. $\dfrac{7}{8}$ $\square$ $-\dfrac{1}{9}$

48. $\dfrac{1}{3}$ $\square$ $-\dfrac{3}{2}$

49. 0 $\square$ $\dfrac{1}{10}$

50. $-\dfrac{8}{7}$ $\square$ 1

51. $-\dfrac{6}{5}$ $\square$ −1

52. −1 $\square$ $-\dfrac{10}{11}$

Objectives 4: Absolute Value

For Exercises 53–64, determine the absolute value. **(See Example 4.)**

53. $|-2|$

54. $|-6|$

55. $|4.5|$

56. $|2.9|$

57. $\left|-\dfrac{5}{2}\right|$

58. $\left|-\dfrac{4}{9}\right|$

59. $|0|$

60. $|6|$

61. $|-3.2|$

62. $|-0.4|$

63. $|21|$

64. $|8|$

65. a. Which is greater, −12 or −8?

   b. Which is greater, $|-12|$ or $|-8|$?

66. a. Which is greater, −14 or −20?

   b. Which is greater, $|-14|$ or $|-20|$?

67. a. Which is greater, 5.2 or 7.8?

    b. Which is greater, $|5.2|$ or $|7.8|$?

69. Which is greater, -5 or $|-5|$?

71. Which is greater, 10 or $|10|$?

68. a. Which is greater, 3.89 or 4.29?

    b. Which is greater, $|3.89|$ or $|4.29|$?

70. Which is greater, -9 or $|-9|$?

72. Which is greater, 256, or $|256|$?

Objective 5: Opposite

For Exercises 73–84, find the opposite. **(See Example 5.)**

73. 5

74. 31

75. -12

76. -25

77. $-\dfrac{1}{6}$

78. $-\dfrac{4}{7}$

79. $\dfrac{2}{11}$

80. $\dfrac{14}{15}$

81. 8.1

82. 9.5

83. -1.14

84. -2.25

For Exercises 85–94, write in symbols, do not simplify.

85. The opposite of 6

86. The opposite of 23

87. The opposite of negative 2

88. The opposite of negative 9

89. The absolute value of 7

90. The absolute value of 11

91. The absolute value of negative 3

92. The absolute value of negative 10

93. The opposite of the absolute value of 14

94. The opposite of the absolute value of 42

For Exercises 95–106, simplify the expression.

95. $-|2|$

96. $-|9|$

97. $-|-5.3|$

98. $-|-6.9|$

99. $-(-15)$

100. $-(-4)$

101. $|-4.7|$

102. $|-9.5|$

103. $-\left|-\dfrac{12}{17}\right|$

104. $-\left|-\dfrac{1}{7}\right|$

105. $-\left(-\dfrac{3}{8}\right)$

106. $-\left(-\dfrac{4}{9}\right)$

section 10.2   Addition of Real Numbers

1. Addition of Integers by Using a Number Line

Addition of real numbers can be visualized on a number line. To do so, we locate the first addend on the number line. Then to add a positive number, we move to the right on the number line. To add a negative number, we move to the left on the number line. This is demonstrated in Example 1.

Objectives

1. Addition of Integers by Using a Number Line
2. Addition of Real Numbers
3. Translating English Phrases to Mathematical Expressions

---

example 1   Using a Number Line to Add Integers

Use a number line to add.

  a. $5 + 3$ **b.** $-5 + 3$ **c.** $5 + (-3)$ **d.** $-5 + (-3)$

Solution:

  a. $5 + 3 = 8$

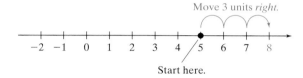

Begin at 5. Then because we are adding *positive* 3, move to the *right* 3 units. The sum is 8.

  b. $-5 + 3 = -2$

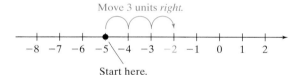

Begin at -5. Then because we are adding *positive* 3, move to the *right* 3 units. The sum is -2.

  c. $5 + (-3) = 2$

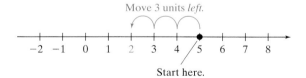

Begin at 5. Then because we are adding *negative* 3, move to the *left* 3 units. The sum is 2.

Skill Practice

Use a number line to add.

1. $3 + 2$
2. $-3 + 2$
3. $3 + (-2)$
4. $-3 + (-2)$

Answers

1. 5   2. -1 3. 1   4. -5

d. $-5 + (-3) = -8$

Move 3 units *left.*

Start here.

Begin at -5. Then since we are adding *negative* 3, move to the *left* 3 units. The sum is -8.

Tip: In Example 1(c) and 1(d), parentheses are inserted for clarity. The parentheses separate the number -3 from the $+$ symbol for addition.

$$5 + (-3) \quad \text{and} \quad -5 + (-3)$$

2. Addition of Real Numbers

It is inconvenient to draw a number line each time we want to add signed numbers. Therefore, we offer two rules for adding real numbers. The first rule is used when the addends have the *same* sign (that is, if the addends are both positive or both negative).

Adding Numbers with the Same Sign

To add two numbers with the same sign, add their absolute values and apply the common sign.

Skill Practice

Add.

5. $-6 + (-8)$

6. $-84 + (-27)$

7. $14 + 31$

example 2 Adding Real Numbers with the Same Sign

Add.

a. $-2 + (-4)$ **b.** $-12 + (-37)$ **c.** $10 + 66$

Solution:

a. $-2 + (-4)$ First find the absolute value of the addends.
                $|-2| = 2$ and $|-4| = 4$.

$= -(2 + 4)$ Add their absolute values and apply the common sign (in this case, the common sign is negative).

Common sign is negative.

$= -6$ The sum is -6.

b. $-12 + (-37)$ First find the absolute value of the addends.
                $|-12| = 12$ and $|-37| = 37$.

$= -(12 + 37)$ Add their absolute values and apply the common sign (in this case, the common sign is negative).

Common sign is negative.

$= -49$ The sum is -49.

Answers

5. -14 6. -111 7. 45

c.   10 + 66   First find the absolute value of the addends.
                         $|10| = 10$ and $|66| = 66$.

= +(10 + 66)   Add their absolute values and apply the common sign (in
                      this case, the common sign is positive).

Common sign is positive.

= 76   The sum is 76.

The next rule helps us add two numbers with different signs.

Adding Numbers with Different Signs

To add two numbers with different signs, subtract the smaller absolute
value from the larger absolute value. Then apply the sign of the number
having the larger absolute value.

Concept Connections

State the sign of the sum.

8. $-9 + 11$

9. $-9 + 7$

example 3   Adding Real Numbers with Different Signs

Add.

a. $2 + (-7)$ **b.** $-6 + 24$ **c.** $-8 + 8$

Skill Practice

Add.

10. $5 + (-8)$

11. $-12 + 37$

12. $-4 + 4$

13. $22 + (-5)$

14. $-88 + (45)$

Solution:

a. $2 + (-7)$ First find the absolute value of the addends.
                         $|2| = 2$ and $|-7| = 7$.

                         Note: The absolute value of -7 is greater than the
                         absolute value of 2. Therefore, the sum is negative.

= -(7 - 2)   Next, subtract the smaller absolute value from the larger
                         absolute value.

└─ Apply the sign of the number with the larger absolute value.

= -5

b. $-6 + 24$ First find the absolute value of the addends.
                         $|-6| = 6$ and $|24| = 24$.

                         Note: The absolute value of 24 is greater than the
                         absolute value of -6. Therefore, the sum is positive.

= +(24 - 6)   Next, subtract the smaller absolute value from the larger
                         absolute value.

└─ Apply the sign of the number with the larger absolute value.

= 18

c. $-8 + 8$ First find the absolute value of the addends.
                         $|-8| = 8$ and $|8| = 8$.

= (8 - 8)   The absolute values are equal. Therefore, their difference is 0.
                         The number zero is neither positive nor negative.

= 0

Example 3(c) illustrates that the sum of a number and its opposite is zero. For
example,

$$-8 + 8 = 0 \qquad -12 + 12 = 0 \qquad 6 + (-6) = 0$$

Answers

8. Positive   9. Negative
10. -3 11. 25   12. 0
13. 17   14. -43

Skill Practice

Simplify.

15. $-8 + 13 + (-18)$

16. $-24 + (-16) + 8 + (-20)$

---

example 4 Applying the Order of Operations

Simplify.

a. $-10 + 4 + (-16)$ **b.** $-30 + (-12) + 4 + (-10) + 6$

Solution:

a. $\underbrace{-10 + 4} + (-16)$ Apply the order of operations by adding from left to right.

$= \underbrace{-6 + (-16)}$

$= -22$

b. $\underbrace{-30 + (-12)} + 4 + (-10) + 6$ Add from left to right.

$= \underbrace{-42 + 4} + (-10) + 6$

$= \underbrace{-38 + (-10)} + 6$

$= \underbrace{-48 + 6}$

$= -42$

Tip: When a string of numbers is added, we can reorder and regroup the addends by using the commutative property and associative property of addition. In particular, we can group all the positive addends, and we can group all the negative addends. This makes the arithmetic easier. For example,

$$-30 + (-12) + 4 + (-10) + 6 = \overbrace{4 + 6}^{\substack{\text{positive} \\ \text{addends}}} + \overbrace{(-30) + (-12) + (-10)}^{\substack{\text{negative} \\ \text{addends}}}$$

$$= 10 + (-52)$$

$$= -42$$

Skill Practice

Add.

17. $-16.8 + 14.3$

18. $-\dfrac{6}{8} + \dfrac{5}{8}$

19. $\dfrac{3}{10} + \left(-\dfrac{3}{5}\right)$

20. $-\dfrac{4}{9} + \left(-\dfrac{5}{6}\right)$

---

example 5 Adding Real Numbers

Add.

a. $-2.73 + 4.81$ **b.** $-\dfrac{6}{11} + \left(-\dfrac{3}{11}\right)$

c. $-\dfrac{4}{3} + \left(-\dfrac{6}{7}\right)$ **d.** $\dfrac{2}{15} + \left(-\dfrac{4}{5}\right)$

Solution:

a. $-2.73 + 4.81$ Find the absolute value of the addends. $|-2.73| = 2.73$ and $|4.81| = 4.81$.

The sum is positive because 4.81 has a greater absolute value than -2.73.

$= +(4.81 - 2.73)$ Subtract the smaller absolute value from the larger absolute value.

└─ Apply the sign of the number with the larger absolute value.

$= 2.08$

Answers

15. -13 16. -52

17. -2.5 18. $-\dfrac{1}{8}$

19. $-\dfrac{3}{10}$ 20. $-\dfrac{23}{18}$

b. $-\dfrac{6}{11} + \left(-\dfrac{3}{11}\right)$ Find the absolute value of the addends.

$$\left|-\dfrac{6}{11}\right| = \dfrac{6}{11} \text{ and } \left|-\dfrac{3}{11}\right| = \dfrac{3}{11}.$$

$= -\left(\dfrac{6}{11} + \dfrac{3}{11}\right)$ Add their absolute values and apply the common sign (in this case, the common sign is negative).

↑ Common sign is negative.

$= -\dfrac{9}{11}$

c. $-\dfrac{4}{3} + \left(-\dfrac{6}{7}\right)$ The least common denominator (LCD) is 21.

$= -\dfrac{4 \cdot 7}{3 \cdot 7} + \left(-\dfrac{6 \cdot 3}{7 \cdot 3}\right)$ Write each fraction with the LCD.

$= -\dfrac{28}{21} + \left(-\dfrac{18}{21}\right)$ Find the absolute value of the addends.

$$\left|-\dfrac{28}{21}\right| = \dfrac{28}{21} \text{ and } \left|-\dfrac{18}{21}\right| = \dfrac{18}{21}.$$

$= -\left(\dfrac{28}{21} + \dfrac{18}{21}\right)$ Add their absolute values and apply the common sign (in this case the common sign is negative).

↑ Common sign is negative.

$= -\dfrac{46}{21}$ The sum is $-\dfrac{46}{21}$.

d. $\dfrac{2}{15} + \left(-\dfrac{4}{5}\right)$ The least common denominator is 15.

$= \dfrac{2}{15} + \left(-\dfrac{4 \cdot 3}{5 \cdot 3}\right)$ Write each fraction with the LCD.

$= \dfrac{2}{15} + \left(-\dfrac{12}{15}\right)$ Find the absolute value of the addends.

$$\left|\dfrac{2}{15}\right| = \dfrac{2}{15} \text{ and } \left|-\dfrac{12}{15}\right| = \dfrac{12}{15}.$$

The absolute value of $-\frac{12}{15}$ is greater than the absolute value of $\frac{2}{15}$. Therefore, the sum is negative.

$= -\left(\dfrac{12}{15} - \dfrac{2}{15}\right)$ Next, subtract the smaller absolute value from the larger absolute value.

↑└ Apply the sign of the number with the larger absolute value.

$= -\dfrac{10}{15}$ Subtract.

$= -\dfrac{2}{3}$ Simplify to lowest terms. $-\dfrac{\overset{2}{\cancel{10}}}{\underset{3}{\cancel{15}}} = -\dfrac{2}{3}$

3. Translating English Phrases to Mathematical Expressions

Recall from Section 1.2 that several key words imply addition: *sum; added to; increased by; more than; plus;* and *total of.*

example 6   Translating to a Mathematical Expression

Translate each phrase to a mathematical expression. Then simplify the expression.

a. The sum of -14.1, 8.7, and 12.9

b. The number $-\dfrac{1}{6}$ added to $\dfrac{2}{3}$

Solution:

a. $\underbrace{-14.1 + 8.7} + 12.9$ Translate the English phrase.

$= \underbrace{-5.4 + 12.9}$ Simplify. Add from left to right.

$= \qquad 7.5$

b. $\dfrac{2}{3} + \left(-\dfrac{1}{6}\right)$ Translate the English phrase.

$= \dfrac{2 \cdot 2}{3 \cdot 2} + \left(-\dfrac{1}{6}\right)$ Write the first fraction with a least common denominator of 6.

$= \dfrac{4}{6} + \left(-\dfrac{1}{6}\right)$ Find the absolute value of the addends.

$\left|\dfrac{4}{6}\right| = \dfrac{4}{6}$ and $\left|-\dfrac{1}{6}\right| = \dfrac{1}{6}$.

The absolute value of $\dfrac{4}{6}$ is greater than the absolute value of $-\dfrac{1}{6}$. Therefore, the sum is positive.

$= +\left(\dfrac{4}{6} - \dfrac{1}{6}\right)$ Subtract the smaller absolute value from the larger absolute value.

⎿ Apply the sign of the number with the larger absolute value.

$= \dfrac{3}{6}$ Subtract.

$= \dfrac{1}{2}$ Simplify to lowest terms.

section 10.2   Practice Exercises

Study Skills Exercise

1. When you are trying to memorize rules for signed numbers, 3×5 cards come in handy. Write the rule you want to memorize on one side and an example of the rule on the other side. Keep these cards handy so that when you have a few minutes (such as waiting for the doctor), you can pull them out and quiz yourself. Write the rules from this section that would be good to put on cards.

Review Exercises

For Exercises 2–8, place the correct symbol ($>$, $<$, or $=$) between the two numbers.

2. $-6 \ \square \ -5$ **3.** $-\dfrac{2}{3} \ \square \ -\dfrac{11}{12}$ **4.** $|-2.4| \ \square \ -|2.4|$ **5.** $|6| \ \square \ |-6|$

6. $0 \ \square \ -0.6$ **7.** $-|-10| \ \square \ 10$ **8.** $-(-2) \ \square \ 2$

Objective 1: Addition of Integers by Using a Number Line

For Exercises 9–20, refer to the number line to add the integers. **(See Example 1.)**

9. $2 + (-4)$ **10.** $5 + (-1)$ **11.** $-3 + 5$ **12.** $-6 + 3$

13. $-4 + (-4)$ **14.** $-2 + (-5)$ **15.** $-3 + 9$ **16.** $-1 + 5$

17. $0 + (-7)$ **18.** $(-5) + 0$ **19.** $-1 + (-3)$ **20.** $-4 + 3$

Objective 2: Addition of Real Numbers

21. Explain the process to add two numbers with the same sign.

For Exercises 22–29, add the numbers with the same sign. **(See Example 2.)**

22. $23 + 12$ **23.** $12 + 3$ **24.** $-70 + (-15)$ **25.** $-40 + (-33)$

26. $-6 + (-10)$ **27.** $-100 + (-24)$ **28.** $23 + 50$ **29.** $44 + 45$

30. Explain the process to add two numbers with different signs.

For Exercises 31–42, add the numbers with different signs. **(See Example 3.)**

31. $75 + (-23)$

32. $26 + (-14)$

33. $-34 + 12$

34. $-88 + 35$

35. $-90 + 66$

36. $-23 + 49$

37. $78 + (-33)$

38. $10 + (-23)$

39. $2 + (-2)$

40. $-6 + 6$

41. $-1.3 + 1.3$

42. $4.5 + (-4.5)$

For Exercises 43–65, simplify. **(See Example 4.)**

43. $12 + (-3)$

44. $-33 + (-1)$

45. $-23 + (-3)$

46. $-5 + 15$

47. $4 + (-45)$

48. $-13 + (-12)$

49. $(-103) + (-47)$

50. $119 + (-59)$

51. $0 + (-17)$

52. $-29 + 0$

53. $-19 + (-22)$

54. $-300 + (-24)$

55. $6 + (-12) + 8$

56. $20 + (-12) + (-5)$

57. $-33 + (-15) + 18$

58. $3 + 5 + (-1)$

59. $7 + (-3) + 6$

60. $12 + (-6) + (-9)$

61. $-10 + (-3) + 5$

62. $-23 + (-4) + (-12) + (-5)$

63. $14 + (-15) + 20 + (-42)$

64. $4 + (-12) + (-30) + 16 + 10$

65. $32 + (-10) + 11 + 13 + (-22)$

For Exercises 66–81, add the real numbers. **(See Example 5.)**

66. $23.9 + 2.1$

67. $10.9 + 6.3$

68. $-34.2 + (-4.1)$

69. $-8.6 + (-12)$

70. $-\dfrac{3}{4} + \left(-\dfrac{5}{4}\right)$

71. $-\dfrac{2}{5} + \left(-\dfrac{1}{10}\right)$

72. $-\dfrac{7}{8} + \left(-\dfrac{1}{4}\right)$

73. $-\dfrac{1}{2} + \left(-\dfrac{7}{12}\right)$

74. $34.8 + (-45)$

75. $90 + (-12.3)$

76. $-23.1 + 24.5$

77. $-12.2 + 10.9$

78. $\dfrac{3}{8} + \left(-\dfrac{3}{16}\right)$ **79.** $\dfrac{1}{3} + \left(-\dfrac{7}{9}\right)$ **80.** $\left(-\dfrac{5}{6}\right) + \dfrac{1}{4}$

81. $\left(-\dfrac{4}{5}\right) + \dfrac{7}{20}$

Objective 3: Translating English Phrases to Mathematical Expressions

82. Name at least two words or phrases that would indicate addition.

For Exercises 83–92, translate to a mathematical expression. Then simplify the expression. **(See Example 6.)**

83. The sum of -23 and 49 **84.** The sum of 89 and -11

85. The total of 3, -10, and 5 **86.** The total of -2, -4, 14, and 20

87. The number -4.2 is added to -2.2 **88.** The number -4.5 is added to -12

89. 8 more than $-\dfrac{1}{4}$ **90.** 2 more than $-\dfrac{1}{3}$

91. $-\dfrac{3}{4}$ increased by 6 **92.** $-\dfrac{1}{5}$ increased by 1

93. At 6:00 A.M. the temperature was $-4°F$. By noon, the temperature had risen by $12°F$. What was the temperature at noon?

94. At noon the temperature was $14°F$. By midnight, the temperature had fallen $10°F$. What was the temperature at midnight?

95. Jorge's checking account is overdrawn. His beginning balance was $-\$56.52$. If he deposits his paycheck for $\$389.81$, what is his new balance?

96. Ellen's checking account balance is $\$23.89$. If she writes a check for $\$40.00$, what will be her balance?

97. Ron bought a new pair of glasses and a bill for $\$320.50$ was sent to his insurance company. If his insurance paid only $\$150$, what is Ron's balance?

98. Savannah was in a minor accident with her car. To repair the damaged car, she paid $\$570.32$. Her insurance company paid for the repair except for a $\$250$ deductible. How much did the insurance company pay?

Expanding Your Skills

99. Find two integers whose sum is -10. Answers may vary.

100. Find two integers whose sum is -14. Answers may vary.

101. Find two integers whose sum is -2. Answers may vary.

102. Find two integers whose sum is 0. Answers may vary.

Calculator Connections

Topic: Adding real numbers on a calculator

To enter negative numbers on a calculator, use the $\boxed{(-)}$ key or the $\boxed{+/-}$ key. To use the $\boxed{(-)}$ key, enter the number the same way that it is written. That is, enter the negative sign first and then the number, such as $\boxed{(-)}$ 5. If your calculator has the $\boxed{+/-}$ key, type the number first, followed by the $\boxed{+/-}$ key. Thus, -5 is entered as 5 $\boxed{+/-}$.

Try entering the expressions below to determine which method your calculator uses.

Expression	Keystrokes	Result
$-10 + (-3)$	$\boxed{(-)}$ 10 $\boxed{+}$ $\boxed{(-)}$ 3 $\boxed{\text{Enter}}$	-13
	or 10 $\boxed{+/-}$ $\boxed{+}$ 3 $\boxed{+/-}$ $\boxed{=}$	-13
$-4.2 + 6.7$	$\boxed{(-)}$ 4.2 $\boxed{+}$ 6.7 $\boxed{\text{Enter}}$	2.5
	or 4.2 $\boxed{+/-}$ $\boxed{+}$ 6.7 $\boxed{=}$	2.5

Calculator Exercises

For Exercises 103–108, add by using a calculator.

103. $302 + (-422)$

104. $-900 + 334$

105. $-23.991 + (-44.23)$

106. $-103.4 + (-229.1)$

107. $23 + (-125) + 912 + (-99)$

108. $891 + 12 + (-223) + (-341)$

section 10.3   Subtraction of Real Numbers

Objectives
1. Definition of Subtraction
2. Subtraction of Rational Numbers
3. Applying the Order of Operations
4. Applications of Subtraction

1. Definition of Subtraction

In Section 10.2, we learned the rules for adding real numbers. Subtraction of real numbers is defined in terms of the addition process. For example, consider the following subtraction problem. The corresponding addition problem produces the same result.

$$6 - 4 = 2 \quad \Leftrightarrow \quad 6 + (-4) = 2$$

In each case, we start at 6 on the number line and move to the *left* 4 units. That is, adding the *opposite* of 4 produces the same result as subtracting 4. This is true in general. To subtract two real numbers, add the opposite of the second number to the first number.

Subtraction of Real Numbers

If a and b are real numbers, then $a - b = a + (-b)$.
Therefore, to perform subtraction, follow these steps:

1. Leave the first number (the minuend) unchanged.
2. Change the subtraction sign to an addition sign.
3. Add the opposite of the second number (the subtrahend).

Concept Connections

Fill in the blank to change subtraction to addition of the opposite.

1. $9 - 3 = 9 + \boxed{}$
2. $-9 - 3 = -9 + \boxed{}$
3. $9 - (-3) = 9 + \boxed{}$
4. $-9 - (-3) = -9 + \boxed{}$

For example,

$$\left.\begin{array}{l} 10 - 4 = 10 + (-4) = 6 \\ -10 - 4 = -10 + (-4) = -14 \end{array}\right\} \text{ Subtracting 4 is the same as adding } -4.$$

$$\left.\begin{array}{l} 10 - (-4) = 10 + (4) = 14 \\ -10 - (-4) = -10 + (4) = -6 \end{array}\right\} \text{ Subtracting } -4 \text{ is the same as adding 4.}$$

example 1   Subtracting Real Numbers

Subtract.

a. $15 - 20$ **b.** $-7 - 12$ **c.** $40 - (-8)$

Solution:

Add the opposite of 20.

a. $15 - 20 = 15 + (-20) = -5$ Rewrite the subtraction in terms of addition.

Change subtraction to addition.

Skill Practice

Subtract.

5. $12 - 19$
6. $-8 - 14$
7. $30 - (-3)$

Answers

1. -3 2. -3 3. 3 4. 3
5. -7 6. -22 7. 33

b. $-7 - 12 = -7 + (-12)$ Rewrite the subtraction in terms of addition.

$ = -19$ Add.

c. $40 - (-8) = 40 + (8)$ Rewrite the subtraction in terms of addition.

$ = 48$ Add.

Recall from Section 1.3 that several key words imply subtraction.

Word or Phrase	Example	In Symbols
a minus b	-15 minus 10	$-15 - 10$
the difference of a and b	the difference of 10 and -2	$10 - (-2)$
a decreased by b	9 decreased by 1	$9 - 1$
a less than b	-12 less than 5	$5 - (-12)$
subtract a from b	subtract -3 from 8	$8 - (-3)$

Skill Practice

Translate to a mathematical expression. Then simplify.

8. The difference of -16 and 4

9. -8 decreased by -9

10. 6 less than 2

example 2   Translating to an Algebraic Expression

Translate each English phrase to a mathematical expression. Then simplify.

a. The difference of -52 and 10

b. -35 decreased by -6

c. 12 less than -8

Solution:

a.   the difference of

$-52 - 10$ Translate: The difference of -52 and 10

$= -52 + (-10)$ Rewrite subtraction in terms of addition.

$= -62$ Add.

b.   decreased by

$-35 - (-6)$ Translate: -35 decreased by -6.

$= -35 + (6)$ Rewrite subtraction in terms of addition.

$= -29$ Add.

c. To translate "12 less than -8," we must *start* with -8 before subtracting 12.

$-8 - 12$ Translate: 12 less than -8.

$= -8 + (-12)$ Rewrite subtraction in terms of addition.

$= -20$ Add.

Answers

8. $-16 - 4$; -20
9. $-8 - (-9)$; 1
10. $2 - 6$; -4

2. Subtraction of Rational Numbers

example 3	Subtracting Rational Numbers

Subtract.

a. $-6.7 - 4.2$ **b.** $\dfrac{9}{4} - \dfrac{7}{4}$ **c.** $\dfrac{3}{10} - \left(-\dfrac{7}{30}\right)$

Solution:

a. $-6.7 - 4.2$

$= -6.7 + (-4.2)$ Rewrite subtraction in terms of addition.

$= -10.9$ Add.

b. $\dfrac{9}{4} - \dfrac{7}{4}$

$= \dfrac{9}{4} + \left(-\dfrac{7}{4}\right)$ Rewrite subtraction in terms of addition.

$= \dfrac{2}{4}$ Add.

$= \dfrac{1}{2}$ Simplify.

c. $\dfrac{3}{10} - \left(-\dfrac{7}{30}\right)$

$= \dfrac{3}{10} + \left(\dfrac{7}{30}\right)$ Rewrite subtraction in terms of addition.

$= \dfrac{3 \cdot 3}{10 \cdot 3} + \left(\dfrac{7}{30}\right)$ The least common denominator is 30.

$= \dfrac{9}{30} + \dfrac{7}{30}$

$= \dfrac{16}{30}$ Add.

$= \dfrac{8}{15}$ Simplify to lowest terms. $\dfrac{\overset{8}{\cancel{16}}}{\underset{15}{\cancel{30}}} = \dfrac{8}{15}$

Skill Practice

Subtract.

11. $-4.6 - 7.1$

12. $-\dfrac{13}{15} + \dfrac{7}{15}$

13. $\dfrac{5}{9} - \left(-\dfrac{5}{6}\right)$

3. Applying the Order of Operations

In Example 4, we perform the order of operations.

example 4	Applying the Order of Operations

Simplify.

a. $-4 - 6 + (-3) - 5 + 8$ **b.** $[3 - (-4)]^2$

Answers

11. -11.7 **12.** $-\dfrac{2}{5}$ **13.** $\dfrac{25}{18}$

Solution:

a. $-4 - 6 + (-3) - 5 + 8$

$= \underbrace{-4 + (-6)} + (-3) + (-5) + 8$ Rewrite all subtractions in terms of addition.

$= \underbrace{-10 + (-3)} + (-5) + 8$ Add from left to right.

$= \underbrace{-13 + (-5)} + 8$

$= \underbrace{-18 + 8}$

$= -10$

b. $[3 - (-4)]^2$ Perform operations inside grouping symbols.

$= [3 + (4)]^2$ Rewrite subtraction in terms of addition.

$= [7]^2$ Add.

$= 49$ Simplify.

4. Applications of Subtraction

example 5   Applying Subtraction of Real Numbers

A helicopter is hovering at a height of 200 ft above the ocean. A submarine is directly below the helicopter 125 ft below sea level. Find the difference in elevation between the helicopter and the submarine.

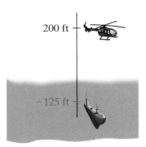

Solution:

$$\begin{pmatrix} \text{Difference between} \\ \text{elevation of helicopter} \\ \text{and submarine} \end{pmatrix} = \begin{pmatrix} \text{Elevation of} \\ \text{helicopter} \end{pmatrix} - \begin{pmatrix} \text{"Elevation" of} \\ \text{submarine} \end{pmatrix}$$

$= 200 \text{ ft} - (-125 \text{ ft})$

$= 200 \text{ ft} + (125 \text{ ft})$ Rewrite as addition.

$= 325 \text{ ft}$

The helicopter and submarine are 325 ft apart.

section 10.3   Practice Exercises

Study Skills Exercise

1. In this section we find that the symbol − has more than one meaning. It can mean minus, opposite, or negative. Write yourself an explanation and example for each of these meanings of −.

Minus: Opposite: Negative:

Review Exercises

For Exercises 2–9, add the numbers.

2. $34 + (-13)$

3. $-34 + (-13)$

4. $-34 + 13$

5. $-\dfrac{5}{9} + \dfrac{7}{12}$

6. $\dfrac{5}{9} + \left(-\dfrac{7}{12}\right)$

7. $-\dfrac{5}{9} + \left(-\dfrac{7}{12}\right)$

8. $3 + (-4.2) + \left(-\dfrac{2}{5}\right) + 5$

9. $\left(-\dfrac{1}{2}\right) + 6.5 + (-8) + 2 + (-4)$

Objective 1: Definition of Subtraction

10. Explain the process to subtract signed numbers.

For Exercises 11–18, rewrite the subtraction problem as an equivalent addition problem. Then simplify.
(See Example 1.)

11. $2 - 9 =$ _____ $=$ ____

12. $5 - 11 =$ _____ $=$ ____

13. $4 - (-3) =$ _____ $=$ ____

14. $12 - (-8) =$ _____ $=$ ____

15. $-3 - 15 =$ _____ $=$ ____

16. $-7 - 21 =$ _____ $=$ ____

17. $-11 - (-13) =$ _____ $=$ ____

18. $-23 - (-9) =$ _____ $=$ ____

For Exercises 19–42, subtract the integers.

19. $35 - (-17)$

20. $23 - (-12)$

21. $-24 - 9$

22. $-5 - 15$

23. $50 - 62$

24. $38 - 46$

25. $-17 - (-25)$

26. $-2 - (-66)$

27. $-8 - (-8)$

28. $-14 - (-14)$

29. $120 - (-41)$

30. $91 - (-62)$

31. $-15 - 19$

32. $-82 - 44$

33. $3 - 25$

34. $6 - 33$

35. $-13 - 13$

36. $-43 - 43$

37. $24 - 25$

38. $43 - 98$

39. $-6 - (-38)$

40. $-75 - (-21)$

41. $-48 - (-33)$

42. $-29 - (-32)$

43. State at least two words or phrases that would indicate subtraction.

44. Is subtraction commutative? For example, does $3 - 7 = 7 - 3$?

For Exercises 45–56, translate each English phrase to a mathematical expression. Then simplify. **(See Example 2.)**

45. 14 minus 23

46. 27 minus 40

47. Subtract 12 from 5.

48. Subtract 10 from 16.

49. The difference of 105 and 110

50. The difference of 70 and 98

51. 320 decreased by -20

52. 150 decreased by 75

53. Subtract 24 from -35.

54. Subtract 189 from 175.

55. 21 less than -34

56. 22 less than -90

Objective 2: Subtraction of Rational Numbers

For Exercises 57–72, subtract the rational numbers. **(See Example 3.)**

57. $5.2 - 13.5$

58. $4.4 - 10.2$

59. $-2.3 - 1.9$

60. $-4.1 - 2.1$

61. $-3.6 - (-9.1)$

62. $-8.9 - (-10.5)$

63. $5.5 - (-2.8)$

64. $11.9 - (-4.3)$

65. $\dfrac{2}{3} - \left(-\dfrac{1}{6}\right)$

66. $\dfrac{5}{9} - \left(-\dfrac{2}{9}\right)$

67. $-\dfrac{3}{10} - \left(-\dfrac{7}{10}\right)$

68. $-\dfrac{5}{8} - \left(-\dfrac{3}{4}\right)$

69. $\dfrac{3}{14} - \dfrac{12}{7}$

70. $\dfrac{2}{5} - \dfrac{9}{10}$

71. $-\dfrac{1}{2} - \dfrac{5}{4}$

72. $-\dfrac{11}{15} - \dfrac{3}{5}$

Objective 3: Applying the Order of Operations

For Exercises 73–84, simplify by using the order of operations. **(See Example 4.)**

73. $2 + 5 - (-3) - 10$

74. $4 - 8 + 12 - (-1)$

75. $-5 + 6 + (-7) - 4 - (-9)$

76. $-2 - 1 + (-11) + 6 - (-8)$

77. $[-2 - (-6)]^2$

78. $[-1 - (-4)]^2$

79. $[-5 - (-6)]^3$

80. $[-3 - (-5)]^3$

81. $25 - 13 - (-40)$

82. $-35 + 15 - (-28)$

83. $5.5 - \left(\dfrac{1}{2} - \dfrac{1}{5}\right)$

84. $-6.8 - \left(\dfrac{2}{5} + \dfrac{3}{10}\right)$

Objective 4: Applications of Subtraction

85. The moon does not have an atmosphere, so temperatures range from $-184°C$ during its night to $214°C$ during its day. Find the difference between the highest temperature on the moon and the lowest temperature. **(See Example 5.)**

86. The temperature at 3:00 P.M. in Chicago was $14°F$. A cold front moved through and by 9:00 P.M. the temperature had dropped by $26°F$. What was the temperature at 9:00 P.M.?

87. In the television show *Jeopardy*, a contestant answers some questions correctly and others incorrectly. The contestant receives the following dollar amounts: $200, $-$600$, $400, $200, $1000, $-$800$. How much did the contestant win overall?

88. For nine holes of golf, Tiger Woods made the following scores: $-1, 0, 0, -1, 0, 1, 0, -2, 0$. What is his total score after nine holes?

For Exercises 89–90, refer to the graph indicating the change in value of a particular stock for one week.

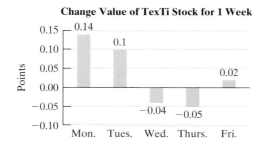

Change Value of TexTi Stock for 1 Week

89. What is the difference between the changes in value of TexTi stock on Monday compared to its value on Wednesday?

90. What is the difference between the changes in value of TexTi stock on Tuesday and its value on Thursday?

91. Ivan owes $320 on his credit card; that is, his balance is $-\$320$. If he charges $55 for a night out, what is his new balance?

92. If Justin's balance on his credit card was $-\$210$ and he made the minimum payment of $25, what is his new balance?

In a statistics class, a student learns that the **range** of a set of data is the difference between the highest and lowest values. That is, range = highest − lowest.

For Exercises 93–94, find the range.

93. Low temperatures (°C) for one week in Anchorage, Alaska: $-4°, -8°, 0°, 3°, -8°, -1°, 2°$

94. Low temperatures (°C) for one week in Fargo, North Dakota: $-6°, -2°, -10°, -4°, -12°, -1°, -3°$

95. Find two integers whose difference is -6. Answers may vary.

96. Find two integers whose difference is -20. Answers may vary.

For Exercises 97–100, write the next three numbers in the sequence.

97. $5, 1, -3, -7,$ ___, ___, ___

98. $-13, -18, -23, -28,$ ___, ___, ___

99. $\dfrac{1}{3}, 0, -\dfrac{1}{3}, -\dfrac{2}{3},$ ___, ___, ___

100. $\dfrac{1}{4}, 0, -\dfrac{1}{4}, -\dfrac{1}{2},$ ___, ___, ___

Expanding Your Skills

For Exercises 101–108, assume $a > 0$ (this means that a is positive) and $b < 0$ (this means that b is negative). Find the sign of each expression.

101. $a - b$

102. $b - a$

103. $|a| + |b|$

104. $|a + b|$

105. $-|a|$

106. $-|b|$

107. $-(a)$

108. $-(b)$

Calculator Connections

Topic: Subtracting real numbers on a calculator

The $\boxed{-}$ is used for subtraction. This should not be confused with the $\boxed{(-)}$ key or $\boxed{+/-}$ key, which is used to enter a negative number.

Expression	Keystrokes	Result
$-7 - 4$	$\boxed{(-)}\ 7\ \boxed{-}\ 4\ \boxed{\text{Enter}}$	-11
	or $7\ \boxed{+/-}\ \boxed{-}\ 4\ \boxed{=}$	-11
$-4.2 - (-6.8)$	$\boxed{(-)}\ 4.2\ \boxed{-}\ \boxed{(-)}\ 6.8\ \boxed{\text{Enter}}$	2.6
	or $4.2\ \boxed{+/-}\ \boxed{-}\ 6.8\ \boxed{+/-}\ \boxed{=}$	2.6

Calculator Exercises

For Exercises 109–114, subtract the real numbers by using a calculator.

109. $-190 - 223$

110. $-288 - 145$

111. $-23.24 - (-90.01)$

112. $-14.93 - (-34.99)$

113. $89.2 - (-23.6)$

114. $104.9 - (-24.8)$

chapter 10 | midchapter review

For Exercises 1–8, add.

1. $5 + (-12)$

2. $8 + (-3)$

3. $-10 + (-4)$

4. $-6 + (-9)$

5. $13 + 4$

6. $2 + 16$

7. $-4 + 11$

8. $-15 + 10$

For Exercises 9–12, find the opposite of the number.

9. 9

10. 3

11. -6

12. -11

For Exercises 13–16, write subtraction as addition by adding the opposite of the second number.

13. $-3 - 9$

14. $-2 - 3$

15. $5 - (-6)$

16. $8 - (-11)$

For Exercises 17–28, subtract.

17. $5 - 12$

18. $8 - 3$

19. $-10 - 4$

20. $-6 - 9$

21. $13 - (-4)$

22. $2 - (-16)$

23. $-4 - (-11)$

24. $-15 - (-10)$

25. $2.01 - 3.99$

26. $-2.5 - (-7.1)$

27. $\left(\dfrac{1}{8}\right) - \left(-\dfrac{5}{4}\right)$

28. $-\dfrac{7}{9} - \dfrac{1}{6}$

Objectives

1. Multiplication of Real Numbers
2. Multiplying Many Factors
3. Exponential Expressions
4. Division of Real Numbers

Concept Connections

1. Write $4(-5)$ as repeated addition.

---

section 10.4   Multiplication and Division of Real Numbers

1. Multiplication of Real Numbers

We know from our knowledge of arithmetic that the product of two positive numbers is a positive number. This can be shown by using repeated addition.

For example: $3(4) = 4 + 4 + 4 = 12$

Now consider a product of numbers with different signs.

For example: $3(-4) = -4 + (-4) + (-4) = -12$ (3 times -4)

These examples suggest that the product of a positive number and a negative number is *negative*.

Now what if we have a product of two negative numbers? To determine the sign, consider the following pattern of products.

$$3 \times -4 = -12$$
$$2 \times -4 = -8$$
$$1 \times -4 = -4$$
$$0 \times -4 = 0$$
$$-1 \times -4 = 4$$
$$-2 \times -4 = 8$$
$$-3 \times -4 = 12$$

The pattern increases by 4 with each row.

The product of two negative numbers is *positive*.

From the first four rows, we see that the product increases by 4 for each row. For the pattern to continue, it follows that the product of two negative numbers must be *positive*.

Multiplication of Real Numbers

1. The product of two real numbers with the same sign is positive.
2. The product of two real numbers with different signs is negative.
3. The product of any real number and zero is zero.

Skill Practice

Multiply.

2. $-2(-6)$ 3. $3 \cdot (-10)$
4. $-14(3)$ 5. $(-12.6)(2.1)$
6. $-\dfrac{6}{5} \cdot \left(-\dfrac{25}{8}\right)$
7. $\dfrac{4}{9} \cdot (-15)$

Answers

1. $-5 + (-5) + (-5) + (-5)$
2. 12   3. -30 4. -42
5. -26.46 6. $\dfrac{15}{4}$ 7. $-\dfrac{20}{3}$

---

example 1   Multiplying Real Numbers

Multiply.

a. $-8(-7)$ **b.** $-5 \cdot 10$ **c.** $(18)(-2)$

d. $(-3.1)(-4.6)$ **e.** $-\dfrac{4}{7} \cdot \dfrac{35}{6}$ **f.** $-12 \cdot \left(-\dfrac{7}{3}\right)$

g. $-\dfrac{3}{4} \cdot 0$

Solution:

a. $-8(-7) = 56$ Same signs. Product is positive.

b. $-5 \cdot 10 = -50$ Different signs. Product is negative.

c. $(18)(-2) = -36$ Different signs. Product is negative.

d. $(-3.1)(-4.6) = 14.26$ Same signs. Product is positive.

e. $-\dfrac{4}{7} \cdot \dfrac{35}{6} = -\dfrac{\overset{2}{\cancel{4}}}{\underset{1}{7}} \cdot \dfrac{\overset{5}{\cancel{35}}}{\underset{3}{6}}$ Multiply fractions.

$\qquad = -\dfrac{10}{3}$ Different signs. Product is negative.

f. $-12 \cdot \left(-\dfrac{7}{3}\right) = -\dfrac{12}{1} \cdot \left(-\dfrac{7}{3}\right)$ Write the whole number as a fraction.

$\qquad = -\dfrac{\overset{4}{\cancel{12}}}{1} \cdot \left(-\dfrac{7}{\underset{1}{3}}\right)$

$\qquad = \dfrac{28}{1}$ Same signs. Product is positive.

$\qquad = 28$

g. $-\dfrac{3}{4} \cdot 0 = 0$ The product of any number and zero is zero.

Recall that the terms *product*, *multiply*, and *times* imply multiplication.

---

example 2   **Translating to an Algebraic Expression**

Translate each English phrase to an algebraic expression. Then simplify.

  a. The product of 7 and -8 **b.** -3 times -11

Solution:

  a. $7(-8)$ Translate: The product of 7 and -8.

$\ \ = -56$ Different signs. Product is negative.

  b. $(-3)(-11)$ Translate: -3 times -11.

$\ \ = 33$ Same signs. Product is positive.

2. Multiplying Many Factors

In each of the following products, we can apply the order of operations and multiply from left to right.

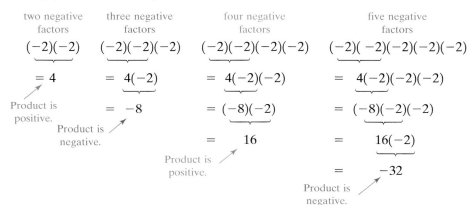

These products indicate the following rules.

- The product of an *even* number of negative factors is *positive*.
- The product of an *odd* number of negative factors is *negative*.

---

example 3 Multiplying Several Factors

Multiply.

a. $(-2)(-5)(-7)$ **b.** $(-4)(2)(-1)(5)$

Solution:

a. $(-2)(-5)(-7)$	This product has an odd number of negative factors.
$= -70$	The product is negative.
b. $(-4)(2)(-1)(5)$	This product has an even number of negative factors.
$= 40$	The product is positive.

---

3. Exponential Expressions

Be particularly careful when evaluating exponential expressions involving negative numbers. An exponential expression with a negative base is written with parentheses around the base, such as $(-3)^4$.

To evaluate $(-3)^4$, the base -3 is multiplied 4 times:

$$(-3)^4 = (-3)(-3)(-3)(-3) = 81$$

If parentheses are *not* used, the expression -3^4 has a different meaning:

- The expression -3^4 has a base of 3 (not -3) and can be interpreted as $-1 \cdot 3^4$. Hence,

$$-3^4 = -1 \cdot (3)(3)(3)(3) = -81$$

- The expression -3^4 can also be interpreted as "the opposite of 3^4." Hence,

$$-3^4 = -(3 \cdot 3 \cdot 3 \cdot 3) = -81$$

---

example 4 Simplifying Exponential Expressions

Simplify.

a. $(-4)^2$ **b.** -4^2 **c.** $(-5)^3$ **d.** -5^3 **e.** $\left(-\dfrac{1}{2}\right)^4$

Solution:

a. $(-4)^2 = (-4)(-4)$	The base is -4.
$= 16$	Multiply.
b. $-4^2 = -(4)(4)$	The base is 4. This is equal to $-1 \cdot 4^2 = -1 \cdot (4)(4)$.
$= -16$	
c. $(-5)^3 = (-5)(-5)(-5)$	The base is -5.
$= -125$	Multiply.

d. $-5^3 = -(5)(5)(5)$ The base is 5. This is equal to
$-1 \cdot 5^3 = -1 \cdot (5)(5)(5)$.

$\quad = -125$ Multiply.

e. $\left(-\dfrac{1}{2}\right)^4 = \left(-\dfrac{1}{2}\right)\left(-\dfrac{1}{2}\right)\left(-\dfrac{1}{2}\right)\left(-\dfrac{1}{2}\right)$ The base is $-\dfrac{1}{2}$.

$\quad = \dfrac{1}{16}$ Multiply.

4. Division of Real Numbers

Recall from Section 2.5 that two numbers are **reciprocals** if their product is 1. In particular, the reciprocal of a negative number must also be a negative number to form a product of 1.

Number	Reciprocal	Product
$\dfrac{2}{3}$	$\dfrac{3}{2}$	$\dfrac{2}{3} \cdot \dfrac{3}{2} = 1$
$-\dfrac{4}{7}$	$-\dfrac{7}{4}$	$-\dfrac{4}{7} \cdot \left(-\dfrac{7}{4}\right) = 1$
-5	$-\dfrac{1}{5}$	$-5 \cdot \left(-\dfrac{1}{5}\right) = 1$

Also recall from Section 2.5 that division can be expressed in terms of multiplication. To divide two real numbers, we multiply the first number (the dividend) by the reciprocal of the second number (the divisor).

$$12 \div (-2) = 12 \cdot \left(-\frac{1}{2}\right) = \frac{\overset{6}{\cancel{12}}}{1} \cdot \left(-\frac{1}{2}\right) = -6$$

multiply

by the reciprocal
of the divisor

Because division of real numbers can be expressed in terms of multiplication, the sign rules that apply to multiplication also apply to division.

> ### Division of Real Numbers
>
> 1. The quotient of two real numbers with the same sign is positive.
> 2. The quotient of two real numbers with different signs is negative.

---

example 5   Dividing Real Numbers

Divide.

a. $50 \div (-5)$ **b.** $\dfrac{-42}{-7}$ **c.** $\dfrac{-39}{3}$

Solution:

a. $50 \div (-5) = -10$ Different signs. The quotient is negative.

Concept Connections

23. If $a \cdot b = 1$, what do you know about a and b?

For each number, give the reciprocal, if possible.

24. $\dfrac{1}{7}$ **25.** $-\dfrac{3}{5}$

26. -6 **27.** 0

Skill Practice

Divide.

28. $-40 \div 10$

29. $\dfrac{-36}{-12}$

30. $\dfrac{18}{-2}$

Answers

23. They are reciprocals.   24. 7
25. $-\dfrac{5}{3}$ 26. $-\dfrac{1}{6}$
27. Not possible   28. -4
29. 3   30. -9

b. $\dfrac{-42}{-7} = 6$ Same signs. The quotient is positive.

c. $\dfrac{-39}{3} = -13$ Different signs. The quotient is negative.

<table><tr><td>

Skill Practice

Divide.

31. $15.96 \div (-3.8)$

32. $-\dfrac{3}{10} \div \left(-\dfrac{21}{2}\right)$

33. $\dfrac{-9}{-5}$

34. $-18 \div 21$

35. $0 \div -2.6$

36. $\dfrac{-9}{0}$

</td><td>

example 6   Dividing Real Numbers

Divide.

a. $-9.45 \div (-2.7)$ **b.** $-\dfrac{4}{15} \div \dfrac{8}{25}$ **c.** $\dfrac{-7}{-4}$

d. $15 \div (-25)$ **e.** $\dfrac{0}{-7}$ **f.** $-6.1 \div 0$

Solution:

a. $-9.45 \div (-2.7) = 3.5$ Same signs. The quotient is positive.

b. $-\dfrac{4}{15} \div \dfrac{8}{25} = -\dfrac{4}{15} \cdot \dfrac{25}{8}$ Multiply by the reciprocal of the divisor.

$= -\dfrac{\overset{1}{4}}{\underset{3}{15}} \cdot \dfrac{\overset{5}{25}}{\underset{2}{8}}$ Multiply fractions.

$= -\dfrac{5}{6}$ Different signs. The quotient is negative.

c. $\dfrac{-7}{-4} = \dfrac{7}{4}$ Same signs. The quotient is positive.

$\dfrac{7}{4}$ or $1\dfrac{3}{4}$ or 1.75 In this example, 7 and 4 share no common factors. Therefore, the fraction cannot be simplified further. However, the quotient may be written in several forms:

d. $15 \div (-25)$ The number 25 does not divide evenly into 15. However, the expression can be written as a fraction and then simplified to lowest terms.

$= \dfrac{15}{-25}$

$= \dfrac{\overset{3}{15}}{\underset{5}{-25}}$ Simplify to lowest terms.

$= -\dfrac{3}{5}$ Different signs. The quotient is negative.

e. $\dfrac{0}{-7} = 0$ Recall from Section 1.6 that 0 divided by any nonzero number is 0.

f. $-6.1 \div 0$ Recall from Section 1.6 that any number divided by zero is undefined.

</td></tr></table>

Answers
31. -4.2 32. $\dfrac{1}{35}$ 33. $\dfrac{9}{5}$
34. $-\dfrac{6}{7}$ 35. 0   36. Undefined

Practice Exercises

Study Skills Exercises

1. Often students learn a rule about signs that states, "Two negatives are a positive." This rule is incomplete and therefore not always true. Note the following combinations of two negatives:

$-2 + (-4)$ the sum of two negatives

$-(-5)$ the opposite of a negative

$-|-10|$ the opposite of an absolute value

$(-3)(-6)$ the product of two negatives

Determine which of the following are negative and which are positive. Then write the rule for multiplying two numbers with the same sign.

$-2 + (-4)$ _____

$-(-5)$ _____

$-|-10|$ _____

$(-3)(-6)$ _____

When multiplying two numbers with the same sign, the answer is _____.

2. Define the key term **reciprocal**.

Review Exercises

For Exercises 3–8, add or subtract as indicated.

3. $14 - (-5)$

4. $-24 - 50$

5. $-33 + (-11)$

6. $-7 - (-23)$

7. $23 - 12 + (-4) - (-10)$

8. $9 + (-12) - 17 - 4 - (-15)$

Objective 1: Multiplication of Real Numbers

For Exercises 9–34, multiply the real numbers. **(See Example 1.)**

9. $-3(5)$

10. $-2(13)$

11. $-12 \cdot 4$

12. $-6 \cdot 11$

13. $-15(3)$

14. $-3(25)$

15. $9(-8)$

16. $8(-3)$

17. $(-1.2)(-3.2)$

18. $(-3.3)(-2.5)$

19. $-6(0.4)$

20. $-8(1.3)$

21. $7(-1.1)$

22. $5(-3.4)$

23. $-14 \cdot 0$

24. $-8 \cdot 0$

25. $\left(-\dfrac{2}{3}\right)\left(-\dfrac{6}{7}\right)$

26. $\left(-\dfrac{8}{9}\right)\left(-\dfrac{3}{4}\right)$

27. $\dfrac{3}{5}\left(-\dfrac{5}{21}\right)$

28. $\dfrac{5}{12}\left(-\dfrac{4}{7}\right)$

29. $6 \cdot \left(-\dfrac{5}{12}\right)$

30. $4 \cdot \left(-\dfrac{1}{16}\right)$

31. $\left(-2\dfrac{3}{5}\right)\left(-1\dfrac{2}{3}\right)$

32. $\left(-3\dfrac{1}{3}\right)\left(-2\dfrac{1}{5}\right)$

33. $\left(-\dfrac{8}{9}\right) \cdot 0$

34. $\left(-\dfrac{2}{11}\right) \cdot 0$

For Exercises 35–40, translate the English phrase to an algebraic expression. Then simplify. **(See Example 2.)**

35. Multiply -3 and -1.

36. Multiply -12 and -4.

37. The product of -5 and 3

38. The product of 9 and -2

39. 1.3 times -3

40. -2.3 times 6

Objective 2: Multiplying Many Factors

For Exercises 41–50, multiply. **(See Example 3.)**

41. $(5)(-2)(4)(-10)$

42. $(-3)(-5)(-2)(4)$

43. $(-11)(-4)(-2)$

44. $(20)(-3)(-1)$

45. $(24)(-2)(0)(-3)$

46. $(3)(0)(-13)(22)$

47. $(-1)(-1)(-1)(-1)(-1)(-1)$

48. $(-1)(-1)(-1)(-1)(-1)(-1)(-1)$

49. $(-1)(-1)(-1)(-1)(-1)$

50. $(-1)(-1)(-1)(-1)$

Objective 3: Exponential Expressions

For Exercises 51–62, simplify. **(See Example 4.)**

51. -10^2

52. -8^2

53. $(-10)^2$

54. $(-8)^2$

55. -3^3

56. -4^3

57. $(-3)^3$

58. $(-4)^3$

59. -0.2^3

60. -0.4^3

61. $\left(-\dfrac{2}{3}\right)^3$

62. $\left(-\dfrac{3}{5}\right)^3$

Objective 4: Division of Real Numbers

For Exercises 63–86, divide the real numbers, if possible. **(See Examples 5–6.)**

63. $\dfrac{-15}{5}$

64. $\dfrac{30}{-6}$

65. $\dfrac{56}{-8}$

66. $\dfrac{-48}{3}$

67. $\dfrac{-25}{-15}$

68. $\dfrac{-6}{-18}$

69. $\dfrac{-14}{-21}$

70. $\dfrac{-81}{-72}$

71. $\dfrac{13}{0}$

72. $\dfrac{-41}{0}$

73. $\dfrac{0}{-2}$

74. $\dfrac{0}{5}$

75. $(-20) \div (-5)$

76. $(-10) \div (-2)$

77. $-0.91 \div -0.7$

78. $-1.3 \div -0.5$

79. $\left(-\dfrac{8}{7}\right) \div \left(-\dfrac{4}{5}\right)$

80. $\left(-\dfrac{2}{5}\right) \div \left(-\dfrac{8}{15}\right)$

81. $\left(-\dfrac{1}{6}\right) \div 0$

82. $\left(\dfrac{2}{11}\right) \div 0$

83. $\left(\dfrac{3}{4}\right) \div (-2)$

84. $\left(\dfrac{4}{7}\right) \div (-12)$

85. $204 \div (-6)$

86. $300 \div (-2)$

For Exercises 87–92, translate the English phrase into a mathematical expression. Then simplify.

87. The quotient of -100 and 20

88. The quotient of 46 and -23

89. -32 divided by -64

90. 108 divided by -24

91. 13 divided into -52

92. -15 divided into -45

Mixed Exercises

For Exercises 93–108, perform the indicated operation.

93. $8 + (-6)$

94. $8 - (-6)$

95. $8(-6)$

96. $8 \div (-6)$

97. $-9 - (-12)$

98. $(-9)(-12)$

99. $-36 \div (-12)$

100. $-36 + (-12)$

101. $(-5)(-4)$

102. $-90 \div (-6)$

103. $0 + (-15)$

104. $0 - (-15)$

105. $\dfrac{1}{3} \div \left(-\dfrac{5}{6}\right)$

106. $\dfrac{1}{3} + \left(-\dfrac{5}{6}\right)$

107. $\dfrac{1}{3} - \left(-\dfrac{5}{6}\right)$

108. $\dfrac{1}{3}\left(-\dfrac{5}{6}\right)$

Expanding Your Skills

109. Which is greater, $(-2)^{50}$ or $(-2)^{51}$?

110. Which is greater, $(-3)^{20}$ or $(-3)^{21}$?

111. Which is greater, $(5)^{40}$ or $(5)^{41}$?

112. Which is greater, $(6)^{10}$ or $(6)^{11}$?

For Exercises 113–118, assume $a > 0$ (this means that a is positive) and $b < 0$ (this means that b is negative). Find the sign of each expression.

113. $a \cdot b$ **114.** $b \div a$ **115.** $|a| \div b$ **116.** $a \cdot |b|$

117. $-a \div (b)$ **118.** $a(-b)$

Calculator Connections

Topic: Multiplying and dividing real numbers on a calculator

Calculator Exercises:

For Exercises 119–122, use a calculator to perform the indicated operations.

119. $(-413)(871)$ **120.** $-6125 \cdot (-97)$

121. $\dfrac{-576{,}828}{-10{,}682}$ **122.** $5{,}945{,}308 \div (-9452)$

Objectives

1. Order of Operations
2. Scientific Notation
3. Converting Decimal Notation to Scientific Notation
4. Converting Scientific Notation to Decimal Notation

section 10.5   Order of Operations and Scientific Notation

1. Order of Operations

The order of operations was first introduced in Section 1.7 and then used throughout the text. The order of operations still applies when simplifying expressions with real numbers.

Order of Operations

1. Perform all operations inside parentheses and other grouping symbols first.
2. Simplify any expressions containing exponents or square roots.
3. Perform multiplication or division in the order that they appear from left to right.
4. Perform addition or subtraction in the order that they appear from left to right.

example 1   Applying the Order of Operations

Simplify.

     a. $-12 - 6(7 - 5)$ **b.** $3^2 - 10^2 \div (-1 - 4)$ **c.** $\dfrac{1}{30} - \left(-\dfrac{1}{3}\right)^2 \cdot \dfrac{3}{5}$

Solution:

a. $-12 - 6(7 - 5)$

$= -12 - 6(2)$ Simplify within parentheses first.

$= -12 - 12$ Multiply before subtracting.

$= -24$ Subtract. *Note:* $-12 - 12 = -12 + (-12) = -24$.

b. $3^2 - 10^2 \div (-1 - 4)$

$= 3^2 - 10^2 \div (-5)$ Simplify within parentheses.
                         Note: $-1 - 4 = -1 + (-4) = -5$.

$= 9 - 100 \div (-5)$ Simplify exponents.
                         Note: $3^2 = 3 \cdot 3 = 9$ and $10^2 = 10 \cdot 10 = 100$.

$= 9 - (-20)$ Divide before subtracting.
                         Note: $100 \div (-5) = -20$.

$= 29$ Subtract. *Note:* $9 - (-20) = 9 + (20) = 29$.

c. $\dfrac{1}{30} - \left(-\dfrac{1}{3}\right)^2 \cdot \dfrac{3}{5}$

$= \dfrac{1}{30} - \dfrac{1}{9} \cdot \dfrac{3}{5}$ Simplify exponents. *Note:* $\left(-\dfrac{1}{3}\right) \cdot \left(-\dfrac{1}{3}\right) = \dfrac{1}{9}$.

$= \dfrac{1}{30} - \dfrac{1}{\overset{}{\underset{3}{9}}} \cdot \dfrac{\overset{1}{3}}{5}$ Multiply fractions.

$= \dfrac{1}{30} - \dfrac{1}{15}$ The least common denominator is 30.

$= \dfrac{1}{30} - \dfrac{1 \cdot 2}{15 \cdot 2}$ Write each fraction with the LCD.

$= \dfrac{1}{30} - \dfrac{2}{30}$

$= \dfrac{1}{30} + \left(-\dfrac{2}{30}\right)$ Write the subtraction in terms of addition.

$= -\dfrac{1}{30}$

Skill Practice

Simplify.

1. $8 - 2(3 - 10)$

2. $8^2 - 2^3 \div (-7 + 5)$

3. $-\dfrac{2}{9} + \left(-\dfrac{5}{6}\right)^2 \cdot \dfrac{2}{5}$

4. $\left(-\dfrac{2}{3}\right)^2 - \left(-\dfrac{1}{6}\right) \cdot \dfrac{5}{2}$

2. Scientific Notation

In Section 1.7, we studied powers of 10. Recall:

$10^6 = 10 \cdot 10 \cdot 10 \cdot 10 \cdot 10 \cdot 10 = 1{,}000{,}000$

$10^5 = 10 \cdot 10 \cdot 10 \cdot 10 \cdot 10 = 100{,}000$

$10^4 = 10 \cdot 10 \cdot 10 \cdot 10 = 10{,}000$

$10^3 = 10 \cdot 10 \cdot 10 = 1000$

$10^2 = 10 \cdot 10 = 100$

$10^1 = 10$

$10^0 = 1$ ⟵——————————————

Notice that as the exponents are decreased by 1, the numbers on the right are one-tenth of the number from the row above.

If we continue this pattern, then 10^0 would equal 1.

Answers

1. 22   2. 68   3. $\dfrac{1}{18}$ 4. $\dfrac{31}{36}$

Fill in the blank.

5. $1000 = 10^{\square}$

6. $\dfrac{1}{1000} = 10^{\square}$

7. $0.01 = 10^{\square}$

8. $100 = 10^{\square}$

$10^{-1} = 0.1$

$10^{-2} = 0.01$

$10^{-3} = 0.001$

$10^{-4} = 0.0001$

$10^{-5} = 0.00001$

$10^{-6} = 0.000001$

If this pattern is continued for negative exponents, we get the fractions $\frac{1}{10}, \frac{1}{100}, \frac{1}{1000}$, and so on. In decimal form, we have $0.1, 0.01, 0.001, \ldots$.

In many applications in mathematics, business, and science, it is necessary to work with very large or very small numbers. For example,

- America's teens spend \$153,000,000,000 yearly in online shopping.
- A piece of paper is 0.002 in. thick.

The value \$153,000,000,000 can be expressed as $1.53 \times 100,000,000,000$

$$= 1.53 \times 10^{11}$$

The value 0.002 can be expressed as 2×0.001

$$= 2 \times 10^{-3}$$

The numbers 1.53×10^{11} and 2×10^{-3} are written in scientific notation. Using scientific notation, we express a number as the product of two factors. One factor is a number greater than or equal to 1 but less than 10. The other factor is a power of 10.

> ### Scientific Notation
>
> A positive number written in scientific notation is written as $a \times 10^{n}$, where a is a number greater than or equal to 1, but less than 10, and n is an integer.

3. Converting Decimal Notation to Scientific Notation

To write a number in scientific notation, follow these guidelines.

Move the decimal point so that its new location is to the right of the first nonzero digit. Count the number of places that the decimal point is moved. Then

1. If the original number is *greater than or equal to 10:*

The exponent for the power of 10 is *positive* and is equal to the number of places that the decimal point was moved.

$$7,500,000 = 7.5 \times 10^{6} \qquad 30,000 = 3 \times 10^{4}$$

Move the decimal point 6 places to the *left.*

Move the decimal point 4 places to the *left.*

2. If the original number is *between 0 and 1:*

The exponent for the power of 10 is *negative.* Its absolute value is equal to the number of places the decimal point was moved.

$$0.000089 = 8.9 \times 10^{-5} \qquad 0.00000004 = 4 \times 10^{-8}$$

Move the decimal point 5 places to the *right.*

Move the decimal point 8 places to the *right.*

Answers

5. 3   6. −3   7. −2   8. 2

example 2	Writing Numbers in Scientific Notation

Write the number in scientific notation.

a. 93,000,000 mi   (the distance between the earth and the sun)

b. 0.000 000 000 753 kg   (the mass of a dust particle)

c. 300,000,000 m/sec   (the speed of light)

d. 0.00017 m   (length of the smallest insect in the world)

Solution:

a. 93,000,000 mi   The number is greater than 10. Move the decimal point left 7 places.

$= 9.3 \times 10^7$ mi   For a number greater than 10, the exponent is *positive*.

b. 0.000 000 000 753 kg   The number is between 0 and 1. Move the decimal point to the right 10 places.

$= 7.53 \times 10^{-10}$ kg   For a number between 0 and 1, the exponent is *negative*.

c. 300,000,000 m/sec   The number is greater than 10. Move the decimal point to the left 8 places.

$= 3 \times 10^8$ m/sec   For a number greater than 10, the exponent is *positive*.

d. 0.00017 m   The number is between 0 and 1. Move the decimal point to the right 4 places.

$= 1.7 \times 10^{-4}$ m   For a number between 0 and 1, the exponent is *negative*.

Skill Practice

Write the numbers in scientific notation.

9. The amount of income brought in by the movie *Shrek* as of November 2004 was $437,200,000.

10. The amount of prize money won by Serena Williams in 2004 was $1,325,000.

11. The time required for light to travel 1 mi is 0.00000534 sec.

12. The diameter of a grain of sand is approximately 0.0025 in.

4. Converting Scientific Notation to Decimal Notation

To convert from scientific notation to decimal notation, follow these guidelines.

1. If the exponent on 10 is *positive*, move the decimal point to the right the same number of places as the exponent. Add zeros as necessary.

2. If the exponent on 10 is *negative*, move the decimal point to the left the same number of places as the exponent. Add zeros as necessary.

example 3	Converting Scientific Notation to Decimal Notation

Convert to decimal notation.

a. 3.52×10^{-5} **b.** 4.6×10^4 **c.** 9×10^{-12} **d.** 1×10^{15}

Solution:

a. $3.52 \times 10^{-5} = 0.0000352$ The exponent is negative. Move the decimal point to the left 5 places.

b. $4.6 \times 10^4 = 46,000$ The exponent is positive. Move the decimal point to the right 4 places.

Skill Practice

Convert to decimal notation.

13. 2.79×10^{-8}

14. 8.603×10^5

15. 6×10^1

16. 1×10^{-6}

Answers

9. 4.372×10^8
10. 1.325×10^6
11. 5.34×10^{-6} sec
12. 2.5×10^{-3} in.
13. 0.0000000279   14. 860,300
15. 60   16. 0.000001

c. $9 \times 10^{-12} = 0.000\,000\,000\,009$ The exponent is negative. Move the decimal point to the left 12 places.

d. $1 \times 10^{15} = 1{,}000{,}000{,}000{,}000{,}000$ The exponent is positive. Move the decimal point to the right 15 places.

---

section 10.5   Practice Exercises

Study Skills Exercises

1. Make a list of resources where you can get help at times when you are not on campus and you need help with your algebra studies (for example, websites, online tutoring, and classmates). Write down Web addresses and phone numbers and keep them handy.

2. Define the key term **scientific notation**.

Review Exercises

For Exercises 3–8, multiply or divide as indicated.

3. $-100 \div (-4)$

4. $\left(-\dfrac{2}{9}\right) \div \left(\dfrac{8}{27}\right)$

5. $10 \cdot \left(-\dfrac{3}{5}\right)$

6. $-2.8(-1.1)$

7. $5.5 \div (-0.5)$

8. $(-1)(-5)(-8)(3)$

9. a. What number must be multiplied by -5 to obtain -35?

   b. What number must be multiplied by -5 to obtain 35?

10. a. What number must be multiplied by -4 to obtain -36?

   b. What number must be multiplied by -4 to obtain 36?

Objective 1: Order of Operations

For Exercises 11–36, simplify by using the order of operations. **(See Example 1.)**

11. $-2 \cdot (3 - 6) + 10$

12. $-4 \div (1 - 3) - 8$

13. $-16 \div (-4) \cdot (-5)$

14. $-12 \cdot (-1) \div 6$

15. $8 - (-3) \cdot (-2)^3$

16. $1 - (-5) \cdot (-3)^2$

17. $12 + (14 - 16)^2 \div (-4)$

18. $-7 + (1 - 5)^2 \div 4$

19. $-48 \div 12 \div (-2)$

20. $-100 \div (-5) \div (-5)$

21. $90 \div (-3) \cdot (-1) \div (-6)$

22. $64 \div (-4) \cdot 2 \div (-16)$

23. $\left[9^2 - (-7)^2\right] \div (-4)$

24. $|(-8)^2 - 5^2| \div (-3)$

25. $2 + 2^3 - |10 - 12|$

26. $14 - 4^2 + 2 - 10$

27. $-6(48 \div 12)^2$

28. $-5(35 \div 5)^2$

29. $\left(-\dfrac{1}{2}\right) \cdot \dfrac{1}{3} \div \dfrac{1}{12}$

30. $\dfrac{2}{9} \div \left(-\dfrac{1}{3}\right) \cdot \dfrac{6}{5}$

31. $\dfrac{1}{6} + \left(-\dfrac{5}{4}\right) \cdot \dfrac{4}{3}$

32. $-\dfrac{3}{8} + \dfrac{5}{24} \div \left(-\dfrac{5}{6}\right)$

33. $\left(-\dfrac{2}{3}\right)^2 - \left(\dfrac{5}{21}\right) \div \dfrac{15}{7}$

34. $\left(-\dfrac{1}{3}\right)^3 + \left(\dfrac{2}{9}\right) \cdot \dfrac{5}{6}$

35. $\dfrac{5}{2} \cdot \left(\dfrac{3}{2}\right)^2 + \left(-\dfrac{1}{2}\right)^3$

36. $\left(-\dfrac{7}{3}\right) \div \left(\dfrac{4}{3}\right)^2 - \left(\dfrac{1}{2}\right)^4$

37. Find the average temperature: $-8°, -11°, -4°, 1°, 9°, 4°, -5°$

38. Find the average temperature: $15°, 12°, 10°, 3°, 0°, -2°, -3°$

39. Find the average score: $-8, -8, -6, -5, -2, -2, 3, 3, 0, -4$

40. Find the average score: $-6, -2, 5, 1, 0, -3, 7, 2, -7, -4$

Objective 2: Scientific Notation

For Exercises 41–48, write each power of 10 in exponential notation.

41. 10,000

42. 100,000

43. 1000

44. 1,000,000

45. 0.001

46. 0.01

47. 0.0001

48. 0.1

For Exercises 49–56, identify which of the expressions are in correct scientific notation. (Answer yes or no.)

49. 43×10^3

50. 82×10^{-4}

51. 6.1×10^{-1}

52. 2.34×10^4

53. 2×10^{10}

54. 8×10^{-5}

55. 0.02×10^4

56. 0.052×10^{-3}

Objective 3: Converting Decimal Notation to Scientific Notation

For Exercises 57–60, convert the number to scientific notation. **(See Example 2.)**

57. For a recent year the national debt was approximately $7,455,000,000,000.

58. The gross national product is approximately $10,533,000,000,000.

59. The diameter of an atom is 0.000 000 2 mm.

60. The length of a flea is 0.0625 in.

For Exercises 61–76, convert to scientific notation.

61. 20,000,000	**62.** 5000	**63.** 8,100,000	**64.** 62,000
65. 0.003	**66.** 0.0009	**67.** 0.025	**68.** 0.58
69. 142,000	**70.** 25,500,000	**71.** 0.0000491	**72.** 0.000116
73. 0.082	**74.** 0.15	**75.** 4920	**76.** 13,400

Objective 4: Converting Scientific Notation to Decimal Notation

For Exercises 77–92, convert to decimal notation. **(See Example 3.)**

77. 6×10^3	**78.** 3×10^4	**79.** 8×10^{-2}	**80.** 2×10^{-5}
81. 4.4×10^{-1}	**82.** 2.1×10^{-3}	**83.** 3.7×10^4	**84.** 5.5×10^3
85. 3.26×10^2	**86.** 6.13×10^7	**87.** 1.29×10^{-2}	**88.** 4.04×10^{-4}
89. 2.003×10^{-6}	**90.** 5.02×10^{-5}	**91.** 9.001×10^8	**92.** 7.07×10^6

Expanding Your Skills

For Exercises 93–96, simplify the expressions containing both fractions and decimals.

93. $\left(\dfrac{1}{2}\right)^2 \div 0.05 + \left(-\dfrac{3}{4} \cdot \dfrac{8}{3}\right)$

94. $-0.8 - \dfrac{19}{20} + \left(\dfrac{4}{5} \div \dfrac{1}{2}\right) - (-0.15)$

95. $2\left(\dfrac{7}{8} - \dfrac{1}{4}\right) - (-1.5)^2$

96. $-2.1 + 4\left(\dfrac{7}{16} - 0.0375\right)$

chapter 10 | summary

section 10.1 Real Numbers and the Real Number Line

Key Concepts

The numbers . . . $-3, -2, -1, 0, 1, 2, 3, . . .$ and so on are called **integers**. The negative integers lie to the left of zero on the number line.

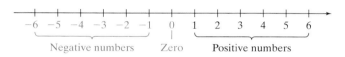

A number that can be written as a ratio of two integers is called a **rational number** (division by zero is excluded).

Irrational numbers are numbers that cannot be written as a ratio of integers.

The set of all rational numbers and irrational numbers together forms the **real numbers**.

The **absolute value** of a number a is denoted $|a|$. The value of $|a|$ is the distance between a and 0 on the number line.

Two numbers that are the same distance from zero on the number line, but on opposite sides of zero are called **opposites**.

The double-negative property states that the opposite of a negative number is a positive number. That is, $-(-a) = a$, for $a > 0$.

Examples

Example 1

The temperature 5° below zero can be represented by a negative number: $-5°$.

Example 2

The following numbers are rational because they can be written as a ratio of two integers.

a. $\frac{1}{2}$ is a ratio of 1 and 2

b. -0.75 is a ratio of -3 and 4

c. -4 is a ratio of -4 and 1

Example 3

a. $|5| = 5$ **b.** $|-13| = 13$ **c.** $|0| = 0$

Example 4

The opposite of 12 is $-(12) = -12$.

Example 5

The opposite of -23 is $-(-23) = 23$.

section 10.2 Addition of Real Numbers

Key Concepts

To add integers by using a number line, locate the first addend on the number line. Then to add a positive number, move to the right on the number line. To add a negative number, move to the left on the number line.

Integers can be added by using following rules:

Adding Numbers with the Same Sign

To add two numbers with the same sign, add their absolute values and apply the common sign.

Adding Numbers with Different Signs

To add two numbers with different signs, subtract the smaller absolute value from the larger absolute value. Then apply the sign of the number having the larger absolute value.

Examples

Example 1

Add $-2 + (-4)$ by using the number line.

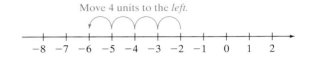
Move 4 units to the *left.*

$-2 + (-4) = -6$

Example 2

a. $5 + 2 = 7$

b. $-5 + (-2) = -7$

Example 3

a. $6 + (-5) = 1$

b. $(-6) + 5 = -1$

section 10.3 Subtraction of Real Numbers

Key Concepts

Subtraction of Real Numbers

If a and b are real numbers, then

$$a - b = a + (-b)$$

To perform subtraction, follow these steps:
1. Leave the first number (the minuend) unchanged.
2. Change the subtraction sign to an addition sign.
3. Add the opposite of the second number (the subtrahend).

Examples

Example 1

a. $3 - 9 = 3 + (-9) = -6$

b. $-3 - 9 = -3 + (-9) = -12$

c. $3 - (-9) = 3 + (9) = 12$

d. $-3 - (-9) = -3 + (9) = 6$

section 10.4 Multiplication and Division of Real Numbers

Key Concepts

Multiplication of Real Numbers

1. The product of two real numbers with the same sign is positive.
2. The product of two real numbers with different signs is negative.
3. The product of any real number and zero is zero.

The product of an *even* number of negative factors is *positive.*

The product of an *odd* number of negative factors is *negative.*

When evaluating an exponential expression, attention must be paid when parentheses are used. That is, $(-2)^4 = (-2)(-2)(-2)(-2) = 16$, while $-2^4 = -1 \cdot (2)(2)(2)(2) = -16$.

Two numbers are **reciprocals** if their product is 1.

Division of Real Numbers

1. The quotient of two real numbers with the same sign is positive.
2. The quotient of two real numbers with different signs is negative.

Division by zero is undefined.
Zero divided by a nonzero number is 0.

Examples

Example 1

a. $-8(-3) = 24$

b. $8(-3) = -24$

c. $-8(0) = 0$

Example 2

a. $(-5)(-4)(-1)(-3) = 60$

b. $(-2)(-1)(-6)(-3)(-2) = -72$

Example 3

a. $\left(-\dfrac{2}{3}\right)^2 = \left(-\dfrac{2}{3}\right)\left(-\dfrac{2}{3}\right) = \dfrac{4}{9}$

b. $-\left(\dfrac{2}{3}\right)^2 = -1 \cdot \left(\dfrac{2}{3}\right)\left(\dfrac{2}{3}\right) = -\dfrac{4}{9}$

Example 4

The reciprocal of -4 is $-\dfrac{1}{4}$ because

$-4 \cdot -\dfrac{1}{4} = 1$

Example 5

a. $\dfrac{42}{-6} = -7$ **b.** $\dfrac{-36}{-9} = 4$

Example 6

a. $\dfrac{-15}{0}$ is undefined.   **b.** $\dfrac{0}{-3} = 0$

section 10.5 Order of Operations and Scientific Notation

Key Concepts

Order of Operations

1. Perform all operations inside parentheses and other grouping symbols first.
2. Simplify any expressions containing exponents or square roots.
3. Perform multiplication or division in the order that they appear from left to right.
4. Perform addition or subtraction in the order that they appear from left to right.

Examples

Example 1

$$-15 - 2(8 - 11)^2 = -15 - 2(-3)^2$$
$$= -15 - 2 \cdot 9$$
$$= -15 - 18$$
$$= -33$$

Example 2

$$\frac{1}{5} \div \left(-\frac{7}{20}\right) \cdot \left(-\frac{9}{2}\right) = \frac{1}{5} \cdot \left(-\frac{20}{7}\right) \cdot \left(-\frac{9}{2}\right)$$
$$= \frac{1}{\overset{1}{5}} \cdot \left(-\frac{\overset{4}{20}}{7}\right) \cdot \left(-\frac{9}{2}\right)$$
$$= -\frac{\overset{2}{4}}{7} \cdot \left(-\frac{9}{\underset{1}{2}}\right)$$
$$= \frac{18}{7}$$

Scientific Notation

A positive number written in **scientific notation** is written as $a \times 10^n$, where a is a number greater than or equal to 1 and less than 10, and n is an integer.

Example 3

The number 130,000 can be written in scientific notation as 1.3×10^5.

The number 0.000102 can be written in scientific notation as 1.02×10^{-4}.

Example 4

The number $3.08 \times 10^3 = 3080$.
The number $2.105 \times 10^{-2} = 0.02105$.

chapter 10 | review exercises

Section 10.1

For Exercises 1–4, write an integer that represents each numerical value.

1. The population of Detroit, Michigan, decreased by 76,704 from 1990 to 2000. (Source: U.S. Census Bureau.)

2. The country's deficit is $5 billion.

3. The temperature rose 15° in one day.

4. Cecelia's bank account earned $10 in interest.

For Exercises 5–8, graph the numbers on the number line.

5. -2 **6.** $-5\frac{1}{3}$ **7.** 0 **8.** 3.8

For Exercises 9–12, determine the opposite and the absolute value for each number.

9. -4 **10.** $-\frac{1}{2}$ **11.** 3.5 **12.** 6

13. Evaluate.

    a. $-(-9)$ **b.** $-|-9|$

14. Evaluate.

    a. $-(1.5)$ **b.** $-|1.5|$

For Exercises 15–20, place the correct symbol, $>$ or $<$, between the two numbers.

15. $-\frac{5}{6} \,\square\, -1$ **16.** $-0.5 \,\square\, 0.5$

17. $|3| \,\square\, -|3|$ **18.** $8 \,\square\, |-9|$

19. $-2.8 \,\square\, -2$ **20.** $\left|-\frac{3}{2}\right| \,\square\, -\frac{1}{2}$

Section 10.2

For Exercises 21–24, add the integers by using the number line.

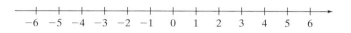

21. $6 + (-2)$ **22.** $-3 + 6$

23. $-3 + (-2)$ **24.** $-3 + 0$

25. State the rule for adding two numbers with the same sign.

26. State the rule for adding two numbers with different signs.

For Exercises 27–34, add the real numbers.

27. $35 + (-22)$ **28.** $-105 + 90$

29. $-29 + (-41)$ **30.** $3.22 + (-4.1)$

31. $-6.5 + (-4.16)$ **32.** $\left(-\frac{7}{4}\right) + \left(\frac{5}{2}\right)$

33. $\left(-\frac{1}{5}\right) + \left(-\frac{7}{10}\right)$ **34.** $2 + \left(-\frac{7}{3}\right)$

For Exercises 35–40, translate each phrase to a mathematical expression. Then simplify the expression.

35. The sum of 23 and -35

36. 57 plus -10

37. The total of -5, -13, and 20

38. -42 increased by 12

39. 3 more than -12

40. -89 plus -22

For Exercises 41–42, add.

41. $-3 + (-10) + 12 + 14 + (-10)$

42. $9 + (-15) + 2 + (-7) + (-4)$

Section 10.3

43. State the steps for subtracting two numbers.

For Exercises 44–51, subtract the real numbers.

44. $4 - (-23)$ **45.** $19 - 44$

46. $-2 - (-24)$ **47.** $-289 - 130$

48. $-2.9 - 4.5$ **49.** $3.8 - 4.5$

50. $\left(-\dfrac{5}{3}\right) - \left(-\dfrac{5}{12}\right)$ **51.** $0 - \left(-\dfrac{20}{21}\right)$

For Exercises 52–55, translate the mathematical statement to an English phrase. Answers will vary.

52. $4 - 6$ **53.** $23 - (-6)$

54. $-2 - 14$ **55.** $-25 - (-7)$

56. The temperature in Fargo, North Dakota, rose from $-6°F$ to $-1°F$. By how many degrees did the temperature rise?

57. Sam's balance in his checking account was $-\$40$, so he deposited 132. What is his present balance?

58. Find the average of the golf scores: $-3, 4, 0, 9,$ $-2, -1, 0, 5, -3$ (These scores are the number of holes above or below par.)

Section 10.4

For Exercises 59–69, multiply or divide as indicated.

59. $6(-3)$ **60.** $\dfrac{-12}{4}$

61. $\dfrac{-900}{-60}$ **62.** $(-7)(-8)$

63. $-2.8 \div 0.04$ **64.** $(-62.6)(2.5)$

65. $\left(-\dfrac{2}{3}\right)\left(-\dfrac{21}{8}\right)$ **66.** $\left(-2\dfrac{1}{8}\right) \div \left(1\dfrac{1}{4}\right)$

67. $\left(-\dfrac{1}{5}\right) \div 0$ **68.** $\dfrac{0}{-5}$

69. $(-1)(-8)(2)(1)(-2)$

For Exercises 70–75, simplify.

70. $(-6)^2$ **71.** -6^2

72. $\left(-\dfrac{3}{4}\right)^3$ **73.** $-\left(\dfrac{3}{4}\right)^3$

74. $(-1)^{10}$ **75.** $(-1)^{21}$

76. What is the sign of the product of three negative factors?

77. What is the sign of the product of four negative factors?

For Exercises 78–81, translate the English phrase into a mathematical expression. Then simplify.

78. The quotient of -45 and -15

79. The product of -4 and 19

80. 30 times -5

81. -136 divided by -8

Section 10.5

For Exercises 82–88, simplify by using the order of operations.

82. $28 \div (-7) \cdot 3 - (-1)$

83. $(-4)^3 \div 8 - (-6)$

84. $\left[10 - (-3)^2\right] \cdot (-11) + 4$

85. $\left[-9 - (-7)\right]^3 \cdot 3 \div (-6)$

86. $18 - (-5)^2 + 14 \div 2$

87. $\left(\dfrac{1}{15}\right) \div \left(-\dfrac{7}{10}\right) \cdot \left(\dfrac{3}{2}\right) + \left(-\dfrac{6}{7}\right)$

88. $\left(-\dfrac{3}{8}\right)^2 - \left(-\dfrac{1}{2}\right)^3$

89. Find the average temperature for one week; $2°, 4°, -6°, -1°, 0°, -4°, -2°$

For Exercises 90–93, convert to scientific notation.

90. 600,000

91. 10,302,000

92. 0.000045

93. 0.009042

For Exercises 94–97, convert to decimal form.

94. 2×10^4

95. 8.7×10^9

96. 1.05×10^{-3}

97. 6.02×10^{-5}

98. The material cost for World War II has been estimated at 1.5×10^{12}. Convert the cost to decimal form.

99. The mass of an electron is 9.11×10^{-31} kg. How would you convert this number to decimal form?

---

chapter 10 | test

1. Write an integer that represents the numerical value.

  a. Dwayne lost $220 during in his last trip to Las Vegas.

  b. Garth Brooks has 26 more platinum albums than Elvis Presley.

For Exercises 2–4, refer to these numbers: $-3, -\frac{3}{5}, 0, \sqrt{7}, 4, -1, \frac{4}{7}, -\pi$

2. List all the numbers that are integers.

3. List all the rational numbers.

4. List all the irrational numbers.

For Exercises 5–10, place the correct symbol, $>$ or $<$, between the two numbers.

5. $-5 \boxed{} -2$

6. $|-5| \boxed{} |-2|$

7. $0 \boxed{} -2.4$

8. $\dfrac{4}{5} \boxed{} -\dfrac{2}{3}$

9. $-|-9| \boxed{} 9$

10. $-|33.1| \boxed{} |-33.1|$

For Exercises 11–16, add or subtract as indicated.

11. $9 + (-14)$

12. $-23 + (-5)$

13. $-4 - (-13)$

14. $-30 - 11$

15. $-15 + 21$

16. $5 - 28$

For Exercises 17–22, multiply or divide as indicated.

17. $6(-12)$

18. $(-11)(-8)$

19. $\dfrac{-24}{-12}$

20. $\dfrac{54}{-3}$

21. $\dfrac{-44}{0}$

22. $(-91)(0)$

23. a. What is the sign of the product of an even number of negative factors?

  b. What is the sign of the product of an odd number of negative factors?

24. Simplify the exponential expressions.

  a. $(-8)^2$ **b.** -8^2 **c.** $(-4)^3$ **d.** -4^3

For Exercises 25–30, translate to a mathematical expression. Then simplify the expression.

25. The product of -3 and -7

26. 8 more than -13

27. Subtract -4 from 18.

28. The quotient of 6 and $-\dfrac{2}{3}$

29. -8.1 increased by 5

30. The total of -3, 15, -6, and -1

For Exercises 31–36, simplify.

31. $-14 + 22 - (-5) + (-10)$

32. $(-3)(-1)(-4)(-1)(-5)$

33. $-20 \div (-2)^2 + (-14)$

34. $12 \cdot (-6) + [20 - (-12)] - 15$

35. $-\dfrac{2}{15} + \left(-\dfrac{20}{21} \cdot \dfrac{7}{5}\right)$

36. $\left(-\dfrac{1}{3}\right)^2 \div \left(\dfrac{5}{6} - \dfrac{1}{9}\right)$

37. Find the average temperature:
$4°, -3°, -1°, 5°, -2°, 0°, 4°$

38. A nanometer is 0.000 000 001 m. Write the number in scientific notation.

39. For a recent year, the number of Internet users in the world was 580,780,000. Write the number in scientific notation.

40. Write the numbers in decimal notation.

    a. 3.0501×10^{10}

    b. 4.009×10^{-4}

---

chapters 1–10 | cumulative review

For Exercises 1–4, add, subtract, or multiply as indicated.

1. $\begin{array}{r} 3490 \\ +123 \\ \hline \end{array}$

2. $\begin{array}{r} 2901 \\ -332 \\ \hline \end{array}$

3. $\begin{array}{r} 23 \\ 34 \\ 98 \\ +22 \\ \hline \end{array}$

4. $\begin{array}{r} 790 \\ \times 24 \\ \hline \end{array}$

5. Write the prime factorization of 720.

6. Write a fraction that represents the shaded region of the figure.

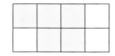

7. On a quiz, Harold missed 3 out of 14 questions. Write a fraction representing the fraction of the quiz questions that he answered *correctly*.

8. Amy has a box that contains 20 oz of snack crackers. How many individual packages will she get if she puts $2\frac{1}{2}$ oz in each package?

9. Find the LCM of 16, 40, and 10.

10. Simplify. $\dfrac{3}{16} + \dfrac{33}{40} - \dfrac{7}{10}$

11. Add. $3\dfrac{3}{5} + 2\dfrac{13}{15}$

12. Subtract. $16\dfrac{1}{2} - 12\dfrac{13}{14}$

13. Round the numbers to the indicated place.

    a. 34.2298 Thousandths   **b.** 9.0314 Tenths

14. Convert cents to dollars. 209.99¢

15. Multiply. 204.55(2.4)   **16.** Divide. 402.5 ÷ 3.5

17. Write a ratio of the shortest side to the longest side of the rectangle.

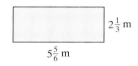

18. A DC-10 aircraft used 9964 gal of fuel in 4 hr. Find the unit rate in gallons per hour.

19. On a map 1 in. represents 6 mi. What is the distance between two cities that measure $3\frac{1}{2}$ in. on the map?

20. Given that the two triangles are similar, find sides x and y.

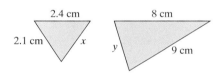

For Exercises 21–23, solve the percent equations.

21. What is 32% of 600?

22. What percent of 300 is 336?

23. 15 is 6% of what?

24. A pair of shoes was discounted 20%. If the original price was $86, what is the sale price?

For Exercises 25–28, convert the units of measure.

25. 2 ft 4 in. = _____ in.

26. 20 qt = _____ gal

27. 60 mL = _____ L

28. 30 oz = _____ lb

29. A car travels 6 mi due north and then turns and travels 8 mi due east. What is the distance of the car from the point of origin? Round to the nearest tenth of a mile.

For Exercises 30–31, find the area.

30. Parallelogram

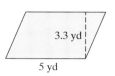

31. Square

32. Find the area A and the circumference C of the circle. Use 3.14 for π.

For Exercises 33–35, use the following data.

The number of miles walked in one day by 10 selected people is given.

4   4   4   3   6   4   6   5   3   4

33. Complete the frequency distribution for the data.

Number of Miles	Tally	Frequency (Number of Walkers)
3		
4		
5		
6		

34. Construct a horizontal bar graph from the frequency distribution in Exercise 33. Label the vertical axis with the number of miles and the horizontal axis with the frequency.

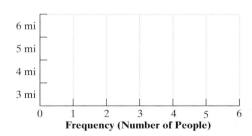

35. What is the mean number of miles walked per day?

36. Refer to the circle graph. If the monthly budget for a small business is $1200, how much will be spent on postage?

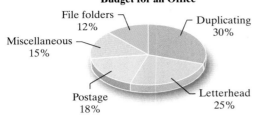

Budget for an Office

File folders
12%

Duplicating
30%

Miscellaneous
15%

Postage
18%

Letterhead
25%

For Exercises 37–40, simplify.

37. $43 - (-12)$

38. $-12 + (-5) - 3 - (-8)$

39. $(-4)^2 - 6^2$

40. $\dfrac{8}{9} \cdot \left(\dfrac{1}{3} - \dfrac{5}{6}\right)^3 \div \left(-\dfrac{2}{3}\right)$

Solving Equations

11

In this chapter we learn how to simplify algebraic expressions by clearing parentheses and combining like terms. Then we move on to solving linear equations and using equations to solve application problems. For example, in Exercise 30 from Section 11.6, we must find the dimensions of an Olympic size swimming pool. The perimeter is given as 150 m, and the width is $\frac{1}{2}$ of its length.

chapter 11 | preview

The exercises in this chapter preview contain concepts that have not yet been presented. These exercises are provided for students who want to compare their levels of understanding before and after studying the chapter. Alternatively, you may prefer to work these exercises when the chapter is completed and before taking the exam.

Section 11.1

1. A DVD costs $17.95. Write an expression that represents the cost of x DVDs.

For Exercises 2–3, evaluate the expression for $a = -3$ and $b = 5$.

2. $a^2 - b$ **3.** $-b^2$

4. Rewrite each expression, using the associative property. Then simplify the expression.

    a. $-3(9x)$

    b. $(y + 10) + 3$

Section 11.2

For Exercises 5–6, combine like terms. Clear parentheses if necessary.

5. $9r + 10s - 3r + s$

6. $8(k - 1) - (4k - 12)$

Section 11.3

For Exercises 7–8, determine whether -5 is a solution to the equation.

7. $6x + 8 = 22$ **8.** $2x + 1 = x - 4$

For Exercises 9–10, solve the equation, using the addition property of equality or the subtraction property of equality.

9. $8 + q = -2$ **10.** $-12 = p - 3$

Section 11.4

For Exercises 11–13, solve the equation, using the multiplication property of equality or the division property of equality.

11. $6w = -15$ **12.** $\dfrac{b}{3} = 7$ **13.** $-\dfrac{2}{9}c = -8$

Section 11.5

For Exercises 14–16, solve the equation.

14. $-5t + 3 = 18$

15. $-9b - 7 = -8b + 1$

16. $-(4n - 10) = 6(n + 2)$

Section 11.6

For Exercises 17–18, write an equation that represents the problem. Then solve the equations.

17. A number increased by 10 is equal to 6 times the number. Find the number.

18. The quotient of a number and 6 is $\frac{2}{3}$. Find the number.

19. In a recent season, Kevin Garnett was the highest-paid NBA basketball player. Shaquille O'Neal made $1.2 million less than Garnett. Together they made $49.2 million. How much did Garnett and O'Neal make individually?

Section 11.7

For Exercises 20–23, plot the points on the rectangular coordinate system.

20. $(1, -1)$

21. $(2, 5)$

22. $(-3, -3)$

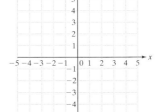

23. $(-4, 0)$

section 11.1   Properties of Real Numbers

Objectives

1. Algebraic Expressions
2. Evaluating Expressions
3. Properties of Real Numbers

1. Algebraic Expressions

We begin our study of algebra with a few key terms. Recall that a **variable** is a letter or symbol that can represent any number. **Constants** have fixed values that never change. Here are some examples of variables and constants.

Variables	Constants
x, y, z, A, l, w	$3, -1, \frac{2}{3}, \pi$

An algebraic **expression** is a collection of variables and constants with algebraic operations such as addition, subtraction, multiplication, and division. Here are some examples of expressions.

$$2x, \quad 4 + y, \quad 3t - 7, \quad \frac{y}{8}$$

Algebraic expressions are often used in applications.

---

example 1   Using Algebraic Expressions in Applications

a. At a discount CD store, each CD costs \$7.99. Suppose n is the number of CDs that a customer buys. Write an expression that represents the cost for n CDs.

b. The length of a rectangle is 5 in. longer than the width, w. Write an expression that represents the length of the rectangle.

c. A rope that is L ft long is to be cut into five pieces of equal length. Write an expression that represents the length of each piece.

Skill Practice

1. Smoked turkey costs \$6.99 per pound. Write an expression that represents the cost of p pounds.

2. The width of a basketball court is 44 ft shorter than its length, l. Write an expression that represents the width.

3. Six tons of gravel is to be carried away by n trucks. Write an expression for the amount carried by each truck. Assume that each truck carries an equal amount of gravel.

Solution:

a. The cost of 1 CD is \$7.99.

The cost of 2 CDs is \$7.99(2) = \$15.98.

The cost of 3 CDs is \$7.99(3) = \$23.97.

The cost of n CDs is \$7.99(n) or simply \$7.99n.

From this pattern, we see that the total cost is the unit cost per CD times the number of CDs.

b. The length of a rectangle is 5 in. more than the width. The phrase *more than* implies addition. Thus, the length (in inches) is represented by

$$\text{Length} = w + 5$$

c. For this scenario, drawing a figure may be helpful (Figure 11-1). The original length L must be cut (divided) into five pieces of equal length. Thus, the length (in feet) of each piece is given by

$$\text{Length of each piece} = \frac{L}{5}$$

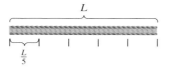

Figure 11-1

Answers

1. \$6.99p 2. $l - 44$ 3. $\dfrac{6}{n}$

2. Evaluating Expressions

The value of an expression depends on the values of the variables within the expression. When we substitute numerical values for the variables within an expression, we call this *evaluating the expression*. We demonstrate this process in Example 2.

example 2   Evaluating Expressions

Evaluate each expression for the given value of the variable(s).

a. $3x^2$ for $x = -5$

b. $4a - 8b$ for $a = -6$, $b = \frac{1}{2}$

Solution:

To evaluate the expressions, we substitute the given number for the variable. We recommend using parentheses so that you can "see" where to insert the given numerical values.

a. $3x^2 = 3(\ \ )^2$ Replace the variable with parentheses.

        $= 3(-5)^2$ Substitute $x = -5$.

        $= 3(25)$ Simplify exponents.

        $= 75$

b. $4a - 8b = 4(\ \ ) - 8(\ \ )$ Replace the variable with parentheses.

             $= 4(-6) - 8(\frac{1}{2})$ Substitute $a = -6$ and $b = \frac{1}{2}$.

             $= -24 - 4$ Multiply from left to right.

             $= -28$

3. Properties of Real Numbers

Several important properties of whole numbers were introduced in Sections 1.2 and 1.5 involving addition and multiplication. These properties also hold for real numbers and are summarized in Tables 11-1 through 11-3.

table 11-1   Commutative Properties of Real Numbers

Property	In Symbols	Examples	Comments/Notes
Commutative property of addition	$a + b = b + a$	$-4 + 7 = 7 + (-4)$ $x + 3 = 3 + x$	The order in which two real numbers are added does not affect the sum.
Commutative property of multiplication	$a \cdot b = b \cdot a$	$(-5)(9) = (9)(-5)$ $8y = y \cdot 8$	The order in which two real numbers are multiplied does not affect the product.

table 11-2   Associative Properties of Real Numbers

Property	In Symbols	Examples	Comments/Notes
Associative property of addition	$(a + b) + c$ $= a + (b + c)$	$(5 + 8) + 1$ $= 5 + (8 + 1)$ $(t + n) + 3$ $= t + (n + 3)$	The manner in which three real numbers are grouped under addition does not affect the sum.
Associative property of multiplication	$(a \cdot b) \cdot c = a \cdot (b \cdot c)$	$-2(3 \cdot 6) = (-2 \cdot 3)(6)$ $3(n \cdot m) = (3 \cdot n)m$	The manner in which three real numbers are grouped under multiplication does not affect the product.

table 11-3   Distributive Property of Multiplication over Addition

Property	In Symbols	Examples	Comments/Notes
Distributive property of multiplication over addition*	$a(b + c)$ $= a \cdot b + a \cdot c$	$2(x + y) = 2x + 2y$ $-3(2 + z)$ $= -3(2) + (-3)z$	The factor outside the parentheses is multiplied by each term within the parentheses.

*Note that the distributive property of multiplication over addition is sometimes referred to as just the *distributive property.*

Example 3 demonstrates the use of the commutative properties.

example 3   Applying the Commutative Properties of Real Numbers

Apply the commutative property of addition or multiplication to rewrite the expression.

a. $6 + p$　　　**b.** $y(7)$　　　**c.** $-5 + n$　　　**d.** xy

Solution:

a. $6 + p = p + 6$　　　Commutative property of addition

b. $y(7) = 7y$　　　Commutative property of multiplication

c. $-5 + n = n + (-5)$　　　Commutative property of addition

　　　$= n - 5$

d. $xy = yx$　　　Commutative property of multiplication

Recall from Section 1.3 that subtraction is not a commutative operation. However, if we rewrite the difference of two numbers $a - b$ as $a + (-b)$, then we can apply the commutative property of addition. For example,

$x - 9 = x + (-9)$　　　Rewrite as addition of the opposite.

　　　$= -9 + x$　　　Apply the commutative property of addition.

Example 4 demonstrates the associative properties of addition and multiplication.

Skill Practice

Use the associative property of addition or multiplication to rewrite the expression. Then simplify the expression.

15. $2(-8w)$

16. $6.2 + (14.7 + x)$

17. $-\dfrac{1}{3}(3y)$

---

example 4   Applying the Associative Properties of Real Numbers

Use the associative property of addition or multiplication to rewrite each expression. Then simplify the expression.

a. $5(7w)$ **b.** $1.2 + (4.5 + y)$ **c.** $-\dfrac{2}{5}\left(-\dfrac{5}{2}z\right)$

Solution:

a. $5(7w) = (5 \cdot 7)w$ Apply the associative property of multiplication.

$\qquad\qquad = 35w$ Simplify.

b. $1.2 + (4.5 + y) = (1.2 + 4.5) + y$ Apply the associative property of addition.

$\qquad\qquad\qquad = 5.7 + y$ Simplify.

c. $-\dfrac{2}{5}\left(-\dfrac{5}{2}z\right) = \left(-\dfrac{2}{5} \cdot -\dfrac{5}{2}\right)z$ Apply the associative property of multiplication.

$\qquad\qquad = 1 \cdot z$ Simplify. Note that the factors within the parentheses are reciprocals.

$\qquad\qquad = z$ Therefore, their product is 1.

Note that in most cases, a detailed application of the associative properties will not be given. Instead the process will be written in one step, such as

$$5(7w) = 35w \qquad 1.2 + (4.5 + y) = 5.7 + y \qquad -\frac{2}{5}\left(-\frac{5}{2}z\right) = z$$

Example 5 demonstrates the use of the distributive property.

Skill Practice

Apply the distributive property.

18. $4(2 + m)$

19. $6(5p - 3q + 1)$

20. $\dfrac{5}{2}\left(4t + \dfrac{1}{3}\right)$

---

example 5   Applying the Distributive Property

Apply the distributive property.

a. $3(x + 4)$ **b.** $2(3y - 5z + 1)$ **c.** $\dfrac{2}{3}\left(6p + \dfrac{1}{4}\right)$

Solution:

a. $3(x + 4) = 3(x) + 3(4)$ Apply the distributive property.

$\qquad\qquad = 3x + 12$ Simplify.

b. $2(3y - 5z + 1) = 2(3y + (-5z) + 1)$ First write the subtraction as addition of the opposite.

$\qquad\qquad = 2(3y + (-5z) + 1)$ Apply the distributive property.

$\qquad\qquad = 2(3y) + 2(-5z) + 2(1)$

$\qquad\qquad = 6y + (-10z) + 2$ Simplify.

$\qquad\qquad \text{or } 6y - 10z + 2$

Answers

15. $[2 \cdot (-8)]w; \ -16w$

16. $(6.2 + 14.7) + x; \ 20.9 + x$

17. $\left(-\dfrac{1}{3} \cdot 3\right)y; \ -y$

18. $8 + 4m$

19. $30p - 18q + 6$

20. $10t + \dfrac{5}{6}$

Tip: In Example 5(b), we rewrote the expression by writing the subtraction as addition of the opposite. Often this step is not shown, and fewer steps are shown overall. For example,

$$2(3y - 5z + 1) = 2(3y) + 2(-5z) + 2(1)$$
$$= 6y - 10z + 2$$

c. $\dfrac{2}{3}\left(6p + \dfrac{1}{4}\right) = \dfrac{2}{3}(6p) + \dfrac{2}{3}\left(\dfrac{1}{4}\right)$ Apply the distributive property.

$= \dfrac{2}{3}\left(\dfrac{6p}{1}\right) + \dfrac{2}{3}\left(\dfrac{1}{4}\right)$ Write the whole number as an improper fraction.

$= \dfrac{2}{\overset{1}{3}}\left(\dfrac{\overset{2}{6p}}{1}\right) + \dfrac{2}{3}\left(\dfrac{1}{\underset{2}{4}}\right)$ Multiply fractions.

$= 4p + \dfrac{1}{6}$ Simplify.

---

example 6   Applying the Distributive Property

Apply the distributive property.

a. $-8(2 - 5y)$ **b.** $-(-4a + b + 3c)$

Solution:

a. $-8(2 - 5y)$

$= -8[2 + (-5y)]$ Write the subtraction as addition of the opposite.

$= -8[2 + (-5y)]$ Apply the distributive property.

$= -8(2) + (-8)(-5y)$

$= -16 + 40y$ Simplify.

b. $-(-4a + b + 3c)$ The negative sign preceding the parentheses indicates that we take the opposite of the expression within parentheses. This is equivalent to multiplying the expression within parentheses by -1.

$= -1 \cdot (-4a + b + 3c)$

$= -1(-4a) + (-1)(b) + (-1)(3c)$ Apply the distributive property.

$= 4a - b - 3c$ Simplify.

Skill Practice

Apply the distributive property.
21. $-4(6 - 10x)$
22. $-(2x - 3y + 4z)$

Tip: Notice that a negative factor outside the parentheses changes the signs of all terms to which it is multiplied.

$$-1 \cdot (-4a + b + 3c)$$
$$= +4a - b - 3c$$

Answers
21. $-24 + 40x$
22. $-2x + 3y - 4z$

section 11.1   Practice Exercises

Study Skills Exercises

1. When beginning a study of algebra, some students do not understand the concept of a variable. A variable is a letter that represents an unknown value. When trying to find what number added to 5 equals -12, we write $5 + x = -12$. Rewrite the following, using variables:

   What number times 4 equals 6? _____

   What number divided by 5 equals 3? _____

2. Define the key terms.

   a. Constant   **b. Expression**   **c. Variable**

Objective 1: Algebraic Expressions

3. Maria needs to buy 8 wine glasses. Write an expression for the cost of 8 glasses at p dollars each. **(See Example 1.)**

4. Jonathan is 4 in. taller than his brother. Write an expression for Jonathan's height if his brother is t in. tall.

5. It takes Perry $\frac{1}{2}$ hr longer than David to mow the lawn. If it takes David l hr to mow the lawn, write an expression for the amount of time it takes Perry to mow the lawn.

6. A sedan travels 6 mph slower than a sports car. Write an expression for the speed of the sedan if the sports car travels v mph.

7. A piece of ribbon is cut into n pieces of equal length. If the length of the ribbon is 4 yd, write an expression for the length of each piece.

8. A party is planned for 20 people. If there is p oz of punch available, write an expression for the amount of punch for each person, assuming that each person drinks an equal amount.

Objective 2: Evaluating Expressions

For Exercises 9–16, evaluate the expression for the given values. **(See Example 2.)**

9. $-6x$ for

   a. $x = 2$

   b. $x = -5$

10. $-2y^2$ for

    a. $y = 3$

    b. $y = -3$

11. $3p + 5q$ for

    a. $p = 2, q = -\dfrac{1}{5}$

    b. $p = -5, q = 0$

12. $9c - 2d$ for

   a. $c = -1, d = \dfrac{1}{2}$

   b. $c = 3, d = -2$

13. $-a^2$ for

   a. $a = -7$

   b. $a = 7$

14. $-b^3$ for

   a. $b = -3$

   b. $b = 3$

15. $-4(r - s)^2$ for

   a. $r = 8, s = 6$

   b. $r = 3, s = -1$

16. $-5(u + v)^2$ for

   a. $u = 10, v = -7$

   b. $u = 0, v = -2$

For Exercises 17–24, evaluate the expression when $x = -2$, $y = \dfrac{2}{3}$, $z = 4$, and $w = -\dfrac{1}{2}$.

17. $6y - 4w^2$

18. $-4x^2 - 2z$

19. $xw + z$

20. $-3yw$

21. $y(x - 4)$

22. $w(-x - 4)$

23. $z^2 - x + 6$

24. $x^3 - w - \dfrac{3}{2}$

For Exercises 25–32, evaluate the expression when $a = 12$, $b = -3$, $c = -2$, and $d = 0$.

25. $a \div (b - d)$

26. $-a \div (2b - d)$

27. $bc \div a$

28. $5b - c$

29. $\dfrac{1}{4}a - b$

30. $-\dfrac{1}{6}a + b$

31. $b^2 - c^2$

32. $b^2 + c^2$

For Exercises 33–38, find the area A, perimeter P, or volume V.

33. $P = 2l + 2w$ for $l = 6$ in. and $w = 2.3$ in. (perimeter of a rectangle)

34. $A = lw$ for $l = \dfrac{3}{2}$ ft and $w = 4$ ft (area of a rectangle)

35. $A = \pi r^2$ for $\pi = \dfrac{22}{7}$ m and $r = \dfrac{7}{2}$ m (area of a circle)

36. $A = \dfrac{1}{2}bh$ for $b = \dfrac{4}{5}$ yd and $h = \dfrac{10}{11}$ yd (area of a triangle)

37. $V = \dfrac{1}{3}\pi r^2 h$ for $\pi = 3.14$, $r = 10$ cm, and $h = 12$ cm (volume of a cone)

38. $V = lwh$ for $l = 2.1$ ft, $w = 3.2$ ft, and $h = 5$ ft (volume of a rectangular solid)

Objective 3: Properties of Real Numbers

For Exercises 39–50, apply the commutative property of addition or multiplication to rewrite each expression. **(See Example 3.)**

39. $5 + w$

40. $t + 2$

41. $-\dfrac{1}{3} + b$

42. $-\dfrac{1}{2} + c$

43. $r(2)$

44. $a(-4)$

45. $t(-s)$

46. $d(-c)$

47. $a - 9$

48. $x - 12$

49. $7 - p$

50. $8 - q$

For Exercises 51–62, apply the associative property of addition or multiplication to rewrite each expression. Then simplify the expression. **(See Example 4.)**

51. $-2(6b)$

52. $-3(2c)$

53. $3 + (8 + t)$

54. $7 + (5 + p)$

55. $-4.2 + (2.5 + r)$

56. $1.1 + (-0.8 + w)$

57. $3(6x)$

58. $9(5k)$

59. $-\dfrac{4}{7}\left(-\dfrac{7}{4}d\right)$

60. $\dfrac{5}{6}\left(\dfrac{6}{5}m\right)$

61. $-9 + (-12 + h)$

62. $-11 + (-4 + s)$

For Exercises 63–78, apply the distributive property. **(See Examples 5–6.)**

63. $4(x + 8)$

64. $5(3 + w)$

65. $-2(p + 4)$

66. $-6(k + 2)$

67. $4(a + 4b - c)$

68. $2(3q - r + s)$

69. $4\left(\dfrac{2}{3} + g\right)$

70. $8\left(\dfrac{5}{6} + m\right)$

71. $-(3 - n)$

72. $-(13 - t)$

73. $-(-a - 8)$

74. $-(-d - 10)$

75. $-(3x + 9 - 5y)$

76. $-(a - 8b + 4c)$

77. $-(-5q - 2s - 3t)$

78. $-(-10p - 12q + 3)$

Mixed Exercises

For Exercises 79–90, apply the appropriate property to simplify the expression.

79. $6(2x)$

80. $-3(12k)$

81. $6(2 + x)$

82. $-3(12 + k)$

83. $-8 + (4 - p)$

84. $3 + (25 - m)$

85. $-8(4 - p)$

86. $3(25 - m)$

87. $\dfrac{5}{9}(9 + y)$

88. $-\dfrac{3}{4}(8 - b)$

89. $\dfrac{5}{9}(9y)$

90. $-\dfrac{3}{4}(8b)$

section 11.2   Simplifying Expressions

Objectives

1. Definition of *Like* Terms
2. Combining *Like* Terms
3. Clearing Parentheses and Combining *Like* Terms

1. Definition of *Like* Terms

An algebraic expression is the sum of one or more terms. A **term** is a constant or the product of a constant and one or more variables. For example, the expression

$$-8x^3 + xy - 40 \qquad \text{can be written as} \qquad -8x^3 + xy + (-40)$$

This expression consists of the terms $-8x^3$, xy, and -40. The terms $-8x^3$ and xy are called **variable terms**, and the term -40 is called a **constant term**.

It is important to distinguish between a term and the factors within a term. For example, the quantity xy is one term, and the values x and y are factors within the term. The constant factor in a term is called the **coefficient** of the term.

Term	Coefficient of the term
$-8x^3$	-8
xy or $1xy$	1
-40	-40

Tip: Variables without a coefficient explicitly written have a coefficient of 1. Thus, the term x is equal to $1x$. The 1 is understood.

Terms are said to be *like* **terms** if they each have the same variables and the corresponding variables are raised to the same powers. For example,

Like Terms	Unlike Terms	
$-4x$ and $6x$	$-4x$ and $6y$	(different variables)
$18ab$ and $4ab$	$18ab$ and $4a$	(different variables)
$7m^2n^5$ and $3m^2n^5$	$7m^2n^5$ and $3mn^5$	(different powers on m)
$5p$ and $-3p$	$5p$ and 3	(different variables)
8 and 10	8 and $10x$	(different variables)

---

example 1   Identifying Terms, Coefficients, and *Like* Terms

a. List the terms of the expression: $1.4x^3 - 6x^2 + x + 5$

b. Identify the coefficient of each term: $1.4x^3 - 6x^2 + x + 5$

c. Which two terms are *like* terms? $-6x$, 5, $-3y$, and $4x$

Skill Practice

Given:

$$-4.8y^5 + 8.1y^2 - \frac{1}{2}y - 8$$

1. List the terms of the expression.
2. List the coefficients of the expression.

Solution:

a. The expression $1.4x^3 - 6x^2 + x + 5$ can be written as
$1.4x^3 + (-6x^2) + x + 5$

Therefore, the terms are $1.4x^3$, $-6x^2$, x, and 5.

b. The coefficients are 1.4, -6, 1, and 5.

c. The terms $-6x$ and $4x$ are *like* terms.

2. Combining *Like* Terms

Two terms may be combined if they are *like* terms. To add or subtract *like* terms, we use the distributive property, as shown in Example 2.

Answers

1. $-4.8y^5$, $8.1y^2$, $-\frac{1}{2}y$, -8
2. -4.8, 8.1, $-\frac{1}{2}$, -8

example 2   Using the Distributive Property to Add and Subtract *Like* Terms

Add or subtract as indicated.

a. $8y + 6y$ **b.** $-15w + 4w - w$

Solution:

a. $8y + 6y = (8 + 6)y$ Apply the distributive property.

$ = 14y$ Simplify.

b. $-15w + 4w - w = -15w + 4w - 1w$ First note that $w = 1w$.

$ = (-15 + 4 - 1)w$ Apply the distributive property.

$ = (-12)w$ Simplify within parentheses.

$ = -12w$

Although the distributive property is used to add and subtract *like* terms, it is tedious to write each step. Observe that adding or subtracting *like* terms is a matter of combining the coefficients and leaving the variable factors unchanged. This can be shown in one step.

$$8y + 6y = 14y \quad \text{and} \quad -15w + 4w - 1w = -12w$$

This shortcut will be used throughout the text.

example 3   Adding and Subtracting *Like* Terms

Simplify by combining *like* terms.

a. $-3x + 8y + 4x - 19 - 10y$ **b.** $\dfrac{2}{5}m + \dfrac{1}{8}n - \dfrac{1}{5}m + \dfrac{3}{8}n$

c. $0.2a - 1.4 + 1.4a - 6b - 2.1$

Solution:

a. $-3x + 8y + 4x - 19 - 10y$

$= -3x + 4x + 8y - 10y - 19$ Arrange *like* terms together.

$= 1x - 2y - 19$ Combine *like* terms.

$= x - 2y - 19$ Note that $1x = x$. Also note that the remaining terms cannot be combined further because they are not *like* terms. The variable factors are different.

b. $\dfrac{2}{5}m + \dfrac{1}{8}n - \dfrac{1}{5}m + \dfrac{3}{8}n$

$= \dfrac{2}{5}m - \dfrac{1}{5}m + \dfrac{1}{8}n + \dfrac{3}{8}n$ Arrange *like* terms together.

$= \dfrac{1}{5}m + \dfrac{4}{8}n$ Combine *like* terms.

$= \dfrac{1}{5}m + \dfrac{1}{2}n$ Simplify fractions.

c. $0.2a - 1.4 + 1.4a - 6b - 2.1$

$= 0.2a + 1.4a - 6b - 1.4 - 2.1$ Arrange *like* terms together.

$= 1.6a - 6b - 3.5$ Combine *like* terms.

3. Clearing Parentheses and Combining *Like* Terms

Notice that when the distributive property is applied, the original parentheses are dropped. This is often called *clearing parentheses*.

---

example 4   Clearing Parentheses and Combining *Like* Terms

Simplify by clearing parentheses and combining *like* terms. $6 - 3(2y + 9)$

Solution:

$6 - 3(2y + 9)$ The order of operations indicates that we must perform multiplication before subtraction.

It is also important to understand that a factor of -3 (not 3) will be multiplied by all terms within the parentheses. To see why, we may rewrite the subtraction in terms of addition of the opposite.

$6 - 3(2y + 9) = 6 + (-3)(2y + 9)$ Rewrite subtraction as addition of the opposite.

$= 6 + (-3)(2y) + (-3)(9)$ Apply the distributive property.

$= 6 + (-6y) + (-27)$ Simplify.

$= -6y + 6 + (-27)$ Arrange *like* terms together.

$= -6y + (-21)$ or $-6y - 21$ Combine *like* terms.

Skill Practice

Simplify.

8. $7 + 4(3x + 6)$

9. $8 - 6(w + 4)$

---

example 5   Clearing Parentheses and Combining *Like* Terms

Simplify by clearing parentheses and combining *like* terms.
$-8(x - 4) + 5(x + 7)$

Solution:

$-8(x - 4) + 5(x + 7)$

$= -8[x + (-4)] + 5(x + 7)$ Rewrite subtraction as addition of the opposite.

$= -8[x + (-4)] + 5(x + 7)$ Apply the distributive property.

$= -8(x) + (-8)(-4) + 5(x) + 5(7)$

$= -8x + 32 + 5x + 35$ Simplify.

$= -8x + 5x + 32 + 35$ Arrange *like* terms together.

$= -3x + 67$ Combine *like* terms.

Skill Practice

Simplify.

10. $-2(t - 3) + 4(t + 2)$

11. $-5(10 - m) - 2(m + 1)$

Answers

8. $12x + 31$ 9. $-6w - 16$
10. $2t + 14$ 11. $3m - 52$

section 11.2   Practice Exercises

Boost *your* GRADE at
mathzone.com!
MathZone

- Practice Problems
- Self-Tests
- NetTutor
- e-Professors
- Videos

Study Skills Exercises

1. Two important concepts in this section are *terms* and *factors*. Consider the expression $2x + 5y$. The quantities $2x$ and $5y$ are terms of the expression. Now consider the expression $2xy$. In this expression, 2, x, and y are factors. Write in your own words the difference between a term and a factor.

2. Define the key terms.

   a. Coefficient   **b. Constant term**   **c. *Like* terms**

   d. Term   **e. Variable term**

Review Exercises

For Exercises 3–8, simplify the expression, using the associative or distributive properties to clear the parentheses.

3. $6(p + 3)$ **4.** $(-7p + 2) + 10$ **5.** $4(-6q)$

6. $-3(t - 2)$ **7.** $13 + (-4 - h)$ **8.** $-(x - 20y - 14z)$

Objective 1: Definition of *Like* Terms

For Exercises 9–16, for each expression, list the terms and identify each term as a variable term or a constant term.

9. $2a + 5b^2 + 6$ **10.** $-5x - 4 + 7y$

11. $8 + 9a$ **12.** $12 - 8k$

13. $4pq - 9p$ **14.** $9t - 8st$

15. $10h^2 - 15 - 4h$ **16.** $w^2 - 3 - 5w$

For Exercises 17–24, identify the coefficients for each term. **(See Example 1.)**

17. $6p - 4q$ **18.** $-5a^3 - 2a$ **19.** $-14h + 12$ **20.** $8x + 9$

21. $x - y$ **22.** $p - q$ **23.** $5t - 8s - 3$ **24.** $6g - 16h - 2$

For Exercises 25–36, determine if the two terms are *like* terms or unlike terms. **(See Example 1.)**

25. $3a, -2a$

26. $8b, 12b$

27. $4x, 4y$

28. $-9k, -9h$

29. $7xy, -3yx$

30. $-5ab, ba$

31. $6a, 13a^2$

32. $20k^3, 3k$

33. $14, 14y$

34. $25x, 25$

35. $17, -32$

36. $8, -22$

Objective 2: Combining *Like* Terms

For Exercises 37–62, combine the *like* terms. **(See Examples 2–3.)**

37. $6rs + 8rs$

38. $4x + 21x$

39. $-4h + 12h$

40. $9p - 13p$

41. $4x^2 + 9 - x^2$

42. $13t^2 - t^2 + 4$

43. $10x - 12y - 4x - 3y$

44. $14a - 5b + 3a - b$

45. $-6k - 9k + 12k$

46. $-11p + 23p - p$

47. $-8uv + 6u + 12uv$

48. $9pq - 9p + 13pq$

49. $6 - 14m - 15 - 2m$

50. $1 - 8n + 5 - 3n$

51. $18 - 3a + 5b - 6a + 2$

52. $13 + w - 5z - 4 + 7w$

53. $-5p^2 + 6p - p^2 + 7 - 8p$

54. $-3q^2 - 10q + q^2 - 15 + 5q$

55. $\frac{1}{2}y + \frac{3}{2}y - \frac{5}{6}$

56. $-\frac{4}{5}p + \frac{2}{5}p + \frac{4}{7}$

57. $\frac{3}{4}a + 3 - \frac{1}{8}a + 6$

58. $\frac{1}{3}b - 4 + \frac{2}{9}b - 4$

59. $2.3x^2 + 4.1x - 5.3x^2 - 6x$

60. $1.2y - 0.4y^2 - 0.3y - 1.5y^2$

61. $4.4 - 0.9a + 3.2$

62. $9.7 - 8.8b - 3.2$

Objective 3: Clearing Parentheses and Combining *Like* Terms

For Exercises 63–86, clear parentheses and combine *like* terms. **(See Examples 4–5.)**

63. $5(t - 6) + 2$

64. $7(a - 4) + 8$

65. $-3(2x + 1) - 13$

66. $-2(4b + 3) - 10$

67. $4 + 6(y - 3)$

68. $11 + 2(p - 8)$

69. $21 - 7(3 - q)$

70. $10 - 5(2 - 5m)$

71. $-3 - (2n + 1)$

72. $-13 - (6s + 5)$

73. $-2(a + 3b) - (4a - 5b)$

74. $-(2m - 7n) - 3(6m - n)$

75. $10(x + 5) - 3(2x + 9)$

76. $6(y - 9) - 5(2y - 5)$

77. $-(12z + 1) + 2(7z - 5)$

78. $-(8w + 5) + 3(w - 15)$

79. $3(w + 3) - (4w + y) - 3y$

80. $2(s + 6) - (8s - t) + 6t$

81. $20a - 4(b + 3a) - 5b$

82. $16p - 3(2p - q) + 7q$

83. $6 - (3m - n) - 2(m + 8) + 5n$

84. $12 - (5u + v) - 4(u - 6) + 2v$

85. $15 + 2(w - 4) - (2w - 5z) + 7z$

86. $7 + 3(2a - 5) - (6a - 8b) - 2b$

Expanding Your Skills

For Exercises 87–94, clear parentheses and combine *like* terms in expressions involving fractions and decimals.

87. $6\left(\frac{1}{2}x - \frac{2}{3}\right) - 4\left(\frac{5}{2}x + \frac{3}{4}\right)$

88. $-12\left(\frac{5}{6}p + \frac{1}{4}\right) + 9\left(\frac{2}{9}p - \frac{1}{3}\right)$

89. $\frac{2}{3}(9y + 6) - \frac{3}{2}(18y - 16)$

90. $-\frac{1}{4}(4w - 8) + \frac{1}{2}(4w + 10)$

91. $10(0.2q - 3) - 100(0.04q - 0.5)$

92. $100(0.14b + 0.2) - 10(1.3b - 4)$

93. $100(1.04a - 2.1b) - 10(21.1a + 0.3b)$

94. $10(-7.2x - y) + 1000(0.023x + 0.004y)$

Objectives

1. Definition of an Equation
2. Addition and Subtraction Properties of Equality

section 11.3   Addition and Subtraction Properties of Equality

1. Definition of an Equation

An **equation** is a statement that indicates that two quantities are equal. The following are equations.

$$x = 7 \qquad z + 3 = 8 \qquad -6p = 18$$

All equations have an equal sign. Furthermore, notice that the equal sign separates the equation into two parts, the left-hand side and the right-hand side. A **solution** to an equation is a value of the variable that makes the equation a true statement. Substituting a solution to an equation for the variable makes the right-hand side equal to the left-hand side.

Equation	Solution	Check	
$x = 7$	7	$x = 7$ $\downarrow$ $7 = 7$ ✓	Substitute 7 for x. Right-hand side equals left-hand side.
$z + 3 = 8$	5	$z + 3 = 8$ $\downarrow$ $5 + 3 = 8$ ✓	Substitute 5 for z. Right-hand side equals left-hand side.
$-6p = 18$	-3	$-6p = 18$ $\downarrow$ $-6(-3) = 18$ ✓	Substitute -3 for p. Right-hand side equals left-hand side.

Concept Connections

Which of the following are equations?

1. $-2w = 6$

2. $4 = t - 3$

3. $x + 8$

example 1   Determining Whether a Number Is a Solution to an Equation

Determine whether the given number is a solution to the equation.

a. $2x - 9 = 3$; 6   **b.** $8 = 8p - 4$; $-\frac{1}{2}$

Solution:

a. $2x - 9 = 3$

$2(6) - 9 \stackrel{?}{=} 3$ Substitute 6 for x.

$12 - 9 \stackrel{?}{=} 3$ Simplify.

$3 = 3$ ✓   Right-hand side equals the left-hand side. Thus, 6 is a solution to the equation $2x - 9 = 3$.

b. $8 = 8p - 4$

$8 \stackrel{?}{=} 8\left(-\frac{1}{2}\right) - 4$ Substitute $-\frac{1}{2}$ for p.

$8 \stackrel{?}{=} -4 - 4$ Simplify.

$8 \neq -8$ Right-hand side does not equal left-hand side. Thus, $-\frac{1}{2}$ is *not* a solution to the equation $8 = -8p - 4$.

Skill Practice

Determine whether the given number is a solution to the equation.

4. $2 + 3x = 23$; 7

5. $-4x + 1 = 9$; 2

In the study of algebra, you will encounter a variety of equations. In this chapter, we focus on a specific type of equation called a linear equation in one variable.

Definition of a Linear Equation in One Variable

Let a and b be real numbers such that $a \neq 0$. A **linear equation in one variable** is an equation that can be written in the form

$$ax + b = 0$$

Answers

1. Equation   2. Equation
3. Not an equation
4. Yes, 7 is a solution
5. No, 2 is not a solution

2. Addition and Subtraction Properties of Equality

Given the equation $x = 3$, we can easily determine that the solution is 3. The solution to the equation $2x + 14 = 20$ is also 3. These two equations are called **equivalent equations** because they have the same solution. However, while the solution to $x = 3$ is obvious, the solution to $2x + 14 = 20$ is not. Our goal in this chapter is to learn how to *solve* equations.

To solve an equation, we use algebraic principles to write an equation such as $2x + 14 = 20$ in an equivalent but simpler form, such as $x = 3$. The addition and subtraction properties of equality are the first tools we will learn to solve an equation.

The Addition and Subtraction Properties of Equality

Let a, b, and c represent algebraic expressions.

1. The **addition property of equality:**   If $a = b$
then $a + c = b + c$

2. The **subtraction property of equality:**   If $a = b$
then $a - c = b - c$

The addition and subtraction properties of equality indicate that adding or subtracting the same quantity to each side of an equation results in an equivalent equation. This is true because if two quantities are increased (or decreased) by the same amount, then the resulting quantities will also be equal (Figure 11-2).

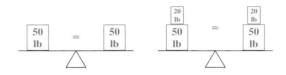

Figure 11-2

example 2   Applying the Addition Property of Equality

Solve the equations and check the solution.

a. $x - 6 = 18$ **b.** $-12 = x - 7$ **c.** $-4.1 + x = 3.8$

Solution:

To solve an equation, the goal is to isolate the variable on one side of the equation. That is, we want to create an equivalent equation of the form $x = $ number. To accomplish this, we can use the fact that the sum of a number and its opposite is zero.

a. $x - 6 = 18$

$x - 6 + 6 = 18 + 6$ To isolate x, add 6 to both sides, because $-6 + 6 = 0$.

$x + 0 = 24$ Simplify.

$x = 24$ The variable is isolated (by itself) on the left-hand side of the equation. The solution is 24.

Check: $x - 6 = 18$ Original equation

$(24) - 6 \overset{?}{=} 18$ Substitute 24 for x.

$18 = 18$ ✓   Right-hand side = left-hand side.

b. $-12 = x - 7$

$-12 + 7 = x - 7 + 7$ To isolate x, add 7 to both sides, because $-7 + 7 = 0$.

$-5 = x + 0$ Simplify.

$-5 = x$ The variable is isolated on the right-hand side of the equation. The solution is -5.

The equation $-5 = x$ is equivalent to $x = -5$.

Check: $-12 = x - 7$ Original equation

$-12 \overset{?}{=} (-5) - 7$ Substitute -5 for x.

$-12 = -12$ ✓   Right-hand side = left-hand side.

c. $-4.1 + x = 3.8$

$-4.1 + 4.1 + x = 3.8 + 4.1$ To isolate x, add 4.1 to both sides, because $-4.1 + 4.1 = 0$.

$0 + x = 7.9$ Simplify.

$x = 7.9$ The solution is 7.9.

Check: $-4.1 + x = 3.8$ Original equation

$-4.1 + (7.9) \overset{?}{=} 3.8$ Substitute 7.9 for x.

$3.8 = 3.8$ ✓

> **Tip:** Notice that the variable may be isolated on *either* side of the equal sign. In Example 2(a), the variable appears on the left. In Example 2(b), the variable appears on the right.

In Example 3, we apply the subtraction property of equality. This indicates that we can subtract the same quantity from both sides of the equation to obtain an equivalent equation.

example 3   Applying the Subtraction Property of Equality

Solve the equations and check.

a. $z + 11 = 14$ **b.** $-8 = 2 + q$

Solution:

a. $z + 11 = 14$

$z + 11 - 11 = 14 - 11$ Subtract 11 from both sides, because $11 - 11 = 0$.

$z + 0 = 3$ Simplify.

$z = 3$ The solution is 3.

Skill Practice

Solve the equation and check the solution.

10. $m + 8 = 21$

11. $-16 = 1 + z$

Answers

10. $m = 13$
11. $z = -17$

Check: $z + 11 = 14$ Original equation

$(3) + 11 \stackrel{?}{=} 14$ Substitute 3 for z.

$14 = 14$ ✓

b. $-8 = 2 + q$

$-8 - 2 = 2 - 2 + q$ Subtract 2 from both sides, because $2 - 2 = 0$.

$-10 = 0 + q$ Simplify.

$-10 = q$ The solution is -10.

Check: $-8 = 2 + q$ Original equation

$-8 \stackrel{?}{=} 2 + (-10)$ Substitute -10 for q.

$-8 = -8$ ✓

In Example 4, we use both the addition and subtraction properties of equality. As you read through each example, remember that you want to isolate the variable.

Skill Practice

Solve the equation.

12. $\dfrac{2}{3} = x + \dfrac{1}{6}$

13. $\dfrac{4}{5} + c = -\dfrac{1}{4}$

14. $y - 2.27 = -9.13$

example 4   Applying the Addition and Subtraction Properties of Equality

Solve the equations.

a. $\dfrac{1}{2} = t - \dfrac{3}{4}$ **b.** $\dfrac{3}{10} + m = -\dfrac{2}{3}$ **c.** $8.54 = p + 1.96$

Solution:

a. $\dfrac{1}{2} = t - \dfrac{3}{4}$ To isolate t, we add $\frac{3}{4}$ because $-\frac{3}{4} + \frac{3}{4} = 0$.

$\dfrac{1}{2} + \dfrac{3}{4} = t - \dfrac{3}{4} + \dfrac{3}{4}$ Add $\frac{3}{4}$ to both sides.

$\dfrac{1 \cdot 2}{2 \cdot 2} + \dfrac{3}{4} = t + 0$ Add the fractions on the left-hand side by first obtaining a common denominator. The LCD is 4.

$\dfrac{2}{4} + \dfrac{3}{4} = t$

$\dfrac{5}{4} = t$ The solution is $\frac{5}{4}$.

Check: $\dfrac{1}{2} = t - \dfrac{3}{4}$ Original equation

$\dfrac{1}{2} \stackrel{?}{=} \dfrac{5}{4} - \dfrac{3}{4}$ Substitute $t = \frac{5}{4}$.

$\dfrac{1}{2} = \dfrac{2}{4}$ ✓

b. $\dfrac{3}{10} + m = -\dfrac{2}{3}$ To isolate m, we subtract $\frac{3}{10}$ because $\frac{3}{10} - \frac{3}{10} = 0$.

$\dfrac{3}{10} - \dfrac{3}{10} + m = -\dfrac{2}{3} - \dfrac{3}{10}$ Subtract $\frac{3}{10}$ from both sides.

Answers

12. $x = \dfrac{1}{2}$ **13.** $c = -\dfrac{21}{20}$

14. $y = -6.86$

$$0 + m = -\frac{2 \cdot 10}{3 \cdot 10} - \frac{3 \cdot 3}{10 \cdot 3}$$

To subtract the fractions, first obtain a common denominator. The LCD is 30.

$$m = -\frac{20}{30} - \frac{9}{30}$$

$$m = -\frac{29}{30}$$

The solution is $-\frac{29}{30}$ and checks in the original equation.

c.

$$8.54 = p + 1.96$$

To isolate p, we subtract 1.96 because $1.96 - 1.96 = 0$.

$$8.54 - 1.96 = p + 1.96 - 1.96$$

Subtract 1.96 from each side.

$$6.58 = p + 0$$

$$6.58 = p$$

The solution is 6.58 and checks in the original equation.

section 11.3   Practice Exercises

Boost *your* GRADE at mathzone.com!

MathZone

• Practice Problems
• Self-Tests
• NetTutor
• e-Professors
• Videos

Study Skills Exercises

1. Up to this point we have been simplifying expressions. We will now begin solving equations. Consider the two lists:

Expressions	Equations
$3x + 2y$	$5x + 2 = 6$
$6(8 + x) + 2$	$2(x - 5) = 14$
$7y$	$7 = y$

Write in your own words the difference between an expression and an equation.

2. Define the key terms.

    a. Addition property of equality   **b. Equation**   **c. Equivalent equations**

    d. Linear equation in one variable   **e. Solution**   **f. Subtraction property of equality**

Review Exercises

For Exercises 3–8, simplify the expression.

3. $-10a + 3b - 3a + 13b$ **4.** $4 - 23y + 11 - 16y$ **5.** $-(-8h + 2k - 13)$

6. $3(-4m + 3) - 12$ **7.** $5z - 8(z - 3) - 20$ **8.** $-(7p - 12) - 10(1 - p) + 6$

Objective 1: Definition of an Equation

For Exercises 9–20, determine whether the given number is a solution to the equation. **(See Example 1.)**

9. $5x + 3 = -2$; -1

10. $3y - 2 = 4$; 2

11. $10 = p - 16$; 26

12. $-14 = q - 1$; -13

13. $-z + 8 = 20$; 12

14. $-7 - w = -10$; -3

15. $6m - 3 = -6$; $-\dfrac{1}{2}$

16. $-12n + 2 = -1$; $\dfrac{1}{4}$

17. $13 = 13 + 6t$; 0

18. $-\dfrac{1}{5} = r - \dfrac{1}{5}$; 0

19. $25 = -5q - 5$; 4

20. $39 = -7p - 4$; 5

Objective 2: Addition and Subtraction Properties of Equality

For Exercises 21–28, fill in the blank with the appropriate number.

21. $13 + (-13) =$ _____

22. $6 +$ _____ $= 0$

23. _____ $+ (-7) = 0$

24. $1 + (-1) =$ _____

25. $3.2 +$ _____ $= 0$

26. _____ $+ (-0.3) = 0$

27. $-\dfrac{3}{8} +$ ____ $= 0$

28. $-\dfrac{1}{6} +$ ____ $= 0$

For Exercises 29–38, solve the equation using the addition property of equality. **(See Example 2.)**

29. $g - 23 = 14$

30. $h - 12 = 30$

31. $-4 + k = 12$

32. $-16 + m = 4$

33. $-18 = n - 3$

34. $-9 = t - 6$

35. $-\dfrac{5}{6} + p = \dfrac{1}{3}$

36. $-\dfrac{3}{4} + q = \dfrac{3}{2}$

37. $k - 4.3 = -1.2$

38. $a - 0.04 = -2.04$

For Exercises 39–44, fill in the blank with the appropriate number.

39. $52 -$ _____ $= 0$

40. $2 - 2 =$ _____

41. $18 - 18 =$ _____

42. _____ $- 15 = 0$

43. _____ $- 100 = 0$

44. $21 -$ _____ $= 0$

For Exercises 45–54, solve the equation using the subtraction property of equality. **(See Example 3.)**

45. $x + 34 = 6$

46. $y + 12 = 4$

47. $17 + b = 20$

48. $5 + c = 14$

49. $-32 = t + 14$

50. $-23 = k + 11$

51. $8.2 = m + 21.8$

52. $16.01 = n + 20.88$

53. $a + \dfrac{3}{5} = -\dfrac{7}{10}$

54. $b + \dfrac{1}{4} = -\dfrac{3}{8}$

Mixed Exercises

For Exercises 55–74, solve the equation by using the appropriate property. **(See Example 4.)**

55. $1 + p = 0$

56. $r - 12 = 13$

57. $-34 + t = -40$

58. $7 + q = 4$

59. $\dfrac{2}{3} = y - \dfrac{5}{12}$

60. $\dfrac{7}{11} = z + \dfrac{3}{11}$

61. $-2.5 = -1.1 + m$

62. $-4.1 = -3.5 + n$

63. $w - 23 = -11$

64. $p - 10 = -9$

65. $x + 21 = 16$

66. $y + 18 = -4$

67. $-2 = a - 15$

68. $-1 = b - 49$

69. $4.01 + p = 3.22$

70. $2.8 + q = 6.1$

71. $t + \dfrac{3}{8} = 2$

72. $r - \dfrac{4}{7} = -1$

73. $27 = z - 22$

74. $109 = x + 49$

Expanding Your Skills

For Exercises 75–80, first simplify each side of the equation. Then solve the equation.

75. $5h - 4h + 4 = 3$

76. $10x - 9x - 11 = 15$

77. $9 + (-2) = 4 + t$

78. $-13 + 15 = p + 5$

79. $3(r - 2) - 2r = 6 + (-2)$

80. $4(k + 2) - 3k = -6 + 9$

Objectives

1. Multiplication and Division Properties of Equality
2. Using the Properties of Equality

section 11.4 Multiplication and Division Properties of Equality

1. Multiplication and Division Properties of Equality

Adding or subtracting the same quantity to both sides of an equation results in an equivalent equation. In a similar way, multiplying or dividing both sides of an equation by the same nonzero quantity also results in an equivalent equation. This is stated formally as the multiplication and division properties of equality.

The Multiplication and Division Properties of Equality

Let a, b, and c represent algebraic expressions.

1. The **multiplication property of equality:** If $a = b$ then $a \cdot c = b \cdot c$
2. The **division property of equality:** If $a = b$ then $\frac{a}{c} = \frac{b}{c}$ (provided $c \neq 0$)

To understand the multiplication property of equality, suppose we start with a true equation such as $10 = 10$. If both sides of the equation are multiplied by a constant such as 3, the result is also a true statement (Figure 11-3).

$$10 = 10$$
$$3 \cdot 10 = 3 \cdot 10$$
$$30 = 30$$

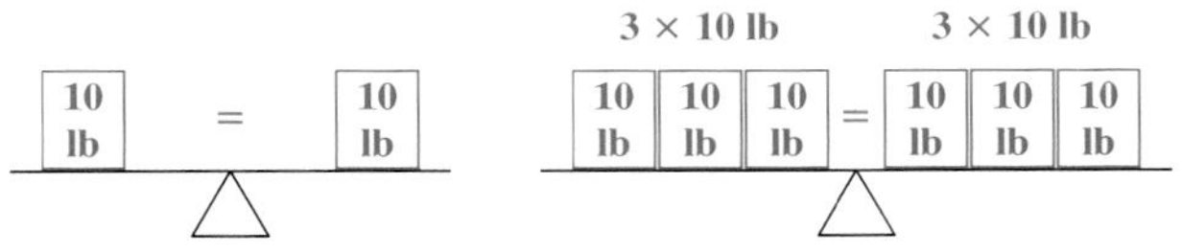

Figure 11-3

To solve an equation in the variable x, the goal is to write the equation in the form $x =$ number. In particular, notice that we desire the coefficient of x to be 1. That is, we want to write the equation as $1 \cdot x =$ number. To solve an equation such as $3x = 12$, we can multiply both sides of the equation by the reciprocal of the x-term coefficient. In this case, multiply both sides by the reciprocal of 3, which is $\frac{1}{3}$.

$$3x = 12$$

$\frac{1}{3} \cdot (3x) = \frac{1}{3} \cdot (12)$ Multiply by the reciprocal of 3, which is $\frac{1}{3}$.

$1 \cdot x = 4$ The coefficient of the x-term is now 1.

$x = 4$ Simplify.

The division property of equality can also be used to solve the equation $3x = 12$ by dividing both sides by the coefficient of the x-term. In this case, divide both sides by 3 to make the coefficient of x equal to 1.

Tip: Recall that the product of a number and its reciprocal is 1. For example:

$$\frac{1}{5} \cdot (5) = 1$$
$$\frac{3}{2} \cdot \frac{2}{3} = 1$$
$$-\frac{7}{2} \cdot \left(-\frac{2}{7}\right) = 1$$

$$3x = 12$$

$$\frac{3x}{3} = \frac{12}{3}$$ Divide by the coefficient of x which is 3.

$$1 \cdot x = 4$$ The coefficient on the x-term is now 1.

$$x = 4$$ Simplify.

example 1   Applying the Multiplication and Division Properties of Equality

Solve the equations by using the multiplication or division property of equality.

a. $10x = 50$ **b.** $28 = -4p$ **c.** $-y = 34$ **d.** $23.18 = 6.1w$

e. $\frac{4}{5}x = -12$ **f.** $-\frac{2}{3}p = -\frac{4}{7}$ **g.** $5 = \frac{d}{8}$

Skill Practice

Solve.

1. $4x = 32$
2. $18 = -2w$
3. $19 = -m$
4. $4.1z = 28.29$
5. $\frac{2}{3}x = 18$
6. $-\frac{5}{9}p = \frac{1}{3}$
7. $\frac{c}{12} = 3$

Solution:

a. $10x = 50$

$$\frac{10x}{10} = \frac{50}{10}$$ To obtain a coefficient of 1 for the x-term, divide both sides by 10.

$$1x = 5$$ Simplify.

$$x = 5$$ Check: $10x = 50$ Original equation

$$\qquad\qquad\qquad 10(5) \stackrel{?}{=} 50 \qquad \text{Substitute } 5 \text{ for } x.$$

$$\qquad\qquad\qquad 50 = 50 \checkmark$$

Tip: In Example 1(a) we could also have multiplied both sides by $\frac{1}{10}$ to obtain a coefficient of 1 for the x-term.

$$\frac{1}{10}(10x) = \frac{1}{10}(50)$$

$$1x = 5$$

b. $28 = -4p$

$$\frac{28}{-4} = \frac{-4p}{-4}$$ To obtain a coefficient of 1 for the x-term, divide both sides by -4. This is also equivalent to multiplying by $-\frac{1}{4}$.

$$-7 = 1p$$ Simplify.

$$-7 = p$$ The solution is -7 and checks in the original equation.

c. $-y = 34$ Note that $-y$ is the same as $-1 \cdot y$.

$$-1y = 34$$

$$\frac{-1 \cdot y}{-1} = \frac{34}{-1}$$ To obtain a coefficient of 1 for the y-term, divide both sides by -1.

$$1y = -34$$ Simplify.

$$y = -34$$ The solution is -34 and checks in the original equation.

Answers

1. $x = 8$ 2. $w = -9$
3. $m = -19$ 4. $z = 6.9$
5. $x = 27$ 6. $p = -\dfrac{3}{5}$
7. $c = 36$

Tip: In Example 1(c), we could have also multiplied both sides by -1 to obtain a coefficient of 1 for y.

$$(-1)(-y) = (-1)34$$

$$y = -34$$

d. $23.18 = 6.1w$

$$\frac{23.18}{6.1} = \frac{6.1w}{6.1}$$ To obtain a coefficient of 1 on the w-term, divide both sides by 6.1.

$$3.8 = 1w$$ Simplify.

$$3.8 = w$$ The solution is 3.8 and checks in the original equation.

e. $\dfrac{4}{5}x = -12$

$$\frac{5}{4} \cdot \frac{4}{5}x = \frac{5}{4}\left(\frac{-12}{1}\right)$$ To obtain a coefficient of 1 for the x-term, multiply by the reciprocal of $\frac{4}{5}$ which is $\frac{5}{4}$.

$$1x = -15$$ Multiply fractions. Note that $\dfrac{5}{4}\left(\dfrac{-\overset{3}{\cancel{12}}}{1}\right) = -15$.

$$x = -15$$ The solution is -15 and checks in the original equation.

f. $-\dfrac{2}{3}p = -\dfrac{4}{7}$

$$-\frac{3}{2}\left(-\frac{2}{3}p\right) = -\frac{3}{2}\left(-\frac{4}{7}\right)$$ To obtain a coefficient of 1 for the p-term, multiply by the reciprocal of $-\frac{2}{3}$ which is $-\frac{3}{2}$.

$$1p = \frac{12}{14}$$ Simplify.

$$p = \frac{6}{7}$$ The solution is $\frac{6}{7}$.

g. $5 = \dfrac{d}{8}$

$$5 = \frac{1}{8}d$$ The expression $\frac{d}{8}$ is equivalent to $\frac{1}{8}d$.

$$8(5) = 8\left(\frac{1}{8}d\right)$$ To obtain a coefficient of 1 on the d-term, multiply both sides by the reciprocal of $\frac{1}{8}$, which is 8.

$$40 = 1d$$ Simplify.

$$40 = d$$ The solution is 40.

Tip: When applying the multiplication or division property of equality to obtain a coefficient of 1 for the variable term, we will generally use the following conventions.

- If the coefficient of the variable term is expressed as a fraction, we will usually multiply both sides by its reciprocal. See Example 1(e)–(g).
- Otherwise, we will divide both sides by the coefficient itself. See Example 1(a)–(d).

2. Using the Properties of Equality

It is important to distinguish between cases where the addition or subtraction property of equality should be used to isolate a variable versus where the multiplication or division property of equality should be used. Compare the equations:

$$4 + x = 12 \quad \text{and} \quad 4x = 12$$

In the first equation, the relationship between 4 and x is addition. Therefore, we want to reverse the process by *subtracting* 4 from both sides. In the second equation, the relationship between 4 and x is multiplication. To isolate x, we reverse the process by *dividing* by 4 or, equivalently, by multiplying by the reciprocal $\frac{1}{4}$.

$$4 + x = 12 \quad\quad \text{and} \quad\quad 4x = 12$$

$$4 - 4 + x = 12 - 4 \quad\quad\quad \frac{4x}{4} = \frac{12}{4}$$

$$x = 8 \quad\quad\quad\quad\quad x = 3$$

In Example 2, we practice distinguishing which property of equality to use.

---

example 2   Solving Linear Equations

Solve the equations.

a. $\dfrac{m}{12} = -3$ **b.** $3.2 = x + 19.5$ **c.** $6 = -4t$ **d.** $y - \dfrac{5}{9} = \dfrac{2}{3}$

Solution:

a. $\dfrac{m}{12} = -3$ The relationship between m and 12 is division. To obtain a coefficient of 1 for the m-term, multiply both sides by 12.

$12\left(\dfrac{m}{12}\right) = 12(-3)$ Multiply both sides by 12.

$m = -36$ Simplify both sides. The solution -36 checks in the original equation.

b. $3.2 = x + 19.5$ The relationship between x and 19.5 is addition. To isolate the x-term, we can subtract 19.5 from both sides because $19.5 - 19.5 = 0$.

$3.2 - 19.5 = x + 19.5 - 19.5$ Subtract 19.5 from both sides.

$-16.3 = x$ Simplify. The solution -16.3 checks in the original equation.

Skill Practice

Solve.

11. $5 = -2p$

12. $z + \dfrac{1}{3} = \dfrac{5}{6}$

13. $\dfrac{1}{3}z = \dfrac{5}{6}$

c. $6 = -4t$ The relationship between t and -4 is multiplication. To obtain a coefficient of 1 on the t-term, we can divide both sides by -4.

$\dfrac{6}{-4} = \dfrac{-4t}{-4}$ Divide both sides by -4.

$-\dfrac{6}{4} = t$

$-\dfrac{3}{2} = t$ Simplify. The solution $-\dfrac{3}{2}$ checks in the original equation.

d. $y - \dfrac{5}{9} = \dfrac{2}{3}$ The relationship between y and $\frac{5}{9}$ is subtraction. Therefore, we can add $\frac{5}{9}$ to both sides to isolate y.

$y - \dfrac{5}{9} + \dfrac{5}{9} = \dfrac{2}{3} + \dfrac{5}{9}$ Add $\frac{5}{9}$ to both sides.

$y = \dfrac{2 \cdot 3}{3 \cdot 3} + \dfrac{5}{9}$ Obtain a common denominator. The LCD is 9.

$y = \dfrac{6}{9} + \dfrac{5}{9}$ Add the fractions.

$y = \dfrac{11}{9}$ The solution $\frac{11}{9}$ checks in the original equation.

Answers

11. $p = -\dfrac{5}{2}$ 12. $z = \dfrac{1}{2}$

13. $z = \dfrac{5}{2}$

---

section 11.4   Practice Exercises

Boost *your* GRADE at mathzone.com!

MathZone

• Practice Problems   • e-Professors
• Self-Tests   • Videos
• NetTutor

Study Skills Exercises

1. One way to know that you really understand a concept is to try to explain it to someone else. In your own words, explain when you would apply the multiplication property of equality or the division property of equality.

2. Define the key terms.

    a. Division property of equality   **b. Multiplication property of equality**

Review Exercises

For Exercises 3–10, solve the equation.

3. $p - 12 = 33$ **4.** $-8 = 10 + k$ **5.** $16 = h - 5$ **6.** $-4 + w = 22$

7. $p - 6 = -19$ **8.** $\dfrac{1}{6} = -\dfrac{11}{6} + m$ **9.** $n + \dfrac{1}{2} = -\dfrac{2}{3}$ **10.** $2.4 + z = -12$

Objective 1: Multiplication and Division Properties of Equality

For Exercises 11–18, fill in the blank with the appropriate number.

11. $3 \cdot \underline{\hspace{1cm}} = 1$ **12.** $-6 \cdot \underline{\hspace{1cm}} = 1$ **13.** $-\dfrac{4}{7} \cdot \underline{\hspace{1cm}} = 1$ **14.** $\dfrac{3}{10} \cdot \underline{\hspace{1cm}} = 1$

15. $-7 \div \underline{\hspace{1cm}} = 1$ **16.** $2 \div \underline{\hspace{1cm}} = 1$ **17.** $5.1 \div \underline{\hspace{1cm}} = 1$ **18.** $-6.8 \div \underline{\hspace{1cm}} = 1$

For Exercises 19–50, solve the equation by using the multiplication or division property of equality. **(See Example 1.)**

19. $14b = -42$ **20.** $-6p = 12$ **21.** $-8k = 56$ **22.** $5y = -25$

23. $-t = -13$ **24.** $-h = -17$ **25.** $\dfrac{2}{3}m = 14$ **26.** $\dfrac{5}{9}n = 40$

27. $\dfrac{b}{7} = -3$ **28.** $\dfrac{a}{4} = -12$ **29.** $-2.8 = -0.7t$ **30.** $-3.3 = -3r$

31. $-\dfrac{u}{2} = -15$ **32.** $-\dfrac{v}{10} = -4$ **33.** $6 = -18w$ **34.** $4 = -32g$

35. $1.3x = 5.33$ **36.** $8.1y = 17.82$ **37.** $\dfrac{5}{4}k = -\dfrac{1}{2}$ **38.** $-\dfrac{11}{12}h = -\dfrac{1}{6}$

39. $0 = \dfrac{3}{8}m$ **40.** $0 = \dfrac{1}{10}n$ **41.** $-\dfrac{9}{4}x = -\dfrac{3}{5}$ **42.** $-\dfrac{15}{14}y = \dfrac{1}{2}$

43. $100 = 5k$ **44.** $95 = 19h$ **45.** $31 = -p$ **46.** $11 = -q$

47. $3p = \dfrac{5}{2}$ **48.** $2q = \dfrac{7}{5}$ **49.** $-4a = 0$ **50.** $-7b = 0$

Objective 2: Using the Properties of Equality

51. In your own words explain how to determine when to use the addition property of equality.

52. In your own words explain how to determine when to use the subtraction property of equality.

53. Explain how to determine when to use the division property of equality.

54. Explain how to determine when to use the multiplication property of equality.

For Exercises 55–78, solve the equation. **(See Example 2.)**

55. $4 + x = -12$

56. $6 + z = -18$

57. $4y = -12$

58. $6p = -18$

59. $q - 4 = -12$

60. $p - 6 = -18$

61. $\dfrac{h}{4} = -12$

62. $\dfrac{w}{6} = -18$

63. $\dfrac{2}{3} + t = 1$

64. $\dfrac{3}{4} + q = 1$

65. $-9a = -12$

66. $-8b = -44$

67. $7 = r - 23$

68. $11 = s - 4$

69. $-\dfrac{y}{3} = 5$

70. $-\dfrac{h}{5} = 1$

71. $2p = \dfrac{5}{6}$

72. $4q = \dfrac{3}{5}$

73. $-\dfrac{3}{7}x = \dfrac{9}{10}$

74. $-\dfrac{2}{11}y = \dfrac{4}{15}$

75. $t - 12.9 = 15$

76. $c - 4.11 = 1.2$

77. $5 + u = 3.2$

78. $3 + v = 1.7$

Expanding Your Skills

For Exercises 79–84, first simplify each side of the equation. Then solve the equation.

79. $5x - 2x = -15$

80. $13y - 10y = -18$

81. $3p + 4p = 25 - 4$

82. $2q + 3q = 54 - 9$

83. $-2(a + 3) - 6a + 6 = 8$

84. $-(b - 11) - 3b - 11 = -16$

Objectives

1. Solving Equations with Multiple Steps
2. Steps to Solve a Linear Equation

section 11.5   Solving Equations with Multiple Steps

1. Solving Equations with Multiple Steps

In Sections 11.3 and 11.4 we studied a one-step process to solve linear equations. We used the addition, subtraction, multiplication, and division properties of equality. In this section we combine these properties to solve equations that require multiple steps. This is shown in Example 1.

example 1   Solving a Linear Equation

Solve. $2x - 3 = 15$

Solution:

Remember that our goal is to isolate x. Therefore, in this equation, we first isolate the *term* containing x. This can be done by adding 3 to both sides.

$$2x - 3 + 3 = 15 + 3 \qquad \text{Add 3 to both sides.}$$

$$2x = 18 \qquad \text{The term containing } x \text{ is now isolated (by itself).}$$
The resulting equation now requires only one step to solve.

$$\frac{2x}{2} = \frac{18}{2} \qquad \text{Divide both sides by 2 to make the coefficient on } x \text{ equal to 1.}$$

$$x = 9 \qquad \text{Simplify. The solution is 9.}$$

Check: $2x - 3 = 15$ Original equation

$2(9) - 3 \overset{?}{=} 15$ Substitute 9 for x.

$18 - 3 = 15 \checkmark$

As Example 1 shows, we will generally apply the addition (or subtraction) property of equality to isolate the variable term first. Then we will apply the multiplication (or division) property of equality to obtain a coefficient of 1 on the variable term.

example 2   Solving Linear Equations

Solve.

   a. $14 = -4y + 8$ **b.** $2z - 9.2 = 2.6$

Solution:

a. $14 = -4y + 8$

$14 - 8 = -4y + 8 - 8$ Subtract 8 from both sides. This will isolate the term containing the variable y.

$6 = -4y$ Simplify.

$\dfrac{6}{-4} = \dfrac{-4y}{-4}$ Divide both sides by -4.

$-\dfrac{6}{4} = y$ Simplify.

$-\dfrac{3}{2} = y$ Simplify to lowest terms.

Check: $14 = -4y + 8$

$14 \overset{?}{=} -4\left(-\tfrac{3}{2}\right) + 8$

$14 = 6 + 8 \checkmark$

b.
$$2z - 9.2 = 2.6$$
$$2z - 9.2 + 9.2 = 2.6 + 9.2$$ Add 9.2 to both sides. This will isolate the term containing the variable z.

$$2z = 11.8$$ Simplify.

$$\frac{2z}{2} = \frac{11.8}{2}$$ Divide both sides by 2.

$$z = 5.9$$ The solution is 5.9 and checks in the original equation.

In Example 3, the variable x appears on both sides of the equation. In this case, apply the addition or subtraction properties of equality to collect the variable terms on one side of the equation and the constant terms on the other side.

example 3   Solving a Linear Equation with Variables on Both Sides

Solve. $4x + 5 = -2x - 13$

Solution:

To isolate x, we must first "move" all x-terms to one side of the equation. For example, suppose we add $2x$ to both sides. This would "remove" the x-term from the right-hand side because $-2x + 2x = 0$. The term $2x$ is then combined with $4x$ on the left-hand side.

$$4x + 5 = -2x - 13$$

$$4x + 2x + 5 = -2x + 2x - 13$$ Add $2x$ to both sides.

$$6x + 5 = -13$$ Simplify. We now have an equation whose form is similar to those presented in Examples 1 and 2. Next, we want to isolate the term containing x.

$$6x + 5 - 5 = -13 - 5$$ Subtract 5 from both sides to isolate the x-term.

$$6x = -18$$ Simplify.

$$\frac{6x}{6} = \frac{-18}{6}$$ Divide both sides by 6 to obtain a coefficient of 1 for the x-term.

$$x = -3$$

Check: $4x + 5 = -2x - 13$

$$4(-3) + 5 \stackrel{?}{=} -2(-3) - 13$$ Substitute -3 for all x's in the equation.

$$-12 + 5 \stackrel{?}{=} 6 - 13$$

$$-7 = -7 \checkmark$$

Tip: Note that the variable may be isolated on either side of the equation. In Example 3, for instance, we could have isolated the x-terms on the right-hand side of the equation.

$4x + 5 = -2x - 13$

$4x - 4x + 5 = -2x - 4x - 13$ Subtract $4x$ from both sides. This "removes" the x-term from the left-hand side.

$5 = -6x - 13$

$5 + 13 = -6x - 13 + 13$ Add 13 to both sides to isolate the x-term.

$18 = -6x$ Simplify.

$\dfrac{18}{-6} = \dfrac{-6x}{-6}$ Divide both sides by -6.

$-3 = x$ This is the same solution as in Example 3.

Often we can simplify both sides of an equation before applying the properties of equality. This is demonstrated in Example 4.

example 4   Solving a Linear Equation by Simplifying First

Solve. $6 - 8y + 3 = y + 3y + 6$

Solution:

Notice that *like* terms can be combined on both sides of the equation first, to make the equation simpler.

$6 - 8y + 3 = y + 3y + 6$

$9 - 8y = 4y + 6$ Combine *like* terms. Note, on the left-hand side $6 + 3 = 9$. On the right-hand side, $y + 3y = 4y$.

$9 - 8y - 4y = 4y - 4y + 6$ We can collect all variable terms on the left-hand side by subtracting $4y$ from both sides.

$9 - 12y = 6$ Simplify.

$9 - 9 - 12y = 6 - 9$ To isolate the y-term on the left, subtract 9 from both sides.

$-12y = -3$ Simplify.

$\dfrac{-12y}{-12} = \dfrac{-3}{-12}$ Divide both sides by -12 to make the coefficient on the y-term 1.

$y = \dfrac{1}{4}$ The solution checks in the original equation.

Skill Practice

Solve.

6. $14 - 3w + 2$
$= 4w + 21 - 2w$

7. $2x - 104 + 8x$
$= 6x + 104 + 2x$

8. $2.2x - 9.6$
$= 0.8x + 3.4 + 0.9x$

Answers

6. $w = -1$ 7. $x = 104$

8. $x = 26$

2. Steps to Solve a Linear Equation

In Examples 1–4, we used multiple steps to solve equations. We also learned how to collect the variable terms on one side of the equation so that the variable could be isolated. The following guidelines summarize the steps to solve a linear equation.

Steps to Solve a Linear Equation in One Variable

1. Simplify both sides of the equation.
   - Clear parentheses if necessary.
   - Combine *like* terms if necessary.
2. Use the addition or subtraction property of equality to collect the variable terms on one side of the equation.
3. Use the addition or subtraction property of equality to collect the constant terms on the *other* side of the equation.
4. Use the multiplication or division property of equality to make the coefficient of the variable term equal to 1.
5. Check the answer in the original equation.

Skill Practice

Solve.

9. $6(z + 4) - 9$
   $= -12 - 3z$
10. $4(x - 2) + 3x$
    $= -2(x + 1) + 6$
11. $2 - y - 4$
    $= 6 - 2(y - 8)$

---

example 5   Solving Linear Equations

Solve.

a. $2(y - 6) + 32 = 8 - 4y$ **b.** $2x + 3x + 2 = -4(3 - x)$

Solution:

a. $2(y - 6) + 32 = 8 - 4y$

$2y - 12 + 32 = 8 - 4y$ **Step 1:** Simplify both sides of the equation. Clear parentheses.

$2y + 20 = 8 - 4y$ Combine *like* terms on the left-hand side. Note that $-12 + 32 = 20$.

$2y + 4y + 20 = 8 - 4y + 4y$ **Step 2:** Add $4y$ to both sides to collect the variable terms on the left.

$6y + 20 = 8$ Simplify.

$6y + 20 - 20 = 8 - 20$ **Step 3:** Subtract 20 from both sides to collect the constants on the right.

$6y = -12$ Simplify.

$\dfrac{6y}{6} = \dfrac{-12}{6}$ **Step 4:** Divide both sides by 6 to obtain a coefficient of 1 on the y-term.

$y = -2$ Simplify.

Answers

9. $z = -3$ 10. $x = \dfrac{4}{3}$

11. $y = 24$

Check: $2(y - 6) + 32 = 8 - 4y$ **Step 5:**   Check the solution in the original equation.

$2(-2 - 6) + 32 \stackrel{?}{=} 8 - 4(-2)$ Substitute -2 for y.

$2(-8) + 32 \stackrel{?}{=} 8 - (-8)$

$-16 + 32 \stackrel{?}{=} 16$

$16 = 16$ ✓   The solution checks.

b. $2x + 3x + 2 = -4(3 - x)$

$5x + 2 = -12 + 4x$ **Step 1:**   Simplify both sides of the equation. On the left, combine *like* terms. On the right, clear parentheses.

$5x - 4x + 2 = -12 + 4x - 4x$ **Step 2:**   Subtract $4x$ from both sides to collect the variable terms on the right.

$x + 2 = -12$ Simplify.

$x + 2 - 2 = -12 - 2$ **Step 3:**   Subtract 2 from both sides to collect the constants on the left.

$x = -14$ **Step 4:**   The coefficient on the x-term is already 1.

Check: $2x + 3x + 2 = -4(3 - x)$ **Step 5:**   Check in the original equation.

$2(-14) + 3(-14) + 2 \stackrel{?}{=} -4[3 - (-14)]$ Substitute -14 for x.

$-28 - 42 + 2 \stackrel{?}{=} -4(17)$

$-70 + 2 \stackrel{?}{=} -68$

$-68 = -68$ ✓   The solution checks.

---

section 11.5   Practice Exercises

Boost *your* GRADE at mathzone.com!

- Practice Problems
- Self-Tests
- NetTutor
- e-Professors
- Videos

Study Skills Exercise

1. When you are solving multistep equations, it is recommended that you write an explanation for each step along the way. On page 742, an equation is solved with each step shown. Your job is to write an explanation for each step to the right of the step.

$$-7x + 2 = 4(x - 5) \qquad \textbf{Explanation}$$

$$-7x + 2 = 4x - 20$$

$$-7x - 4x + 2 = 4x - 4x - 20$$

$$-11x + 2 = -20$$

$$-11x + 2 - 2 = -20 - 2$$

$$-11x = -22$$

$$\frac{-11x}{-11} = \frac{-22}{-11}$$

$$x = 2$$

Review Exercises

For Exercises 2–8, solve the equation.

2. $4c = -\dfrac{1}{3}$ **3.** $\dfrac{1}{3}b = -4$ **4.** $-\dfrac{1}{5} + t = \dfrac{6}{5}$ **5.** $-\dfrac{3}{8} = w + \dfrac{1}{4}$

6. $-p = -\dfrac{7}{10}$ **7.** $-8h = 0$ **8.** $5 + q = 0$

Objective 1: Solving Equations with Multiple Steps

For Exercises 9–26, solve the equation. **(See Examples 1–2.)**

9. $3m + 2 = 14$ **10.** $-2n + 5 = -15$ **11.** $-8c - 12 = 36$ **12.** $5t - 1 = -11$

13. $1 = -4z + 21$ **14.** $-4 = -3p + 14$ **15.** $9 = 12x - 7$ **16.** $-7 = 5y - 8$

17. $3.4 - 2d = 8.2$ **18.** $2.9 - 4g = 23.3$ **19.** $-0.57 = 15h + 16.23$ **20.** $1.9 = 8k + 4.06$

21. $\dfrac{b}{3} - 12 = -9$ **22.** $\dfrac{c}{5} + 2 = 4$ **23.** $-9 = \dfrac{w}{2} - 3$ **24.** $-16 = \dfrac{t}{4} - 14$

25. $3x + \dfrac{1}{2} = \dfrac{5}{4}$ **26.** $9z - \dfrac{3}{8} = \dfrac{9}{16}$

For Exercises 27–44, solve the equation. **(See Examples 3–4.)**

27. $8 + 4b = 2 + 2b$ **28.** $2w + 10 = 5w - 5$ **29.** $7 - 5t = 3t - 2$

30. $4 - 2p = 8 + 5p$ **31.** $4 - 3d = 5d - 4$ **32.** $-3k + 14 = -4 + 3k$

33. $12p = 3p + 21$ **34.** $2x + 10 = 4x$ **35.** $-z - 2 = -2z$

36. $9y = -y + 25$

37. $p - 1 + \dfrac{1}{4}p = 2 + \dfrac{3}{4}p$

38. $\dfrac{4}{3} + \dfrac{2}{3}q = -\dfrac{5}{3} - q - \dfrac{1}{3}$

39. $4 + 2a - 7 = 3a + a + 3$

40. $4b + 2b - 7 = 2 + 4b + 5$

41. $-8w + 8 + 3w = 2 - 6w + 2$

42. $-12 + 5m + 10 = -2m - 10 - m$

43. $6y + 2y - 2 = 14 + 3y - 12$

44. $-7t - 20 + 7 = -7 - 3t$

Objective 2: Steps to Solve a Linear Equation

For Exercises 45–58, solve the equation. **(See Example 5.)**

45. $3n - 4(n - 1) = 16$

46. $4p - 3(p + 2) = 18$

47. $9q - 5(q - 3) = 5q$

48. $6h - 2(h + 6) = 10h$

49. $2(1 - m) = 5 - 3m$

50. $3(2 - g) = 12 - g$

51. $-4(k - 2) + 14 = 3k - 20$

52. $-3(x + 4) - 9 = -2x + 12$

53. $3z - 9 = 3(5z - 1)$

54. $4y - 9 = 8(y - 2)$

55. $6w + 2(w - 1) = 14 - (3w + 1)$

56. $-3t - 3(t - 4) = 2 - (2t - 1)$

57. $6(u - 1) + 5u + 1 = 5(u + 6) - u$

58. $2(2v + 3) + 8v = 6(v - 1) + 3v$

---

chapter 11 | midchapter review

1. a. Simplify the expression. $16x - 5(x + 3) - 4x$

 b. Simplify the expression. $2 - (3x + 1) - 10$

 c. Solve the equation.
 $16x - 5(x + 3) - 4x = 2 - (3x + 1) - 10$

2. a. Simplify the expression. $20 + 3(2y - 5) - 3$

 b. Simplify the expression. $7 - (3y + 1) - 3y$

 c. Solve the equation.
 $20 + 3(2y - 5) - 3 = 7 - (3y + 1) - 3y$

For Exercises 3–18, identify as an expression or an equation. If it is an expression, simplify. If it is an equation, solve.

3. $5x + 3x - 12$

4. $9t - 5 = 22$

5. $6(h - 3) = 2h - 14$

6. $4(k - 1) + 6k + 4$

7. $8(w - 1) - 6w = -8$

8. $4b - 6 - 4(b - 2)$

9. $m + \dfrac{1}{4} - \dfrac{1}{8}$

10. $m + \dfrac{1}{4} = \dfrac{1}{8}$

11. $-3(y + 1) + 2$

12. $-3(y + 1) = 2$

13. $7 - t = 2(t - 1)$ **14.** $7 - t + 2(t - 1)$ **15.** $2x + 4x + 1 = 0$ **16.** $2x + 4x + 1$

17. $5 + 3p - 2$ **18.** $5 + 3p - 2 = 0$

Objectives

1. Problem-Solving Flowchart
2. Translating Verbal Statements into Equations
3. Applications of Linear Equations

section 11.6   Applications and Problem Solving

1. Problem-Solving Flowchart

Linear equations can be used to solve many real-world applications. In Section 1.8, we introduced guidelines for problem solving. In this section, we expand on these guidelines to enable us to write equations to solve applications. Consider the problem-solving flowchart.

Problem-Solving Flowchart for Word Problems

Step 1 | Read the problem completely. | • Familiarize yourself with the problem. Identify the unknown, and if possible estimate the answer.

Step 2 | Assign labels to unknown quantities. | • Identify the unknown quantity or quantities. Let a variable represent one of the unknowns. Draw a picture and write down relevant formulas.

Step 3 | Write an equation in words. | • Verbalize what quantities must be equal.

Step 4 | Write a mathematical equation. | • Replace the verbal equation with a mathematical equation using x or some other variable.

Step 5 | Solve the equation. | • Solve for the variable by using the steps for solving linear equations.

Step 6 | Interpret the results and write the final answer in words. | • Once you have obtained a numerical value for the variable, recall what it represents in the context of the problem. Can this value be used to determine other unknowns in the problem? Write an answer to the word problem *in words*.

2. Translating Verbal Statements into Equations

We begin solving word problems with practice translating between an English sentence and an algebraic equation. First, spend a minute to recall some of the key words that represent addition, subtraction, multiplication, and division. See Table 11-4.

table 11-4

Addition: $a + b$	Subtraction: $a - b$
the sum of a and b	the difference of a and b
a plus b	a minus b
b added to a	b subtracted from a
b more than a	a decreased by b
a increased by b	b less than a
the total of a and b	

Multiplication: $a \cdot b$	Division: $a \div b$
the product of a and b	the quotient of a and b
a times b	a divided by b
a multiplied by b	b divided into a
	the ratio of a and b
	a over b
	a per b

example 1   Translating Sentences into Mathematical Equations

A number decreased by 7 is 12. Find the number.

Solution:

	Step 1: Read the problem completely.
Let x represent the number.	**Step 2:** Label the variable.
A number decreased by 7 is 12.	**Step 3:** Write the equation in words.
$\quad x \qquad - \qquad 7 = 12$	**Step 4:** Translate to a mathematical equation.
$x - 7 = 12$	**Step 5:** Solve the equation.
$x - 7 + 7 = 12 + 7$	Add 7 to both sides.
$x = 19$	
The number is 19.	**Step 6:** Interpret the answer in words.

Skill Practice

1. 6 subtracted from a number is -22. Find the number.

Answer

1. The number is -16.

2. 8 added to twice a number is 22. Find the number.

example 2   Translating Sentences into Mathematical Equations

Two times the sum of a number and 8 results in 38.

Solution:

Step 1: Read the problem completely.

Let x represent the number.   **Step 2:** Label the variable.

Two times the sum of a number and 8 results in 38.   **Step 3:** Write the equation in words.

$$\underbrace{2 \cdot}_{\text{two times}} \underbrace{(x + 8)}_{\substack{\text{the sum of a} \\ \text{number and 8}}} = \underbrace{38}_{38} \xleftarrow{\text{results in}}$$

Step 4: Translate to a mathematical equation.

$$2(x + 8) = 38$$ **Step 5:** Solve the equation.

$$2x + 16 = 38$$ Clear parentheses.

$$2x + 16 - 16 = 38 - 16$$ Subtract 16 from both sides.

$$2x = 22$$ Simplify.

$$\frac{2x}{2} = \frac{22}{2}$$ Divide both sides by 2.

$$x = 11$$

The number is 11.   **Step 6:** Interpret the answer in words.

3. Applications of Linear Equations

In Examples 3–6, we solve application problems by using linear equations.

3. A piece of cable 92 ft long is to be cut into two pieces. One piece must be 3 times longer than the other. How long should each piece be?

example 3   Applying a Linear Equation to Carpentry

A carpenter must cut an 8-ft board into two pieces to build a brace for a picnic table. If one piece is to be 4 times as long as the other piece, how long should each piece be?

Solution:

Step 1: Read the problem completely.

We can let x represent the length of either piece. However, if we choose x to be the length of the shorter piece, then the longer piece has to be $4x$ (4 times as long).

Let x = length of the shorter piece.
Then $4x$ = length of the longer piece.

Step 2: Label the variables. Draw a picture.

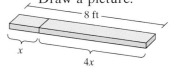

Answers

2. The number is 7.
3. One piece should be 23 ft, and the other should be 69 ft long.

$$\begin{pmatrix} \text{Length of} \\ \text{one piece} \end{pmatrix} + \begin{pmatrix} \text{Length of} \\ \text{other piece} \end{pmatrix} = \begin{pmatrix} \text{Total} \\ \text{length} \end{pmatrix}$$

$$x \quad + \quad 4x \quad = \quad 8$$

Step 3: Write an equation in words.

Step 4: Write a mathematical equation.

$$x + 4x = 8$$

$$5x = 8$$

$$\frac{5x}{5} = \frac{8}{5}$$

$$x = \frac{8}{5} \text{ or } 1.6$$

Step 5: Solve the equation.

Combine *like* terms.

Divide both sides by 5.

Recall that x represents the length of the shorter piece. Therefore, the shorter piece is 1.6 ft. The longer piece is given by $4x$ or $4(1.6 \text{ ft}) = 6.4 \text{ ft}$.

The pieces are 1.6 and 6.4 ft.

Step 6: Interpret the results in words.

Tip: It is good practice to verify that an answer is reasonable. In Example 2, the two boards should total 8 ft. We have 1.6 ft + 6.4 ft = 8 ft, as desired.

---

example 4   Applying a Linear Equation

One model of a Panasonic high-definition plasma TV sells for $1500 more than a certain model made by Planar. The combined cost for these two models is $8500. Find the cost for each model. (Source: *Consumer Reports.*)

Solution:

Step 1: Read the problem completely.

Let x represent the cost of the Planar TV.

Step 2: Label the variables.

Then $x + 1500$ represents the cost of the Panasonic.

$$\begin{pmatrix} \text{Cost of} \\ \text{Planar} \end{pmatrix} + \begin{pmatrix} \text{Cost of} \\ \text{Panasonic} \end{pmatrix} = \begin{pmatrix} \text{Total} \\ \text{cost} \end{pmatrix}$$

$$x \quad + \quad x + 1500 \quad = 8500$$

Step 3: Write an equation in words.

Step 4: Write a mathematical equation.

$$x + x + 1500 = 8500$$

$$2x + 1500 = 8500$$

$$2x + 1500 - 1500 = 8500 - 1500$$

Step 5: Solve the equation.

Combine *like* terms.

Subtract 1500 from both sides.

Skill Practice

4. A kit of cordless 18-volt tools made by Craftsman costs $310 less than a similar kit made by DeWalt. The combined cost for both models is $690. Find the cost for each model. (Source: *Consumer Reports.*)

Answer
4. The Craftsman model costs $190, and the DeWalt model costs $500.

$$2x = 7000$$

$$\frac{2x}{2} = \frac{7000}{2} \qquad \text{Divide both sides by 2.}$$

$$x = 3500$$

The Planar model TV costs $3500.
The Panasonic model is represented by $x + 1500 = \$3500 + \$1500 = \$5000$.

Tip: In Example 4, we could have let x represent the cost of *either* the Planar model TV or the Panasonic model.

Suppose we had let x represent the cost of the Panasonic model. Then $x - 1500$ is the cost of the Planar model (the Planar model is *less* expensive).

$$\left(\begin{matrix}\text{Cost of} \\ \text{Planar}\end{matrix}\right) + \left(\begin{matrix}\text{Cost of} \\ \text{Panasonic}\end{matrix}\right) = \left(\begin{matrix}\text{Total} \\ \text{cost}\end{matrix}\right)$$

$$x - 1500 + \qquad x \qquad = 8500$$

$$2x - 1500 = 8500$$

$$2x - 1500 + 1500 = 8500 + 1500$$

$$2x = 10{,}000$$

$$x = 5000$$

Therefore, the Panasonic model costs $5000 as expected.
The Planar model costs $x - 1500$ or $\$5000 - \$1500 = \$3500$.

Skill Practice

5. The perimeter of a tennis court is 228 ft. If the length is 6 ft longer than twice the width, find the dimensions of the court.

example 5   Applying a Linear Equation to Geometry

The perimeter of the soccer field at Giants Stadium is 338 m. If the length is 37 m longer than the width, find the dimensions of the field.

Solution:

	Step 1: Read the problem completely.
Let w represent the width. Then $w + 37$ represents the length.	**Step 2:** Label the variables and write down relevant formulas. Draw a figure to help you.

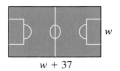
$w + 37$

Recall that the formula for the perimeter of a rectangle is given by $P = 2l + 2w$.

Step 3: The perimeter formula for a rectangle can be used as the equation.

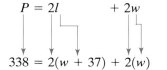

$$338 = 2(w + 37) + 2(w)$$

Step 4: Write a mathematical equation, using the labeled variables.

Answer

5. The length is 78 ft, and the width is 36 ft.

$$338 = 2w + 74 + 2w$$

Step 5: Solve the equation. Clear parentheses.

$$338 = 4w + 74$$

Combine *like* terms.

$$338 - 74 = 4w + 74 - 74$$

Subtract 74 from both sides.

$$264 = 4w$$

$$\frac{264}{4} = \frac{4w}{4}$$

Divide by 4 on both sides.

$$66 = w$$

Step 6: Interpret the answer in words.

The width is 66 m.

The length is given by $w + 37 = 66\text{ m} + 37\text{ m} = 103\text{ m}$.

---

example 6   Using a Linear Equation in a Consumer Application

Joanne has a cellular phone plan in which she pays $39.95 per month for 450 min of air time. Additional minutes beyond 450 are charged at a rate of $0.40 per minute. If Joanne's bill comes to $87.95, how many minutes did she use beyond 450 min?

Solution:

Step 1: Read the problem.

Let x represent the number of minutes beyond 450.

Step 2: Label the variable.

Then $0.40x$ represents the cost for x additional minutes.

$$\begin{pmatrix}\text{Monthly}\\\text{fee}\end{pmatrix} + \begin{pmatrix}\text{Cost of}\\\text{additional minutes}\end{pmatrix} = \begin{pmatrix}\text{Total}\\\text{cost}\end{pmatrix}$$

Step 3: Write an equation in words.

$$39.95 \quad + \quad\quad 0.40x \quad\quad = \quad 87.95$$

Step 4: Write a mathematical equation.

$$39.95 + 0.40x = 87.95$$

Step 5: Solve the equation.

$$39.95 - 39.95 + 0.40x = 87.95 - 39.95$$

Subtract 39.95.

$$0.40x = 48.00$$

$$\frac{0.40x}{0.40} = \frac{48.00}{0.40}$$

Divide by 0.40.

$$x = 120$$

Joanne talked for 120 min beyond her allotted 450 min.

Answer

6. DJ charged $9400.

section 11.6   Practice Exercises

Study Skills Exercise

1. In solving an application it is very important first to read and understand what is being asked in the problem. One way to do this is to read the problem several times. Another is to read it out loud so you can hear yourself. Still another is to try to rewrite the problem in your own words. Which of these methods do you think will help you in understanding an application?

Review Exercises

For Exercises 2–8, solve the equation.

2. $3t - 15 = -22$

3. $\dfrac{b}{5} - 5 = -14$

4. $2x + 22 = 6x - 2$

5. $4(r + 4) - 12 = 18 - r$

6. $-(y - 9) + 5(y + 3) = -2(3y + 7)$

7. $4.4p - 2.6 = 1.2p - 5$

8. $\dfrac{2}{3}w - \dfrac{1}{6} = \dfrac{1}{3}w + \dfrac{5}{6}$

Objective 2: Translating Verbal Statements into Equations

For Exercises 9–24, **a.** Write an equation that represents the problem and **b.** Solve the problem. **(See Examples 1–2.)**

9. The quotient of a number and 3 is -8. Find the number.

10. The difference of -2 and a number is -14. Find the number.

11. A number subtracted from -30 results in 42. Find the number.

12. A number increased by 13 results in -100. Find the number.

13. The total of 30 and a number is 13. Find the number.

14. Sixty is -5 times a number. Find the number.

15. Five less than the quotient of a number and 4 is equal to -12. Find the number.

16. Eight decreased by the product of a number and 3 is equal to 5. Find the number.

17. One-half increased by a number is 4. Find the number.

18. Five-thirds decreased by a number is 1. Find the number.

19. The product of -12 and a number is the same as the sum of the number and 26. Find the number.

20. The difference of a number and 16 is the same as the product of the number and -3. Find the number.

21. Ten times the total of a number and 5.1 is 56. Find the number.

22. Three times the difference of a number and 5 is 15. Find the number.

23. The product of 3 and a number is the same as 10 less than twice the number.

24. Six less than a number is the same as 3 more than twice the number.

Objective 3: Applications of Linear Equations

For Exercises 25–44, solve the problem by using the problem-solving flowchart found on page 744. **(See Examples 3–6.)**

25. Felicia has a piece of ribbon that is 4 ft long. She wants to cut the ribbon to make two pieces so that one piece is twice as long as the other. How long should each piece be?

26. A pipe, 6 m long, needs to be cut into two sections. If one section must be 3 times longer than the other, how long must each section be?

27. In the 1990s the musical group Metallica had 6 fewer hits than the group Boyz II Men. If together the groups had 26 hits, how many hits did each group have?

28. A two-piece set of luggage costs $150. If sold individually, the large bag costs $40 more than the small bag. What are the individual costs for each bag?

29. The perimeter of a rectangular soccer field is 460 yd. The length is 30 yd longer than the width. What are the dimensions of the field?

30. The width of an Olympic size swimming pool is one-half its length. If the perimeter is 150 m, what are the dimensions of the pool?

31. A cellular phone company charges $49.95 each month, which includes 500 min. For minutes used beyond the first 500, the charge is $0.25 per minute. Jim's bill came to $62.45. How many minutes over 500 min did he use?

32. U-Rent-It car company charges $19 per day, which includes 200 mi of driving. If a person travels more than 200 mi, a fee of $0.30 per mile is charged for each mile over 200 mi. If Mr. Cain's bill came to $62.50, how many miles did he travel over 200 mi?

33. In Superbowl XXXVII (37) the Tampa Bay Buccaneers won the game with 6 points more than twice the number of points earned by their opponents, the Oakland Raiders. If there was a total of 69 points in the game, how many points did each team score?

34. In the first Superbowl in 1967, the Green Bay Packers won the game with 5 points more than 3 times the points earned by their opponents, the Kansas City Chiefs. If there was a total of 45 points in the game, how many points did each team score?

35. An apartment complex charges a refundable security deposit when renting an apartment. The deposit is $350 less than the monthly rent. If Charlene paid a total of $950 for her first month's rent and security deposit, how much is her monthly rent? How much is the security deposit?

36. José buys a shirt and pants for $71.90. If the shirt costs $20 less than the pants, what is the cost of the shirt? What is the cost of the pants?

37. Stefan is paid a salary of $480 a week at his job. When he works overtime, he receives $18 an hour. If his weekly paycheck came to $588, how many hours of overtime did he put in that week?

38. Victoria is paid a salary of $480 a week at her job. She worked 8 hr of overtime during the holidays, and her weekly paycheck came to $672. What is her overtime pay per hour?

39. Raul signed up for his classes for the spring semester. He signed up for 4 credit-hours more in the spring than he took in the fall. If he has a total of 28 hr in the two semesters, how many hours did he take in the fall? How many hours did he sign up for in spring?

40. A computer with a monitor costs $899. If the computer costs $241 more than the monitor, what is the price of the computer? What is the price of the monitor?

41. Mercedes had a calling card worth $20. On this card, long-distance calls cost $0.05 more per minute than local calls. If Mercedes made 35 min of local calls and 52 min of long-distance calls, what was she paying per minute for each type of call?

42. Ann-Marie got two offers for jobs as a salesperson. One job pays a weekly salary of $300, and the other pays on commission. If commission pays 24% of sales, how much merchandise must Ann-Marie sell each week to match the weekly salary?

43. New York is connected on the Internet to 10 more countries than London, UK. Together, New York and London are connected to 132 countries. How many countries are connected to London?

44. In 2005, recording artist Kanye West received 2 more Grammy nominations than Alicia Keys. Together they received 18 nominations. How many nominations did each receive?

---

section 11.7 Rectangular Coordinate System (Optional)

Objectives

1. Rectangular Coordinate System
2. Plotting Points in a Rectangular Coordinate System
3. Graphing Ordered Pairs in an Application

1. Rectangular Coordinate System

In Section 9.1 we created line graphs that illustrated how one variable relates to another. For example, Table 11-5 represents the daily revenue for the movie *Ladder 49* for two weeks in October 2004. In table form, the information is difficult to picture and interpret. However, Figure 11-4 shows a scatter diagram of these data. From the graph we can speculate that October 1–3 and 8–10 were weekends because revenue was up. We can also see that as time went on, revenue dropped.

table 11-5

Date	Day Number	Revenue ($ Millions)
Oct. 1	1	7.6
Oct. 2	2	9.4
Oct. 3	3	5.3
Oct. 4	4	1.5
Oct. 5	5	1.6
Oct. 6	6	1.3
Oct. 7	7	1.4
Oct. 8	8	3.8
Oct. 9	9	5.9
Oct. 10	10	3.7
Oct. 11	11	1.8
Oct. 12	12	0.9
Oct. 13	13	0.8
Oct. 14	14	0.9

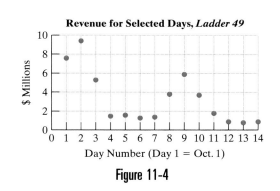

Figure 11-4

Figure 11-4 represents the variables of time and revenue. In general, to picture two variables, we use a graph with two number lines drawn at right angles to each other (Figure 11-5). This forms a **rectangular coordinate system**. The horizontal line is called the **x-axis**, and the vertical line is called the **y-axis**. The point where the lines intersect is called the **origin**. On the x-axis, the numbers to the right of the origin are positive, and the numbers to the left are negative. On the y-axis, the numbers above the origin are positive, and the numbers below the origin are negative. The x- and y-axes divide the graphing area into four regions called **quadrants**.

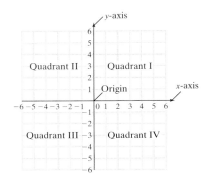

Figure 11-5

2. Plotting Points in a Rectangular Coordinate System

Points graphed in a rectangular coordinate system are defined by two numbers as an **ordered pair** (x, y). The first number (called the **x-coordinate** or first coordinate) is the horizontal position from the origin. The second number (called the **y-coordinate** or second coordinate) is the vertical position from the origin. Example 1 shows how points are plotted in a rectangular coordinate system.

Plot the points and state the quadrant in which the point lies.

1. $(-2, 3)$

2. $(4, 1)$

3. $(3, -1)$

4. $(-5, -4)$

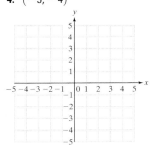

Answers

1. Quadrant II   2. Quadrant I
3. Quadrant IV   4. Quadrant III

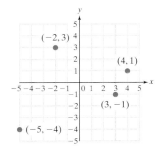

example 1 Plotting Points in a Rectangular Coordinate System

Plot the points.

a. $(3, 4)$ **b.** $(-4, 2)$ **c.** $(2, -4)$ **d.** $(-5, -3)$

Solution:

a. The ordered pair $(3, 4)$ indicates that $x = 3$ and $y = 4$.

Because the x-coordinate is positive, start at the origin and move 3 units in the positive x direction (to the right). Then, because the y-coordinate is positive, move 4 units in the positive y direction (upward). Draw a dot at the final location. See Figure 11-6.

The point $(3, 4)$ is located in Quadrant I.

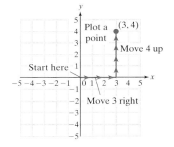

Figure 11-6

b. The ordered pair $(-4, 2)$ indicates that $x = -4$ and $y = 2$.

Because the x-coordinate is *negative*, we start at the origin and move 4 units in the *negative* x direction (left). Then, because the y-coordinate is positive, move 2 units in the positive y direction (upward). Draw a dot at the final location. See Figure 11-7.

The point $(-4, 2)$ is located in Quadrant II.

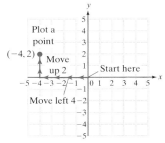

Figure 11-7

c. The ordered pair $(2, -4)$ indicates that $x = 2$ and $y = -4$.

$$\begin{array}{cc} \downarrow & \downarrow \\ x & y \end{array}$$

Because the x-coordinate is positive, start at the origin and move 2 units in the positive x direction (to the right). Then, because the y-coordinate is *negative*, move 4 units in the *negative* y direction (downward). Draw a dot at the final location. See Figure 11-8.

The point $(2, -4)$ is located in Quadrant IV.

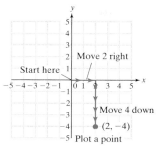

Figure 11-8

Tip: Notice that changing the order of the x- and y-coordinates changes the location of the point. In Example 1(b), the point $(-4, 2)$ is in Quadrant II. In Example 1(c), the point $(2, -4)$ is in Quadrant IV (Figure 11-8). This is why points are represented by *ordered* pairs. The order is important.

d. The ordered pair $(-5, -3)$ indicates that $x = -5$ and $y = -3$.

$$\begin{array}{cc} \downarrow & \downarrow \\ x & y \end{array}$$

Because the x-coordinate is *negative*, start at the origin and move 5 units in the *negative* x direction (to the left). Then, because the y-coordinate is *negative*, move 3 units in the *negative* y direction (downward). Draw a dot at the final location. See Figure 11-9.

The point $(-5, -3)$ is located in Quadrant III.

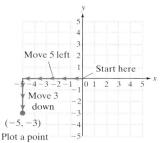

Figure 11-9

In Example 2, we plot points whose coordinates are fractions, decimals, or zero.

example 2   Plotting Points in a Rectangular Coordinate System

Plot the points.

a. $(-4.6, -3.8)$ **b.** $\left(-\dfrac{5}{2}, \dfrac{13}{3}\right)$ **c.** $(3, 0)$ **d.** $(0, -4)$

Solution:

a. The point $(-4.6, -3.8)$ is located 4.6 units to the left and 3.8 units down from the origin. The point is in Quadrant III. See Figure 11-10.

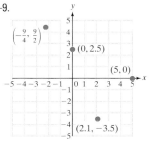

b. The improper fraction $-\frac{5}{2}$ can be written as the mixed number $-2\frac{1}{2}$. The fraction $\frac{13}{3}$ can be written as $4\frac{1}{3}$. Therefore, the ordered pair $(-\frac{5}{2}, \frac{13}{3})$ can be written as $(-2\frac{1}{2}, 4\frac{1}{3})$. The point is located $2\frac{1}{2}$ units to the left and $4\frac{1}{3}$ units up from the origin. The point is in Quadrant II. See Figure 11-10.

c. In the ordered pair $(3, 0)$, the y-coordinate is zero. Therefore, we move neither upward nor downward from the origin. This indicates that the point is on the x-axis. The point is located on the x-axis, 3 units to the right of the origin. See Figure 11-10.

Figure 11-10

d. In the ordered pair $(0, -4)$, the x-coordinate is zero. Therefore, we move to neither the left nor the right of the origin. This indicates that the point is on the y-axis. The point is located on the y-axis 4 units down from the origin. See Figure 11-10.

In Example 3, we practice reading ordered pairs from a graph.

Skill Practice

Estimate the coordinates of the points in the graph.

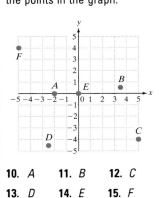

| 10. | A | 11. | B | 12. | C |
| 13. | D | 14. | E | 15. | F |

example 3   Reading Ordered Pairs from a Graph

Estimate the coordinates of points A, B, C, D, E, and F from Figure 11-11.

Solution:

Point A is at $(-5, -1)$.

Point B is at $(-4, 4)$.

Point C is at $(2\frac{1}{2}, 0)$.

Point D is at $(5, 2)$.

Point E is at $(3\frac{1}{2}, -1\frac{1}{2})$.

Point F is at $(0, 0)$.

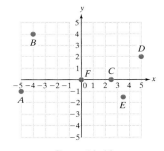

Figure 11-11

3. Graphing Ordered Pairs in an Application

Graphing ordered pairs in an application is essentially the same process as creating a scatter plot.

example 4   Graphing Ordered Pairs in an Application

The data given in Table 11-6 represent the gas mileage for a subcompact car for various speeds. Let x represent the speed of the car, and let y represent the corresponding gas mileage.

a. Write the table values as ordered pairs.

b. Graph the ordered pairs.

Answers

10. $(-2, 0)$ 11. $(3.5, 0.5)$
12. $(5, -4)$ 13. $(-2.5, -4.5)$
14. $(0, 0)$ 15. $(-5, 4)$

table 11-6

Speed *x* (mph)	Gas Mileage *y* (mpg)
25	27
30	29
35	33
40	34
45	35
50	33
55	30
60	27
65	24
70	21

Solution:

a. The ordered pairs are given by (25, 27), (30, 29), (35, 33), (40, 34), (45, 35), (50, 33), (55, 30), (60, 27), (65, 24), and (70, 21).

b.

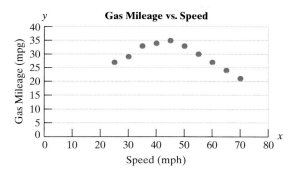

From the graph we see that the most efficient speed is approximately 45 mph. That is where the gas mileage is at its peak.

Skill Practice

For Exercises 16–17, refer to the data values given in the table representing the number of arrests made by a police department for one week.

16. Write the table values as ordered pairs.

17. Graph the ordered pairs.

Day Number *x*	Number of Arrests *y*
1	25
2	12
3	16
4	15
5	20
6	40
7	50

Answers

16. (1, 25), (2, 12), (3, 16), (4, 15), (5, 20), (6, 40), (7, 50)

17.

---

section 11.7   Practice Exercises

Boost *your* GRADE at mathzone.com!

MathZone

- Practice Problems
- Self-Tests
- NetTutor
- e-Professors
- Videos

Study Skills Exercises

1. You should not wait until the last minute to study for any test in math. It is particularly important for the final exam that you begin your review early. Check all the activities that you could do to help you study for the final.

☐ Rework all old exams.

☐ Rework the chapter tests for the chapters you covered in class.

☐ Do the cumulative review exercises at the end of Chapters 2–11.

2. Define the key terms.

    a. Origin   **b. Quadrant**   **c. Rectangular coordinate system**   **d. *x*-axis**

    e. *x*-coordinate   **f. *y*-axis**   **g. *y*-coordinate**   **h. Ordered pair**

Review Exercises

For Exercises 3–6, write an equation that represents the problem. Then solve the problem.

  3. A number less than 9 results in twice the number. Find the number.

  4. The total of 6, -11, and twice a number is the same as the number decreased by 8. Find the number.

  5. A number divided by -6 is 13. Find the number.

  6. The number 30 subtracted from a number is the same as the product of the number and -4. Find the number.

Objective 1: Rectangular Coordinate System

Use the words in Exercises 7–13 to label the rectangular coordinate system.

  7. *x*-axis   **8.** *y*-axis   **9.** Origin

10. Quadrant I   **11.** Quadrant II   **12.** Quadrant III

13. Quadrant IV

Objective 2: Plotting Points in a Rectangular Coordinate System

For Exercises 14–18, plot the points on the rectangular coordinate system. **(See Example 1.)**

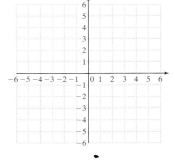

14. $(-1, 4)$ **15.** $(4, -1)$ **16.** $(2, 2)$

17. $(3, -3)$ **18.** $(-5, -2)$

19. Explain how to plot the point $(-1.8, 3.1)$.

20. Explain how to plot the point $\left(\dfrac{15}{2}, \dfrac{15}{7}\right)$.

For Exercises 21–26, plot the points on the rectangular coordinate system. **(See Example 2.)**

21. $(-0.6, 1.1)$ **22.** $(2.3, 4.9)$ **23.** $(-1.4, 4.1)$

24. $(5.1, -3.8)$ **25.** $(0.9, -1.1)$ **26.** $(-3.3, -4.6)$

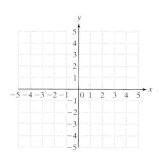

For Exercises 27–32, plot the points on the rectangular coordinate system. **(See Example 2.)**

27. $\left(\dfrac{1}{2}, \dfrac{5}{2}\right)$ **28.** $\left(-\dfrac{10}{3}, \dfrac{8}{3}\right)$ **29.** $\left(-\dfrac{13}{6}, -1\right)$

30. $\left(\dfrac{24}{5}, -\dfrac{8}{5}\right)$ **31.** $\left(\dfrac{13}{4}, \dfrac{29}{8}\right)$ **32.** $\left(\dfrac{15}{7}, -\dfrac{17}{6}\right)$

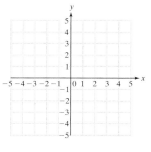

For Exercises 33–38, plot the points on the rectangular coordinate system. **(See Example 2.)**

33. $(0, -3)$ **34.** $(2, 0)$ **35.** $(4, 0)$

36. $(0, 5)$ **37.** $(0, 1)$ **38.** $(-4, 0)$

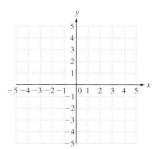

For Exercises 39–50, for each point, identify which quadrant or which axis it lies on.

39. $(32, -44)$ **40.** $(-12, 25)$ **41.** $(-10, -5)$ **42.** $(100, 82)$

43. $(54.9, 0)$ **44.** $(0, -23.33)$ **45.** $\left(0, \dfrac{55}{17}\right)$ **46.** $\left(-\dfrac{3}{4}, 0\right)$

47. $(-27, 3)$ **48.** $(5, -75)$ **49.** $(35, 66)$ **50.** $\left(-\dfrac{2}{17}, -\dfrac{1}{50}\right)$

For Exercises 51–58, estimate the coordinate of points *A, B, C, D, E, F, G,* and *H.* **(See Example 3.)**

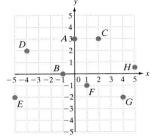

51. *A*   **52.** *B*   **53.** *C*

54. *D*   **55.** *E*   **56.** *F*

57. *G*   **58.** *H*

Objective 3: Graphing Ordered Pairs in an Application

For Exercises 59–66, write the table values as ordered pairs. Then graph the ordered pairs.

59. The table gives the average monthly temperature in Anchorage, Alaska, for one year. Let January represent $x = 1$, February represent $x = 2$, etc. **(See Example 4.)**

Month x	Temperature y (°C)
1	−7
2	−3
3	1
4	6
5	12
6	17
7	18
8	18
9	14
10	6
11	−1
12	−7

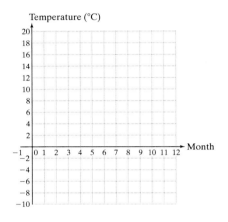

60. The table gives the average monthly temperature in Moscow for one year. Let January represent $x = 1$, February represent $x = 2$, etc.

Month x	Temperature y (°C)
1	−6
2	−5
3	1
4	11
5	18
6	22
7	24
8	22
9	15
10	8
11	1
12	−4

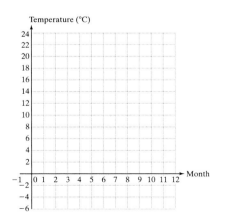

61. The table gives the depreciation amount each year of a new car purchased for $20,000.

Year After Purchase x	Amount of Depreciation y ($)
1	5,000
2	6,800
3	8,384
4	10,010
5	11,509
6	12,782

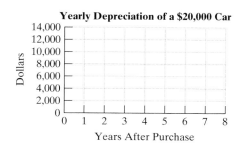

62. The table gives the value each year after purchase of a used car bought for $10,000.

Year After Purchase x	Value of Car y ($)
1	13,200
2	11,352
3	9,649
4	8,201
5	6,971
6	5,925

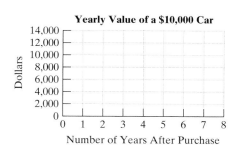

63. The table gives the sales for yogurt and yogurt drinks for 6 years. The dollar amounts are in millions. (Source: Information Resources Inc.)

Year x	Sales y ($ Millions)
1997	1736
1998	1789
1999	1896
2000	1104
2001	2268
2002	2393

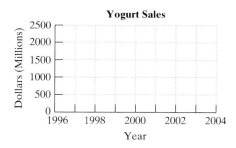

64. The table gives the sales for milk for 6 years. The dollar amounts are in millions. (Source: Information Resources Inc.)

Year x	Sales y ($ Millions)
1997	9,762
1998	9,882
1999	10,486
2000	10,618
2001	10,939
2002	10,929

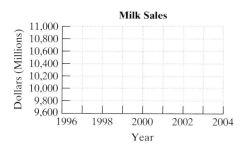

65. The table gives the average snowfall in Juneau, Alaska, for each month of the year. Let January represent $x = 1$, February represent $x = 2$, etc.

Month x	Snowfall y (in.)
1	26.8
2	19.6
3	14.4
4	2.8
5	0
6	0
7	0
8	0
9	0
10	1.1
11	11.7
12	21.8

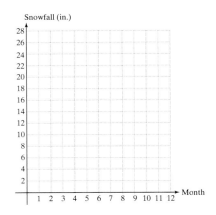

66. The table gives the average rainfall in Seattle, Washington, for each month of the year. Let January represent $x = 1$, February represent $x = 2$, etc.

Month x	Rainfall y (mm)
1	136.7
2	88.3
3	81.2
4	53.3
5	38.1
6	32.3
7	13.2
8	22.3
9	42.0
10	76.4
11	118.5
12	113.8

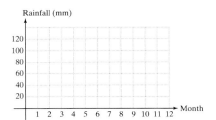

chapter 11 | summary

section 11.1 Properties of Real Numbers

Key Concepts

An algebraic **expression** is a collection of **variables** and **constants** with algebraic operations such as addition, subtraction, multiplication, and division.

To evaluate an expression, first replace the variable with parentheses. Then insert the values and simplify, using the order of operations.

The Properties of Real Numbers

Commutative property of addition:

$a + b = b + a$

Commutative property of multiplication:

$a \cdot b = b \cdot a$

Associative property of addition:

$(a + b) + c = a + (b + c)$

Associative property of multiplication:

$(a \cdot b) \cdot c = a \cdot (b \cdot c)$

Distributive property of multiplication over addition:

$a(b + c) = a \cdot b + a \cdot c$

Examples

Example 1

$3x + 8$ is an algebraic expression, x is a variable, and 8 is a constant.

Example 2

Evaluate $4x - 5y$ for $x = -\dfrac{1}{2}$ and $y = 3$.

$$4x - 5y = 4(\ \ ) - 5(\ \ )$$
$$= 4\left(-\frac{1}{2}\right) - 5(3)$$
$$= -2 - 15$$
$$= -17$$

Example 3

$5 + w = w + 5$ is an example of the commutative property of addition.

Example 4

$3(4y) = (3 \cdot 4)y$ is an example of the associative property of multiplication.

Example 5

$-2(7 + t) = -2(7) + (-2)t$ is an example of the distributive property of multiplication over addition.

section 11.2 **Simplifying Expressions**

Key Concepts

An algebraic expression is the sum of one or more terms. A **term** is a constant or the product of a constant and one or more variables. If a term contains a variable, it is called a **variable term**. A number multiplied by the variable is called a **coefficient**. A term with no variable is called a **constant term**.

Terms that have exactly the same variable factors are called *like* **terms**.

Like terms can be combined by applying the distributive property.

To simplify an expression, first clear parentheses by using the distributive property. Arrange *like* terms together. Then combine *like* terms.

Examples

Example 1

In the expression $12x + 3$,

$12x$ is a variable term

12 is the coefficient of the term $12x$

The term 3 is a constant term.

Example 2

$5h$ and $-2h$ are *like* terms because the variable part h is the same.

$6t$ and $6t^2$ are not *like* terms because the variable parts t and t^2 are not exactly the same.

Example 3

$$3x + 15x - 7x = (3 + 15 - 7)x$$
$$= 11x$$

Example 4

Simplify: $3(k - 4) - (6k + 10) + 14$

$3(k - 4) - (6k + 10) + 14$

$= 3k - 12 - 6k - 10 + 14$ Clear parentheses.

$= 3k - 6k - 12 - 10 + 14$ Arrange *like* terms together.

$= -3k - 8$ Combine *like* terms.

section 11.3 Addition and Subtraction Properties of Equality

Key Concepts

An **equation** is a statement that indicates that two quantities are equal.

A **solution** to an equation is a value of the variable that makes the equation a true statement.

Definition of a Linear Equation in One Variable

Let a and b be real numbers such that $a \neq 0$. A **linear equation in one variable** is an equation that can be written in the form $ax + b = 0$.

Two equations that have the same solution are called **equivalent equations**.

The Addition and Subtraction Properties of Equality

Let a, b, and c represent algebraic expressions.

1. The **addition property of equality**:
   If $a = b$
   then $a + c = b + c$

2. The **subtraction property of equality**:
   If $a = b$
   then $a - c = b - c$

Examples

Example 1

$3x + 4 = 6$ is an equation while $3x + 4$ is an expression.

Example 2

The number -4 is a solution to the equation $5x + 7 = -13$ because when we substitute -4 for x, we get a true statement.

$$5(-4) + 7 = -13$$
$$-20 + 7 = -13$$
$$-13 = -13 \checkmark$$

Example 3

The equation $5x + 7 = -13$ is equivalent to the equation $x = -4$ because they both have the same solution of -4.

Example 4

To solve the equation $t - 12 = -3$, use the addition property of equality.

$$t - 12 = -3$$
$$t - 12 + 12 = -3 + 12$$
$$t = 9$$

Example 5

To solve the equation $-1 = p + 2$, use the subtraction property of equality.

$$-1 = p + 2$$
$$-1 - 2 = p + 2 - 2$$
$$-3 = p$$

section 11.4 Multiplication and Division Properties of Equality

Key Concepts

<u>The Multiplication and Division Properties of Equality</u>

Let a, b, and c represent algebraic expressions.

1. The **multiplication property of equality**:
   If $a = b$
   then $a \cdot c = b \cdot c$

2. The **division property of equality**:
   If $a = b$
   then $\dfrac{a}{c} = \dfrac{b}{c}$ (provided $c \neq 0$)

When applying the multiplication or division properties of equality to obtain a coefficient of 1 for the variable term, we will generally use the following conventions.

- If the coefficient of the variable term is expressed as a fraction, multiply both sides by its reciprocal.
- Otherwise, divide both sides by the coefficient itself.

We can determine which property to use to solve an equation as follows: Note the operation on the variable. Then use the property of equality of the inverse operation.

Examples

Example 1

Solve. $3a = -18$

$$\frac{3a}{3} = \frac{-18}{3}$$ Divide both sides by 3.

$$a = -6$$

Example 2

Solve. $-\dfrac{3}{5}b = 9$

$$\left(-\frac{5}{3}\right)\left(-\frac{3}{5}\right)b = \left(-\frac{5}{3}\right) \cdot 9$$ Multiply both sides by the reciprocal of $-\frac{3}{5}$.

$$b = -15$$

Example 3

Solve. $\dfrac{w}{2} = -11$

$$2\left(\frac{w}{2}\right) = 2(-11)$$ Multiply both sides by 2.

$$w = -22$$

Example 4

Solve. $4x = 20$ and $4 + x = 20$

$$4x = 20 \qquad\qquad 4 + x = 20$$

$$\frac{4x}{4} = \frac{20}{4} \qquad\qquad 4 - 4 + x = 20 - 4$$

$$x = 5 \qquad\qquad\qquad x = 16$$

section 11.5 Solving Equations with Multiple Steps

Key Concepts

<u>Steps to Solve a Linear Equation in One Variable</u>

1. Simplify both sides of the equation.
   - Clear parentheses if necessary.
   - Combine *like* terms if necessary.
2. Use the addition or subtraction property of equality to collect the variable terms on one side of the equation.
3. Use the addition or subtraction property of equality to collect the constant terms on the *other* side of the equation.
4. Use the multiplication or division property of equality to make the coefficient of the variable term equal to 1.
5. Check the answer in the original equation.

Examples

Example 1

Solve. $3y + 2 = -4y + 3$

$$3y + 4y + 2 = -4y + 4y + 3 \qquad \text{Add } 4y.$$

$$7y + 2 = 3 \qquad \text{Simplify.}$$

$$7y + 2 - 2 = 3 - 2 \qquad \text{Subtract 2.}$$

$$7y = 1 \qquad \text{Simplify.}$$

$$\frac{7y}{7} = \frac{1}{7} \qquad \text{Divide by 7.}$$

$$y = \frac{1}{7} \qquad \text{Simplify.}$$

Example 2

Solve. $4(x - 3) + 1 = -(2x + 6) + 5x$

$$4x - 12 + 1 = -2x - 6 + 5x \qquad \text{Clear parentheses.}$$

$$4x - 11 = 3x - 6 \qquad \text{Combine } like \text{ terms.}$$

$$4x - 3x - 11 = 3x - 3x - 6 \qquad \text{Subtract } 3x.$$

$$x - 11 = -6 \qquad \text{Simplify.}$$

$$x - 11 + 11 = -6 + 11 \qquad \text{Add 11.}$$

$$x = 5 \qquad \text{Simplify.}$$

section 11.6 Applications and Problem Solving

Key Concepts

Problem-Solving Flowchart for Word Problems

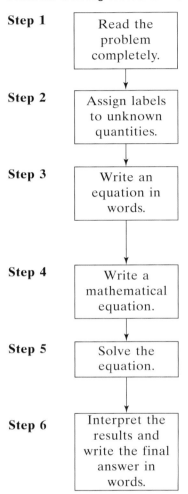

Step 1 — Read the problem completely.

Step 2 — Assign labels to unknown quantities.

Step 3 — Write an equation in words.

Step 4 — Write a mathematical equation.

Step 5 — Solve the equation.

Step 6 — Interpret the results and write the final answer in words.

Examples

Example 1

Subtract 5 times a number from 14. The result is -6. Find the number.

Let n represent the unknown number.

$$\underbrace{14}_{\text{from 14}} \quad \underbrace{- \; 5n}_{\text{subtract } 5n} \quad \underbrace{= \; -6}_{\text{result is } -6}$$

$$14 - 14 - 5n = -6 - 14$$

$$-5n = -20$$

$$\frac{-5n}{-5} = \frac{-20}{-5}$$

$$n = 4$$

The number is 4.

Example 2

An electrician needs to cut an 8-ft wire into two pieces so that one piece is 4 times as long as the other. How long should each piece be?

Let x represent one piece of the wire.
The other piece will be $4x$ ft long.

The two pieces added will be 8 ft.

$$x + 4x = 8$$

$$5x = 8$$

$$\frac{5x}{5} = \frac{8}{5}$$

$$x = \frac{8}{5} \quad \text{or} \quad 1\frac{3}{5}$$

One piece of wire is $1\frac{3}{5}$ ft.
The other piece is $4(1\frac{3}{5} \text{ ft}) = 6\frac{2}{5}$ ft.

section 11.7 Rectangular Coordinate System (Optional)

Key Concepts

The **rectangular coordinate system** is a means to plot points in two dimensions.

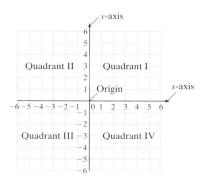

Examples

Example 1

Plot the points on the coordinate system.

$(0, -3)\ (2, 1)\ (3, -2)\ (-4, 1)\ (-3, -3)\ (5, 0)$

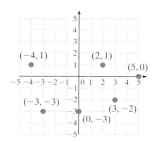

---

chapter 11 | review exercises

Section 11.1

1. a. Michael is 8 years older than his sister. Write an expression for Michael's age if his sister is a years old.

  b. What is Michael's age if his sister is 35 years old?

2. Evaluate the expression $-\left(\dfrac{4}{x}\right)^2$ for

  a. $x = 3$ **b.** $x = 4$

  c. $x = 2$ **d.** $x = -2$

3. Evaluate the expression $-2(x + y)^2$ for $x = 6$ and $y = -9$.

4. Find the volume V for a sphere with $r = \frac{3}{2}$ m and $\pi = \frac{22}{7}$, using the formula $V = \frac{4}{3}\pi r^3$

5. Apply the commutative property of addition or multiplication to rewrite the expression.

  a. $t - 5$ **b.** $h \cdot 3$

6. Apply the associative property of addition or multiplication to rewrite the expression. Then simplify the expression.

  a. $-4(2 \cdot p)$ **b.** $(m + 10) - 12$

7. Apply the distributive property. $3(2b + 5)$

8. Apply the distributive property.
  $-(-4k + 8k - 12)$

Section 11.2

For Exercises 9–10, list the terms and identify the term as a variable term or a constant term. Then identify the coefficient.

9. $3a^2 - 5a + 12$

10. $-6xy - y + 2x + 1$

For Exercises 11–14, determine if the two terms are *like* terms or *unlike* terms.

11. $5t^2, 5t$ **12.** $4h, -2h$

13. $21, -5$ **14.** $-8, -8k$

For Exercises 15–18, combine *like* terms. Clear parentheses if necessary.

15. $6y + 8x - 2y - 2x + 10$

16. $12a - 5 + 9b - 5a + 14$

17. $4(u - 3v) - 5u + v$

18. $-5(p + 4) + 6(p + 1) - 2$

Section 11.3

For Exercises 19–20, determine if -3 is a solution to the equation.

19. $5x + 10 = -5$

20. $-3(x - 1) = -9 + x$

21. Explain how to decide whether to use the addition property of equality or the subtraction property of equality to solve an equation.

For Exercises 22–30, solve the equation, using either the addition or the subtraction property of equality.

22. $r + 23 = -12$

23. $k - 3 = -15$

24. $10 = p - 4$

25. $21 = q - 3$

26. $4.1 + m = 5.2$

27. $-3.1 + n = 1.9$

28. $a - \dfrac{2}{3} = \dfrac{5}{6}$

29. $b + \dfrac{1}{5} = -\dfrac{9}{10}$

30. $\dfrac{3}{4} + h = 2$

Section 11.4

For Exercises 31–42, solve the equation by using either the multiplication or the division property of equality.

31. $4d = -28$

32. $-3c = -12$

33. $\dfrac{t}{2} = -13$

34. $\dfrac{p}{5} = 7$

35. $-\dfrac{4}{5}y = -16$

36. $\dfrac{2}{3}x = -14$

37. $1.4 = -0.7m$

38. $-3.6 = 0.9n$

39. $\dfrac{1}{3}w = \dfrac{3}{7}$

40. $-\dfrac{1}{4}s = \dfrac{2}{3}$

41. $-42 = -7p$

42. $51 = -3b$

Section 11.5

For Exercises 43–54, solve the equation.

43. $9x + 7 = -2$

44. $8y - 3 = 13$

45. $45 = 6m - 3$

46. $-25 = 2n - 1$

47. $4 = \dfrac{3}{5}m - 2$

48. $\dfrac{5}{8}p - 1 = 14$

49. $5x + 12 = 4x - 16$

50. $-4t - 2 = -3t + 5$

51. $6(w - 2) + 15 = 3(w + 3) - 2$

52. $-4(h - 5) + h = 7(h + 1) - 5$

53. $-(5a + 3) - 3(a - 2) = 24 - a$

54. $-(4b - 7) = 2(b + 3) - 4b + 13$

Section 11.6

For Exercises 55–58, **a.** write an equation that represents the problem and **b.** solve the problem.

55. The product of a number and -6 results in the sum of the number and 2. Find the number.

56. The total of -9, 3, and twice a number is -2. Find the number.

57. One-third decreased by a number is 2. Find the number.

58. The number 2 less than the quotient of a number and 8 is $\frac{1}{4}$. Find the number.

59. Actor Tom Cruise has starred in 5 fewer films than Tom Hanks. If together they have starred in 65 films, how many films did Tom Hanks and Tom Cruise star in individually?

60. A company offering Internet access charges only $6.95 per month but charges $0.25 per minute for each minute over 100 min online. Marty's bill was for $16.95. How many minutes was he online over the first 100 min?

61. The top of a rectangular table has length 8 in. less than twice the width. If the perimeter of the table is 224 in., find the width and length of the table.

Section 11.7

For Exercises 62–69, estimate the coordinates of points $A, B, C, D, E, F, G,$ and H.

62. A **63.** B

64. C **65.** D

66. E **67.** F

68. G **69.** H

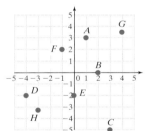

For Exercises 70–75, plot the points on the rectangular coordinate system.

70. $(3, 4)$ **71.** $(-3, 0)$

72. $(2, -2)$ **73.** $(-5, 4)$

74. $(0, 4)$ **75.** $(-1, -1)$

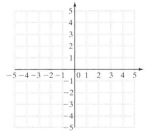

76. The table gives the average monthly temperature in Neepawa, Manitoba, Canada, for one year. Let January represent $x = 1$, February represent $x = 2$, etc. Write the table values as ordered pairs. Then graph the ordered pairs.

Month x	Temperature y (°C)
1	−17
2	−15
3	−7
4	2
5	11
6	16
7	19
8	19
9	12
10	6
11	−4
12	−13

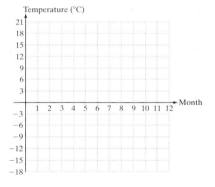

chapter 11 | test

1. A high school student sells magazine subscriptions at \$19.95 each. Write an expression that represents the total amount made for m magazines.

2. Evaluate the expression for $x = 4$ and $y = -1$.
   $-x^2 + y^2$

3. Find the area A in the geometry formula $A = 2lw + 2lh + 2wh$ for $l = 3$ ft, $w = 2.5$ ft, and $h = 1.75$ ft (surface area of a rectangular solid).

For Exercises 4–8, name the property demonstrated. Choose from the commutative property of addition, commutative property of multiplication, associative property of addition, associative property of multiplication, and distributive property of multiplication over addition.

4. $-5(9x) = (-5 \cdot 9)x$ 5. $-5x + 9 = 9 - 5x$

6. $-3 + (u + v) = (-3 + u) + v$

7. $-4(b + 2) = -4b - 8$ 8. $g(-6) = -6g$

For Exercises 9–12, simplify the expressions.

9. $4(a + 9) - 12$

10. $-3(6b) + 5b + 8$

11. $14y + 2(y - 9) + 21$

12. $2 + (5 - w) + 3(-2w)$

13. Explain the difference between an expression and an equation.

14. Identify the following as either an expression or an equation.

    a. $4x + 5$ b. $4x + 5 = 2$

    c. $-3(p + 5)$ d. $2(q - 3) = 6$

    e. $2(q - 3) + 6$ f. $6(4y + 3) + (y - 2)$

For Exercises 15–26, solve the equation.

15. $-6x = 12$ 16. $-6 + x = 12$

17. $\dfrac{x}{-6} = 12$ 18. $12x = -6$

19. $12 = -3p + 9$ 20. $1.8m = 0.36$

21. $p + \dfrac{1}{16} = -\dfrac{3}{4}$ 22. $-\dfrac{5}{12} = -\dfrac{5}{6}n$

23. $5h - 2 = -h + 16$ 24. $\dfrac{x}{7} = -12$

25. $-2(q - 5) = 6q + 10$

26. $-(4k - 2) - k = 2(k - 6)$

27. The product of -2 and a number is the same as the total of 15 and the number. Find the number.

28. The United States has 0.4 million more than 3 times the number of home satellite antennas as the United Kingdom. Together the two countries have 21.2 million antennas. How many antennas does each country have?

29. Sela's electric bill was \$52.45. She pays a monthly fee of \$5.95 and is charged \$0.075 per kWh (kilowatthour). How many kilowatthours did she use that month?

For Exercises 30–35, plot the points on the rectangular coordinate system.

30. $(0, -3)$ 31. $(4, -4)$

32. $(1.5, 5)$ 33. $(1, 0)$

34. $(-2, -4)$ 35. $(-3, 1)$

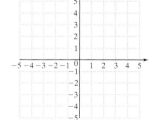

For Exercises 36–39, identify the quadrant in which the point lies.

36. $(100, -59)$

37. $\left(\dfrac{1}{6}, \dfrac{3}{20}\right)$

38. $\left(-26.22, -\dfrac{6}{5}\right)$

39. $(-239, 0.05)$

40. This table gives the gas mileage for a VW Beetle at different speeds. Write the table values as ordered pairs. Then graph the ordered pairs.

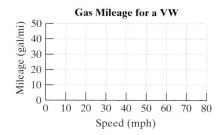

Speed x (mph)	Gas Mileage y (mpg)
30	33
40	35
50	40
60	37
70	30

Gas Mileage for a VW

---

chapters 1–11 | cumulative review

1. Identify the place value of the underlined digit.

    a. 34,9<u>1</u>1   **b.** 2<u>0</u>9,001   **c.** 5,<u>9</u>01,888

For Exercises 2–4, round the number to the indicated place value.

2. 45,921; thousands

3. 1,285,000; ten-thousands

4. 25,449; hundreds

5. Divide 39,190 by 46. Identify the dividend, divisor, whole part of the quotient, and remainder.

6. Identify each number as prime or composite.

    a. 59   **b.** 91   **c.** 39

7. Write the reciprocal of the number $\dfrac{8}{23}$.

8. Multiply. $\dfrac{2}{13} \cdot \dfrac{39}{5}$ **9.** Divide. $\dfrac{21}{10} \div \dfrac{75}{8}$

10. Simplify. $\dfrac{1300}{10{,}000}$

11. Add and subtract. $\dfrac{9}{25} - \dfrac{1}{10} + \dfrac{4}{15}$

12. Find the area of the triangle.

13. Convert to an improper fraction. $3\dfrac{7}{8}$

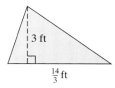

14. Convert to a mixed fraction. $\dfrac{28}{5}$

15. Simplify. Write the answer as a mixed number. $2\dfrac{1}{4} + 5\dfrac{5}{6} \cdot 1\dfrac{1}{10}$

For Exercises 16–20, simplify.

16. $3.1(4.5)$ **17.** $0.08 \div (0.16)$

18. $23.991 + 3.2 + 4.03$ **19.** $78.002 - 34.25$

20. $4.2 - (2.0 - 1.2)^2$

21. Sarah cleans three apartments in a weekend. The apartments have 5, 6, and 4 rooms, respectively. If she earns $300 for the weekend, how much does she make per room?

For Exercises 22–24, solve the proportion.

22. $\dfrac{2}{9} = \dfrac{1.8}{p}$ **23.** $\dfrac{n}{1\frac{1}{2}} = \dfrac{8}{15}$ **24.** $\dfrac{24}{15} = \dfrac{m}{35}$

25. One type of cat treats, Kitty Treats, contains 2 oz and sells for $1.84. Cat Goodies sells for $2.25 but contains 2.5 oz. Find the unit prices to determine the better buy.

For Exercises 26–30, complete the table.

	Decimal	Fraction	Percent
26.			15%
27.		$\dfrac{1}{8}$	
28.	1.1		
29.		$\dfrac{2}{9}$	
30.			0.2%

31. A 1-gal jug of fruit punch will be poured into 6-oz cups. How many complete cups can be filled?

32. A Formula One Grand Prix race was won with a time of 1:41:45. Convert this to minutes.

For Exercises 33–35, convert the metric measurements.

33. 680 cc = _____ L

34. 3.2 km = _____ m

35. 45 g = _____ cg

36. What is the area for a round table with a 4-in. diameter? Use 3.14 for π.

37. A page in a book is $8\frac{1}{2}$ in. by 10 in. However, there is a $\frac{1}{2}$-in. margin around the edges of the page. How much area can be used for printed material?

38. Find the volume of a can of paint if the can is a cylinder with radius 6.5 in. and height 7.5 in. Use 3.14 for π, and round the answer to the nearest whole unit.

39. Find the circumference C of the circle. Use $\pi = 3.14$.

40. Find the length of side x.

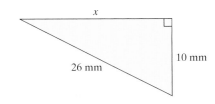

For Exercises 41–43, refer to the list of data depicting the number of turkey subs sold each day for two weeks at Subs-R-Us.

15	12	8	5	6	12	10
20	7	5	8	9	11	12

41. Find the mean.

42. Find the median.

43. Find the mode.

44. The annual amounts of overtime for Ormond Beach firefighters for 7 years are presented in the table. Construct a bar graph with horizontal bars by following these steps.

Year	Overtime ($)
1997–1998	273,000
1998–1999	260,000
1999–2000	241,000
2000–2001	284,000
2001–2002	375,000
2002–2003	633,000
2003–2004	748,000

a. On the horizontal scale, make tick marks, starting from 0 to $800,000 in increments (steps) of 100,000.

b. On the vertical scale, write the year. Begin at the bottom with the year 1997–1998.

c. Construct the bar graph. The length of each bar corresponds to the amount of overtime for that year.

Overtime Pay for Ormond Beach Firefighters

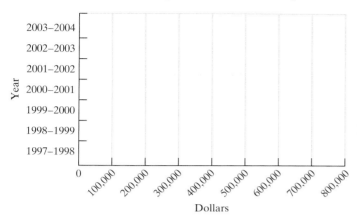

45. The circle graph represents activities of high school students after school.

After-School Activites

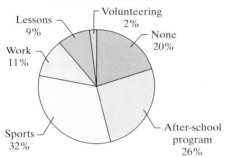

a. If a school has about 520 students, how many are enrolled in sports?

b. If a school has about 650 students, how many do not participate in any after-school activities?

For Exercises 46–49, perform the indicated operations.

46. $-129 - (-132)$ **47.** $16 \div (-4) \cdot 3$

48. $4(5 - 11) + (-1)$ **49.** $5 - 23 + 12 - 3$

50. a. Write the number 3,001,000 in scientific notation.

    b. Write 4×10^{-5} in decimal notation.

For Exercises 51–52, simplify the expression.

51. $-2(x + 14) + 15 - 3x$

52. $3y - (5y + 6) - 12$

For Exercises 53–54, solve the equation.

53. $4p + 5 = -11$

54. $9(t - 1) - 7t + 2 = t - 15$

55. Plot the following points on the rectangular coordinate system.

    a. $(-3, 5)$ **b.** $(0, 0)$

    c. $(-2, -2)$ **d.** $(3, 4)$

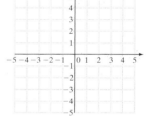

STUDENT ANSWER APPENDIX

Chapter 1

Chapter 1 Preview, p. 2

1. Ten-thousands place **3.** One hundred thirty is less than two hundred forty-four. **5.** 262 **7.** 766 **9.** 930 **11.** 248 **13.** 105 R2 **15.** 0 **17.** 140 m² **19.** 6 **21.** 187 mi

Section 1.1 Practice Exercises, pp. 7–10

3. 7: ones; 5: tens; 4: hundreds; 3: thousands; 1: ten-thousands; 2: hundred-thousands; 8: millions **5.** Tens **7.** Ones **9.** Hundreds **11.** Thousands **13.** Hundred-thousands **15.** Billions **17.** Ten-thousands **19.** Millions **21.** Ten-millions **23.** Billions **25.** 5 tens + 8 ones **27.** 5 hundreds + 3 tens + 9 ones **29.** 5 hundreds + 3 ones **31.** 1 ten-thousand + 2 hundreds + 4 tens + 1 one **33.** 2 millions + 6 thousands + 4 ones **35.** 524 **37.** 150 **39.** 1,906 **41.** 85,007 **43.** Answers will vary. **45.** Ones, thousands, millions, billions **47.** Two hundred forty-one **49.** Six hundred three **51.** Nine thousand, five hundred thirty-five **53.** Thirty-one thousand, five hundred thirty **55.** One hundred thousand, two hundred thirty-four **57.** Twenty thousand, three hundred twenty **59.** One thousand, three hundred seventy-seven **61.** 6,005 **63.** 672,000 **65.** 1,484,250 **67.**

69. 10 **71.** 4 **73.** 8 is greater than 2, or 2 is less than 8 **75.** 3 is less than 7, or 7 is greater than 3 **77.** < **79.** > **81.** < **83.** > **85.** < **87.** < **89.** False **91.** 99 **93.** There is no greatest whole number. **95.** 7 **97.** 964

Section 1.2 Practice Exercises, pp. 18–23

3. 3 hundreds + 5 tens + 1 one **5.** 1 hundred + 7 ones **7.** 4012 **9.**

+	0	1	2	3	4	5	6	7	8	9
0	0	1	2	3	4	5	6	7	8	9
1	1	2	3	4	5	6	7	8	9	10
2	2	3	4	5	6	7	8	9	10	11
3	3	4	5	6	7	8	9	10	11	12
4	4	5	6	7	8	9	10	11	12	13
5	5	6	7	8	9	10	11	12	13	14
6	6	7	8	9	10	11	12	13	14	15
7	7	8	9	10	11	12	13	14	15	16
8	8	9	10	11	12	13	14	15	16	17
9	9	10	11	12	13	14	15	16	17	18

11. Addends: 2, 8; sum: 10 **13.** Addends: 11, 10; sum: 21 **15.** Addends: 5, 8, 2; sum: 15 **17.** 74 **19.** 58 **21.** 48 **23.** 19 **25.** 588 **27.** 798 **29.** 237 **31.** 198 **33.** 84 **35.** 115 **37.** 937 **39.** 850 **41.** 41 **43.** 29 **45.** 1003 **47.** 836 **49.** 24,004 **51.** 13,121 **53.** 21 + 30 **55.** 13 + 8 **57.** 23 + (9 + 10) **59.** (41 + 3) + 22 **61.** The sum of any number and 0 is that number. **a.** 423 **b.** 25 **c.** 67 **63.** 100 + 42; 142 **65.** 23 + 81; 104 **67.** 76 + 2; 78 **69.** 1320 + 448; 1768 **71.** 78 + 12 + 22; 112 **73.** For example: The sum of 33 and 15 **75.** For example: 15 added to 70 **77.** For example: The total of 11, 41, and 53 **79.** 521 deliveries **81.** 423 mi **83.** 1,206,655 athletes **85.** $245 **87.** 821,024 nonteachers **89.** 13,538 participants **91.** 60 in. **93.** 35 m **95.** 26 ft **97.** 360 ft

Calculator Connections 1.2, p. 23

99. 908,788 **101.** 21,491,394 **103.** 121,480,019 votes

Section 1.3 Practice Exercises, pp. 29–33

3. 1151 **5.** 899 **7.** 0 < 10 **9.** Minuend: 12; subtrahend: 8; difference: 4 **11.** Minuend: 21; subtrahend: 12; difference: 9 **13.** Minuend: 9; subtrahend: 6; difference: 3 **15.** 18 + 9 = 27 **17.** 27 + 75 = 102 **19.** 5 **21.** 3 **23.** 6 **25.** 45 **27.** 61 **29.** 1126 **31.** 321 **33.** 10,004 **35.** 1103 **37.** 34,331 **39.** 17 **41.** 49 **43.** 104 **45.** 521 **47.** 23 **49.** 4764 **51.** 1303 **53.** 2217 **55.** 378 **57.** 722 **59.** 30,941 **61.** 5,662,119 **63.** 78 − 23; 55 **65.** 78 − 6; 72 **67.** 422 − 100; 322 **69.** 1090 − 72; 1018 **71.** 50 − 13; 37 **73.** For example: 93 minus 27 **75.** For example: Subtract 85 from 165. **77.** The expression 7 − 4 means 7 minus 4, yielding a difference of 3. The expression 4 − 7 means 4 minus 7 which results in a difference of −3. (This is a mathematical skill we have not yet learned.) **79.** $33 **81.** 55 more hits **83.** 8 plants **85.** 9190 yd **87.** 13 m **89.** 10 yd **91.** 7748 **93.** 107,489

Calculator Connections 1.3, p. 33

95. 4,447,302 **97.** 49,408 mi² **99.** The difference in land area between Colorado and Rhode Island is 102,673 mi².

Section 1.4 Practice Exercises, pp. 38–41

3. 26 **5.** 5007 **7.** Ten-thousands **9.** If the digit in the tens place is 0, 1, 2, 3, or 4, then change the tens and ones digits to 0. If the digit in the tens place is 5, 6, 7, 8, or 9, increase the digit in the hundreds place by 1 and change the tens and ones digits to 0.
11. 340 **13.** 730 **15.** 9400 **17.** 8500 **19.** 10,000 **21.** 3000
23. 35,000 **25.** 109,000 **27.** 490,000 **29.** $317,000,000
31. 22,000 words **33.** 160 **35.** 180 **37.** 500 **39.** 100
41. 70,000 more women **43.** $151,000,000 **45.** $13,500,000
47. a. 1994; $3,600,000 **b.** 1995; $3,200,000 **49.** Massachusetts; 79,000 students **51.** 71,000 students **53.** Answers may vary.
55. 10,000 mm **57.** 440 in.

Midchapter Review, p. 41

1. 1033 **2.** 3381 **3.** 101 **4.** 9542 **5.** Thousands **6.** Hundreds
7. Hundred-thousands **8.** Millions **9.** 15,000 **10.** 4,100,000
11. 268 **12.** 222 **13.** 21 **14.** 123

Section 1.5 Practice Exercises, pp. 50–55

3. 1,010,000 **5.** 5400 **7.** 6 × 5; 30 **9.** 3 × 9; 27
11. Factors: 6, 4; product: 24 **13.** Factors: 13, 42; product: 546
15. Factors: 3, 5, 2; product: 30 **17.** For example: 5 × 12; 5 · 12; 5(12) **19.** d **21.** e **23.** c **25.** 8 × 14
27. (6 × 2) × 10 **29.** (5 × 7) + (5 × 4) **31.** 144 **33.** 52 **35.** 655
37. 1376 **39.** 11,280 **41.** 23,184 **43.** 378,126 **45.** 448 **47.** 1632
49. 864 **51.** 2431 **53.** 6631 **55.** 19,177 **57.** 186,702
59. 21,241,448 **61.** 24,000 **63.** 2,100,000 **65.** 72,000,000
67. 36,000,000 **69.** 60,000,000 **71.** 2,400,000,000 **73.** $1000
75. $780,000 **77.** 375 lb **79.** $1665 **81.** 287,500 sheets
83. 372 miles **85.** 276 ft^2 **87.** 5329 cm^2 **89.** 128 ft^2
91. a. 2552 in.^{2} **b.** 42 windows **c.** 107,184 in.^{2}

Calculator Connections 1.5, p. 55

93. 7,665,000,000 bbl **95.** $215,644,000

Section 1.6 Practice Exercises, pp. 63–66

3. 4944 **5.** 1253 **7.** 664,210 **9.** 902 **11.** 9; the dividend is 72; the divisor is 8; the quotient is 9 **13.** 8; the dividend is 64; the divisor is 8; the quotient is 8 **15.** 5; the dividend is 45; the divisor is 9; the quotient is 5 **17.** You cannot divide a number by zero (the quotient is undefined). If you divide zero by a number (other than zero), the quotient is always zero. **19.** 15 **21.** 0 **23.** Undefined
25. 1 **27.** Undefined **29.** 0 **31.** 2 × 3 = 6, 2 × 6 ≠ 3
33. Multiply the quotient and the divisor to get the dividend.
35. 13 **37.** 41 **39.** 486 **41.** 409 **43.** 952 **45.** 822
47. Correct **49.** Correct **51.** Correct **53.** Incorrect; 25 R3
55. 7 R5 **57.** 10 R2 **59.** 27 R1 **61.** 197 R2 **63.** 42 R4
65. 1557 R1 **67.** 751 R6 **69.** 835 R2 **71.** 479 R9 **73.** 43 R19
75. 308 **77.** 1259 R26 **79.** 229 R96 **81.** 302 **83.** 497 ÷ 71; 7
85. 877 ÷ 14; 62 R9 **87.** 42 ÷ 6; 7 **89.** 14 classrooms
91. 5 cases; 8 cans left over **93.** 52 mph **95.** 22 lb **97.** Yes, they can all attend if they sit in the second balcony.
99. 1200 ÷ 20 = 60; approximately 60 words per minute

Calculator Connections 1.6, p. 66

101. 234 **103.** $180 billion

Section 1.7 Practice Exercises, pp. 71–74

3. True **5.** False **7.** True **9.** 9^4 **11.** 2^7 **13.** 3^6 **15.** 4^4 · 2^3
17. 8 · 8 · 8 · 8 **19.** 4 · 4 · 4 · 4 · 4 · 4 · 4 **21.** 8 **23.** 9
25. 27 **27.** 125 **29.** 32 **31.** 81 **33.** 1 **35.** 1 **37.** The number 1 raised to any power equals 1. **39.** 1000 **41.** 100,000 **43.** 2

45. 6 **47.** 10 **49.** 0 **51.** No, addition and subtraction should be performed in the order in which they appear from left to right.
53. 26 **55.** 1 **57.** 49 **59.** 3 **61.** 2 **63.** 55 **65.** 8 **67.** 45
69. 24 **71.** 4 **73.** 25 **75.** 4 **77.** 81 **79.** 18 **81.** 0 **83.** 5
85. 6 **87.** 48 **89.** 109 **91.** 18 **93.** 38¢ per pound **95.** 121 mm per month **97.** 24 **99.** 70

Calculator Connections 1.6, pp. 74–75

101. 24,336 **103.** 248,832 **105.** 8028 **107.** 66,049 **109.** 35

Section 1.8 Practice Exercises, pp. 80–85

3. 71 + 14; 85 **5.** 2 · 14; 28 **7.** 102 − 32; 70 **9.** 10 · 13; 130
11. 24 ÷ 6; 4 **13.** For example: sum, added to, increased by, more than, plus, total of **15.** For example: difference, minus, decreased by, less, subtract **17.** The whole screen has 12,096 pixels. **19.** There will be 9 gal used. **21.** Denali is 6074 ft higher than White Mountain Peak. **23.** Jeannette will pay $29,560 for one year.
25. The Insight can go 1320 mi. **27.** There will be 120 classes of Beginning Algebra. **29.** The maximum capacity is 3150 seats.
31. There will be $36 left in Gina's account. **33.** The total bill was $154,032. **35. a.** Latayne will receive $48. **b.** She can buy 6 CDs.
37. Michael Jordan scored 33,454 points with the Bulls. **39. a.** The difference was 8,410,500 passengers. **b.** There was a total of 143,306,500 passengers. **41. a.** The distance is 360 mi. **b.** 14 in. represents 840 mi. **43.** 104 boxes will be filled completely with 2 books left over. **45. a.** Marc needs five $20 bills. **b.** He will receive $16 in change. **47.** Jackson's monthly payment was $390.
49. Each trip will take 2 hr. **51.** Perimeter **53.** It will cost $1650.
55. The cost is $1020. **57.** He earned $520.

Chapter 1 Review Exercises, pp. 92–96

1. Ten-thousands **3.** 92,046
5. 3 millions + 4 hundred-thousands + 8 hundreds + 2 tens
7. Two hundred forty-five **9.** 3602

11.

13. True **15.** Addends: 105, 119; sum: 224 **17.** 71 **19.** 17,410
21. a. Commutative property **b.** Associative property
c. Commutative property **23.** 44 + 92; 136 **25.** 23 + 6; 29
27. 45,797 thousand seniors **29.** Minuend: 14; subtrahend: 8; difference: 6 **31.** 26 **33.** 121 **35.** 31,019 **37.** 38 − 31; 7
39. 251 − 42; 209 **41.** 71,892,438 tons **43.** 2336 thousand visitors
45. 9,330,000 **47.** 1500 **49.** 163,000 m^3 **51.** Factors: 33, 40; product: 1320 **53.** c **55.** d **57.** b **59.** 52,224 **61.** $429 **63.** 7; divisor: 6, dividend: 42, quotient: 7 **65.** 3 **67.** Undefined
69. Multiply the quotient and the divisor to get the dividend.
71. 58 **73.** 52 R3 **75.** 9)‾108 (12) **77. a.** 4 T-shirts **b.** 5 hats
79. 2^4 · 5^3 **81.** 256 **83.** 1,000,000 **85.** 12 **87.** 15 **89.** 55
91. $89 **93. a.** The Cincinnati Zoo has 13,000 more animals than the San Diego Zoo. **b.** The San Diego Zoo has 50 more species than the Cincinnati Zoo. **95.** He will receive $19,600,000 per year.

Chapter 1 Test, pp. 97–98

1. a. Hundreds **b.** Thousands **c.** Millions **d.** Ten-thousands
2. a. 4,065,000 **b.** Twenty-one million, three hundred twenty-five thousand **c.** Twelve million, two hundred eighty-seven thousand
d. 729,000 **e.** Eleven million, four hundred ten thousand
3. a. 14 > 6 **b.** 72 < 81 **4.** 129 **5.** 328 **6.** 113 **7.** 227
8. 2842 **9.** 447 **10.** 21 R9 **11.** 546 **12.** 8103 **13.** 20
14. 1,500,000,000 **15.** 336 **16.** 0 **17.** Undefined **18. a.** The associative property of multiplication; the expression shows a change in grouping. **b.** The commutative property of

multiplication; the expression shows a change in the order of the factors. **19. a.** 4900 **b.** 12,000 **c.** 8,000,000 **20.** There were approximately 1,430,000 people. **21.** 4 **22.** 24 **23.** 48 **24.** Jennifer has a higher average of 29. Brittany has an average of 28. **25.** There were 20,099 foreign adoptions in 2002. **26. a.** 12,392 subscribers **b.** The largest increase was between the years 1999 and 2000. **27.** The North Side Fire Department is the busiest with an average of 5 calls per week. **28.** 156 mm **29.** Perimeter: 350 ft; area: 6016 ft^2 **30.** 4,560,000 m^2

Chapter 2

Chapter 2 Preview, p. 100

1. a. Proper **b.** Improper **c.** Improper **3.** $5\frac{4}{7}$ **5.** $\frac{3}{8}$
7. 1, 3, 5, 9, 15, 45 **9.** a **11.** $\frac{1}{10}$ **13.** 5 cm^2 **15.** $\frac{9}{16}$ **17.** 120 ft^2
19. 17 **21.** $6\frac{1}{2}$

Section 2.1 Practice Exercises, pp. 107–111

3. $\frac{3}{4}$ **5.** $\frac{5}{9}$ **7.** $\frac{1}{6}$ **9.** $\frac{3}{8}$ **11.** $\frac{3}{4}$ **13.** Numerator: 2; denominator: 3
15. Numerator: 12; denominator: 11 **17.** $6 \div 1$; 6 **19.** $2 \div 2$; 1
21. $0 \div 3$; 0 **23.** $2 \div 0$; undefined **25.** $\frac{2}{5}$ **27.** $\frac{10}{21}$ **29.** Proper
31. Improper **33.** Improper **35.** $\frac{5}{2}$ **37.** $\frac{12}{4}$ **39.** $\frac{7}{4}$; $1\frac{3}{4}$
41. $\frac{13}{8}$; $1\frac{5}{8}$ **43.** $\frac{7}{4}$ **45.** $\frac{38}{9}$ **47.** $\frac{24}{7}$ **49.** $\frac{29}{4}$ **51.** $\frac{137}{12}$ **53.** $\frac{171}{8}$
55. $\frac{133}{16}$ **57.** $\frac{269}{20}$ **59.** 19 **61.** 7 **63.** $4\frac{5}{8}$ **65.** $7\frac{4}{5}$ **67.** $2\frac{7}{10}$
69. $5\frac{7}{9}$ **71.** $12\frac{1}{11}$ **73.** $3\frac{5}{6}$ **75.** $7\frac{5}{9}$ **77.** $8\frac{1}{8}$ **79.** $44\frac{1}{7}$ **81.** $1056\frac{1}{5}$
83. $810\frac{3}{11}$
85.
87.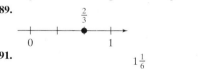
89.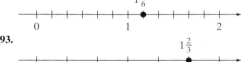
91.

93.

95. False **97.** True

Calculator Connections 2.1, p. 111

99. $\frac{8586}{407}$ **101.** $\frac{5399}{112}$

Section 2.2 Practice Exercises, pp. 116–119

3. $\frac{8}{12}$, $\frac{4}{12}$ **5.** $\frac{5}{4}$, $\frac{3}{4}$ **7.** $\frac{7}{12}$; proper **9.** $4\frac{3}{5}$ **11.** For example, $2 \cdot 4$ and $1 \cdot 8$ **13.** For example, $4 \cdot 6$ and $2 \cdot 2 \cdot 2 \cdot 3$ **15.** For example, $8 \cdot 4$ and $2 \cdot 2 \cdot 2 \cdot 2 \cdot 2$

17.

Product	36	42	30	15	81
Factor	12	7	30	15	27
Factor	3	6	1	1	3
Sum	15	13	31	16	30

19. A whole number is divisible by 2 if it is an even number. **21.** A whole number is divisible by 3 if the sum of its digits is divisible by 3. **23. a.** No **b.** Yes **c.** Yes **d.** No **25. a.** Yes **b.** Yes **c.** No **d.** No **27. a.** Yes **b.** Yes **c.** No **d.** No **29. a.** Yes **b.** No **c.** Yes **d.** Yes **31. a.** No **b.** No **c.** No **d.** No **33.** Yes **35.** There are two whole numbers that are neither prime nor composite, 0 and 1. **37.** False **39.** 2, 3, 5, 7, 11, 13, 17, 19, 23, 29, 31, 37, 41, 43, 47 **41.** Prime **43.** Composite **45.** Composite **47.** Prime **49.** Neither **51.** Composite **53.** Prime **55.** Composite **57.** No, 9 is not a prime number. **59.** Yes **61.** $2 \cdot 5 \cdot 7$ **63.** $2 \cdot 2 \cdot 5 \cdot 13$ or $2^2 \cdot 5 \cdot 13$ **65.** $3 \cdot 7 \cdot 7$ or $3 \cdot 7^2$ **67.** $2 \cdot 3 \cdot 23$ **69.** $2 \cdot 2 \cdot 2 \cdot 7 \cdot 11$ or $2^3 \cdot 7 \cdot 11$ **71.** Prime **73.** 1, 2, 3, 6, 12 **75.** 1, 2, 4, 8, 16, 32 **77.** 1, 3, 9, 27, 81 **79.** 1, 2, 3, 4, 6, 8, 12, 16, 24, 48 **81.** No **83.** Yes **85.** Yes **87.** No **89.** Yes **91.** No **93.** Yes **95.** No

Section 2.3 Practice Exercises, pp. 124–127

3. $5 \cdot 29$ **5.** $2 \cdot 2 \cdot 23$ or $2^2 \cdot 23$ **7.** $5 \cdot 17$ **9.** $3 \cdot 5 \cdot 13$
11. 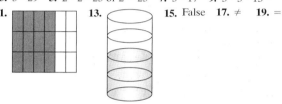 **13.** **15.** False **17.** ≠ **19.** =
21. = **23.** ≠ **25.** $\frac{1}{2}$ **27.** $\frac{1}{3}$ **29.** $\frac{9}{5}$ **31.** $\frac{5}{4}$ **33.** $\frac{4}{5}$ **35.** 1
37. 2 **39.** 1 **41.** $\frac{3}{4}$ **43.** 3 **45.** $\frac{7}{10}$ **47.** $\frac{13}{15}$ **49.** $\frac{77}{39}$ **51.** $\frac{2}{5}$
53. $\frac{5}{2}$ **55.** $\frac{3}{4}$ **57.** $\frac{5}{3}$ **59.** $\frac{21}{11}$ **61.** $\frac{17}{100}$ **63.** Heads: $\frac{5}{12}$; tails: $\frac{7}{12}$
65. a. $\frac{3}{22}$ **b.** $\frac{9}{22}$ **67. a.** Jonathan: $\frac{5}{7}$; Jared: $\frac{6}{7}$ **b.** Jared sold the greater fractional part. **69. a.** Raymond: $\frac{10}{11}$; Travis: $\frac{9}{11}$
b. Raymond read the greater fractional part.
71. For example, $\frac{6}{8}$, $\frac{9}{12}$, $\frac{12}{16}$ **73.** For example, $\frac{6}{9}$, $\frac{4}{6}$, $\frac{2}{3}$

Midchapter Review, p. 127

1. A fraction in which the numerator is greater than or equal to the denominator is an improper fraction. **2.** A fraction in which the numerator is less than the denominator is a proper fraction. **3.** $\frac{2}{3}$ is a proper fraction and $\frac{3}{2}$ is an improper fraction.
4. **5.** **6. a.** 1, 2, 4, 8, 16
b. 1, 3, 7, 9, 21, 63 **c.** 1, 2, 4, 5, 8, 10, 20, 40 **d.** 1, 2, 3, 5, 6, 10, 15, 30 **7. a.** $2 \cdot 2 \cdot 2 \cdot 2$ or 2^4 **b.** $3 \cdot 3 \cdot 7$ or $3^2 \cdot 7$ **c.** $2 \cdot 2 \cdot 2 \cdot 5$ or $2^3 \cdot 5$ **d.** $2 \cdot 3 \cdot 5$ **8.** $\frac{8}{15}$ **9.** $\frac{2}{3}$ **10.** $\frac{5}{4}$ **11.** $\frac{5}{4}$ **12.** $\frac{3}{2}$ **13.** $\frac{5}{6}$ **14.** $2\frac{1}{2}$
15. $2\frac{3}{4}$ **16.** $1\frac{1}{3}$

Section 2.4 Practice Exercises, pp. 135–139

3. Numerator: 10; denominator: 14; $\frac{5}{7}$

5. Numerator: 25; denominator: 15; $\frac{5}{3}$

7. Numerator: 7200; denominator: 90,000; $\frac{2}{25}$

9. **11.** $\frac{1}{8}$ **13.** 6 **15.** $\frac{3}{16}$ **17.** $\frac{14}{81}$ **19.** $\frac{24}{35}$

21. $\frac{8}{11}$ **23.** $\frac{24}{5}$ **25.** $\frac{65}{36}$ **27.** $\frac{2}{15}$ **29.** $\frac{5}{8}$ **31.** $\frac{35}{4}$ **33.** $\frac{8}{3}$ **35.** $\frac{4}{5}$

37. 8 **39.** 12 **41.** $\frac{30}{7}$ **43.** 10 **45.** $\frac{5}{3}$ **47.** $\frac{3}{8}$ **49.** 24

51. a. $\frac{1}{1000}$ **b.** $\frac{1}{10,000}$ **c.** $\frac{1}{1,000,000}$ **53.** $\frac{1}{16}$ **55.** $\frac{64}{27}$ **57.** 8

59. $\frac{1}{900}$ **61.** 3 **63.** $\frac{21}{2}$

65. 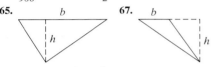 **67.** **69.** 90 in.^{2}

71. 4 yd^2 **73.** $\frac{8}{3}$ or $2\frac{2}{3}$ mm^2 **75.** 8 m^2 **77.** $\frac{23}{32}$ ft^2 **79.** 36 m^2

81. 24 m^2 **83.** The cost is $13,750. **85.** $\frac{1}{10}$ of the sample has

O negative blood. **87.** Nancy spends $\frac{9}{4}$ or $2\frac{1}{4}$ hr a day.

89. Frankie mowed 960 yd^2. He has 480 yd^2 left to mow.

91. a. $\frac{4}{49}$ **b.** $\frac{2}{7}$ **93.** $\frac{1}{10}$ **95.** $\frac{3}{2}$ **97.** $\frac{2}{81}$ **99.** They are the same.

Calculator Connections 2.4, p. 139

101. $\frac{555}{5356}$ **103.** $\frac{275,152}{225}$

Section 2.5 Practice Exercises, pp. 145–149

3. $\frac{18}{5}$ **5.** 2 **7.** $\frac{5}{3}$ **9.** 1 **11.** 1 **13.** $\frac{8}{7}$ **15.** $\frac{9}{10}$ **17.** $\frac{1}{4}$

19. No reciprocal exists. **21.** $\frac{1}{3}$ **23.** Multiplying **25.** $\frac{8}{25}$ **27.** $\frac{35}{26}$

29. $\frac{35}{9}$ **31.** 5 **33.** 1 **35.** $\frac{21}{2}$ **37.** 20 **39.** 16 **41.** $\frac{3}{5}$ **43.** $\frac{1}{4}$

45. $\frac{90}{13}$ **47.** 20 **49.** $\frac{154}{3}$ **51.** $\frac{7}{2}$ **53.** $\frac{5}{36}$ **55.** 8 **57.** $\frac{2}{5}$ **59.** $\frac{40}{3}$

61. 2 **63.** $\frac{55}{56}$ **65.** $\frac{3}{2}$ **67.** $\frac{2}{3} \cdot 6$ multiplies $\frac{2}{3}$ by $\frac{6}{1}$, and $\frac{2}{3} \div 6$

multiplies $\frac{2}{3}$ by $\frac{1}{6}$. So $\frac{2}{3} \cdot 6 = 4$ and $\frac{2}{3} \div 6 = \frac{1}{9}$. **69.** 3 **71.** $\frac{7}{6}$

73. $\frac{7}{32}$ **75.** $\frac{9}{400}$ **77.** 49 **79.** 18 **81.** 24 cups of juice

83. Li wrapped 54 packages. **85.** The stack will be 12 in. high.
87. a. 27 commercials in 1 hr **b.** 648 commercials in 1 day
89. a. She plans to sell $\frac{3}{4}$ acre. **b.** She will keep $\frac{3}{2}$ or $1\frac{1}{2}$ acres.
91. She can prepare 14 samples. **93. a.** Ricardo's mother will
pay $16,000. **b.** Ricardo will have to pay $8000. **c.** He will have
to finance $216,000. **95.** 12 ft, because $\frac{5}{2} \cdot 12 = 30$.

Calculator Connections 2.5, p. 149

97. $\frac{40,002}{851}$ **99.** $\frac{681}{23,968}$

Section 2.6 Practice Exercises, pp. 154–156

3. $\frac{26}{9}$ **5.** $\frac{12}{11}$ **7.** $\frac{2}{9}$ **9.** $\frac{17}{5}$ **11.** $\frac{11}{7}$ **13.** $\frac{77}{6}$ **15.** $\frac{39}{4}$ **17.** $7\frac{2}{5}$

19. $1\frac{2}{3}$ **21.** 38 **23.** $27\frac{2}{3}$ **25.** $1\frac{4}{5}$ **27.** $72\frac{1}{2}$ **29.** 0 **31.** $7\frac{1}{2}$

33. $2\frac{4}{25}$ **35.** $2\frac{5}{8}$ **37.** $\frac{34}{55}$ **39.** $4\frac{5}{12}$ **41.** $2\frac{6}{17}$ **43.** 2 **45.** 0 **47.** 17

49. $4\frac{2}{3}$ **51.** $1\frac{3}{4}$ **53.** $2\frac{3}{8}$ **55.** Tabitha earned $38. **57. a.** 7 weeks
old **b.** $8\frac{1}{2}$ weeks old **59. a.** Lucy earned $72 more than Ricky.
b. Together they earned $922. **61.** 2 **63.** $5\frac{1}{3}$ **65.** 0 **67.** $1\frac{1}{6}$

69. 0 **71.** $1\frac{1}{2}$ **73.** Undefined **75.** The total cost is $168.

Chapter 2 Review Exercises, pp. 162–165

1. $\frac{1}{2}$ **3. a.** $\frac{5}{3}$ **b.** Improper **5.** $\frac{7}{15}$ **7.** $\frac{7}{6}$ or $1\frac{1}{6}$ **9.** $\frac{57}{5}$ **11.** $5\frac{2}{9}$

13., 15.

```
        13   10
         8    5
+--+--+--+--●--+--●--+--+--+--+--+--+--+--+--+--+
0         1         2         3         4
```

17. $60\frac{11}{13}$ **19.** 55, 140, 260, 1200 **21.** Prime **23.** Neither

25. $2 \cdot 2 \cdot 2 \cdot 2 \cdot 2 \cdot 2$ or 2^6 **27.** $2 \cdot 2 \cdot 3 \cdot 3 \cdot 5 \cdot 5$ or $2^2 \cdot 3^2 \cdot 5^2$

29. 1, 2, 4, 5, 8, 10, 16, 20, 40, 80 **31.** = **33.** $\frac{2}{7}$ **35.** $\frac{7}{3}$ **37.** 2

39. $\frac{7}{10}$ **41. a.** $\frac{3}{5}$ **b.** $\frac{2}{5}$ **43.** $\frac{32}{9}$ **45.** 15 **47.** $\frac{12}{7}$ **49.** $\frac{1}{625}$

51. $\frac{1}{17}$ **53.** $A = lw$ **55.** $\frac{10}{3}$ or $3\frac{1}{3}$ m^2 **57.** Maximus requires $\frac{7}{2}$

or $3\frac{1}{2}$ yd of lumber. **59.** There are 300 Asian American students.

61. There are 750 Caucasian male students. **63.** 1 **65.** $\frac{1}{7}$ **67.** 6

69. Multiplying **71.** $\frac{7}{5}$ **73.** $\frac{1}{6}$ **75.** 14 **77.** $\frac{4}{5}$ **79.** $\frac{1}{52}$

81. $18 \div \frac{2}{3}$; 27 **83.** Amelia earned $576. **85.** Yes. $9 \div \frac{3}{8} = 24$ so
he will have 24 pieces, which is more than enough for his class.
87. $23\frac{2}{3}$ **89.** $22\frac{1}{2}$ **91.** $1\frac{1}{2}$ **93.** $4\frac{1}{2}$ **95.** $\frac{3}{5}$ **97.** It will take $3\frac{1}{8}$ gal.

Chapter 2 Test, pp. 166–167

1. a. $\frac{5}{8}$ **b.** Proper **2. a.** $\frac{7}{3}$ **b.** Improper **3.** $\frac{11}{2}$; $5\frac{1}{2}$ **4.** $\frac{7}{7}$ is an
improper fraction because the numerator is greater than or equal to
the denominator. **5. a.** $3\frac{2}{3}$ **b.** $\frac{34}{9}$

6.
```
+--+--+--+--●--+--+--+--+--+--+
0          1/2          1
```
7.
```
+--+--+--+--+--+--+--●--+--+--+
0                   3/4       1
```

8.
```
+--+--+--+--+--+--●--+--+--+--+--+
0                7/12            1
```

9.
```
                    13
                     5
+--+--+--+--+--+--+--+--●--+--+--+--+--+--+--+--+
0         1         2         3         4
```

10. a. Composite **b.** Neither **c.** Prime **d.** Neither **e.** Prime **f.** Composite **11. a.** 1, 2, 4, 5, 8, 10, 16, 20, 40, 80 **b.** $2 \cdot 2 \cdot 2 \cdot 2 \cdot 5$ or $2^4 \cdot 5$ **12. a.** Add the digits of the number. If the sum is divisible by 3, then the original number is divisible by 3. **b.** Yes. **13. a.** No **b.** Yes **c.** Yes **d.** No **14.** = **15.** ≠ **16.** $\frac{10}{7}$ or $1\frac{3}{7}$ **17.** $\frac{6}{7}$ **18. a.** Christine; $\frac{3}{5}$; Brad: $\frac{4}{5}$ **b.** Brad has the greater fractional part completed. **19.** $\frac{19}{69}$ **20.** $\frac{25}{2}$ or $12\frac{1}{2}$ **21.** $\frac{4}{9}$ **22.** $\frac{1}{2}$ **23.** $\frac{4}{15}$ **24.** $\frac{3}{4}$ **25.** $\frac{4}{35}$ **26.** $9\frac{3}{5}$ **27.** $\frac{13}{12}$ **28.** $\frac{44}{3}$ or $14\frac{2}{3}$ cm^2 **29.** $20 \div \frac{1}{4}$ **30.** 48 quarter-pounders **31.** 5 dogs are female pure breeds. **32.** They can build on a maximum of $\frac{2}{5}$ acre.

Chapters 1 and 2 Cumulative Review, pp. 167–168

1.

	Height (ft)	
Mountain	Standard Form	Words
Mt. Foraker (Alaska)	17,400	Seventeen thousand, four hundred
Mt. Kilimanjaro (Tanzania)	19,340	Nineteen thousand, three hundred forty
El Libertador (Argentina)	22,047	Twenty-two thousand, forty-seven
Mont Blanc (France-Italy)	15,771	Fifteen thousand, seven hundred seventy-one

2. 1430 **3.** 139 **4.** 214,344 **5.** 24 **6.** 1863 **7.** 18 R2 **8.** 120,000,000,000 **9.** 184 **10.** 6 **11.** 22 **12.** 16 **13.** 4 **14.** d **15.** c **16.** b **17.** e **18.** a **19. a.** $\frac{4}{7}$ **b.** $\frac{7}{3}$ or $2\frac{1}{3}$ **20. a.** Proper **b.** Improper **c.** Improper **21. a.** 1, 2, 3, 5, 6, 10, 15, 30 **b.** $2 \cdot 3 \cdot 5$ **22. a.** $\frac{12}{7}$ or $1\frac{5}{7}$ **b.** $\frac{2}{5}$ **23.** $\frac{119}{171}$ **24.** $\frac{5}{6}$ **25.** Yes. $\frac{8}{13} \cdot \frac{5}{16} = \frac{5}{26}$ and $\frac{5}{16} \cdot \frac{8}{13} = \frac{5}{26}$ **26.** Yes. $\left(\frac{1}{2} \cdot \frac{2}{9}\right) \cdot \frac{5}{3} = \frac{1}{9} \cdot \frac{5}{3} = \frac{5}{27}$ and $\frac{1}{2} \cdot \left(\frac{2}{9} \cdot \frac{5}{3}\right) = \frac{1}{2} \cdot \left(\frac{10}{27}\right) = \frac{5}{27}$ **27.** $\frac{6}{25}$ **28.** $\frac{11}{9}$ or $1\frac{2}{9}$ m^2 **29.** 50 ft^2 **30.** $\frac{3}{40}$ of the students are males from out of state.

Chapter 3

Chapter 3 Preview, p. 170

1. $\frac{13}{2}$ **3.** $\frac{7}{4}$ **5.** 72 **7.** $\frac{8}{15}, \frac{11}{20}, \frac{7}{12}, \frac{3}{5}$ **9.** $\frac{11}{16}$ **11.** $\frac{3}{2}$ **13.** $2\frac{1}{14}$ **15.** $\frac{7}{9}$ **17.** Area: 201 ft^2

Section 3.1 Practice Exercises, pp. 175–178

3. 8 ft **5.** 20 m **7.** 7 fourths

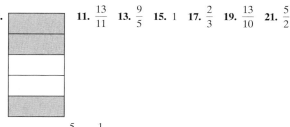

9.

11. $\frac{13}{11}$ **13.** $\frac{9}{5}$ **15.** 1 **17.** $\frac{2}{3}$ **19.** $\frac{13}{10}$ **21.** $\frac{5}{2}$

23. Bethany has $\frac{5}{2}$ or $2\frac{1}{2}$ cups of bleach and water mixture.

25. 11 baskets **27.** 6 fifths

29.

31. $\frac{3}{8}$ **33.** $\frac{3}{2}$ **35.** 2 **37.** $\frac{2}{3}$ **39.** $\frac{2}{5}$ **41.** $\frac{1}{4}$

43. $\frac{1}{4}$ g is left. **45.** $\frac{3}{2}$ **47.** $\frac{12}{5}$ **49.** 1 **51.** $\frac{4}{5}$ **53.** $\frac{5}{2}$ **55.** $\frac{9}{5}$ **57.** $\frac{5}{3}$ **59.** $\frac{16}{7}$ **61.** $\frac{8}{27}$ **63.** $\frac{13}{3}$ **65.** There was $\frac{1}{2}$ gal left over. **67.** He used $\frac{3}{8}$ L. **69.** $\frac{12}{7}$ or $1\frac{5}{7}$ m **71.** $\frac{7}{2}$ or $3\frac{1}{2}$ in.

73. a. Thilan walked $5\frac{1}{2}$ mi total. **b.** He walked an average of $\frac{11}{12}$ mi per day. **75.** Perimeter: 2 ft; area: $\frac{15}{64}$ ft^2 **77.** Perimeter: $\frac{70}{3}$ or $23\frac{1}{3}$ yd; area: $\frac{286}{9}$ or $31\frac{7}{9}$ yd^2 **79.** $\frac{3}{5} + \frac{2}{5}$; 1 **81.** $\frac{11}{15} - \frac{8}{15}$; $\frac{1}{5}$

Section 3.2 Practice Exercises, pp. 184–188

3. $\frac{1}{2}$ **5.** $\frac{5}{3}$ **7.** 6 **9. a.** 48, 72, 240 **b.** 4, 8, 12 **11. a.** 72, 360, 108 **b.** 6, 12, 9 **13.** 5, 10, 15, 20, 25 **15.** 14, 28, 42, 56, 70 **17.** 16, 32, 48, 64, 80 **19.** 50 **21.** 48 **23.** 120 **25.** $2 \cdot 2 \cdot 2 \cdot 3$ **27.** $2 \cdot 2 \cdot 2 \cdot 5$ **29.** $2 \cdot 2 \cdot 3 \cdot 3$ **31.** 72 **33.** 60 **35.** 75 **37.** 540 **39.** 60 **41.** 120 **43.** 210 **45.** 240 **47.** 180 **49.** 32 **51.** 120 **53.** 180 **55.** 80 **57.** 72 **59.** 84 **61.** 360 **63.** 120 **65.** 180 **67.** The shortest length of floor space is 60 in. **69.** It will take 84 months (7 years) for the planets to be aligned again. **71.** $\frac{14}{21}$ **73.** $\frac{10}{16}$ **75.** $\frac{12}{16}$ **77.** > **79.** < **81.** = **83.** $\frac{7}{8}$ **85.** $\frac{2}{3}, \frac{3}{4}, \frac{7}{8}$ **87.** $\frac{1}{4}, \frac{5}{16}, \frac{3}{8}$ **89.** $\frac{13}{12}, \frac{17}{15}, \frac{4}{3}$ **91.** The greatest amount is $\frac{2}{3}$ lb of turkey. The least amount is $\frac{3}{5}$ lb of ham. **93.** a and b

Section 3.3 Practice Exercises, pp. 194–197

3. $\frac{12}{14}$ **5.** $\frac{14}{21}$ **7.** $\frac{25}{5}$ **9.** $\frac{8}{4}$ **11.** $\frac{80}{100}$ **13.** $\frac{5}{40}$ **15.** $\frac{19}{16}$ **17.** $\frac{1}{4}$ **19.** $\frac{1}{4}$ **21.** $\frac{83}{42}$ **23.** $\frac{3}{8}$ **25.** $\frac{1}{3}$ **27.** $\frac{19}{10}$ **29.** $\frac{5}{8}$ **31.** $\frac{25}{8}$ **33.** $\frac{8}{3}$ **35.** $\frac{17}{3}$ **37.** $\frac{2}{7}$ **39.** $\frac{89}{100}$ **41.** $\frac{1}{100}$ **43.** $\frac{391}{1000}$ **45.** $\frac{101}{120}$ **47.** $\frac{23}{60}$ **49.** $\frac{38}{35}$ **51.** $\frac{7}{12}$ **53.** $\frac{37}{28}$ **55.** $\frac{4}{3}$ **57.** $\frac{1}{36}$ **59.** $\frac{13}{125}$ **61.** $\frac{23}{24}$ **63.** Inez added $\frac{9}{8}$ or $1\frac{1}{8}$ cup. **65.** Yes. There will be $\frac{3}{20}$ oz left. **67.** The job did not get completed. There is still $\frac{1}{40}$ of the job left. **69. a.** $\frac{13}{36}$ **b.** $\frac{23}{36}$ **71.** $\frac{13}{5}$ or $2\frac{3}{5}$ ft **73.** Perimeter: 3 ft **75.** b

Calculator Connections 3.3, p. 197

77. $\dfrac{5925}{8036}$ **79.** $\dfrac{1013}{1674}$

Midchapter Review, p. 198

1. Addition and subtraction   **2.** To subtract fractions with the same denominator, subtract the numerators and keep the denominator. Simplify the answer to lowest terms, if possible.
3. Rewrite the fractions so that they have a common denominator. Then add the numerators and keep the common denominator. Simplify the answer to lowest terms, if possible.   **4.** To multiply fractions, multiply the numerators and multiply the denominators. Simplify the product to lowest terms, if possible. To divide fractions, multiply the first fraction by the reciprocal of the second fraction.

Simplify to lowest terms, if possible. **5.** $\dfrac{9}{5}$ **6.** $\dfrac{14}{25}$ **7.** $\dfrac{4}{9}$ **8.** $\dfrac{5}{3}$

9. $\dfrac{13}{28}$ **10.** $\dfrac{3}{14}$ **11.** $\dfrac{50}{27}$ **12.** $\dfrac{5}{9}$ **13.** $\dfrac{5}{9}$ **14.** $\dfrac{7}{4}$ **15.** 2 **16.** 5

17. $\dfrac{1}{140}$ **18.** $\dfrac{91}{90}$ **19.** $\dfrac{44}{15}$ **20.** $\dfrac{29}{48}$

Section 3.4 Practice Exercises, pp. 204–208

3. $\dfrac{13}{6}$ **5.** $\dfrac{13}{7}$ **7.** $\dfrac{1}{2}$ **9.** $7\dfrac{4}{11}$ **11.** $15\dfrac{3}{7}$ **13.** $15\dfrac{9}{16}$ **15.** $10\dfrac{13}{15}$

17. 5 **19.** 2 **21.** 15 **23.** 22 **25.** $3\dfrac{1}{5}$ **27.** $8\dfrac{2}{3}$ **29.** $11\dfrac{1}{6}$ **31.** $6\dfrac{1}{3}$

33. $14\dfrac{1}{2}$ **35.** $23\dfrac{1}{8}$ **37.** $19\dfrac{17}{48}$ **39.** $9\dfrac{7}{8}$ **41.** $42\dfrac{2}{7}$ **43.** $11\dfrac{3}{5}$ **45.** $2\dfrac{2}{15}$

47. $12\dfrac{1}{6}$ **49.** $2\dfrac{5}{14}$ **51. a.** $\dfrac{3}{3}$ **b.** $\dfrac{5}{5}$ **c.** $\dfrac{12}{12}$ **d.** $\dfrac{6}{6}$ **53.** $11\dfrac{1}{2}$ **55.** $1\dfrac{3}{4}$

57. $7\dfrac{13}{14}$ **59.** $3\dfrac{1}{6}$ **61.** $2\dfrac{7}{9}$ **63.** $2\dfrac{3}{17}$ **65.** $6\dfrac{5}{14}$ **67.** $7\dfrac{7}{24}$ **69.** $6\dfrac{2}{15}$

71. $\dfrac{11}{16}$ **73.** $9\dfrac{7}{36}$ **75.** $\dfrac{29}{32}$ **77.** $10\dfrac{20}{21}$ **79.** $\dfrac{32}{35}$ **81.** $7\dfrac{13}{72}$ **83.** $7\dfrac{3}{4}$ in.

85. The index finger is longer.   **87.** The total distance is $24\dfrac{7}{24}$ mi.

89. The water rose $6\dfrac{5}{8}$ in.   **91.** The total distance is $13\dfrac{19}{20}$ mi.

93. He worked $5\dfrac{5}{12}$ hr more on Monday.   **95.** $2\dfrac{2}{3}$ **97.** $2\dfrac{1}{6}$

Section 3.5 Practice Exercises, pp. 214–219

3. $12\dfrac{2}{9}$ **5.** $2\dfrac{2}{3}$ **7.** $3\dfrac{13}{36}$ **9.** $\dfrac{67}{13}$ **11.** $\dfrac{39}{10}$ **13.** $5\dfrac{4}{5}$ **15.** $1\dfrac{11}{19}$ **17.** $3\dfrac{2}{3}$

19. $7\dfrac{1}{2}$ **21.** $4\dfrac{2}{7}$ **23.** 13 **25.** $\dfrac{13}{25}$ **27.** $\dfrac{8}{15}$ **29.** $14\dfrac{7}{12}$ **31.** $1\dfrac{3}{7}$

33. a. The difference is $\dfrac{3}{10}$ sec. **b.** The average is $3\dfrac{3}{5}$ sec.

35. a. The total weight loss is 51 lb. **b.** The average is $8\dfrac{1}{2}$ lb.

c. The difference is $6\dfrac{1}{2}$ lb. **37.** The stock dropped $\$3\dfrac{7}{8}$.

39. George will receive $26,750. **41.** Each piece is $3\dfrac{13}{16}$ ft.

43. $2\dfrac{1}{4}$ lb of cheese was eaten. **45.** 20 loaves can be made.

47. The new rate is $7\dfrac{1}{4}$ points. **49.** Stephanie will need $11\dfrac{1}{4}$ yd for the dresses. **51.** Wilma has $1\dfrac{1}{12}$ lb left. **53.** Joan saves $152\dfrac{1}{2}$ gal.

55. She needs $15\dfrac{1}{3}$ ft more. **57.** The perimeter is 100 in.

59. The total area of the four shutters is $26\dfrac{8}{9}$ ft^2. **61.** The area is $212\dfrac{1}{2}$ ft^2. **63. a.** The area is $247\dfrac{1}{2}$ m^2. **b.** The perimeter is 65 m.

65. $152\dfrac{3}{4}$ m^2

Chapter 3 Review Exercises, pp. 224–226

1. 8 books **3.** 12 mi **5.** Fractions with the same denominators are considered like fractions. **7.** $\dfrac{3}{2}$ **9.** $\dfrac{1}{2}$ **11.** $\dfrac{9}{7}$ **13.** 3 **15.** $\dfrac{3}{4}$

17. $\dfrac{11}{13}$ **19.** 12 in. or 1 ft **21. a.** 7, 14, 21, 28 **b.** 13, 26, 39, 52
c. 22, 44, 66, 88 **23. a.** 1, 2, 4, 5, 10, 20, 25, 50, 100 **b.** 1, 5, 13, 65
c. 1, 2, 5, 7, 10, 14, 35, 70 **25.** 150 **27.** 420 **29.** They will meet on the 12th day. **31.** $\dfrac{63}{35}$ **33.** > **35.** $\dfrac{8}{15}, \dfrac{72}{105}, \dfrac{7}{10}, \dfrac{27}{35}$ **37.** $\dfrac{29}{100}$ **39.** $\dfrac{1}{2}$

41. $\dfrac{43}{20}$ **43.** $\dfrac{1}{34}$ **45.** $\dfrac{17}{40}$ **47.** $\dfrac{1}{15}$ **49. a.** $\dfrac{35}{4}$ or $8\dfrac{3}{4}$ m

b. $\dfrac{315}{128}$ or $2\dfrac{59}{128}$ m^2 **51.** $11\dfrac{11}{63}$ **53.** $2\dfrac{5}{8}$ **55.** $3\dfrac{1}{24}$ **57.** $12\dfrac{5}{14}$ **59.** $3\dfrac{2}{5}$

61. $63\dfrac{15}{16}$ **63.** Estimate: 8 Exact: $8\dfrac{5}{18}$ **65.** Estimate: 50 Exact: $50\dfrac{9}{40}$ **67.** Corry drove a total of $8\dfrac{1}{6}$ hr. **69.** $12\dfrac{2}{5}$ **71.** $\dfrac{4}{27}$

73. 12 **75.** The appraised value is $144,000.

Chapter 3 Test, p. 227

1. $\dfrac{7}{5}$ **2.** $\dfrac{1}{2}$ **3.** When subtracting like fractions, keep the same denominator and subtract the numerators. When multiplying fractions, multiply the denominators as well as the numerators.
4. a. 24, 48, 72, 96 **b.** 1, 2, 3, 4, 6, 8, 12, 24 **c.** $2 \cdot 2 \cdot 2 \cdot 3$ or $2^3 \cdot 3$
5. 240 **6.** $\dfrac{35}{63}$ **7.** $\dfrac{33}{63}$ **8.** $\dfrac{36}{63}$ **9.** $\dfrac{11}{21}, \dfrac{5}{9}, \dfrac{4}{7}$ **10.** $\dfrac{9}{16}$ **11.** $\dfrac{49}{27}$ **12.** $\dfrac{1}{3}$

13. $\dfrac{2}{3}$ **14.** $17\dfrac{3}{8}$ **15.** $2\dfrac{7}{11}$ **16.** $60\dfrac{5}{12}$ **17.** $1\dfrac{1}{2}$ **18.** $\dfrac{25}{6}$ or $4\dfrac{1}{6}$

19. 7 **20.** $\dfrac{12}{295}$ **21.** $\dfrac{10}{3}$ or $3\dfrac{1}{3}$ **22.** 1 lb is needed. **23.** The Ford Expedition can tow 8950 lb. **24.** Area: $25\dfrac{2}{25}$ m^2; perimeter: $20\dfrac{1}{5}$ m

25. Justin has $10,500 for cabinets. **26. a.** The difference is $4\dfrac{2}{3}$ ft.

b. The average is $7\dfrac{3}{8}$ ft.

Chapters 1–3 Cumulative Review, pp. 228–229

1. Twenty-three million, four hundred thousand, eight hundred six
2. 96 **3.** 48 **4.** 1728 **5.** 3 **6.** 1,500,000,000 **7.** $4^2 \cdot 5^4 \cdot 8^2$
8. 36 **9.** 17, 19, 23, 29, 31 **10.** $2 \cdot 5 \cdot 7$ **11.** Numerator: 21;
denominator: 17 **12.** $\dfrac{5}{8}$ **13.** $\dfrac{17}{22}$ had pepperoni and $\dfrac{5}{22}$ did not have pepperoni. **14. a.** Improper **b.** Proper **c.** Improper **15.** b
16. a. Composite **b.** Composite **c.** Prime
17. $2 \cdot 2 \cdot 2 \cdot 3 \cdot 3 \cdot 5$ or $2^3 \cdot 3^2 \cdot 5$ **18.** $\dfrac{1}{5}$ **19.** $\dfrac{3}{8}$ **20.** $\dfrac{4}{7}$ **21.** $\dfrac{3}{4}$

22. $\dfrac{33}{16}$ **23.** $\dfrac{2}{5}$ **24.** $\dfrac{305}{22}$ or $13\dfrac{19}{22}$ **25.** $\dfrac{26}{17}$ or $1\dfrac{9}{17}$ **26.** $\dfrac{10}{3}$ or $3\dfrac{1}{3}$

27. The distance around is approximately 88 cm. **28.** $4\dfrac{1}{3}$ yd

29. $\dfrac{63}{8}$ or $7\dfrac{7}{8}$ m^2 **30. a.** The difference in the intensity is $1\dfrac{3}{10}$.

b. The average intensity is $6\dfrac{7}{10}$.

Chapter 4

Chapter 4 Preview, p. 232

1. Forty-five and three hundredths　**3. a.** $<$　**b.** $>$　**5.** 74.001
7. \$1296.37　**9.** 23,889　**11.** 63.48 cm^2　**13.** 124.$\overline{6}$
15.

	Decimal form	Fraction form
a.	0.55	$\frac{11}{20}$
b.	$1.\overline{8}$	$1\frac{8}{9}$
c.	$2.\overline{3}$	$2\frac{1}{3}$ or $\frac{7}{3}$
d.	4.35	$4\frac{7}{20}$
e.	$0.\overline{27}$	$\frac{3}{11}$

17. 3.31

Section 4.1 Practice Exercises, pp. 239–243

3. 100　**5.** 10,000　**7.** $\frac{1}{100}$　**9.** $\frac{1}{10,000}$　**11.** Tenths　**13.** Hundredths
15. Tens　**17.** Ten-thousandths　**19.** Thousandths　**21.** Ones
23. Seven-tenths　**25.** Nineteen hundredths　**27.** Fifty-one
thousandths　**29.** Four and twenty-six hundredths　**31.** Three and
four-tenths　**33.** Twenty-one and five-tenths　**35.** Seven and three
hundred thirty-eight thousandths　**37.** One and two hundred one
ten-thousandths　**39.** $1\frac{9}{10}$　**41.** $4\frac{4}{5}$　**43.** $\frac{3}{4}$　**45.** $\frac{9}{20}$　**47.** $32\frac{181}{200}$
49. $\frac{5}{2}$　**51.** $\frac{113}{20}$　**53.** $\frac{73}{5}$　**55.** $\frac{2133}{100}$　**57.** $<$　**59.** $>$　**61.** $>$　**63.** $<$
65. a, c, d　**67.** 12.4, 12.46, 12.49, 12.5　**69.** 0.0499, 0.04999, 0.05001,
0.4999, $\frac{5}{10}$　**71.** 148.124, 148.148, 148.295, 148.466　**73.** 49.9
75. 33.42　**77.** 9.096　**79.** 21.0　**81.** 16.80　**83.** 7.1　**85.** a

	Number	Hundreds	Tens	Tenths	Hundredths	Thousandths
87.	349.2395	300	350	349.2	349.24	349.240
89.	79.0046	100	80	79.0	79.00	79.005

91. Five and twenty-three hundredths dollars　**93.** Fifteen and
three-hundredths dollars　**95.** Twenty-one and thirteen hundredths
dollars　**97.** 0.279

Section 4.2 Practice Exercises, pp. 248–253

3. b, c　**5.** a, d　**7.** 42.31　**9.** 1.0　**11.** 63.2　**13.** 8.951　**15.** 15.991
17. 79.8005　**19.** 31.0148　**21.** 62.6032　**23.** 100.414　**25.** 128.44
27. 82.063　**29.** 14.24　**31.** 3.68　**33.** 12.32　**35.** 5.677　**37.** 1.877
39. 57.368　**41.** 21.6646　**43.** 14.765　**45.** 159.558　**47.** 15.347
49. 6.581　**51.** 19.912　**53.** 10.3327　**55. a.** 321.724 days
b. 156.73 days　**57. a.** The water is rising 1.7 in./hr.　**b.** At 1:00 P.M.
the level will be 11 in.　**c.** At 3:00 P.M. the level will be 14.4 in.
59.

Check No.	Description	Credit	Debit	Balance
				\$ 245.62
2409	Electric bill		\$ 52.48	193.14
2410	Groceries		72.44	120.70
2411	Department store		108.34	12.36
	Payroll	\$1084.90		1097.26
2412	Restaurant		23.87	1073.39
	Transfer from savings	200		1273.39

61. The new price is \$2.158.　**63.** The pile containing the two
nickels and two pennies is higher.　**65.** $x = 8.9$ in.; $y = 15.4$ in.;

the perimeter is 98.8 in.　**67.** $x = 2.075$ ft; $y = 2.59$ ft; the perimeter
is 22.17 ft.　**69.** 27.2 mi　**71.** 7 mm

Calculator Connections 4.2, pp. 253–254

73. 26.6 million　**75.** 1831.6 million

Section 4.3 Practice Exercises, pp. 260–264

3. 1000　**5.** 0.01　**7.** 100　**9.** 30　**11.** 7000　**13.** 0.2　**15.** 0.04
17. 0.4　**19.** 3.6　**21.** 8　**23.** 0.18　**25.** 17.6　**27.** 37.35　**29.** 4.176
31. 4.736　**33.** 2.891　**35.** 114.88　**37.** 2.248　**39.** 0.00144
41. The decimal point will move to the right 2 places.　**43. a.** 51
b. 510　**c.** 5100　**d.** 51,000　**45.** 216.3　**47.** 18,220　**49.** 59.32
51. The decimal point will move to the left 1 place.　**53.** 0.933
55. 0.05403　**57.** 0.00005　**59.** 96,700,000　**61.** 16,000
63. \$20,549,000,000　**65.** 324¢　**67.** 6134¢　**69.** 37¢　**71.** \$3.47
73. \$20.41　**75.** \$0.34　**77. a.** \$1　**b.** \$1.50　**79.** The total cost is
\$24.29.　**81.** The bill was \$294.50.　**83.** \$2.81 can be saved.
85. 0.00115 km^2　**87.** The area is 333 ft^2.　**89.** $(0.2)^2 = 0.04$, which
is not equal to 0.4.　**91.** 0.16　**93.** 1.69　**95.** 0.001　**97. a.** 0.09
b. 0.3　**99.** 0.1　**101.** 0.6

Calculator Connections 4.3, pp. 264–265

103. 1914.0625　**105.** \$1991.25 is saved.　**107.** The yearly total is
\$1183.08.

Midchapter Review, p. 265

1. 223.04　**2.** 12,304　**3.** 23.04　**4.** 1.2304　**5.** 123.03　**6.** 123.05
7. 10.72　**8.** 10.92　**9.** 1.082　**10.** 20.82　**11.** 108.2　**12.** 0.82
13. a. 7.191　**b.** 7.191　**c.** Yes　**14.** The commutative property of
addition　**15. a.** 11.4768　**b.** 11.4768　**c.** Yes　**16.** The
commutative property of multiplication　**17.** 0.02484　**18.** 84.31
19. 4.5579　**20.** 0.12291　**21.** 67.032　**22.** 1.672

Section 4.4 Practice Exercises, pp. 274–277

3. 5280　**5.** 3.776　**7.** 2.02　**9.** 0.9　**11.** 0.18　**13.** 0.53　**15.** 21.1
17. 1.96　**19.** 2.55　**21.** 0.035　**23.** 16.84　**25.** 5.$\overline{3}$　**27.** 3.1$\overline{6}$
29. 2.1$\overline{5}$　**31.** 2.$\overline{54}$　**33.** 56　**35.** 2.975　**37.** 208.$\overline{3}$　**39.** 48.5
41. a. 2.4　**b.** 2.44　**c.** 2.444　**43. a.** 1.9　**b.** 1.89　**c.** 1.889
45. a. 3.6　**b.** 3.63　**c.** 3.626　**47.** 0.26　**49.** 14.8　**51.** 20.667
53. 35.67　**55.** The decimal point will move to the left 2 places.
57. 0.03923　**59.** 9.802　**61.** 0.00027　**63.** 0.00102　**65.** The
decimal point will move to the right 1 place.　**67.** 503　**69.** 9.92
71. 3200　**73.** 12,340　**75.** Unreasonable; 32 miles per gallon
77. Unreasonable; \$340.00　**79.** The monthly payment is \$42.50.
81. 65 balls per match.　**83.** Babe Ruth's batting average was 0.342.
85. 47.265　**87.** b, d

Calculator Connections 4.4, p. 277

89. \$886　**91.** 265 people per square mile

Section 4.5 Practice Exercises, pp. 283–286

3. 0.39　**5.** 0.0071　**7.** $\frac{1}{625}$　**9.** $\frac{1}{8}$　**11.** $\frac{16}{100}$; 0.16　**13.** $\frac{38}{1000}$; 0.038
15. $\frac{8}{10}$; 0.8　**17.** 0.875　**19.** 0.3125　**21.** 5.25　**23.** 1.2　**25.** 0.75
27. 2.35　**29.** 7.45　**31.** 0.88　**33.** 3.$\overline{8}$　**35.** 0.$\overline{46}$　**37.** 0.5$\overline{27}$
39. 0.5$\overline{4}$　**41.** 0.1$\overline{26}$　**43.** 0.$\overline{123}$　**45.** 0.$\overline{142857}$　**47.** 0.$\overline{076923}$
49. 0.9　**51.** 0.71　**53.** 1.2　**55.** Multiply the numerator and
denominator by 25 to make the denominator a power of 10.
$\frac{1 \cdot 25}{4 \cdot 25} = \frac{25}{100} = 0.25$; or divide $4\overline{)1.00}^{\,0.25}$　**57. a.** 0.$\overline{1}$　**b.** 0.$\overline{2}$　**c.** 0.$\overline{4}$

d. $0.\overline{5}$ If we memorize that $\frac{1}{9} = 0.\overline{1}$, then $\frac{2}{9} = 2 \cdot \frac{1}{9} = 2 \cdot 0.\overline{1} = 0.\overline{2}$, and so on. **59. a.** 0.375 **b.** 0.625 **c.** 0.875
61.

	Decimal Form	Fraction Form
a.	0.45	$\frac{9}{20}$
b.	1.625	$1\frac{5}{8}$ or $\frac{13}{8}$
c.	$0.\overline{7}$	$\frac{7}{9}$
d.	$0.\overline{45}$	$\frac{5}{11}$

63.

	Decimal Form	Fraction Form
a.	$0.\overline{3}$	$\frac{1}{3}$
b.	2.125	$2\frac{1}{8}$ or $\frac{17}{8}$
c.	$0.8\overline{63}$	$\frac{19}{22}$
d.	1.68	$\frac{42}{25}$

65.

Stock	Symbol	Closing Price (\$) (Decimal)	Closing Price (\$) (Fraction)
Corning	GLW	12.38	$12\frac{19}{50}$
Walgreen	WAG	34.95	$34\frac{19}{20}$
Brookstone	BKST	19.50	$19\frac{1}{2}$
SnapOn	SNA	33.44	$33\frac{11}{25}$

67. = **69.** < **71.** > **73.** < **75.** = **77.** <
79. $\frac{1}{10}, 0.\overline{1}, \frac{1}{5}$

81. $1.75, 1.\overline{7}, 1.8$

83. $\frac{9}{9} = 1$ **85.** 7

Section 4.6 Practice Exercises, pp. 294–298

3. 313.72 **5.** $\frac{107}{27}$ **7.** $\frac{5}{4}$ **9.** 6.96 **11.** 8.77 **13.** 25.75 **15.** 6.25
17. 2 **19.** 12.98 **21.** 67.35 **23.** 25.05 **25.** 23.4 **27.** 1.28
29. 10.83 **31.** 2.84 **33.** $0.93\overline{5}$ **35.** $4.4\overline{3}$ **37. a.** Professor McGonagal makes \$21.75 per hour overtime. **b.** She earns \$797.50. **39.** Jorge will be charged \$98.75. **41.** She has 24.3 g left for dinner. **43.** Caren should get \$4.77 in change. **45.** Duncan's average is 78.75. **47.** The average snowfall per month is 14.54 in.
49. a. The stock increased by \$1.07. **b.** Melanie made \$214.00.
51. Answers will vary. **53. a.** 29.8 **b.** Overweight **55.** 3.475
57. 0.52

Calculator Connections 4.6, p. 299

59. a. Marty will have to finance \$120,000. **b.** There are 360 months in 30 years. **c.** He will pay \$287,409.60 **d.** He will pay \$167,409.60 in interest. **61.** Each person will get approximately \$13,410.10.

Chapter 4 Review Exercises, pp. 306–309

1. The 3 is in the tens place, 2 is in the ones place, 1 is in the tenths place, and 6 is in the hundredths place. **3.** Five and seven-tenths
5. Fifty-one and eight thousandths **7.** $4\frac{4}{5}$ **9.** $\frac{13}{10}$ **11.** >
13. 4.07, 4.24, 4.40, 4.41, 4.48 **15.** 34.890 **17.** a, b **19.** 49.743
21. 5.45 **23.** 197.96 **25.** 7.809 **27.** $x = 4.5$ in., $y = 5.07$ in.; the perimeter is 201 in. **29.** 8.19 **31.** 264.44 **33.** 85,490 **35.** 0.9201
37. 28,000,000 **39. a.** \$2.34 **b.** \$0.55 **41. a.** Eight batteries cost \$7.96 on sale. **b.** A customer can save \$2.03. **43.** Area = 940 ft², perimeter = 127 ft **45.** 17.1 **47.** $4.1\overline{3}$ **49.** 27
51.

	8.6	52.52	0.409
Tenths	8.7	52.5	0.4
Hundredths	8.67	52.53	0.41
Thousandths	8.667	52.525	0.409
Ten-thousandths	8.6667	52.5253	0.4094

53. 11.97 **55.** 9.0234 **57.** 260 **59.** $\frac{6}{10}$; 0.6 **61.** $\frac{54}{1000}$; 0.054
63. 3.52 **65.** 0.4375 **67.** $1.52\overline{7}$ **69.** $0.\overline{153846}$ **71.** 0.87 **73.** 2.83
75. $1\frac{2}{3}$ **77.** $5\frac{7}{9}$ **79.** > **81.** < **83.** 0.713 **85.** 125.6 **87.** 25.12
89. Marvin must drive 34 mi more.

Chapter 4 Test, pp. 310–311

1. a. Tens place **b.** Hundredths place **2.** Five hundred nine and twenty-four thousandths **3.** $1\frac{13}{50}; \frac{63}{50}$ **4.** 0.465, 0.486, 0.522, 0.550
5. b is correct. **6.** 52.832 **7.** 21.29 **8.** 126.45 **9.** 5.08 **10.** 1.22
11. 12.2243 **12.** $120.\overline{6}$ **13.** 439.81 **14. a.** 61.4°F **b.** 1.4°F
15. a. 50,500,000 votes **b.** 51,000,000 votes **c.** The difference is approximately 500,000 in favor of Al Gore. **16.** When dividing by 10, move the decimal point 1 place to the left. When multiplying by 10, move the decimal point 1 place to the right. **17.** When dividing by 0.01, move the decimal point 2 places to the right. When multiplying by 0.01, move the decimal point 2 places to the left.
18. 4.592 **19.** 57,923 **20.** 8012 **21.** 0.002931 **22. a.** 67.5 in.²
b. 75.5 in.² **c.** 157.3 in.² **23.** 75 cents equals 75¢ which is the same as \$0.75. The number .75¢ means seventy-five hundredths of a cent.
24.

Year	Decimal	Fraction
1984	41.02 sec	$41\frac{1}{50}$ sec
1988	39.10	$39\frac{1}{10}$
1992	40.33	$40\frac{33}{100}$
1994	39.25	$39\frac{1}{4}$

25. $3.2, 3\frac{1}{2}, 3.\overline{5}$

26. 9.57 **27.** 13.65
28. a. Faulkner walked 19.8 mi. **b.** The average is 2.8 mi.

Chapters 1–4 Cumulative Review Exercises, pp. 311–312

1. 14 **2.** 4039 **3.** 4840 **4.** 3872 **5.** 2,415,000 **6.** Dividend: 4530; divisor: 225; whole-number part of the quotient: 20; remainder: 30 **7.** To check a division problem, multiply the whole-number part of the quotient and the divisor. Then add the remainder to get the dividend. That is, $20 \times 225 + 30 = 4530$.
8. The difference between sales for Wal-Mart and Sears is

$181,956 million. **9.** $\frac{6}{55}$ **10.** $\frac{4}{7}$ **11.** $\frac{49}{100}$ **12.** 2 **13.** $\frac{2}{3}$ **14.** 0

15. There is $9000 left. **16.** $\frac{2}{5}$ **17.** $\frac{97}{100}$ **18.** $\frac{38}{11}$ **19.** $\frac{33}{7}$ **20.** $\frac{3}{2}$

21. Area: $\frac{15}{64}$ ft²; perimeter: 2 ft **22.** The average is $1\frac{3}{16}$ km.

23. 174.13 **24.** 668.79 **25.** 75.275 **26.** 16 **27.** 339.12 **28.** 46.48
29. a. 3.75248 **b.** 3.75248 **c.** Commutative property of multiplication
30.

Bone	Length (in.) (Decimal)	Length (in.) (Mixed Number)
Femur	19.875	$19\frac{7}{8}$
Fibula	15.9375	$15\frac{15}{16}$
Humerus	14.375	$14\frac{3}{8}$
Innominate bone (hip)	7.5	$7\frac{1}{2}$

Chapter 5

Chapter 5 Preview, p. 314

1. $3:11, \frac{3}{11}$ **3.** $\frac{13}{20}$ **5.** $\frac{26}{1}$ **7.** 492.5 mi/hr or mph **9.** $1.05 per
pound for the 8-lb bag, $1.25 per pound for the 4-lb bag, and $0.65
per pound for the 40-lb bag. The best buy is the 40-lb bag.
11. a. $2.05 **b.** $0.1025/yr **13.** $\frac{22}{14} = \frac{33}{21}$ **15.** $x = 0.5$ **17.** There
are approximately 96,900 births. **19.** The tree is 14.4 ft tall.

Section 5.1 Practice Exercises, pp. 319–322

3. $5:6$ and $\frac{5}{6}$ **5.** 11 to 4 and $\frac{11}{4}$ **7.** $1:2$ and 1 to 2 **9. a.** $\frac{3}{2}$
b. $\frac{2}{3}$ **c.** $\frac{3}{5}$ **11. a.** $\frac{21}{52}$ **b.** $\frac{21}{31}$ **13.** $\frac{2}{3}$ **15.** $\frac{1}{5}$ **17.** $\frac{4}{1}$ **19.** $\frac{11}{5}$
21. $\frac{6}{5}$ **23.** $\frac{1}{2}$ **25.** $\frac{3}{2}$ **27.** $\frac{6}{7}$ **29.** $\frac{8}{9}$ **31.** $\frac{7}{1}$ **33.** $\frac{1}{8}$ **35.** $\frac{4}{3}$ **37.** $\frac{1}{15}$
39. a. $\frac{6}{16} = \frac{3}{8}$ **b.** $\frac{\frac{1}{2}}{1\frac{1}{3}} = \frac{3}{8}$ **41.** $\frac{41}{5}$ **43.** $\frac{1}{12}$ **45.** $\frac{15}{32}$ **47.** $\frac{20}{61}$
49. $\frac{2}{3}$ **51.** $\frac{1}{4}$

Section 5.2 Practice Exercises, pp. 327–330

3. $3:5$ and $\frac{3}{5}$ **5.** $\frac{4}{3}$ **7.** $\frac{9}{17}$ **9.** $\frac{33}{37}$ **11.** $\frac{44 \text{ ft}}{5 \text{ sec}}$ **13.** $\frac{7 \text{ blooms}}{3 \text{ plants}}$
15. $\frac{112 \text{ words}}{5 \text{ min}}$ **17.** $\frac{1 \text{ in.}}{3 \text{ hr}}$ **19.** $\frac{7 \text{ plants}}{11 \text{ ft}}$ **21.** $\frac{25 \text{ students}}{2 \text{ advisers}}$
23. 113 mi/day **25.** 96 km/hr **27.** $55 per payment **29.** $0.38/lb
31. $256,000 per person **33.** 0.069 sec/m **35.** $0.050 per oz
37. $0.995 per liter **39.** $52.50 per tire **41.** $5.417 per bodysuit
43. 305,000 vehicles/year **45. a.** $415/year **b.** $135/year
c. Private colleges **47.** Cheetah: 29 m/sec; antelope: 24 m/sec. The
cheetah is faster. **49.** The larger can is $0.041 per ounce, and the
smaller can is $0.051 per ounce. The larger can is the better buy.

Calculator Connections 5.2, pp. 331–332

51. a. 9.9 wins/yr **b.** 8.6 wins/yr **c.** Shula **53. a.** $0.18
b. $0.14 **c.** $0.08; The best buy is Irish Spring. **55. a.** $0.299 per
ounce **b.** $0.208 per ounce; The best buy is the 4-pack of 6-oz cans
for $4.99. **c.** $0.332 per ounce

Midchapter Review, p. 332

1. A ratio is a comparison of two quantities with the same units. A
rate is a comparison of two quantities with different units.

2. A rate needs to have the units written. **3.** $\frac{14 \text{ mi}}{3 \text{ hr}}$ **4.** $\frac{21}{10}$
5. $\frac{13}{24}$ **6.** $\frac{9 \text{ bushels}}{2 \text{ trees}}$ **7.** $\frac{36 \text{ plants}}{5 \text{ yd}}$ **8.** $\frac{8}{3}$ **9.** $\frac{2}{3}$ **10.** $\frac{3 \text{ gal}}{100 \text{ mi}}$
11. $\frac{1}{4}$ **12.** $\frac{4 \text{ m}}{3 \text{ sec}}$ **13.** $\frac{25 \text{ tiles}}{2 \text{ ft}^2}$ **14.** $\frac{5 \text{ students}}{1 \text{ teacher}}$

Section 5.3 Practice Exercises, pp. 338–340

3. $\frac{1}{15}$ **5.** $\frac{3 \text{ apples}}{1 \text{ pie}}$ **7.** $\frac{22 \text{ mi}}{3 \text{ gal}}$ **9.** $\frac{4}{16} = \frac{5}{20}$ **11.** $\frac{25}{15} = \frac{10}{6}$
13. $\frac{2}{3} = \frac{4}{6}$ **15.** $\frac{30}{25} = \frac{12}{10}$ **17.** $\frac{\$6.25}{1 \text{ hr}} = \frac{\$187.50}{30 \text{ hr}}$ **19.** $\frac{1 \text{ in.}}{7 \text{ mi}} = \frac{5 \text{ in.}}{35 \text{ mi}}$
21. No **23.** Yes **25.** Yes **27.** Yes **29.** Yes **31.** Yes **33.** No
35. 2 **37.** 5 **39.** 8 **41.** 0.6 **43.** Yes **45.** No **47.** $x = 4$
49. $x = 3$ **51.** $p = 75$ **53.** $n = 12$ **55.** $t = 12$ **57.** $y = 36$
59. $x = 12$ **61.** $m = \frac{15}{2}$ or $7\frac{1}{2}$ or 7.5 **63.** $k = 30$ **65.** $h = 2.5$

Section 5.4 Practice Exercises, pp. 346–349

3. $=$ **5.** $\neq$ **7.** $n = \frac{20}{3}$ or $6\frac{2}{3}$ or $6.\overline{6}$ **9.** $k = 6$ **11.** $y = 4.9$
13. Pam can drive 610 mi on 10 gal of gas. **15.** 78 kg of crushed
rock will be required. **17.** The actual distance is about 80 mi.
19. There were 55 Republicans. **21.** Heads would come up about
315 times. **23.** There would be approximately 3 for a 9-inning
game. **25.** Pierre can buy 743.4 €. **27. a.** 340 women would be
expected. **b.** 160 men would be expected. **29.** There are
approximately 357 bass in the lake. **31.** There are approximately
4000 bison in the park. **33.** $x = 24$ cm, $y = 36$ cm **35.** $x = 1$ yd,
$y = 10.5$ yd **37.** The flagpole is 12 ft high. **39.** The platform is
2.4 m tall. **41.** $x = 17.5$ in. **43.** $x = 6$ ft, $y = 8$ ft **45.** $x = 21$ ft;
$y = 21$ ft; $z = 53.2$ ft

Calculator Connections 5.4, p. 350

47. There were approximately 166,005 crimes committed.
49. Approximately 15,400 women would be expected to have
breast cancer.

Chapter 5 Review Exercises, pp. 354–357

1. 5 to 4 and $\frac{5}{4}$ **3.** $8:7$ and 8 to 7 **5. a.** $\frac{4}{5}$ **b.** $\frac{5}{4}$ **c.** $\frac{5}{9}$ **7.** $\frac{4}{1}$
9. $\frac{2}{5}$ **11.** $\frac{9}{2}$ **13.** $\frac{4}{3}$ **15. a.** This year's enrollment is 1520 students.
b. $\frac{4}{19}$ **17.** $\frac{1}{5}$ **19.** $\frac{4 \text{ hot dogs}}{9 \text{ min}}$ **21.** $\frac{650 \text{ tons}}{9 \text{ ft}}$ **23.** All unit rates
have a denominator of 1, and reduced rates may not. **25.** 33 mi/hr
or mph **27.** 90 times/sec **29.** $0.599 per ounce **31.** $0.050 per
bag **33.** The difference is about 12¢ per roll or $0.12 per roll.
35. $0.057/yr **37.** $\frac{16}{14} = \frac{12}{10\frac{1}{2}}$ **39.** $\frac{5}{3} = \frac{10}{6}$ **41.** $\frac{\$11}{1 \text{ hr}} = \frac{\$88}{8 \text{ hr}}$
43. No **45.** Yes **47.** Yes **49.** No **51.** $x = 4$ **53.** $b = 3$
55. $h = 13.6$ **57.** The human equivalent is 84 years. **59.** Alabama
has approximately 4,500,000 people. **61.** $x = 10$ in.,
$y = 62.1$ in. **63.** $x = 1.6$ yd, $y = 1.8$ yd

Chapter 5 Test, pp. 357–358

1. 25 to 521, $25:521$, $\frac{25}{521}$ **2. a.** $\frac{17}{23}$ **b.** $\frac{17}{6}$ **3.** $\frac{9}{7}$ **4.** $\frac{3}{1}$ **5.** $\frac{5}{8}$
6. a. $\frac{21}{125}$ **b.** $\frac{9}{125}$ **c.** The poverty ratio was greater in New Mexico.

7. a. $\frac{\frac{1}{2}}{1\frac{1}{2}} = \frac{1}{3}$ **b.** $\frac{30}{90} = \frac{1}{3}$ **8.** $\frac{85 \text{ mi}}{2 \text{ hr}}$ **9.** $\frac{10 \text{ lb}}{3 \text{ weeks}}$ **10.** $\frac{1 \text{ g}}{2 \text{ cookies}}$
11. 21.45 g/cm³ **12.** 2.3 oz/lb **13.** $0.22 per ounce **14.** $0.50 per ring **15.** Generic: $0.05/tablet; Aleve: $0.08/capsule. The generic pain reliever is the better buy. **16.** They form equal ratios or rates.
17. $\frac{42}{15} = \frac{28}{10}$ **18.** $\frac{20 \text{ pages}}{12 \text{ min}} = \frac{30 \text{ pages}}{18 \text{ min}}$ **19.** $\frac{\$15}{1 \text{ hr}} = \frac{\$75}{5 \text{ hr}}$ **20.** No
21. $p = 35$ **22.** $x = 12.5$ **23.** $n = 5$ **24.** $y = 6$ **25.** It will take 7.5 min. **26.** There are 80 Republicans. **27.** There are approximately 27 fish in her pond. **28.** $x = 1\frac{1}{2}$ mi, $y = 8$ mi **29.** $x = 21$ cm

Chapters 1–5 Cumulative Review, pp. 359–360

1. Five hundred three thousand, forty-two **2.** Approximately 1400 **3.** 22,600,000 **4.** 22 R 3 **5.** Multiply the divisor and the whole-number part of the quotient. Then add the remainder to get the dividend. $16(22) + 3 = 355$ **6.** 6

7.  **8.** $\frac{7}{5}$ **9.** $\frac{39}{14}$ **10.** $\frac{9}{25}$

11. Bruce has $4\frac{1}{2}$ in. of sandwich left. **12.** 2 **13.** $\frac{35}{9}$ **14.** $\frac{9}{13}$ **15.** Emil needs $13\frac{1}{12}$ ft of wallpaper border. **16.** It sold $61\frac{11}{16}$ acres, and $20\frac{9}{16}$ acres is left. **17.** There are 59 ninths. **18.** One thousand four and seven hundred one thousandths. **19.** 28.057 **20.** $\frac{109}{25}$
21. 4392.3 **22.** 2.379 **23.** 130.9 cm **24.** $\frac{212}{221}$ or 212 : 221
25. $\frac{13}{1}$ **26.** 5.525 in./month **27.** 125 people/mi² **28. a.** Yes **b.** No **29.** $x = 4.5$ **30.** Jim can drive 100 mi on 4 gal.

Chapter 6

Chapter 6 Preview, p. 362

1. 43% **3.** 50% **5.** $0.7; \frac{7}{10}$ **7.** $0.004; \frac{1}{250}$ **9.** $33.\overline{3}\%$ or $33\frac{1}{3}\%$
11. 150% **13.** 45% **15.** 0.25% **17.** 200 **19.** 155.6%
21. $27,686 **23.** $2800 in interest is earned in 4 years.

Section 6.1 Practice Exercises, pp. 368–371

3. 48% **5.** 50% **7.** 25% **9.**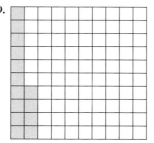

11. 5% **13.** 26% **15.** $\frac{13}{100}$ **17.** $\frac{21}{25}$ **19.** $\frac{1}{4}$ **21.** $\frac{7}{20}$ **23.** $\frac{23}{20}$ or $1\frac{3}{20}$ **25.** $\frac{7}{4}$ or $1\frac{3}{4}$ **27.** $\frac{1}{200}$ **29.** $\frac{1}{400}$ **31.** $\frac{2}{3}$ **33.** $\frac{49}{200}$
35. Replace the % symbol by $\times$ 0.01 (or $\div$ 100). **37.** 0.72
39. 0.66 **41.** 0.129 **43.** 0.4105 **45.** 2.01 **47.** 1.265 **49.** 0.1625
51. 0.622 **53.** 25% **55.** 100% **57.** 150% **59.** d **61.** b **63.** a

65. d **67.** b **69.** c **71.** $0.138; \frac{69}{500}$ **73.** $0.043; \frac{43}{1000}$ **75.** $0.2; \frac{1}{5}$
77. $0.35; \frac{7}{20}$ **79.** $40\% = 0.4$ or $\frac{2}{5}$; $42\% = 0.42$ or $\frac{21}{50}$; $59\% = 0.59$ or $\frac{59}{100}$; $73\% = 0.73$ or $\frac{73}{100}$

Section 6.2 Practice Exercises, pp. 377–380

3. $\frac{13}{10}$ or $1\frac{3}{10}$ **5.** $\frac{1}{200}$ **7.** $0.06\overline{3}$ **9.** 0.003 **11.** 162% **13.** 26%
15. 125% **17.** 77% **19.** Write the whole number as a fraction by writing the number over 1. Then multiply the numerators and multiply the denominators. Simplify the fraction to lowest terms.

21. a. 0.17 **b.** 17% **23. a.** $\frac{37}{100}$ **b.** 37% **25.** 27% **27.** 19%
29. 175% **31.** 12.4% **33.** 0.6% **35.** 101.4% **37.** 71%
39. 95% **41.** 87.5% or $87\frac{1}{2}\%$ **43.** 81.25% or $81\frac{1}{4}\%$ **45.** $83.\overline{3}\%$ or $83\frac{1}{3}\%$ **47.** $44.\overline{4}\%$ or $44\frac{4}{9}\%$ **49.** 25% **51.** 10% **53.** $66.\overline{6}\%$ or $66\frac{2}{3}\%$ **55.** 175% **57.** 135% **59.** $122.\overline{2}\%$ or $122\frac{2}{9}\%$
61. $166.\overline{6}\%$ or $166\frac{2}{3}\%$ **63.** 42.9% **65.** 7.7% **67.** 45.5%
69. 86.7% **71.** The fraction $\frac{1}{2} = 0.5$ and $\frac{1}{2}\% = 0.5\% = 0.005$.
73. $25\% = 0.25$ and $0.25\% = 0.0025$ **75.** a, c **77.** a, c
79.

	Fraction	Decimal	Percent
a.	$\frac{1}{4}$	0.25	25%
b.	$\frac{23}{25}$	0.92	92%
c.	$\frac{3}{20}$	0.15	15%
d.	$\frac{8}{5}$ or $1\frac{3}{5}$	1.6	160%
e.	$\frac{1}{100}$	0.01	1%
f.	$\frac{1}{200}$	0.005	0.5%

81.

	Fraction	Decimal	Percent
a.	$\frac{7}{50}$	0.14	14%
b.	$\frac{87}{100}$	0.87	87%
c.	1	1	100%
d.	$\frac{1}{3}$	$0.\overline{3}$	$33.\overline{3}\%$ or $33\frac{1}{3}\%$
e.	$\frac{1}{500}$	0.002	0.2%
f.	$\frac{19}{20}$	0.95	95%

83. $1.4 > 100\%$ **85.** $0.052 < 50\%$

Section 6.3 Practice Exercises, pp. 387–392

3. 55% **5.** 0.06% **7.** 250% **9.** $\frac{5}{8}$ **11.** $\frac{77}{100}$ **13.** 0.003 **15.** Yes
17. No **19.** Yes **21.** 45 **23.** 48 **25.** Amount: 12; base: 20; $p = 60$ **27.** Amount: 99; base: 200; $p = 49.5$ **29.** Amount: 50; base: 40; $p = 125$ **31.** $\frac{10}{100} = \frac{12}{120}$ **33.** $\frac{80}{100} = \frac{72}{90}$
35. $\frac{104}{100} = \frac{21,684}{20,850}$ **37.** 0.2 **39.** 108 employees **41.** 560
43. Pedro pays $20,160 in taxes. **45.** Jesse Ventura received approximately 762,200 votes. **47.** 36 **49.** 230 lb **51.** 1350
53. Albert makes $1600 per month. **55.** Amiee has a total of 35 e-mails. **57.** 35% **59.** 120% **61.** 87.5% **63.** She answered 72.5% correctly. **65.** 20% **67.** 26.7% **69.** 70 mm of rain fell in August. **71.** Approximately 1900 freshmen were admitted.
73. Smith had approximately 47.2% completion of three-point shots. **75. a.** 106 five-person households own a dog. **b.** 23 three-person households own a dog. **77.** 73 were Chevys. **79.** There were 180 total vehicles. **81.** $331.20 **83.** $11.60 **85.** $6.30

Section 6.4 Practice Exercises, pp. 397–401

3. Divided both sides of the equation by 26 to get $x = 2.5$.
5. $x = 4$ **7.** $x = 187.5$ **9.** $x = (0.35)(700)$; $x = 245$
11. $(0.55)(900) = x$; $x = 495$ **13.** $x = (0.33)(600)$; $x = 198$
15. 50% equals one-half of the number. So divide the number by 2.
17. $2 \times 14 = 28$ **19.** $\frac{1}{2} \times 40 = 20$ **21.** There is 3.84 oz of sodium
hypochlorite in household bleach. **23.** Marino completed
approximately 5015 passes. **25.** $18 = 0.4x$; $x = 45$
27. $0.92x = 41.4$; $x = 45$ **29.** $3.09 = 1.03x$; $x = 3$ **31.** There were
1175 subjects tested. **33.** At that time, the population was about
280 million. **35.** 13% **37.** 108% **39.** 0.5% **41.** 17%
43. $x \cdot 480 = 120$; $x = 25\%$ **45.** $666 = x \cdot 740$; $x = 90\%$
47. $x \cdot 300 = 375$; $x = 125\%$ **49.** 70% of the hot dogs were sold.
51. a. There are 80 total employees. **b.** 12.5% missed 3 days of
work. **c.** 75% missed 1 to 5 days of work. **d.** 62.5% missed at
least 4 days of work. **53.** There were 35 million total hospital stays
that year. **55.** Approximately 12.6% of Florida's panthers live in
Everglades National Park. **57.** 416 parents would be expected to
have started saving for their children's education. **59.** 15.6 min of
commercials would be expected. **61.** 6,350,000 people ages 25–34
made over $10/hr. **63.** There are a total of 16,000,000 workers in
the 16–24 age group. **65. a.** 200 beats per minute. **b.** Between
120 and 170 beats per minute. **67. a.** Answers will vary.
b. Answers will vary.

Midchapter Review, p. 401

1. 41% **2.** 75% **3.** $33\frac{1}{3}\%$ **4.** 100% **5.** c **6.** b **7.** Greater
than **8.** Less than **9.** Greater than **10.** Greater than
11. 3000 **12.** 24% **13.** 4.8 **14.** 15% **15.** 70 **16.** 36

Section 6.5 Practice Exercises, pp. 409–413

3. 12 **5.** 28 **7.** 81 **9.** 24 **11.** 115%

	Cost of Merchandise	Sales Tax Rate	Amount of Tax	Total Cost
13.	$ 56.00	6%	$ 3.36	$ 59.36
15.	$212.00	7%	$14.84	$ 226.84
17.	$ 55.00	6%	$ 3.30	$ 58.30

19. The total bill is $71.66. **21.** The tax rate is 7%. **23.** The price
is $44.50.

	Total Sales	Commission Rate	Amount of Commission
25.	$ 20,000.00	5%	$ 1000.00
27.	$125,000.00	8%	$ 10,000.00
29.	$ 5400.00	10%	$ 540.00

31. Zach made $3360 in commission. **33.** Rodney's commission
rate is 15%. **35.** Her sales were $1,400,000.
37. Jeff's commission totaled $5810.00.

	Original Price	Discount Rate	Amount of Discount	Sale Price
39.	$175.00	15%	$ 26.25	$ 148.75
41.	$900.00	30%	$270.00	$630.00
43.	$ 110.00	30%	$ 33.00	$ 77.00
45.	$ 58.40	40%	$ 23.36	$ 35.04

47. a. The discount is $55. **b.** The discounted yearly membership
will cost $495. **49.** The discount rate is 20%. **51.** The set of
dishes is not free. After the first discount, the price was 50% or
one-half of $112, which is $56. Then the second discount is 50% or
one-half of $56, which is $28. **53.** The discount is $47.00, and the
discount rate is 20%.

	Original Price	Markup Rate	Amount of Markup	Retail Price
55.	$ 92.00	5%	$ 4.60	$ 96.60
57.	$110.00	8%	$ 8.80	$118.80
59.	$ 325.00	30%	$ 97.50	$422.50
61.	$ 45.00	20%	$ 9.00	$ 54.00

63. a. The markup is $27.00. **b.** The retail price is $177.00.
c. The total price is $189.39. **65.** The markup rate is 25%.
67. The markup rate is 54%.

Section 6.6 Practice Exercises, pp. 416–418

3. a. The total price will be $68.25. **b.** The total price with the 20%
discount would be $54.60. **c.** Chris will save $13.65. **5.** Katie's
commission is $31.50. **7.** Multiply the decimal by 100% by moving
the decimal point 2 places to the right and attaching the % sign.
9. 5% **11.** 12% **13. a.** Increase **b.** 11 **15. a.** Decrease
b. 10 **17. a.** Decrease **b.** 9 **19. a.** Increase **b.** 12 **21.** c
23. 75% **25.** 75% **27.** a **29.** 5% **31.** 15%

Calculator Connections 6.6, pp. 418–420

33. 97% **35.** 4% **37.** 10% **39.** 37.5%

	Country	Population in 2000 (Millions)	Population in 2005 (Millions)	Change (Millions)	Percent Increase or Decrease
41.	Mexico	100.3	110.8	10.5	10.5% increase
43.	Bulgaria	8.15	8.11	0.04	0.5% decrease

	Item	Value in 1995	Value in 2000	Change	Percent Increase or Decrease
45.	Number of unemployed people in the U.S.	7.4 million	5.6 million	1.8 million	24.3% decrease
47.	U.S. federal debt	$4.9 trillion	$5.7 trillion	$0.8 trillion	16.3% increase

Section 6.7 Practice Exercises, pp. 426–430

	U.S. National Parks	Visitors in 2000 (Thousands)	Visitors in 2002 (Thousands)	Change	Percent Increase or Decrease
3.	Bryce Canyon, UT	1099	886	213	19% decrease
5.	Great Basin, NV	81	86	5	6% increase
7.	Dry Tortugas, FL	84	80	4	5% decrease

9. Interest: $240; Total Amount: $4240 **11.** Interest: $576; Total Amount: $5376 **13.** Interest: $2761.97; Total Amount: $8991.97
15. a. $350 **b.** $2850 **17. a.** $48 **b.** $448 **19.** $12,360
21. $5625
23.

Year	Interest	Total
1	$20.00	$520.00
2	20.80	540.80
3	21.63	**562.43**

25. There are 6 total compound periods. **27.** There are 24 total compound periods. **29.** $6365.40; $365.40 **31. a.** $8960
b. $8998.91 **c.** $38.91 **33.** A = total amount in the account; P = principal; r = annual interest rate; n = number of compounding periods per year; t = time in years

Calculator Connections 6.7, p. 430

35. Total Amount: $6230.91 **37.** Total Amount: $6622.88
39. Total Amount: $10,934.43 **41.** Total Amount: $16,019.47

Chapter 6 Review Exercises, pp. 437–441

1. 75% **3.** 125% **5.** b, c **7.** f **9.** a **11.** c **13.** e **15.** f
17. d **19.** $\frac{21}{50}$; 0.42 **21.** 0.0615 **23.** $\frac{183}{2000}$ **25.** 17% **27.** 80%
29. 12% **31.** 0.5% **33.** 87.5% **35.** 20%

	Fraction	Decimal	Percent
37.	$\frac{9}{20}$	0.45	45%
39.	$\frac{3}{50}$	0.06	6%
41.	$\frac{9}{1000}$	0.009	0.9%

43. Amount: 67.50; base: 150; $p = 45$ **45.** Amount: 30.24; base: 144; $p = 21$ **47.** $\frac{6}{8} = \frac{75}{100}$ **49.** $\frac{840}{420} = \frac{200}{100}$ **51.** 6
53. 12.5% **55.** 39 **57.** Approximately 11 people would be no-shows. **59.** Victoria spends 40% on rent.
61. $0.18 \cdot 900 = x; x = 162$ **63.** $18.90 = x \cdot 63; x = 30\%$
65. $30 = 0.25 \cdot x; x = 120$ **67.** The original price is $68.00.
69. Elaine can consume 720 fat calories. **71.** The sales tax is $14.28.
73. a. The tax is $0.54. **b.** The tax rate is 8%. **75.** The commission rate was approximately 10.6%. **77.** Sela will earn $75 that day. **79.** The discount is $8.69. The sale price is $20.26.
81. The markup rate is 30%. **83. a.** Increase **b.** 25% **85.** 47.5%
87. 700% **89.** Interest: 12,224; Total Amount: $11,424
91. Jean-Luc will have to pay $2687.50.

93.

Year	Interest	Total
1	$240.00	$6240.00
2	249.60	6489.60
3	259.58	**6749.18**

95. Total Amount: $995.91 **97.** Total Amount: $16,976.32

Chapter 6 Test, pp. 441–443

1. 22% **2.** 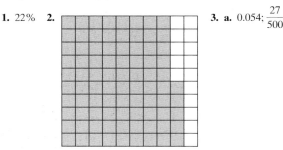 **3. a.** 0.054; $\frac{27}{500}$
b. 0.009; $\frac{9}{1000}$ **c.** 1.70; $\frac{17}{10}$ **4. a.** $\frac{1}{100}$ **b.** $\frac{1}{4}$ **c.** $\frac{1}{3}$ **d.** $\frac{1}{2}$ **e.** $\frac{2}{3}$
f. $\frac{3}{4}$ **g.** 1 **h.** $\frac{3}{2}$ **5.** 0.006; $\frac{3}{500}$ **6.** 0.099; $\frac{99}{1000}$ **7.** Multiply the fraction by 100%. **8.** 60% **9.** 0.4% **10.** 175% **11.** 71.4%
12. Multiply the decimal by 100%. **13.** 32% **14.** 5.2% **15.** 130%
16. 0.6% **17.** 36 **18.** 19.2 **19.** 350 **20.** 200 **21.** 90%
22. 50% **23. a.** 730 mg **b.** 98.6% **24.** 390 m³ **25.** 420 m³
26. a. The amount of sales tax is $2.10. **b.** The sales tax rate is 7%. **27.** Charles will earn $610. **28.** The discount rate of this product is 60%. **29.** 37.5% **30. a.** $1200 **b.** $6200
31. $23,397.17 **32.** $31,268.76

Chapters 1–6 Cumulative Review, pp. 443–444

1. Millions place **2. a.** 3,539,245 **b.** Eight hundred thirty thousand **c.** Four thousand, seven hundred **d.** 401,044
3. 3,488,200 **4.** 87 **5.** 3185 **6.** 11 **7. a.** Improper
b. Improper **c.** Proper **d.** Proper **8.** $\frac{4}{3}$ or $1\frac{1}{3}$ **9.** 24
10. $\frac{3}{2}$ or $1\frac{1}{2}$ **11.** $\frac{2}{3}$ **12.** $\frac{15}{32}$ yd² **13.** 9 km **14.** $\frac{473}{1000}$ **15.** $24\frac{3}{10}$
16. $\frac{459}{2}$ or $229\frac{1}{2}$ in.² **17. a.** 18, 36, 54, 72 **b.** 1, 2, 3, 6, 9, 18
c. $2 \cdot 3^2$ **18. a.** $\frac{5}{2}$ **b.** $\frac{5}{6}$ **19.** 0.375 **20.** $1.\overline{3}$ **21.** $0.\overline{7}$ **22.** 0.75
23. 65.3% **24.** 42.1% **25.** 0.085 **26.** 8500 **27.** 8.5 **28.** 850,000
29. $p = 20$ **30.** $p = 3.75$ **31.** $p = 6\frac{1}{2}$ **32.** $p = 27$ **33.** It will take $2\frac{1}{2}$ hr. **34.** The unit price is $0.25 per ounce. **35.** It will take about 7.2 min. **36.** The DC-10 flew 514 mph. **37.** The increase will be about 370%. **38. a.** 4.6 million **b.** 0.046 million people per year or 46,000 people per year **39.** Kevin will have $15,080. **40.** There is $91,473.02 paid in interest.

Chapter 7

Chapter 7 Preview, p. 446

1. 18 ft **3.** $3\frac{1}{3}$ min **5.** 5 qt **7.** 9 ft **9.** 5903 g **11.** 5600 cL
13. 250 dm **15.** 42.6 ft **17.** 6.3 qt **19.** 22.2 lb **21.** 77°F
23. 5000 ft·lb **25.** 583,500 ft·lb **27.** 5500 $\frac{\text{ft·lb}}{\text{sec}}$

Section 7.1 Practice Exercises, pp. 452–456

3. 1 mi   **5.** = 3 ft   **7.** $\frac{1}{3}$ yd   **9.** 6 ft   **11.** 72 in.   **13.** 10,560 ft

15. 8 yd   **17.** $\frac{3}{4}$ ft   **19.** $\frac{1}{3}$ mi   **21. b**   **23. a**   **25.** $\frac{3}{8}$ ft   **27.** 42 in.

29. $2\frac{1}{4}$ mi   **31.** 18 ft   **33.** $4\frac{2}{3}$ yd   **35.** 563,200 yd   **37.** $4\frac{3}{4}$ yd

39. 72 in.   **41.** 50,688 in.   **43.** $\frac{1}{10}$ mi   **45. a.** 76 in.   **b.** $6\frac{1}{3}$ ft

47. a. 8 ft   **b.** $2\frac{2}{3}$ yd   **49.** 7'7"   **51.** 6 ft   **53.** 11'2"   **55.** 3 ft 4 in.

57. 4'4"   **59.** 8 ft 10 in.   **61.** 28 ft   **63.** 3'2"   **65.** 6 ft 1 in.

67. $5\frac{1}{2}$ ft   **69.** 18 pieces of border are needed.   **71.** The plumber used 7'2" of pipe.   **73.** 7 ft is left over.   **75.** The cable is $0.50 per foot.   **77.** The total length is 46'.   **79.** 6 yd^2 **81.** 720 in.^{2}

Section 7.2 Practice Exercises, pp. 461–463

	Object	in.	ft	yd	mi
3.	Length of a hallway	144 in.	12 ft	4 yd	
5.	Height of a tree	216 in.	18 ft	6 yd	
7.	Perimeter of a backyard	1,800 in.	150 ft	50 yd	

9. 2 pt   **11.** 16 oz   **13.** 365 days   **15.** 4 qt   **17.** 1 hr   **19.** 730 days

21. $1\frac{1}{2}$ hr   **23.** 3 min   **25.** 3 days   **27.** 1 hr   **29.** 80.5 min

31. 175.25 min   **33.** Gil ran for 5 hr 35 min.   **35.** The total time is 1 hr 34 min.   **37.** 2 lb   **39.** 4000 lb   **41.** 64 oz

43. $1\frac{1}{2}$ tons or 1.5 tons   **45.** 10 lb 8 oz   **47.** 8 lb 2 oz   **49.** 6 lb 8 oz

51. The total weight is 312 lb 8 oz.   **53.** The truck will have to make 2 trips.   **55.** 2 c   **57.** 24 qt   **59.** 16 c   **61.** $\frac{1}{2}$ gal   **63.** 16 fl oz

65. 6 tsp   **67.** Yes, 3 c is 24 oz, so the 48-oz jar will suffice.
69. The unit price for the 24-fl-oz jar is about $0.112 per ounce, and the unit price for the 1-qt jar is about $0.103 per ounce; therefore the 1-qt jar is the better buy.

	Object	fl oz	c	pt	qt	gal
71.	Bottle of canola oil	32 fl oz	4 c	2 pt	1 qt	0.25 gal
73.	Laundry detergent	128 fl oz	16 c	8 pt	4 qt	1 gal
75.	Bottle of Gatorade	16 fl oz	2 c	1 pt	0.5 qt	0.125 gal
77.	Bottle of spring water	8 fl oz	1 c	0.5 pt	0.25 qt	0.0625 gal
79.	Jug of maple syrup	64 fl oz	8 c	4 pt	2 qt	0.5 gal

Section 7.3 Practice Exercises, pp. 469–473

3. 1.25 mi   **5.** 3 lb   **7.** 1440 min   **9.** 56 oz   **11. b, f, g**   **13.** 3.2 cm or 32 mm   **15.** 2.1 cm or 21 mm   **17. a.** 5 cm   **b.** 2 cm   **c.** 14 cm   **d.** 10 cm^2 **19. a.** 4 cm   **b.** 4 cm   **c.** 16 cm   **d.** 16 cm^2 **21. a**
23. d   **25. d**   **27.** $\frac{1 \text{ km}}{1000 \text{ m}}$ **29.** $\frac{1 \text{ m}}{100 \text{ cm}}$ **31.** $\frac{1 \text{ m}}{10 \text{ dm}}$ **33.** 2.43 km
35. 10.3 m   **37.** 5 dam   **39.** 4000 m   **41.** 431 dam   **43.** 0.3328 km
45. 3.45 dam   **47.** 250 m   **49.** 4.003 dam   **51.** 700 cm   **53.** 2091 cm
55. 2.538 km   **57.** 0.269 km   **59.** No, she needs 1.04 m of molding.
61. It will take 13 tiles.   **63.** 150 cm or 1.5 m   **65.** 3 dm^2
67. 41,000 cm^2

Section 7.4 Practice Exercises, pp. 478–483

	Object	mm	cm	m	km
3.	Distance between Orlando and Miami	670,000,000	67,000,000	670,000	670
5.	Length of a screw	25	2.5	0.025	0.000025
7.	Thickness of a dime	1.35	0.135	0.00135	0.00000135
9.	World record in men's long jump as of the year 2000	2,450	245	2.45	0.00245

11. Centigram   **13.** Kilogram   **15.** Dekagram   **17.** 0.1 g   **19.** 0.01 g
21. 0.001 g   **23.** 0.539 kg   **25.** 2500 g   **27.** 33.4 mg   **29.** 0.09 hg
31. 0.45 kg

	Object	mg	cg	g	kg
33.	Bag of cat food	1,580,000	158,000	1580	1.58
35.	Can of tuna	170,000	17,000	170	0.17
37.	Box of raisins	425,000	42,500	425	0.425
39.	Dose of acetaminophen	325	32.5	0.325	0.000325

41. <   **43.** >   **45.** =   **47.** <   **49.** Cubic centimeter   **51.** 3.2 L
53. 700 cL   **55.** 0.42 dL   **57.** 64 mL   **59.** 40 cc

	Object	mL	cL	L	kL
61.	1 Tablespoon	15	1.5	0.015	0.000015
63.	Bottle of vinegar	355	35.5	0.355	0.000355
65.	Bottle of soda pop	2,000	200	2	0.002
67.	Capacity of a cooler	37,700	3,770	37.7	0.0377

69. c   **71. b**   **73. c, d**   **75.** 11.2014 dm   **77.** 0.6 g   **79.** 0.019 kL
81. Stacy gets 9.45 g per week.   **83.** The price is $0.50 per liter.
85. A 6-pack contains 4.26 L.   **87.** 5.25 g of the drug would be given in 1 wk.   **89.** 520 mg of sodium per 1-qt bottle   **91.** 2 mL
93. 3.3 metric tons   **95.** 10,900 kg   **97.** 20 μg   **99.** 50 μg

Midchapter Review, p. 483

1. 9 qt   **2.** 2.2 m   **3.** 12 oz   **4.** 300 mL   **5.** 4 yd   **6.** 6030 g
7. 4.5 m   **8.** $\frac{3}{4}$ ft   **9.** 2640 ft   **10.** 3 tons   **11.** 4 qt   **12.** $\frac{1}{2}$ T
13. 0.021 km   **14.** 6.8 cg   **15.** 36 cc   **16.** 4 lb   **17.** 4.322 kg
18. 5000 mm   **19.** 2.5 c   **20.** 8.5 min

Section 7.5 Practice Exercises, pp. 489–492

3. d, f   **5. b, e**   **7. c, f**   **9. b, g**   **11. a.** Numerator
b. Denominator   **13. d**   **15. c**   **17. b**   **19.** 5.1 cm   **21.** 8.8 yd
23. 122 m   **25.** 1.6 mi   **27.** 168 g   **29.** 8.9 lb   **31.** 0.5 oz
33. 0.135 kg ≈ 0.1 kg   **35.** 5.7 L   **37.** 4 fl oz   **39.** 32 fl oz
41. 18 mi is about 28.98 km. Therefore the 30-km race is longer than 18 mi.   **43.** The box of sugar costs $0.100 per ounce, and the packets cost $0.118 per ounce. The 2-lb box is the better buy.
45. 97 lb is approximately 43.65 kg.   **47.** The price is approximately $6.08 per gallon.   **49.** A hockey puck is 1 in. thick.   **51.** Mario weighs about 222 lb.   **53.** 45 cc is 1.5 fl oz.   **55.** 77°F   **57.** 20°C
59. 86°F   **61.** 7232°F   **63.** It is a hot day. The temperature is 95°F.
65. $F = \frac{9}{5}C + 32 = \frac{9}{5} \cdot 100 + 32 = 9(20) + 32 = 180 + 32 = 212$

67. The Navigator weighs approximately 2.565 metric tons.
69. The average weight of the blue whale is approximately 240,000 lb.

Section 7.6 Practice Exercises, pp. 495–499

	Description	°F	°C
3.	Heat of oven for baking cookies	350°F	176.7°C
5.	Temperature of a typical winter day in Modesto, California	41°F	5°C

7. 22,800 ft·lb   **9.** 1200 ft·lb   **11.** 15,000 ft·lb   **13.** 31,120,000 ft·lb
15. 10,892,000 ft·lb   **17.** 5 Btu   **19.** 41 Btu   **21.** 96,472,000 ft·lb
23. 70,020,000 ft·lb   **25.** $\frac{2}{3}$ hr or 0.67 hr   **27.** $\frac{5}{4}$ hr or 1.25 hr
29. $\frac{12}{5}$ hr or 2.4 hr   **31.** 933 Cal   **33.** 770 Cal   **35.** 304 Cal
37. 295 Cal   **39.** $25\ \frac{\text{ft·lb}}{\text{sec}}$ **41.** $100\ \frac{\text{ft·lb}}{\text{sec}}$ **43.** $400\ \frac{\text{ft·lb}}{\text{sec}}$ **45.** 1 hp
47. 9 hp   **49.** 2.8 hp   **51.** $220,000\ \frac{\text{ft·lb}}{\text{sec}}$ **53.** $302,500\ \frac{\text{ft·lb}}{\text{sec}}$
55. $167,750\ \frac{\text{ft·lb}}{\text{sec}}$ **57. a.** 22,500 Wh   **b.** 22.5 kWh   **c.** $2.48
59. $6.39

Chapter 7 Review Exercises, pp. 507–510

1. 4 ft   **3.** 3520 yd   **5.** $1\frac{1}{3}$ mi   **7.** 72 in.   **9.** 9 ft 3 in.   **11.** 2′10″
13. 21′   **15.** 2 ft 1 in.   **17.** $7\frac{1}{2}$ ft   **19.** 3 days   **21.** 80 oz   **23.** $1\frac{1}{2}$ c
25. $1\frac{3}{4}$ tons   **27.** 0.5 hr   **29.** $\frac{3}{4}$ lb   **31.** 144.5 min   **33.** 375 lb will
go to each location.   **35.** 5.5 cm by 3.5 cm   **37. b**   **39. c**
41. 520 mm   **43.** 2338 m   **45.** 3.4 m   **47.** 0.004 dam   **49.** 1200 dm
51. The difference is 3688 m.   **53.** 610 cg   **55.** 3.212 g   **57.** 50 mg
59. 0.3 L   **61.** 8.3 L   **63.** 22.5 cL   **65.** Perimeter: 6.5 m; area:
2.5 m²   **67.** The difference is 64.8 kg.   **69.** There is 1.2 cc or 1.2 mL
of fluid left.   **71.** 15.75 cm   **73.** 5 oz   **75.** 1.04 m   **77.** 74.53 mi
79. 45 cc   **81.** The difference in height is 38.2 cm.   **83.** The total
amount of cough syrup is approximately 0.42 L.
85. $C = \frac{5}{9}(F - 32)$ **87.** $F = \frac{9}{5}C + 32$ **89.** 15,400 ft·lb
91. 11,670,000 to 23,340,000 ft·lb   **93.** 16,000 Btu   **95.** She will
burn 200 Cal more by walking briskly.   **97.** $100\ \frac{\text{ft·lb}}{\text{sec}}$ **99.** 0.5 hp

		hp	ft·lb/sec
101.	Dodge Viper	450	247,500
103.	Chevrolet Corvette	345	189,750

Chapter 7 Test, pp. 510–511

1. c, d, g, j   **2.** f, h, i   **3.** a, b, e   **4.** $8\frac{1}{3}$ yd   **5.** 5.5 tons   **6.** 10 mi
7. 10 oz of liquid   **8.** 20 min   **9.** 9′   **10.** 4′2″   **11.** He lost 7 oz.
12. 19 ft 7 in.   **13.** 75.25 min   **14.** 2.4 cm or 24 mm   **15. c**
16. 1.158 km   **17.** 15 mL   **18. a.** Cubic centimeters
b. 235 cc   **c.** 1000 cc   **19.** 41,100 cg   **20.** 7 servings   **21.** 2.1 qt
22. 109 yd   **23.** 2.8 mi   **24.** 2929 m   **25.** 50.8 cm tall and 96.52-cm
wingspan   **26.** 11 lb   **27.** 190.6°C   **28.** 35.6°F   **29.** 1100 Cal
30. 270 ft·lb   **31.** 77,800,000 ft·lb   **32.** $2475\ \frac{\text{ft·lb}}{\text{sec}}$

Chapters 1–7 Cumulative Review, pp. 512–513

1. a. 2000   **b.** 42,100   **2.** 56 cm   **3.** 180 cm²   **4.** 4   **5. a.** Ford
Motor Company spends the most. That amount is $7400 million or
$7,400,000,000.   **b.** The difference between IBM and Motorola is
$302 million or $302,000,000.   **c.** The total amount spent is $26,917
million or $26,917,000,000.   **6.** $\frac{6}{39}$ **7.** The number 32,542 is not
divisible by 3 because the sum of the digits (16) is not divisible by
3.   **8.** 2 · 2 · 3 · 3 · 3   **9.** 540 in.²   **10.** $\frac{1}{4}$ of the recipe would call
for $\frac{3}{4}$ c of oatmeal. This is less than 1 c so Keesha does have enough.
11. 10   **12.** $\frac{7}{5}$ **13.** $9\frac{1}{2}$ **14.** $18\frac{8}{9}$ **15.** $2\frac{6}{17}$ **16.** $3\frac{5}{6}$

	Fraction	Decimal
17.	$\frac{1}{3}$	$0.\overline{3}$
18.	$\frac{9}{20}$	0.45
19.	$\frac{5}{4}$	1.25
20.	$\frac{7}{2}$	3.5
21.	$\frac{3}{8}$	0.375
22.	$\frac{1}{25}$	0.04

23. a. $\frac{6}{5}$ **b.** $\frac{6}{11}$ **24. a.** $6100   **b.** $610 per year   **c.** $4700
d. $470 per year   **e.** Men   **25.** 90 cars   **26.** 6.7 beds per nurse
27. No, because $\frac{6}{8} \neq \frac{2}{3}$.   **28.** 80%   **29.** $x = \frac{16}{3}$ yd   **30.** 2290 trees
31. 27 people   **32.** 6%   **33.** $15,000 in sales   **34.** $1020 in interest
35. 5.8 kg   **36.** 12.9 lb   **37.** 182.9 cm   **38.** 6 ft   **39.** 7 pt   **40.** 3.3 L

Chapter 8

Chapter 8 Preview, p. 516

1. The complement is 68° and the supplement is 158°.   **3.** False
5. True   **7.** 30 mm   **9.** 0.48 mi²   **11.** 119 m²   **13. c**   **15. a**

Section 8.1 Practice Exercises, pp. 522–527

3. A line extends forever in both directions. A line segment
is a portion of a line between two endpoints.
5. Ray   **7.** Point   **9.** Line   **11.** For example:

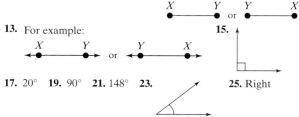

13. For example:

15.

17. 20°   **19.** 90°   **21.** 148°   **23.**   **25.** Right
27. Obtuse   **29.** Acute   **31.** Straight   **33.** 10°   **35.** 63°   **37.** 60.5°
39. 1°   **41.** 100°   **43.** 53°   **45.** 142.6°   **47.** 1°   **49.** No   **51.** Yes
53. A 90° angle   **55.**

57.

59. $m(\angle a) = 41°$; $m(\angle b) = 139°$; $m(\angle c) = 139°$

61. $m(\angle a) = 26°$; $m(\angle b) = 112°$; $m(\angle c) = 26°$; $m(\angle d) = 42°$
63. The two lines are perpendicular. **65.** a, c or b, h or e, g or f, d
67. a, e or f, b **69.** $m(\angle a) = 55°$; $m(\angle b) = 125°$;
$m(\angle c) = 55°$; $m(\angle d) = 55°$; $m(\angle e) = 125°$; $m(\angle f) = 55°$;
$m(\angle g) = 125°$ **71.** True **73.** True **75.** False **77.** True
79. True **81.** 70° **83.** 90° **85. a.** 48° **b.** 48° **c.** 132°
87. 180° **89.** 120°

Section 8.2 Practice Exercises, pp. 533–538

3. Yes **5.** No **7.** No **9.** $m(\angle a) = 54°$ **11.** $m(\angle b) = 78°$
13. $m(\angle a) = 60°$, $m(\angle b) = 80°$ **15.** $m(\angle a) = 40°$, $m(\angle b) = 72°$
17. c, f, g **19.** b, d **21.** b, c, e, g **23.** 7 **25.** 49 **27.** 16 **29.** 4
31. 6 **33.** 36 **35.** 81 **37.** 9

	Square Root	Estimate	Calculator Approximation (Round to 3 Decimal Places)
	$\sqrt{50}$	is between 7 and 8	7.071
39.	$\sqrt{10}$	is between __3__ and __4__	3.162
41.	$\sqrt{116}$	is between __10__ and __11__	10.770
43.	$\sqrt{5}$	is between __2__ and __3__	2.236

45. 20.682 **47.** 1116.244 **49.** 0.7 **51.** 0.748 **53.** $c = 5$ m
55. $b = 12$ yd **57.** Leg = 10 ft **59.** Hypotenuse = 40 in. **61.** The
brace is 20 in. long. **63.** The height is 9 km. **65.** The car is 25 mi
from the starting point. **67.** 24 m **69.** 30 km **71.** $c = 5$ in.;
perimeter = 28 in. **73.** Perimeter = 72 ft

Calculator Connections 8.2, p. 538

75. $b = 21$ ft **77.** Hypotenuse = 11.180 mi **79.** Leg = 18.439 in.
81. The diagonal length is 1.41 ft.

Section 8.3 Practice Exercises, pp. 546–552

3. An isosceles triangle has two sides of equal length. **5.** An
acute triangle has all acute angles. **7.** An obtuse triangle has an
obtuse angle. **9.** A quadrilateral is a polygon with four sides.
11. A trapezoid has one pair of opposite sides that are parallel.
13. A rectangle has four right angles. **15.** True **17.** False
19. 80 cm **21.** 260 mm **23.** 10.7 m **25.** 10 ft 6 in. **27.** 5 ft
29. $x = 550$ mm; $y = 3$ dm; perimeter = 26 dm or 2600 mm
31. 280 ft of rain gutters is needed. **33.** 576 yd² **35.** 54 m²
37. 656 in.² **39.** 18.4 ft² **41.** 12.375 ft² **43.** 217.54 ft²
45. 280 mm² **47.** 60 in.² **49.** The area to be carpeted is 382.5 ft².
The area to be tiled is 13.5 ft². **51.** The area is 276 m². **53.** The
area is 1.625 ft². **55.** The area is increased by 9 times.

Midchapter Review, p. 552

1. Perimeter **2.** Perimeter **3.** Area **4.** Area **5.** Area = 25 ft²;
perimeter = 20 ft **6.** Area = 12 m² or 120,000 cm²;
perimeter = 14 m or 1400 cm **7.** Area = 0.473 km² or 473,000 m²;
perimeter = 3.24 km or 3240 m **8.** Area = 6 yd²;
perimeter = 12 yd **9.** Area = 88 in.²; perimeter = 40 in.

Section 8.4 Practice Exercises, pp. 557–561

3. 1260 cm² **5.** 630 cm² **7.** Yes. Since a rectangle is a special
type of parallelogram (one that contains four right angles), the area
formula for a parallelogram applies to a rectangle. **9.** 12 in.
11. 3 m **13.** 5.6 km **15.** 4 in. **17.** 7.5 km **19.** 8.3 m **21.** c
23. π is the circumference divided by the diameter. That is, $\pi = \frac{C}{d}$.
25. 12.56 m **27.** 62.8 cm **29.** 13.188 cm **31.** 15.7 km
33. 18.84 cm **35.** 14.13 in. **37.** 6.908 cm **39.** 154 m²
41. 346.5 cm² **43.** 491 mm² **45.** 121 ft² **47.** 2.72 ft² **49.** 55.04 in.²
51. 18.28 in.² **53.** 113.04 mm² **55. a.** \$1051.75 **b.** \$9377.28
57. 2826 ft² **59. a.** 81.64 in. **b.** 147 times **61.** 69,080 in. or 5757 ft
63.

Diameter	Cost	Area	Cost per in.²
8 in.	\$ 6.50	50.24 in.²	\$ 0.129
12 in.	12.40	113.04 in.²	0.110

The 12-in. is the better buy.

Section 8.5 Practice Exercises, pp. 566–570

3. $C = 25.12$ in; $A = 50.24$ in.² **5.** $C = 18.84$ m; $A = 28.26$ m²
7. b, d **9.** Area = 1 ft²; volume = 1 ft³ **11.** Area = 1 km²;
volume = 1 km³ **13.** 2.744 cm³ **15.** 48 ft³ **17.** 12.56 mm³
19. 235.5 cm³ **21.** 113.04 yd³ **23.** 452.16 ft³ **25.** 289 in.³
27. 314 ft³ **29.** 10 ft³ **31.** 32 ft³ **33.** 39.8 in.³ **35.** $\frac{11}{36}$ ft³ or
0.306 ft³ or 528 in.³ **37.** 109.3 in.³ **39.** 450 ft³
41. 267,946,666,667 mi³ **43.** 84.78 in.³

Chapter 8 Review Exercises, pp. 576–580

1. d **3.** c **5.** $\overrightarrow{SR}$ or $\overrightarrow{SQ}$ **7.** S **9.** The measure of an acute angle
is between 0° and 90°. **11.** The measure of a straight angle is 180°.
13. a. 58° **b.** 148° **15.** 60° **17.** 175° **19.** b **21.** True **23.** False
25. True **27.** $m(\angle x) = 40°$ **29.** An obtuse triangle has one
obtuse angle. **31.** A right triangle has a right (90°) angle. **33.** An
isosceles triangle has two sides of equal length and two angles of
equal measure. **35.** 5 **37.** 10 **39.** 3.742 **41.** 2.236 **43.** The
sum of the squares of the legs of a right triangle equals the square
of the hypotenuse. **45.** $c = 20$ ft **47.** They both have sides of
equal length, but a square also has four right angles. **49.** A square
is a rectangle with four sides of equal length. **51.** 90 cm **53.** 56 mi
55. 42 ft **57.** 20 in.² **59.** 7056 ft² **61.** 90 mm **63.** 22.5 mm
65. $C = 50.24$ m; $A = 200.96$ m² **67.** $C = 440$ in.; $A = 15,400$ in.²
69. 134.88 in.² **71.** 5.57 yd² **73.** 226.08 ft³ **75.** 37.68 km³
77. 113 in.³ **79.** 28,500 in.³

Chapter 8 Test, pp. 580–582

1. Yes **2.** Yes **3.** Yes **4.** No **5.** Yes **6.** No **7.** No
8. Yes **9.** No **10.** No **11.** Obtuse **12.** Acute **13.** Acute
14. Obtuse **15.** Right **16.** Acute **17.** Straight **18.** Acute
19. $m(\angle x) = 125°$, $m(\angle y) = 55°$ **20.** They are each 45°. **21.** 49°
22. 180° **23.** $m(\angle A) = 80°$ **24.** 12 ft **25.** 100 m **26.** d **27.** c
28. f **29.** b **30.** a **31.** e **32.** 96 in. **33.** 3 rolls are needed.
34. The area is 72 in.². **35.** The area of the rectangular pizza is
96 in.² The area of the round pizza is approximately 113.04 in.² The
round pizza is larger by about 17 in.² **36.** The volume is about
151 ft³. **37.** The volume is 1260 in.³ **38.** The volume is 2002 cm³.

Chapters 1–8 Cumulative Review, pp. 582–583

1. 3835 **2.** 0 **3.** Undefined **4.** 666,000 **5.** 2,511,000
6.

7. There is $10\frac{1}{2}$ oz left. **8.** 18

9. $\frac{1}{18}$ **10.** $\frac{6}{5}$ **11.** 60 **12.** $\frac{67}{60}$ **13.** $\frac{67}{60}$ **14.** $16\frac{1}{2}$ **15.** $\frac{46}{9}$

16. Four glasses cost $47.96. **17.** Geraldo will save the cost of one shirt which is $13.49.

	Fraction	Decimal
18.	$\frac{3}{8}$	0.375
19.	$\frac{2}{9}$	$0.\overline{2}$
20.	$\frac{1}{50}$	0.02

21. $\frac{2}{3}$ **22.** $n = 37.35$

23. 17 pizzas **24. a.** 0.58 **b.** 0.52 **c.** O'Neal **25.** $3436 per hour **26.** 52.8 **27.** 72 **28.** 130% **29.** 20% markup **30.** 16% discount **31.** 10 ft **32.** Yes, $4\frac{1}{2}$ ft is 54 in. **33.** There is a total of $1\frac{1}{4}$ c or 10 fl oz of liquid. **34.** 100 kph $\approx$ 62 mph

35. 165,000 $\frac{\text{ft} \cdot \text{lb}}{\text{sec}}$ **36.** 13.3 m **37.** 11 ft **38.** 1256 cm^2 **39.** 3 yd^2 or 27 ft^2 **40.** 452 in.^{3}

Chapter 9

Chapter 9 Preview, p. 586

1. a. New York **b.** California **3. a.** 398 Hispanic students **b.** 94 Asian/Pacific Islander students **5.** $\frac{13}{20}$

Section 9.1 Practice Exercises, pp. 592–599

3. Mt. Everest **5.** Mt. Kosciusko; Australia **7.** 11,200 ft **9.** 2.5 yr **11.** 1.7 yr **13.** Women

15.

	Attends Church	Does Not Attend Church
With children	7	4
Without children	4	3

17.

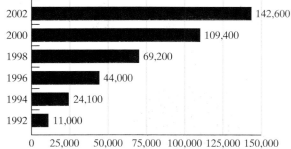

Number of Cellular Phone Subscriptions in the United States by Year

Year	
2002	142,600
2000	109,400
1998	69,200
1996	44,000
1994	24,100
1992	11,000

0   25,000   50,000   75,000   100,000   125,000   150,000

19. a. The health care industry has 219,400 new jobs, which is the greatest number.

b.

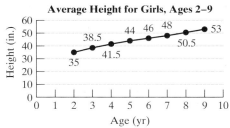

Number of New Jobs for Selected Industries

Health Care 219,400; Temporary Help 212,000; Construction 173,000; Food Service 167,600; Retail 78,600

21. a. Each computer icon represents 10% of adults who access the Internet. **b.** About 60% **c.** International news **23. a.** About 2.25 million **b.** About 2 million **25.** 8.6% **27.** The trend for men over 65 in the labor force shows a significant decrease until 1980 and then levels off. **29.** For example: 10.5% **31.** In the year 1990 the least number of SUVs was sold. There were 98 million sold. **33.** 2 million **35.** 2000–2002

37.

Average Height for Girls, Ages 2–9

(Height in. vs Age yr) 35, 38.5, 41.5, 44, 46, 48, 50.5, 53

39. There is 630 mg of sodium in one container. **41.** The daily value of cholesterol is about 294 mg.

Section 9.2 Practice Exercises, pp. 602–608

3. There are 72 data. **5.** 9–12 **7. a.** The class of 65–67 has the most values. **b.** There are 20 values represented in the table. **c.** Of the professors, 25% retire when they are 68 to 70 years old.

Class Intervals (Age Group)	Tally	Frequency (Number of Professors)
56–58	II	2
59–61	I	1
62–64	I	1
65–67	IIII II	7
68–70	IIII	5
71–73	IIII	4

9. a. The 12.0–13.9 class has the highest frequency. **b.** There are 16 data values represented in the table. **c.** Of the customers, 12.5% purchase 18 to 19.9 gal of gas.

Class Intervals (Amount Purchased)	Tally	Frequency (Number of Customers)
8.0–9.9	IIII	4
10.0–11.9	I	1
12.0–13.9	IIII	5
14.0–15.9	IIII	4
16.0–17.9		0
18.0–19.9	II	2

11. 1. Whenever possible, make the classes the same width. 2. The classes should not overlap. That is, a data value should

belong to one and only one class.   **3.** In general, we usually create a frequency distribution with between 5 and 15 classes.
13. The class widths are not the same.   **15.** There are too few classes.   **17.** The class intervals overlap. For example, it is unclear whether the data value 5 should be placed in the first class or the second class.

19.

Class Interval (Weight, lb)	Frequency (Number of Females)
115–124	6
125–134	2
135–144	5
145–154	3
155–164	3
165–174	0
175–184	1

21.

Class Interval (Amount, $)	Frequency (Number of Customers)
0–49	3
50–99	4
100–149	7
150–199	1
200–249	5

23.

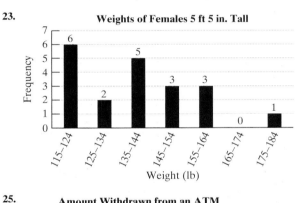

25.

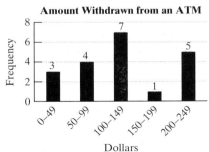

27.

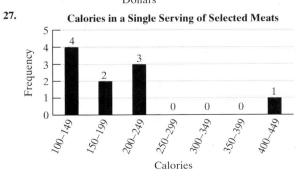

29.

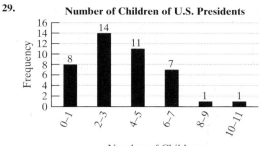

Section 9.3 Practice Exercises, pp. 613–616

3. 20,400 troops   **5.** 300 more troops   **7.** Approximately 11.3%
9. There are 5.5 times more troops from the United Kingdom than from Ukraine.   **11.** There are 20 million viewers represented.
13. There are 1.8 times as many viewers who watch *The Young and the Restless* as *Guiding Light.*   **15.** Of the viewers, 18% watch *General Hospital.*   **17.** There are 960 Latina CDs.   **19.** There are 640 CDs that are either classical or jazz.   **21.** There were 9 Super Bowls played in Louisiana.   **23.** There were 2 Super Bowls played in Georgia.   **25.**   **27.**

29.   **31.**

33.

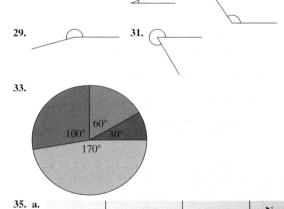

35. a.

	Expenses	Percent	Number of Degrees
Tuition	$9000	75%	270°
Books	600	5%	18°
Housing	2400	20%	72°

b.

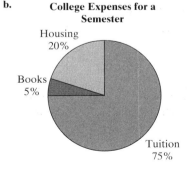

37.

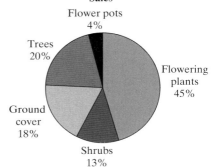

Sunshine Nursery Distribution of Sales

Flower pots 4%
Trees 20%
Flowering plants 45%
Ground cover 18%
Shrubs 13%

Midchapter Review, pp. 616–617

1.

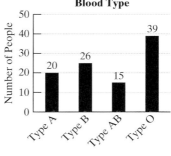

Number of People Surveyed by Blood Type

Type A 20, Type B 26, Type AB 15, Type O 39

2.

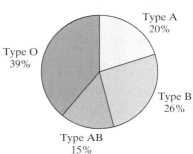

Percent by Blood Type

Type A 20%
Type B 26%
Type AB 15%
Type O 39%

3.

Mortgage Rates

6.8%, 5.8%, 6.0%, 5.9%, 5.6%, 5.3%
July, September, November, January, March, May

4.

Mortgage Rates

6.8%, 5.8%, 6.0%, 5.9%, 5.6%, 5.3%
July, September, November, January, March, May

5.

Grade Point Average	Frequency
2.50–2.74	1
2.75–2.99	5
3.00–3.24	7
3.25–3.49	6
3.50–3.74	1
3.75–3.99	3

6.

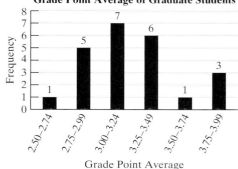

Grade Point Average of Graduate Students

1, 5, 7, 6, 1, 3
2.50–2.74, 2.75–2.99, 3.00–3.24, 3.25–3.49, 3.50–3.74, 3.75–3.99

Grade Point Average

Section 9.4 Practice Exercises, pp. 623–628

3. 6 **5.** 4 **7.** 15.2 **9.** 8.76 in. **11.** 5.8 hr **13. a.** 397 Cal
b. 386 Cal **c.** There is only an 11-Cal difference in the means.
15. a. 86.5% **b.** 81% **c.** The low score of 59% decreased Zach's
average by 5.5%. **17.** 17 **19.** 110.5 **21.** 52.5 **23.** 3.93 deaths
per 1000 **25.** 58 years old **27.** 3.1 million albums **29.** 4
31. No mode **33.** 21, 24 **35.** 39 **37.** 5.2% **39.** These data are
bimodal: $2.49 and $2.51.

41.

Age (yr)	Number of Students	Product
16	2	32
17	9	153
18	6	108
19	3	57
Total:	20	350

The mean age is approximately 17.5 years.
43. The weighted mean is about 26 students initially enrolled in
each class. **45.** 2.38 **47.** 2.77

Section 9.5 Practice Exercises, pp. 632–637

3. Mean: 17.2; median: 16; no mode **5.** Mean: 8.875; median: 8.5;
mode: 8 **7.** Mean: 82%; median: 88.5%; mode: 88%
9. {1, 2, 3, 4, 5, 6, 7, 8, 9, 10} **11.** {2, 3, 4, 5, 6, 7, 8, 9, 10, 11, 12}

13. For example: {2} That is, a 2 comes up when the die is rolled.
15. c, d, g, h   **17.** $\frac{2}{6} = \frac{1}{3}$ **19.** $\frac{3}{6} = \frac{1}{2}$ **21.** $\frac{2}{8} = \frac{1}{4}$ **23.** 0
25. An impossible event is one in which the probability is 0.
27. 1   **29.** $\frac{12}{52} = \frac{3}{13}$ **31.** $\frac{12}{16} = \frac{3}{4}$ **33. a.** $\frac{3}{20}$ **b.** $\frac{9}{40}$ **c.** 30%
35. a. $\frac{21}{60} = \frac{7}{20}$ **b.** 50%   **37. a.** $\frac{7}{29}$ **b.** $\frac{11}{29}$ **c.** 62%   **39.** $1 - \frac{2}{7} = \frac{5}{7}$
41. $100\% - 1.2\% = 98.8\%$ **43. a.** $\frac{312}{530} = \frac{156}{265}$ **b.** $\frac{14}{206} = \frac{7}{103}$

Chapter 9 Review Exercises, pp. 643–645

1. Godiva   **3.** Blue Bell has 2 times more sodium than Edy's Grand.   **5.** 374 acres   **7.** The difference is 4 acres.   **9.** 1 house represents 10,000 detached single-family houses.   **11.** About 6.0 ten-thousands or 60,000 houses.   **13.** 1999   **15.** Decreasing
17.

Class Intervals (Age)	Frequency
18–21	4
22–25	5
26–29	4
30–33	3
34–37	1
38–41	1
42–45	2

19. There are 24 types of subs.   **21.** $\frac{1}{3}$ of the subs do not contain beef.   **23.** Mean: 17.5; median: 18; mode: 20   **25.** The median is 20,562 seats.   **27.** {blue, green, brown, black, gray, white}
29. a, c, d, e, g   **31.** 0

Chapter 9 Test, pp. 646–648

1.

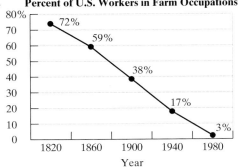

World's Major Producer's of Primary Energy (Quadrillions of Btu)

2. a. The year 1820 had the greatest percent of workers employed in farm occupations. This was 72%.
b.

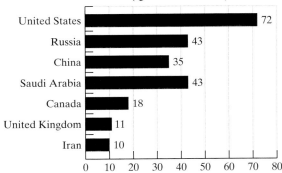

Percent of U.S. Workers in Farm Occupations

c. It appears that 10% of U.S. workers were employed in farm occupations in the year 1960.   **3. a.** $1000   **b.** $4500   **c.** February
4.

Number of Days Late	Tally	Number of Employees
0	III	3
1	II	2
2	IIII	5
3	IIII I	6
4	IIII	5
5	III	3
6	I	1
7	III	3
8	II	2

5.

Number of Minutes Used Monthly	Tally	Frequency
51–100	IIII I	6
101–150	II	2
151–200	III	3
201–250	II	2
251–300	IIII	4
301–350	III	3

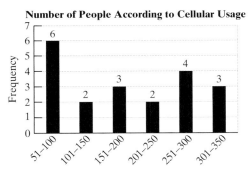

Number of People According to Cellular Usage

6. a. 66 people would have carpet.   **b.** 40 people would have tile.
c. 270 people would have something other than linoleum.
7. a. 19,173 ft   **b.** 19,340 ft   **c.** There is no mode.   **8.** Mean: $14.60; median: $15   **9.** The mode is $14.97.   **10. a.** {1, 2, 3, 4, 5, 6, 7, 8}
b. $\frac{1}{8}$ **c.** $\frac{4}{8} = \frac{1}{2}$ **d.** $\frac{2}{8} = \frac{1}{4}$ **11. a.** $\frac{4}{14} = \frac{2}{7}$ **b.** $1 - \frac{2}{7} = \frac{5}{7}$
12. 3.09

Chapters 1–9 Cumulative Review, pp. 648–651

1. a. Millions   **b.** Ten-thousands   **c.** Hundreds   **2.** 12,645
3. $700 \times 1200 = 840,000$ **4.** Divisor: 23; dividend: 651; quotient: 28; remainder: 7   **5.** $\frac{3}{8}$ **6.** $\frac{2}{3}$ **7.** $\frac{5}{2}$ **8.** $\frac{3}{2}$ **9.** $\frac{1}{3}$ **10.** $\frac{37}{100}$ **11.** 2
12. $\frac{1}{6}$ **13.**

Stock	Yesterday's Closing Price ($)	Increase/ Decrease	Today's Closing Price ($)
RylGold	13.28	0.27	13.55
NetSolve	9.51	−0.17	9.34
Metals USA	14.35	0.10	14.45
PAM Transpt	18.09	0.09	18.18
Steel Tch	21.63	−0.37	21.26

14. 6841.2 **15.** 6.8412 **16.** 68,412 **17. a.** 1,900,000 **b.** 95,000 people per year **18.** Quick Cut Lawn Company's rate is 0.55 hr per customer. Speedy Lawn Company's rate is 0.5 hr per customer. Speedy Lawn Company is faster. **19.** 125 min or 2 hr 5 min **20.** $x = 5$ m, $y = 22.4$ m **21.** 122 people **22.** 17.02 million **23.** 65% **24.** $1404 **25.** 29 in. **26.** 18 qt **27.** 9 yd 1 ft **28.** 9.64 km or 9640 m **29.** 2 lb 11 oz **30.** Obtuse **31.** Right **32.** Acute **33.** Area: 8 ft² **34.** 66 m³

35.

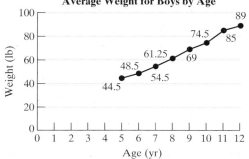

Average Weight for Boys by Age

36. a. 0.5% **b.** 31.9% **c.** 59.0%

37.

Number of Videos	Tally	Number of Customers
0	卌	5
1	卌 卌	10
2	卌 II	7
3	IIII	4
4	III	3
5		0
6	I	1

38. {yellow, blue, red, green} **39.** $\frac{1}{4}$ **40.** $\frac{3}{4}$

Chapter 10

Chapter 10 Preview, p. 654

1. a. 3,740,000 **b.** −2,446,000 **3. a.** $\frac{4}{3}$ **b.** 4.2 **5.** −8 **7.** 0
9. −2.3 + 8; 5.7 **11.** −34 + 6; −28 **13.** −16.2 **15.** 23
17. −2 − 16; −18 **19.** 19 **21.** 0 **23.** 81 **25.** 2
27. a. 4.502×10^9 **b.** 2.2301×10^{-4}

Section 10.1 Practice Exercises, pp. 659–662

3. −86 m **5.** $3800 **7.** −$500 **9.** −14 lb **11.** 140,000

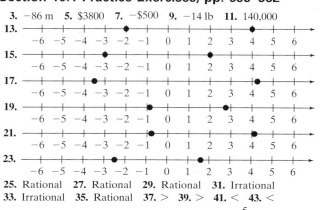

25. Rational **27.** Rational **29.** Rational **31.** Irrational
33. Irrational **35.** Rational **37.** > **39.** > **41.** < **43.** <
45. > **47.** > **49.** < **51.** < **53.** 2 **55.** 4.5 **57.** $\frac{5}{2}$

59. 0 **61.** 3.2 **63.** 21 **65. a.** −8 **b.** |−12| **67. a.** 7.8 **b.** |7.8|
69. |−5| **71.** Neither, they are equal. **73.** −5 **75.** 12 **77.** $\frac{1}{6}$
79. −$\frac{2}{11}$ **81.** −8.1 **83.** 1.14 **85.** −6 **87.** −(−2) **89.** |7|
91. |−3| **93.** −|14| **95.** −2 **97.** −5.3 **99.** 15 **101.** 4.7
103. −$\frac{12}{17}$ **105.** $\frac{3}{8}$

Section 10.2 Practice Exercises, pp. 669–672

3. > **5.** = **7.** < **9.** −2 **11.** 2 **13.** −8 **15.** 6 **17.** −7
19. −4 **21.** To add two numbers with the same sign, add their absolute values and apply the common sign. **23.** 15 **25.** −73
27. −124 **29.** 89 **31.** 52 **33.** −22 **35.** −24 **37.** 45 **39.** 0
41. 0 **43.** 9 **45.** −26 **47.** −41 **49.** −150 **51.** −17 **53.** −41
55. 2 **57.** −30 **59.** 10 **61.** −8 **63.** −23 **65.** 24 **67.** 17.2
69. −20.6 **71.** −$\frac{1}{2}$ **73.** −$\frac{13}{12}$ **75.** 77.7 **77.** −1.3 **79.** −$\frac{4}{9}$
81. −$\frac{9}{20}$ **83.** −23 + 49; 26 **85.** 3 + (−10) + 5; −2
87. −2.2 + (−4.2); −6.4 **89.** −$\frac{1}{4}$ + 8; $\frac{31}{4}$ **91.** −$\frac{3}{4}$ + 6; $\frac{21}{4}$
93. 8°F **95.** $333.29 **97.** −$170.50 **99.** For example: −12 + 2
101. For example: −1 + (−1)

Calculator Connections 10.2, p. 672

103. −120 **105.** −68.221 **107.** 711

Section 10.3 Practice Exercises, pp. 677–680

3. −47 **5.** $\frac{1}{36}$ **7.** −$\frac{41}{36}$ **9.** −4 **11.** 2 + (−9); −7 **13.** 4 + 3; 7
15. −3 + (−15); −18 **17.** −11 + 13; 2 **19.** 52 **21.** −33
23. −12 **25.** 8 **27.** 0 **29.** 161 **31.** −34 **33.** −22 **35.** −26
37. −1 **39.** 32 **41.** −15 **43.** *Minus, difference, decreased, less than, subtract from* **45.** 14 − 23; −9 **47.** 5 − 12; −7
49. 105 − 110; −5 **51.** 320 − (−20); 340 **53.** −35 − 24; −59
55. −34 − 21; −55 **57.** −8.3 **59.** −4.2 **61.** 5.5 **63.** 8.3
65. $\frac{5}{6}$ **67.** $\frac{2}{5}$ **69.** −$\frac{3}{2}$ **71.** −$\frac{7}{4}$ **73.** 0 **75.** −1 **77.** 16 **79.** 1
81. 52 **83.** 5.2 **85.** 398°C **87.** The contestant won $400.
89. The difference is 0.18 point. **91.** His new balance is −$375.
93. The range is 3° − (−8°) = 11°. **95.** For example, 4 − 10
97. −11, −15, −19 **99.** −1, −$\frac{4}{3}$, −$\frac{5}{3}$ **101.** Positive **103.** Positive
105. Negative **107.** Negative

Calculator Connections 10.3, p. 680–681

109. −413 **111.** 66.77 **113.** 112.8

Midchapter Review, p. 681

1. −7 **2.** 5 **3.** −14 **4.** −15 **5.** 17 **6.** 18 **7.** 7 **8.** −5
9. −9 **10.** −3 **11.** 6 **12.** 11 **13.** −3 + (−9) **14.** −2 + (−3)
15. 5 + (6) **16.** 8 + (11) **17.** −7 **18.** 5 **19.** −14 **20.** −15
21. 17 **22.** 18 **23.** 7 **24.** −5 **25.** −1.98 **26.** 4.6 **27.** $\frac{11}{8}$
28. −$\frac{17}{18}$

Section 10.4 Practice Exercises, pp. 687–690

3. 19 **5.** −44 **7.** 17 **9.** −15 **11.** −48 **13.** −45 **15.** −72
17. 3.84 **19.** −2.4 **21.** −7.7 **23.** 0 **25.** $\frac{4}{7}$ **27.** −$\frac{1}{7}$

29. $-\dfrac{5}{2}$ or $-2\dfrac{1}{2}$ **31.** $\dfrac{13}{3}$ or $4\dfrac{1}{3}$ **33.** 0   **35.** $-3(-1)$; 3
37. $-5 \cdot 3$; -15 **39.** $1.3(-3)$; -3.9 **41.** 400   **43.** -88 **45.** 0
47. 1   **49.** -1 **51.** -100 **53.** 100   **55.** -27 **57.** -27
59. -0.008 **61.** $-\dfrac{8}{27}$ **63.** -3 **65.** -7 **67.** $\dfrac{5}{3}$ **69.** $\dfrac{2}{3}$
71. Undefined   **73.** 0   **75.** 4   **77.** 1.3   **79.** $\dfrac{10}{7}$ **81.** Undefined
83. $-\dfrac{3}{8}$ **85.** -34 **87.** $-100 \div 20$; -5 **89.** $-32 \div (-64)$; $\dfrac{1}{2}$
91. $-52 \div 13$; -4 **93.** 2   **95.** -48 **97.** 3   **99.** 3   **101.** 20
103. -15 **105.** $-\dfrac{2}{5}$ **107.** $\dfrac{7}{6}$ **109.** $(-2)^{50}$ **111.** $(5)^{41}$
113. Negative   **115.** Negative   **117.** Positive

Calculator Connections 10.4, p. 690

119. $-359{,}723$ **121.** 54

Section 10.5 Practice Exercises, pp. 694–696

3. 25   **5.** -6 **7.** -11 **9. a.** 7   **b.** -7 **11.** 16   **13.** -20
15. -16 **17.** 11   **19.** 2   **21.** -5 **23.** -8 **25.** 8   **27.** -96
29. -2 **31.** $-\dfrac{3}{2}$ **33.** $\dfrac{1}{3}$ **35.** $\dfrac{11}{2}$ **37.** $-2°$ **39.** -2.9 **41.** 10^{4}
43. 10^{3} **45.** 10^{-3} **47.** 10^{-4} **49.** No   **51.** Yes   **53.** Yes   **55.** No
57. $\$7.455 \times 10^{12}$ **59.** 2×10^{-7} mm   **61.** 2×10^{7} **63.** 8.1×10^{6}
65. 3×10^{-3} **67.** 2.5×10^{-2} **69.** 1.42×10^{5} **71.** 4.91×10^{-5}
73. 8.2×10^{-2} **75.** 4.92×10^{3} **77.** 6000   **79.** 0.08   **81.** 0.44
83. 37,000   **85.** 326   **87.** 0.0129   **89.** 0.000002003   **91.** 900,100,000
93. 3   **95.** -1

Chapter 10 Review Exercises, pp. 701–703

1. $-76{,}704$ **3.** $15°$ **5. & 7.**

```
  +——+——+——+——●——+——+——●——+——+——+——+——+——+——→
 −6  −5  −4  −3  −2  −1   0   1   2   3   4   5   6
```

9. 4, 4   **11.** $-3.5, 3.5$ **13. a.** 9   **b.** -9 **15.** $>$ **17.** $>$
19. $<$ **21.** 4   **23.** -5 **25.** To add two numbers with the same
sign, add their absolute values and apply the common sign.   **27.** 13
29. -70 **31.** -10.66 **33.** $-\dfrac{9}{10}$ **35.** $23 + (-35)$; -12
37. $-5 + (-13) + 20$; 2   **39.** $-12 + 3$; -9 **41.** 3   **43.** 1. Leave
the first number (the minuend) unchanged.   2. Change the
subtraction sign to an addition sign.   3. Add the opposite of the
second number (the subtrahend).   **45.** -25 **47.** -419 **49.** -0.7
51. $\dfrac{20}{21}$ **53.** For example: 23 minus negative 6   **55.** For example:
Subtract -7 from -25.   **57.** Sam's balance is now $92.   **59.** -18
61. 15   **63.** -70 **65.** $\dfrac{7}{4}$ **67.** Undefined   **69.** -32 **71.** -36
73. $-\dfrac{27}{64}$ **75.** -1 **77.** Positive   **79.** $-4 \cdot 19$; -76
81. $-136 \div (-8)$; 17   **83.** -2 **85.** 4   **87.** -1 **89.** $-1°$
91. 1.0302×10^{7} **93.** 9.042×10^{-3} **95.** 8,700,000,000
97. 0.0000602   **99.** In the number 9.11, move the decimal point 31
places to the left (inserting 30 zeros).

Chapter 10 Test, pp. 703–704

1. a. $-\$220$ **b.** 26   **2.** $-3, 0, 4, -1$ **3.** $-3, -\dfrac{3}{5}, 0, 4, -1, \dfrac{4}{7}$
4. $\sqrt{7}, -\pi$ **5.** $<$ **6.** $>$ **7.** $>$ **8.** $>$ **9.** $<$ **10.** $<$
11. -5 **12.** -28 **13.** 9   **14.** -41 **15.** 6   **16.** -23 **17.** -72
18. 88   **19.** 2   **20.** -18 **21.** Undefined   **22.** 0   **23. a.** Positive
b. Negative   **24. a.** 64   **b.** -64 **c.** -64 **d.** -64 **25.** $-3(-7)$; 21

26. $-13 + 8$; -5 **27.** $18 - (-4)$; 22   **28.** $6 \div \left(-\dfrac{2}{3}\right)$; -9
29. $-8.1 + 5$; -3.1 **30.** $-3 + 15 + (-6) + (-1)$; 5   **31.** 3
32. -60 **33.** -19 **34.** -55 **35.** $-\dfrac{22}{15}$ **36.** $\dfrac{2}{13}$ **37.** $1°$
38. 1×10^{-9} m   **39.** 5.8078×10^{8} **40. a.** 30,501,000,000
b. 0.0004009

Chapters 1–10 Cumulative Review, pp. 704–706

1. 3613   **2.** 2569   **3.** 177   **4.** 18,960   **5.** $2 \cdot 2 \cdot 2 \cdot 2 \cdot 3 \cdot 3 \cdot 5$ or
$2^{4} \cdot 3^{2} \cdot 5$ **6.** $\dfrac{5}{8}$ **7.** Harold got $\dfrac{11}{14}$ of the quiz correct.   **8.** Amy
will have 8 packages.   **9.** 80   **10.** $\dfrac{5}{16}$ **11.** $6\dfrac{7}{15}$ **12.** $3\dfrac{4}{7}$
13. a. 34.230   **b.** 9.0   **14.** $\$2.0999$ **15.** 490.92   **16.** 115   **17.** $\dfrac{2}{5}$
18. The aircraft used 2491 gal/hr.   **19.** 21 mi   **20.** $x = 2.7$ cm;
$y = 7$ cm   **21.** 192   **22.** 112%   **23.** 250   **24.** The sale price is
$\$68.80$.   **25.** 28 in.   **26.** 5 gal   **27.** 0.06 L   **28.** $1\dfrac{7}{8}$ lb   **29.** The
distance is 10 mi.   **30.** 16.5 yd^{2} **31.** $5\dfrac{1}{16}$ m^{2} **32.** $A = 7.065$ km^{2};
$C = 9.42$ km
33.

Number of Miles	Tally	Frequency (Number of Walkers)
3	‖	2
4	‖‖‖	5
5	‖	1
6	‖	2

34.

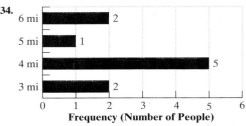

35. 4.3 mi   **36.** $\$216$ **37.** 55   **38.** -12 **39.** -20 **40.** $\dfrac{1}{6}$

Chapter 11

Chapter 11 Preview, p. 708

1. $\$17.95x$ **3.** -25 **5.** $6r + 11s$ **7.** -5 is not a solution
9. $q = -10$ **11.** $w = -\dfrac{5}{2}$ **13.** $c = 36$ **15.** $b = -8$
17. $n + 10 = 6n$; the number is 2   **19.** Garnett made $25.2 million
and O'Neal made $24 million.
21., 23.

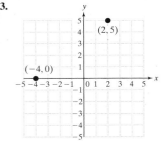

Section 11.1 Practice Exercises, pp. 714–716

3. $8p$ **5.** $l + \dfrac{1}{2}$ **7.** $\dfrac{4}{n}$ **9. a.** -12 **b.** 30   **11. a.** 5   **b.** -15

13. a. -49 **b.** -49 **15. a.** -16 **b.** -64 **17.** 3   **19.** 5   **21.** -4

23. 24   **25.** -4 **27.** $\dfrac{1}{2}$ **29.** 6   **31.** 5   **33.** $P = 16.6$ in.

35. $A = \dfrac{77}{2}\,\text{m}^2$ **37.** $V = 1256\,\text{cm}^3$ **39.** $w + 5$ **41.** $b + (-\frac{1}{3})$ or $b - \frac{1}{3}$ **43.** $2r$ **45.** $-st$ **47.** $-9 + a$ **49.** $-p + 7$

51. $(-2 \cdot 6)b$; $-12b$ **53.** $(3 + 8) + t$; $11 + t$

55. $(-4.2 + 2.5) + r$; $-1.7 + r$ **57.** $(3 \cdot 6)x$; $18x$

59. $\left(-\dfrac{4}{7} \cdot -\dfrac{7}{4}\right)d$; d **61.** $(-9 + (-12)) + h$; $-21 + h$ **63.** $4x + 32$

65. $-2p - 8$ **67.** $4a + 16b - 4c$ **69.** $\dfrac{8}{3} + 4g$ **71.** $-3 + n$

73. $a + 8$ **75.** $-3x - 9 + 5y$ **77.** $5q + 2s + 3t$ **79.** $12x$

81. $12 + 6x$ **83.** $-4 - p$ **85.** $-32 + 8p$ **87.** $5 + \dfrac{5}{9}y$ **89.** $5y$

Section 11.2 Practice Exercises, pp. 720–722

3. $6p + 18$ **5.** $-24q$ **7.** $9 - h$ **9.** $2a$: variable term; $5b^2$: variable term; 6: constant term   **11.** 8: constant term; $9a$: variable term

13. $4pq$: variable term; $-9p$: variable term   **15.** $10h^2$: variable term; -15: constant term; $-4h$: variable term   **17.** 6, -4 **19.** -14, 12

21. 1, -1 **23.** 5, -8, -3 **25.** *Like* terms   **27.** Unlike terms

29. *Like* terms   **31.** Unlike terms   **33.** Unlike terms   **35.** *Like* terms   **37.** $14rs$ **39.** $8h$ **41.** $3x^2 + 9$ **43.** $6x - 15y$ **45.** $-3k$

47. $4uv + 6u$ **49.** $-16m - 9$ **51.** $-9a + 5b + 20$

53. $-6p^2 - 2p + 7$ **55.** $2y - \dfrac{5}{6}$ **57.** $\dfrac{5}{8}a + 9$ **59.** $-3x^2 - 1.9x$

61. $-0.9a + 7.6$ **63.** $5t - 28$ **65.** $-6x - 16$ **67.** $6y - 14$

69. $7q$ **71.** $-2n - 4$ **73.** $-6a - b$ **75.** $4x + 23$ **77.** $2z - 11$

79. $-w - 4y + 9$ **81.** $8a - 9b$ **83.** $-5m + 6n - 10$ **85.** $12z + 7$

87. $-7x - 7$ **89.** $-21y + 28$ **91.** $-2q + 20$ **93.** $-107a - 213b$

Section 11.3 Practice Exercises, pp. 727–729

3. $-13a + 16b$ **5.** $8h - 2k + 13$ **7.** $-3z + 4$ **9.** -1 is a solution.

11. 26 is a solution.   **13.** 12 is not a solution.

15. $-\dfrac{1}{2}$ is a solution.   **17.** 0 is a solution.   **19.** 4 is not a solution.

21. 0   **23.** 7   **25.** -3.2 **27.** $\dfrac{3}{8}$ **29.** $g = 37$ **31.** $k = 16$

33. $n = -15$ **35.** $p = \dfrac{7}{6}$ **37.** $k = 3.1$ **39.** 52   **41.** 0   **43.** 100

45. $x = -28$ **47.** $b = 3$ **49.** $t = -46$ **51.** $m = -13.6$

53. $a = -\dfrac{13}{10}$ **55.** $p = -1$ **57.** $t = -6$ **59.** $y = \dfrac{13}{12}$

61. $m = -1.4$ **63.** $w = 12$ **65.** $x = -5$ **67.** $a = 13$

69. $p = -0.79$ **71.** $t = 1\dfrac{5}{8}$ **73.** $z = 49$ **75.** $h = -1$

77. $t = 3$ **79.** $r = 10$

Section 11.4 Practice Exercises, pp. 734–736

3. $p = 45$ **5.** $h = 21$ **7.** $p = -13$ **9.** $n = -\dfrac{7}{6}$ **11.** $\dfrac{1}{3}$ **13.** $-\dfrac{7}{4}$

15. -7 **17.** 5.1   **19.** $b = -3$ **21.** $k = -7$ **23.** $t = 13$

25. $m = 21$ **27.** $b = -21$ **29.** $t = 4$ **31.** $u = 30$ **33.** $w = -\dfrac{1}{3}$

35. $x = 4.1$ **37.** $k = -\dfrac{2}{5}$ **39.** $m = 0$ **41.** $x = \dfrac{4}{15}$ **43.** $k = 20$

45. $p = -31$ **47.** $p = \dfrac{5}{6}$ **49.** $a = 0$ **51.** If the operation between

a number and a variable is subtraction, use the addition property to isolate the variable.   **53.** If the operation between a number and a variable is multiplication, use the division property to isolate the variable.   **55.** $x = -16$ **57.** $y = -3$

59. $q = -8$ **61.** $h = -48$ **63.** $t = \dfrac{1}{3}$ **65.** $a = \dfrac{4}{3}$ **67.** $r = 30$

69. $y = -15$ **71.** $p = \dfrac{5}{12}$ **73.** $x = -\dfrac{21}{10}$ **75.** $t = 27.9$

77. $u = -1.8$ **79.** $x = -5$ **81.** $p = 3$ **83.** $a = -1$

Section 11.5 Practice Exercises, pp. 741–743

3. $b = -12$ **5.** $w = -\dfrac{5}{8}$ **7.** $h = 0$ **9.** $m = 4$ **11.** $c = -6$

13. $z = 5$ **15.** $x = \dfrac{4}{3}$ **17.** $d = -2.4$ **19.** $h = -1.12$ **21.** $b = 9$

23. $w = -12$ **25.** $x = \dfrac{1}{4}$ **27.** $b = -3$ **29.** $t = \dfrac{9}{8}$ **31.** $d = 1$

33. $p = \dfrac{7}{3}$ **35.** $z = 2$ **37.** $p = 6$ **39.** $a = -3$ **41.** $w = -4$

43. $y = \dfrac{4}{5}$ **45.** $n = -12$ **47.** $q = 15$ **49.** $m = 3$ **51.** $k = 6$

53. $z = -\dfrac{1}{2}$ **55.** $w = \dfrac{15}{11}$ **57.** $u = 5$

Midchapter Review, pp. 743–744

1. a. $7x - 15$ **b.** $-3x - 9$ **c.** $x = \dfrac{3}{5}$ **2. a.** $6y + 2$

b. $-6y + 6$ **c.** $y = \dfrac{1}{3}$ **3.** Expression; $8x - 12$ **4.** Equation; $t = 3$

5. Equation; $h = 1$ **6.** Expression; $10k$ **7.** Equation; $w = 0$

8. Expression; 2   **9.** Expression; $m + \dfrac{1}{8}$ **10.** Equation; $m = -\dfrac{1}{8}$

11. Expression; $-3y - 1$ **12.** Equation; $y = -\dfrac{5}{3}$

13. Equation; $t = 3$ **14.** Expression; $t + 5$ **15.** Equation; $x = -\dfrac{1}{6}$

16. Expression; $6x + 1$ **17.** Expression; $3p + 3$

18. Equation; $p = -1$

Section 11.6 Practice Exercises, pp. 750–753

3. $b = -45$ **5.** $r = \dfrac{14}{5}$ **7.** $p = -0.75$ **9. a.** $\dfrac{x}{3} = -8$ **b.** The number is -24.   **11. a.** $-30 - x = 42$ **b.** The number is -72.

13. a. $30 + x = 13$ **b.** The number is -17.   **15. a.** $\dfrac{x}{4} - 5 = -12$

b. The number is -28.   **17. a.** $\dfrac{1}{2} + x = 4$ **b.** The number is $\dfrac{7}{2}$ or

$3\dfrac{1}{2}$.   **19. a.** $-12x = x + 26$ **b.** The number is -2.

21. a. $10(x + 5.1) = 56$ **b.** The number is 0.5.

23. a. $3x = 2x - 10$ **b.** The number is -10.   **25.** The pieces are $1\dfrac{1}{3}$ ft and $2\dfrac{2}{3}$ ft long.   **27.** Metallica had 10 hits while Boyz II Men had 16 hits.   **29.** The soccer field is 100 yd by 130 yd.   **31.** Jim used 50 min over the 500 min.   **33.** Tampa Bay had 48 points and Oakland had 21 points.   **35.** Charlene's rent is $650 a month with a security deposit of $300.   **37.** Stefan worked 6 hr of overtime.

39. Raul took 12 hr in fall and signed up for 16 hr in spring.

41. Local calls are $0.20 per minute, and long-distance calls are $0.25 per minute.   **43.** London is connected to 61 countries on the Internet.

Section 11.7 Practice Exercises, pp. 757–762

3. $9 - x = 2x$; The number is 3.   **5.** $\dfrac{a}{-6} = 13$; the number is -78

7., 9., 11., 13.

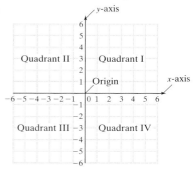

15., 17.

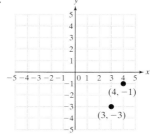

19. First move to the left 1.8 units from the origin. Then go up 3.1 units. Place a dot at the final location. The point is in Quadrant II.

21., 23., 25.

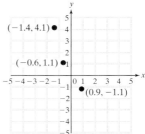

27., 29., 31

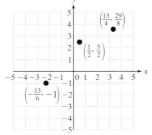

33., 35., 37.

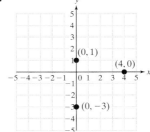

39. Quadrant IV   **41.** Quadrant III   **43.** x-axis   **45.** y-axis
47. Quadrant II   **49.** Quadrant I   **51.** $(0, 3)$ **53.** $(2, 3)$

55. $(-5, -2)$ **57.** $(4, -2)$ **59.** $(1, -7), (2, -3), (3, 1), (4, 6), (5, 12),$
$(6, 17), (7, 18), (8, 18), (9, 14), (10, 6), (11, -1), (12, -7)$

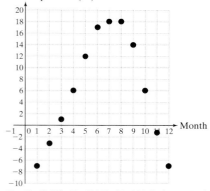

61. $(1, 5000), (2, 6800), (3, 8384), (4, 10{,}010), (5, 11{,}509), (6, 12{,}782)$

63. $(1997, 1736), (1998, 1789), (1999, 1896), (2000, 1104), (2001, 2268),$
$(2002, 2393)$

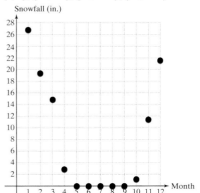

65. $(1, 26.8), (2, 19.6), (3, 14.4), (4, 2.8), (5, 0), (6, 0), (7, 0), (8, 0),$
$(9, 0), (10, 1.1), (11, 11.7), (12, 21.8)$

Chapter 11 Review Exercises, pp. 769–771

1. a. $a + 8$ **b.** 43 years old   **3.** -18 **5. a.** $-5 + t$ **b.** $3h$
7. $6b + 15$ **9.** $3a^2$ is a variable term with coefficient 3; $-5a$ is a
variable term with coefficient -5; 12 is a constant term with
coefficient 12   **11.** Unlike terms   **13.** *Like* terms

15. $6x + 4y + 10$ **17.** $-u - 11v$ **19.** -3 is a solution. **21.** If a constant is being added to the variable term, use the subtraction property. If a constant is being subtracted from a variable term, use the addition property. **23.** $k = -12$ **25.** $q = 24$ **27.** $n = 5$

29. $b = -\dfrac{11}{10}$ **31.** $d = -7$ **33.** $t = -26$ **35.** $y = 20$

37. $m = -2$ **39.** $w = \dfrac{9}{7}$ **41.** $p = 6$ **43.** $x = -1$

45. $m = 8$ **47.** $m = 10$ **49.** $x = -28$ **51.** $w = \dfrac{4}{3}$

53. $a = -3$ **55.** $-6x = x + 2$; $x = -\dfrac{2}{7}$

57. $\dfrac{1}{3} - x = 2$; $x = -\dfrac{5}{3}$ **59.** Tom Hanks starred in 35 films, and Tom Cruise starred in 30 films. **61.** The width is 40 in. and the length is 72 in. **63.** $(2, 0)$ **65.** $(-4, -2)$ **67.** $(-1, 2)$

69. $\left(-3, -3\dfrac{1}{3}\right)$ **71., 73., 75.**

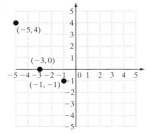

Chapter 11 Test, pp. 772–773

 1. $19.95m$ **2.** -15 **3.** $A = 34.25 \text{ ft}^2$ **4.** Associative property of multiplication **5.** Commutative property of addition **6.** Associative property of addition **7.** Distributive property of multiplication over addition **8.** Commutative property of multiplication **9.** $4a + 24$ **10.** $-13b + 8$ **11.** $16y + 3$ **12.** $7 - 7w$ **13.** An expression is a collection of terms. An equation has an equal sign that indicates that two expressions are equal. **14. a.** Expression **b.** Equation **c.** Expression **d.** Equation **e.** Expression **f.** Expression **15.** $x = -2$ **16.** $x = 18$ **17.** $x = -72$ **18.** $x = -\dfrac{1}{2}$ **19.** $p = -1$ **20.** $m = 0.2$ **21.** $p = -\dfrac{13}{16}$ **22.** $n = \dfrac{1}{2}$ **23.** $h = 3$ **24.** $x = -84$ **25.** $q = 0$ **26.** $k = 2$ **27.** The number is -5. **28.** The U.S. has 16 million antennas, and the United Kingdom has 5.2 million antennas. **29.** Sela used 620 kWh.
30.–35.

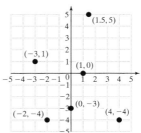

36. Quadrant IV **37.** Quadrant I **38.** Quadrant III **39.** Quadrant II

40. $(30, 33)$, $(40, 35)$, $(50, 40)$, $(60, 37)$, $(70, 30)$

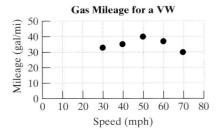

Chapters 1–11 Cumulative Review, pp. 773–775

 1. a. Hundreds **b.** Ten-thousands **c.** Hundred-thousands **2.** 46,000 **3.** 1,290,000 **4.** 25,400 **5.** Dividend is 39,190; divisor is 46; quotient is 851; remainder is 44. **6. a.** Prime **b.** Composite **c.** Composite **7.** $\dfrac{23}{8}$ **8.** $\dfrac{6}{5}$ **9.** $\dfrac{28}{125}$ **10.** $\dfrac{13}{100}$ **11.** $\dfrac{79}{150}$ **12.** 7 ft^2 **13.** $\dfrac{31}{8}$ **14.** $5\dfrac{3}{5}$ **15.** $8\dfrac{2}{3}$ **16.** 13.95 **17.** 0.5 **18.** 31.221 **19.** 43.752 **20.** 3.56 **21.** Sarah makes $20 per room. **22.** $p = 8.1$ **23.** $n = \dfrac{4}{5}$ **24.** $m = 56$ **25.** Kitty Treats costs $0.92 per ounce. Cat Goodies costs $0.90 per ounce. Cat Goodies is the better buy.

	Decimal	Fraction	Percent
26.	0.15	$\dfrac{3}{20}$	15%
27.	0.125	$\dfrac{1}{8}$	12.5%
28.	1.1	$\dfrac{11}{10}$	110%
29.	$0.\overline{2}$	$\dfrac{2}{9}$	$22.\overline{2}\%$
30.	0.002	$\dfrac{1}{500}$	0.2

31. 21 cups can be filled. **32.** 101.75 min **33.** 0.68 L **34.** 3200 m **35.** 4500 cg **36.** 12.56 in.² **37.** The area is 67.5 in.² **38.** 995 in.³ **39.** $C = 18.84$ yd **40.** 24 mm **41.** 10 **42.** 9.5 **43.** 12 **44.**

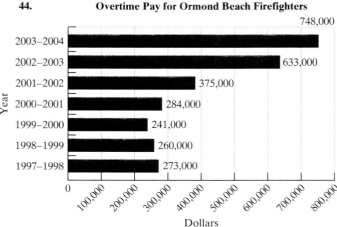

45. a. Approximately 166 students   **b.** 130 students   **46.** 3
47. -12 **48.** -25 **49.** -9 **50. a.** 3.001×10^6 **b.** 0.00004
51. $-5x - 13$ **52.** $-2y - 18$ **53.** $p = -4$ **54.** $t = -8$
55.

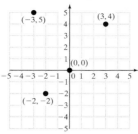

Applications Index

BIOLOGY/HEALTH/LIFE SCIENCES

BUSINESS AND ECONOMICS

CONSTRUCTION AND DESIGN

CONSUMER APPLICATIONS

COOKING

DISTANCE/SPEED/ TIME

EDUCATION AND SCHOOL

Index